AGRICULTURAL MECHANICS

Fundamentals and Applications

7th Editi

AGRICULTURAL MECHANICS

Fundamentals and Applications

7th Edition

Ray V. Herren, Ed.D.

CENGAGE

Australia • Brazil • Mexico • Singapore • United Kingdom • United States

Agricultural Mechanics: Fundamentals and Applications, 7E Precision Exams Edition
Ray V. Herren

SVP, GM Skills & Global Product Management:
    Jonathan Lau

Product Director: Matt Seeley

Product Manager: Nicole Robinson

Executive Director, Development: Marah Bellegarde

Senior Content Developer: Jennifer Starr

Product Assistant: Nicholas Scaglione

Vice President, Marketing Services: Jennifer Ann Baker

Marketing Manager: Abigail Hess

Senior Production Director: Wendy Troeger

Production Director: Patty Stephan

Senior Content Project Manager: Betsy Hough

Digital Delivery Lead: Jim Gilbert

Design Director: Jack Pendleton

Cover image credits: Irrigation Equipment: © Steven
    Ringler/Shutterstock.com

Credits for top row of images: (top left image) Saw:
    © Cengage Learning 2015; (top middle image) Electrical
    Outlet: © Brandon Blinkenberg/Shutterstock.com;
    (top right image) Linkage: © iStockphoto/francv

DESIGN IMAGE CREDITS
Unit Opener Unit 01— © iStockphoto/Simply Creative
Photography
Unit 02: © iStockphoto/asterix0597
Units 03-06: Copyright © 2015 Cengage Learning®.
Unit 07: © iStockphoto/flyfloor
Unit 08: © iStockphoto/Zmiy
Units 09-16: Copyright © 2015 Cengage Learning®.
Unit 17: © iStockphoto/Prill Mediendesign & Fotografie
Unit 18: © iStockphoto/Ruslan Dashinsky
Unit 19: © iStockphoto/taonga
Unit 20: © iStockphoto/Silvia Jansen
Units 21-22: Copyright © 2015 Cengage Learning®.
Unit 23: © iStockphoto/GlenJ
Unit 24: Copyright © 2015 Cengage Learning®.
Unit 25: © Larry Jeffus
Unit 26: © iStockphoto/kentarus
Unit 27: Copyright © 2015 Cengage Learning®
Unit 28: © iStockphoto/kai813
Unit 29: © iStockphoto/Janice Richard
Unit 30: Copyright © 2015 Cengage Learning®
Unit 31: © iStockphoto/Banks Photos
Unit 32: © iStockphoto/daseaford
Unit 33: © iStockphoto/Fertnig
Unit 34: © iStockphoto/Gary Alvis
Unit 35: © iStockphoto/Living Images
Unit 36: © iStockphoto/emesilva
Unit 37: © iStockphoto/Branislav
Unit 38: © iStockphoto/cenix
Unit 39: © iStockphoto/Banks Photos
Unit 40: © iStockphoto/Carole Gomez
Unit 41: © iStockphoto/iacona
Unit 42: Copyright © 2015 Cengage Learning®
Unit 43: © iStockphoto/Charles Schug

© 2019, 2015 Cengage

ALL RIGHTS RESERVED. No part of this work covered by the copyright herein may be reproduced or distributed in any form or by any means, except as permitted by U.S. copyright law, without the prior written permission of the copyright owner.

For product information and technology assistance, contact us at
Cengage Customer & Sales Support, 1-800-354-9706
For permission to use material from this text or product,
submit all requests online at **www.cengage.com/permissions**.
Further permissions questions can be e-mailed to
permissionrequest@cengage.com

Library of Congress Control Number: 2013943110

ISBN: 978-1-337-91870-1

Cengage
20 Channel Center Street
Boston, MA 02210
USA

Cengage is a leading provider of customized learning solutions with employees residing in nearly 40 different countries and sales in more than 125 countries around the world. Find your local representative at **www.cengage.com**.

Cengage products are represented in Canada by Nelson Education, Ltd.

To learn more about Cengage platforms and services, register or access your online learning solution, or purchase materials for your course, visit **www.cengage.com**.

Notice to the Reader
Publisher does not warrant or guarantee any of the products described herein or perform any independent analysis in connection with any of the product information contained herein. Publisher does not assume, and expressly disclaims, any obligation to obtain and include information other than that provided to it by the manufacturer. The reader is expressly warned to consider and adopt all safety precautions that might be indicated by the activities described herein and to avoid all potential hazards. By following the instructions contained herein, the reader willingly assumes all risks in connection with such instructions. The publisher makes no representations or warranties of any kind, including but not limited to, the warranties of fitness for particular purpose or merchantability, nor are any such representations implied with respect to the material set forth herein, and the publisher takes no responsibility with respect to such material. The publisher shall not be liable for any special, consequential, or exemplary damages resulting, in whole or part, from the readers' use of, or reliance upon, this material.

Printed in the United States of America
Print Number: 04   Print Year: 2021

TABLE OF CONTENTS AT A GLANCE

CONTENTS

PREFACE

Agricultural mechanics is one of the most widely taught courses in agricultural education programs. It is taught in all 50 states and is considered to be among the most useful courses taught. Like any applied science, this discipline is constantly changing as new advances are made and new techniques and equipment are put into use. *Agricultural Mechanics: Fundamentals and Applications* has for many years been a standard text for students studying agricultural mechanics. It grew out of the need for an easy-to-read, easy-to-understand, and highly illustrated text on modern agricultural mechanics for high school and post-secondary programs. This edition addresses the specific needs of students enrolled in agriscience, production agriculture, ornamental horticulture, agribusiness, agricultural mechanics, and natural resources programs. All of these areas require basic knowledge and skills in mechanics to succeed in a career path. The text starts with very basic and general information, such as career opportunities, and then provides instruction on basic mechanical skills and applications.

This seventh edition focuses on additional technical information, as needed, throughout. The text is consistent in format, easy to use for individualized instruction, easy to teach from, and simple for substitute teachers. It also provides easy methods to assess student progress. Each unit is part of a section, and each includes (1) a statement of objective, (2) competencies to be developed, (3) a list of new terms, (4) a materials list, (5) highly illustrated text material, (6) student activities, and (7) self-evaluation. This edition contains updated relevant websites that provide more information on the unit topics. All new terms are carefully defined in the text as well as in the glossary.

The appendices include 40 project plans with bills of materials and construction procedures. The projects were carefully selected to match the skills covered in the text. Plans include some projects that have become classics in the field, some that cover targeted enterprises, and some that are new and innovative. The projects were also chosen because they are used in high school agriculture/agribusiness, agriscience, or related programs.

The appendices also include 36 tables containing information for estimating, planning, selecting, purchasing, and building in agricultural mechanics. The project plans, tables, glossary, and index provide unique reference materials that, for many users, are alone worth the price of the text.

AGRICULTURAL MECHANICS
New and Enhanced Content for Seventh Edition

The *Precision Exam Edition* of *Agricultural Mechanics: Fundamentals and Applications, 7E* combines current top-notch content with new information aligned to Precision Exams' **Agricultural Systems Technology I** exam, part of the **Agriculture, Food and Natural Resources** Career Cluster. The **Agriculture, Food and Natural Resources** pathway connects industry with skills taught in the classroom to help students successfully transition from high school to college and/or career. For more information on how to administer the **Agricultural Systems Technology I** exam or any of the 170+ exams available to your students, contact your local NGL/Cengage Sales Consultant.

Features that remain in this enhanced text, and continue to engage and educate students, include the following:

- Updated statistics dealing with agricultural mechanics.
- Expanded coverage of safety using power hand tools, stationary power equipment, and other areas.
- Updated Relevant Websites at the end of each unit provide clearer search information in the event web links change or become out of date.
- Integrates agricultural mechanization with plant, animal, and environmental sciences to provide students with a broad view of the world of agriculture.
- Every unit combines theory with practice.
- Mathematical skill development is emphasized throughout.
- Each unit has been updated to include the latest information on agricultural mechanics.
- Over 350 new and modernized illustrations and photographs offer improved detail and modern equipment related to small engines and power mechanics, plumbing, planning and constructing agricultural structures, and more. New photos depict actual high school students working in the agricultural mechanics workshop.

UNIT SPECIFIC UPDATES
Units 1 through 3

- Many of the photos in these units have been replaced with updated images that depict modern equipment, practices, and professionals in the field of agricultural mechanics.

Unit 4: Personal Safety in Agricultural Mechanics

- New photographs related to workplace safety, including guards or shields on moving parts, safety switches, first aid stations, radiation hazards, wearing eye protection, coveralls, and work boots

Unit 5: Reducing Hazards in Agricultural Mechanics

- New images related to electrical fires, fire extinguisher safety, and fire classification

Unit 6: Shop Cleanup and Organization

- New photos of a clean work shop and a dust collector/dust collection ducts

Unit 7: Hand Tools, Fasteners, and Hardware

- No changes aside from Relevant Websites

Unit 8: Layout Tools and Procedures

- New images of layout tools, stainless steel tools, aluminum tools, plastic tools
- New image showing how patterns can be made from fiber board or card stock

- New images illustrating CAD (computer-aided design) and 3-dimensional designs

Unit 9: Selecting, Cutting, and Shaping Wood

- New images of green lumber, rough lumber, a kerf, and files
- New images of drying lumber (stacking to air dry and drying in kilns)

Unit 10: Fastening Wood

- Expanded discussion of using biscuit joints
- New photo of deck screws
- New photo and information on lag screws

Unit 11: Finishing Wood

- Updates related to types of fillers and wood-finishing products
- New photos of proper sanding technique, applying clear finishes, applying an oil finish, applying polyurethane, applying lacquer, and applying penetrating wood stains

Unit 12: Identifying, Marking, Cutting, and Bending Metal

- New images of structural steel, iron, cast iron, wrought iron, and the carbon content of steel, stainless steel, aluminum, copper, and lead

Unit 13: Fastening Metal

- New image and information on taps and dies
- Updated images of soldering

Unit 14: Portable Power Tools

- New photos of portable power tools (cordless drill, handsaw, belt sander, finishing sander, saber saw, reciprocating saw) and a ground-fault circuit

Unit 15: Woodworking with Power Machines

- New photos related to safe operation of machinery using safety guards, eye and hearing protection; using a bandsaw, jigsaw blades, a contractor's saw, table saw blade guard, and anti-kickback

device; blades with carbide tips; using a radial-arm saw, a miter saw, a sliding compound miter saw, proper standing position when feeding a board into a planer, etc.
- New information related to a newer type of locking table saw that protects against finger/hand injuries
- Updated safety information related to sawing and cross cutting
- New paragraph and photo illustrating careless use of a table saw

Unit 16: Adjusting and Maintaining Power Woodworking Equipment

- No changes aside from Relevant Websites

Unit 17: Metalworking with Power Machines

- New photos of a grinding wheel, wire wheels, a tool rest, and an abrasive cutoff saw
- New safety information and photo related to grinders

Unit 18: Sketching and Drawing Projects

- New images providing updated examples of sketching and drawing projects

Units 19 and 20

- No changes aside from Relevant Websites

Unit 21: Repairing and Reconditioning Tools

- New images of various tools, including types of farm and household tools, metal tools, a mushroomed head, a wedge, an improperly installed handle, and a split handle

Unit 22: Sharpening Tools

- New images of a hand stone, bench stones, sharpening a rotary mower blade
- New information and an image of an overheated tool with damaged temper

- New information and images related to various types of knife sharpeners
- New image depicting the last step in sharpening a wood chisel (lay the blade on its back and remove the wire edge)
- New information and images depicting the differences between a dull drill bit and a sharp drill bit

Unit 23: Using Gas Welding Equipment

- New information about acetylene gas including precautions when using

Unit 24: Cutting with Oxyfuels and Other Gases

- New images related to cutting torches and the brazing process
- Updated photo of plasma arc cutting

Unit 25: Brazing and Welding with Oxyacetylene

- No changes aside from Relevant Websites

Unit 26: Selecting and Using Arc Welding Equipment

- New photos of a welding machine
- New information about proper storage of electrodes along with images of corroded electrodes

Unit 27: Arc Welding Mild Steel and GMAW/GTAW Welding

- New photo related to proper positioning of an electrode
- Updated photos and images related to welding equipment and personal protective equipment
- New photos of plasma arc welding and automated/robotic welding

Unit 28: Preparing Wood and Metal for Painting

- New photo related to primer
- Updated photos of using wood preservatives and removing loose paint

Unit 29: Selecting and Applying Painting Materials

- Updated information about enamels including interior and exterior enamels along with a new photo
- Updated information about paint color and color matching, including a new photo
- Expanded information about different types of paintbrushes and when to use each kind, including a new photo
- Updated photos depicting painting with brushes, applying paint with a roller, and an HVLP spray gun

Unit 30: Fundamentals of Small Engines

- New photos of early engines powered by steam; use of eye and ear protection and proper shoes when using small engine equipment
- New images of a modern two-cycle engine, piston rings, and a type of governor
- New information about gasoline additives
- New information and images related to setting the gap using a spark plug gauge when replacing spark plugs

Unit 31: Small Engine Maintenance and Repair

- Minor updates related to additives and gasoline volatility
- New photos of spark plugs—one in good condition and one in need of replacement; setting the gap using a spark plug gauge
- New photo and reference to lubricating linkages
- Updated illustrations throughout this unit better depicting changing oil, the flywheel key, pulling a flywheel, checking valve stem clearance, and other tasks related to small engine maintenance and repair
- New photo and information related to OHC (overhead cam)

Unit 32: Diesel Engines and Tractor Maintenance

- New photos of a diesel engine; inventor Rudolf Diesel; modern diesel engines; checking oil; a hydrometer; a dirty radiator/grill to illustrate importance of regular cleaning; proper

maintenance of tires (tire gauge); cleaning grease fittings; and batteries
- New illustration of an injector pump
- Updated illustrations of oil flow through an oil filter; the function of the radiator; properly adjusting tension on drive belt
- New information about air filtering systems in the fuel system including a new series of photos related to cleaning a modern dry filter
- New photos related to hydraulic system maintenance including use of a dipstick to monitor fluid levels; servicing fluid filters; and disconnecting quick couplings

Unit 33: Electrical Principles and Wiring Materials

- Expanded information and a new photo related to fluorescent lighting
- Updated images illustrating magnetism, including a new photo of a strong magnet
- New photos of the commutator and the armature, and a modern electric motor
- New photos or nonmetallic sheathed cable, armored cable, and conduit

Unit 34: Installing Branch Circuits

- New photos related to new work installations; wiring boxes; wire nuts; using a wire stripper; connecting wire to a fixture

Unit 35: Electronics in Agriculture

- Updated photos of modern electronic devices; resistors; analog and digital meters; an oscilloscope; integrated circuit packages; water analysis equipment; diode packages; and microcomputers

Unit 36: Electric Motors, Drives, and Controls

- New image of a modern electric motor
- Updated illustrations depicting a power generation and distribution system and the movement of electrical current from the local transformer to a building's distribution panel

Unit 37: Plumbing

- Most illustrations in this unit updated to modernize the fixtures and equipment shown, including the steps for soldering or sweating copper tubing, the steps involved in cementing plastic pipe, plastic vent and drain systems, common pipe fittings and valves, parts of a washer-type faucet, and parts of float valve and flush valve assemblies
- New and updated photos depicting a pipe wrench and pipe vise; a tube cutter; using a flaring tool; and tools used for cutting plastic pipe

Unit 38: Irrigation Technology

- Several new illustrations including updated depictions of irrigation wells and plant rooting pattern and moisture sensor placement

Unit 39: Hydraulic, Pneumatic, and Robotic Power

- Updated illustrations of automotive braking systems and hydraulic systems

Unit 40: Concrete and Masonry

- New photos related to the history of masonry; sizes of aggregates; slag; recycling concrete; sand being flushed with water to remove clay and silt; and a broom finish
- Updated illustrations related to proper form construction and pouring and floating concrete

Unit 41: Planning and Constructing Agricultural Structures

- No changes aside from updates to Relevant Websites

Unit 42: Aquaculture, Greenhouse, and Hydroponics Structures

- Updated photos of net pens; mechanical aerators; fish feeders; maintenance of dissolved oxygen levels; greenhouse production; types of greenhouses; fans; cooling pad; hydroponic crops; and trickle irrigation

Unit 43: Fence Design and Construction

- No changes aside from Relevant Websites

EXTENSIVE TEACHING/LEARNING PACKAGE

The complete supplement package was developed to achieve two goals:

1. To assist students in learning the essential information needed to continue their exploration into the exciting field of agricultural mechanics
2. To assist instructors in planning and implementing their instructional programs for the most efficient use of time and other resources

LAB MANUAL TO ACCOMPANY AGRICULTURAL MECHANICS, 7TH EDITION

ISBN-13: 978-1-28505-901-3

This comprehensive workbook tests students' knowledge and reinforces learning of text content. Job sheets for each unit include an objective, tools and materials needed, and a procedure with short-answer questions, procedural checklists, image labeling activities, or other activities meant to reinforce comprehension of unit content.

MINDTAP FOR AGRICULTURAL MECHANICS: FUNDAMENTALS AND APPLICATIONS, 7E PRECISION EXAMS EDITION

The MindTap for *Agricultural Mechanics: Fundamentals and Applications, 7E* Precision Exams Edition features an integrated course offering a complete digital experience for the student and teacher. This MindTap is highly customizable and combines an enhanced, interactive ebook along with a multitude of engaging activities and assignments including powerpoint, videos, matching, image labeling, crossword puzzles, job sheets, and auto-graded quizzing to enable students to directly analyze and apply what they are learning and allow teachers to measure skills and outcomes with ease.

CLASSMASTER CD-ROM TO ACCOMPANY AGRICULTURAL MECHANICS, 7TH EDITION

ISBN-13: 978-1-28505-899-3

This technology supplement provides the instructor with valuable resources to simplify the planning and implementation of the instructional program. It has been expanded for this edition to include the following support materials:

- A **PDF Instructor's Manual** provides objectives, competencies, glossary terms with definitions, and answers to the end-of-unit question. New to this edition, the instructor's manual now includes lesson plan outlines for each unit with suggested class activities, a correlation guide to the National AFNR Career Cluster Standards, and additional suggested resources.
- A **PDF Lab Manual Instructor's Guide** provides answers to lab manual exercises and additional guidance for the instructor. It has been expanded for this edition to include more answer keys for the lab exercises found in the lab manual.
- A **computerized test bank** created in ExamView® makes generating tests and quizzes a snap. Expanded to include 1,500+ questions with different question formats to choose from, you can create customized assessments for your students with the click of a button. Add your own unique questions and print rationales for easy class preparation.
- **Instructor support slide presentations that can be customized** in PowerPoint® format focusing on key points for each chapter. There is a slide

presentation for each unit available to accompany the textbook.

- New! An **Image Library** with all of the illustrations from the textbook can be used in slide presentations or as part of classroom discussion.
- New! **Correlation guides** include content mapping to STEM, National AFNR Career Cluster Standards, and National Science Standards.

INSTRUCTOR COMPANION WEBSITE

NEW! The instructor companion website provides online access to many of the instructor support materials provided on the ClassMaster CD-ROM, including the Instructor's Manual, Lab Manual Instructor's Guide, computerized test bank files, correlation guides, and support slides. To access the available materials, sign up for a faculty account at login.cengage.com. Add the core textbook to your bookshelf using the 13-digit ISBN that appears on the back cover of the textbook.

ABOUT THE AUTHOR

Dr. Ray V. Herren grew up on a diversified farm near Carbon Hill, Alabama. He earned a B.S. in Agricultural Education at Auburn University, an M.S. in Agribusiness at Alabama A&M University, and an Ed.D. in Vocational Education at Virginia Tech University. He taught high school agriculture in Gaylesville, Alabama, for 9 years and was on the faculty at Oregon State University for 5 years. He is currently Professor Emeritus in the Department of Agricultural Leadership, Education, and Communication at the University of Georgia. He has traveled extensively throughout the world and has authored and coauthored several books.

Dr. Herren has been involved with agricultural mechanics almost all of his life. From practical experience on the farm to teaching preservice and in-service courses, he is well versed in most areas of technical agricultural mechanics. As a hobby, he is an accomplished woodworker.

ACKNOWLEDGMENTS

The author wishes to thank the following for their help in preparing the seventh edition: Ross Hargett for his help in reviewing the text materials and Dr. Frank Flanders, the University of Georgia, for his help with computer problems and his encouragement. Additionally, we extend our thanks and appreciation to the students and staff at Habersham Central High School in Mt. Airy, Georgia, including Jonathan Stribling, CTAE Director, and Kyle Dekle, Teacher; Clay T. Corey, BOCES Instructor, and the Saratoga Springs, New York BOCES location; and Tom Stock, Stock Studios, for their time and participation in photoshoots that provide many of the photos used in this textbook.

The author and Cengage Learning wish to thank the many individuals who provided encouragement and assistance during the preparation of this text. Reviewers who contributed their expertise to the seventh edition of *Agricultural Mechanics: Fundamentals and Applications* include:

Gary Blankenship
Hinckley Big Rock High School
Sheridan, IL

Michael Crim
Agriscience Instructor
Ridge Spring-Monetta High School
Monetta, SC

Al Garner, MS
Agricultural Education Instructor
NAAE and GVATA
Dublin, GA

Marshall Gerbitz
Agriculture Instructor
Okeechobee High School
Okeechobee, FL

Donna Moeller
Instructor
Steinbrenner High School
Lutz, FL

Ted O'Neil
Keene ISD
Keene, TX

Nikki Reed
Danbury High School
Danbury, TX

Mark Rose
Robinson High School
Robinson, TX

HOW TO USE THIS TEXTBOOK

Agricultural Mechanics: Fundamentals and Applications has been carefully designed to enhance the study of mechanics and technology in agriscience programs. For best results, you may want to become familiar with the features incorporated into this text and accompanying learning tools.

Each unit begins with the following tools:

- a core *objective* explains the purpose of the unit;
- *competencies to be developed* list specific goals to meet as you read and review the unit;
- a *materials list* identifies items you will need to complete the unit; and
- a list of *Terms to Know* identifies key vocabulary to master.

SPECIAL FEATURES

- Hundreds of new full-color photographs provide up-to-date visual references for the procedures described throughout the textbook.

- Procedures are highlighted throughout the textbook in order to highlight the basic steps in performing specific tasks covered in each unit.

- Notes emphasize important items you should be aware of before proceeding with a unit.

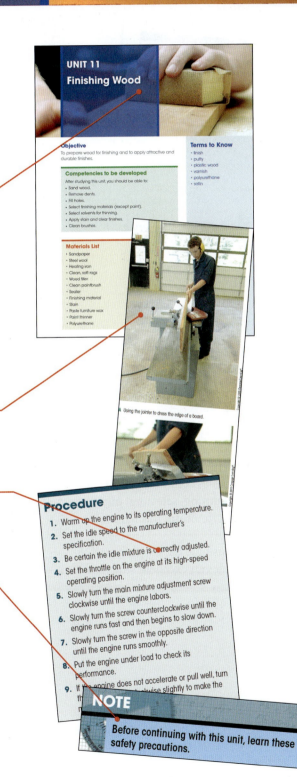

CAUTION!

Fuses and circuit breakers are used to protect circuits from damage and fire. Never destroy this protection by installing fuses ~~or~~ breakers with a ~~...~~mmended for

line from one end to the other on the face of the board. Use a block plane or a smoothing plane to make a chamfer. A chamfer is a tapered cut along the edge of a board.

16. Discuss your workmanship in the preceding procedures with your instructor. Save the board for an activity in Unit 10.

Relevant Web Sites

American Hardwood Export Council
www.ahec.org/index.asp

Softwood Export Council
http://softwood.org

Self-Evaluation

A. **Multiple Choice.** Select the best answer.

1. Grain in lumber is caused by
   a. the age of the board
   b. annual rings
   c. special drying techniques
   d. the stain

2. Lumber is graded according to its
   a. appearance and soundness
   b. color and species
   c. strength and durability
   d. cost and length

3. A crosscut handsaw with very coarse teeth would have how many teeth per inch?
   a. 6
   b. 10
   c. 12
   d. 14

4. A crosscut handsaw with very fine teeth would have how many teeth per inch?
   a. 6
   b. 8
   c. 14
   d. 20

5. The wood removed by a saw blade leaves an opening called a
   a. bevel
   b. channel
   c. chamfer
   d. kerf

6. The backsaw gets its name from
   a. its use as a backup tool
   b. its fine teeth
   c. its stiff back
   d. its original use in making chair backs

7. A suitable tool for cutting curves is the
   a. compass saw
   b. coping saw
   c. keyhole saw
   d. all of these

8. The brace and bit has been replaced by the
   a. drill press
   b. coping saw
   c. screwdriver bit
   d. portable electric drill

9. Lumber for construction should be dried to about
   a. 2%
   b. 15%
   c. 35%
   d. 20%

10. The teeth on a file cut
    a. best when oiled
    b. only soft materials
    c. only on the backward stroke
    d. only on the forward stroke

- Caution! features indicate safety precautions you need to be aware of.

- End of unit review materials include:
  - a unit summary;
  - student activities, including supervised agricultural experiences;
  - a list of relevant websites for further information and research; and
  - *Self-Evaluations,* which allow you to review the unit content using multiple choice, matching, and completion question.

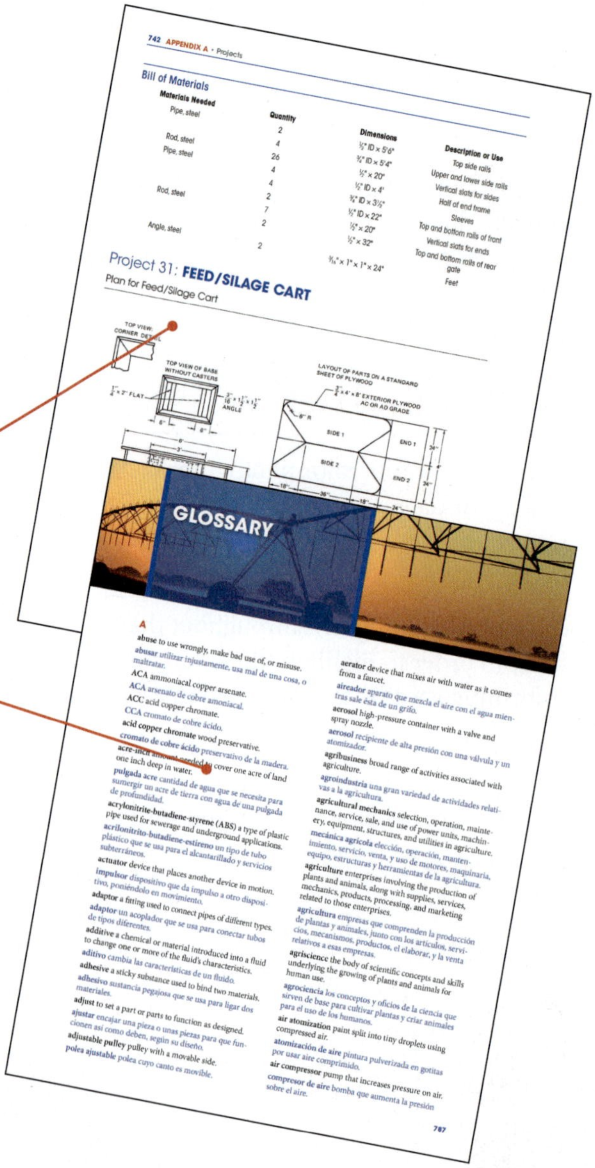

742 APPENDIX A • Projects

Bill of Materials

Materials Needed

Pipe, steel	Quantity	Dimensions	Description or Use
Rod, steel	2		
Pipe, steel	4	½" ID × 5'6"	Top side rails
	26	½" ID × 5'4"	Upper and lower side rails
	4	½" × 20"	Vertical slats for sides
Rod, steel	4	½" ID × 4"	Half of end frame
	2	½" ID × 3½"	Sleeves
Angle, steel	7	½" ID × 22"	Top and bottom rails of front
	2	½" × 20"	Vertical slats for ends
		½" × 32"	Top and bottom rails of rear gate
	2	¾" × 1" × 1" × 24"	Feet

Project 31: FEED/SILAGE CART

Plan for Feed/Silage Cart

GLOSSARY

A

abuse to use wrongly, make bad use of, or misuse.
abusar utilizar injustamente, usa mal de una cosa, o maltratar.

ACA ammoniacal copper arsenate.
ACA arsenato de cobre amoniacal.

ACC acid copper chromate.
CCA cromato de cobre ácido.

acid copper chromate wood preservative.
cromato de cobre ácido preservativo de la madera.

acre-inch amount of water to cover one acre of land one inch deep in water.
pulgada acre cantidad de agua que se necesita para sumergir un acre de tierra con agua de una pulgada de profundidad.

acrylonitrile-butadiene-styrene (ABS) a type of plastic pipe used for sewerage and underground applications.
acrilonitrilo-butadieno-estireno un tipo de tubo plástico que se usa para el alcantarillado y servicios subterráneos.

actuator device that places another device in motion.
impulsor dispositivo que da impulso a otro dispositivo, poniéndolo en movimiento.

adaptor a fitting used to connect pipes of different types.
adaptor un acoplador que se usa para conectar tubos de tipos diferentes.

additive a chemical or material introduced into a fluid to change one or more of the fluid's characteristics.
aditivo cambia las características de un fluido.

adhesive a sticky substance used to bind two materials.
adhesivo sustancia pegajosa que se usa para ligar dos materiales.

adjust to set a part or parts to function as designed.
ajustar encajar una pieza o unas piezas para que funcionen así como deben, según su diseño.

adjustable pulley pulley with a movable side.
polea ajustable polea cuyo canto es movible.

aerator device that mixes air with water as it comes from a faucet.
aireador aparato que mezcla el aire con el agua mientras sale ésta de un grifo.

aerosol high-pressure container with a valve and spray nozzle.
aerosol recipiente de alta presión con una válvula y un atomizador.

agribusiness broad range of activities associated with agriculture.
agroindustria una gran variedad de actividades relativas a la agricultura.

agricultural mechanics selection, operation, maintenance, service, sale, and use of power units, machinery, equipment, structures, and utilities in agriculture.
mecánica agrícola elección, operación, mantenimiento, servicio, venta, y uso de motores, maquinaria, equipo, estructuras y utensilios de la agricultura.

agriculture enterprises involving the production of plants and animals, along with supplies, services, mechanics, products, processing, and marketing related to these enterprises.
agricultura empresas que comprenden la producción de plantas y animales, junto con los artículos, servicios, mecanismos, productos, el elaborar, y la venta relativos a esas empresas.

agriscience the body of scientific concepts and skills underlying the growing of plants and animals for human use.
agrociencia los conceptos y oficios de la ciencia que sirven de base para cultivar plantas y criar animales para el uso de los humanos.

air atomization paint split into tiny droplets using compressed air.
atomización de aire pintura pulverizada en gotitas por usar aire comprimido.

air compressor pump that increases pressure on air.
compresor de aire bomba que aumenta la presión sobre el aire.

787

OTHER RESOURCES AVAILABLE

- The appendices offer numerous project plans with bill of materials and construction procedures as well as tables containing information for estimating, planning, selecting, purchasing, and building in agricultural mechanics.

- An English–Spanish glossary with over 425 definitions provides an excellent study aid and reference.

EXPLORING CAREERS IN AGRICULTURAL MECHANICS

UNIT 1
- Mechanics in the World of Agriculture

UNIT 2
- Career Options in Agricultural Mechanics

Objective

To determine how mechanical skills, concepts, and principles are used in agriculture and related occupations.

Competencies to be developed

After studying this unit, you should be able to:

- Define *agriculture* and *agricultural mechanics*.
- Define *occupation* and describe an occupational cluster.
- Describe the role of mechanics and mechanical applications in society.
- Demonstrate knowledge of contributions made by mechanical application to the development of agriculture.
- Name eight inventors of important agricultural machines.

Materials List

- Pencil
- Paper
- Encyclopedias

Terms to Know

- agriculture
- agriscience
- renewable natural resources
- agricultural mechanics
- efficiency

To many people, the term **agriculture** refers to production agriculture—farming or the production of plants and animals. However, agriculture is a very broad industry that includes not only producing plants and animals, but also all the related supplies, services, mechanics, products, processing, and marketing related to producing plants and animals, and keeping the environment sound.

Agriculture is a very complex industry. The industry produces plant and animal products from which thousands of commodities are made. Because every person and many industries depend upon agriculture, it is said to be a basic industry. Some products of agriculture are food, oils, fiber, lumber, ornamental trees and shrubs, flowers, leather, fertilizers, feed, seed, and more. Basic agricultural products form the raw materials for many items of everyday living.

Fabrics for clothing, curtains, and floor coverings are made from oils such as corn oil, soybean oil, and cottonseed oil. Plastics of all kinds are also made from vegetable oils. Products from animals are used to make materials such as glue, leather, and paint. Many medicines come from plants and animals. The manufacture of automobiles, furniture, airplanes, radios, stereos, and computers depends on agriculture for certain raw materials. The construction of homes, boats, and factories depends on agriculture for lumber, fiber, and other basic commodities. Most dwellings in America are surrounded by lawn, shrubs, or other plants for beautification. These are also agricultural commodities.

Agriculture is, indeed, a basic industry upon which all people depend. It is the backbone of the American society. Even though the number of farms has dramatically decreased over the past 150 years, the amount of agricultural production has increased manifold. For example, in 1940 there were about 6.5 million U.S. farms and one farmer fed about 19 people. Today there are a little over 2.2 million U.S. farms. Of these, 98 percent are family farms with only about 2 percent corporate operations. The families who live on and operate our farms and ranches comprise only about 2 percent of the U.S. population. The producers have become so efficient that one farmer feeds about 155 people (Figure 1-1). A lot of different factors contributed to this increase in efficiency. Scientific research that developed better varieties, more efficient fertilization programs, more effective insect and disease control, and an overall better understanding of how plants and animals reproduce has been a tremendous factor. However, just as important is the development of agricultural mechanics that brought

FIGURE 1-1 The average American farmer produces enough food to feed about 155 people.

electricity to all rural areas and developed the marvelous machines and implements that made human labor much more efficient.

The term **agriscience** refers to the science involved with the industry of agriculture. Agriculture is a science that involves most of the knowledge we process concerning the growth and reproduction of plants and animals. In fact, when one considers all the vast knowledge we have in biology, there are only three applications: medicine, ecology, and agriculture. Agriculture plays a vital role in both medicine and ecology, so our knowledge in biology is applied in agriculture.

Renewable natural resources are considered to be a part of agriculture. These are the resources provided by nature that can be replaced or renewed. Examples of such resources are our forests; fish in our streams, lakes, and oceans; and our wildlife such as game animals. At one time most natural resources were not managed, and humans relied on nature to replenish the resources. As our population began to grow, nature could not replenish natural resources as fast as they were being consumed. A large industry began to develop that helps to manage our resources. Most U.S. forestland is now managed—in terms of planting and cultivating trees, regulating harvesting, and managing fire-control efforts (Figure 1-2). Large hatcheries now produce young fish that are released into the wild. Both commercial and sport fishing are closely regulated to prevent the depletion of fish populations. Game animals are closely regulated to prevent overhunting and to control population numbers to prevent overpopulation (Figure 1-3). Just think of all the industry that is needed to support the management and enjoyment of our natural resources!

FIGURE 1-2 Managed forests are a renewable resource.

Think of all the industry that goes with fishing—everything from the production of fishing lures, rods and reels to fishing boats.

American farmers are among the world's most efficient businesspeople. Much of this efficiency and success is due to the invention, development, and

FIGURE 1-3 Game animals account for a significant part of our natural-resource industry.

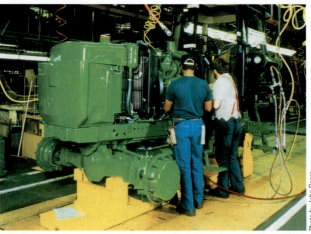

FIGURE 1-4 Workers in agricultural mechanics are responsible for the design, manufacture, testing, sales, and service of farm machinery.

maintenance of modern farm machinery. Also, consider all the machinery needed to plant, cultivate, and harvest trees (Figure 1-4). Consider all that goes into the construction and maintenance of buildings that house agricultural products, equipment, supplies, and machinery. A large industry is built around agricultural construction (Figure 1-5).

FIGURE 1-5 Agricultural construction is an important part of agricultural mechanics.

THE ROLE OF AGRICULTURAL MECHANICS

Mechanical applications are found throughout agriculture. A few examples of people whose occupations involve **agricultural mechanics** are:

- the engineer who designs tractors and other farm and ranch machines (Figure 1-6)
- the forester who keeps chainsaws and other equipment going
- the builder who constructs processing plants, farm buildings, and aquaculture facilities
- the electrician who installs climate controls, silo unloaders, and milling equipment
- the soil conservationist who constructs terraces to control erosion
- the hardware store employee who obtains and stocks repair parts for agricultural tools and machines (Figure 1-7)
- the specialist who creates air-conditioning and refrigeration systems in processing and storage facilities
- the specialist who designs and installs drainage and irrigation systems for fields, turf, landscaping, and golf courses (Figure 1-8)
- the lawn equipment service mechanic who repairs lawn tractors
- the welder who repairs farm machinery
- the mechanic who keeps diesel trucks and machines in good repair

Processing plants for field crops, livestock, poultry, fruits, and vegetables all use machinery. Machines

FIGURE 1-7 Machinery parts can be ordered by computer.

FIGURE 1-8 Agricultural mechanics is needed in urban areas. Ball fields, parks, and golf courses rely on machinery to install and maintain turf, plants, and landscaping.

require designers, engineers, operators, maintenance and repair personnel, and construction workers. Even people with jobs in finance, publishing, and communications may need some knowledge of mechanics when their assignments deal with agriculture. All are likely to use computers and computer applications in their work.

THE INFLUENCE OF MECHANIZATION

At the birth of the United States in 1776, more than 90 percent of the American colonists were farmers, yet many of General Washington's troops at Valley Forge died for lack of food and clothing. Today, less than 10 percent of all Americans work in production

FIGURE 1-6 Engineers are needed to design machinery.

agriculture, yet there are generally food surpluses in America. The ratio of farm workers to nonfarm workers in America approximately reversed in the last 230 years. In 1776, the farm-to-nonfarm ratio was approximately 9:1. Today, the ratio of production agricultural workers to the remaining population is approximately 1:9. Mechanization has played a major role in this rise in production efficiency. **Efficiency** means the ability to produce with a minimum waste of time, energy, and materials.

America provided the inventors for many of the world's most important agricultural machines. For example, grain has always been a basic food for humans. For thousands of years people have planted, cultivated, and harvested grain to eat whole or to make bread. One of the most labor-intensive jobs was that of cutting the plants and threshing out the grain. This was done from prehistoric times until the early part of the 1800s. Cyrus McCormick invented the reaper in 1834 to cut small grain crops. This machine used horse power to pull the reaper that cut the mature grain plants. This allowed one person to do the work of many people and the labor was easier. This still left the job of threshing out the wheat. Later, a machine called a combine was invented, which both cut and threshed the grain in the field. The name *combine* was adopted because the machine combined the job of cutting and threshing. Today, one modern combine operator can cut and thresh as much grain in one day as 100 people could cut and bundle in one day in the 1830s (Figure 1-9).

Two inventions had a profound influence on the settling of this country. The first was by a man named John Deere, who developed a steel plow that replaced an iron plow invented by Thomas Jefferson. The plow, invented by Deere in 1837, allowed farmers to break up the tough sod that previously had prevented pioneers from cultivating the rich prairie soils. Previously plowshares (the parts that cut through the soil) were made either of wood or cast iron. Both of these materials were so heavy that the plow could not be drawn through the sod. Deere constructed plowshares out of steel that was both lighter and tough enough to cut through the sod. Settlers inhabited all of the Midwest and Plains region where so much of our food is grown today. From this beginning, John Deere became one of the world's leading manufacturers of agricultural and construction implements (Figure 1-10).

The other invention that affected settlement was by Eli Whitney, who developed a machine, called a gin (short for "engine"), to remove seeds from cotton. Prior to his invention in 1793, seeds were removed from cotton by hand. The seeds could be separated from the cotton, but it was a time-consuming task. Another problem was the type of cotton grown. Cotton with loose seeds was the Sea Island or

FIGURE 1-9 A combine cuts and threshes grain in one operation.

© Orientaly/Shutterstock.com

Photo by John Deere

FIGURE 1-10 John Deere invented a plow that could cut through the heavy sod of the Midwest.

long-staple variety that would grow only along the coast of Georgia and the Carolinas. Upland cotton would grow anywhere in the South, but the seeds of this type were almost impossible to separate by hand. Whitney's gin not only saved labor, but it also opened up the entire southern portion of the nation for cultivation of upland cotton.

The power for agricultural machines was first supplied by horses, oxen, and mules. Later, as heavier and more complex machines were developed, power was supplied by steam. Steam engines had proven to be very effective in powering boats, ships, and trains. They provided more power than animals, but they were extremely heavy and cumbersome on land (Figure 1-11). The internal combustion engine made machines tremendously efficient. Machinery equipped with these relatively lightweight, powerful engines replaced the steam-powered machines,

revolutionizing the production of food and fiber. Today, most of the machinery used in agricultural production is based on the internal combustion engine.

Perhaps no invention has had more impact on agriculture and on the lives of people than the invention of refrigeration. Prior to its invention, produce and meats had to be sold fresh, and their shelf life was very short. With mechanical refrigeration, meats and produce could not only be stored much longer, but could also be transported long distances. Refrigerated railcars and trucks allowed livestock, fruits, and vegetables to be produced in one part of the country and shipped across the country to large cities. For the first time in history, people could have fresh meats and vegetables year-round.

The machines discussed here are only a few of the ones that have revolutionized agriculture. Our entire modern agricultural industry is based on complex, efficient machinery that performs crucial roles, all the way from the research laboratories that develop agricultural inputs to the businesses that harvest, store, process, and transport commodities to the consumer. All of these thousands of different types of machines and implements that keep the agricultural industry going rely on a complex support system that designs, produces, installs, maintains, and repairs them. Agriculture has become highly mechanized in all of the developed countries of the world. In undeveloped countries, many engineers, teachers, and technicians have sought simple, tough, and reliable small machines to improve agriculture. In such countries, America's highly developed, complex, computerized, expensive machinery will not do, in part because these countries lack people trained for the variety of agricultural mechanics jobs that are needed to support agriculture in the United States.

Many features, such as rubber tires, have been standard equipment on American farms since the 1930s. Yet a machine with rubber tires is useless if a tire gets damaged and repair services are not available. This is the case in most undeveloped countries in Central and South America, Asia, and Africa. Much of the world cannot compete with American agriculture because the related agricultural products and services are not available to support farm workers.

The efficiency of American agriculture will increase in the future as computer-controlled machines and robotics play an important part. It is exciting to envision the changes in store for agricultural mechanization in the twenty-first century.

© Alan Egginton/Shutterstock.com

FIGURE 1-11 Heavy, bulky steam engines began to replace animals as a power source in the 1800s. This antique tractor has been restored.

SUMMARY

Agricultural mechanics has been fundamental to the development of the agricultural industry in this country. Much of the tremendous increase in the efficiency of the American producers is due to innovations in mechanics. The wiring of buildings to supply power, the repairing of engines and equipment, the laying of pipes for water supplies, and constructing buildings are only a few examples of mechanics in agriculture. As further advances are made, the role of mechanics in agriculture will be as prominent in the future as it has been in the past.

Student Activities

1. Define the Terms to Know in this unit.
2. Interview a cooperative extension specialist for agricultural resources in your county or city. Ask the specialist to describe the different agricultural or agriculturally related jobs that people do in your locality.
3. Look up "inventors" or "inventions" in an encyclopedia. Pick out the inventions that relate to agriculture, and report your findings to the class.
4. Select three or five classmates to join you in a debate on the role of agriculture in society. One team should support the position that agriculture is the backbone of society. The opposing team should support the notion that it is not.
5. Consider an everyday product such as bread, milk, leather gloves, or a corsage for Mother's Day. Trace the production, processing, and marketing of the item from its source to its sale as a finished product. List points along the way where agricultural mechanics is involved.

Relevant Web Sites

About.com's web page on Inventors, Agriculture, and Farm Innovations
http://inventors.about.com/library/inventors/blfarm.htm

Delaware Agricultural Museum & Village
www.agriculturalmuseum.org

National FFA Online, Teacher's Workroom
http://web.missouri.edu/~schumacherl/natcon.html

Online Encyclopedia Britannica
www.britannica.com

Self-Evaluation

A. Multiple Choice. Select the best answer.

1. The production of plants and animals and the provision and management of related supplies, services, mechanics, products, processing, and marketing defines
    a. horticulture
    b. renewable natural resources
    c. agricultural mechanics
    d. agriculture

2. Agriscience is
    a. the same as agricultural mechanics
    b. limited to the sale of agricultural products
    c. business stemming from agriculture
    d. the science that is behind agricultural production

3. Examples of renewable natural resources are
    a. oil, gas, and coal
    b. fish, trees, and wildlife
    c. rubber, steel, and water
    d. air, soil, and minerals

4. Over the past 230 years, the number of American farms has
    a. decreased
    b. increased
    c. remained about the same
    d. not been measured

5. Agricultural mechanics stems mostly from
    a. physics
    b. biology
    c. medicine
    d. horticulture

6. Agricultural products come from
    a. soil and coal
    b. plants and animals
    c. iron ore and aluminum
    d. atomic fuel

7. Products of agriculture include
    a. leather seat covers
    b. paint
    c. flower arrangements
    d. all of these

8. Agricultural mechanics includes the occupation of
    a. garden tractor repairperson
    b. automobile mechanic
    c. pile driver
    d. systems analyst

9. Mechanization of agriculture has resulted in
    a. decreased soil production
    b. decreased farm expenses
    c. increased production efficiency
    d. increased numbers of farm workers

10. Cyrus McCormick invented the
    a. steel plow
    b. cotton gin
    c. milking machine
    d. reaper

B. Matching. Match the items in column I with those in column II.

Column I

1. efficiency
2. agricultural mechanics
3. Thomas Jefferson
4. loose seeds
5. John Deere
6. steam engines
7. Eli Whitney
8. mechanic
9. technician
10. trade

Column II

a. one who is specifically trained to perform tasks having to do with a machine, a mechanism, or machinery

b. the iron plow

c. the selection, operation, maintenance, servicing, selling, and use of power units, machinery, equipment structures, and utilities used in agriculture

d. the ability to produce with minimum waste of time, energy, and materials

e. Sea Island cotton

f. heavy and cumbersome

g. specific kinds of work or business, especially those that require skilled mechanical work

h. the cotton gin

i. a mechanic who uses high technology

j. the steel plow

C. **Completion.** Fill in the blanks with the word or words that make the following statements correct.

1. Today a single combine operator can cut and thresh as much grain in one day as _____ people could cut and bundle in the same amount of time in the 1830s.

2. The average U.S. farmer today feeds about _____ people per year.

3. Early inventors such as Eli Whitney and John Deere could be considered workers in the area of _____.

4. Most of the machinery used today in agricultural production is based on the _____.

5. All jobs and types of work in the field of agriculture/agribusiness and renewable natural resources make up an _____ _____.

D. **Brief Answer.** Briefly answer the following questions.

1. It can be said that agriculture is the backbone of American society. Explain.

2. Compare the ratio of farm workers to nonfarm workers in 1776 with the farm-to-nonfarm ratio today. What has made today's food surpluses possible in spite of this reversal?

3. Identify two inventions that had a profound influence on the settling of this country, and explain the importance of each.

4. Explain the impact of refrigeration on agriculture and society.

5. In the past, *agriculture* meant to farm or to grow plants or animals. Why is such a simple definition not possible today?

UNIT 2
Career Options in Agricultural Mechanics

Objective

To determine how skills in agricultural mechanics may be used to earn a good living.

Competencies to be developed

After studying this unit, you should be able to:

- List the major divisions in the agricultural cluster of occupations.
- Identify occupations in agriculture that require mechanical skills.
- Describe the relationship between mechanical applications and success in certain agricultural occupations.
- Conduct an in-depth study of one or more jobs in agricultural mechanics.
- Establish tentative personal goals for using agricultural mechanics skills.

Materials List

- A computerized, microfiche, or other career information system
- *Occupational Outlook Handbook*

Terms to Know

- off-the-farm agricultural jobs
- supervised agricultural experience
- occupational division
- FFA
- agribusiness and agricultural production
- 4-H
- Boy Scouts of America

Young people today have a wide variety of options when planning a career (Figure 2-1). It is never too early to begin thinking about what you want to do for a living. Earning enough money to make a comfortable living is important; however, it is also important to choose a career that is enjoyable and rewarding. Just think of all the time you will be spending on the job. If you do not enjoy what you do, the quality of life will be much reduced. With the wide array of all the different career choices and the opportunities available, you should be able to find an enjoyable career.

Copyright © 2015 Cengage Learning®.

FIGURE 2-1 The agriculture industry provides interesting careers in production, management, science, education, finance, communication, government, conservation, and mechanics.

FIGURE 2-2 The serenity of rural areas attracts many people to careers in agriculture.

Agriculture may be the area with the greatest opportunity for a satisfying career since it encompasses so many different areas of interest. Whether you like the serenity of living in a rural setting or you prefer the bustle of suburban living, there are careers in agriculture that may suit you (Figure 2-2).

AGRICULTURE IS NUMBER ONE

For many years agriculture has been America's number one employer. While some labor analysts may say this is no longer the case, there are literally many millions of jobs in the field of agriculture. While the number of farmers and ranchers has decreased, the need for support people has opened up many job opportunities. Any way the job market is analyzed, agriculture accounts for a very high percentage of America's total workforce. A large proportion of these jobs are off-the-farm agricultural jobs. **Off-the-farm agricultural jobs** are those jobs requiring agricultural skills but not regarded as farming or ranching. The efficiency of this unique mix of farm and nonfarm agricultural jobs permits the average American to use only about 9.8 percent of his or her total disposable income on food with about 5.7 percent spent on food for home and about 4.1 percent spent on food away from home at restaurants and other prepared food outlets This is the lowest of any nation in the world.

The opportunities in agriculture are found at all levels of employment. The first level of employment is laborer; this level requires the least amount of preparation. A few laborer jobs require less than a high school diploma. Additional training is needed for the better-paying jobs. Other job levels include semiskilled, skilled, managerial,

and professional. Some professional jobs, such as veterinarian and scientist, require a doctor's degree. The doctor's degree requires 7 or more years of college education. It is important to make a career choice as early as possible so the proper training can be obtained. Career goals should be set while in high school. High school courses, **supervised agricultural experience**, work experience abroad, and future schooling can then help prepare for the chosen career. Jobs are available in agricultural occupations at every work level (Figure 2-3).

A Agricultural engineers and soil and water technicians seek better ways to use water wisely.

B Machinery repair is a necessary part of modern agriculture.

FIGURE 2-3 Many occupations are available in the field of agriculture. (*continued*)

C Agriculture provides many jobs for those who prefer to work outdoors.

D Salespeople are needed to work in agricultural machinery dealerships.

E Food processors hire many people to repair and maintain equipment.

FIGURE 2-3 (*continued*)

Students enrolled in an agricultural education course have the edge on others who seek careers in agriculture. Agricultural education programs prepare the student specifically for the world of work, as well as for college. With so many areas inside the broad field of agriculture, choices must be narrowed. Choosing a division within agriculture generally permits more specializing and better job opportunities. The largest career areas in agriculture are in marketing and sales, science, and engineering (Figure 2-4). There are also many other areas to choose from. This suggests an excellent employment outlook for agricultural education graduates.

AGRICULTURAL DIVISIONS

A number of U.S. government agencies have worked together to classify occupations. The National Center for Educational Statistics publishes *A Classification of*

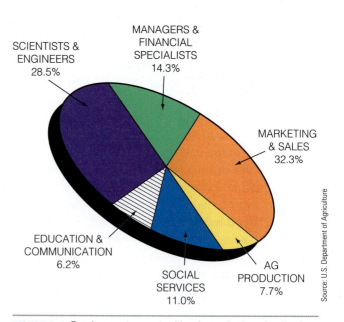

FIGURE 2-4 Employment opportunities for agriculture/agribusiness, as reported by the USDA.

Instructional Programs. This book arranges all occupations into occupational clusters and divisions. An **occupational division** is a group of occupations or jobs within a cluster that requires similar skills. All jobs in agriculture are in one agricultural cluster grouped under the headings of (1) Agribusiness and Agricultural Production, (2) Agricultural Sciences, and (3) Renewable Natural Resources.

Divisions in Agribusiness and Agricultural Production

The **agribusiness and agricultural production** area contains eight divisions. They are:

- agricultural business and management
- agricultural mechanics
- agricultural production
- agricultural products and processing
- agricultural services and supplies
- horticulture
- international agriculture
- agribusiness and agricultural production

Notice that agricultural mechanics is one of the divisions. This division is absolutely vital to all aspects of agriculture. Can you imagine what the industry would be like without equipment to plant, cultivate, and harvest crops? Without mechanics, produce could not be transported to processing plants. The processing plants could not run, and supermarkets could not be stocked. Within the designing, building, operating, and maintaining of the machines, there are many job opportunities. For example, farm equipment operator is a part of agricultural mechanics. Many young people find satisfaction in a career operating heavy equipment (Figure 2-5). There are many interesting jobs available in agriculture that use mechanical applications. Figure 2-6 shows how solar energy can be used to help power agricultural buildings. Figure 2-7 shows career areas in agriculture.

Divisions in Agricultural Sciences

The agricultural sciences have six divisions. They are:

1. agricultural sciences, general
2. animal sciences
3. food sciences
4. plant sciences
5. soil sciences
6. agricultural sciences

Photo by Scott Bauer, USDA NRCS

FIGURE 2-5 Many people make careers out of operating the massive machinery used in agricultural work.

© iStockphoto/fotolinchen

FIGURE 2-6 Solar panels can help create energy for farm use. Note solar panels on the top of the building.

Notice that each division under agricultural sciences has an *s* on "science." This is because there are numerous sciences under each division.

American agriculture is based on scientific knowledge. Production, management, and mechanics all rely on information obtained through the scientific approach. Figure 2-8 shows satellite technology being used to predict the weather.

Divisions in Renewable Natural Resources

The renewable natural resources area contains seven divisions:

1. renewable natural resources, general
2. conservation and regulation
3. fishing and fisheries

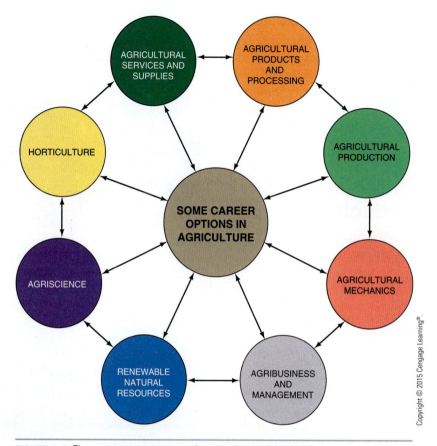

FIGURE 2-7 There are many career options to choose from in agriculture. Each area of agriculture relies on many others to be successful. For example, a greenhouse cannot be successful without people knowledgeable in mechanics and business.

4. forestry production and processing
5. forestry and related sciences
6. wildlife management
7. renewable natural resources

There are many jobs listed under the renewable natural resources division that require agricultural mechanics.

FIGURE 2-8 Numerous scientific advances play important roles in agriculture. This photo shows a satellite image that is used to predict weather patterns.

CAREER SELECTION

Agriculture teachers, guidance counselors, and librarians are sources of information regarding careers in the field of agriculture. Figure 2-9 provides a breakdown of the various divisions in agriculture.

There are specific job titles in agricultural mechanics. They are classified under the following categories:

- agricultural mechanics, general
- agricultural electrification, power, and controls
- agricultural mechanics, construction, and maintenance skills
- agricultural power machinery
- agricultural structures, equipment, and facilities
- soil and water mechanical practices
- agricultural mechanics, other

Figure 2-10 shows some examples of jobs that are included in these classifications.

Farmers, ranchers, greenhouse operators, pesticide applicators, veterinarians, wildlife officers—all are better at their jobs if they have agricultural mechanics skills. Even those who simply use buildings, equipment, or materials find mechanical skills help them solve problems.

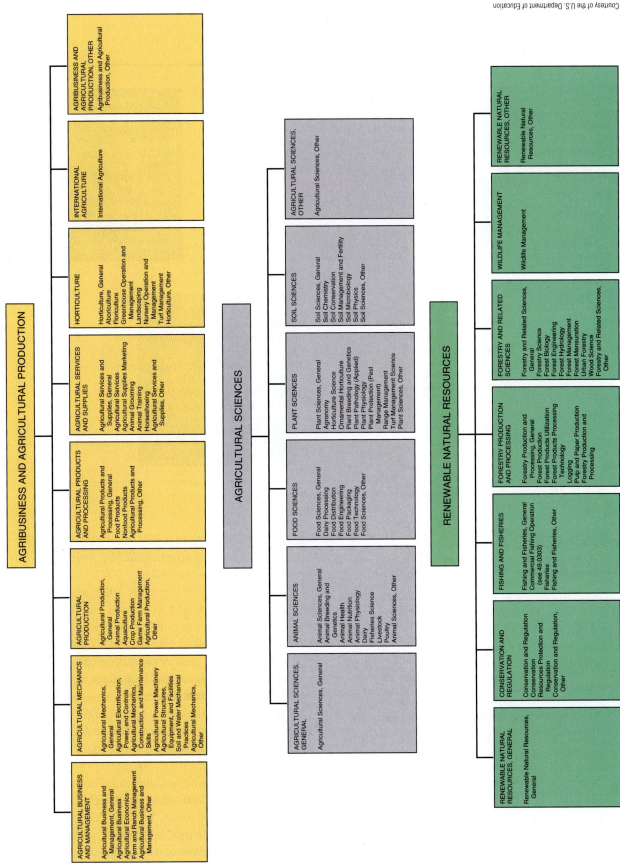

Courtesy of the U.S. Department of Education

FIGURE 2-9 There are many instructional programs in agriculture.

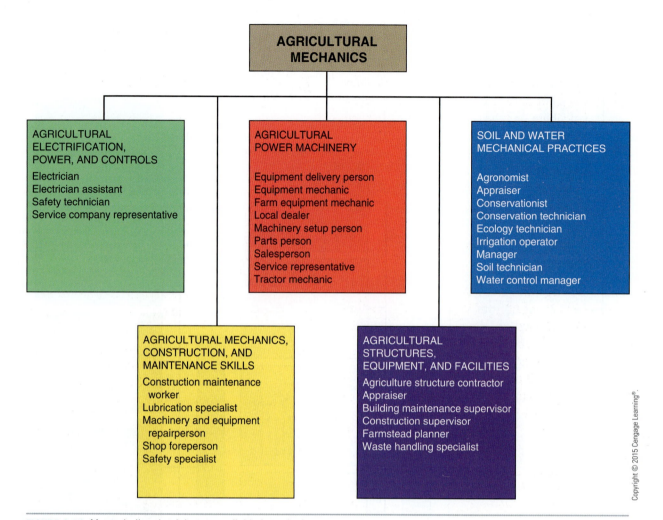

FIGURE 2-10 Many challenging jobs are available in agriculture.

Copyright © 2015 Cengage Learning®.

AGRICULTURAL EDUCATION PROGRAMS

As students plan courses in agricultural education, they should get as much experience in agricultural mechanics as possible. Such experiences should be stimulating and rewarding. They may occur in the classroom, shop, greenhouse, school farm, at home, in supervised agricultural experiences, or FFA.

The Smith-Hughes Act of 1917 provided funding for agriculture in high schools. Through these programs, students in rural areas had the opportunity to learn the latest scientific principles in agricultural production. Teachers began visiting students after school to help with the student projects. This was the beginning of the SAE. This program component is explained in detail in Appendix C at the back of this text. The third component, the Future Farmers of America (FFA), completed the three parts of the program: classroom/lab, SAE, and FFA.

The National FFA is an organization for students studying agricultural education in high school. This organization is active in all 50 states and Puerto Rico. Over 450,000 students participate in activities and programs sponsored by the FFA.

This organization began in 1928 when students studying agriculture met in Kansas City for the purpose of establishing a national association. Prior to this, many states had created similar organizations for rural young people. The state organizations began to compete in livestock judging and looked for a place to hold a national competition. The American Royal in Kansas City agreed to host the event. This brought the students together for the first time, and they decided there was a need for a national organization. In 1928, the FFA came into existence.

The FFA was patterned after the Future Farmers of Virginia. This state association was developed by Henry Groseclose, Walter Newman, Edmund Magill, and Harry Sanders. They saw the need to give rural youth opportunities to develop social graces and leadership skills. Ceremonies for the FFA were designed after the ceremonies of the Freemasons, and the awards were patterned after the Boy

Scouts. The colors selected for the organization are corn gold and national blue.

The FFA emblem was designed by Henry Groseclose. It features a cross section of an ear of corn that represents the nationwide scope of FFA. The rising sun represents a new era in agriculture, and the plow represents labor and tillage of the soil. The owl stands for wisdom that is necessary for progress. The eagle at the top symbolizes freedom and the ability fulfills to go as far as our abilities take us.

The FFA has passed many milestones since 1928. In 1965, the FFA merged with the New Farmers of America (NFA). The NFA was an organization for black youths studying agriculture. In 1969, girls were allowed to join the organization. In 1988, the name was changed from the Future Farmers of America to the National FFA Organization. That same year the program name was changed from Vocational Agriculture to Agricultural Education. The name change was reflected on the FFA emblem.

Today, the National FFA and its state chapters are thriving and continue to educationally enrich the lives of academic students nationwide and uphold the mission described in the FFA Creed:

The FFA Creed

I believe in the future of agriculture, with a faith born not of words but of deeds—achievements won by the present and past generations of agriculturists; in the promise of better days through better ways, even as the better things we now enjoy have come to us from the struggles of former years.

I believe that to live and work on a good farm, or to be engaged in other agricultural pursuits, is pleasant as well as challenging; for I know the joys and discomforts of agricultural life and hold an inborn fondness for those associations which, even in hours of discouragement, I cannot deny.

I believe in leadership from ourselves and respect from others. I believe in my own ability to work efficiently and think clearly, with such knowledge and skill as I can secure, and in the ability of progressive agriculturists to serve our own and the public interest in producing and marketing the product of our toil.

I believe in less dependence on begging and more power in bargaining; in the life abundant and enough honest wealth to help make it so—for others as well as myself; in less need for charity and more of it when needed; in being happy myself and playing square with those whose happiness depends upon me.

I believe that American agriculture can and will hold true to the best traditions of our national life and that

I can exert an influence in my home and community which will stand solid for my part in that inspiring task.

By E.M. Tiffany. Adopted at the Third National FFA Convention. Revised at the 38th and 63rd Conventions.

To that end, each state chapter includes a program of activities (POA) at the local level, which defines chapter goals, outlines steps in order to meet those goals, and acts as a written guide for events in the upcoming year pertaining to committee members, including administrators, advisory committees, alumni members, and other stakeholders. A well-planned POA includes divisions, or the types of activities a chapter conducts, and quality standards, which provide guidance for the planning and execution of those activities.

Chapter officers are a vital component of ensuring the success of chapter activities. These officers include the president, who presides over meetings, appoints committees, coordinates activities, and represents the chapter; the vice president, who works closely with the president to achieve chapter goals and may assume duties of president if necessary; the secretary, who is responsible for many administrative duties; the treasurer, who oversees the finances of the chapter and maintains treasury records; the reporter, who participates in public relation activities; the sentinel, who assists the president in official meetings and other activities; and the advisor, who supervises activities year-round and acts a leader and guide for other FFA members and stakeholders. Other chapter officers may include a historian, parliamentarian, and chaplain.

To learn more about the National FFA Organization, refer to the *Official FFA Manual and FFA Student Handbook*. Visit www.ffa.org.

In FFA, students may participate in agricultural mechanics or tractor operation contests. There are proficiency awards for farm and home safety and agricultural mechanics. Proficiency award winners receive significant monetary awards, depending on the level of the award. Additionally, agricultural mechanics may be used as a speech topic in any organization.

Members of **4-H** conduct many projects using agricultural mechanics skills. Some of these projects are wood science, electricity, tractor safety and maintenance, and automotive and small engines. The **Boy Scouts of America** has over 115 merit-badge areas, of which at least 25 involve agricultural mechanics skills. Some of these are camping, drafting, electricity, energy, engineering, farm mechanics, forestry, gardening, home repairs, landscape architecture, machinery, metalwork, plumbing, soil and water conservation, and woodwork.

Areas of Career Opportunities in American Agriculture

FIGURE 2-11 Areas of career opportunities in American agriculture.

MEETING THE CHALLENGE

Students having trouble choosing a specific career within agriculture should examine Figure 2-11. This may help them set career goals. Most importantly, students should talk with the teacher, prospective employers, parents, grandparents, counselors, and workers to learn more about appropriate careers. Answering the following questions may be helpful in gaining insight into a career choice.

- Do you want to feel pride in a job well done?
- Do you want to help others by doing a job for them they cannot do themselves?

Copyright © 2015 Cengage Learning®

The following is a list of questions that could be used when interviewing people about their occupation. This list is not complete and is intended to be used as a guide for developing questions.

1. Why did you pick this job?
2. How did you get started in your occupation?
3. How did you choose your place of training?
4. What educational, training, and other qualifications are there for the job?
5. If you should wish to change jobs, would the training contribute in any way?
6. Do you think this job would have a good future for me?
7. How could I get started in this career?
8. What is the salary range of this occupation?
9. What could a beginning person expect to make?
10. What are the fringe benefits?
11. Do you get paid vacations?
12. Do you have medical insurance?
13. Is there any chance of layoff? If so, how often?
14. What sort of planning does this business have for retirement?
15. What do you or don't you like about your job?
16. What are the advantages?
17. What are the disadvantages?
18. What are the hours and working conditions?
19. Do you ever have to work holidays? If so, which ones?
20. Do you ever work on weekends?
21. Is there a special uniform you must wear, or are you free to wear what you want? Does the company provide the uniform or does the employee?
22. What tools do you need?
23. Do you have to provide your own tools and equipment?
24. What are the physical requirements?
25. What do you do in this occupation?
26. How much traveling is involved?
27. What kinds of people do you work with?
28. Are there chances for advancement?
29. What are your responsibilities?
30. Do you belong to a union?
31. What's a typical day like for you in this job?
32. Is there any on-the-job training?
33. Has there ever been a time when you couldn't stand your job? If so, why and when?
34. Do you have to move if the company does?
35. What work experience did you have before you started to work in this occupation?
36. Who depends on your work? Upon whom do you depend?
37. Are there opportunities for advancement in this job? If so, what are the requirements for advancement?
38. How does your job affect your personal life?
39. What kinds of people do you meet?
40. Do you work mainly with people or things?
41. Do you work a lot with ideas?
42. Does your job offer opportunities to be creative?
43. Are people with your skills usually needed—even when business may be bad?
44. Is your work at all seasonal?
45. Could you briefly describe the personal qualities a person would need to do your job—strength, height, agility, ability to think rapidly, ability to make decisions, ability to deal with other people, etc.?
46. Would you recommend this kind of work for your children?
47. How do you spend your time after work?
48. If you could have any job in the world, what would you like to be?
49. Do you still go to school for special training?
50. When are people promoted? When are people fired?

FIGURE 2-12 Some questions for career exploration interviews.

Copyright © 2015 Cengage Learning®

- Do you want to keep machinery operating efficiently, from soil preparation through harvesting?
- Do you want to work on engines, build buildings, operate equipment, or install electrical equipment?
- Are you interested in welding, hydraulics, soil management, water management, or use of tools?

Agricultural mechanics is an area that can lead to great specialization, or it may provide supplemental skills for most agriculture-related careers. It can lead to business and career opportunities locally or it can lead to international opportunities. Agricultural mechanics skills may be helpful in many career areas, or may be absolutely necessary, depending on the career choice. Many areas in agricultural mechanics offer certification or licensing that requires added training. For example, the welding industry offers nine different areas in which you can be certified. These include Certified Welder, Certified Welding Instructor, Certified Robotic Arc Welder and more. Electricians and plumbers can earn a license that tells potential employers they are qualified.

Students planning a career in this field need to get prepared. To get prepared, students should:

- visit with people who have jobs in agriculture that use mechanical skills
- prepare a list of questions to ask workers about their jobs (Figure 2-12)
- talk to as many people as possible about agricultural mechanics
- plan to learn every skill possible in school, at home, and on the job
- learn what and why in the classroom
- learn how through shop and laboratory activities

You may plan to go into business for yourself and be an employer. Or, you may wish to work as an employee. Either way, it will be helpful to develop good work and management skills. The experts say one needs the proper schooling, good grades, a serious work attitude, good communication skills, and leadership competencies to do well in today's competitive job market and business world.

SUMMARY

The industry of agriculture has many career opportunities. The areas are broad and diverse, ranging from production agriculture to sales and service. The most widely available jobs in the industry will be in the sales and service segments. Although the number of individuals needed in production agriculture is declining, many areas of sales and service are growing. Agriculture will always be needed to supply food and fiber for people. Because of this, there will always be careers for people in agriculture. Due to the wide range of opportunities, there is probably a career for you that will make the best use of your interests and abilities.

Student Activities

1. Define the Terms to Know in this unit.
2. Check the *Dictionary of Occupational Titles* (www.occupationalinfo.org) for agricultural mechanics job titles that interest you.
3. Ask your school guidance counselor to help you use the career information systems available in your school. Ask about computerized and/or microfiche systems. Use these or other systems to find information on three or more job titles of your choice in agricultural mechanics.
4. Use these sources and the *Occupational Outlook Handbook* (www.bls.gov/ooh) to prepare a report on the job in agricultural mechanics that interests you the most. Include the following parts in your report: (a) job title, (b) outlook for the future, (c) working conditions,(d) personal and educational requirements for success, (e) certification requirements, (f) special advantages of the job, and (g) hazards and other disadvantages.
5. Shadow a person who owns his or her own business in agricultural mechanics. Examples include small engine repair shop, tractor dealership, tool sharpening business, construction business that builds barns and other agricultural business, and a fence building business.

6. Write a report telling how owning a business differs from just working as an employee. Also prepare a list of competencies that an entrepreneur should have.

7. Consult the Yellow Pages of your local phone book or use an Internet search engine. Record the names, addresses, and phone numbers of all agencies that hire people in the job of most interest to you.

8. Use the classified ads section of a large newspaper. Mark all jobs mentioned in which a knowledge of agricultural mechanics would be helpful.

9. Visit a business that hires people with agricultural mechanics skills. Ask the employer and some workers about the working conditions, travel requirements, employee benefits, skills needed, salary, and other items of interest to you.

10. Choose a specialty area such as welding, electricity, or plumbing. Research requirements for certification in that field.

Relevant Web Sites

National FFA Organization, web page on Career Development Events (CDE)
www.ffa.org/programs/awards/cde/Pages/default.aspx

Boy Scouts of America
www.scouting.org

National 4-H Council
www.4-h.org

Dictionary of Occupational Titles
www.occupationalinfo.org

U.S. Department of Labor, Bureau of Labor Statistics, Occupational Outlook Handbook
www.bls.gov/ooh

Self-Evaluation

A. Multiple Choice. Select the best answer.

1. America's number one employer is
   a. agriculture
   b. chemicals
   c. oil
   d. steel

2. The estimated number of agriculture-related jobs in the United States is
   a. around 800
   b. in the thousands
   c. many millions
   d. not significant

3. The first level of employment in agriculture is
   a. skilled
   b. semiskilled
   c. professional
   d. laborer

4. The National Center for Educational Statistics classifies agricultural jobs by the number(s)
   a. 01
   b. 02
   c. 03
   d. all of these

5. The average American uses only _____ percent of his or her income for food.
   a. 15
   b. 5
   c. 9.8
   d. 20

6–10. Categorize each of the following occupational divisions within its area.
   Areas:
   a. Agribusiness and Agricultural Production
   b. Agricultural Sciences
   c. Renewable Natural Resources

Occupational Divisions:

    6. food sciences _____

    7. wildlife management _____

    8. agricultural mechanics _____

    9. forestry and related sciences _____

    10. horticulture _____

B. Matching. Match the job classifications in column I with the job titles in column II.

Column I

    1. agricultural electrification, power, and controls

    2. agricultural mechanics, construction, and maintenance skills

    3. agricultural power machinery

    4. agricultural structures, equipment, and facilities

    5. soil and water mechanical practices

Column II

    a. farmstead planner

    b. electrician assistant

    c. machinery setup person

    d. safety specialist

    e. ecology technician

    f. tractor mechanic

    g. water control manager

    h. agriculture structure contractor

    i. machinery and equipment repair person

    j. safety technician

C. Completion. Fill in the blanks with the word or words that will make the following statements correct.

    1. In terms of the number of people employed, agriculture ranks number _____.

    2. Organizations such as _____, _____, and _____ help young people develop agricultural mechanics skills.

    3. A group of occupations or jobs within a cluster that require similar skills is called an _____ _____.

    4. Agricultural will always be needed to supply _____ and _____.

    5. The number of individuals needed in production agriculture is _____, but many areas of sales and service are _____.

D. Brief Answer. Briefly answer the following questions.

    1. Why is it a good idea to set one's career goals while in high school?

    2. According to experts, what does one need to succeed in today's competitive job market and business world?

    3. What are some things a student should do to prepare for a career in agricultural mechanics?

Supervised Agricultural Experiences (SAEs) in Agricultural Mechanics

Supervised agricultural experiences are teacher-supervised, individualized, hands-on, student-developed projects that give students real-life experience.

SAEs in Agricultural Mechanics may involve Student Ownership of an individual agricultural mechanics enterprise. Examples may include:

- a tool-sharpening business
- a lawnmower repair shop
- a welding shop

Placement in an agricultural mechanics–related job. Examples may include:

- working at a tractor repair shop
- working as an electrician's aid
- working in an electric motor repair shop

Completing an agricultural mechanics–related research project. Examples may include:

- testing the strength of different concrete mixtures
- experimenting with the decay resistance of different types of wood
- researching the effect of pipe size on water pressure and flow

USING THE AGRICULTURAL MECHANICS SHOP

UNIT 3
- Shop Orientation and Procedures

UNIT 4
- Personal Safety in Agricultural Mechanics

UNIT 5
- Reducing Hazards in Agricultural Mechanics

UNIT 6
- Shop Cleanup and Organization

UNIT 3
Shop Orientation and Procedures

Objective

To recognize major work areas and use safe procedures when working in an agricultural mechanics shop.

Competencies to be developed

After studying this unit, you should be able to:

- Identify major work areas in an agricultural mechanics shop.
- State school policies regarding shop procedures.
- State general safety precautions regarding the shop.
- Inform parents of shop policies and procedures.
- Sign a shop policy and procedures statement and a physical problems statement.
- Have parents sign a shop policy and procedures statement and a physical problems statement.
- Submit a signed policy and procedures statement and physical problems statement to the teacher.

Materials List

- Tape
- Drawing paper
- Pencil, eraser, and ruler
- Shop policies and procedures statement form
- Physical problems statement form

Terms to Know

- hands-on experience
- electricity
- scale drawing
- plan reading
- farmstead layout
- concrete
- supervised agricultural experience
- safe
- efficient
- policy
- procedure

THE SHOP AS A PLACE TO LEARN

The agricultural mechanics shop is a wonderful place to learn. Such shops may be found in schools, on farms, and in other agricultural businesses. Some people have shops in their basement or in special buildings. Good lighting, adequate electrical power, proper ventilation, and grounded electrical outlets are essential for safety and efficiency in all shop areas. A sturdy bench and vise are basics for the shop (Figure 3-1). Some people have tools and equipment that are easy to move from place to place. Large equipment like table saws and band saws can be put on rolling bases that allow movement of the saw and have locking wheels that prevent it from rolling when in use. Large toolboxes on rollers are used to store tools. The box can be easily moved to work on a project (Figure 3-2).

Sometimes students and teachers refer to the agricultural mechanics shop in school as the ag mech shop

FIGURE 3-2 A large toolbox may hold frequently used tools, fasteners, and lubricants. This toolbox is built on rollers for portability.

or simply the ag shop. Students learn many things that help them now and in the future. Important skills for now include the use of basic hand tools and power equipment (Figure 3-3). Most teenagers and young adults find many uses for tools, such as repair of bicycles, automobiles, stereos, appliances, garden equipment, and engines. Agriculture students use tools to help them in their agricultural projects and agricultural experiences.

The modern agricultural mechanics shop has facilities for the serious student to learn skills that are useful for a lifetime. Many adults who are agricultural education graduates learned mechanical skills when in high school. They now use these skills at home, in their businesses, and for leisure.

FIGURE 3-1 A sturdy bench and vise are basics for a shop.

FIGURE 3-3 An agricultural mechanics instructor demonstrates the proper use of a table saw.

Agricultural Shop Operations

Most agricultural mechanics shop operations fall under several broad categories:

- selection, care, and correct use of hand tools and power equipment
- woodworking and carpentry
- use of sheet metal (metalworking)
- handling and fastening structural steel, including welding
- pipe selection and fastening (pipe fitting)
- rope work, such as tying knots and splicing
- machinery maintenance and repair
- painting and finishing
- electrical wiring, motors, and applications
- hydraulic and pneumatic applications

Many of these items deal with the use of materials for constructing equipment or buildings used in agriculture. Therefore, a large amount of space in a shop is used for material storage.

Agricultural Power Machinery

The agricultural mechanics shop should have large, open spaces for project construction and machinery repair (Figure 3-4). Students typically build projects to develop skills. They also bring equipment from their homes, farms, or places of employment and repair it as a way of learning through hands-on experiences (Figures 3-5 and 3-6). **Hands-on experience** simply means learning by doing an operation rather than just reading or talking about it. It is very important,

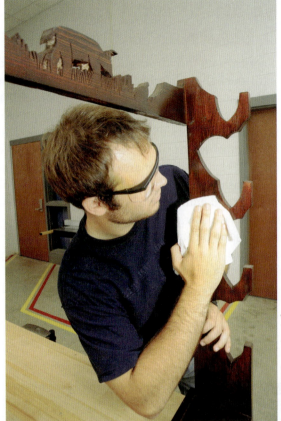

FIGURE 3-5 Many skills and techniques are required to build this gun rack.

FIGURE 3-6 An agriculture education student completes a tractor tune-up.

FIGURE 3-4 The agricultural mechanics shop must have large, open areas for students to work on large projects.

however, that reading, discussion, and demonstrations precede hands-on experiences for most activities.

Agricultural power machinery experiences include the selection, management, adjustment, operation, maintenance, and repair of:

- gas and diesel engines
- trucks
- tractors
- field machines
- feed-handling equipment
- crop storage and crop-handling equipment
- special machines such as those used in horticulture
- other mechanical devices used in agriculture

Agricultural Electrification

American agriculture is a very efficient industry. Because of this efficiency, the average American spends only 9.8 percent of income on food. American agriculture is more efficient than that of most other countries. This is due in part to the use of energy to operate machines.

Electricity is a form of energy that can produce light, heat, magnetism, and chemical changes. By using the energy of magnetism, engineers have developed electric motors and controls. Similarly, they create high-voltage current to provide spark for gasoline engines.

The agriculture student can benefit much from a knowledge of electricity and its use. This includes the use of electricity in the home, on the farm, and in all agricultural settings. Knowledge of electricity includes the selection, installation, and maintenance of wiring and electrical equipment. Basic wiring skills are useful throughout the farmstead and the home (Figure 3-7). Many of these skills may be learned in the agricultural mechanics shop.

Agricultural Buildings and Equipment

Many people in agriculture are around buildings and equipment in their work. This is true whether the person is an aquaculturalist, crop farmer, nursery operator, greenhouse operator, agricultural banker, or agriculture teacher. Many agricultural businesses are family businesses. Many of the construction and maintenance procedures on buildings and equipment are done by the owner, family members, or employees (Figure 3-8). Therefore, agriculturists should have a thorough knowledge of building construction and repair and maintenance of equipment.

© iisafx. Image from BigStockPhoto.com

FIGURE 3-7 Basic electrical wiring skills are useful throughout the farmstead and the home.

Scale Drawing and Plan Reading. Most agricultural mechanics projects are too large to be drawn actual size. Therefore, drawings must be scaled down (reduced) to make the dimensions proportionately smaller so the project can be planned on paper. **Scale drawing** is object representation on paper and uses a smaller dimension to represent a larger dimension, e.g., ¼ inch on the paper equals 1 foot on the object, or ¼" = 1'. This means each foot of the actual object is represented by ¼ inch in the drawing. The dimension ratio used must be indicated on the drawing. The actual dimensions of the object are stated on the drawing.

Copyright © 2015 Cengage Learning®.

FIGURE 3-8 Equipment operators are frequently called upon to maintain and repair their machinery.

Scale drawings can also be made of a part or object that is too small to show details clearly. The actual dimensions of the part or object are increased on the drawing. A scale drawing that states 2" = 1" means that the each inch of the actual object is represented by 2 inches in the drawing.

Students build projects from plans developed by others, or they design their own projects themselves. Either way, they need to know about scale drawing and plan reading.

Plan reading simply means using the scale drawing to build the project from the information given. The ability to do scale drawing and plan reading is basic to all construction (Figure 3-9).

Farmstead Layout. **Farmstead layout** refers to the efficient arrangement of buildings on a farm. Although arrangement has some relationship to construction, students typically study farmstead layout in the classroom rather than in a shop. Mistakes on a plan can be corrected quickly and easily. However, once construction is completed, mistakes are difficult and expensive to correct.

Functional Requirements of Buildings. Functional skills are needed to plan, install, or repair water systems, irrigation equipment, sprinkler systems, drainage pipe, and sewage systems. This requires the agriculturalist to be a junior engineer or a jack-of-all-trades to keep operating costs down.

Concrete. **Concrete** is a mixture of portland cement, water, sand, and coarse aggregate (stones). Concrete is used in agriculture for patios, landscaping features, feeding floors, walks, roads, bridges, holding tanks, and other projects. Concrete can be poured to make objects or structures of nearly any shape. It is hard, durable, and resistant to abuse. Hence, a lot of concrete is used in agriculture (Figure 3-10).

Soil and Water Management

Mechanical skills needed for soil and water management are generally developed in the classroom or field. These include land leveling, land measurement, mapping, drainage, irrigation, terracing, contouring, and strip cropping (Figure 3-11). However, some of the

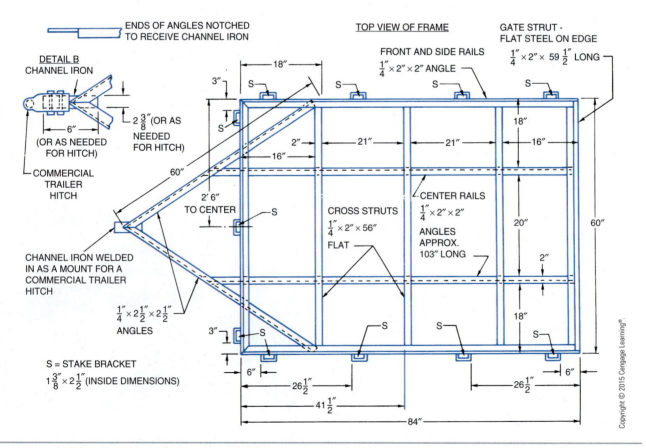

Copyright © 2015 Cengage Learning®.

FIGURE 3-9 A plan or blueprint provides workers with information on materials, dimensions, and details of construction.

FIGURE 3-10 After pouring concrete, the worker must smooth the surface.

basic shop skills are useful in learning soil and water management practice.

Briefly, the major work areas in a modern agricultural mechanics shop are:

- tool storage
- materials storage
- woodworking
- finishing
- metalworking (including welding)
- electricity
- machinery repair
- spray painting

FIGURE 3-11 Bulldozers are commonly used to construct terraces to control the flow of water on slopes.

In some schools, one or more of these areas are combined, such as tool storage and materials storage. In other schools, these areas are subdivided. For example, metalworking may be in three or four areas, such as cold metal, hot metal, sheet metal, and welding.

Work areas in the shop are generally not divided by definite barriers. Major walkways are often identified with strips of vinyl tape or painted lines. This helps students identify work areas and keep walkways clear of tools and materials.

Large, open areas should be available inside the agricultural mechanics shop for students to build large projects and repair machinery that they use in their supervised agricultural experience.

The term **supervised agricultural experience** (SAE) refers to activities the student does outside the agricultural class or laboratory to develop agricultural knowledge and skills (Figures 3-12 and 3-13). Refer to Appendix C for more information on the SAE program.

FIGURE 3-12 Students can learn occupational skills such as parts sales outside the classroom through a supervised agricultural experience (SAE).

FIGURE 3-13 Students built this project as part of their supervised agricultural experience (SAE).

Many school agricultural mechanics shops have fenced-in areas with concrete or blacktopped surfaces. Such areas are located just outside the shop and greatly increase the amount of space available for students to develop valuable shop skills (Figure 3-14).

POLICIES FOR SAFE PROCEDURES IN THE SHOP

The school shop is similar to well-equipped shops used by contractors, farmers, or other agricultural businesses. Such home or farm shops must be equipped with basic hand tools, power equipment, and materials storage. Most procedures recommended for safe operation of a school shop also apply to home shops. Students are encouraged to think in terms of both situations as they study agricultural mechanics shop procedures and organization.

Shop Size

The school shop must be big enough for 20 or more students to work at one time. Home and farm shops generally need work areas for only several people at one time. To be safe and efficient, the shop must be large enough to meet reasonable standards for space for each person in the shop. **Safe** means free from harm or danger. **Efficient** means able to produce with a minimum of time, energy, and expense. Even with the recommended amount of space per pupil, learning

FIGURE 3-14 Large wood and metal projects are frequently constructed in areas outside the school shop.

opportunities and personal safety are decreased without careful management.

Proper Instruction

Students must be properly instructed in the correct and safe way to use each tool. To avoid crowding, which creates serious hazards, students must be spread throughout the shop. The instructor generally has students working in different shop areas during the shop period.

Safety Policies and Procedures

Every school shop should have policies and procedures to improve instruction and promote student safety. A **policy** is a plan of action or a way of management. A **procedure** is a method of doing things or a particular course of action. Figure 3-15 gives an example of

SHOP POLICIES AND PROCEDURES

Safe Conduct and Dress. I AGREE TO:

1. Purchase school insurance or bring a note signed by my parent(s) verifying that I have suitable personal accident and medical insurance.
2. Inform my teacher of any allergies or handicaps before using the shop.
3. Occupy my assigned place at the beginning of the period.
4. Wear safety glasses at all times when in the shop, use special goggles as needed for handling chemicals and caustic materials, and wear shields or helmet when conducting hazardous operations.
5. Wear prescribed protective clothing and not wear long neckties, loose sleeves, or other loose clothing.
6. Practice general cleanliness and orderliness at all times.
7. Not wear finger rings when handling molten metal or working on electrical systems.
8. Not loiter in the shop.
9. Not throw objects in the shop.
10. Avoid "horseplay" at all times in the shop.
11. Report all accidents including minor cuts, scratches, and splinters to the instructor promptly.
12. Remain in the shop or classroom at all times except when excused by the instructor.
13. Help with shop cleanup and storage duties until the job is finished.

Safe Machine Use. I AGREE TO:

1. Never operate power equipment unless specifically authorized.
2. Never operate machines unless the guards are in place.
3. Use all tools for the purpose intended and in the approved manner as taught by the instructor.
4. Stand to the side of grinding wheels, buffers, and blades while the machine is gaining speed, and out of the fire path when furnaces and other burners are being started.
5. Remove keys from chucks when adjustments are complete.
6. Fasten all work securely before drilling, milling, sanding, and other such operations.
7. Avoid talking to or otherwise distracting others using machines or doing hazardous activities.
8. Not leave machines unattended, nor repair, oil, clean, or adjust them while they are running.
9. Not force machines beyond the capacity for which they were designed.
10. Report to the teacher all electrical equipment that is not properly grounded or otherwise safe to use.
11. Report to the teacher all tools and machinery in need of repair, and any hazards that I observe.

Safe Materials Handling. I AGREE TO:

1. Obtain the teacher's permission and read the label before using any pesticides or other chemicals.
2. Handle, use, and store pesticides and other chemicals properly.
3. Not use gasoline as a cleaner or solvent, and to handle all fuels in the prescribed ways.
4. Not permit water to be poured into acid or molten metal, nor permit grease or oil to come into contact with molten metal.
5. Handle and store all soiled materials, rags, paints, solvents, and flammable materials in the approved manner.
6. Lift or move heavy objects in the approved manner only.

I have read the above rules and discussed them in class with my instructor. I realize they are for my protection and I will do all I can to see that they are enforced. I will observe all precautions given by my instructor or others assigned to supervise my participation in school and school-related activities.

Signed _____ / _____

               student's signature   date

_____ / _____

               parent's signature   date

Copyright © 2015 Cengage Learning®.

FIGURE 3-15 A shop policies and procedures statement.

a good policies and procedures statement for an agricultural mechanics shop. The instructor may add to or take away from this list according to established requirements in the school. It is suggested that the student discuss every item in the instructor's policy and procedures list in detail with the instructor and parents. Students and parents should sign the statement to indicate complete understanding. The signed statements may then be filed with other student records at the school.

Another important procedure to help protect students in the shop is to fill out and sign an allergies and physical problems statement (Figure 3-16). Parents should sign these statements also to ensure proper medical treatment can be given in case of accident or other emergency.

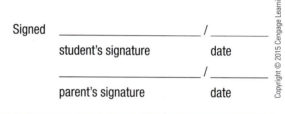

ALLERGIES AND PHYSICAL PROBLEMS STATEMENT

The agriculture teachers and others responsible for providing treatment for me in case of an emergency should be aware of the following:

1. I am allergic to _____

_____

2. I should not be given the following medication

_____

3. I have the following physical or personal problems

_____

4. My family physician is _____

5. My physician's phone number is _____

6. My parents' names are _____

7. My parents' phone numbers are _____

Signed   _____ / _____

              student's signature   date

           _____ / _____

              parent's signature   date

Copyright © 2015 Cengage Learning®.

FIGURE 3-16 An allergies and physical problems statement.

SUMMARY

The agricultural mechanics shop is a valuable asset to any agricultural enterprise. It should be well planned, laid out properly, and well maintained. Areas should be divided into major work areas, and specific tools should be stored in the appropriate place. Essential to any mechanics shop is a well-drawn plan for safety. Proper storage, marking, and equipping for safety should be fundamental to any shop procedure.

Student Activities

1. Define each of the Terms to Know in this unit.
2. Measure the length, width, and other major dimensions of your agricultural mechanics shop. How many square feet are available per pupil in your class?
3. Make (or ask your teacher to provide) a drawing of your shop and any fenced-in area that may be used by students. Label the tool storage, lumber storage, metal storage, woodworking, metalworking, plumbing, electrical, and project areas.
4. List five projects that other agriculture students have constructed in the shop.
5. List five projects that you would like to make in the shop.
6. Using Figure 3-15 as a guide, create a poster for your shop wall reminding students of appropriate work habits and safety measures in the shop.
7. Read, discuss with your parents, sign, and obtain your parents' signatures on the policies and procedures statement for your shop. Return the signed document to your teacher.
8. Fill out, sign, and have your parents sign a personal allergies and physical problems statement for you. Return the signed statement to your teacher.

Self-Evaluation

A. Multiple Choice. Select the best answer.

1. The agricultural mechanics shop is also referred to as the
   a. ag mechanics shop
   b. ag mech shop
   c. ag shop
   d. all of these

2. The agricultural mechanics shop is a good place for serious students to learn skills that are useful
   a. at home
   b. in their businesses
   c. for leisure
   d. all of these

3. Agricultural mechanics includes
   a. woodworking and carpentry
   b. metalworking and welding
   c. pipe fitting and irrigation
   d. all of these

4. *Hands-on experience* means
   a. made by hand tools only
   b. a process of learning by doing
   c. a procedure requiring many people to help
   d. a wasteful method of education

5. Large open spaces are needed in agricultural mechanics shops as compared with other shops for
   a. safe operation of stationary power equipment
   b. meeting fire code regulations
   c. storing materials for the school custodians
   d. student project work

6. Safety in the shop depends on
   a. students
   b. teacher(s)
   c. school shop designers
   d. all of these

7. Safety in the shop depends on
   a. students staying in assigned areas
   b. students wearing proper clothing
   c. use of safety glasses by all persons in the shop
   d. all of these

8. Which of the following enhances student safety in the shop?
   a. shop cleanliness and orderliness
   b. proper instruction
   c. machines that are kept in adjustment
   d. all of these

B. Matching. Match the definitions in column I with the correct term in column II.

Column I

1. energy source for agriculture

2. construction, repair, and maintenance

3. irrigation and drainage

4. woodworking, metalworking, welding, and pipe fitting

5. efficient arrangement of buildings

6. portland cement, water, sand, and stone

7. selection, management, adjustment, operation, maintenance, and repair

8. object representation on paper

Column II

a. agricultural shop operations

b. agricultural power machinery

c. agricultural electrification

d. agricultural buildings and equipment

e. scale drawing

f. farmstead layout

g. concrete

h. soil and water management

C. Completion. Fill in the blanks with the word or words that make the following statements correct.

1. Activities the student does outside the agricultural class or laboratory to develop agricultural skills are referred to as the _____ _____ _____ or SAE.

2. The use of a scale drawing to build a project from the information given is called _____ _____.

3. Every school shop should have _____ and _____ to improve instruction and promote student safety.

4. Agriculturalists should have a thorough knowledge of building _____, and repair and _____ of equipment.

5. _____ is a form of energy that can produce light, heat, magnetism, and chemical changes.

UNIT 4

Personal Safety in Agricultural Mechanics

Objective

To interpret safety colors and codes, protect the body against injury, and work safely in agricultural mechanics settings.

Competencies to be developed

After studying this unit, you should be able to:

- State how to create a safe place to work.
- Recognize hazards in agricultural mechanics.
- List the types of parts and areas identified by various safety colors.
- Describe what each safety color means.
- Select appropriate protective clothing and devices for personal protection.

Materials List

- Examples of shop protective clothing and devices

Terms to Know

- safety
- safety color
- focal color
- noise intensity
- noise duration
- decibel

Students sometimes grow weary of teachers, parents, and others who constantly remind them to work and play safely. Yet, far more people are injured each year than need be to carry on the work of society (Figure 4-1).

Most people react to injury with anger at themselves or others who were responsible. This is because many injuries could have been avoided. People are willing to accept severe pain and hardship if they must, but they are resentful if they could have escaped the loss by being more careful.

Injury and disability are troublesome problems in society. Not only does the victim suffer, but others suffer as well. Friends of the injured are frightened and inconvenienced. Parents, guardians, spouses, and others frequently must take off work for visits to the doctor's office or hospital. They often forfeit income and must pay for extra expenses. Therefore, it is well worth the time to create safe places to work, and to learn safety in everything. **Safety** means freedom from accidents. Some common causes of accidents are indicated in Figure 4-2.

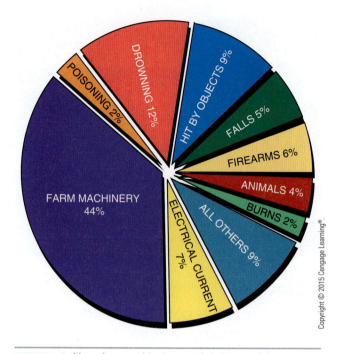

FIGURE 4-3 Many farm accidents are related to machinery.

Many accidents leave the victim partially or totally disabled. This carries a lifetime of regret. Many accidents are fatal. This results in needless loss of loved ones and bereavement to those left behind. Obviously, the thinking person will make every reasonable effort to work safely. This helps avoid the hardships caused by accidents. Accidents in farm populations may involve machinery, drownings, firearms, falls, falling objects, burns, and other dangers (Figure 4-3).

SAFETY IN THE WORKPLACE

Work in agricultural mechanics involves extensive contact with tools and machinery. Therefore, workers should be especially aware of the hazards that exist and take special precautions as needed. By taking the following precautions, a safer workplace can be created:

- Install all electrical devices according to the *National Electrical Code*®.
- Install all machinery according to the manufacturer's specifications.
- Keep all tools and equipment adjusted or fitted according to specifications.
- Use tools and equipment skillfully.
- Provide proper storage for tools, materials, fuels, chemicals, and waste materials.
- Keep work areas clean and free of tools, materials, grease, and dirt.
- Keep moving parts properly shielded (Figure 4-4).

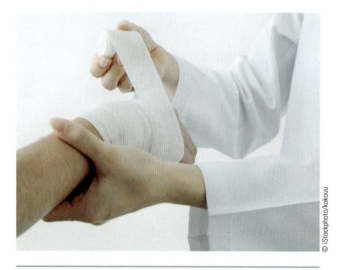

FIGURE 4-1 Far too many people are injured every year.

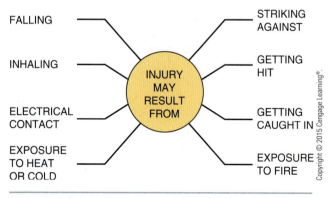

FIGURE 4-2 Some common causes of accidents.

is determined by the employer, but it is defined and driven by worker-safety legislation and enforced by state and federal agencies.

Employers frequently focus on two related approaches for keeping employees safe from harm in the workplace. The work area is carefully evaluated for safety concerns. This assessment is followed by correcting hazardous conditions and targeting safety training to employees based upon how potential hazards intersected with their specific duties. Every employee should be trained in all aspects of the work environment in which he or she might reasonably be expected to be engaged.

The second approach to employee safety is to provide basic first-aid training to each employee and specialized training such as Emergency Medical Technician (EMT) training to enough individuals to ensure that trained personnel are located nearby whenever the workplace is in use. Specialty training goes beyond basic first aid to provide expertise for responding to serious trauma. For example, such training might include the use of a heart defibrillator (functions to reset heart rhythm back to normal).

Both basic first-aid training and the more advanced levels of emergency response training are provided by professional instructors who are approved to offer certifications to workers who successfully complete training programs. Certifications from an accredited training organization may be required by state and federal governments to protect workers. In addition, the issue of liability for employee safety is a major concern of employers. Work environment inspections and training for emergency responses are important ways to limit employer liability and increase worker safety.

No amount of safety training can take the place of people who know what to do when an injury accident occurs. Life-threatening accidents and incidents can, and do, happen wherever people are located. For example, safety training is not likely to prevent a heart attack or stroke from happening to someone who is at work, but an employee who is on-site with first-aid training could save a life. For this reason, many employers provide first-aid training to at least some of their employees.

One of the most important things you can do when an incident occurs is to call for professional help first. It is critical that medical help arrives as soon as possible, so the person who administers first aid should make sure that someone calls for help immediately. First aid is limited care that lasts only until more qualified help arrives. Once a first-aid procedure is started, this level of care should be continued until the patient can be placed in the care of medical personnel such as a doctor or paramedic.

Copyright © 2015 Cengage Learning®.

FIGURE 4-4 Guards or shields should be kept on all moving parts. Note the guard on the saw blade.

- Manage all situations to avoid the likelihood of falling objects.
- Avoid areas where objects may fall.
- Avoid the flight path of objects that could be thrown by machines.
- Protect eyes, ears, face, feet, and other parts of the body with protective clothing and devices.
- Move slowly enough to avoid creating hazards to self and others.
- Read and heed all precautions.

These precautions help make the workplace safe. However, each person should have insurance to cover any personal injury or property damage that may occur.

First Aid

First aid is care given to a victim immediately following an injury or the onset of illness. First-aid training is a critical need in every work setting because such incidents often occur without warning. First-aid policy

Basics of First Aid

Before rendering aid to a distressed person, it is important to quickly and accurately assess the situation. Do not place yourself in danger by entering an unsafe area too quickly. For example, electrical shock, hazardous materials, danger of collapse, unsafe machinery, and other hazards may claim the lives of those who fail to evaluate the cause of injury or illness to someone else.

Pause to ask questions and to observe danger signs. Moving someone with a neck or spine injury may cripple them for life or even contribute to death. Of course, there are times when an injured person must be removed to a place of safety (fire danger, dangerous fumes). Keep in mind that we should avoid causing greater harm to the patient.

Once the decision has been made to administer first aid, one must prioritize the order of care:

A. Airway—make sure the airway is not obstructed.
B. Breathing—determine whether breathing has stopped and administer aid if there is a breathing problem.
C. Circulation—identify and treat sources of bleeding and symptoms of shock.
D. Disabling injury—check for burns, broken bones, or similar injuries.

A person who is not breathing must be given resuscitation immediately. Current practice is to perform cardiopulmonary resuscitation (CPR). It would be a mistake to first treat the accident victim by immobilizing a broken bone before restoring the circulation of oxygen-rich blood to the brain and heart. It is critical to remember that the greatest threat to life should be treated first.

Once the injury victim is breathing and bleeding has been stopped, other injuries should be cared for while you wait for professional help. Make the patient as comfortable as possible, doing what you can to calm the patient and ease his or her pain. Do not leave the patient alone; keep someone with him or her until medical help arrives.

First-Aid Kits. Every shop and lab should have a first-aid kit to treat minor injuries. Injuries that are serious such as cuts that require stitches should be treated by trained medical personnel. The kit should be readily accessible in an area of the shop where there is nothing around the kit that would hinder access. The kit should be enclosed so that no dust or any other foreign material can get inside. Make sure you understand the fundamentals of first aid before treating any wound. The contents should include:

- an antiseptic such as alcohol to sterilize the wound
- sterile water to wash out the wound
- adhesive bandages of various sizes
- a roll of sterile gauze
- adhesive tape
- scissors
- antibiotic ointment
- burn gel for treating minor burns
- tweezers to remove splinters

SAFETY COLORS

National organizations have worked together to develop a safety color-coding system for shops. The American Society of Agricultural Engineers and the Safety Committee of the American Vocational Association have published such a code. In developing the code, these agencies drew upon materials published by the American National Standards Institute (ANSI), the U.S. Department of Transportation (DOT), the National Safety Council (NSC), and the Occupational Safety and Health Administration (OSHA).

Colors in the coding system are used to:

- alert people to danger or hazards
- help people locate certain objects
- make the shop a pleasant place to work
- promote cleanliness and order
- help people react quickly to emergencies

Each color or combination of colors conveys a specific message, based on a standard code (Figure 4-5).

Students need to memorize the message conveyed by each color; then the safety message will be

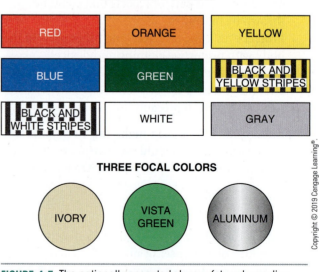

NINE SAFETY COLORS

RED	ORANGE	YELLOW
BLUE	GREEN	BLACK AND YELLOW STRIPES
BLACK AND WHITE STRIPES	WHITE	GRAY

THREE FOCAL COLORS

IVORY   VISTA GREEN   ALUMINUM

Copyright © 2019 Cengage Learning®.

FIGURE 4-5 The nationally accepted shop safety color-coding system uses nine safety colors and three focal colors to designate objects and areas.

FIGURE 4-6 Safety switches are colored red.

FIGURE 4-7 First-aid stations are marked in green.

understood in the shop. Shops must be properly painted for the color-coding system to make its proper contribution to shop safety and efficiency.

The following descriptions show how each **safety color** is used to convey a safety message and help students work safely in the shop. The colors are listed in Figure 4-5, and used as follows:

Red = Danger. Red is used to identify areas or items of danger or emergency, such as safety switches and fire equipment (Figure 4-6).

Orange = Warning. Orange is used to designate machine hazards, such as edges and openings. Orange is also used as background for electrical switches, levers, and controls.

Yellow = Caution. Yellow, like the amber traffic light, means be cautious. It is used to identify parts of machines, such as wheels, levers, and knobs that control or adjust the machine. Yellow and black stripes are used in combination to mark stairs, protruding objects, and other stationary hazards.

Blue = Information. Blue is used for signs if a warning or caution is intended. Such signs are made of white letters on blue background and carry messages such as "OUT OF ORDER" or "DO NOT OPERATE."

Green = Safety. Safety green is a special shade of green and indicates the presence of safety equipment, safety areas, first aid, and medical practice (Figure 4-7).

Black and Yellow Diagonal Stripes = Radioactivity. A yellow sign with a black fan-like symbol in the center is designated as the marking for radiation hazards (Figure 4-8).

FIGURE 4-8 Radioactive areas are marked with a yellow sign with a fan-like symbol in the center.

White. White is used to mark off traffic areas. White arrows indicate the direction of traffic. White lines also mark work areas around objects in the shop. Yellow may be used in place of white to mark areas and lanes.

White and Black Stripes. White and black in alternate stripes or checkers are traffic markings. An example of such use is to mark traffic-stopping barricades.

Gray. Gray is used on floors of work areas in the shop. It is a restful color and provides good contrast for other safety colors. It is used to paint body areas of machines and may be used on the table tops if painting is desired.

FOCAL COLORS

The nationally accepted shop safety color-coding system includes three focal colors. A **focal color** is used to draw attention to large items such as machines, cabinets, and floors. The focal colors provide contrast for the safety colors and create pleasant surroundings for people using the shop. The focal colors are ivory, vista green, and aluminum.

Ivory. Ivory is used to highlight or improve visibility of certain items. These items include tool storage chests, table edges, and freestanding vises and anvils.

Vista Green. Vista green is a special shade of green. It is used to paint bodies of machines, cabinets, and stationary tools such as vises. It is regarded as a pleasing color and contrasts with the safety colors.

Aluminum. Aluminum is used on waste containers such as those for scrap wood, scrap metal, and rags.

Some of these colors are used in combination with other colors to mark pipes, hoses, and vents. Such markings identify the material in the lines as oxygen, natural gas, compressed air, air at atmospheric pressure, and water.

A properly color-coded shop is attractive and pleasant. The coloring system helps guide the user safely through many enjoyable experiences in agricultural mechanics.

PROTECTIVE CLOTHING AND DEVICES

Most work situations in agricultural mechanics require some type of body protection. The eyes, ears, hands, arms, feet, and legs are easily injured (Figure 4-9). The best protection against injury is to prevent the accident

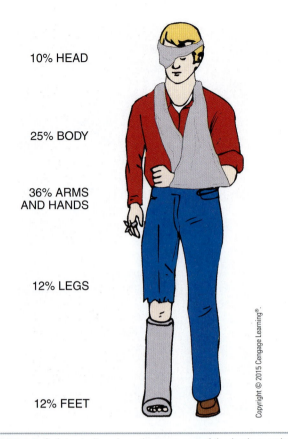

10% HEAD

25% BODY

36% ARMS AND HANDS

12% LEGS

12% FEET

Copyright © 2015 Cengage Learning®.

FIGURE 4-9 Safety studies show that any part of the body may be injured in the shop.

from happening. The next best approach is to protect the body where injury may occur. Both lines of defense are needed to minimize injuries.

Safety Glasses and Face Shields

The face and eyes are regarded as the most critical parts of the body to be protected. This is because the eyes are so easily damaged and the face is easily disfigured. Flying objects striking the head can easily cause blindness or result in death. Acids, caustic chemicals, fertilizers, pesticides, solvents, molten metal, and hot water are all dangerous materials. Those who work around these materials should shield themselves against the possibility of accidental contact with them.

Safety Glasses and Goggles. Safety glasses and goggles offer minimum eye protection and are the first line of defense for the eyes (Figure 4-10). Glasses and goggles should be the approved type with special impact-resistant lenses and side shields. They should fit the user's face and be kept clean for proper visibility. Single-piece goggles and cup goggles are used where special eye protection is needed against chemicals, flying

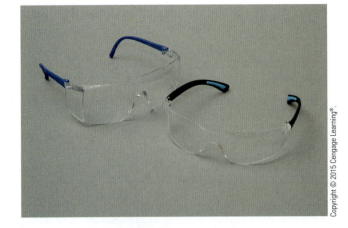

A

FIGURE 4-11 Wearing eye protection on your head does no good at all.

B

FIGURE 4-10 Safety glasses, goggles, or face shields (A) should be worn when working around materials, processes, and machines (B).

objects, or damaging light. Special shaded lenses are needed for work in welding.

Many states have special laws requiring students to wear safety glasses or cover goggles upon entering and participating in shops and laboratories in the school.

Note that eye protection is worthless unless it covers the eyes. Wearing them on your head does no good at all (Figure 4-11). Some states have laws requiring approved face shields when an individual is:

- participating in, or exposed to, the immediate vicinity where hot molten metal or solder is being prepared, poured, or used in any form
- participating in, or exposed to, the immediate vicinity where milling, turning, sawing, shaping, grinding, sanding, cutting, or stamping of any solid materials is taking place
- participating in, or exposed to, the immediate vicinity where heat treating, tempering, or kiln firing of any material is taking place

Approved cup goggles, helmets, or hand shields are required when in areas of gas welding, electric arc welding, or weld flash exposure. Every shop should be equipped with an eyewash station. If foreign material gets into the eyes, they can be flooded with water to dislodge the material (Figure 4-12). Prompt action can prevent damage to the eyes from chemicals or other foreign matter.

Students must always wear safety glasses as minimum protection against eye injury. When students are in special hazard areas, special eye and face shields must be used (Figure 4-13).

Hair Restraints. Serious accidents can occur if long hair becomes tangled in drill presses, saws, or other turning equipment. When hair is long or loose, it should be contained by wearing one of the following:

- a woolen hat
- a headband
- a hardhat
- a hairnet

FIGURE 4-12 An eyewash station uses water to flush foreign material from a person's eyes. In emergency situations, prompt action may help a person avoid serious injury.

Protective Clothing

No part of the body is safe from injury in shop accidents. Suitable protective clothing that fits properly helps to prevent or reduce injuries. There should be no cuffs, strings, or ties for turning machinery to catch. Clothing should be fire resistant and provide protection from scrapes and abrasions. Protective clothing must be easily cleaned. Keeping clothing clean keeps clothing more fire resistant.

Coveralls. Coveralls are the most versatile and all-around item of clothing for agricultural mechanics (Figure 4-14). Coveralls protect the arms, body, and legs. They can be buttoned or zipped to the neck for maximum protection. Coveralls should fit well and be easily removable. Coveralls are safer if they do not have sleeve or pant cuffs. Cuffs are hazardous because flying sparks can be caught and trapped, resulting in the garment catching fire. Bulky cuffs can catch in machinery and other objects.

Coveralls have the advantage of many pockets for pencil, pad, small tools, and objects the worker uses. The pockets should be covered with snap-down flaps to provide a smooth body cover that will not catch sparks and other objects. Coveralls should have an elastic waist, sleeves that button, and flaps that cover all buttons and zippers.

FIGURE 4-13 Clear face shields protect the entire face.

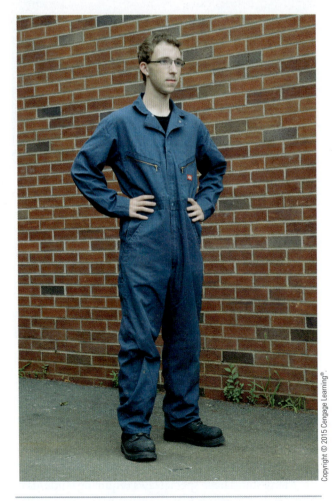

FIGURE 4-14 Coveralls provide the best protection when working in the shop.

When jackets, pants, or shirts are worn instead of coveralls, it is important to select clothing that fits snugly and has the same safety qualities as coveralls.

Aprons. Some shop teachers permit students to wear aprons in the shop or laboratory. Heavy cloth or leather aprons provide good protection for the front of the body and upper legs. Aprons are recommended only for limited shop work at benches or as additional covering over coveralls.

Vinyl or rubber aprons should be worn when liquids are used. Aprons are economical to buy and easy to store.

Shop Coats. Shop coats have the benefits of aprons plus additional body protection. The arms, body, and upper legs are protected for most work where the operator is standing. The shop coat is easy to put on. Therefore, it is used frequently by instructors and others who need to put on and remove protective clothing frequently throughout the day.

Footwear. Leather shoes with steel toes are recommended when working in the shop and when using machinery (Figure 4-15). Many workers prefer shoes with 6-inch tops; others prefer the higher, boot-type shoes. Leather is the preferred material for footwear because of its strength, durability, and comfort. The ability of leather to breathe explains why it is more healthful, cool, and comfortable than other materials. Farmers, ranchers, foresters, carpenters, plumbers, and all who handle heavy or hot objects need sturdy shoes. The popular soft, lightweight vinyl or canvas shoes that many wear for casual activities are not safe for agricultural work.

Rubber boots are needed when working in water or using pesticides. Rubber is more durable than leather in extremely wet conditions and offers better

FIGURE 4-16 Leather gloves provide good protection for the hands.

protection against wet feet. Rubber will not absorb pesticides and can be washed free of such materials.

Gloves. Gloves are used to keep the hands warm as well as to protect them from excessive abrasion, heat, liquids, or chemicals. Cloth gloves are suitable for warmth, but leather is needed where protection from heat, abrasion, or impact is needed (Figure 4-16). Only rubber or vinyl gloves are suitable where liquids or chemicals are involved.

Hardhats. Hardhats are needed when working where objects are above head level or flying objects could be encountered (Figure 4-17). Hardhats are made from special lightweight and impact-resistant materials. It is important that hardhats be approved by the Occupational Safety and Health Administration.

Masks and Respirators. Masks that cover the nose and mouth are needed to filter out particles of dust or spray paint (Figure 4-18). Such materials irritate

FIGURE 4-15 Leather shoes with steel-reinforced toes provide excellent protection for the feet.

FIGURE 4-17 Hardhats are essential protective gear in work environments where falling objects or other potential sources of head injury are a possibility.

FIGURE 4-19 Two types of hearing protectors are ear plugs (left) and earmuffs (right).

the nostrils and sinuses. Continuous inhaling of dust leads to lung diseases such as black lung and cancer. Effective dust masks are not expensive and should be worn when sanding, painting, welding, mixing soil,

FIGURE 4-18 Face masks should be worn when dust is encountered in the work area.

shoveling grain, or whenever dust is encountered. When working around inhalable fumes such as spray painting and applying pesticides, you should wear an approved respirator that will prevent the entrance of toxic fumes.

Respirators that cover the nose and mouth and contain special filters are needed for certain jobs. When using pesticides, it is important to use the specific type of respirator recommended by the pesticide manufacturer.

Earmuffs and Ear Plugs. Ear protection is recommended when working in certain types and levels of noise (Figure 4-19). Equipment such as the radial arm saw, planer, router, chain saw, tractor, and lawnmower can produce noise that may damage the ears and cause a hearing impairment.

Earmuffs or ear plugs are recommended when the intensity, frequency, or duration of noise reaches certain levels. **Noise intensity** refers to the energy in the sound waves. **Noise duration** refers to the length of time a person is exposed to a sound.

Distance has a great effect on sound pressure or intensity. A person standing 5 feet from a machine can reduce the sound pressure to 25 percent by moving away another 5 feet. The **decibel** (dB) is the standard unit of sound. The Occupational Safety and Health Administration (OSHA) has established a 90-dB noise level for an 8-hour period as the maximum safe limit. An 85-dB limit is safer. The sound level in decibels and duration of time is illustrated in Figure 4-20. A sound-level meter is used to determine noise levels.

Time is an important factor bearing on the effect of noise on hearing. The ears can stand loud noises for a few minutes. That same noise may damage the ears if exposed for longer periods of time.

DURATION OF TIME PERMITTED AT VARIOUS SOUND LEVELS

Duration per Day (in Hours)	Sound Level (in dB)
8	90
6	95
3	97
2	100
1½	102
1	105
½	110
¼ or less	115
none	over 115

DECIBEL (dB) LEVELS OF COMMON SOUNDS AT TYPICAL DISTANCE FROM SOURCE

Sound Level (in dB)	Common Sound
0	Acute threshold of hearing
15	Average threshold of hearing
20	Whisper
30	Leaves rustling, very soft music
40	Average residence
60	Normal speech, background music
70	Noisy office, inside auto traveling at 60 mph
80	Heavy traffic, window air conditioner
85	Inside acoustically insulated protective tractor cab in field
90	OSHA limit—hearing damage from excessive exposure to noise above 90 dB
100	Noisy tractor, power mower, all-terrain vehicle, snowmobile, motorcycle, inside subway car, chain saw
120	Thunderclap, jackhammer, basketball crowd, amplified rock music
140	Threshold of pain—shotgun, near jet taking off, 50-hp siren (100')

FIGURE 4-20 The Occupational Safety and Health Administration (OSHA) has established regulations governing the maximum safe levels of exposure to noise, in decibels, for specified periods of time. Workers should not be exposed to higher levels of noise for the time durations given without wearing hearing protection.

Source: Occupational Safety & Health Administration Regulations.

SUMMARY

Tools and machines in an agricultural mechanics shop are dangerous if improperly used. Each shop must be properly designed and organized to provide safety to workers. Each person using the shop should be aware of the proper use of all tools and machines and the inherent dangers of each. By using the proper precautions and observing all the rules of safety, injuries can be avoided.

Student Activities

1. Define the Terms to Know in this unit.

2. Interview the school nurse or medical assistant to determine what kinds of accidents have occurred in the school.

3. Search recent newspapers or the Internet for articles on home, shop, farm, or work accidents, and discuss them in class. Classify each accident or incident described in the articles as preventable or not preventable.

4. Make a bulletin board using newspaper clippings about home, shop, farm, or work accidents.

5. Make safety posters depicting the causes of accidental injury.

6. Contact your Cooperative Extension Service or other agencies and request current accident information regarding your community.

7. Study every machine and work area in your home shop or school shop, and list all hazards that should be corrected. If at home, correct them; if at school, assist your teacher in correcting them.

8. Examine your school agricultural mechanics shop for proper use of safety colors and focal colors. Ask your teacher what plans there are to correct any incorrectly painted areas.

9. Survey your agricultural mechanics shop for the following items: safety glasses for all students, goggles, face shields, welding helmets, protective clothing, respirators, hearing protectors, and fire extinguishers. Ask your teacher to acquire any of the items not on hand.

Relevant Web Sites

National Fire Protection Association
www.nfpa.org

The Essential Health & Safety Manual
www.osha-occupational-health-and-safety.com

Self-Evaluation

A. Multiple Choice. Select the best answer.

1. Accidents among farm workers most often involve
   a. burns
   b. drowning
   c. falls
   d. machinery

2. For safety purposes, moving parts on machines should be
   a. labeled
   b. oiled
   c. painted
   d. shielded

3. Color coding is used in the shop to
   a. alert people to dangers and hazards
   b. make the shop a pleasant place to work
   c. help people react quickly to emergencies
   d. all of these

4. Which of the following is **not** regarded as a major type of accident that causes injury?
   a. assault and battery
   b. electrical contact
   c. falling
   d. inhaling

5. The national organization(s) that helped to develop safety color coding is/are the
   a. American Society of Agricultural Engineers
   b. American Vocational Association
   c. National Safety Council
   d. all of these

6. The safety color used to identify wheels, levers, or knobs that control or adjust machines is
   a. red
   b. yellow
   c. orange
   d. none of these

7. Fire equipment and safety switches are indicated by the color
   a. orange
   b. purple
   c. red
   d. bright green

8. The number of safety colors in the shop color-coding system is
   a. nine
   b. eight
   c. seven
   d. four

9. The number of focal colors in the shop color-coding system is
   a. one
   b. two
   c. three
   d. four

10. Suitable eye protection must be worn when working with
    a. chemicals
    b. grinding machinery
    c. welding equipment
    d. all of these

11. Protective clothing used in the shop must
    a. be fire resistant
    b. fit properly
    c. be clean
    d. all of these

12. The best item of protective clothing for agricultural workers is
    a. an apron
    b. a shop coat
    c. jeans
    d. coveralls

13. The length of time a person is exposed to sound is called
    a. noise intensity
    b. noise duration
    c. decibels
    d. sound pressure

14. Hearing damage may occur if excessively exposed to noise above
    a. 30 decibels
    b. 60 decibels
    c. 75 decibels
    d. 90 decibels

B. **Matching.** Match the meaning in column I with the correct color code in column II.

Column I

1. indicates warning
2. indicates danger
3. indicates information
4. indicates radioactivity
5. indicates caution
6. indicates safety
7. indicates traffic areas
8. indicates traffic markings

Column II

a. yellow
b. red
c. green
d. blue
e. black and yellow stripes
f. orange
g. black and white stripes
h. white

C. **Completion.** Fill in the blanks with the word or words that make the following statements correct.

1. Many accidents leave the victim partially or totally _____.

2. All electrical devices should be installed according to the _____ _____ Code.

3. All machinery should be installed according to the _____ specifications.

4. The standard unit of sound is the _____.

5. Minimum eye protection is provided by wearing _____ _____ or _____.

6. The hazards of breathing dust and paint spray can be reduced by wearing a _____.

7. The body parts most often injured in shop accidents are the _____ and _____.

8. When working with objects above head level, a _____ should be worn.

D. **Brief Answer.** Briefly answer the following questions.

1. What are considered the most critical parts of the body to be protected? Why?

2. The colors in the safety color-coding system for shops serve what five general purposes?

3. According to OSHA (Occupational Safety and Health Administration), how long can a worker safely use a power mower without using ear protection? A jackhammer? (See Figure 4-20.)

4. What are some general considerations one should keep in mind when selecting protective clothing for use in the shop environment?

Objective

To recognize and reduce hazards in agricultural mechanics settings, and to react effectively in case of fire or other emergencies.

Competencies to be developed

After studying this unit, you should be able to:

- Reduce hazards in agricultural mechanics.
- State the three conditions necessary for combustion.
- Match appropriate types of fire extinguishers to each class of fire.
- Use a fire extinguisher.
- Interpret labels on hazardous materials.
- Describe appropriate action in case of fire, accident, or other emergency.

Materials List

- Different types of fire extinguishers
- Labels from agricultural chemicals

Terms to Know

- fire triangle
- fuel
- combustion
- heat
- oxygen
- extinguished
- slow-moving vehicle (SMV)
- cardiopulmonary resuscitation (CPR)

ire, slow-moving vehicles, highway crossings, and chemicals create unique hazards in agricultural mechanics. Each may hold dangers to the workers and to others in the area. Fortunately, there are ways to reduce hazards and take action quickly if accidents occur.

REDUCING FIRE HAZARDS

The discovery of fire and how to create it was one of man's most important achievements. Fire is used to heat homes, cook food, generate electricity, melt ore to refine metals, heat metals to bend and form them, and cut metals. Yet, fire has not really been tamed. Fire breaks out of control at unexpected times, causing injury and loss of property and lives. Burns are probably the most painful of all injuries.

It is known how to prevent uncontrolled fire, except in the case of volcanic eruptions and lightning strikes. Even lightning can be directed to a safe ground and thus reduce its destruction. The important thing is that most losses from fire can be prevented; however, it requires attention and knowledge of how fire works (Figure 5-1).

The Fire Triangle

To produce fire, three components must be present at the same time. These three components are fuel, heat, and oxygen. They are known as the **fire triangle** (Figure 5-2).

Fuel is any combustible material that will burn. "Combustible" comes from the word **combustion**, which means "to burn." Common fuels are gasoline, kerosene, diesel fuel, wood, paper, acetylene, and

FIGURE 5-1 Most losses from fire can be prevented.

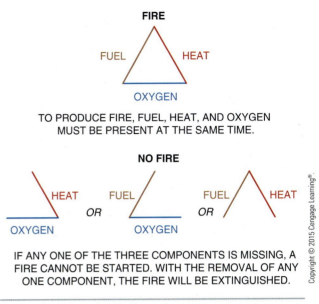

TO PRODUCE FIRE, FUEL, HEAT, AND OXYGEN MUST BE PRESENT AT THE SAME TIME.

IF ANY ONE OF THE THREE COMPONENTS IS MISSING, A FIRE CANNOT BE STARTED. WITH THE REMOVAL OF ANY ONE COMPONENT, THE FIRE WILL BE EXTINGUISHED.

Copyright © 2015 Cengage Learning®.

FIGURE 5-2 If any one of the three components of the fire triangle is missing, a fire cannot be started. If any component is removed from an existing fire, the fire will go out.

propane. Most materials will burn if they are made hot enough in the presence of oxygen.

Heat is simply a type of energy that causes the temperature to rise. If the temperature of a room is changed from 50 degrees to 70 degrees, it is done by using heat. Remember, most materials will burn if they are made hot enough in the presence of oxygen.

Oxygen is a gas in the atmosphere. It is not a fuel, but must be present for fuels to burn. Oxygen is nearly always present except in airtight conditions. This fact is important to remember in fire safety and control.

Preventing Fires in Agricultural Mechanics

If any one of the three components of the fire triangle (fuel, heat, oxygen) is eliminated, fire will be prevented from starting; or it will be stopped if it has started. Therefore, to prevent, control, or stop fires:

- store fuels in approved containers
- store fuels away from other materials that burn easily
- store materials in areas that are cooler than their combustion temperature
- use fire only in safe surroundings
- put out fires by removing one or more elements in the fire triangle

The prevention of fire goes hand-in-hand with safe use of equipment and efficient management of work areas.

For instance, the proper use of a gas cutting torch decreases the likelihood of fire resulting from its use. Proper storage of materials decreases the chance of fire and keeps materials readily available when needed. Clean work areas prevent people from slipping or tripping, and damage to parts or projects. A clean work area also decreases the chance of a fire. Special paint booths provide a clean area for paint jobs and also decrease the likelihood of fire.

EXTINGUISHING FIRES

Fires are **extinguished** or put out by adding water or other materials to cool them, covering them to cut off the oxygen, or removing the fuel. This could mean wrapping a person whose clothing is on fire with a blanket. It could mean stopping fire in a field by raking grass and leaves out of the path of the fire or throwing soil on the fire to smother it. Fire at a gas torch or hose may be stopped by shutting off the gas at the cylinder. A burning container of paper may be extinguished by cooling it with water from a hose or bucket.

Classes of Fires

To effectively and safely put out a fire with a fire extinguisher, the class of fire must be known. This is determined by the material and the surroundings as follows:

- **Class A—Ordinary Combustibles.** Ordinary combustibles include wood, papers, and trash. Class A combustibles do not include any item in the presence of electricity or any type of liquid.
- **Class B—Flammable Liquids.** Flammable liquids include fuels, greases, paints, and other liquids, as long as they are not in the presence of electricity.
- **Class C—Electrical Equipment.** Class C fires involve the presence of electricity (Figure 5-3).
- **Class D—Combustible Metals.** Combustible metals are metals that burn. Burning metals are very difficult to extinguish. Only Class D extinguishers will work on burning metals.

Fire classification is based on how to safely and cheaply extinguish each type of material (Figure 5-4). Water is generally the cheapest material to use in fire control, but it may not be safe or effective. A firefighter can be electrocuted if the stream of water hits exposed electrical wires, plugs, appliances, or controls. Water is not suitable on fires involving petroleum products because the fuel floats to the top of the water and continues to burn.

© Thomas Skjaeveland/Shutterstock.com

FIGURE 5-3 Electrical fires are Class C fires and require the use of a non-water extinguisher.

Types of Fire Extinguishers

The proper fire extinguisher can put out a fire within seconds. However, such results occur only if the fire is extinguished when it first bursts into flames. The key is the proper extinguisher, used immediately, and in the proper way. This combination may make the difference between a mere frightful moment or a multimillion-dollar fire loss with serious injuries and death.

Students should learn to recognize extinguishers by their type (Figure 5-5). Common types of extinguishers are:

- water with pump or gas pressure (used for Class A fires)
- carbon dioxide (CO_2) gas (used for Class B and C fires)
- dry chemical (used for Class A, B, and C fires)
- blanket (used for smothering fires on humans or animals)

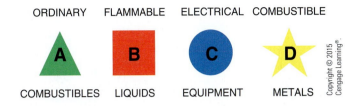

ORDINARY | FLAMMABLE | ELECTRICAL | COMBUSTIBLE

A | **B** | **C** | **D**

COMBUSTIBLES | LIQUIDS | EQUIPMENT | METALS

Copyright © 2015 Cengage Learning®.

FIGURE 5-6 *Standardized symbols on fire extinguishers indicate the type of fire that they are designed to put out.*

© Jonas San Luis/Shutterstock.com

FIGURE 5-4 Extinguishers are marked according to the class of fires on which they will safely work.

Copyright © 2015 Cengage Learning®.

FIGURE 5-5 Types of fire extinguishers.

Extinguishers are marked according to the class or classes of fires on which they will safely work. Fire extinguisher labels contain standardized symbols to help the reader act quickly in an emergency (Figure 5-6). The symbols are as follows:

- Green triangle for Class A, ordinary combustibles
- Red square for Class B, flammable liquids
- Blue circle for Class C, electrical equipment
- Yellow star for Class D, combustible metals

Location of Fire Extinguishers

The location of fire extinguishers is very important. A few seconds lost in looking for the right type of extinguisher could allow a fire to rage out of control. Class A extinguishers should be placed in areas where Class A fires are likely to occur, Class B extinguishers should be placed in areas where Class B fires are likely to occur, and so on. Placing a water-type extinguisher in an area where an electrical fire is likely to occur is of little value. The extinguisher should not be used on an electrical fire.

Extinguishers should be placed in clean, dry locations near exits within easy reach. The extinguisher should be hung on the wall so the top of the extinguisher is not more than 3½ to 5 feet above the floor. The bottom of the extinguisher should be at least 4 inches above the floor. The extinguisher should be positioned so it can be removed quickly. Also, if possible, it should be near a fire alarm (Figure 5-7).

Everyone should be familiar with the locations and use of all types of extinguishers.

Using Fire Extinguishers

Generally, fire extinguishers are held upright and operated by a lever. However, some types are activated by inverting the tank, causing chemicals to mix inside the container. Sometimes, the lever is blocked by a pin to prevent accidental discharge (Figure 5-8). Before operating an extinguisher, the instructions on the container

FIGURE 5-7 A fire extinguisher should be hung on the wall in such a way that the top of the extinguisher is not more than 3½ to 5 feet above the floor. Ideally, it should be near a fire alarm.

FIGURE 5-9 Instructions are usually printed on the extinguisher.

should be read (Figure 5-9). For most extinguishers, a pin is pulled and a lever pressed.

Before discharging an extinguisher, the firefighter should move to within 6 to 10 feet of the fire and direct the extinguisher nozzle toward the base of the fire (Figure 5-10). The extinguisher will be empty in a matter of seconds, so any material that misses the fuel at the base of the fire will be wasted.

Extinguishers should be carefully maintained. Tragedy can occur if an extinguisher is needed and it will not function properly. Over time the charge may be lost due to leaking or deterioration of the contents. Make it a habit to frequently check the gauge on top of the extinguisher to make sure it is fully charged (Figure 5-11). A monthly inspection of all fire extinguishers should be made to ensure that the extinguishers are usable in case of an emergency (Figure 5-12).

Always think before acting. Call for help immediately. Be sure any fire is completely out before leaving the area. The local fire department is the best source of help on fire safety and prevention.

FIGURE 5-8 Extinguishers have pins that prevent accidental discharge. The pin is removed prior to use.

FIGURE 5-10 The extinguisher should be aimed at the base of the fire.

FIGURE 5-11 Frequently check the extinguisher's gauge to make sure it is fully charged.

MONTHLY FIRE EXTINGUISHER CHECK

- See that the proper class extinguisher is in the area of fire class risk.
- See that the extinguisher is in its place.
- See that there is no obvious mechanical damage or corrosive condition to prevent safe reliable operation.
- Examine or read visual indicators (safety seals, pressure indicators, gauges) to make certain the extinguisher has not been used or tampered with.
- Check nameplate for readability and lift or weigh extinguisher to provide reasonable assurance extinguisher is fully charged.
- Examine nozzle opening for obstruction. If equipped with shut-off type nozzle at the end of the hose, check the handle for free movement.

CHECKLIST

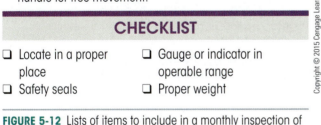

- ❑ Locate in a proper place
- ❑ Safety seals
- ❑ Gauge or indicator in operable range
- ❑ Proper weight

FIGURE 5-12 Lists of items to include in a monthly inspection of fire extinguishers.

SIGNS OF DANGER

There are many signs to warn of possible hazards. Stop, yield, caution, and crossing signs alert drivers to dangers on the highway. Danger, no trespassing, condemned, and keep-out signs warn of dangers around old buildings.

On the farm and in other agricultural mechanics settings, warning signs are found on machinery. Such signs may say not to remove a shield, not to overspeed,

FIGURE 5-13 Manufacturers place safety warning labels on their machinery to alert operators to danger. Following the instructions can help prevent personal injury or property damage.

to keep bolts tight, or to position a part in a certain way (Figure 5-13).

The Slow-Moving Vehicle Emblem

A very important sign for safety on the highway is the **slow-moving vehicle (SMV)** emblem. This is a reflective-type emblem consisting of an orange triangle with a red strip on each of the three sides (Figure 5-14).

FIGURE 5-14 The slow-moving vehicle (SMV) emblem is commonly used on farm and industrial equipment, road trucks, and other vehicles whose maximum speed is less than 25 miles per hour.

ACME PESTICIDE

Controls Insects and Diseases
Fruit and Ornamental Trees

ACTIVE INGREDIENTS		
Captan (n-trichloromethythio-4-cyclohexene 1,2-dicarboximide)		9.00%
Related Derivatives		0.25%
Malathion (0,0-dimethyl dithiophosphate of diethyl mercatosuccinate)		4.00%
Carbaryl (1-naphthyl N-methylcarbamate)		10.00%
INERT INGREDIENTS		76.75%
Contains Petroleum Distillate	Total	100.00%

Consumer Chemicals, Nowhere, Georgia	EPA Reg. No. 010-1133	EPA Est. No. 010XXX

DANGER
Keep out of reach of children

STATEMENT OF PRACTICAL TREATMENT--Flush eyes with flowing water and call a physician. Wash from skin with plenty of soap & water. If redness or irritation persists, get medical attention. If swallowed, drink a large quantity of milk, egg whites, gelatin solutions, or large quantities of water. Avoid alcohol. If inhaled: Remove victim to fresh air. If not breathing, give artificial respiration, preferably by mouth and call an ambulance.

Note To Physician: Carbaryl is a modest cholinesterase inhibitor. Atropine is antidotal Emergency medical information, 1-800-XXX-XXXX   .

PRECAUTIONARY STATEMENTS--

HAZARDS TO HUMANS AND DOMESTIC ANIMALS--DANGER-- Can cause irreversible eye damage. Harmful if swallowed. May cause allergic skin reactions. Do not get in eyes. Wear goggles or face shield when handling to avoid contact with eyes or skin or clothing. Also, chemical resistant gloves should be worn. Also wear long pants and long-sleeved shirt and apply with the wind to your back. Wash nondisposable gloves thoroughly with soap and water before removing. Clothing worn while handling this product must be laundered separately from other clothing before reusing.

ENVIRONMENTAL HAZARDS--This pesticide is toxic to fish, aquatic invertebrates, and aquatic life stages of amphibians. Do not allow drift and runoff from the application or runoff from the cleaning of equipment or disposal of equipment be released in a manner that will contaminate water resources. This product is highly toxic to bees exposed to direct treatment on blooming crops or weeds. Do not apply this product or allow it to drift to blooming crops or weeds if bees are visiting the treatment area.

PHYSICAL OR CHEMICAL HAZARDS: Do not use or store near heat or open flame. Do not store below 32*F.

DIRECTIONS FOR USE

It is a violation of Federal law to use this product in a manner inconsistent with its labeling.

Acme Insecticide and Fungicide is a complete concentrate containing fungicide, aphicide, miticide, scalicide, and spreader-sticker. Mixes with water instantly. Designed especially for home gardens to protect fruit and ornamental trees from the listed insects and diseases.

SHAKE PRODUCT THOROUGHLY BEFORE USING. Contains micronic particles which settle upon standing and require reblending by agitation. Apply during a cool, calm period, preferably early morning or evening. The sprayer should be occasionally agitated (shaken) to keep spray particles in suspension during application.

CONTROLS

Insects: Aphids, apple maggot, bagworm, black cutworm, bud moth, cherry fruit fly and worm, codling moth, plum cucurlio, mites, oriental fruit moth, pear slugs, psylla, red banded leafroller, scale (Forbes, Putnam San Jose), and tent caterpillars.
Diseases: Bitter rot, black spot, black rot, blossom blight, botrytis blossom end rot, downy mildew, fly speck, frog eye, leaf spot, scab, and sooty blotch.

MIX 1.5 TABLESPOONS PER GALLON OF WATER. Begin applications when insects or disease symptoms first appear or conditions favor their development and repeat at weekly intervals or as necessary to maintain control. Therefore a preventive spray schedule is recommended. Select still periods for application (early morning or evening) to reduce waste by blow away and blow back. Spray in early morning or in the evening to avoid direct sunlight. Do not apply through any type of irrigation equipment.

STORAGE AND DISPOSAL

Storage: Keep pesticide in original container. Do not put concentrate or dilute solution into food or drink container. Avoid contamination of feed and foodstuffs. Store in a cool, dry place, preferably in a locked storage area. PRODUCT DISPOSAL: Empty container by use. CONTAINER: Do not reuse empty jug. Rinse thoroughly before discarding in trash.

NOTICE: Buyer assumes all responsibilities for safety and use not in accordance with directions.

Net Contents: 1 quart

Copyright © 2015 Cengage Learning®.

FIGURE 5-15 Agricultural pesticides and other hazardous chemicals have labels that warn the user of any hazards associated with the material.

It glows brightly when only a small amount of light hits it. Therefore, it is generally the first item to be seen on a vehicle. When operators of fast-moving vehicles know the meaning of the emblem, they have time to slow down before running up on slow vehicles. SMV emblems are required on all vehicles that travel a maximum of 25 miles per hour on public roads. It is important to have an SMV emblem on every piece of machinery on the highway. Drivers of automobiles and trucks always must be prepared to slow down when the SMV emblem comes into view.

Package Labels

All commercial products have a label to provide the user with certain information. Food product labels tell about nutrient content. Repair parts come with instructions for installation. Paint labels name the components, and clothing labels tell how they are to be cleaned.

Labels on hazardous products may be a matter of life or death in the event of accident. Common products such as kerosene and turpentine are poisonous if taken internally. Agricultural pesticides are designed to be poisonous to pests; they may cause illness or death to humans if misused. If used according to the instructions on the label, pesticides are generally safe.

It is very important to keep materials in their original containers. Original containers are the correct type of container. They carry the label that describes the product, its hazards, and procedures to use in case of an emergency with the product.

Pesticide labels are legal documents. Label directions are required by law to be accurate and complete. Labels should be read again before handling, opening, mixing, using, or disposing of pesticides.

Pesticide labels have at least 16 different items of information. Study the pesticide label shown in Figure 5-15 to see how the 16 items of information are displayed on the label. Pesticides should always be stored in a locked cabinet or area (Figure 5-16).

EMERGENCIES OR ACCIDENTS

Quick action of the correct type can change the outcome of an emergency from a tragedy to simply a frightening moment. If fire should break out at school:

- notify the teacher
- keep everyone calm

FIGURE 5-16 Pesticides should always be stored in a locked cabinet or other locked, secure area.

- set off the fire alarm
- call the fire department
- clear the area
- use fire extinguishers if this seems logical under the circumstances

If an injury occurs, quick action is in order. Such action must be correct and based on good thinking; otherwise, additional injury may result. For many emergencies, the following procedure should be helpful.

Procedure

1. Call or send for help. Most communities have emergency services that can be reached by dialing 911 on any phone.
2. Tell the emergency operator that you need police or ambulance.

NOTE

Some communities now have ambulance service and medical helicopters standing by that can reach an accident scene with paramedics and emergency equipment in minutes. These services often make the difference between life and death.

3. Do not move the victim unnecessarily.
4. Try to arouse the victim by talking.

(continued)

Copyright © 2015 Cengage Learning®

Procedure, *continued*

5. Treat for shock. Keep the victim lying down. Elevate the victim's feet 8 to 10 inches if there are no signs of bone fractures or head or back injuries. Place a blanket under and over the victim to maintain body heat. If the victim complains of being thirsty, moisten a clean cloth and wet the victim's lips, tongue, and inside of the mouth.

6. If the victim is bleeding, stop the bleeding by wrapping or pressing clean cloth or gauze directly on the wound.

7. If the victim is not breathing, clear the air passage and find someone to administer rescue breathing or cardiopulmonary resuscitation. **Cardiopulmonary resuscitation (CPR)** is a first-aid technique to provide oxygen to the body and circulate blood when breathing and heartbeat stop. CPR training is given to many teachers and students. The technique requires a special course. To save a victim's life, CPR must be started within 4 to 6 minutes of drowning, electrocution, suffocation, smoke inhalation, gas poisoning, or heart attack.

8. Do not move the victim if broken bones are suspected, unless problems with breathing, bleeding, or other life-threatening factors exist.

SUMMARY

Shops often contain many materials that are combustible, and many shop operations can create sparks to ignite the materials. Proper precautions must be taken to ensure that fires do not occur. Not only must fire extinguishers be readily available, but they must also be of the correct type and must be properly charged. All hazardous materials should be labeled and correctly stored. By planning and reducing hazards, accidents and injuries can be prevented.

Qualified people should be called to give medical aid to accident victims. In schools, there are specific procedures for teachers and students to follow in case of injury or other emergency. Each student should learn these specific policies and procedures.

Student Activities

1. Define the Terms to Know in this unit.

2. List the types of fire extinguishers needed for the classroom, shop, and laboratory. Check each fire extinguisher that is present for state of charge, type, proper hanger, and appropriate location. Ask your teacher to correct all problems regarding fire extinguishers.

3. Ask your teacher to arrange for a representative of the local fire department to visit your class and demonstrate appropriate fire safety practices.

4. Install slow-moving vehicle (SMV) emblems on pieces of school equipment that need them.

5. Install SMV emblems on pieces of equipment at home that need them.

6. As an FFA chapter project, conduct an SMV emblem sale and campaign to increase the use of the emblem in your community.

7. List on your class whiteboard all the signs of danger you and your classmates can name.

Relevant Web Sites

Learn CPR, a free public service supported by the University of Washington School of Medicine
http://depts.washington.edu/learncpr

U.S. Fire Administration, Fire Safety for Kids, Parents, and Teachers
http://usfa.fema.gov/kids

CDC web page on machine safety
www.cdc.gov/niosh/topics/machine

Self-Evaluation

A. Multiple Choice. Select the best answer.

1. Which is **not** part of the fire triangle?
   a. fuel
   b. combustion
   c. oxygen
   d. heat

2. A commonly used fuel is
   a. acetylene
   b. acetone
   c. oxygen
   d. magnesium

3. Fire can always be prevented or stopped by eliminating
   a. combustible gases in the area
   b. congestion in the shop
   c. improper storage of fuels
   d. any item in the fire triangle

4. Fire hazards associated with painting can be reduced by
   a. using a spray gun instead of a brush
   b. using newspaper to protect bench surfaces
   c. using a special paint booth
   d. painting with several people in the area

5. Effective fire control techniques include
   a. cooling a fire with water
   b. wrapping a blanket around a person whose clothes are on fire
   c. raking dead leaves and grass away from an advancing fire
   d. all of these

6. Fires are classified according to
   a. materials involved and techniques that safely extinguish them
   b. size and duration of the fire
   c. season of the year when the fire occurs
   d. the amount of material being burned

7. A green triangle on a fire extinguisher means the extinguisher can be used to put out burning
   a. metals
   b. liquids
   c. wood
   d. electrical wires

8. Most fire extinguishers will discharge when
   a. the pin is pulled and the lever is pressed
   b. the extinguisher is inverted
   c. either (a) or (b), depending on the extinguisher
   d. none of these

9. SMV means
   a. small mechanical vehicle
   b. stop! moving vehicle
   c. slow-moving vehicle
   d. none of these

10. SMV emblems are required when
    a. vehicles are standing
    b. vehicles travel 25 miles per hour or slower
    c. vehicles travel 30 miles per hour or slower
    d. vehicles travel faster than 30 miles per hour

11. Pesticide labels are
    a. legal documents
    b. used only on insecticides
    c. used primarily on powdered chemicals
    d. generally written in two or more languages

B. **Matching.** Match the class of fire in column I with the burning material in column II.

Column I

   1. Class A fire

   2. Class B fire

   3. Class C fire

   4. Class D fire

Column II

   a. paper and wood

   b. flammable liquids

   c. combustible metals

   d. electrical equipment

C. **Matching.** Match the type of extinguisher in column I with the type of fire for which it is used in column II.

Column I

   1. dry chemical extinguisher

   2. water extinguisher

   3. carbon dioxide gas extinguisher

   4. foam extinguisher

Column II

   a. Class B and C fires

   b. Class A fires

   c. Class A and B fires

   d. Class A, B, and C fires

d. **Brief Answer.** Briefly answer the following questions.

   1. Name four items of information provided on pesticide labels.

   2. Name four signs that may be seen along highways that alert people of possible hazards.

   3. Name three warning signs often found on machinery.

   4. Describe the slow-moving vehicle emblem.

   5. Name five things to do if a fire breaks out.

   6. Name six emergency procedures used for accident victims.

   7. Explain how to properly extinguish a fire using a fire extinguisher.

UNIT 6
Shop Cleanup and Organization

Objective

To work cooperatively with classmates to clean the shop efficiently and to store all tools and materials properly.

Competencies to be developed

After studying this unit, you should be able to:

- Use shop-cleaning equipment properly.
- Clean benches, machines, and floors.
- Store materials properly.
- Store tools properly.
- Do assigned tasks.
- Work cooperatively with others.

Materials List

- Bench brush
- Floor broom
- Dust mop
- Vacuum cleaner
- Rag can
- Scrap wood box
- Scrap metal can
- Shop cleanup wheel chart
- Shop cleanup assignment sheet

Terms to Know

- silhouettes
- flammable materials cabinets
- vertical racks
- floor brooms
- dust mops
- bench brushes
- scoop shovels
- dust pans
- cleanup wheel
- cleanup skills checklist
- cleanup assignment sheet

A CLEAN AND ORDERLY SHOP

The agricultural mechanics shop can be a place where it is a pleasure to work and learn skills or it can be a place of frustration and danger. Much of the difference is in how the shop is equipped and maintained. A clean, orderly environment is much safer and pleasant to work in than a dirty, cluttered shop (Figure 6-1). Both students and instructors have a responsibility to make sure the shop is kept clean and orderly.

Some positive indicators of a properly cleaned shop are as follows:

- A signal is given to stop work and start cleanup at a specified time. A whistle is effective as a cleanup signal.
- Every student helps with cleanup.
- Benches are cleared and clean.
- Machines are clean.
- Paintbrushes and spray equipment are properly cleaned and stored.
- Solvents, paints, and greases are properly stored.
- Tools are in their places.
- Lumber, metal, and other construction materials are stored.
- Projects and related materials are in approved places.
- Floor is clean and trash is in containers.
- Cabinets and storage areas are locked.
- Every job is checked for completeness.
- Every student is evaluated according to the quality of his or her cleanup contribution.
- Sinks and restrooms are clean and orderly.
- Students are waiting in an orderly manner for dismissal by the teacher.

REASONS FOR KEEPING THE SHOP CLEAN

A quick and efficient cleanup procedure is important to the safety of students. It adds greatly to the success of an agricultural mechanics shop program. There are good reasons for cleaning the shop after each class, every day. Some of these are related to personal safety, some to learning efficiency, and some to student comfort and convenience. A properly organized shop cleanup procedure is important for the following reasons:

- Each student's projects and possessions are stored properly. When projects are stored properly, they do not interfere with the work of other students and are not damaged by other students using the shop. Projects or project parts may be stored in drawers, lockers, storage cabinets, storage rooms, or fenced areas. In well-managed shops, project parts may be stored in containers by, on, or under large projects such as tractors, wagons, or machinery.
- All project parts are stored together. This enables students to see if any items need to be brought the next day to continue the project.
- Shop spaces are cleared, so other classes can safely use the areas.
- Tools are returned to their proper places. Tools should be mounted on panels over colored outlines of each tool (Figure 6-2). These outlines are called **silhouettes**; they make it easy to check for missing tools at the end of each class. Tools can also be easily checked to see if they are in need of repair.
- Each student learns to put tools and materials in their proper places and can expect to find them quickly when needed. This eliminates lost time looking for tools and materials.

FIGURE 6-1 A clean, orderly shop is more pleasant and safe to work in.

FIGURE 6-2 Tools should be mounted on panels in an orderly manner. Note how the use of silhouettes makes it easy to see when tools are missing.

- Paint materials and equipment are cleaned and stored to avoid wasted materials and ruined finishing materials.
- The hazards of fire and explosion are reduced by proper storage of materials.
- Students learn cooperation and teamwork.

EQUIPMENT AND CONTAINERS USEFUL FOR SHOP CLEANUP

It is important to have enough cleanup equipment and materials on hand so all students can participate in cleaning and storage activities. Each student must do his or her part to make the cleanup easier for everyone. It is also important to have storage containers or racks for all types of materials used in the shop (Figure 6-3). Many shops have excellent commercial **flammable materials cabinets**. Flammable liquids, such as grease, oil, and solvents, are stored in these cabinets, which are made of steel and close automatically in the presence of fire (Figure 6-4). Such cabinets must meet all safety requirements.

Racks for lumber and metal provide safe and convenient storage for these materials (Figure 6-5). **Vertical racks** permit the storage of both long and short items. On such racks, materials can be reached with little moving of other items.

Many items of equipment are necessary to clean a shop quickly and efficiently and to store materials safely. These items include the following:

- **floor brooms**
- floor **dust mops**
- **bench brushes**
- shop vacuum cleaner(s)
- dust collection and chip removal system
- metal cans for rag storage
- large metal trash cans
- storage cabinets for combustible materials
- **scoop shovels** and **dust pans** to pick up dirt and trash
- solvents for cleaning up grease and oil spills
- sawdust to absorb liquids

FIGURE 6-3 Storage racks and containers are needed for different types of materials.

FIGURE 6-4 Flammable liquids such as grease, oil, and solvents can be stored in special cabinets that close automatically if a fire occurs.

FIGURE 6-5 Racks for lumber, steel rods, structure steel, and pipe make good use of wall space.

- commercial material to sprinkle on the floor to control dust
- clean rags
- storage cabinets for tools and hardware
- storage racks for lumber and metal
- steel cans for metal used for practice welding
- suitable containers for scrap wood
- cabinets, lockers, fenced areas, and the like for project storage

The soft-bristled brush and shop vacuum cleaner are the standard tools for removing dirt, sawdust, and trash from benches and machines (Figure 6-6). The floor broom and dust mop are important floor-cleaning equipment. The dust pan and standard scoop shovels are commonly used to move the trash from the floor to the trash can.

Dust Collection Systems

Dust is a major problem in most shops, particularly those that are used for woodworking. The use of woodworking machines creates a lot of sawdust that can cover most areas of the shop. This can cause not only a cleaning problem, but safety and health concerns as well. Fine particles in the air are not healthy to breathe and can cause a fire hazard. Many shops are equipped with dust collection systems. These systems consist of a large centrally located vacuum with ducts running to various machines and areas in the shop (Figure 6-7). As a machine creates dust (such as sawdust or sanding dust), the dust is pulled from around the machine and transported to a collection bin (Figure 6-8). Also, vents may be located in the floor for collecting dust swept from the floor. These systems have to be properly maintained by cleaning

FIGURE 6-6 The shop vacuum cleaner makes cleanup easier and does not stir up dust.

the ducts, making sure they are not clogged, and emptying the storage bin periodically. Particular attention should also be paid to the type of ducts used. Only ducts approved for dust collection systems should be used. Proper grounding must be installed because the movement of particles through the duct can build up static electricity that can cause shock or present a fire hazard.

TECHNIQUES FOR EFFECTIVE CLEANING

Effective cleaning skills cannot be taken for granted; they must be developed. Many students go through the motions of cleaning, but many do not do a good job. Shop cleanup tasks include:

- removing tools and materials from benches and floor before cleaning

FIGURE 6-8 This table saw has a dust collection duct over the blade to collect sawdust.

- putting all trash in suitable metal containers
- placing trash containers in the proper place
- storing all cleanup equipment properly
- cleaning all sink areas and picking up paper towels
- helping others finish the cleanup

SHOP CLEANUP SYSTEMS

Organization is the key to a clean shop. Without good organization, students who see the need for a clean and orderly shop soon lose their willingness to put the shop in good order. This is natural since the value of teamwork and fair play is learned at a very young age. Therefore, a system that involves every student on an equal basis is needed. Several systems have been developed to get the job done.

All-Pitch-In Method

For lack of a better term, the most simple system of shop cleanup is called the "all-pitch-in" method. With this system, the teacher announces cleaning time verbally, or by whistle, bell, or other device. Students start by putting their own materials away. They then do cleanup, arrangement, or storage tasks according to their knowledge, maturity, personal cleanup habits, and commitment to the program. This system generally fails because students lack the knowledge and skills necessary to do a good job at all cleanup tasks. Students and teacher soon discover that certain important cleanup tasks are left undone. This is due to lack of organization.

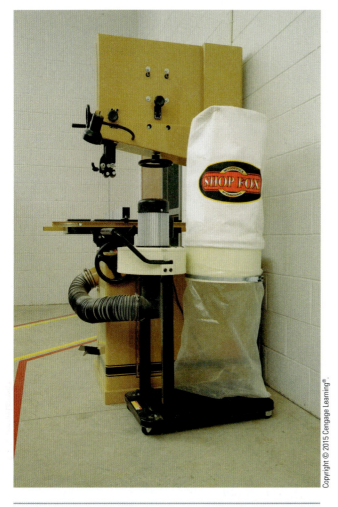

FIGURE 6-7 A dust collector consists of a large vacuum with ducts running to various machines.

- cleaning all paintbrushes
- cleaning high areas such as racks, machines, and bench tops
- cleaning the shop, starting at its far sides and ends and working toward the trash-collecting area(s)
- using brushes and brooms in short strokes and lifting intermittently (brushes and brooms should be tapped against the surface or floor frequently to shake out dirt particles)
- using a commercial dust-absorbing material, if available
- using a vacuum cleaner to clean machines whenever possible (the vacuum cleaner is very desirable because it does not create dust)
- putting oily rags in closed metal containers
- using sawdust or commercial materials to absorb liquids such as oil spills
- sweeping fine waste particles into floor drops of dust collection systems, if provided

Cleanup Wheel Method

Many shops use a **cleanup wheel** (Figure 6-9). This system uses a chart shaped like a wheel. The teacher specifies all cleanup tasks in equal sections on the outer section of the wheel chart. Students are placed in groups or given a group number. Either names or group numbers are placed in the inner section of the wheel chart. Each group of students is assigned the task or tasks listed in line with their name or group number on the wheel chart. The tasks remain stationary at the edge of the chart. The wheel itself can be periodically rotated so each group has a chance to do all cleanup tasks.

For the cleanup wheel method to work, some kind of **cleanup skills checklist** is needed. It is important that a student who is respected by his or her classmates and is a good judge of achievement be given this job. A checklist with a format that works well is shown in Figure 6-10. The checklist should be placed in a prominent location so students can see how they are being evaluated. Students with zeroes (0) and ones (1) should be encouraged to improve their cleanup skills.

When using a checklist, the foreperson's name is listed along with all others. During the shop cleanup time, the foreperson carries the checklist on a clipboard and evaluates all students except himself or herself. The teacher then evaluates how well the foreperson has done the evaluations.

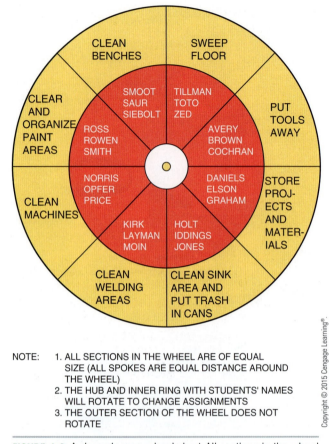

NOTE: 1. ALL SECTIONS IN THE WHEEL ARE OF EQUAL SIZE (ALL SPOKES ARE EQUAL DISTANCE AROUND THE WHEEL)
2. THE HUB AND INNER RING WITH STUDENTS' NAMES WILL ROTATE TO CHANGE ASSIGNMENTS
3. THE OUTER SECTION OF THE WHEEL DOES NOT ROTATE

FIGURE 6-9 A shop cleanup wheel chart. All sections in the wheel are of equal size. To change assignments, the teacher rotates the inner circle containing the students' names. The outer ring of the wheel does not rotate.

SHOP CLEANUP SKILLS CHECKLIST

Key:

3 = Done well
2 = Done satisfactorily
1 = Done poorly
0 = Job not done
ab = Student is absent

DATE _____ 4/4

SHOP FOREPERSON _____ *Walker*

DATE AND RATING

NAME	3/5	3/6	3/7	3/8	3/9	3/12	3/13	3/14	3/15	3/16	AVERAGE SCORE	COMMENTS
Avery	3	3	2	3	3	3	3	3	3	3	2.9	
Saur	3	2	1	1	1	ab	0	0	1	1	1.0	
Cochran	1	3	0	3	3	3	3	3	3	3	2.5	
Opfer	ab	3	3	2	2	2	2	2	3	2	2.1	
Walker (Foreperson)	*JW*	*JW*	*JW*	*JW*	*JW*	*JW*	*JW*	*JW*	*JW*	*JW*		

FIGURE 6-10 A shop cleanup skills checklist. In this example, the teacher has initialed the box by the foreperson's name, indicating that the teacher observed the general conditions of the shop and approved the job done by the foreperson.

Copyright © 2015 Cengage Learning®.

This method helps students see how well they are developing their cleanup skills. It also helps develop management and supervisory skills. The checklist has only enough columns for a limited number of shop periods. New cleanup tasks should be assigned to students every two weeks. This ensures that everyone will do all jobs sometime during the year and learn how to do each one in a skillful manner.

Both the teacher and the students must realize that shop cleanup skills are valuable. Future employment demands such skills be developed. Therefore, every effort should be made to teach and learn procedures and tasks used in effective cleanup operations.

Assignment Sheet Method

The assignment sheet method may be used in place of the cleanup wheel and the shop cleanup skills checklist (Figure 6-11). One important advantage of the assignment sheet is that as many students as necessary can be assigned to any given task to get it done. The equal-sized groups necessary for the cleanup wheel are not needed for the assignment sheet. A disadvantage of assignment sheets is the time required to periodically reassign students.

With the shop **cleanup assignment sheet**, the teacher and class develop a list of cleanup tasks. Class members are then assigned to these tasks in numbers needed to get the job done. For instance, if it takes two people to sweep off benches, four to clean the welding areas, and five to sweep the floor, the appropriate number of students can be assigned to each task. The teacher decides who will do the tasks although students can volunteer for various tasks. Assignments may be changed as often as desired. A new shop cleanup assignment sheet must be prepared each time a rotation is made.

A student foreperson and the check-off system are used to evaluate cleanup skills. The check-off, or evaluation, process is the same as described for the shop cleanup wheel method.

Choosing a System

The shop cleanup system used is largely up to the teacher and students. Each method requires a cooperative attitude on the part of all. Each student must do the assigned

---

SHOP CLEANUP ASSIGNMENT SHEET

Key:
3 = Done well
2 = Done satisfactorily
1 = Done poorly
0 = Job not done
ab = Student is absent

DATE _____

SHOP FOREPERSON _____

Task	Person(s) Responsible	Date and Rating							Average Score	Comments
	Date:									
Sweep benches	Avery									
	Saur									
Clean welding areas	Opfer									
	Elson									
	Graham									
Sweep floor	Holt									
	Iddings									
	Jones									
	Kirk									
Foreperson	Mozier									

FIGURE 6-11 A shop cleanup assignment sheet. This sheet resembles a shop cleanup skills checklist except that it includes a task column.

Copyright © 2015 Cengage Learning®.

task every day or the system will not work. This is because others must do the job for the negligent students. Jobs of absent students must also be covered by others assigned to the same task. To avoid this problem, several students may be designated as substitutes to do the tasks assigned to others who are absent on any given day.

The following must be provided if a shop cleanup system is to work well:

- assignment of every task to some person
- fair assignments based on desirability of the task and effort required to do it

- rotation of assignments so all students learn all tasks
- cooperation by all parties
- genuine honesty in evaluating performance
- a record of individual performance
- a clean and safe shop by the end of each period

By working together, agricultural mechanics students can enjoy the benefits of a clean and safe shop. This provides satisfaction, good workmanship, cooperative effort, and personal gain.

SUMMARY

A clean and organized shop is essential to any agricultural mechanics project. Knowing where to find tools, supplies, and materials will save time and be useful in maintaining the proper inventory of tools and materials. A shop that is cluttered and disorganized will not only be unsafe, but will also hinder the proper maintenance of tools and equipment. A disciplined approach to daily cleaning and organizing will save time and effort in the long run and help ensure that accidents are prevented.

Student Activities

1. Define the Terms to Know in this unit.
2. List the items in the school agricultural mechanics shop that are used for the storage of:
   - lumber and metal
   - fasteners such as nails and screws
   - flammable liquids
   - waste materials
   - tools
3. Examine the tools and equipment in your school agricultural mechanics shop that are used to clean the shop. Learn to use each item to clean thoroughly.
4. Check all containers used in your agricultural mechanics shop to see if they are of the proper type and properly labeled. Report your findings to the teacher.
5. Help improve the cleanup equipment and facilities in your shop.
6. Volunteer for the shop-cleaning job of your choice.

Relevant Web Sites

Texas A&M University, Department Of Chemistry, Flammable and Hazardous Materials information
www.chem.tamu.edu/safety/flammable_haz.html

Ryvac Engineering Company, Dust Collection and Vacuum Systems information
www.ryvac.com

Self-Evaluation

A. Multiple Choice. Select the best answer.

1. Oily rags should be stored in a
   a. cardboard box
   b. plastic bag
   c. wooden box
   d. closed, metal can

2. A clean, organized shop reduces the chance of
   a. fire
   b. lost tools
   c. damage to projects
   d. all of these

3. Brushes and brooms work better if pushed
   a. in a continuous path
   b. and lifted intermittently
   c. back and forth
   d. in long strokes

4. Sawdust is useful in shop cleanup to
   a. absorb liquids on the floor
   b. reduce dust in the trash container
   c. condition bristles on floor brooms
   d. none of these

5. A recommended material for cleaning grease from the floor is
   a. water
   b. gasoline
   c. solvents
   d. sawdust

6. The foreperson's job in the cleanup process is
   a. supervision
   b. reward
   c. evaluation
   d. assigning jobs

7. The best item for cleaning nongreasy machines is a/an
   a. rag
   b. brush
   c. air gun
   d. vacuum cleaner

8. The shop-cleaning method that gives the best control over the cleanup process is the
   a. all-pitch-in method
   b. cleanup wheel method
   c. assignment sheet method
   d. honor system method

9. The main advantage of the shop cleanup assignment sheet over the shop cleanup wheel is the
   a. flexibility in assigning students to tasks
   b. ease in reassigning tasks
   c. use of a checklist for evaluations
   d. use of a foreperson for evaluations

10. Rotating shop cleanup duties
    a. enables everyone to learn the various cleaning tasks
    b. promotes fairness in assigning undesirable tasks
    c. involves every student on an equal basis
    d. all of these

B. Completion. Fill in the blanks with the word or words that make the following statements correct.

1. A quick and efficient cleanup procedure is important to the _____ of the students.

2. The key to a clean shop is _____.

3. Any shop cleanup method requires a _____ _____ on the part of all involved.

4. Tools should be mounted on panels over colored outlines of each tool, called _____.

5. In the presence of _____, a steel flammable materials cabinet automatically closes.

6. Effective cleaning skills cannot be taken for granted; they must be _____.

C. Brief Answer. Briefly answer the following questions.

1. List fifteen signs of a properly cleaned shop.

2. List seven reasons for using a properly organized cleanup procedure.

3. List fifteen tasks involved in shop cleanup.

4. Name two cleaning tools that are used to move trash from the floor to the trash cans.

5. What is the best tool to use for cleaning machines?

HAND WOODWORKING AND METALWORKING

Hand Tools, Fasteners, and Hardware

Objective

To identify and correctly spell hand tools, hardware, and fasteners that are commonly used in agricultural mechanics.

Competencies to be developed

After studying this unit, you should be able to:

- Describe how tools are classified.
- Name the major tool categories, according to use.
- Classify and correctly spell hand tools commonly used in agricultural mechanics.
- Identify and correctly spell commonly used screws, nails, and bolts.
- Select screws, nails, and bolts for various uses.
- Identify and correctly spell important items of hardware.

Materials List

- Hand tools commonly used in agricultural mechanics
- Common types of screws, nails, and bolts
- Items of hardware commonly used in agricultural mechanics

Terms to Know

- tool
- hand tool
- power tool
- layout tool
- saw
- boring tool
- driving tool
- hammer
- pliers
- wrench
- chisel
- punch
- clamp
- wrecking tool
- digging tool
- fastener
- laminate
- adhesive
- nail
- diameter
- shank
- head
- penny
- galvanized
- screw
- threads

continued

The correct use of hand tools is fundamental in agricultural mechanics. Large and efficient power tools are used to do most of the work in society. However, hand tools are used to do the small jobs and to do the work where large machines cannot function.

The use of hand tools is basic and essential for work in design, construction, maintenance, and repair. Hand tools are used by all who construct buildings, install landscape structures, wire computers, or repair tractors. In fact, all people in society use hand tools or pay others to do jobs that require their use. A **tool** is any instrument used in doing work. A **hand tool** generally refers to any tool operated by hand that makes use of muscle power rather than electricity or other power sources (Figure 7-1). This is contrasted with a **power tool**, which is operated by some source of power other than human power. However, note that some people may classify such tools as a power hand drill as a hand tool.

TOOL CLASSIFICATIONS

There are various methods of classifying tools. Therefore, the student needs to be prepared to recognize tools by their various classifications. Tools may be classified according to who uses them, e.g., carpenters' tools, masons' tools, mechanics' tools, or machinists' tools. Tools from all such trades are used in agriculture. Therefore, classification of tools by use or function is more meaningful to workers in agricultural mechanics.

FIGURE 7-1 Hand tools are instruments that use human muscle power to do work.

Terms to Know (continued)

- square head
- hex head
- slotted-head screw
- Phillips-head screw
- Allen screw
- drywall screw
- machine bolt
- cap screw
- carriage bolt
- stove bolt
- plow bolt
- nut
- washer
- machine screw
- hardware
- hinge
- butt hinges
- strap hinge
- T hinge
- screw hook and strap hinge
- hasp
- gussets
- flush plate

The following discussion outlines some of the various types of hand tools commonly used in the agricultural mechanics lab. It is essential that you learn the proper name of tools to be able to communicate correctly about tools. One of the common ways to misname a tool is to call it by a brand name. The following outline and images should help you to learn the proper classification and name of a particular tool. For example, locking pliers are sometimes called Vise Grip® pliers because of the popularity of that brand. Many of the specific tools outlined here will be explained in other chapters that deal with their use.

Layout Tools (L)

A **layout tool** is a tool used to measure or mark wood, metal, or other materials (Figure 7-2). Rules, squares, levels, chalk lines, measuring tapes, and plumb bobs are among some of the tools classified as layout tools. These tools are used to measure and mark materials before cutting or shaping is done. Measuring and marking are done when laying out work so other functions can follow according to plan.

Saws

A **saw** is a tool used to cut wood or other materials. This permits the user to shape the material (Figure 7-3). There are several different types of saws that are operated by hand, and each has a specific purpose. These purposes can range from the simple sawing of a board using a handsaw to the cutting of dovetail joints using a dovetail saw.

Boring Tools (B)

A **boring tool** is a tool used to make holes or change the size or shape of holes. Boring tools include bits, drills, reams, and the devices used to turn them (Figure 7-4). In addition to many common hand tools, this group includes a large variety of specialized tool bits for power machines. There is a wide variety of different types and sizes of drill bits. Some are for drilling holes in wood, some for metal, and some for masonry such as concrete.

Hammers

Another group of tools classified by their use is hammers that are used as **driving tools** (Figure 7-5). Hammers rely on their weight and speed to provide force to move an object when it is hit. A hammer is used for moving a sliding belt tightener, striking a cold **chisel** to cut a rivet, forcing nails into wood, or driving stakes into the soil. Some examples of hammers are sledges, claw hammers, peen hammers, and non-marring hammers such as rubber mallets.

Pliers

Very common tools used to grip wood, metal, plastic, and other materials are known as **pliers** (Figure 7-6). Pliers are used to hold material while other tools are used to cut, shape, modify, or turn threaded items like screws. Pliers are also used to grip objects such as bolts or pieces of wire. They may be used to bend or shape such objects as needed. Pliers come in an array of types for different uses. They range from the commonly used slip-joint pliers to locking pliers to wire-crimping pliers.

Wrenches

A very basic hand tool group is referred to as wrenches. Engine and machinery mechanics as well as people doing everyday household maintenance tasks make frequent use of wrenches. Wrenches are used to turn nuts, bolts, or screws. **Wrenches** may include tools such as boxed-end, open-end, combination, socket, and Allen wrenches (Figure 7-7).

Chisels and Punches

A category of tools that are usually used by striking them with a hammer is chisels and punches. Chisels are used to cut metal, masonry products, or wood depending on the type of chisel used. **Punches** are used to align holes to insert a pin, to make a dent in metal to start a drill, set a nail below the surface in wood, and many other uses (Figure 7-8).

Screwdrivers

Screwdrivers are found in every toolbox and are basic tools used on a daily basis. There are several different types of screwdrivers. Straight-bit screwdrivers turn traditional slot-headed screws; Phillips screwdrivers turn screws with a cross-shaped opening in the head of the screw. A newer type of screwdriver is called a torx that drives a screw with a star-shaped or groove-shaped head. Nut runners are also included in this category because of their close resemblance to and similar function of a screwdriver. These turn small bolts or nuts (Figure 7-9).

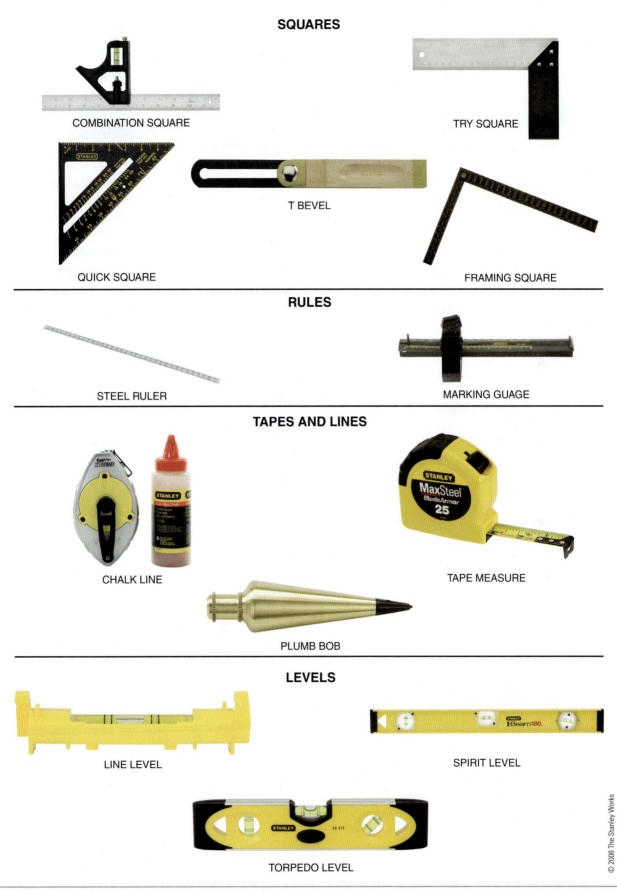

SQUARES

COMBINATION SQUARE

TRY SQUARE

QUICK SQUARE

T BEVEL

FRAMING SQUARE

RULES

STEEL RULER

MARKING GUAGE

TAPES AND LINES

CHALK LINE

TAPE MEASURE

PLUMB BOB

LEVELS

LINE LEVEL

SPIRIT LEVEL

TORPEDO LEVEL

© 2008 The Stanley Works

FIGURE 7-2 Layout tools.

SAWS

BACKSAW

COPING SAW

DRYWALL SAW

HACKSAW

HANDSAW

KEYHOLE SAW

DOVETAIL SAW

MINI HACKSAW

FOLDING POCKET SAW

MITER SAW

© 2008 The Stanley Works

FIGURE 7-3 Cutting tools.

CUTTERS

AVIATION CUTTERS

CARPET KNIFE

RAZOR BLADE SCRAPER

NAIL NIPPERS

SIDE CUTTERS

SNAP-OFF BLADE KNIFE

SPORT UTILITY KNIFE

TIN SNIPS

UTILITY KNIFE

HAWK BILL KNIFE

HOBBY KNIFE

© 2008 The Stanley Works

FIGURE 7-3 (*Continued*)

BORING TOOLS

BRAD-POINT BIT

COUNTERSINK BIT

PLUG CUTTER

FORSTNER BIT

DRILL-BIT EXTENSION

SPADE-BORE BIT

TWIST BIT

© 2008 The Stanley Works

FIGURE 7-4 Boring tools.

HAMMERS

BALL-PEEN HAMMER

BLACKSMITH'S HAMMER

CURVED CLAW HAMMER

DEAD-BLOW HAMMER

RUBBER HAMMER

SHINGLING HATCHET

SLEDGEHAMMER

STRAIGHT CLAW HAMMER

MASON'S HAMMER

DRILLING HAMMER

TACK HAMMER

© 2008 The Stanley Works

FIGURE 7-5 Hammers.

PLIERS

DIAGONAL OR SIDE-CUTTING PLIERS

GROOVE-JOINT PLIERS

LINEMAN'S PLIERS

LOCKING PLIERS

NAIL NIPPERS

SLIP-JOINT PLIERS

LONG NOSE OR NEEDLENOSE PLIERS

WELDING PLIERS

WIRE-CRIMPING PLIERS

© 2008 The Stanley Works

FIGURE 7-6 Pliers.

Clamps

Clamps are used to draw and hold objects together. As you can imagine, they have a very wide range of uses and come in many different sizes, shapes, and types to fit different needs. Vises are also included in this group (Figure 7-10).

Wrecking Tools

Some tools are designed for taking apart materials such as walls and other structures that are to be demolished. Wrecking tools are used as pry bars to pry apart boards that are nailed together. Obviously, the larger the boards the larger the wrecking tool that is needed. Most of these tools can also be used to remove nails (Figure 7-11).

Digging Tools (D)

Many people working in agriculture use a group of tools known as digging tools (designated by the capital letter **D**). A digging tool is any device used to turn up, loosen, or remove earth (Figure 7-12). Digging tools include shovels, mattocks, hoes, rakes, posthole diggers, and garden trowels.

WRENCHES

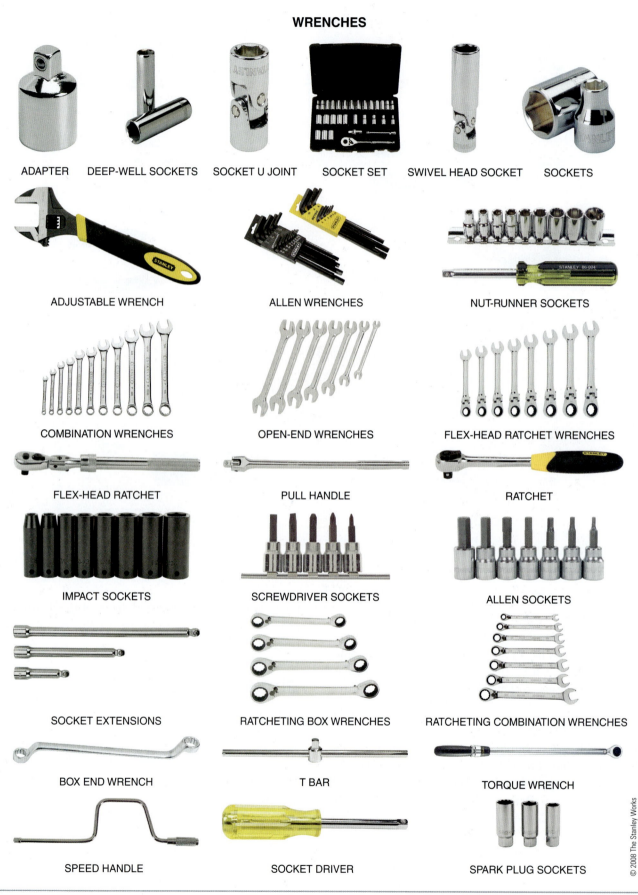

ADAPTER   DEEP-WELL SOCKETS   SOCKET U JOINT   SOCKET SET   SWIVEL HEAD SOCKET   SOCKETS

ADJUSTABLE WRENCH   ALLEN WRENCHES   NUT-RUNNER SOCKETS

COMBINATION WRENCHES   OPEN-END WRENCHES   FLEX-HEAD RATCHET WRENCHES

FLEX-HEAD RATCHET   PULL HANDLE   RATCHET

IMPACT SOCKETS   SCREWDRIVER SOCKETS   ALLEN SOCKETS

SOCKET EXTENSIONS   RATCHETING BOX WRENCHES   RATCHETING COMBINATION WRENCHES

BOX END WRENCH   T BAR   TORQUE WRENCH

SPEED HANDLE   SOCKET DRIVER   SPARK PLUG SOCKETS

© 2008 The Stanley Works

FIGURE 7-7 Wrenches.

CHISELS AND PUNCHES

COLD CHISEL   CENTER PUNCH   FLOORING CHISEL

MASON'S CHISEL   PRICK PUNCH   DRIFT PUNCH

WOOD CHISELS   NAIL SETS   CHISELS AND PUNCHES

PIN PUNCH   BRICK SET   SCRAPER

© 2008 The Stanley Works

FIGURE 7-8 Chisels and punches.

FASTENERS

A **fastener** is any device used to hold two or more pieces of material together or in place. The carpenter creates a building out of individual pieces by using fasteners such as nails. By using fasteners such as bolts and nuts, the manufacturer assembles a machine from many parts. Fasteners are used to hang pictures and laminate panels. **Laminate** means to fasten two or more flat pieces together with an adhesive. An **adhesive** is a sticky substance such as glue. The most common fasteners used in agricultural

mechanics are nails, screws, bolts, nuts, washers, and machine screws.

Nails

A **nail** is a fastener that is driven into the material it holds. There are many types of nails classified generally by their use or form (Figure 7-13). When buying nails, the purchaser must know the use for the nail, the desired length, and the desired thickness or diameter. **Diameter** refers to the distance

SCREWDRIVERS

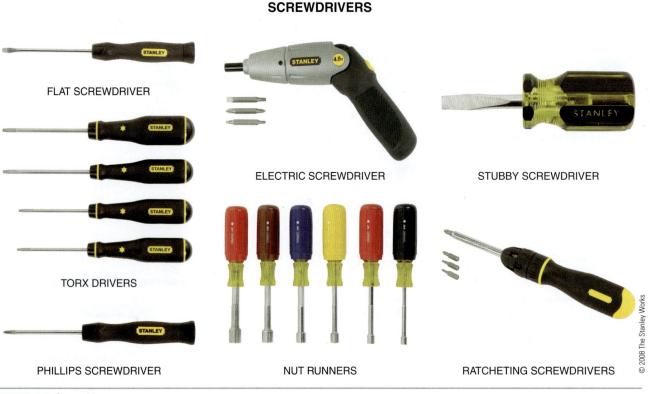

FLAT SCREWDRIVER

TORX DRIVERS

PHILLIPS SCREWDRIVER

ELECTRIC SCREWDRIVER

NUT RUNNERS

STUBBY SCREWDRIVER

RATCHETING SCREWDRIVERS

© 2008 The Stanley Works

FIGURE 7-9 Screwdrivers.

across the center of a round circle or object. The **shank** is the long stem part of a nail or screw; the **head** is the enlarged part on top. Probably the most familiar nail is the common nail. It has a fairly thick shank and medium-sized head. Nails that are flat and tapered are called cut nails. These are cut from steel and used for wood flooring, or they may be hardened and used as fasteners in masonry materials.

Nail Lengths. The unit of measure used to designate the length of most nails is the **penny** (Figure 7-14). The symbol for penny is the lower-case letter d. The term penny was originally used to indicate the number of English pennies needed to purchase 100 nails of a given size. The lengths of common nails, box nails, finishing nails, cut nails, and spikes are designated by penny. The size ranges for these nails are:

- common nail—2d to 60d
- box nail—2d to 40d
- finishing nail—2d to 20d
- cut nail—2d to 20d
- spike—16d to 12 inches

Uses for Nails. Nails vary in thickness of shank and diameter of head according to their use. If the material being held is soft, a large-headed nail is needed. Otherwise, the material will pull over the head of the nail. If the material being held is heavy, a thick, strong shank is required. Following are some important types of nails and their uses:

- common nail—used for general construction; nailing sheeting, shiplap, and board fencing
- cut nail—used for nailing tongue-and-groove flooring; if hardened, used for nailing in masonry materials
- box nail—used for light household construction; nailing siding on buildings; nailing into the end grain of boards
- finishing nail—used for interior finishing of buildings; trim, cabinet, and furniture work, when countersinking is needed
- shingle nail—used for nailing wood and shingles
- roofing nail—used for nailing rolled roofing and composition shingles
- plasterboard nail—used to attach plasterboard to studs in buildings
- hinge nail—used to fasten hinges on doors and cabinets

CLAMPS

FIGURE 7-10 Clamps.

- duplex nail—used for construction of forms for concrete work and for nailing insulators on wooden posts for electric fencing
- wire staple—used in wire fence construction
- lead-head nail—used for nailing galvanized steel roofing and siding

Improved Nails. Changes in nail forms have come about with the development of new types of building materials. Some new materials require special fastening devices. For instance, soft insulating boards should be nailed with special nails having large, square heads.

© 2008 The Stanley Works

WRECKING TOOLS

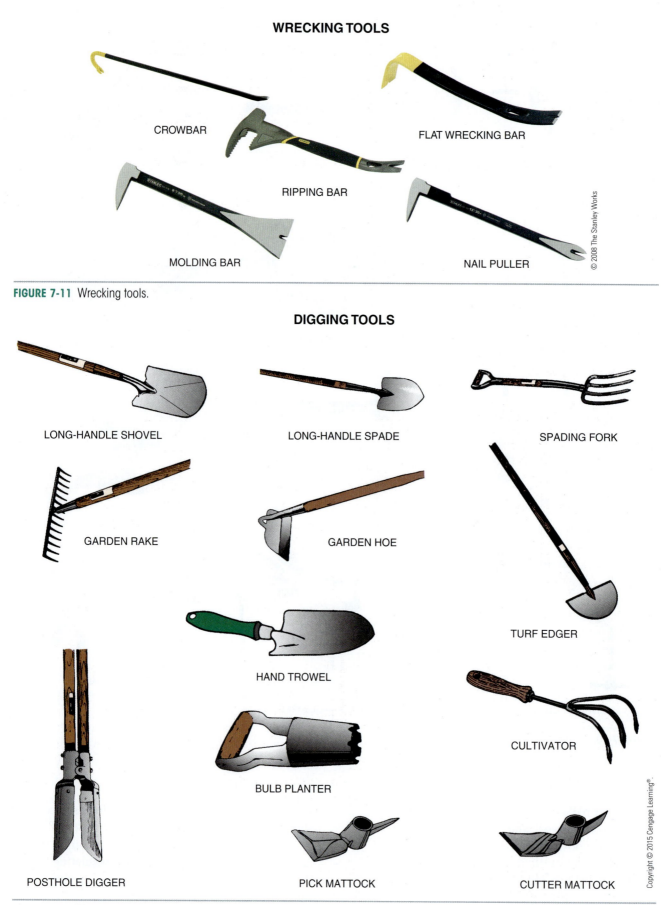

CROWBAR

RIPPING BAR

FLAT WRECKING BAR

MOLDING BAR

NAIL PULLER

© 2008 The Stanley Works

FIGURE 7-11 Wrecking tools.

DIGGING TOOLS

LONG-HANDLE SHOVEL

LONG-HANDLE SPADE

SPADING FORK

GARDEN RAKE

GARDEN HOE

TURF EDGER

HAND TROWEL

CULTIVATOR

POSTHOLE DIGGER

BULB PLANTER

PICK MATTOCK

CUTTER MATTOCK

Copyright © 2015 Cengage Learning®.

FIGURE 7-12 Digging tools.

STAPLE   HINGE PLASTERBOARD ROOFING

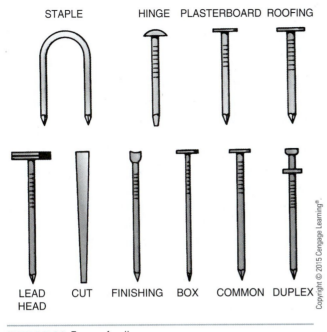

LEAD
HEAD   CUT   FINISHING   BOX   COMMON   DUPLEX

FIGURE 7-13 Types of nails.

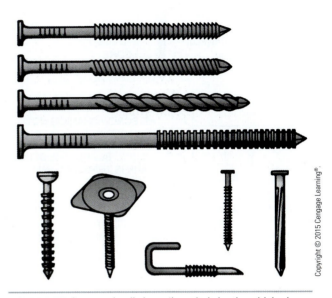

FIGURE 7-15 Improved nails have threaded shanks, which give them more holding power than regular nails.

2d 3d 4d 6d 8d 10d 12d 16d 20d 30d 40d 50d 60d

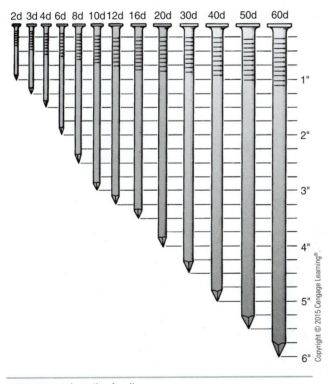

1"

2"

3"

4"

5"

6"

FIGURE 7-14 Length of nails.

Improved nails, sometimes referred to as thread nails, are basically the same as regular nails. They differ in that a portion of the nail shank is threaded with annular or helical threads (Figure 7-15). The indentations in the threads along the nail shank provide grooves into which the wood fibers can expand. Thus, both friction and sheer resistance make it very difficult to remove the nail. Nails that must be driven into very hard substances are heat treated to make them hard.

Nails are generally made of steel or aluminum. Steel nails will rust and must be galvanized if exposed to moisture. **Galvanized** means coated with zinc. Steel nails can be coated with a varnishlike cement to make them hold wood together more securely. Nails used with aluminum should be made of aluminum. This is necessary since steel nails will rust and galvanized nails will react with and destroy the aluminum that they touch.

Screws

A **screw** is a fastener with threads that bite into the material it fastens. The term **threads** refers to grooves of even shape and taper that wrap continuously around a shank or hole. Screws are made for use in wood, sheet metal, plastic, and any material solid enough to hold them. Screws generally cut threads into the material to which they hold (Figure 7-16).

Classifications of Screws. Screws can be classified according to the material they hold. Wood screws have threads designed to bite into wood fibers, which draw the screw into the wood when turned. Sheet metal screws have threads that are wide enough to

Copyright © 2015 Cengage Learning®.

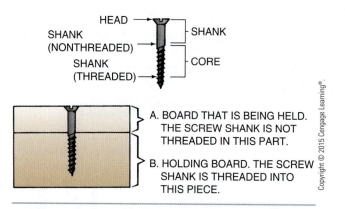

FIGURE 7-16 Screws hold and resist pulling out by threading into the holding material (B). The screw must not be threaded in the material being held (A). This permits material A to draw tight against material B.

permit the thin metal to fit between the ridges of the threads. Cap screws are designed to thread into thick metal that has matching threads cut into the metal. Lag screws, also called lag bolts, have very coarse threads designed for use in structural timber or lead-wall anchors.

Screws can be classified according to the metal they are made from or the finish used. Steel screws may be coated with a blued, galvanized, cadmium, nickel, chromium, or brass finish. Solid brass screws are rustproof and are used where severe moisture problems destroy coated or plated screws.

Screws can be classified by the shape of their heads (Figure 7-17). Flat-head screws have tapered heads designed to fit into a hole countersunk in the material.

The result is a screw head that is flush with the surface. Round-head screws have heads that are half-round with a slot in the upper part. These heads or pan-head screws look like an upside-down frying pan. They are used on sheet metal. A Phillips-head screw has a flat head with slots in the head that look like a plus sign (+). Lag screws have unslotted heads that are either square or hex-shaped. (A **square head** has four equal sides; a **hex head** has six equal sides.) Lag screws are turned with a wrench instead of a screwdriver to obtain more leverage.

Screws can be classified by the type of tool needed to turn them. A **slotted-head screw** requires a standard screwdriver. A **Phillips-head screw** requires a screwdriver shaped like a plus sign. A six-sided, hex, or Allen wrench is used to turn **Allen screws**.

Other types of screws are designed to be driven with a power drill or power screwdriver. These screws, which must be harder to withstand the force needed to turn them, usually have a Phillips head, but they might have a square-shaped notch in the head for accepting a square drive bit. A common type of screw designed for power drivers is called a deck screw. Deck screws come in lengths of 1 to 3½ inches and have the same diameter size as regular wood screws. They are galvanized and are used to install outdoor decking. They are also used in a variety of woodworking applications where a power driver is used.

Size of Screws. The size of the screw is specified by the diameter and length of shank. Diameter is expressed by gauge numbers that run from 2 through 24.

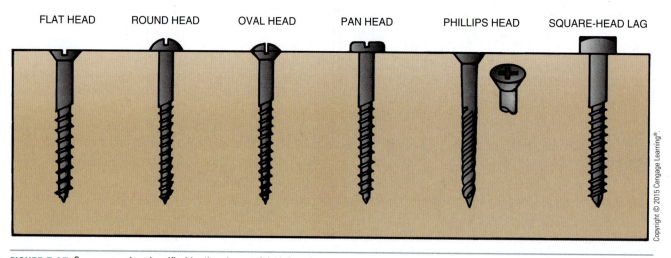

FIGURE 7-17 Screws can be classified by the shape of their heads.

Numbers 6, 8, and 10 are common sizes. Common lengths range from ¼ inch through 4 inches (Figure 7-18).

Drywall Screws. The **drywall screw** has a flat Phillips head, with a very thin shank. It is made from very hard and tough steel that can be driven with a power driver or drill. Screws can be purchased with black or galvanized finish and flat or pan heads and have threads that run the entire length of the screw. They are commonly used to install drywall but are also used as wood screws in many woodworking projects. They can be driven with power tools without

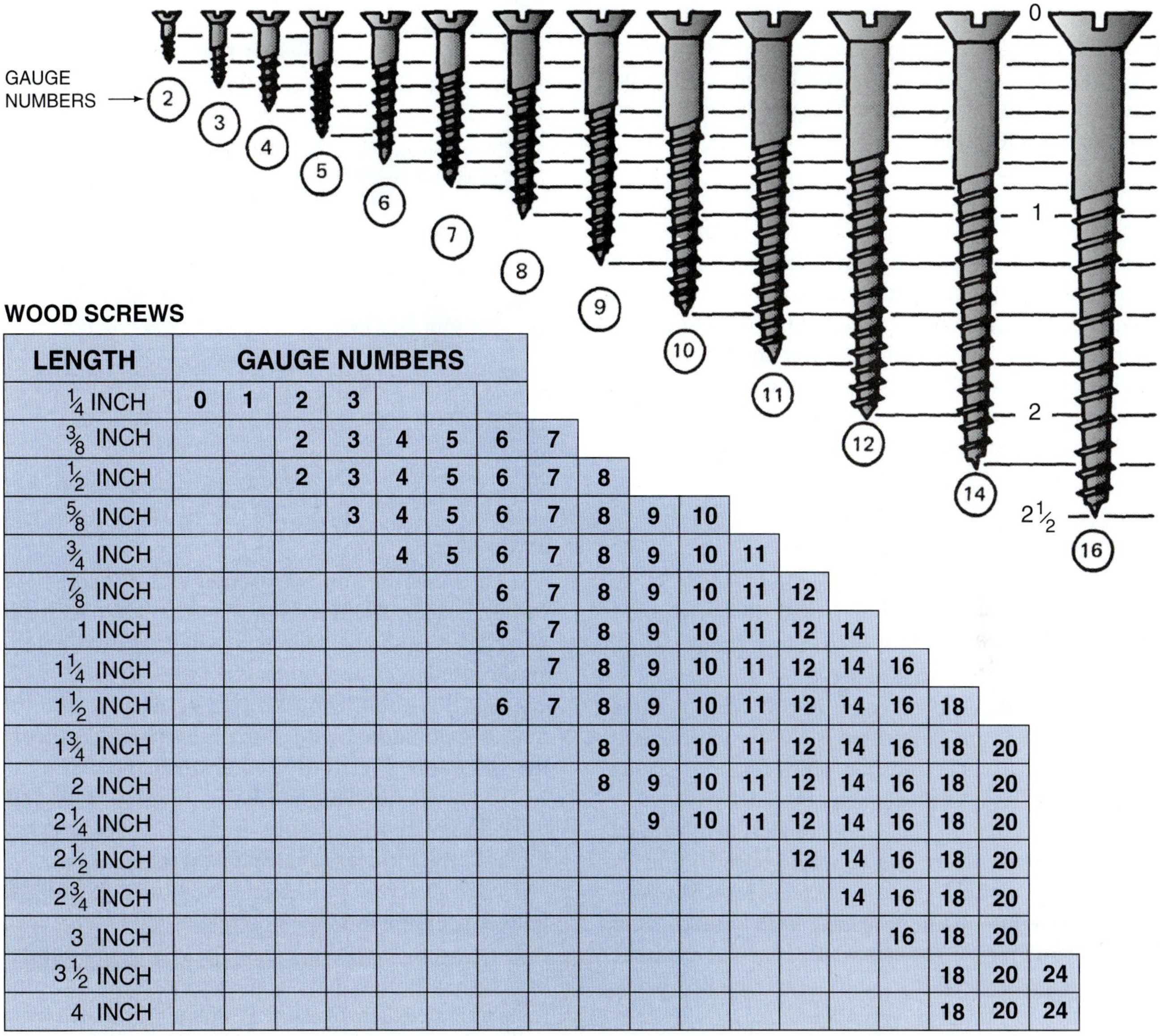

WOOD SCREWS

LENGTH	0	1	2	3	4	5	6	7	8	9	10	11	12	14	16	18	20	24
¼ INCH	0	1	2	3														
⅜ INCH			2	3	4	5	6	7										
½ INCH			2	3	4	5	6	7	8									
⅝ INCH				3	4	5	6	7	8	9	10							
¾ INCH					4	5	6	7	8	9	10	11						
⅞ INCH							6	7	8	9	10	11	12					
1 INCH							6	7	8	9	10	11	12	14				
1¼ INCH								7	8	9	10	11	12	14	16			
1½ INCH							6	7	8	9	10	11	12	14	16	18		
1¾ INCH									8	9	10	11	12	14	16	18	20	
2 INCH									8	9	10	11	12	14	16	18	20	
2¼ INCH										9	10	11	12	14	16	18	20	
2½ INCH													12	14	16	18	20	
2¾ INCH														14	16	18	20	
3 INCH															16	18	20	
3½ INCH																18	20	24
4 INCH																18	20	24

When you buy screws specify (1) length, (2) gauge number, (3) type of head (flat, round, or oval), (4) material (steel, brass, bronze, etc.), and (5) finish (bright, steel blued, cadmium, nickel, or chromium-plated).

FIGURE 7-18 A screw's gauge number refers to the diameter of its head, while a screw's length is expressed in inches. Wood screws can be purchased in various combinations of length and gauge number. Drywall and deck screws can also be purchased in a variety of lengths and diameters.

Copyright © 2015 Cengage Learning®.

making drilled holes as needed with standard screws (Figure 7-18)

Deck Screws.

Outdoor decks are generally installed using long flat-head screws called deck screws. These screws are hardened and made with a Phillips head so they can be driven using a power drill or power driver. They can be galvanized, specially coated, or made of stainless steel to prevent corrosion. Like drywall screws, they are used in many different types of woodworking projects. They differ from drywall screws, in that they have a shank where the threads do not go the entire length of the screw. This allows wood to be tightly drawn together.

Square and Torx Head Screws.

Some screws are designed for use with a power drill or power driver. The head is indented with a square-shaped or star-shaped hole that a power bit fits into. This design provides for a sure fit between the driver and the screw to minimize slippage. These screws are generally harder and come in a wide variety of sizes and styles designed for both wood and metal.

Bolts

A bolt is a fastener with a threaded nut (Figure 7-19). Types of bolts that are often used in agricultural mechanics include machine bolts, carriage bolts, stove bolts, plow bolts, and special bolts.

The threads on bolts come in two types—coarse and fine. They are designated UNC for coarse and UNF for fine. The U stands for "unified," and N stands for "national." A $\frac{3}{8}$-inch-diameter thread designated as $\frac{3}{8} \times 24$ means that the bolt has 24 threads per inch and is a fine thread. A bolt $\frac{3}{8} \times 16$ would have 16 threads per inch and would be a coarse thread. Most automotive engine bolts are fine-threaded. Both types come in standard and metric sizes.

Bolts are generally straight, with round, square, or hex heads. However, some are bent, modified, or shaped for special uses. These include the eye bolt, hook or J bolt, U bolt, turnbuckle, and plow bolt. A lag screw, when used with an expansion plug in a masonry wall, is known as an expansion bolt. A toggle is a collapsible winged nut that opens when inserted into a hollow wall. A toggle with bolt is called a toggle bolt.

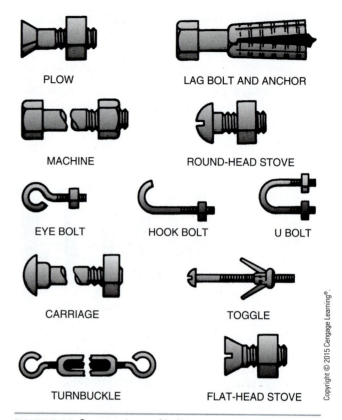

PLOW

LAG BOLT AND ANCHOR

MACHINE

ROUND-HEAD STOVE

EYE BOLT

HOOK BOLT

U BOLT

CARRIAGE

TOGGLE

TURNBUCKLE

FLAT-HEAD STOVE

Copyright © 2015 Cengage Learning®.

FIGURE 7-19 Common types of bolts.

The **machine bolt** is a fastener with a square head or hex head with threads on just the last 1 inch (or more) of the shank. Machine bolts are used extensively in engines and other machines. Special hardened machine bolts with extra strength and toughness have heads marked with short lines that point from the shoulders toward the center of the head (Figure 7-20). Three lines designate a very tough bolt, and six lines designate a very, very tough bolt. The tougher the bolt, the more expensive it is to manufacture.

The **cap screw** looks like a machine bolt with the following exceptions. A cap screw usually is threaded over its entire length, generally is 2 inches or shorter in length, and may have a special head, such as one requiring an Allen wrench. It threads into an object rather than a nut.

The **carriage bolt** is a fastener with a round head over square shoulders and is used with wood. The shoulders are drawn down into the wood to prevent the bolt from turning. The round, low profile of the head makes it almost flush with the surface of the wood.

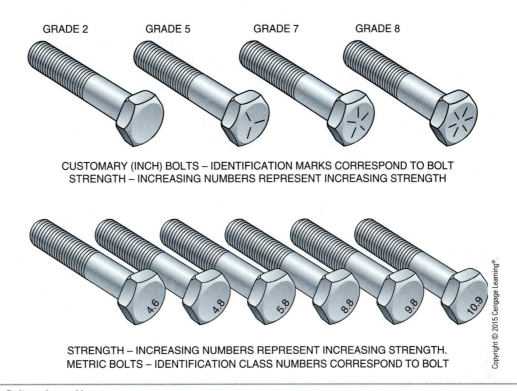

GRADE 2   GRADE 5   GRADE 7   GRADE 8

CUSTOMARY (INCH) BOLTS – IDENTIFICATION MARKS CORRESPOND TO BOLT
STRENGTH – INCREASING NUMBERS REPRESENT INCREASING STRENGTH

STRENGTH – INCREASING NUMBERS REPRESENT INCREASING STRENGTH.
METRIC BOLTS – IDENTIFICATION CLASS NUMBERS CORRESPOND TO BOLT

Copyright © 2015 Cengage Learning®.

FIGURE 7-20 Bolt grade markings.

Carriage bolts are used for construction of wagon bodies, feeders, doors, and other wooden projects.

The **stove bolt** is a round-headed bolt with a straight screwdriver slot. It is threaded the entire length. When purchased, stove bolts generally come with square nuts. They are generally available in sizes up to ⅜ inch diameter and up to 6 inches in length.

The **plow bolt** has a square tapered head. When placed in a hole designed for a plow bolt, the head is flush with the surface. The plow bolt is used to hold shares and other parts in place on tillage implements.

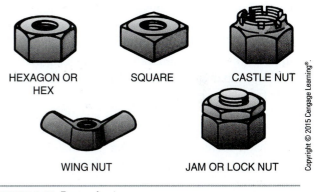

HEXAGON OR   SQUARE   CASTLE NUT
HEX

WING NUT   JAM OR LOCK NUT

Copyright © 2015 Cengage Learning®.

FIGURE 7-21 Types of nuts.

Nuts

A **nut** is a device with a threaded hole (Figure 7-21). Nuts are the movable part of bolts that are used to fasten two items together. Nuts may be square (with four sides) or hexagonal (with six sides). Hexagonal is also simply referred to as hex. Wing nuts have winged extensions for tightening with the fingers.

Special nuts are available with slots on one side to permit the use of cotter pins or keys to lock the nut in place. Another technique used to lock a nut in place is to tighten a second nut against the first. The second nut is then referred to as a lock nut. Some nuts are said to be self-locking. These have special design features that make them difficult to turn on or off a bolt. Like bolts, nuts are also graded according to their strength (Figure 7-22).

Washers

A **washer** is a flat device with a hole in the center and is used as part of a fastener (Figure 7-23). Flat washers are used to prevent bolt heads or nuts from

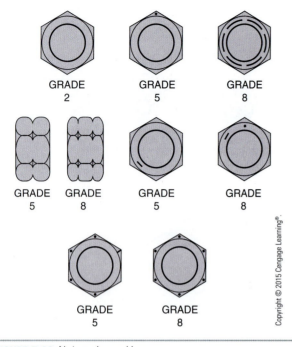

FIGURE 7-22 Nut grade markings.

DIAMETER NUMBER	THREADS PER INCH
4	40
6	32
6	40
8	32
8	36
10	24
10	32
12	24
12	28
¼*	20
¼	28

*Same as ¼-inch stove bolt or cap screw with standard threads

FIGURE 7-24 Machine screw sizes and threads.

penetrating material. Special lock washers are used to prevent nuts or bolts from loosening due to vibration and use. Lock washers may be split and made with spring steel to exert back pressure when pressed. They may also have spring steel fingers that bite the surfaces that press against them. The spring pressure or biting capability of lock washers prevents nuts or bolts from loosening.

Washers are designated by the size of the bolt for which they are designed. For instance, a washer with a hole that is correct for fitting on a ⅜-inch bolt is called a ⅜-inch washer. Washers are sold by the piece or by the pound.

Machine Screws

A **machine screw** is a small bolt with a hex nut. Machine screws range in size from very tiny up to less than ¼ inch in diameter, and up to several inches in length. The diameter of machine screws is expressed by a number (Figure 7-24). The most common machine screws range from number 4 (very small) to number 12 (nearly ¼ inch in diameter).

Threads on machine screws are designated by the number per inch. Finer threads have more threads per inch than coarse threads and are less likely to vibrate loose. On the other hand, coarse threads permit greater ease of starting and greater speed in applying a nut.

When buying machine screws, the diameter by number, the threads per inch, and the length must be specified. For example, a 4–40 × ½ machine screw is one with a number 4 diameter, 40 threads per inch, and ½ inch long.

HARDWARE

The term **hardware** is used in agricultural mechanics for special fasteners. These include hinges, brackets, plates, and miscellaneous metal objects. Hardware is used to support doors, lock windows, secure sideboards on trucks, and anchor roofs. Whenever two objects need to be connected, hardware is generally used. Hardware, as the term suggests, is hard because it is usually made from steel or brass.

FLAT WASHER

EXTERNAL LOCK WASHER

SPLIT LOCK WASHER

COUNTERSUNK LOCK WASHER

FIGURE 7-23 Types of washers.

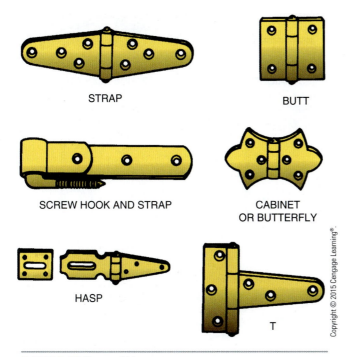

STRAP

BUTT

SCREW HOOK AND STRAP

CABINET
OR BUTTERFLY

HASP

T

Copyright © 2015 Cengage Learning®.

FIGURE 7-25 Types of hinges.

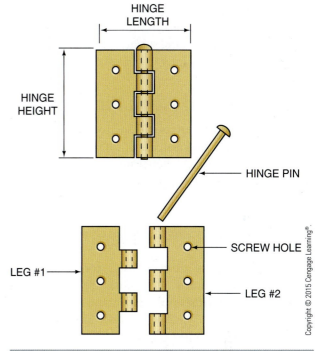

HINGE
LENGTH

HINGE
HEIGHT

HINGE PIN

SCREW HOLE

LEG #1

LEG #2

Copyright © 2015 Cengage Learning®.

FIGURE 7-26 A hinge consists of three parts: two legs and a pin. The hinge pin interlocks leg 1 with leg 2. Hinges with removable pins may be purchased for use on doors intended for occasional removal. Many hinge pins, however, are not removable.

Hinges

A **hinge** is an object that pivots and permits a door or other object to swing back and forth or up and down. Small hinges used on cabinets and furniture are referred to as cabinet hinges. Hinges are classified according to their type and size. Some common types of hinges are butt hinges, strap hinges, T hinges, screw hook and strap hinges, and hasps (Figure 7-25). All hinges consist of two parts plus a pin. The parts are fastened to the stationary object and the moving object. The pin is then inserted to permit movement between the parts (Figure 7-26).

Butt Hinges. **Butt hinges** are used to mount doors in a butted position. This means the door and the strip it is mounted on are flush and form a smooth surface. Butt hinges are fairly short in length as compared to their height or the length of their pins. Butt hinges may be mounted on the surface or may be mounted in a concealed position between the door and its mounting strip. If the hinges are concealed or out of sight, they must be set in the material and have countersunk screw holes with flat-head screws.

Some surface-mounted butt hinges are designed to be attractive. An example is the butterfly hinge. It is so named because it is shaped like a butterfly in flight.

Strap Hinges. **Strap hinges** are much longer than they are high. Their long, thin, straplike appearance gives them their name. The strap hinge is used where the hinge must reach across a long surface; it provides additional support for the door. This feature is useful where the door or its mounting strip is not very strong. Strap hinges are also useful where excessive weight or pressure is used on a door.

T Hinges. The **T hinge** is made by combining the features of the butt hinge and the strap hinge. One leg of the hinge is short and high, like a butt hinge. The other leg is narrow and long, like a strap hinge. The T hinge is useful where the mounting strip for a door is thin but strong, and the door is flimsy and needs extra hinge support.

Screw Hook and Strap Hinges. The **screw hook and strap hinge** is also called a gate hinge. Two special screw hooks are screwed into a post or doorjamb. The two straps are then mounted in the correct positions on the door. The door with hinge straps mounted is set or hung on the screw hooks. This completes the installation.

Hasps. The **hasp** is not really a hinge but is constructed like one and is sometimes classified with hinges.

It is used to hold doors closed. One part of the hasp is attached to the door. The hinged part fits over a metal loop attached to the mating surface. A padlock or other device is placed in the loop to hold the door closed.

Brackets and Flush Plates

Hardware suppliers can provide many types of brackets, plates, door closers, door catches, and other hardware items for construction and maintenance. Special devices called **gussets** are available to reinforce joints in wood and metal. Wall brackets and special mounting channels with prefinished shelving are some hardware items that make construction fast and easy.

Some common devices used for strengthening corners of frames and doors are flush plates (Figure 7-27). A **flush plate** is made of metal about $\frac{1}{16}$ inch thick and ½ inch wide. They are drilled and have countersunk holes to receive flat-head screws. Flush plates are available as straight, tee, inside, or flush corners.

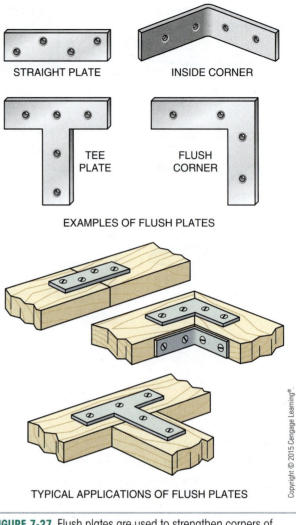

STRAIGHT PLATE   INSIDE CORNER

TEE PLATE   FLUSH CORNER

EXAMPLES OF FLUSH PLATES

TYPICAL APPLICATIONS OF FLUSH PLATES

Copyright © 2015 Cengage Learning®.

FIGURE 7-27 Flush plates are used to strengthen corners of frames and doors.

SUMMARY

The beginning point for developing most agricultural mechanics skills is learning the proper tools and hardware. The most basic of tools are those powered by humans rather than by electricity or another power source. Learning the names and uses of these tools is the first step toward achieving the skill level needed to carry out projects. Likewise, learning the basic types of hardware and fasteners is also essential.

Student Activities

1. Define the Terms to Know in this unit.

2. Study individual tools provided by the instructor and classify them according to their use.

3. Prepare a bulletin board or computer document with captions showing the major classes of tools: layout, cutting, boring, driving, holding, turning, digging, and other. Cut pictures of tools from old issues of magazines and catalogs or copy and paste from the Internet, and place them under their caption on the bulletin board.

4. Examine the tool storage area in your agricultural mechanics department. On a sheet of paper, list all tools that you see. Place a letter after each tool to indicate its classification as follows: L = Layout; C = Cutting; B = Boring; Dr = Driving; H = Holding; T = Turning; D = Digging; and O = Other.

5. Make up a board with samples of various types of nails attached; properly label the nails.

6. Make up a board with samples of screws and bolts attached; properly label them.

7. Visit a hardware store and examine various nails, screws, bolts, and other fasteners. Ask the store manager for manufacturers' charts or tables that describe fasteners.

8. Ask your instructor if there are some activities you can do to improve the arrangement and storage of nails, screws, bolts, and other fasteners in the agricultural mechanics department.

Relevant Web Sites

The Home Depot
www.homedepot.com

New Tool News
www.newtoolnews.com

Self-Evaluation

A. Multiple Choice. Select the best answer.

1. The use of hand tools is
   a. for those who cannot afford power tools
   b. for a limited number of highly specialized jobs
   c. primarily for engine and machinery mechanics
   d. the foundation of agricultural mechanics

2. Tools are generally classified according to
   a. use
   b. color
   c. construction
   d. origin

3. An example of a layout tool is/are
   a. claw hammer
   b. outside calipers
   c. handsaw
   d. plug cutter

4. Saws are classified as
   a. kerf tools
   b. push tools
   c. flexing tools
   d. cutting tools

5. Taps and dies are classified as
   a. holding tools
   b. digging tools
   c. cutting tools
   d. turning tools

6. Wrenches are classified as
   a. turning tools
   b. digging tools
   c. other tools
   d. cutting tools

7. The lowercase letter *d* is used to designate sizes of
   a. lumber
   b. screws
   c. nails
   d. bolts

8. Which is not the name of a type of nail?
   a. lumber
   b. plasterboard
   c. roofing
   d. duplex

9. The term *improved* means a nail
   a. is made of copper
   b. is easy to remove
   c. has a thick shank
   d. holds better

10. Screws are classified
    a. according to the material they hold
    b. by the metal from which they are made or the finish used
    c. by the shape of their heads
    d. by all of these

11. The number of a screw refers to its
    a. diameter
    b. length
    c. head type
    d. use

12. The difference between a bolt and a screw is
    a. a bolt has threads
    b. a bolt has a nut
    c. a screw has a slotted head
    d. a screw is suitable for use in wood

13. A bolt used in wood that has a round head over square shoulders is
    a. a stove bolt
    b. a machine bolt
    c. a carriage bolt
    d. none of these

14. How many sides are there on a hexagon nut?
    a. four
    b. six
    c. eight
    d. twelve

15. A 4–40 × ½ machine screw
    a. has 4 threads per inch
    b. has 40 threads per inch
    c. is 4 inches long
    d. comes four to the package

16. Which is not a type of hinge?
    a. butt
    b. strap
    c. T
    d. N

17. The hinge that contains a feature from each of the two other hinges is the
    a. butt hinge
    b. strap hinge
    c. T hinge
    d. N hinge

18. The best hinge to use if it is not to be seen is the
    a. butt hinge
    b. strap hinge
    c. T hinge
    d. N hinge

19. The best hinge for a very large and extra-heavy door or gate is the
    a. butt hinge
    b. T hinge
    c. N hinge
    d. screw hook and strap hinge

20. Corners of frames and doors may be strengthened by using a
    a. hasp
    b. butt hinge
    c. flush plate
    d. hook and eye bolt

B. **Brief Answer.** Briefly answer the following questions.

1. What is a hand tool?
2. Name and correctly spell six layout tools.
3. Name and correctly spell six cutting tools.
4. Name and correctly spell six boring tools.
5. Name and correctly spell six driving tools.
6. Name and correctly spell six holding tools.
7. Name and correctly spell six turning tools.
8. Name and correctly spell six digging tools.

C. **Identification.** Name, correctly spell, and classify each of the following tools.

1.

2.

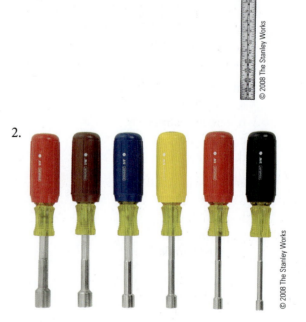

© 2008 The Stanley Works

3.

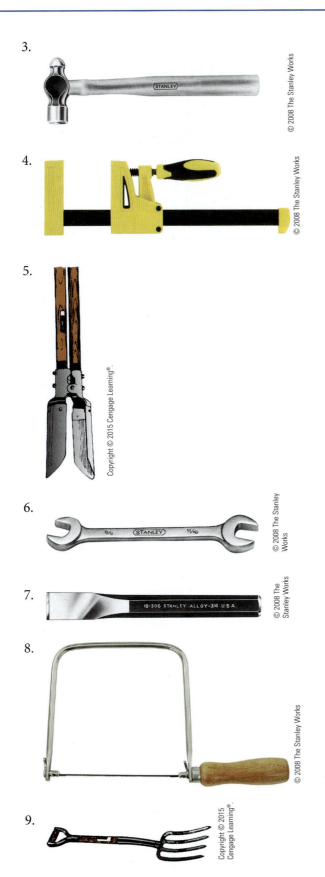

4.

5.

6.

7.

8.

9.

© 2008 The Stanley Works

© 2008 The Stanley Works

Copyright © 2015 Cengage Learning®.

© 2008 The Stanley Works

© 2008 The Stanley Works

© 2008 The Stanley Works

Copyright © 2015 Cengage Learning®.

D. Identification. Name, correctly spell, and classify each of the following fasteners.

1.

2.

3.

4.

5.

Copyright © 2015 Cengage Learning®.

Copyright © 2015 Cengage Learning®.

Copyright © 2015 Cengage Learning®.

Copyright © 2015 Cengage Learning®.

Copyright © 2015 Cengage Learning®.

6.

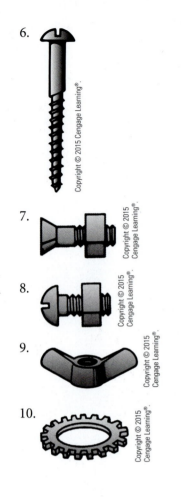

7.

8.

9.

10.

E. **Identification.** Name, correctly spell, and classify each of the following items of hardware.

1.

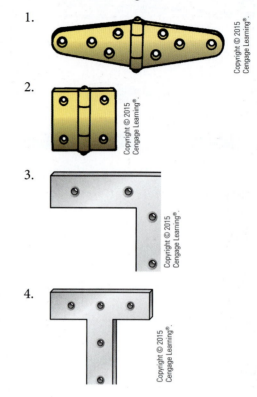

2.

3.

4.

Objective

To select and use appropriate layout tools and procedures for woodworking and metalworking.

Competencies to be developed

After studying this unit, you should be able to:

- Select appropriate tools for layout procedures in woodworking and metalworking.
- Use layout tools correctly and accurately.
- Make and use a pattern.
- Lay out wood and metal for cutting and shaping.

Materials List

- Examples of tools of poor quality that have been bent or broken with normal use
- A meter stick and a 12-inch ruler
- 12-foot tape (with English and metric numbers preferred)
- Steel or wooden scale
- Folding rule
- Inside calipers
- Outside calipers
- A variety of gauges (sheet metal, wire, drill, feeler, and spark plug)
- Framing square
- Combination square

continued

Terms to Know

- layout
- gradations
- scale
- aluminum
- plastic
- craftsperson
- English system
- U.S. customary system
- linear
- metric system
- meter
- millimeters
- kilo
- tape
- folding rule
- caliper
- gauge
- framing square
- heel
- try square
- combination square
- plumb
- miter
- sliding T bevel
- bevel
- right triangle
- dividers

continued

Terms to Know (Continued)

- scribers
- level
- spirit level
- laser
- line
- chalk line
- plumb line
- plumb bob
- pattern

Materials List (Continued)

- Try square
- Sliding T bevel
- Dividers and scribers
- Carpenter's level and line level
- Plumb line
- Chalk line
- Tag board and scissors
- String for 12-foot lines
- Hammer and three slender 12-inch stakes

Layout tools are used to guide the worker when planning or constructing projects (Figure 8-1). The term **layout** means to prepare a pattern for future operations. A truck driver uses a map to determine the correct route between two points. The map requires the use of many roads, numerous turns, and changes in direction to get to the destination. The situation is similar when workers design and construct projects, machinery, equipment, or buildings. They place marks on their work and then cut, shape, dig, or fasten according to the marks.

Many workers must use layout tools in performing their work. These workers include plumbers, electricians, designers, engineers, surveyors, farmers, landscapers, builders, and many others. The student will do well to learn how to use layout tools carefully. Layout tools have many uses.

When laying out work, it is important to choose the right tool. A second important factor is to use the tool correctly. For many jobs, there are several tools that will work, but usually one works better than others. For instance, a 1-foot ruler or yardstick may be used to measure a house that is 60 feet long. However, a 100-foot tape is quicker and more accurate.

© iStockphoto/Lisa McDonald

FIGURE 8-1 Layout tools are used to guide the worker in planning and constructing projects.

MATERIALS USED FOR LAYOUT TOOLS

Measuring devices are generally made of steel, aluminum, wood, plastic, or cloth. Steel tools, especially stainless steel, are generally more durable than those made of other materials (Figure 8-2). When numbers and lines, called **gradations**, are stamped into steel tools, they stay readable for a long time. Numbers and gradations make up a **scale** on measuring tools. Another advantage of steel is that it bends without breaking. Steel can withstand the rough use found at construction sites. Steel is the first choice of material for calipers, dividers, and other slender tools that must measure very accurately and not bend easily.

Aluminum is the second choice of material for use in many layout tools. Aluminum is a very tough metal that is light and durable. It is used extensively in levels and squares (Figure 8-3).

Wood is a cheap, relatively soft, and light material. It is good for handles of layout tools. Wood is not used extensively in tools because it breaks easily and does not wear well. Also, wood may absorb moisture and expand. When it dries out, it contracts. Any distortion of layout tools can be a problem. An exception is the wooden folding rule, which is compact, useful, and popular. However, it is very easily broken if not handled correctly.

Plastic is a term used for a group of materials made from chemicals and molded into objects. Some plastics are tough and light and can be made into useful layout tools. A major advantage is they are relatively cheap and are light and easy to use. A major disadvantage of plastics is their tendency to melt if touched by a hot object or flame. They may also be damaged by solvents found in shops. Some layout tools such as rulers, protractors, squares, and angles are made of plastic (Figure 8-4). Their use is quite popular for limited jobs where durability is not a factor.

Cloth is a material used in some 50-foot tapes. It is cheap and lightweight. Cloth tapes are not very accurate as they stretch under stress and moisture. Any stretching of a measuring device reduces its accuracy.

Cost Versus Quality

Cheap tools are seldom a bargain. A cheap tool is likely to be inaccurate and unsatisfactory after little use. Good tools may seem expensive; however, they last a lifetime if used properly. High-quality tools are an excellent investment. Craftspeople use only good tools. A **craftsperson** is a skilled worker. Good tools permit the worker to become a good craftsperson if proper

FIGURE 8-2 Stainless steel tools are generally the most durable.

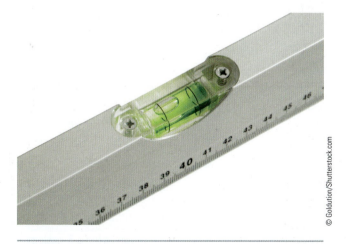

FIGURE 8-3 Aluminum is light, tough, and durable. Many levels are made from aluminum.

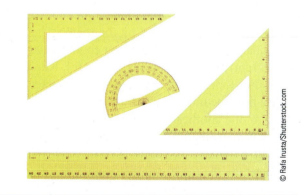

FIGURE 8-4 Plastic is used to make rulers, protractors, squares, and triangles.

techniques are used. On the other hand, even skilled workers cannot do good work with poor tools.

SYSTEMS OF MEASUREMENT

The **English system** has always been the standard system of measurement in the United States. This system is now called the **U.S. customary system** (Figure 8-5). It uses the inch, the foot, the yard, the rod, and the mile as units of linear measure. **Linear** simply means in a line. The English system uses units of various sizes, based on their similarity to a certain familiar object or part of the human body, such as the foot or the hand. Even today, horses are said to be so many "hands" high.

There are 12 inches in a foot, 3 feet in a yard, 16 ½ feet in a rod, and 5,280 feet in a mile. The English system does not seem very logical.

In the early times, people could function satisfactorily with approximate units of measure such as a hand measure for the height of a horse. However, as the technological age evolved, measurements had to be standardized. Can you imagine the results if the human foot was used as the method for measuring when constructing buildings?

Today, the foot and most other linear units of measure are very precise lengths. If layout tools are constructed properly, a foot is the same length on all tools. Variations may be observed on poor-quality tools and on tools that measure long distances, such as tapes. The longer the instrument, the greater the problem of inaccuracy due to stretching, expansion, and contraction. The foot is represented by one mark (') and the inch by two marks ("). Hence, 2'6" means 2 feet and 6 inches.

The **metric system** of measurement has always been used for scientific work in the United States (Figure 8-6). Only recently has the metric system been used for nonscientific work. Linear measurements in the metric system are based on the **meter**. The meter (m) equals 39.37 inches. The meter contains 100 centimeters (cm) and 1,000 **millimeters** (mm). **Kilo** means 1000. One thousand meters equals 1 kilometer (km). Notice how the units of measure in the metric system relate to one another by multiples of ten. The metric system is mathematically logical and easy to use.

The millimeter, the centimeter, the meter, and the kilometer are the most used metric units of linear measurement in agricultural mechanics. The inch, the foot, the yard, the rod, and the mile are the most useful units of linear measurement in the U.S. customary system.

Liquid Measure

2 cups	1 pint
2 pints	1 quart
4 quarts	1 gallon
31½ gallons	1 barrel
2 barrels	1 hogshead
7½ gallons of water	1 cu. ft.
U.S. Gallon	231 cu in

Avoirdupois Weight

16 ounces	1 pound
100 pounds	1 hundredweight
20 hundredweight	1 ton

Dry Measure

2 pints	1 quart
8 quarts	1 peck
4 pecks	1 bushel

1 bushel contains 2150.42 cubic inches or approximately 1¼ cubic feet.

Square Measure

144 square inches	1 square foot
9 square feet	1 square yard
30¼ square yards	1 square rod
160 square rods	1 acre (43,560 sq. ft.)
640 acres	1 square mile
1 square mile	1 section

Weight

Gram	15.432 grains
Gram	.0353 ounce
Kilogram	2.2046 lb
Kilogram	.0011 ton (short)
Metric ton	1.1025 ton (short)
Grain	.064 gram
Ounce	28.35 grams
Pound	453.5 grams
Ton (Short)	907.18 kilograms
Ton (Short)	907 metric tons
Ton (Short)	2,000 lb

FIGURE 8-5 The English system, now known as the U.S. customary system, has long been the standard system of weights and measures in the United States.

Copyright © 2015 Cengage Learning®.

Linear Measure *continued*

12 inches..1 foot
3 feet .... 1 yard
5½ yards (16½ feet)......................................1 rod
320 rods .... 1 mile
1,760 yards (5,280 feet) .... 1 mile

Cubic Measure

1,728 cubic inches..1 cu. ft.
27 cubic feet .... 1 cu. yd.
128 cu. ft. (8′ × 4′ × 4′) ....1 cord
1′ × 1′ × 1″...1 bd. ft.

Liquid Data

1 gallon of water weighs 8.34 pounds
1 gallon of milk weighs 8.6 pounds
1 cubic foot of water is equal to 7.48 gallons
1 inch of rainfall amounts to 27,154 gallons per acre

FIGURE 8-5 (*continued*)

The U.S. customary system is still the most used system in the United States. However, the metric system is increasing in use and many layout tools use both systems (Figure 8-7).

The Inch as a Unit of Measurement

The inch is the traditional unit of measurement for woodworking and metalworking in the United States. It must be divided into smaller units to be useful for most applications. Some fine rules or scales may have as many as 32 marks per inch. Each mark is $\frac{1}{32}$ of an inch apart. One-sixteenth inch is more commonly used as the smallest unit on a rule. Lines of different lengths are used to show ½, ¼, and ⅛ of an inch on many measuring devices, such as rules and squares (Figure 8-8).

The Millimeter as a Unit of Measurement

The millimeter (mm) is slightly smaller than $\frac{1}{16}$ of an inch. It is a very convenient unit for linear measurement without using fractions. It has the advantage of being $\frac{1}{1000}$ of a meter and $\frac{1}{10}$ of a centimeter. This means 1 meter plus 250 millimeters equals 1¼

meters. It is more convenient to write 1¼ meter using a decimal rather than a fraction. Using the decimal, 1¼ meters reads 1.250 meters. Similarly, 1½ meters reads 1.500 meters, or simply 1.5 meters. On many metric rules, each centimeter division contains ten marks to represent millimeters (Figure 8-9). Centimeters can be changed to millimeters by multiplying by 10. Simply adding a 0 to the centimeter's value obtains the millimeter value, e.g., 4 centimeters is 40 millimeters.

When working on small projects, measurements may be made entirely in millimeters to avoid the use of fractions or decimals. If a board is 1.5 meters in length, this value can be written as 1,500 millimeters. To change meters to millimeters, multiply by 1,000.

MEASURING TOOLS

Measuring tools include tapes, rules, calipers, and other devices used to determine specific distances. They are used to measure length, width, height, depth, thickness, spacing, and clearance.

When measuring, the last number on the scale is read and then the fraction of inches or number of millimeters is added, depending on the scale. If very exact measurements are necessary, a thin tape is used or the measuring device is turned up on its edge to get an exact reading.

Tapes

The term **tape** refers to flexible measuring devices that roll into a case. Tapes range in length from several feet to hundreds of feet and are made of steel, cloth, or fiber materials. Some tapes are self-retracting; they have a spring inside to wind up the tape when released. Some have locks to hold the tape out while in use. Some have a button on the side to control the movement of the tape.

Most tapes have a hook on the end that slides a distance equal to the thickness of the hook. This permits the tape to provide accurate measurements whether hooked to an object or pushed against a surface. Figure 8-10 shows various ways of measuring using a tape measure. Students must handle tapes carefully as they are easily broken. Tapes break if pulled out too far, forced back into the case, or bent excessively.

Folding Rules

A **folding rule** is a rigid rule, 2 to 8 feet in length. It can be folded for handling and storage. At one time the folding

A Comparison of the International Metric System and the U.S. Customary System of Measurement

1 Centimeter	= 0.3937 inch	1 Kilometer	= 1000 meters	1 Gallon	= 3.785 liters
1 Inch	= 2.54 centimeters	1 Kilometer	= 0.62137 mile	1 Gram	= 15.43 grains
1 Foot	= 30.48 centimeters	1 Sq. Centimeter	= 0.155 sq. inch	1 Ounce	= 28.35 grams
1 Meter	= 39.37 inches	1 Sq. Decimeter	= 100 cu. centimeters	1 Kilogram	= 1000 grams
1 Meter	= 100 centimeters	1 Cu. Centimeter	= 0.061 cu. inch	1 Kilogram	= 2.205 pounds
1 Meter	= 1.094 yards	1 Cu. Decimeter	= 1000 cu. centimeters	1 Pound	= 7000 grains
1 Meter	= 1000 millimeters	1 Cu. Meter	= 100 liters	1 Pound	= 0.4536 kilogram
1 Millimeter	= 0.001 meter	1 Fluid Ounce	= 29.54 milliliters	1 Kilogram	= 1000 milliters
1 Yard	= 0.9144 meter	1 Liter	= 1000 cu. centimeters	1 Kilogram	= 1 liter
1 Mile	= 1609.344 meters	1 Liter	= 1.057 quarts		

Metric Prefixes

Names of multiples of Metric units are formed by adding a prefix to "meter," "gram," or "liter." These prefixes are used also with units other than these three metric ones.

tera	- T	10^{12}	1,000,000,000,000.	ONE TRILLION					
tera	- T	10^{12}	1,000,000,000,000.	ONE TRILLION	—	-	10^{0}	1.	ONE
giga	- G	10^{9}	1,000,000,000.	ONE BILLION	deci	- d	10^{-1}	0.1	ONE TENTH
mega	- M	10^{6}	1,000,000.	ONE MILLION	centi	- c	10^{-2}	0.01	ONE HUNDREDTH
kilo	- k	10^{3}	1,000.	ONE THOUSAND	milli	- m	10^{-3}	0.001	ONE THOUSANDTH
hecto	- h	10^{2}	100.	ONE HUNDRED	*micro	- μ	10^{-6}	0.000001	ONE MILLIONTH
deka	- dk	10^{1}	10.	TEN	nano	- n	10^{-9}	0.000000001	ONE BILLIONTH
—	-	10^{0}	1.	ONE	pico	- p	10^{-12}	0.000000000001	ONE TRILLIONTH

*The special case of one millionth of a meter is called a micron.

Conversion Table

Multiply	To Obtain	Multiply	To Obtain	Multiply	To Obtain
Bushels by	Cu. Feet	Cu. Yards by 764600	Cu. Centimeters	Meters by 100	Centimeters
Bushels by	Pecks	Cu. Yards by 22	Bushels	Meters by 3.281	Feet
Bushels by 0.04545	Cu. Yards	Cu. Yards by 27	Cu. Feet	Meters by 39.37	Inches
Centimeters by 0.3937	Inches	Cu. Yards by 46.656	Cu. Inches	Meters by 0.01	Kilometers
Centimeters by 0.01	Meters	Cu. Yards by 0.7646	Cu. Meters	Meters by 1000	Millimeters
Centimeters by 10	Millimeters	Cu. Yards by 202.0	Gallons	Meters by 1.094	Yards
Cu. Centimeters by 0.00003531	Cu. Feet	Cu. Yards by 764.6	Liters	Microns by 0.00001	Meters
Cu. Centimeters by 0.06102	Cu. Inches	Cu. Yards by 1616	Pints (Liq.)	Miles by 160900	Centimeters
Cu. Centimeters by 0.00001	Cu. Meters	Cu. Yards by 807.9	Quarts (Liq.)	Miles by 5280	Feet
Cu. Centimeters by 0.000001308	Cu. Yards	Fathoms by 6	Feet	Miles by 1.609	Kilometers
		Feet by 30.48	Centimeters	Miles by 1760	Yards

FIGURE 8-6 Units of measure in the metric system, and factors for converting U.S. customary system units to metric system units and vice versa.

Copyright © 2015 Cengage Learning®

Conversion Table *continued*

Multiply	To Obtain	Multiply	To Obtain	Multiply	To Obtain
Cu. Centimeters by 0.0002642	Gallons	Feet by 12	Inches	Miles per Hr. by 44.70	Centimeters per Sec
Cu. Centimeters by 0.001	Liters	Feet by 0.3048	Meters	Miles per Hr. by 88	Feet per Min.
Cu. Centimeters by 0.002113	Pints (Liq.)	Feet by 1/3	Yards	Miles per Hr. by 1.467	Feet per Sec.
Cu. Centimeters by .001057	Quarts (Liq.)	Gallons by 3785	Cu. Centimeters	Miles per Hr. by 1.609	Kilometers per Hr.
Cu. Feet by .0002832	Cu. Centimeters	Gallons by 0.1337	Cu. Feet	Miles per Hr. by 0.8684	Knots
Cu. Feet by 1728	Cu. Inches	Gallons by 231	Cu. Inches	Miles per Hr. by 26.82	Meters per Min.
Cu. Feet by 0.02832	Cu. Meters	Gallons by .003785	Cu. Meters	Millimeters by 0.03937	Inches
Cu. Feet by 0.03704	Cu. Yards	Gallons by .004951	Cu. Yards	Ounces by 2	Tablespoons (Liq.)
Cu. Feet by 7.48052	Gallons	Gallons by 3.785	Liters	Ounces by 6	Teaspoons (Liq.)
Cu. Feet by 28.32	Liters	Gallons by 8	Pints (Liq.)	Ounces by 3	Tablespoons (Dry)
Cu. Feet by 59.84	Pints (Liq.)	Gallons by 4	Quarts (Liq.)	Ounces by 9	Teaspoons (Dry)
Cu. Feet by 29.92	Quarts (Liq.)	Gallons Water by 8.3453	Lbs. of Water	Ounces by 28.349527	Grams
Cu. Feet by 1.25	Bushels	Grams by 0.03527	Ounces	Ounces by 0.9115	Ounces (Troy)
Cu. Inches by 16.39	Cu. Centimeters	Grams by 0.03215	Ounces (Troy)	Ounces (Fluid) by 1.805	Cu. Inches
Cu. Inches by 0.0005787	Cu. Feet	Grams by .002205	Pounds	Pounds by 16	Ounces
Cu. Inches by 0.00001639	Cu. Meters	Inches by 2.540	Centimeters	Pounds of Water by 0.01602	Cu. Feet
Cu. Inches by 0.00002143	Cu. Yards	Kilometers by 100000	Centimeters	Pounds of Water by 27.68	Cu. Inches
Cu. Inches by 0.004329	Gallons	Kilometers by 3281	Feet	Pounds of Water by 0.1198	Gallons
Cu. Inches by 0.01639	Liters	Kilometers by 1000	Meters	Tablespoons (Liq.) by 0.5	Ounces
Cu. Inches by 0.03463	Pints (Liq.)	Kilometers by 0.6214	Miles	Tablespoons (Dry) by 0.3333	Ounces
Cu. Inches by 0.01732	Quarts (Liq.)	Kilometers by 1094	Yards	Tablespoons by 3	Teaspoons
Cu. Meters by 10000	Cu. Centimeters	Liters by 1000	Cu. Centimeters	Teaspoons (Liq.) by 0.1666	Ounces
Cu. Meters by 35.31	Cu. Feet	Liters by 0.03531	Cu. Feet	Teaspoons (Dry) by 0.1111	Ounces
Cu. Meters by 61.023	Cu. Inches	Liters by 61.02	Cu. Inches	Teaspoons by 0.3333	Tablespoons
Cu. Meters by 1.308	Cu. Yards	Liters by .01	Cu. Meters	Temp (C)+17.78 by 1.8	Temp (F)
Cu. Meters by 264.2	Gallons	Liters by .001308	Cu. Yards	Temp (F)−32 by 5/9	Temp (C)
Cu. Meters by 1000	Liters	Liters by 0.2642	Gallons	Tons (Long) by 2240	Pounds
Cu. Meters by 2113	Pints (Liq.)	Liters by 2.113	Pints (Liq.)		
Cu. Meters by 1057	Quarts (Liq.)	Liters by 1.057	Quarts (Liq.)		
		T (in) by W (ft) by L (ft) Board Feet			

FIGURE 8-6 *(continued)*

rule was the standard measuring device used by carpenters. However, the compactness and length of modern tapes make them preferable to the folding rule for many jobs. Folding rules are made of wood, plastic, or metal.

Good-quality wooden folding rules have spring-loaded buttons to keep them open; these rules are called spring-joint rules. A good wooden folding rule is made of tough wood and is easy to read. Some folding rules have a sliding metal insert in one end (Figure 8-11). The insert permits accurate inside measurements and depth measurements.

Special care must be used in opening and closing folding rules. A careless twist can cause the rule to break. However, careful workers, who oil the joints periodically, may use the same folding rule day after day for many years without breakage.

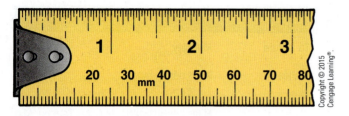

FIGURE 8-7 This tape displays units of measure from both the U.S. customary system and the metric system. The 1, 2, 3 values refer to inches. The 20, 30, 40 values refer to millimeters.

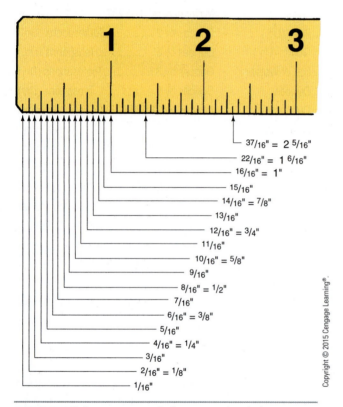

37/16" = 2 5/16"
22/16" = 1 6/16"
16/16" = 1"
15/16"
14/16" = 7/8"
13/16"
12/16" = 3/4"
11/16"
10/16" = 5/8"
9/16"
8/16" = 1/2"
7/16"
6/16" = 3/8"
5/16"
4/16" = 1/4"
3/16"
2/16" = 1/8"
1/16"

FIGURE 8-8 Rules, squares, and other measuring devices are marked to show halves, quarters, eighths, and sixteenths of an inch. Often, vertical lines of different heights represent these various intervals.

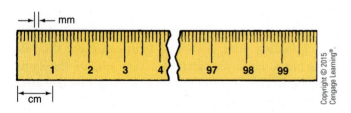

FIGURE 8-9 The marks that appear between each centimeter division on a rule represent millimeters. A meter is equal to 100 centimeters or 1,000 millimeters.

MEASURING THE CIRCUMFERENCE OF A ROUND OBJECT

TAKING A MEASUREMENT STARTING AT A NAIL

TAKING A MEASUREMENT STARTING FROM THE END OF AN OBJECT

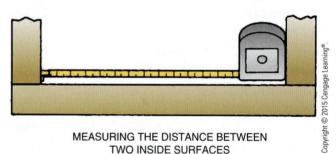

MEASURING THE DISTANCE BETWEEN TWO INSIDE SURFACES

FIGURE 8-10 Tapes can be used to measure in various ways. The hook provided on one end of the tape measure allows accurate measurement whether it is hooked onto an object or pushed against a surface. A slot where the hook fastens to the tape provides a sliding action that allows for the thickness of the hook.

Scales and Bench Rules

The term *scale* as used here means a rigid steel or wooden measuring device (Figure 8-12). Scales are generally 1 to 3 feet in length and ¾ to 1 inch in width. Wooden scales are sometimes called *bench rules*. They are about ¼ inch thick. Metal scales are relatively thin and, therefore, better than wooden scales where very accurate work is required.

Scales and bench rules are handy in the shop. They are not used much on other jobs since they cannot be folded.

Copyright © 2015 Cengage Learning®

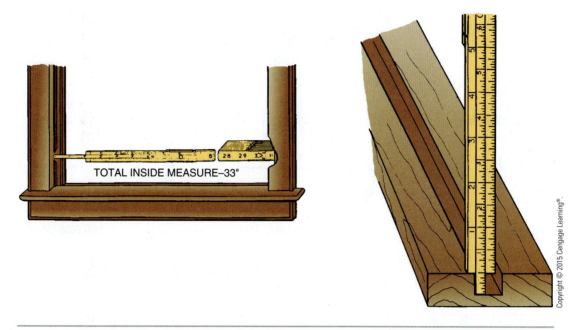

TOTAL INSIDE MEASURE–33"

FIGURE 8-11 The insert in the end of a folding ruler makes the ruler useful for taking accurate inside measurements.

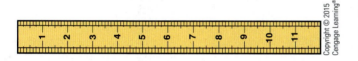

FIGURE 8-12 Wooden bench rules and steel scales are useful in the shop. They are rarely used on job sites, however, because they cannot be folded.

Calipers

A **caliper** is an instrument used to measure the diameter or thickness of an object. There are inside calipers and outside calipers (Figure 8-13). Inside calipers are so named because they measure inside distances, such as the interior diameter of a cylinder. Outside calipers are used to measure the outside of round objects, such as pipes.

With the most economical caliper, a rule or scale may be needed to measure its settings. Other calipers have their own dimension scales. Another type of caliper has a dial to indicate very small differences in the size of objects measured.

Gauges

A **gauge** is a device used to determine thickness, gap in space, diameter of materials, or pressure of flow. Examples are sheet-metal gauges, wire gauges, drill gauges, spark plug gauges, feeler gauges, and pressure gauges.

SQUARES AND THEIR USES

Squares are used to guide the builder. A square is a device used to draw angles for cutting and check the cuts for accuracy. The squares most often used are the framing square, try square, combination square, and the sliding T bevel.

Framing Square. The **framing square** is a flat square with a body and tongue. It is usually made of steel. The body is 24 inches long and the tongue is 16 inches long. The **heel** is the place where the tongue and body meet to form a 90-degree angle.

The framing square is also called a carpenter's square or a steel square. It may contain several tables. One table found on many framing squares helps the user to calculate the board feet in a piece of lumber. Another table may help the user to lay out rafters. Special booklets and chapters in advanced texts provide more information on the use of these tables.

Framing squares have measurements that start at the inside of the heel and run outward along the body and the tongue. Inside measurements of the tongue are 1 to 14 inches. Inside measurements of the body are 1 to 22 inches. Outside measurements of the tongue are 1 to 16 inches. Another set of measurements starts at the outside of the heel and runs out of the body and tongue. Outside measurements of the body are 1 to 24 inches. It is important to study the scales carefully

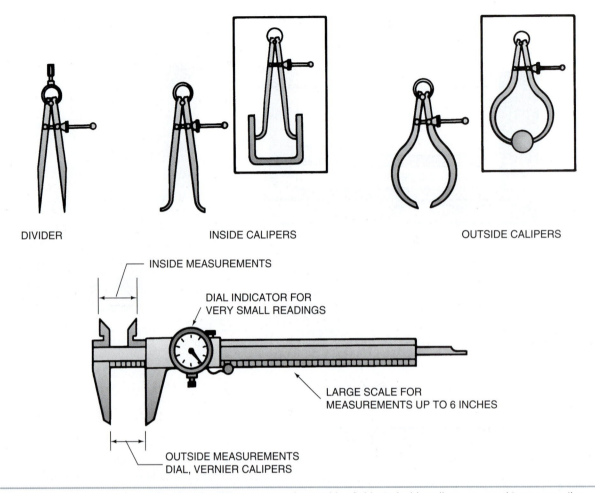

DIVIDER   INSIDE CALIPERS   OUTSIDE CALIPERS

INSIDE MEASUREMENTS

DIAL INDICATOR FOR
VERY SMALL READINGS

LARGE SCALE FOR
MEASUREMENTS UP TO 6 INCHES

OUTSIDE MEASUREMENTS
DIAL, VERNIER CALIPERS

Copyright © 2015 Cengage Learning®.

FIGURE 8-13 Outside calipers are used to measure the distance across the outside of objects. Inside calipers are used to measure the interiors of objects. Vernier calipers provide very precise measurements.

before using a framing square. Figure 8-14 shows how the tongue and the body are used to measure different materials.

Try Square. The **try square** is so named because it is used to try or test the accuracy of cuts that have been made. It is also used to mark lines on boards in preparation for cutting. Try squares have wood, steel, or plastic handles and steel blades.

The try square is a good tool for marking lines on boards up to 12 inches wide. Small try squares are handy to mark lines across the edge or end of boards. They may be used to draw both 90- or 45-degree lines on boards (Figure 8-15).

Combination Square. The **combination square** is so named because it combines many tools into one (Figure 8-16). It contains a bubble that permits the tool

to be used as a level. Another bubble permits it to be used for plumbing objects. An object is **plumb** when it is vertical to the axis of the earth or in line with the pull of gravity. In other words, when something is straight up and down, it is said to be plumb.

The combination square contains a removable blade that can be used as a steel scale. When the blade is in place, it may be used as a depth gauge. The tool is very handy as an ordinary square and as a 45-degree miter square. (A **miter** is an angle.) Special heads make the tool useful for locating the center of round objects. The combination square is probably the most used square because of its many functions.

Sliding T Bevel. The **sliding T bevel** is a device used to lay out angles. It is also called a *bevel square*. A **bevel** is a sloping edge. If the corner is cut from a board, a bevel is created (Figure 8-17). To use a sliding

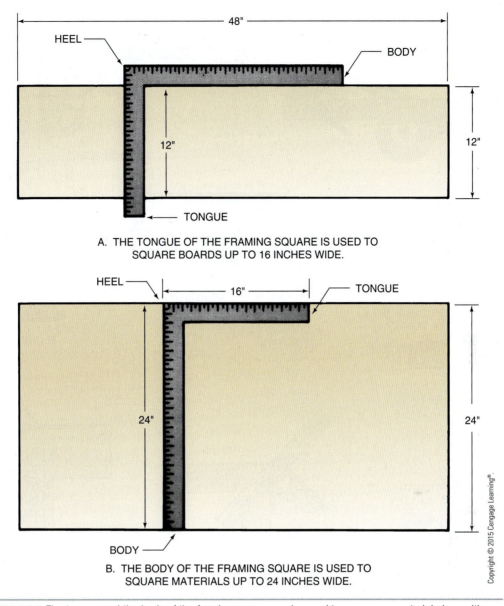

A. THE TONGUE OF THE FRAMING SQUARE IS USED TO SQUARE BOARDS UP TO 16 INCHES WIDE.

B. THE BODY OF THE FRAMING SQUARE IS USED TO SQUARE MATERIALS UP TO 24 INCHES WIDE.

FIGURE 8-14 The tongue and the body of the framing square can be used to measure a material along either their inside or their outside edge.

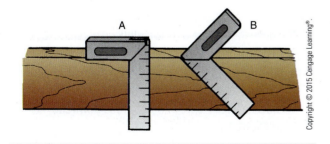

FIGURE 8-15 The try square can be used to draw 90-degree lines across the face, edge, or ends of a board. For marking a 90-degree angle, the try square is positioned as shown in (A). For marking a 45-degree angle, the 45-degree shoulder of the try square is used as in (B).

T bevel, first loosen the wing nut or thumb screw. Then adjust the blade to the desired angles and length, and retighten.

Using a Square to Mark a Board

A square is used when marking boards to be sawed exactly square or at 90 degrees to the edge. When using a square to mark a board, the following procedure is used:

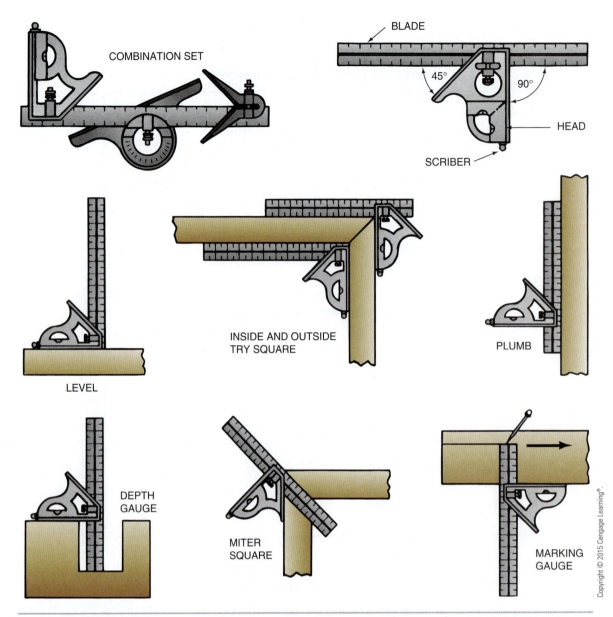

FIGURE 8-16 The combination square unites many tools into one. It can work as a plumb, a level, a scriber, a scale, a depth gauge, a marking gauge, a miter gauge, a miter square, and a regular square.

Procedure

1. Measure and mark the length of board desired. The mark should be thin and at a 90-degree angle to the board's edge.

2. Place the handle of the square firmly against the edge of the board.

3. Move the square until the blade is against the mark.

4. Hold a sharp pencil or knife against the blade and draw a line through the mark and across the board (Figure 8-18). The board is now ready to be sawed.

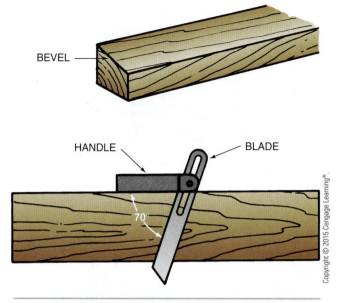

FIGURE 8-17 The sliding T bevel is used to lay out bevels and other angles.

FIGURE 8-18 Step 4 of the procedure for using a square to mark boards for cutting: Hold the square against the mark, and draw a line through the mark and across the board.

SQUARING LARGE AREAS

When working large areas, such as laying out a building, care must be taken when squaring corners. An error as small as 1 degree will cause serious construction problems. One way to establish 90-degree corners is to create a right triangle with sides that are 3 feet, 4 feet, and 5 feet in length. Any right triangle with sides that are multiples of 3 feet, 4 feet, and 5 feet will work just as well. A **right triangle** is a three-sided figure with one angle of 90 degrees.

To lay out a 90-degree corner for a large area, use 6, 8, and 10 feet, respectively, and the following steps.

Procedure

1. Drive a thin stake at the point of the right angle or corner.

2. Lay the heel of a framing square against the stake.

3. Attach a piece of string to the stake, and stretch it along the line of one leg of the square. Stake the other end of the string, and call this line A.

4. Attach a second string at the point of the right angle, and stretch it along the other leg of the square. Call this line B.

5. Measure along line A, a distance of 6 feet. Tie a piece of string at this 6-foot mark.

6. Measure along line B and tie a piece of string at the 8-foot mark.

7. Measure the distance from the string tied on line A and the string tied on line B. The two points should be exactly 10 feet apart. If they are not, move either line A or B in or out until the distance is exactly 10 feet (within $\frac{1}{16}$ of an inch). The angle will then be square or 90 degrees (Figure 8-19).

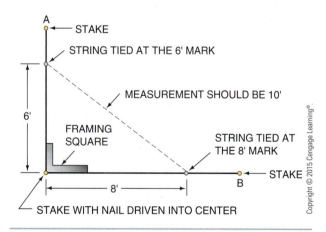

FIGURE 8-19 The 6, 8, 10 method is useful for establishing a perfectly square corner in a large area. The framing square provides a means to position the strings at approximately a right angle to each other. Measurements of 6 feet, 8 feet, and 10 feet provide greater accuracy in placing lines A and B to create an angle that is exactly 90 degrees.

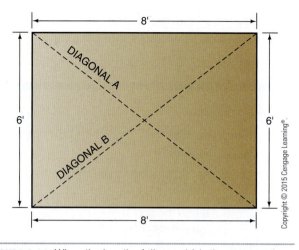

Copyright © 2015 Cengage Learning®.

FIGURE 8-20 When the length of diagonal A is the same as the length of diagonal B, all four corners of this rectangle are square. To meet the definition of a rectangle, the two long sides must be of equal length and the two short sides must be of equal length.

A useful procedure for determining if four corners of a rectangle are square is to measure the distance diagonally from corner to corner. If all four corners are square, the diagonals will be equal in length (Figure 8-20). To be considered equal, two measurements must be exactly the same.

DIVIDERS AND SCRIBERS

Dividers and **scribers** are instruments used to make circles and other curved lines. They are also used to transfer equal measurements from a scale or piece of material to another item. Dividers and scribers are very similar. One difference is that the scriber has one steel leg and a pencil for the other leg. The divider has two sharp steel legs (Figure 8-21).

LEVELS

A **level** is a device used to determine if an object has the same height at two or more points. To determine this, a spirit level is generally used. A **spirit level** consists of alcohol in a sealed, curved tube with a small air space or bubble mounted in a bar of wood or aluminum. When both ends of the level are exactly the same height, the bubble flows to the high point of the tube. This point is indicated by the bubble fitting exactly between two lines marked on the glass. If the tube is mounted crossways in the bar, the level may be used to plumb an object.

Levels may be made to attach to a string or line stretched between two distant points. This is called a "line level." When the string is pulled tight, the level indicates when both ends are the same height (Figure 8-22).

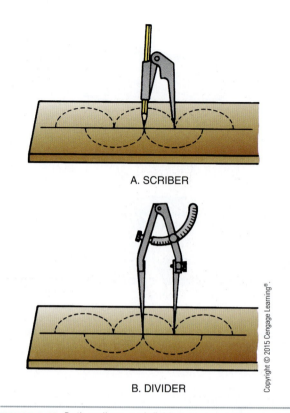

A. SCRIBER

B. DIVIDER

Copyright © 2015 Cengage Learning®.

FIGURE 8-21 Both scribers and dividers are used to mark circles. Scribers have one steel leg and a pencil attachment. Dividers have two steel legs, which can be used to mark metal.

Line levels are useful for leveling the corners when building with block and setting stakes to gauge the depth of concrete.

A newer type of level uses a **laser** beam that directs a strong beam of light across a space. The beam is level and level marks can be made at several points and at varying distances. These new tools are very accurate and easy to use. They are used for a variety of jobs such as squaring tiles, where straight lines are necessary (Figure 8-23). Not only do the levels project a red line that is level but also a line that is straight. This combination can be a tremendous benefit in such tasks as hanging pictures on the wall where they must be not only level but straight (Figure 8-24).

LINES

Strong cotton or nylon string is used as a guide when laying block or flooring, cutting rafters, applying ceiling tiles, and other jobs. A **line** is any thin material stretched tightly between two or more points.

A **chalk line** is a cord with chalk applied by rubbing or by drawing the line through powdered chalk. When stretched and snapped, the line leaves a thin trail of chalk to mark straight lines up to 50 feet.

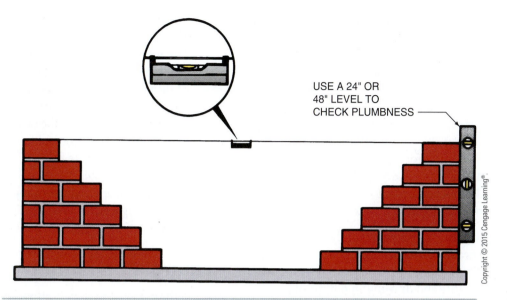

USE A 24" OR
48" LEVEL TO
CHECK PLUMBNESS

Copyright © 2015 Cengage Learning®

FIGURE 8-22 A level attached to a string is called a line level. It is used to check whether two points are of equal height. The corners of the walls shown in this drawing are 16 feet apart. A carpenter's level may be used to check the wall for plumbness.

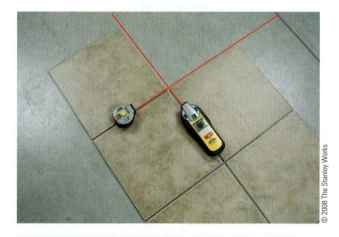

© 2008 The Stanley Works

FIGURE 8-23 Laser levels can be used to lay out straight lines.

© 2008 The Stanley Works

A **plumb line** is a string with a round and pointed piece of metal called a **plumb bob** attached. The plumb line hangs perfectly vertical. It is useful when building greenhouses, sheds, barns, and other buildings to check plumbness of walls and other structures.

FIGURE 8-24 Laser levels project a line that is both straight and level.

USING PATTERNS

Layout tools and instruments can be used to create almost any object. However, much time may be saved by copying the shapes of objects or designs using a **pattern**. A pattern is a model or guide used to create a project. The pattern can be made using layout tools or can be copied from a drawing or a project. One advantage in using a pattern is that the pattern can be saved and used again.

To build a wall, a carpenter may cut one 2 × 4-inch piece of lumber exactly 7 feet 6 inches long with both ends square. This piece will then be used as a pattern to mark and saw several dozen pieces exactly like the pattern.

In the agricultural mechanics shop, students may wish to develop simple woodworking skills by making a decorative cutting board as a project. The teacher may have a pattern made of a thin piece of high density fiber board or card stock cut in the shape that is to be cut. The fiber board is laid on a piece of wood and the outline of the pattern is drawn. Within minutes, the project is ready to be cut.

Individual designs for a project or project part can be created. When creating designs, it is always wise to begin by making a pattern out of paper. Several techniques may be used to make a pattern:

- **Tracing a pattern.** If a picture of the desired design is available, the pattern can be made by tracing. This is done by placing thin tracing paper over the picture and drawing an outline of the object. The tracing paper is then glued on to a thin (1/8 inch) piece of high density fiber board using spray adhesive. The tracing can then be cut using a jigsaw or band saw (Figure 8-25). Be sure to cut just outside the line and sand the edge to the line.
- **Using carbon paper.** Carbon paper may be placed between a picture and a piece of heavy paper.

The picture is then outlined with a ballpoint pen and the image is transferred to the heavy paper. The paper is then cut into a pattern.

- **Drawing a pattern.** A pattern can be made by using a pencil, ruler, compass, or other drawing instruments. A pattern can also be drawn freehand without instruments. If the object is to be the same on both the left and the right, the paper should be folded in half to make the pattern symmetrical. One-half of the object is then drawn. By cutting the doubled paper, a pattern is obtained with both halves being exactly alike. This procedure is especially helpful if the pattern has curved lines.

Enlarging a Pattern

If a pattern is smaller than desired, an enlarged version can be made. First, a grid of vertical and horizontal lines is drawn over the original pattern or a copy of the pattern. The lines must all be spaced the same distance apart. One-quarter inch spacings work well for objects smaller than a standard sheet of paper.

A similar grid is then drawn on a piece of heavy drawing paper, but the space between lines in the grid is increased. This increase is in proportion to the desired increase in size of the pattern. For example, suppose the original is a picture in a book that is 6 inches high and 4 inches wide. If the pattern is to be twice as large, the desired height would be 12 inches and the width 8 inches. In this case, the distance between the lines of the grid in the pattern is doubled. Suppose the picture has identical left and right sides, such as a Christmas tree. After the grid on the paper is prepared, the paper is folded in half. Observe how the lines of the original picture cut across the grid (Figure 8-26). An outline of half of the tree is drawn onto the new grid, using the corresponding lines and squares of the original as a guide for drawing on the pattern. The pattern is then cut out and unfolded. It should be a perfect enlargement of the original.

An alternate method is to use an overhead or opaque projector to project an enlarged pattern onto paper. The image is then marked with a pencil or pen.

Computer-Aided Design. The most modern layout tool is the computer. Data such as dimensions and area are entered into the computer. Also points can be plotted that will assist in using the data entered. The program called CAD for computer-assisted design will develop a detailed drawing of the project. The program can be used for such tasks as determining the optimum use of space in rooms, buildings, cabinets, and other storage and work areas. Also CAD can calculate a bill of materials and a total cost for the project.

FIGURE 8-25 Patterns can be made of high density fiber board on card stock.

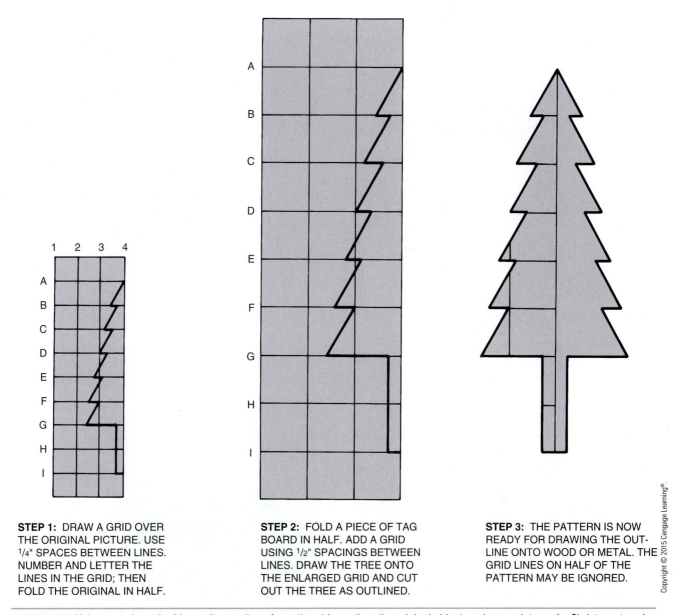

STEP 1: DRAW A GRID OVER THE ORIGINAL PICTURE. USE 1/4" SPACES BETWEEN LINES. NUMBER AND LETTER THE LINES IN THE GRID; THEN FOLD THE ORIGINAL IN HALF.

STEP 2: FOLD A PIECE OF TAG BOARD IN HALF. ADD A GRID USING 1/2" SPACINGS BETWEEN LINES. DRAW THE TREE ONTO THE ENLARGED GRID AND CUT OUT THE TREE AS OUTLINED.

STEP 3: THE PATTERN IS NOW READY FOR DRAWING THE OUTLINE ONTO WOOD OR METAL. THE GRID LINES ON HALF OF THE PATTERN MAY BE IGNORED.

Copyright © 2015 Cengage Learning®.

FIGURE 8-26 Using an enlarged grid permits creation of a pattern bigger than the original object, such as a picture of a Christmas tree. In this example, the resulting pattern is double the size of the original picture.

CAD has evolved tremendously over the past 20 years; even the acronym itself has changed meaning. Originally, CAD stood for "computer-aided drafting"; but as CAD evolved, the name has changed meaning to "computer-aided design" to reflect the broader range of tasks that CAD now performs. There are additional variations in the CAD name—CAM (computer-aided manufacturing), CAAD (computer-aided architectural design), CAID (computer-aided instructional design), and AutoCAD (CAD software).

While CAD has been known by many names, the basic principle remains the same. CAD is simply the use of computers to assist in the design of products (Figure 8-27). The greatest benefit of CAD technology is that designers and engineers are able to develop a product on their computers without actually manufacturing it. The computer-generated design can then be saved for future editing and revisions, which saves the designer time in redrawing the product each time changes need to be made. CAD has become the predominant method of design for architects, agricultural engineers, builders, mechanical engineers, electrical engineers, software and hardware developers, and a cadre of other engineers and designers.

CAD technology has moved to the forefront of drafting and design since 1982, when the company that developed AutoCAD released the first CAD software compatible with home computers. However, the

FIGURE 8-27 CAD is the use of computers to assist in the design of products. This cabinet plan was designed on a computer.

technology had been utilized as early as the 1960s in the aerospace and automotive arenas and evolved from two-dimensional drawings similar to those done on paper into multilayered three-dimensional graphics that can predict how the product will perform once it is produced. Many companies are now producing CAD software packages to meet the needs of various niche markets.

The original CAD software was capable of producing only two-dimensional drawings similar to what can be drafted on paper, much like blueprints. Current CAD software is capable of creating designs in 3-dimension (Figure 8-28). It is also capable of much more such as predicting the fluid dynamics of products, rotating graphics on a three-dimensional axis, and even animating working components. One of the most immediate benefits of CAD programming is that products that are "drawn" on the computer can be saved to a disk and carried to and loaded directly onto production equipment, thus greatly reducing the amount of time needed to move a product from paper to production.

Each CAD system requires the designer to utilize different hardware and software components to meet his or her needs. Perhaps the ones most useful for the agricultural mechanics lab are the programs that draw in simple three-dimensional format and can be used to design wood or metal projects. One such program called Sketchup® is free and downloadable from Google.

FIGURE 8-28 Modern computer programs can design projects in 3-dimension.

Most agricultural mechanics design needs that you will encounter in school can be created with this program.

With the advent of CAD, entire professions have disappeared; teams of drafters are now replaced by a single skilled CAD operator. While CAD has evolved from the drafting boards of the 1950s to the desktops of the new millennium, the basic goal is the same: for designers to create and communicate ideas as efficiently as possible. The current trends suggest that CAD systems will continue to shorten the amount of time it takes for an idea to go from someone's mind to being an actual product.

SUMMARY

Without standard units of measurement, projects would be difficult to complete and almost impossible to reproduce. Whether working with the U.S. customary system or the metric system, a person must know the basic units of measurements and how to properly divide the units into smaller components. Layout tools use units of measurements that allow people to draw plans and to transfer dimensions on the plan to materials used on a project. Learning to measure properly is a skill that takes practice. No project can be properly done until this skill is mastered.

Student Activities

1. Define each of the Terms to Know in this unit.

2. Obtain a steel tape, a steel or wooden scale, a combination square, a try square, and a framing square. Examine these tools and record the information needed to complete the following table.

Tool	Basic Units of Measure	Marks per Unit	Length
a. Steel tape	inch/ foot	16 per inch	12 inches
b. Steel or wooden scale	_____	_____	_____
c. Combination square	_____	_____	_____
d. Try square	_____	_____	_____
e. Framing square	_____	_____	_____

3. Using a ruler, draw ten straight lines, each 5 inches long, on a piece of paper. Place short marks across each line at distances exactly as follows:
   a. Line 1: Distance between marks = 3 inches
   b. Line 2: Distance between marks = 4½ inches
   c. Line 3: Distance between marks = 2⅛ inches
   d. Line 4: Distance between marks = 2¾ inches
   e. Line 5: Distance between marks = 3⅞ inches
   f. Line 6: Distance between marks = $^{15}/_{16}$ inches
   g. Line 7: Distance between marks = 7 centimeters
   h. Line 8: Distance between marks = 84 millimeters
   i. Line 9: Distance between marks = 7.3 centimeters
   j. Line 10: Distance between marks = 7 centimeters and 3 millimeters

4. Use a tape to measure the length and width of your classroom, your desktop, and the classroom door.

5. Examine a folding rule. Unfold and refold the rule being careful to swing each section open and closed, using the spring joint as a hinge.

6. Examine inside calipers and outside calipers. Measure the inside and outside diameter of a round object, such as a can, pipe fitting, or piece of pipe.

7. Obtain a square of your choice and a board that is 6 inches to 10 inches wide. Draw a line across the board that is straight and square to the edge of the board.

8. Obtain a 12-inch, 24-inch, or 48-inch level.
   a. Place the level on the windowsill in the classroom, shop, or other room. Is the window level?
   b. With the level resting on the windowsill, lift the right end of the level. Which way did the bubble move?
   c. Lift the left end of the level. Which way did the bubble move?
   d. Place the level in a vertical position against the classroom or shop door. Is the door plumb? Do all of the bubbles in the level read alike?

e. With the level still in a vertical position against the door, hold the bottom of the level against the door but move the top away from the door. Which way did the bubble move?

f. Hold the top of the level against the door and move the bottom away. Which way did the bubble move?

9. Use a combination square to:
   a. draw a 90-degree line on a board
   b. draw a 45-degree line on a board
   c. measure the width of a board
   d. measure the depth of a groove
   e. test the levelness of a bench

10. Create a 90-degree corner in the lawn using the 6', 8', 10' method. You will need three stakes; a hammer; two lengths of string, each 12 feet long; two string ties; a framing square; and a 12-foot tape.

11. Use a scriber to create a 4-inch circle on a piece of paper. Use a divider to create a 3-inch circle on a scrap piece of wood or metal.

12. Prepare a pattern for some small object you would like to make out of wood or metal.

13. Design a project using the CAD program and the appropriate software. Enter the dimensions and shape of your project. Be sure to include a scale. Print the results and make modifications as needed.

Relevant Web Sites

Museum of Woodworking Tools
www.antiquetools.com

CAD/CAM/CAE Design Information Network
www.cadinfo.net

Self-Evaluation

A. Multiple Choice. Select the best answer.

1. On most rules and tapes used in the shop, the shortest line represents
   a. ⅛ inch
   b. ¹⁄₁₆ inch
   c. ¹⁄₆₄ inch
   d. 1 inch

2. Of the following, the smallest unit of metric measurement is
   a. meter
   b. millimeter
   c. kilometer
   d. centimeter

3. Tapes generally break due to
   a. frequent use
   b. use of spring retractors
   c. use outdoors
   d. forcing the tape back into the case

4. Wooden scales are sometimes called
   a. depth gauges
   b. bench rules
   c. marking gauges
   d. scribers

5. The best tool for measuring the outside diameter of a pipe in its middle section is
   a. a try square
   b. tape
   c. calipers
   d. dividers

B. Matching. Match the word or phrase in column I with the correct word or phrase in column II.

Column I

1. high-quality tools
2. U.S. customary system
3. metric system
4. millimeter
5. centimeter
6. meter
7. kilometer
8. framing square
9. combination square
10. plumb line
11. line level
12. 90 degrees
13. body, tongue, and heel
14. sliding T bevel
15. 6', 8', 10'
16. scriber
17. glass tubes and alcohol
18. tracing paper
19. pattern enlargement
20. makes a circle

Column II

a. m
b. cm
c. mm
d. board-foot table
e. km
f. good investment
g. yards, feet, inches
h. check high structures
i. meters, centimeters, millimeters
j. includes a sliding scale and a level
k. requires tight string
l. used to lay out angles
m. right triangle
n. dividers
o. bubble
p. grid
q. framing square
r. right angle
s. contains a pencil
t. pattern making

C. Completion. Fill in the blanks with the word or words that make the following statements correct.

1. When laying out work, it is important to choose the right tool and use the tool _____.

2. The materials tools are made of include _____, _____, _____, _____, and _____.

3. Tools used to guide workers when cutting, sawing, drilling, shaping, or fastening are called _____ tools.

4. The official system of measurement for non-scientific purposes in the United States is the _____ _____ system.

5. The metric system of measurement is regarded as a logical system because its units of measure relate to each other in multiples of _____.

6. Use the following drawing and read the rule to the nearest ⅛ inch.

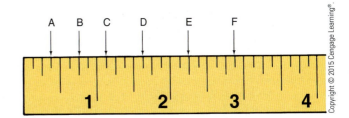

A. _____
B. _____
C. _____
D. _____
E. _____
F. _____

Copyright © 2015 Cengage Learning®.

7. Use the following drawing and read the rule to the nearest ⅟₁₆ inch.

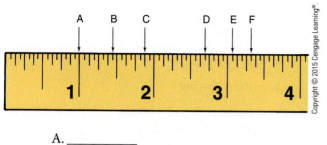

Copyright © 2015 Cengage Learning®.

A. _____
B. _____
C. _____
D. _____
E. _____
F. _____

D. Brief Answer. Briefly answer the following questions.

1. Greater inaccuracy in longer measuring tools is due to what?

2. Give two advantages of layout tools made of steel.

3. When is an object said to be "plumb"?

4. What four aspects might a gauge be used to determine?

5. Briefly describe two methods one might use to enlarge a pattern that is smaller than desired if CAD (computer-aided design) is unavailable.

6. What useful measuring technique might be used to determine if four corners of a rectangle are square?

Selecting, Cutting, and Shaping Wood

Objective

To select wood and to use tools and procedures in cutting and shaping wood.

Competencies to be developed

After studying this unit, you should be able to:

- Name and correctly spell twelve species of lumber that may be used in agricultural mechanics shops.
- Select lumber for agricultural projects.
- Determine the most useful hand tools for cutting and shaping wood.
- Cut and shape wood.

Materials List

- Twelve specified species of lumber
- Planed and rough lumber
- Dried and green lumber
- Samples of S2S, S4S, and sanded lumber
- Samples of standard cuts of lumber
- Combination square
- Handsaws of various kinds
- Boring tools
- Planes
- Files
- Miter box
- ¾" wood chisel
- Sandpaper

Terms to Know

- wood
- annual rings
- species
- softwood
- hardwood
- kiln
- warping
- planer
- nominal
- handsaw
- crosscut
- rip
- grain
- crosscut saws
- ripsaws
- kerf
- backsaw
- miter box
- coping saw
- compass saw
- keyhole saw
- bore
- drill
- file
- rasps
- file card
- dado
- rabbet
- perpendicular

The agricultural mechanics student should know some basic information about wood and its source (Figure 9-1). **Wood** is the hard, compact fibrous material that comes from the stems and branches of trees. Nutrients flow up and down a tree trunk in bundles of tubelike tissues called xylem and phloem that are formed just inside the bark of trees each year. It is the yearly addition of new bundles that causes a tree to have annual rings. **Annual rings** are patterns caused by the hardening and disuse of the bundles; they

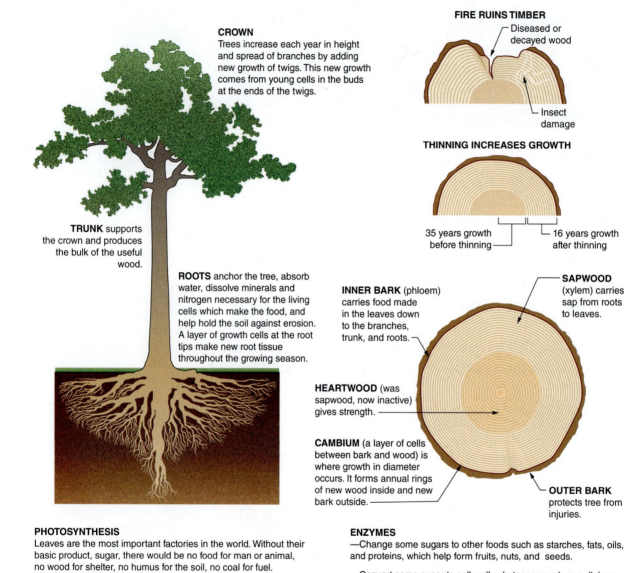

CROWN
Trees increase each year in height and spread of branches by adding new growth of twigs. This new growth comes from young cells in the buds at the ends of the twigs.

TRUNK supports the crown and produces the bulk of the useful wood.

ROOTS anchor the tree, absorb water, dissolve minerals and nitrogen necessary for the living cells which make the food, and help hold the soil against erosion. A layer of growth cells at the root tips make new root tissue throughout the growing season.

FIRE RUINS TIMBER
Diseased or decayed wood
Insect damage

THINNING INCREASES GROWTH
35 years growth before thinning
16 years growth after thinning

INNER BARK (phloem) carries food made in the leaves down to the branches, trunk, and roots.

SAPWOOD (xylem) carries sap from roots to leaves.

HEARTWOOD (was sapwood, now inactive) gives strength.

CAMBIUM (a layer of cells between bark and wood) is where growth in diameter occurs. It forms annual rings of new wood inside and new bark outside.

OUTER BARK protects tree from injuries.

PHOTOSYNTHESIS
Leaves are the most important factories in the world. Without their basic product, sugar, there would be no food for man or animal, no wood for shelter, no humus for the soil, no coal for fuel.

Inside each leaf, millions of green-colored, microscopic "synthetic chemists" (chloroplasts) manufacture sugar, they trap radiant energy from sunlight for power. Their raw materials are carbon dioxide from the air and water from the soil. Oxygen, a by-product, is released. This fundamental energy-storing, sugar-making process is called photosynthesis.

What happens to this leaf-made sugar in a tree? With the aid of "chemical specialists" (enzymes), every living cell—from root tips to crown top—goes to work on the sugar. New products result. Each enzyme does a certain job, working with split-second timing and in harmony with the others. In general, they break down sugar and recombine it with nitrogen and minerals to form other substances.

ENZYMES
—Change some sugars to other foods such as starches, fats, oils, and proteins, which help form fruits, nuts, and seeds.

—Convert some sugar to cell-wall substances such as cellulose, wood, and bark.

—Make some of the sugar into other substances which find special uses in industry. Some of these are rosin and turpentine from southern pines, syrup from maples, chewing gum from chicle trees and spruces, tannin from hemlocks, oaks, and chestnuts.

—Use some of the sugar directly for energy in the growing parts of the tree—its buds, cambium layer, and root tips.

TRANSPIRATION
Transpiration is the release of water-vapor from living plants. Most of it occurs through the pores (stomates) on the underside of the leaves. Air also passes in and out.

FIGURE 9-1 How a tree grows.

Source: USDA, Forest Service, "How a Tree Grows," FS-32, May 1970, Reviewed and Approved for reprinting October 2006.

form the grain in wood. Annual rings can be seen by looking at the end of a tree trunk or tree stump.

CHARACTERISTICS OF WOOD AND LUMBER

It is important to know the characteristics of wood from various species of trees (Figure 9-2). The word *species* means plants or animals with the same permanent characteristics. Some woods are hard and some are relatively soft. Some woods resist rotting better than others. Some are very attractive while other woods are plain. Some woods are stronger and tougher than others.

Wood can be classified according to the type of tree that produced the lumber. Trees are divided into two broad groups—softwood and hardwood. These terms can often be misleading. For example, southern pine is classified as a softwood, but it is actually harder than poplar, which is classified as a hardwood. The classification is based on whether the tree sheds its leaves in the winter. These deciduous trees are classified as hardwoods. Examples of hardwood trees are oak, maple, poplar, walnut, and ash. Evergreen trees or conifers are classified as softwood. Examples are spruce, fir, cedar, hemlock, and pine.

Many of the hardwoods such as oak, maple, cherry, mahogany, and walnut are best suited for furniture

manufacturing. Softwoods such as pine and spruce are used extensively for construction purposes, such as framing. Selecting the proper type of wood for a specific purpose is essential to any successful project.

Lumber Grades

Hardwood. Hardwood lumber is graded according to the amount of usable, clear lumber in a board. *Clear* refers to wood that has no knots or other obvious defects. Knots are the result of limbs that grow from a tree. When the lumber is sawed, the limbs appear as knots in the wood. Knots generally make the lumber weaker, and are difficult to cut and mill.

The highest grade of hardwood is FAS, which stands for "first and seconds." A board that is graded FAS must be at least 8 feet long and 6 inches wide. It must also have a minimum area of 83 percent that is clear. The next-highest grade is FAS-1. This grade indicates that only one side of the board meets the specifications of FAS (Figure 9-3A through C). Sometimes these two grades are sold as one grade called Face or Better. The grade that is about average for most of the lumber sold is called Number 1 Common (1C), which is a lower grade than FAS. The following chart outlines the specifications established for the different grades of hardwood lumber.

Species	Hardness	Known For	Some Major Uses
Birch	Hard	Surface veneer for panels	Cabinets and doors
Cedar, red	Medium	Pleasant odor	Furniture, chests, and closet linings
Cherry	Hard	Red grain	Fine furniture
Cypress	Medium	Rot resistance	Structural material in wet places
Fir and Hemlock	Soft	Light, straight, strong	Construction framing, siding, sheathing
Locust, black	Hard	Rot resistance	Fence posts
Mahogany	Medium	Reddish color	Fine furniture
Maple	Hard	Light grain	Floors, bowling alleys, durable furniture
Oak	Hard	Toughness, strength	Floors, barrels, wagon bodies, feeders, farm buildings
Pine, white	Soft	Easy to work, straight	Shelving, siding, trim
Pine, yellow	Medium	Wear resistance, tough	Floors, stairs, trim
Redwood	Soft	Excellent rot resistance	Yard posts, fences, patios, siding
Walnut, black	Hard	Brown grain	Fine furniture
Willow, black	Soft	Brown grain, easy to work, walnut look	Furniture

FIGURE 9-2 Characteristics of common woods.

Copyright © 2015 Cengage Learning®.

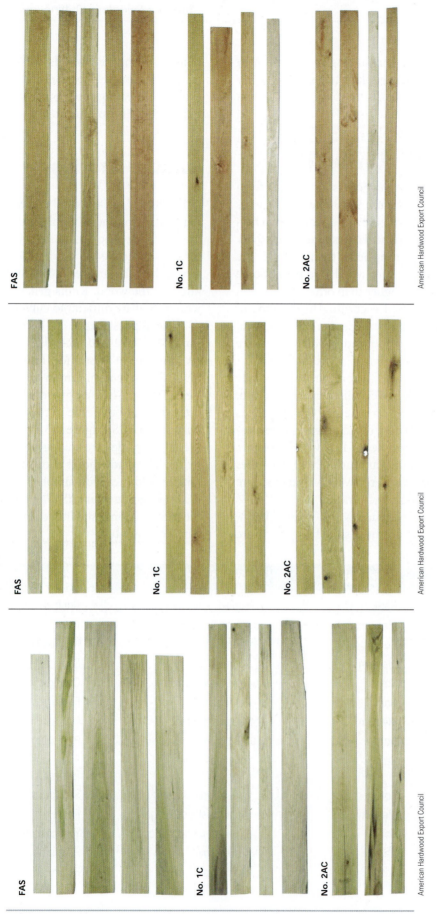

FAS   No. 1C   No. 2AC

American Hardwood Export Council

FAS   No. 1C   No. 2AC

American Hardwood Export Council

FAS   No. 1C   No. 2AC

American Hardwood Export Council

FIGURE 9-3 Hardwood lumber is graded according to how much usable, clear lumber each board contains.

Softwood. Most softwood lumber is classified as "yard lumber" or lumber commonly found at lumber yards in building supply stores. Yard lumber is broken down into three broad grades: *select*, *common*, and *dimension*. These grades are labeled according to use.

Select is used for purposes where appearance (see the following table) is important, such as face boards on a house or the trim around siding. All defects must be so minor that they are covered when the board is finished or painted. Select is further graded as Grade A Select, Grade B Select, Grade C Select, and Grade D Select.

Grade	Minimum Board Length	Minimum Board Width	Minimum Cutting Size	Minimum Area of Clear Cuttings
FAS	8'	6"	4" × 5' or 3" × 7'	83%
1C	4'	3"	4" × 2' or 3" × 3'	67%
2C	4'	3"	3" × 4'	50%
3AC	4'	3'	3" × 4'	33%

Molding and trim are also classified as *paint grade* or *stain grade*. Paint grade is used for trim work that is to be painted. It is made up of short pieces of molding that are joined together to make longer lengths. Stain grade is more expensive and is usually of one-piece construction. This grade is used where the trim is stained and coated with a clear finish that shows the grain in the wood.

Common is usually used for construction purposes where appearance is not a factor. This grade is further divided into Number 1 Common, Number 2 Common, and so on through Number 5 Common. The lowest-quality lumber would be Number 5 Common (Figure 9-4).

Dimension is recommended for purposes where strength is needed. The grade is based on the straightness, strength, and rigidity of the board. Common uses include rafters, studs, joists, and general framing. This grade is divided into Number 1 Dimension, Number 2 Dimension, and Number 3 Dimension, according to specific usage.

Product	Grade	Character of Grade and Typical Uses
Finish	B&B	Highest recognized grade of finish. Generally clear, although a limited number of pin knots permitted. Finest quality for natural or stain finish.
	C	Excellent for painted or natural finish where requirements are less exacting. Reasonably clear but permits limited number of surface checks and small tight knots.
	C&Btr	Combination of B&B and C grades; satisfies requirements for high-quality finish.
	D	Economical, serviceable grade for natural or painted finish.
Boards S4S	No. 1	High quality with good appearance characteristics. Generally sound and tight-knotted. Largest hole permitted is ⅟₁₆". A superior product suitable for wide range of uses, including shelving, form, and crating lumber.
	No. 2	High-quality sheathing material, characterized by tight knots. Generally free of holes.
	No. 3	Good, serviceable sheathing, usable for many applications without waste.
	No. 4	Admit pieces below No. 3 that can be used without waste or contain usable portions at least 24" in length.
Dimension		
Structural light framing 2" to 4" thick 2" to 4" wide	Select Structural	High quality, relatively free of characteristics that impair strength or stiffness. Recommended for uses where high strength, stiffness, and good appearance are required.
	Dense Select Structural	
	No. 1	Provide high strength; recommended for general utility and construction purposes.
	No. 1 Dense	Good appearance, especially suitable where exposed because of the knot limitations.
	No. 2	Although less restricted than No. 1, suitable for all types of construction.
	No. 2 Dense	Tight knots.
	No. 3	Assigned design values meet wide range of design requirements. Recommended for general construction purposes where appearance is not a controlling factor. Many pieces included in this grade would qualify as No. 2 except for single limiting characteristic. Provides high-quality, low-cost construction.

(continued)

Courtesy Southern Forest Products Association

FIGURE 9-4 Southern pine softwood lumber grades.

Product	Grade	Character of Grade and Typical Uses, *continued*
Studs 2" to 4" thick 2" to 6" wide 10' and shorter	Stud	Stringent requirements as to straightness, strength, and stiffness adapt this grade to all stud uses, including load-bearing walls. Crook restricted in 2" × 4"-8' to ¼", with wane restricted to ⅓ of thickness.
Structural joists and planks 2" to 4" thick 5" and wider	Select Structural Dense Select	High quality, relatively free of characteristics that impair strength or stiffness.
	Structural	Recommended for uses where high strength, stiffness, and good appearance are required.
	No. 1 No. 1 Dense No. 2 No. 2 Dense	Provide high strength; recommended for general utility and construction purposes. Good appearance; especially suitable where exposed because of the knot limitations. Although less restricted than No. 1, suitable for all types of construction. Tight knots.
	No. 3 No. 3 Dense	Assigned stress values meet wide range of design requirements. Recommended for general construction purposes where appearance is not a controlling factor. Many pieces included in this grade would qualify as No. 2 except for single limiting characteristic. Provides high-quality, low-cost construction.
Light framing 2" to 4" thick 2" to 4" wide	Construction	Recommended for general framing purposes. Good appearance, strong, and serviceable.
	Standard Utility	Recommended for same uses as Construction grade, but allows larger defects. Recommended where combination of strength and economy is desired. Excellent for blocking, plates, and bracing.
	Economy	Usable lengths suitable for bracing, blocking, bulkheading, and other utility purposes where strength and appearance are not controlling factors.
Appearance framing 2" to 4" thick 2" and wider	Appearance	Designed for uses such as exposed-beam roof systems. Combines strength characteristics of No. 1 with appearance of C&Btr.
Timbers 5" × 5" and larger	No. 1 SR No. 1 Dense SR No. 2 SR No. 2 Dense SR	No. 1 and No. 2 are similar in appearance to corresponding grades of 2" dimension. Recommended for general construction uses. SR in grade name indicates Stress Rated.
Structural lumber	Dense Str. 86 Dense Str. 72 Dense Str. 65	Premier structural grades from 2" through and including timber sizes. Provides some of the highest design values in any softwood species with good appearance.

FIGURE 9-4 (*continued*)

Sawing Methods

Lumber can also be broadly classified according to the way the lumber was sawed from the log. Plain sawn (sometimes called *slash sawn*) means that the lumber was cut from the log by making parallel passes with the saw straight through the log. The grain in the lumber appears to be wide apart. Most structural-grade lumber is sawed in this manner. Quartersawn lumber is cut by dividing the log into quarters and sawing at an angle to the center axis of the log (Figure 9-5). Some hardwoods are cut in this manner because the grain of the lumber is more attractive. Most lumber that goes into fine furniture is quartersawn. This type of lumber is more stable because it has less tendency to warp or twist (Figure 9-6).

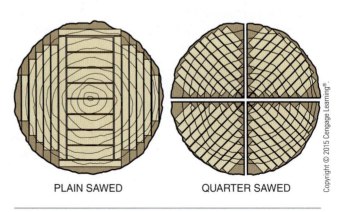

PLAIN SAWED   QUARTER SAWED

Copyright © 2015 Cengage Learning®.

FIGURE 9-5 Lumber may be produced by plain-sawing or by quarter-sawing logs.

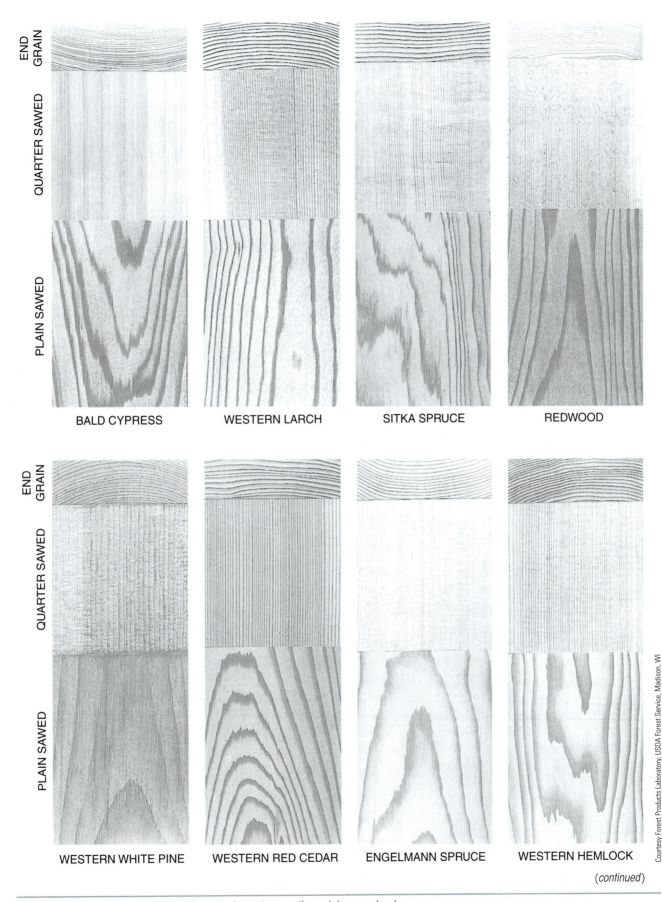

END GRAIN

QUARTER SAWED

PLAIN SAWED

BALD CYPRESS WESTERN LARCH SITKA SPRUCE REDWOOD

WESTERN WHITE PINE WESTERN RED CEDAR ENGELMANN SPRUCE WESTERN HEMLOCK

(continued)

Courtesy Forest Products Laboratory, USDA Forest Service, Madison, WI

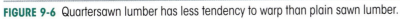

FIGURE 9-6 Quartersawn lumber has less tendency to warp than plain sawn lumber.

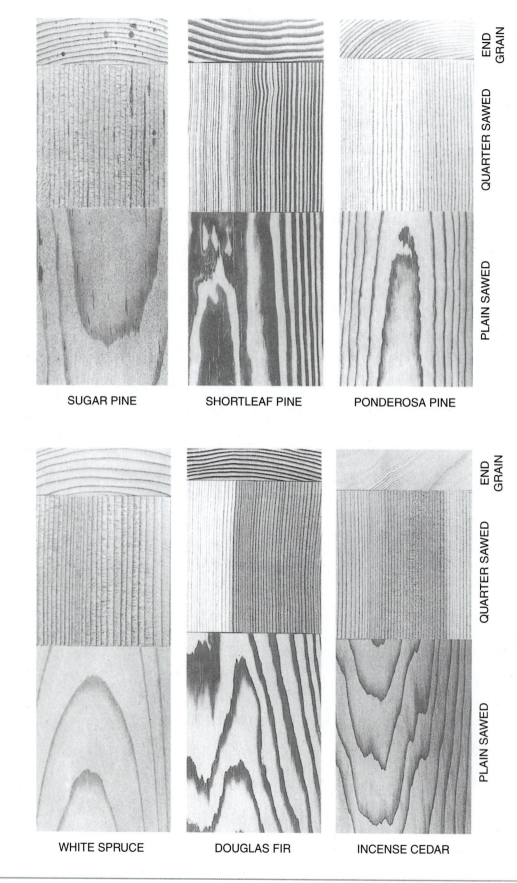

END GRAIN

QUARTER SAWED

PLAIN SAWED

SUGAR PINE

SHORTLEAF PINE

PONDEROSA PINE

WHITE SPRUCE

DOUGLAS FIR

INCENSE CEDAR

FIGURE 9-6 (continued)

Lumber Curing

For most uses, lumber has to be properly cured. When lumber is first sawed from a tree, it is said to be green (Figure 9-7). This means that it has a very high moisture content. For example, a freshly sawn oak board measuring 1 inch thick, 8 inches wide, and 3 feet long may have a gallon of water bound inside. This moisture content is different from lumber that has absorbed moisture from rainfall. Moisture from freshly sawn lumber is bound chemically within the cells of the wood and this moisture has to be removed slowly in a curing process. Green lumber may have from 120 to 130 percent moisture content. This is calculated by dividing the weight of completely dry lumber by the weight of the lumber when it was green. However, the moisture content of lumber is usually determined by the use of a moisture meter. Moisture meters are small, relatively inexpensive electronic devices. They work by the use of probes that are inserted into the wood. A small electric current is passed through the wood from one probe to the other. The amount of electrical flow passing through the wood indicates the moisture content of the wood. The more moisture, the more current that can pass through the wood (Figure 9-8).

Lumber used for construction purposes needs to be dried to about 15 percent moisture. As green lumber dries out, it shrinks and can distort. If improperly dried lumber is used in a building or a project, in a few weeks the lumber may crack, shrink to a smaller size, or otherwise distort. Lumber used for furniture and other inside projects needs to be dried to about 6 to 8 percent moisture. The heated area inside a house can cause the lumber to dry out. Therefore, the lumber needs to be as close to the humidity of the room as possible.

The curing process begins with the lumber being stacked so that air can move through the stack. This is accomplished by laying sticks about 1" × 1" at right

FIGURE 9-8 A moisture meter reads the amount of moisture in lumber by measuring the amount of current allowed to pass between the probes.

angles to the lumber and stacking the lumber on the "stacking sticks" (Figure 9-9). The lumber is laid in rows so that the sides are close but not touching. This method of stacking allows air to move through the boards. Many layers of lumber are added and the top is covered (or the lumber is stacked in a shed). As a general rule, lumber must be air-cured for 1 year for every inch of thickness.

FIGURE 9-9 To air dry, lumber is stacked so the boards do not touch. This allows air to move through the stacks.

FIGURE 9-7 When lumber is first sawed, it is said to be green.

FIGURE 9-10 After air drying, the lumber is placed in a kiln to remove more moisture. Note the steam coming from the kiln.

FIGURE 9-11 Lumber that comes from the sawing process at the mill is called rough lumber.

In other words, a board 2 inches thick would need to be cured for 2 years before being placed in a kiln.

Air drying will not remove a sufficient amount of moisture, so the lumber is placed in a kiln to remove additional moisture (Figure 9-10). A **kiln** is a huge oven that heats the boards at a steady rate to remove the moisture slowly. The lumber may stay in the kiln for a few days or a few weeks, depending on the moisture content and the type of wood. If the lumber is cured too quickly, it will dry unevenly and this will cause splitting, **warping**, and/or cracking.

Lumber Finish

Lumber is generally purchased with all sides and edges smooth. The **planer** is a machine that cuts lumber down to an exact size and leaves it smooth. Lumber may be purchased in the following four categories:

- **Rough:** not planed; delivered as it comes from the sawmill; width and thickness varies from piece to piece (Figure 9-11)
- **S2S:** surfaced two sides; all pieces are the same thickness; edges are not planed; widths vary
- **S4S:** surfaced four sides; the sides and edges are planed to exact dimensions
- **Sanded:** width and thickness are exact on all pieces; all surfaces are sanded.

Standard Lumber Sizes

Rough pieces of lumber, as they come from a sawmill, are not all the same size. What is called a 2 × 4 may be 2¼" × 4 or 2¼" × 4¼". When a 2 × 4 is planed, the actual size is only 1½" × 3½". The name given to cuts of lumber reflects the **nominal** or approximate size. The real or actual sizes are ½ inch less than the nominal sizes (Figure 9-12).

STANDARD LUMBER SIZES

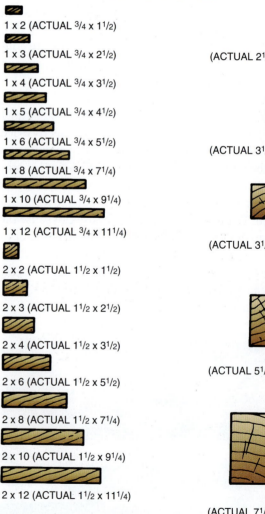

1 x 2 (ACTUAL ³/₄ x 1¹/₂)

1 x 3 (ACTUAL ³/₄ x 2¹/₂)

1 x 4 (ACTUAL ³/₄ x 3¹/₂)

1 x 5 (ACTUAL ³/₄ x 4¹/₂)

1 x 6 (ACTUAL ³/₄ x 5¹/₂)

1 x 8 (ACTUAL ³/₄ x 7¹/₄)

1 x 10 (ACTUAL ³/₄ x 9¹/₄)

1 x 12 (ACTUAL ³/₄ x 11¹/₄)

2 x 2 (ACTUAL 1¹/₂ x 1¹/₂)

2 x 3 (ACTUAL 1¹/₂ x 2¹/₂)

2 x 4 (ACTUAL 1¹/₂ x 3¹/₂)

2 x 6 (ACTUAL 1¹/₂ x 5¹/₂)

2 x 8 (ACTUAL 1¹/₂ x 7¹/₄)

2 x 10 (ACTUAL 1¹/₂ x 9¹/₄)

2 x 12 (ACTUAL 1¹/₂ x 11¹/₄)

3 x 4
(ACTUAL 2¹/₂ x 3¹/₂)

4 x 4
(ACTUAL 3¹/₂ x 3¹/₂)

4 x 6
(ACTUAL 3¹/₂ x 5¹/₂)

6 x 6
(ACTUAL 5¹/₂ x 5¹/₂)

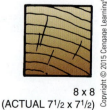

8 x 8
(ACTUAL 7¹/₂ x 7¹/₂)

FIGURE 9-12 The actual width and thickness of planed lumber are smaller than their nominal size.

CUTTING AND SHAPING WOOD WITH SAWS

Wood has become quite expensive; therefore, much care should be taken when measuring and cutting. A good rule is to measure twice and cut once. If a piece of wood must be recut because of an error, it generally means a new piece of wood must be used; or it may mean all pieces must be recut and the project made smaller.

Saws are used to cut boards to length and width. Saws are also used to cut curves, make holes, and cut panels to size. The more common saws are the handsaw, backsaw, coping saw, and compass saw. The compass saw is very similar to the keyhole saw.

Handsaws

The **handsaw** is used to cut across boards or to rip boards and panels. The word **crosscut** means to cut across the grain of a board. The word **rip** means to cut along the length of the board or with the grain. **Grain** in a board refers to lines caused by the annual rings in the tree.

The teeth of a handsaw determine how the saw should be used (Figure 9-13). Teeth cut and filed to a point are designed to cut across the grain of boards. Such saws are called **crosscut saws**. Teeth filed to a knifelike edge are designed to cut with the grain. These saws are called **ripsaws**.

Both crosscut and rip handsaws may be purchased in sizes ranging from 20 to 28 inches in length. The shorter saws are easier to use by smaller people. Shorter saws are frequently the choice for saws with small teeth designed for finer cuts.

Saws may be purchased with various tooth sizes. These range from 6 to 14 teeth per inch on handsaws. The tooth size is designated by the number of tooth points per inch. For example, an 8-point saw has 8 large teeth per inch of blade and is said to be a coarse saw. A 12-point saw has 12 small teeth per inch and is considered a fine saw. When sawing a board, the saw removes wood and leaves an opening called a **kerf** (Figure 9-14).

When sawing wood, the material should rest on a solid work area or be held in a vise. Most carpenters use sawhorses to support the material being sawed. The following procedure is recommended when using a crosscut saw or a ripsaw.

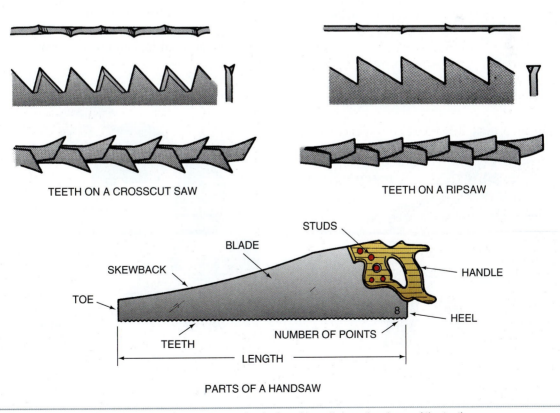

TEETH ON A CROSSCUT SAW

TEETH ON A RIPSAW

STUDS

BLADE

SKEWBACK

TOE

HANDLE

HEEL

TEETH

NUMBER OF POINTS

LENGTH

PARTS OF A HANDSAW

Copyright © 2015 Cengage Learning®.

FIGURE 9-13 Handsaws may be designed for crosscutting or for ripping, depending on the shape of the teeth.

FIGURE 9-14 The opening left by the saw is called a kerf.

4. Grasp the saw handle in the right hand. The saw can be better controlled by placing the right forefinger along the outside of the handle.

5. Place the left hand on the board and use the thumbnail against the saw as a guide to start the cut.

6. Place the heel of the saw on the mark at the edge of the board and pull it toward your body to start the cut. Stand so your eyes are in line with the cut.

7. After the cut or kerf is started, push the saw forward, applying a light, downward pressure. Since the saw is designed to cut on the downward stroke only, apply no pressure on the return stroke. Operate the saw at a 45-degree angle (60-degree angle when ripping with a ripsaw) to the surface of the board (Figure 9-15).

8. Use long, slow strokes to complete the cut. The last few strokes should be very slow with no pressure on the saw.

Procedure

1. Select a piece of 1" × 6" lumber, several feet long.

2. Use a square and draw a thin pencil mark across the board 1 inch from the end.

3. Place the stock (board) on a sawhorse with the marks over the end to the right. This assumes the operator is right-handed (reverse if left-handed). Hold the stock down on the sawhorse with the left knee.

Backsaws

A **backsaw** is similar to a crosscut handsaw. It has very, very fine teeth and a stiff metal back from which it gets its name. These features make it an excellent tool for making very accurate cuts in smooth lumber. The backsaw may be placed in a tool called a miter box to guide the saw (Figure 9-16). A **miter box** is a device used to cut molding and other narrow boards at any

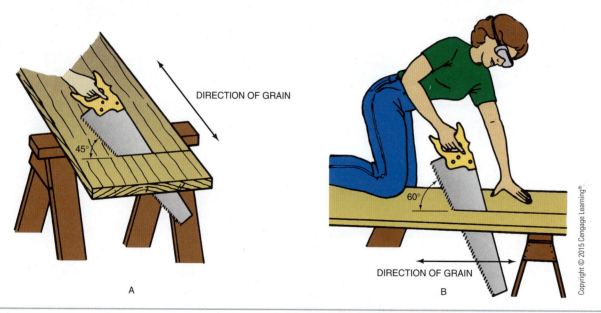

FIGURE 9-15 When using a crosscut saw, hold the saw at a 45-degree angle (A). When using a ripsaw, hold the saw at a 60-degree angle (B). Start all cuts with short strokes, and then use full, even strokes to accomplish the cut. The saw kerf should be on the waste (scrap) side of the wood.

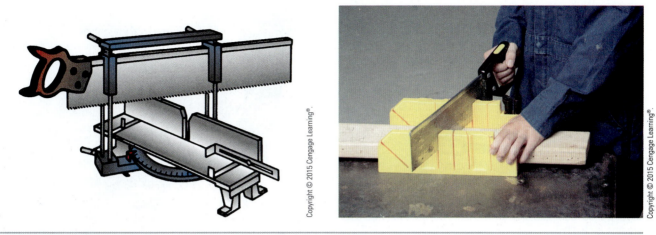

FIGURE 9-16 A backsaw can be used alone or with a miter box. It is excellent for making very fine and accurate cuts.

desired angle. An adjustable miter box is operated by simply squeezing a lever and swinging the saw to the desired angle. The lever is released and the saw locks at that angle. The wood is inserted, held firmly, and sawed.

Coping Saws

The **coping saw** has a very thin and narrow blade supported by a spring-steel frame. This design permits the blade to be removed from the frame (Figure 9-17). This makes it possible to use the saw to cut large holes or other shapes that are totally surrounded by wood. The coping saw is useful for cutting any kind of irregular, curved cuts in wood or other soft materials.

Compass and Keyhole Saws

The **compass saw** and **keyhole saw** are designed for making cuts starting from a hole (Figure 9-18). A hole (about 1 inch in diameter) is bored and the slim blade of the saw is inserted into the hole. The operator saws outward and follows the lines of the desired cut. The compass saw is similar to the keyhole saw, but its blade is wider at the base.

CUTTING AND SHAPING WOOD WITH BORING AND DRILLING TOOLS

The words **bore** and **drill** are often used to mean the same thing—to make a hole. However, the word *boring* generally refers to low speed, whereas drilling is a high-speed operation. The most popular hand

tools for boring and drilling were at one time the brace and bit, the hand drill, and the push drill. Most of these have been replaced by the portable electric hand drill.

FIGURE 9-17 A coping saw is used to cut irregular shapes out of wood. The user can remove the blade from the frame, insert it into a hole, and then reattach it to the frame, in order to cut an inside hole.

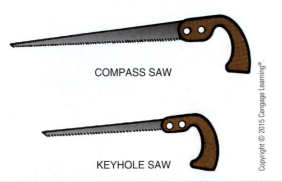

FIGURE 9-18 Compass and keyhole saws are used to make cuts starting from a hole.

CUTTING A BOARD TO WIDTH

The procedure for cutting a board to a specified width involves several tools and steps. The following procedure is used when cutting boards to width. The board used as an example in the procedure is to be cut to a width of 4 inches.

Procedure

1. Set a combination square so its blade extends 4 inches out from the handle.

2. Place the handle of the square against a smooth edge of the board.

3. Place a sharp pencil at the end of the blade.

4. Move the pencil and the square along the entire length of the board making a line (Figure 9-19). The pencil should be held straight up and down, and the angle should not be changed at any point along the way.

5. Reset the square at $4\frac{1}{16}$ inches, and draw a second line.

Using Surforms

A relatively new tool, called a Surform, is available. Surform is a trademark of Stanley Tool Works. This tool may be used instead of a plane or file (Figure 9-20). It consists of several styles of handles and blades. The blades have sharp ridges that cut wood, plastic, or body fillers. Small curls of material are produced much like a plane.

FIGURE 9-19 When marking the width of a board for cutting, the worker moves the pencil and the combination square along the board together.

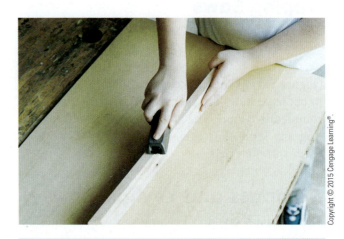

FIGURE 9-20 A Surform tool can be used in place of a plane or file to shape the edges of boards.

Surforms are handy since they do not need to be adjusted or sharpened. The blade is simply replaced with a new blade. Surforms come in a variety of shapes. Some are useful for cutting material to form straight and flat surfaces. Others are useful for rounding edges and smoothing curves. A variety of surfaces can be produced by simply changing the direction of stroke (Figure 9-21). Surforms will remove large amounts of material if pushed at a 45-degree angle to the material. Pushing the tool in a straight line in the direction of the work produces a fine, smooth cut.

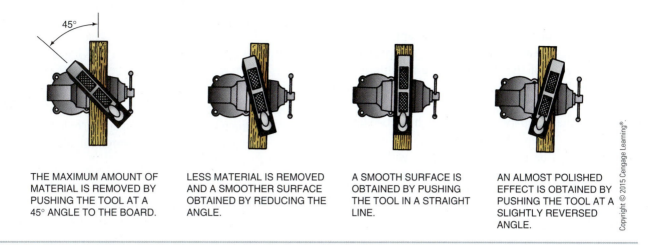

THE MAXIMUM AMOUNT OF MATERIAL IS REMOVED BY PUSHING THE TOOL AT A 45° ANGLE TO THE BOARD.

LESS MATERIAL IS REMOVED AND A SMOOTHER SURFACE OBTAINED BY REDUCING THE ANGLE.

A SMOOTH SURFACE IS OBTAINED BY PUSHING THE TOOL IN A STRAIGHT LINE.

AN ALMOST POLISHED EFFECT IS OBTAINED BY PUSHING THE TOOL AT A SLIGHTLY REVERSED ANGLE.

FIGURE 9-21 By changing the direction of the Surform's stroke, a worker can produce various surfaces. For maximum removal of material, the worker should push the tool at a 45-degree angle to the board. To remove less material and produce a smoother surface, the worker simply reduces the angle. Pushing the tool in a straight line produces a smooth surface, and pushing it at a slightly reversed angle yields an almost polished effect.

Using Files

Various files are used to dress down and finish off curved cuts on boards. Files are handy for smoothing edges and shaping materials to odd shapes (Figure 9-22). A **file** may be flat, round, half-round, square, or three-sided. Files have teeth that cut wood or metal. Files designed for cutting wood and soft metals have coarse teeth. Those designed for cutting steel have smaller teeth. Most files are hard enough to cut either wood or metal. The rasp is a type of file with very coarse teeth. It is designed to be used only on wood and other soft materials. **Rasps** are good for making rough cuts when a lot of material must be removed.

When filing wood, the file must not be pushed across the edge of the board as this would tear down the corner and leave it ragged. Best results are obtained when the file is pushed lengthwise along the board. File teeth cut only when the file is pushed forward. Therefore, it is advisable to apply pressure on the forward stroke only. The file should be lifted on the return stroke. The teeth of the file may become filled with the material being filed. Material may be removed and the file cleaned by tapping the handle up and down on a bench top. A special wire brush called a **file card** may also be used to clean files (Figure 9-23).

A file has a sharp end called the *tang*. The tang is designed to go into a wooden or plastic handle. Files should not be used without handles. Serious injury to the hand can result from using a file without a handle. A handle can be installed by driving the handle on the tang using a soft mallet. Another method is to place the tang in the handle and tap the handle up and down on the bench top until the tang is totally hidden in the handle.

FIGURE 9-22 Files can be used to shape wood.

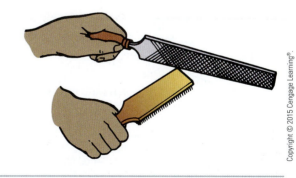

FIGURE 9-23 A file card is a wire brush used to clean material from the teeth of a file.

Copyright © 2015 Cengage Learning®.

Using Sandpaper

A small piece of sandpaper can be wrapped around a file for the finishing touches. The file provides a thin, rigid core of the desired shape to support the sandpaper. The sandpaper is wrapped with the rough surface out.

To use the sandpaper and file, pinch the two lightly at both ends. Then, simply move the two as a unit. Sandpaper can be pushed across edges without splitting the corner of boards. Medium sandpaper should be used first, and then very fine sandpaper can be used to finish.

CUTTING DADOS AND RABBETS

A **dado** is a square or rectangular groove in a board. The purpose of the dado is to receive the end or edge of another board to make a dado joint. A **rabbet** is a cut or groove made at the end or edge of a board to receive another board. The result is a rabbet joint when the two are fastened together. Dado and rabbet joints are used wherever strong 90-degree joints are needed (Figure 9-24). To make a dado, the following procedure is used. (A rabbet is made using a similar procedure.)

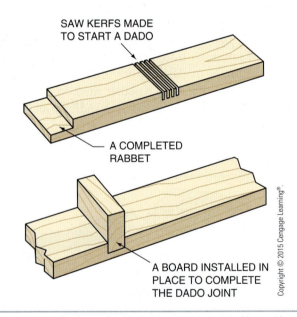

SAW KERFS MADE TO START A DADO

A COMPLETED RABBET

A BOARD INSTALLED IN PLACE TO COMPLETE THE DADO JOINT

Copyright © 2015 Cengage Learning®.

FIGURE 9-24 Dados and rabbets are grooves created by making multiple saw cuts and finishing with a wood chisel and file. A groove made in the end or on the edge of a board to help create a perpendicular joint is called a rabbet. A groove made elsewhere in the board to help create a similar joint is called a dado.

Procedure

1. Using a combination square, mark the face of the board with two lines ¾ inch apart to indicate the position and width of the dado.

2. Use a square to mark lines down across the edge of the board with lines perpendicular to the lines on the face of the board. **Perpendicular** means at a 90-degree angle.

3. Set the blade on the combination square to extend for a distance equal to one-half the thickness of the board. The distance will be ⅜ inch for a board that is ¾ inch thick.

4. Lay the blade of the square against the edge of the board, and mark the bottom of the dado by drawing a line at the end of the blade.

5. Mark the depth and sides of the dado on the other edge of the board.

6. Carefully saw four kerfs into the dado, stopping with the kerf exactly at the bottom of the dado. This can be seen on both edges of the board. A backsaw or miter box saw should be used for this operation.

7. Use a wood chisel to remove the wood between the kerfs (Figure 9-25). Be careful to chisel the wood to a uniform depth at the bottom of the kerf.

8. Smooth the bottom of the dado with a flat file. Be careful not to rock the file. Rocking will round the edges at the bottom of the dado and make a poor joint.

Copyright © 2015 Cengage Learning®./Ray Herren.

FIGURE 9-25 A wood chisel can be used to remove wood from between kerfs when making a dado.

SUMMARY

Completing quality projects made of wood requires a thorough understanding of wood's properties and grades. Each type of project involves selecting the type of wood that has the characteristics demanded by the uses of the project. Wood is graded according to the intended use and the appearance of the wood. Also, wood must be properly cured and dried to the appropriate moisture level. Once the proper type and grade of wood is selected, various hand tools can be used to work the wood.

Student Activities

1. Define the Terms to Know in this unit.

2. Examine the stump of a recently cut tree. Count the number of rings in the stump. How old was the tree? Examine the distance between the rings. Wide distances indicate a year of good growth, perhaps plenty of rain.

3. Examine samples of oak, walnut, maple, cherry, mahogany, white pine, red cedar, poplar, fir, redwood, and cypress. Record the color and degree of hardness of each.

4. Learn to identify and correctly spell the various species of wood.

5. Examine lumber that is warped and has split ends. Try to explain the reason for the warping.

6. Describe the steps in preparing lumber, from log to sanded board.

7. Measure the width and thickness of five different types of finished lumber. Record the actual size and the nominal size of each.

8. Obtain a board 18 inches long and 5½ inches wide. Softwood is recommended. Place five lines across the board ½ inch apart, near the end and square to the edge. Use a crosscut saw to saw off the piece nearest the end. Check the cut with a try square. Is it exactly square when checked in both directions? Saw and check each of the other four cuts, trying to make perfect cuts. Report the results to your instructor.

9. Use a combination square and mark a line ½ inch from the edge of a 1- to 2-foot-long board. Rip the board on the side of the line closest to an edge. Check the cut for straightness and squareness. What are the results?

10. Use a Surform tool to improve the edge of the board sawed in activity 9. Now is the edge straight? Is it square?

11. Set a compass at one-half the width of the board used in activity 10. Place the steel point on the board so that the pencil can be swung around and just touch both edges and an end of the board. Swing the compass in an arc and mark a rounded end on the board. Saw the end round with a coping saw and finish it with a file or Surform.

12. Using the board from activity 11, bore a ½-inch hole with a power hand drill at the mark left by the steel compass point.

13. Using the board from activity 12, remove ½ inch from the nonrounded end by using a miter box. Cut a rabbet in the end that is ¾ inch deep and ¾ inch wide.

14. Measure and mark the board used in activity 13 at 3 inches and again at 3¾ inches from one end. Cut a dado ⅜ inch deep and ¾ inch wide.

15. Set a combination square so that the blade extends ¼ inch. Choose one of the four corners of the board you used in activity 14. Place the square and draw a line on the edge of the board ¼ inch down from the corner you chose. The line should run the entire length of the board. Reposition the square and draw a

line from one end to the other on the face of the board. Use a block plane or a smoothing plane to make a chamfer. A chamfer is a tapered cut along the edge of a board.

16. Discuss your workmanship in the preceding procedures with your instructor. Save the board for an activity in Unit 10.

Relevant Web Sites

American Hardwood Export Council
www.ahec.org/index.asp

Softwood Export Council
http://softwood.org

Self-Evaluation

A. Multiple Choice. Select the best answer.

1. Grain in lumber is caused by
   a. the age of the board
   b. annual rings
   c. special drying techniques
   d. the stain

2. Lumber is graded according to its
   a. appearance and soundness
   b. color and species
   c. strength and durability
   d. cost and length

3. A crosscut handsaw with very coarse teeth would have how many teeth per inch?
   a. 6
   b. 10
   c. 12
   d. 14

4. A crosscut handsaw with very fine teeth would have how many teeth per inch?
   a. 6
   b. 8
   c. 14
   d. 20

5. The wood removed by a saw blade leaves an opening called a
   a. bevel
   b. channel
   c. chamfer
   d. kerf

6. The backsaw gets its name from
   a. its use as a backup tool
   b. its fine teeth
   c. its stiff back
   d. its original use in making chair backs

7. A suitable tool for cutting curves is the
   a. compass saw
   b. coping saw
   c. keyhole saw
   d. all of these

8. The brace and bit has been replaced by the
   a. drill press
   b. coping saw
   c. screwdriver bit
   d. portable electric drill

9. Lumber for construction should be dried to about
   a. 2%
   b. 15%
   c. 35%
   d. 20%

10. The teeth on a file cut
    a. best when oiled
    b. only soft materials
    c. only on the backward stroke
    d. only on the forward stroke

11. The groove cut across the end of a board to receive another board is called a
    a. dado
    b. rabbet
    c. miter
    d. ratchet

12. A file handle is necessary for
    a. operator protection
    b. file functioning
    c. compliance with the law
    d. none of these

B. Matching. Match the type of wood in column I with its major use in column II.

Column I

1. red cedar
2. cypress
3. fir and hemlock
4. black locust
5. mahogany
6. maple
7. oak
8. white pine
9. yellow pine
10. redwood
11. black walnut
12. birch

Column II

a. fence posts
b. reddish, fine furniture
c. barrels, wagon bodies, farm buildings
d. shelving, siding, trim
e. floors, stairs, trim
f. excellent rot resistance; patios
g. structural material in wet places
h. brown, fine furniture
i. floors and bowling alleys
j. chests and closet linings
k. construction framing, siding, sheathing
l. surface veneer for cabinets and doors

C. Completion. Fill in the blanks with the word or words that will correctly complete the statement.

1. A piece of lumber that is planed and measures 1½" × ½" is called a _____ by _____.

2. A ripsaw is designed to cut _____ the grain, and the crosscut saw is designed to cut _____ the grain.

3. The compass saw is used to cut a board starting at a _____.

4. Two advantages of Surform tools over planes are that Surforms do not need to be _____ or _____.

5. *Clear* refers to lumber with no _____.

D. Brief Answer. Briefly answer the following questions.

1. Explain why "measure twice, cut once" is a good rule when cutting wood.
2. What does the designation S2S for lumber mean?
3. Explain the difference between boring and drilling.
4. Explain how green lumber is cured.

UNIT 10
Fastening Wood

Objective

To use nails, screws, bolts, and glue in assembling wood.

Competencies to be developed

After studying this unit, you should be able to:
- Drive nails.
- Set screws.
- Use bolts.
- Use glue.
- Assemble wood parts.

Materials List

- Samples of common wood joints (butt, lap, dado, rabbet, and dovetail)
- Common and finishing nails: 4d, 8d, 10d
- Scraps of 2 × 4s hardwood and softwood
- Claw hammers
- Nail set
- Stapling gun and staples
- Flat-head wood screws (#8 × 1¼")
- Hand drill and drill-bit set; countersink
- Push drill and drill bits
- Bit brace and auger bit (⅜" recommended)
- Standard screwdriver to fit #8 screws
- Hinge set with screws

continued

Terms to Know

- joint
- butt joint
- lap joint
- member
- assembly
- dado joint
- rabbet joint
- miter joint
- dovetail joint
- biscuits
- setting
- nail set
- toe nail
- end nailing
- flat nailing
- clinch
- staple
- countersink
- shank hole
- clearance hole
- pilot hole
- anchor hole
- glue
- dowel
- doweling jig

Materials List, *continued*

- Liquid glue
- Assortment of dowel pins
- Glue
- Clamps
- Clean, soft rags

Wood and wood products are attractive, strong, and long lasting. They are used for framing, siding, walls, doors, cabinets, counters, and furniture. Wood is used extensively on farms for construction of buildings, fences, and machinery such as wagons and trailers. Wood is easy to fasten. Common materials for fastening wood include nails, screws, bolts, and glue.

TYPES OF JOINTS

When fastening wood, a **joint** that is strong enough to do the job should be chosen. A joint is the union of two materials. Joints may be secured with nails, screws, bolts, or glue. The most popular joints are butt, lap, dado, rabbet, miter, and dovetail (Figure 10-1).

The **butt joint** is formed by placing two pieces end-to-end or edge-to-edge in line or at a 90-degree angle. Butt joints may be strengthened by

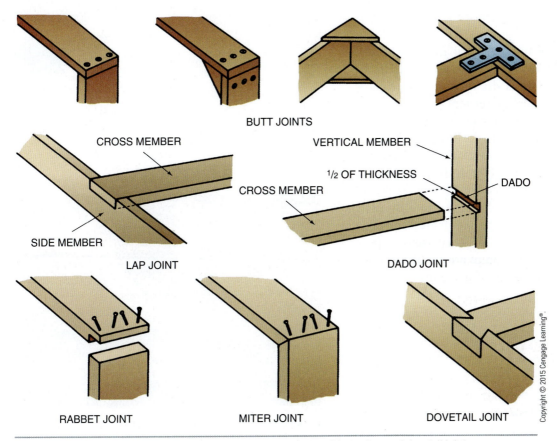

FIGURE 10-1 Popular types of wood joints include butt, lap, dado, rabbet, miter, and dovetail joints.

applying thin wood or metal plates at corners or across flat surfaces where parts meet.

The **lap joint** is formed by fastening one member face-to-face on another member of an assembly. **Member** means piece; **assembly** means pieces fastened together. The members may be offset or may be cut so the two form a flat surface. In other words, the members are set into each other. Lap joints are stronger than butt joints.

The **dado joint** is (a rectangular groove) cut in a board. The end or edge of another board is then inserted in this groove.

The **rabbet joint** is similar to a dado except that it occurs at the end or edge of a board.

The **miter joint** is formed by cutting the ends of two pieces of lumber at a 45-degree angle. The two pieces are then joined to form a 90-degree angle.

The **dovetail joint** is formed by interlocking parts of two pieces. This is one of the strongest joints. However, it is also the most difficult to make. Dovetail joints are used extensively in high-quality and expensive furniture.

Using Biscuit Joints

A relatively new method of joinery is the use of wood **biscuits** (Figure 10-2). These biscuits are thin, oval-shaped pieces of wood that are fitted and glued into slots cut into the edge of boards that are being joined. The boards to be joined are laid out as they are to be joined and a short line is drawn across the boards (Figure 10-2A). A special power tool is used to align and cut the slots so that the biscuits will fit together. A mark on the tool is aligned with the line on each board. The trigger is pressed and a rotary cutter extends and cuts a slot in the board (Figure 10-2B). Once all the slots are cut, the thin wooden "biscuit" is coated with glue (Figure 10-2C) and inserted into the slots (Figure 10-2D). The edges of the boards are also coated with glue. The boards are then realigned using the pencil marks (Figure 10-2E). Clamps are applied to draw the joint tight (Figure 10-2F). This joint is extremely strong and durable.

FASTENING WITH NAILS

Nailing is the fastest way to fasten wood. However, it is the least rigid and has the least strength compared to the other methods of fastening wood. Carpenters use nails to build frames of houses, attach siding, and trim out house interiors. Farmers and others in agricultural

A The first step in making a biscuit joint is to lay out the boards as they are to be joined. Using a square, draw a line across both pieces.

B Align the mark on the board with the mark on the biscuit cutter. Press the start trigger and cut the slot. Do this for both boards.

C Apply glue to both sides of the biscuit.

FIGURE 10-2 Joining wood with biscuits.

D Press the biscuit into the slot in one of the boards.

E Align the other board and press it onto the biscuit.

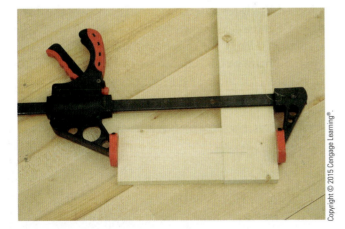

F Clamp the boards together and let it sit for at least two hours.

FIGURE 10-2 (continued)

mechanics find that proper nailing techniques are a very useful skill. A few nails, carefully placed and carefully driven, make quick and effective repairs. Where extensive nailing is done, electric and air-driven automatic nailing machines speed the process.

Driving Nails

Nails can be driven into flat surfaces easily if the material is soft. To drive a nail, the following procedure is used.

Procedure

1. Hold the nail between the thumb and index finger, and place the point of the nail on the material (Figure 10-3A). The fingers should be placed high on the nail to permit them to be knocked free rather than smashed if the hammer accidentally hits them.

2. Hold a claw hammer in the other hand with the hand near the end of the handle (Figure 10-3B).

3. Keep your eyes focused where you want the hammer to hit.

4. Tap the nail with the hammer until it stands up itself.

5. Use the wrist and arm to deliver firm blows, using the weight of the hammer to drive the nail.

6. Drive the nail until the head is flush with the wood and the two pieces of wood are tightly fastened.

Match the hammer to the size of the nail. Common weights of claw hammer heads are 7, 13, 16, and 20 ounces. The 7-ounce hammer is excellent for driving brads, tacks, and small finishing nails. The 13-ounce hammer is good for light general nailing. For most farm applications, the 16- or 20-ounce hammer is preferred because it is heavy enough to drive large nails and spikes. Splitting of lumber is decreased when the end of the nail is filed to create a blunt end.

Pulling Nails

Claw hammers and ripping bars are used to pull nails and to rip boards from surfaces. When pulling a nail with a claw hammer, place a block under the hammer to prevent breaking the hammer handle (Figure 10-4).

Setting Nails

When using finishing nails, the head of the nail is hidden in the wood. This is done by **setting** the nail. **Setting** a nail

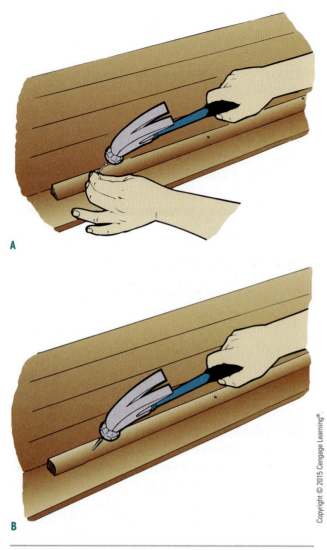

A

B

FIGURE 10-3 Proper procedure for driving a nail. Start a nail by holding it with one hand and tapping it lightly with the hammer held in your other hand (A). Drive the nail with long, even strokes; the hammer handle should be parallel with the work as the head strikes the nail (B).

means driving the head below the surface (Figure 10-5). To set a nail, the nail is driven with a hammer until the head touches the wood. A **nail set** is then used to drive the head below the surface about ¹⁄₁₆ inch. A nail set is a punchlike tool that has a cupped end rather than the flat or pointed end of a punch.

Toe Nailing

When two large pieces of wood must be fastened at right angles, they may be toe-nailed. To **toe nail** means to drive a nail at an angle near the end of one piece and into the face of another piece (Figure 10-6). Toe nailing is done extensively in framing using 2 × 4 or 2 × 6 lumber and 8d common nails.

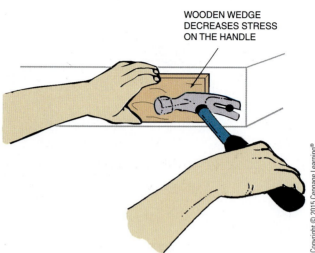

WOODEN WEDGE DECREASES STRESS ON THE HANDLE

FIGURE 10-4 When using a claw hammer to pull a nail, the worker places a block under the hammer's head to avoid breaking the hammer's handle.

End Nailing

End nailing is done by nailing through the thickness of one piece of lumber and into the end of another piece (Figure 10-6). The nail is driven with the grain into the end of the receiving piece. Common nails or spikes are used for end nailing; box nails can be used for end nailing thin materials. End nailing has poor holding power. The technique is most useful when weight is permanently resting on the assembly. When end-nailing small projects, it is advisable to use glue as well.

Flat Nailing

Flat nailing means to fasten two flat pieces to each other. One flat piece may also be nailed to a thicker piece. When two thin pieces are nailed together, the nails must be clinched. To **clinch** means bending the nail over and driving the flattened end down into the wood (Figure 10-6). If splitting is likely, the nail should be clinched across the grain. Clinching results in a very strong nailed joint. To clinch nails, use the following procedure.

Procedure

1. Choose a common nail that is about 1 inch longer than the thickness of the two boards.

2. Drive the nail through the boards until the head is flush with the surface of the wood. Be careful to support the wood so there is a space under the

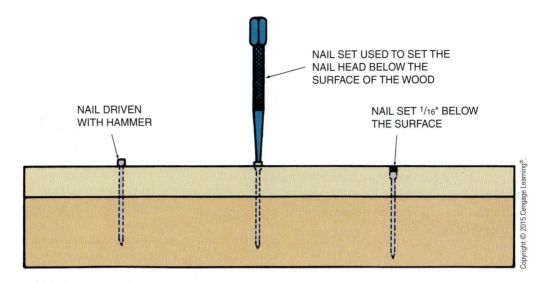

Copyright © 2015 Cengage Learning®

FIGURE 10-5 A nail set is used to sink finishing nails so that their heads are below the wood's surface. The resulting holes can then be filled in with putty or other filler before the wood is finished.

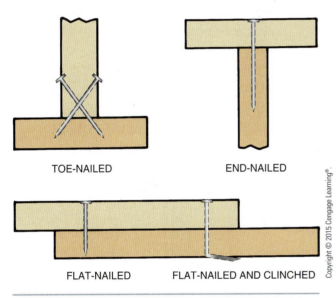

TOE-NAILED   END-NAILED

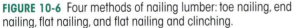

FLAT-NAILED   FLAT-NAILED AND CLINCHED

Copyright © 2015 Cengage Learning®

FIGURE 10-6 Four methods of nailing lumber: toe nailing, end nailing, flat nailing, and flat nailing and clinching.

area where the nail comes through. If a space is not provided, the nail will damage the surface or the support, or the pieces will be nailed to the bench or sawhorse.

3. Turn the two boards over and place the head of the driven nail on a hard surface. A steel vise works well. When nailing large assemblies, a heavy sledge hammer can be held against the nail head.

4. Hit the end of the nail sideways until it is bent flat to the surface.

5. Drive the nail until it is embedded in the wood.

Stapling

A **staple** is a piece of wire with both ends sharpened and bent to form two parallel legs. Staples are generally used to attach fencing or electrical wiring to wood. Staples are driven until they just touch the wire. If they are driven too tight, they will cause a short in electrical wire. It is also possible to weaken or break fence wire if staples are driven tightly against the wire.

Staples are also used to fasten ceiling tile and back panels on furniture. This type of staple is driven with a staple gun.

FASTENING WITH SCREWS

Screws hold better than nails. They are installed with a screwdriver, which is a turning tool with a straight tip, Phillips tip, or special tip. The Phillips screwdriver end is shaped like a plus sign (+). Screws may be driven quickly and easily with power screwdrivers or variable speed drills. Screws are used extensively in doors, windows, wall systems, and furniture. The flat-head screw is the one used more extensively in woodworking. Its head is countersunk and flush with the surface after it is installed. The Phillips-head is preferred because it is easier to keep the screwdriver bit in the head of the screw.

Preparing Wood for Flat-Head Screws

A hole must be drilled in the wood before a screw is inserted. Holes made in wood to receive screws must be the correct diameter and depth; otherwise, the screw will not hold properly. Screws have a head, shank, and core with threads; all require exact holes (Figure 10-7). The flat-head must have a hole with tapered sides that fit. The hole for the head is called the **countersink**. The shank requires a hole that will permit it to drop through. The hole for the shank is called a **shank hole** or **clearance hole**. The core requires a hole small enough to permit the threads to screw into the wood. The hole for the core is called a **pilot hole** or **anchor hole**. When drilling holes for screws, it is recommended that the drill sizes be determined using a drill size chart (Figure 10-8).

Attaching Wood with Screws

When using screws to fasten two pieces of wood together, the following procedure is recommended.

Procedure

1. Measure the thickness of the material to be attached.

2. If possible, use a screw three times as long as the thickness of the board being attached. If the screw will reach all the way through the second board, use a shorter screw.

3. Determine the appropriate screw spacing, and mark the spots for all screws.

4. The diameter of the screws should look in balance with the spacing. Close screws should be smaller in diameter. The most frequently used screw sizes are numbers 6, 8, 9, and 10.

5. Use a chart to determine the drill size for the shank and pilot holes. An alternative method is to hold different drills under the screw shank until you have one that appears to be the same diameter. For the pilot hole, the drill must be the size of the screw core, not the diameter of the threads.

6. Insert the pilot-hole drill into a hand drill chuck. Adjust the length of the exposed drill until it equals the length of the screw. The bit may be too long. If so, mark the drill with a sliver of masking tape or use a piece of wooden dowel to limit the exposed length of the drill.

7. Place two pieces of wood together and drill the pilot hole through both pieces of wood.

8. Install the shank-hole drill in a hand drill. Use the shank-hole drill to enlarge the hole in the first piece.

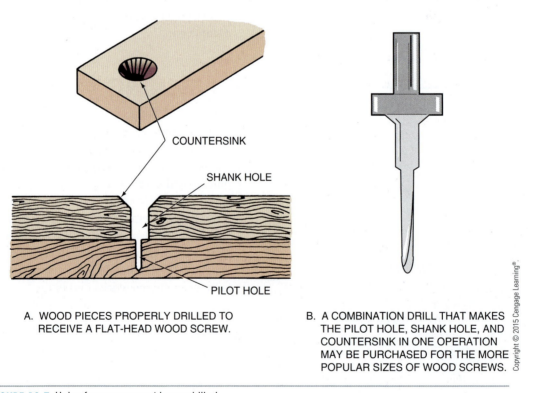

COUNTERSINK

SHANK HOLE

PILOT HOLE

A. WOOD PIECES PROPERLY DRILLED TO RECEIVE A FLAT-HEAD WOOD SCREW.

B. A COMBINATION DRILL THAT MAKES THE PILOT HOLE, SHANK HOLE, AND COUNTERSINK IN ONE OPERATION MAY BE PURCHASED FOR THE MORE POPULAR SIZES OF WOOD SCREWS.

Copyright © 2015 Cengage Learning®.

FIGURE 10-7 Holes for screws must be predrilled.

Screw Gauge	0	1	2	3	4	5	6	8	10	12	14	16	18
Shank Hole	1/16	5/64	3/32	7/64	7/64	1/8	9/64	11/64	3/16	7/32	1/4	17/64	19/64
Pilot Hole	1/64	1/32	1/32	3/64	3/64	1/16	1/16	5/64	3/32	7/64	7/64	9/64	9/64

Copyright © 2015 Cengage Learning®.

FIGURE 10-8 Chart of drill sizes suitable for screws of different gauges.

> **NOTE**
>
> The practice of enlarging a hole with a drill is acceptable in wood only. If attempted in metal, it may damage the bit.

9. Use a countersink tool to create a countersink that exactly fits the screw head. The screw should drop through the piece to be attached. The head of the screw should be perfectly level with the surface of the board.

> **NOTE**
>
> Steps 7, 8, and 9 can be done in one operation using a combination drill designed for this purpose.

10. Screw the two pieces together. Use a screwdriver that fits the screw slot. Turn the screw until it is snug. Do not overtighten.
11. Drill pilot holes for all additional screws.
12. Drill the shank holes for all additional screws. The depth of the shank hole is very critical. If it is drilled deeper than the thickness of the board, the screw will not hold properly.
13. Countersink all holes.
14. Set all screws.

> **NOTE**
>
> A project looks better if the screws are evenly tightened and all of the screw slots are aligned.

Installing Round-Head Screws

The procedure for using round-head screws is similar to the procedure for using flat-head screws. Countersinking is simply eliminated. When tightening screws, screws should not be forced. If all holes are the proper depth and diameter, the screw should be easy to turn. In hardwood, it may be necessary to drill the pilot hole 1/64 inch larger than the chart size. Screws are easier to turn if the threads are lubricated with soap, wax, oil, or water.

If the screwdriver slips out of the screw slot, the following should be checked:

- be sure the screwdriver fits the screw (Figure 10-9)
- be sure the holes are the proper depth and diameter
- lubricate the screw threads
- check the condition of the tip of the screwdriver—reshape if necessary

A standard screwdriver tip should have a flat end with square edges. If the tip is rounded, it should be reshaped with a file or another screwdriver should be used.

If a Phillips screwdriver end is damaged or the end does not fit the screw properly, another screwdriver should be used. Damaged or worn Phillips screwdrivers should be discarded or reshaped for other purposes.

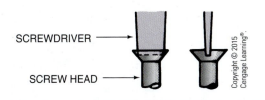

SCREWDRIVER

SCREW HEAD

Copyright © 2015 Cengage Learning®.

FIGURE 10-9 The screwdriver tip should fit snugly into the slot on the screw's head. The tip of the screwdriver must be square so that it fits the slot exactly.

A modern method of installing screws is using a portable drill or power screwdriver. Screws can be efficiently and quickly installed using these tools. Special screws are usually required for driving with a power driver. Most woodworkers use deck screws either with a Phillips head or a head with a square slot. These screws are designed for use with a power driver and are composed of harder metal that helps prevent the slot in the head from stripping out (Figure 10-10).

Power drivers come with an adjustment below the chuck that allows the operator to set the amount of torque (Figure 10-11). *Torque* refers to the amount of twisting or turning power. If the tool has too much torque, the screw can be easily stripped and the screw will not hold. On the other hand, if there is too little torque, the screw cannot be driven in completely. Trial and error is usually required to get the torque adjustment correct.

Lag screws are a type of screw used on large or thick boards (Figure 10-12). They usually range from

FIGURE 10-12 Lag screws are used on large or thick boards.

FIGURE 10-10 Most woodworkers use deck screws.

FIGURE 10-11 Power drivers can be adjusted to provide the proper amount of torque.

¼ inch to ¾ inch in diameter and have varying lengths. These bolts are used in places such as fastening timbers together to build a pole barn.

Attaching Metal with Screws

Metal objects, such as hinges, require short screws. This is due to the thinness of the material being attached. When setting screws to attach hinges, the following procedure is recommended.

Procedure

1. Hold the door in place.
2. Hold the hinge in place.
3. Draw a circle in one hole of the hinge.
4. Make a pilot hole with a push drill, a hand drill, or a pointed awl pushed by hand.

NOTE

The shank hole is already drilled in the hinge.

5. Install one screw.
6. Repeat the process on the other end of the hinge.
7. Install the second hinge using two screws.
8. Check the door for proper closing and clearance.
9. Reset one or more screws if necessary.
10. Install all remaining screws.

FASTENING WITH BOLTS

Bolts can be the strongest method of fastening wood—with the possible exception of glue. The strength of bolts and the addition of large washers make them especially useful at high stress points. To install bolts, use the following procedure.

Procedure

1. Select the type of head needed. The carriage bolt is designed specifically for wood. It leaves a fairly smooth head.
2. Drill a hole in both members the size of the bolt.
3. Place a washer on the bolt if a machine bolt is being used.
4. Insert the bolt.
5. Add a flat washer and nut.
6. Tighten the nut until the members are tight.

Lock washers are not needed when using bolts in wood. The wood creates back pressure on the nut. When tightening bolts, crushing the wood fibers or drawing the head below the surface of the wood should be avoided.

FASTENING WITH GLUE

Gluing is used extensively as a fastener of wood. **Glue** is a sticky liquid used to hold things together. It is used in the manufacture of plywood and particleboards. It is used extensively in manufacturing furniture. Even the rafters of wooden barns and large gymnasiums require glue to laminate the pieces into arched shapes. Glue is often used in combination with nails, screws, or bolts to provide extrastrong joints.

There are many kinds of glue. Some examples are resorcinol, urea, polyvinyl, epoxy, contact cement, casein, and animal glues. The most common type of glue used with wood is aliphatic resin glue (sometimes called *yellow carpenter's glue*). It comes in colors ranging from yellow to tan and brown depending on the brand. This glue forms a very tight, durable bond between pieces of wood. This type of glue comes in a range of grades according to moisture resistance (Figure 10-13). In Figure 10-13, the glue in the white bottle on the left is used for applications where light materials are glued together and little moisture is encountered. To make a proper glue joint, the glue must be pressed tightly into the cells of the wood. This is usually accomplished by using clamps on the wood.

Another important adhesive for agricultural uses is epoxy. Epoxy produces a strong, waterproof bond on wood, plastics, ceramics, metals, and other materials. Epoxy materials are also available for difficult masonry repairs. The disadvantage of epoxy is that it is supplied as two materials. These materials must be mixed in the correct proportions and used promptly once mixed.

Using Aliphatic Resin Glue

A properly glued wood joint will be as strong as the wood itself. When gluing two pieces of wood together with aliphatic resin glue, the following procedure is used.

FIGURE 10-13 These are four grades of glue according to their resistance to moisture. The one on the left is least resistant and the one on the far right is most resistant.

Copyright © 2015 Cengage Learning®/Ray Herren

Procedure

1. Cut the two pieces of wood so all surfaces match perfectly.

2. Put the two dry pieces of wood together to check that the fit is good. Modify the pieces to make them fit, if needed.

3. Drill all screw or bolt holes.

4. Obtain all clamps that may be needed.

5. Apply a small bead of glue on the mating surfaces of both pieces of wood.

6. Spread the glue evenly with a flat object.

7. Put the two pieces together and nail, screw, bolt, or clamp the material (Figure 10-14).

8. Check the joint and project to be certain every part is correctly aligned.

9. Retighten all clamps.

10. Remove all glue runs when slightly dry with a putty knife or wood chisel (Figure 10-15).

11. Wipe all glue marks a second time using a clean wet rag. Finally, wipe the wood with a dry rag.

NOTE

If all glue is not removed, the wood will not stain evenly. If too much water is used in the cleanup process, the grain of the wood will get rough and will need sanding before finishing.

12. Clamps may be removed after $\frac{1}{2}$ hour. For better results, leave the clamps on for about 12 hours.

Glue will not adhere to paint, grease, or wax. Therefore, it is best to cut, file, sand, or otherwise clean the areas to be glued.

Glue guns that melt a glue stick and apply a thick substance that does not require clamping are very useful for arts and crafts projects.

USING DOWEL PINS

A **dowel** is a round piece of wood. Dowels are generally sold in 36-inch lengths. They range from $\frac{1}{16}$ inch to 1 inch or larger in diameter. Popular sizes are $\frac{1}{8}$, $\frac{3}{16}$, $\frac{1}{4}$, $\frac{5}{16}$, $\frac{3}{8}$, $\frac{7}{16}$, $\frac{1}{2}$, $\frac{5}{8}$, $\frac{3}{4}$, $\frac{7}{8}$, and 1 inch. An important use of

FIGURE 10-14 Bar clamps can be used to draw the pieces of wood together until the glue sets. To avoid squeezing too much glue out of the joint and leaving it glue-starved, the worker should tighten the clamps to no more than a moderate level of pressure.

FIGURE 10-15 When the correct amount of glue is used, only small droplets of excess glue will be squeezed from the joint. These droplets can be removed with a putty knife or wood chisel after the glue dries.

Copyright © 2015 Cengage Learning® /Ray Herren

dowels is to strengthen glue joints by extending from one part to the other.

Dowels are also sold in lengths of about 2 inches. These lengths have tapered ends that make them easy to use when gluing parts together. They have grooves that permit glue to make an excellent bond between the dowel and the board (Figure 10-16). For jointing boards ¾ inches thick, use a dowel ⅜ inches in diameter. In a process similar to using a biscuit joiner, the pieces of wood are laid out and marked using a square and pencil (Figure 10-17A). A **doweling jig** is used to align the mark on the jig and the boards (Figure 10-17B). A drill is used to make the holes for the dowels (Figure 10-17C). The dowels are coated with glue and inserted into the holes (Figure 10-17D). The other board is then aligned and pressed onto the dowels. A clamp is applied and let sit for at least 2 hours (Figure 10-17E).

1B

2B Align the marks on the boards with the mark on the doweling jigs.

FIGURE 10-16 Dowels have grooves to permit glue to form a strong bond.

A Align the boards and draw lines using a square and pencil.

C A drill is used to make the holes for the dowels.

FIGURE 10-17 Joining wood with dowels. *(continued)*

D The dowels are coated with glue and inserted into the holes.

E The other board is pressed onto the dowels and a clamp is applied.

FIGURE 10-17 *(continued)*

SUMMARY

Humans have been fastening pieces of wood together for many centuries. Over the years different methods have been used ranging from basic nailing to the use of modern glues. Each method has a specific use as well as a proper procedure and technique. Selecting the correct method and using it properly will give the desired results. Improper use and/or methods will result in improperly joined wood.

Student Activities

1. Define the Terms to Know in this unit.

2. Obtain two pieces of scrap wood 2 × 4 or 2 × 6, each about 12 inches long, cut from pine, spruce, hemlock, redwood, poplar, or other softwood. Place one piece on top of the other to double their thickness. Practice driving 4d, 8d, and 10d common nails and finishing nails. Try different sizes of hammers.

3. Drive several nails part way into a piece of wood. Place a block of wood (about two-thirds the height of the nail) beside a nail. Place the hammer claws under the head of the nail with the hammer head resting on the block. Pull the nail. Practice this skill, being careful to use wooden blocks to avoid breaking the hammer handle. Why does the use of a block under the hammer reduce strain on the handle?

4. Repeat activity 3 using oak, locust, maple, or other hardwood. What differences do you observe? Nails are easier to drive into hardwood if a hole smaller than the nail is drilled through the first piece.

5. Use a nail set to drive the heads of several finishing nails below the surface.

6. Practice toe-nailing a 2 × 4 to a 2 × 4.

7. Study some pieces of furniture and identify butt, lap, dado, rabbet, and dovetail joints.

8. Attach two small boards together using glue and two flat-head screws. Be careful to provide the correct countersink, shank hole, and pilot hole.

9. Take out the board you used to cut a dado in Unit 9. Cut a ¾-inch-thick piece of wood several inches wide to fit into the dado groove. Glue and clamp the board to form a dado joint. Save this board for an activity in Unit 11.

10. Make a sanding block with a hold-down cleat like the one shown in Figure 10-17.

Relevant Web Sites

Humboldt Woodworkers Guild, Wood Types and Techniques
www.woodguild.com/demo/woodt.html

Technologystudent.com, Types of Screws information page
www.technologystudent.com/joints/screws1.htm

Self-Evaluation

A. Multiple Choice. Select the best answer.

1. Two pieces of wood joined together is called
   a. an angle
   b. a joint
   c. a lap
   d. a splice

2. The tool used to push nail heads below the surface of wood is a nail
   a. driver
   b. press
   c. punch
   d. set

3. The strongest nailing method is
   a. clinching
   b. end nailing
   c. flat nailing
   d. toe nailing

4. An advantage of epoxy glue is
   a. ease of use
   b. no mixing
   c. waterproof bond
   d. none of these

5. The most difficult joint to make, but one of the strongest, is the
   a. miter joint
   b. butt joint
   c. dado joint
   d. dovetail joint

6. The _____-ounce hammer is good for light, general nailing.
   a. 32
   b. 13
   c. 20
   d. 40

7. The glue most commonly used with wood is _____.
   a. aliphatic resin
   b. epoxy
   c. contact cement
   d. urea

B. Matching. Match the words or phrases in column I that best match the word or phrase in column II.

Column I

1. groove in end of board
2. interlocking extensions
3. groove in middle of board
4. face-to-face
5. edge-to-edge
6. flat head of screw
7. smooth stem of screw
8. threaded core of screw
9. combination drill
10. shank hole for No. 8 screw
11. shank hole for No. 10 screw
12. pilot hole for No. 8 screw
13. pilot hole for No. 10 screw

Column II

a. butt joint
b. dado joint
c. dovetail joint
d. lap joint
e. rabbet joint
f. three holes at once
g. countersink
h. shank hole
i. pilot hole
j. $1\frac{1}{64}$ inch
k. $\frac{3}{64}$ inch
l. $\frac{5}{64}$ inch
m. $\frac{3}{32}$ inch

C. Completion. Fill in the blanks with the word or words that make the following statements correct.

1. The strength of bolts and the addition of large washers make them especially useful at _____.

2. When fastening two pieces of wood together with screws, it is best to use a screw that is _____ times as long as the thickness of the board being attached, if possible.

3. For the pilot hole, the drill must be the size of the _____, not the diameter of the threads.

4. An important use of _____ is to strengthen glue joints by extending from one part to the other.

5. To make a proper glue joint, _____ are usually used to press glue tightly into the cells of the wood.

D. Brief Answer. Briefly answer the following questions.

1. If the screwdriver slips out of the screw slot, what four things should be checked?

2. List the steps for installing bolts.

3. What two problems can occur if glue is not appropriately cleaned from a glued wood joint?

4. In woodworking, what are biscuits, and how are they used?

5. List the steps for setting screws to attach hinges.

UNIT 11
Finishing Wood

Objective
To prepare wood for finishing and to apply attractive and durable finishes.

Competencies to be developed
After studying this unit, you should be able to:
- Sand wood.
- Remove dents.
- Fill holes.
- Select finishing materials (except paint).
- Select solvents for thinning.
- Apply stain and clear finishes.
- Clean brushes.

Materials List
- Sandpaper
- Steel wool
- Heating iron
- Clean, soft rags
- Wood filler
- Clean paintbrush
- Sealer
- Finishing material
- Stain
- Paste furniture wax
- Paint thinner
- Polyurethane

Terms to Know
- finish
- putty
- plastic wood
- varnish
- polyurethane
- satin

No matter how well a project is planned, designed, and constructed, if the finish is not properly applied the results will not be good. Wood objects and buildings that were properly finished have remained durable and beautiful for centuries. There are many ways to finish wood to protect the surface. Many have a stain or dye applied to enhance the color and to make the grain more enhanced. A top coating of a clear finish is usually applied. Wood can be truly beautiful if finished properly. Most often poor finishes are due to improperly prepared wood before the finish is applied. The secret to an attractive wood project is to prepare the wood for finishing. This means removing dents, leveling high spots, removing glue stains, and sanding to a smooth surface. Applying the finish too thickly, using the wrong finish, or allowing dust in the finish can also give poor results.

This unit covers the finishing of wood with stain and clear finishes. Painting is covered in another unit.

PREPARING WOOD FOR FINISHES

Soft wood is easily damaged until a finish is applied. A **finish** is a chemical layer that protects the surface of a material. Lumber should always be handled carefully. Walking on clean boards or laying them on dirty benches or floors should be avoided. Sanded boards should be protected with cloth or paper. Pencil marks should be made lightly and then erased. If pencil marks are sanded, they become embedded in the wood.

Removing Dents

If wood is accidentally dented, it may be repaired unless the wood fibers are cut. The procedure is as follows.

Procedure

1. Heat a soldering copper to about 400°F or a clothes iron to its hottest setting.
2. Place a damp, soft, clean cloth over the dent.
3. Apply the hot iron to the cloth until steam rises from the cloth (Figure 11-1).
4. Let it steam for about 5 seconds, but do not burn the wood.
5. Remove the iron.
6. Remove the cloth.
7. Examine the dent. The steam should have caused the compressed wood fibers to expand. The previous dent should be level.
8. If there is still a small dent, redampen the cloth and repeat steps 2 through 7.
9. Sand the area when the surface dries.

Filling Holes

A word of caution is in order about filling holes. Some students expect to correct mistakes with filling materials. It is very difficult to restore the looks of a project after bad cuts are made or it has been inadequately sanded. In this regard, wood filling materials work well only in holes where dowels or nails are set below the surface, or where a puncture or blemish is small in diameter.

Holes made by installing finishing nails may be filled with several different types of filler. These come in a variety of colors and types. Some are applied to bare wood and sanded smooth. Others are applied when the wood is stained and before the final finish is applied. This allows the matching of the color of the filler to the color of the stain.

Copyright © 2015 Cengage Learning®.

FIGURE 11-1 To remove a dent from wood, lay a wet cloth over the dent and press the flat area of a hot iron on the cloth. The steam from the water in the cloth will cause the compressed wood fibers to expand, eliminating the dent.

To use most compounds to fill holes, the procedure is as follows:

Procedure

1. Seal the wood with a sealer or primer.
2. Take a small amount of **putty** or compound from the can.
3. If you wish to color the material, work in some oil-based tinting pigment to obtain the desired color.
4. Place a bit of the material on a putty knife tip and press it into the hole (Figure 11-2A).
5. Smooth and level the material by wiping the surface firmly with a flexible putty knife or your finger (Figure 11-2B).
6. Permit the material to firm up overnight.
7. Sand the compound smooth and level with the board.
8. Use a tack cloth to remove dust.
9. Apply the desired finishing materials to the wood.

Plastic Wood. **Plastic wood** is a filler product that is soft when it comes from a sealed can but dries very quickly. It has the advantage of becoming very hard in a matter of minutes. Plastic wood cannot be colored. However, it can be purchased in colors to match various woods such as pine, oak, walnut, and mahogany. To apply plastic wood, the following procedure is recommended.

Procedure

1. Select a can or tube of plastic wood of the color to match the finished color of the wood.
2. Put a small amount of plastic wood on the tip of a screwdriver and close the can or tube.
3. Quickly press the plastic wood into the hole.
4. Smooth the surface by drawing the flat part of the screwdriver blade back over the plastic wood.
5. Add more plastic wood and smooth out if necessary.
6. Recheck to be sure the can or tube is tightly closed.
7. Scrape off any plastic wood on the surface around the hole.
8. After 3 to 5 minutes, sand the area thoroughly with fine sandpaper. Otherwise, every spot touched by plastic wood will leave an ugly stain on the wood.
9. Apply the desired finishing materials to the wood.

Putty Stick. A more recent product for filling holes is the putty stick. Putty sticks may be purchased in dozens of shades to match any color of wood finish. The procedure for using a putty stick follows.

A Place a bit of the filling material on the hole, and press it in.

B Smooth the filler with a putty knife.

FIGURE 11-2 Holes in wood may be filled with putty, glazing compound, or plastic wood.

Material	Recommended Use	Special Comments
Flint	Painted or pitchy surfaces; jobs where clogging is a problem	Low cost; paperback abrasive wears off quickly; use where paper clogs easily
Garnet	Hand sanding of wood	Durable; low cost; paper backed
Emery	Hand sanding of metal to remove rust or smooth rough areas	Expensive; cloth backed; disintegrates when wet
Aluminum oxide	Machine sanding of wood or metal; finishing of bronze and steel	Fast cutting; long lasting
Silicone carbide (wet and dry)	Wet: remove automotive-type finishes; feather edges of chipped paint. Dry: finish brass, copper, aluminum; sand plastics, glass, and ceramics	Has a waterproof backing to permit rinsing of paint from the abrasive; very effective when used with water; durable; expensive
Steel wool	Remove old finishes; smooth and polish fine wood finishes; polish brass	Coarse and medium grades work well with paint removers. Very fine is used to cut the gloss and provide a dead smooth surface after applying wood finishes.

FIGURE 11-3 Guidelines for selecting sanding materials and steel wool.

Copyright © 2015 Cengage Learning®.

Procedure

1. Apply a stain (if desired) and a final finish on the wood.
2. Select a putty stick to match the finish of the wood.
3. Rub the stick back and forth over the hole until the hole is filled with putty.
4. Smooth the putty so it is level with the surface of the wood.
5. Rub the area with a cloth or paper towel to remove all excess putty. The job is then complete, since all finishes are applied before using the putty stick, and putty is a good, protective finish.

Selecting the Finish

Wood products must be carefully prepared before applying stains and other finishing materials. Paint and opaque stains require less sanding than penetrating stains and clear finishes. The type of finish must be selected before the project is completed.

If expensive wood with an attractive grain is used, a penetrating stain with a clear finish is desirable. If the grain is not attractive, an opaque stain or paint can be used. This requires less time than preparing the wood for a finish that lets the grain show.

Sanding

All edges and ends of boards should be worked down with a file or other cutting tool to the desired shape.

The wood is then sanded with medium sandpaper (100-grit) and finally with fine (150-grit) or very fine (180- to 220-grit) sandpaper. Figure 11-3 gives guidelines for selecting sanding materials.

For best results, always do the sanding with the grain; otherwise, ugly sanding marks will be left in the wood (Figure 11-4).

Faces of boards should be sanded with a power sander or with sandpaper wrapped around a block of wood. This helps keep the surface of the wood level. It also spreads the use across a large area of the sandpaper. A sanding block with a beveled edge permits sanding in corners. Again, start with medium sandpaper and end with very fine. After final sanding, use a tack cloth to wipe the surface to remove any dust prior to applying the finish. A tack cloth is a cloth with a sticky surface that will pick up dust when rubbed across the wood.

FIGURE 11-4 Sanding should always be done with the grain of the wood.

© iStockphoto/Jon Whitney

APPLYING CLEAR FINISHES

Clear finishes protect the wood and let the beauty and color of the wood show through (Figure 11-5). Basically, there are five types of clear finishes: oil, varnish, polyurethane, lacquer, and shellac. Each one has advantages and disadvantages, and the usage, type of desired finish, and durability needs determine which type to use.

The procedures given here for applying finishes are general instructions. The student is reminded to always read the label and follow the specific directions given with the product. Also, always apply the finish in a dust-free environment. Any dust that settles on the uncured finish will be rough and unsightly.

Oil Finishes

One of the oldest methods used for finishing wood is an oil finish. Linseed oil, which is an oil extruded from flaxseed, can be rubbed into the surface of wood to

FIGURE 11-6 An oil finish is applied with a lint-free cloth.

provide a semiclear finish. This method has been used for centuries. Today, modern oil finishes are a blend of linseed oil and other ingredients that help the finish stay clear and durable. Oil finishes are easy to apply. The oil is applied using a soft, lint-free cloth (Figure 11-6). The oil is rubbed into the wood and allowed to dry. After the first coat has dried, the surface should be lightly sanded with 220-grit or finer sandpaper until the surface is smooth. Wipe the surface free of dust using a tack cloth. Apply another coat and allow it to dry. Several coats should be applied because the oil penetrates the surface of the wood. This is an advantage because the finish becomes a part of the wood and not just a layer on top of the surface of the wood. The disadvantages are that the finish tends to become darker over a period of time and the wood may need more coats of oil as the finish dries out.

Varnish

Varnish is a finish composed of a variety of different oils. Varnish has been used for a long time and for many years was a standard in the furniture industry. Varnish does not adhere very well to bare wood, so a sealer coat is applied prior to the varnish. For a long time shellac was used as a sealer coat. This sealer had a tendency to turn dark and was relatively expensive. Today, shellac has been replaced with a newer sealer called sanding sealer. This sealer is easy to apply with a brush and dries very quickly. As with any sealer, it causes the grain to rise and the surface must be sanded lightly when dry. Two top coats of varnish are then applied. Varnish has the advantage of being easy to apply, but it has the disadvantage of being softer than most of the more modern finishes. Also, it tends to turn dark and become soft after years of use.

FIGURE 11-5 Clear finishes protect the wood and let the beauty and color of the wood show through.

Polyurethane

Polyurethane is a clear, durable, water-resistant finish. It requires no separate sealer. It seals the pores of the wood and provides a very durable finish. This finish is ideal for use on wood floors because of its durability (Figure 11-7). Polyurethane can be purchased in a high-gloss form or **satin** form. The satin form looks like a hand-rubbed varnish finish without the need for steel wooling or sanding between coats. A satin finish has a low sheen and is regarded by many as more desirable than a very shiny or glossy finish. Polyurethane dries quickly and is applied like varnish. Certain types of polyurethane may be wiped on with a cloth. This gives a superior, even, smooth finish. Two or three coats are recommended. Once dried, polyurethane is not easily removed even with chemical paint and varnish removers. Polyurethane comes in two types: oil-based and water-based.

Paint thinner or mineral spirits can be used as thinners and brush cleaners for oil-based polyurethane just as water-based polyurethane can be thinned and cleaned with water. Oil-based polyurethane provides a finish that gradually turns a warm amber color. Water-based polyurethane usually remains clearer as the finish and the project ages.

Lacquer

Lacquer is a clear finish that requires its own special thinner. It seals wood and provides a good protective finish. Lacquer and lacquer thinners are identified by their banana-like odor. Lacquer thinner softens paint, so it is important not to apply lacquer over other finishes.

Lacquer is popular because it is a clear, hard finish that dries quickly. This finish is best applied with a sprayer because the finish dries so quickly (Figure 11-8).

FIGURE 11-7 Polyurethane is an ideal finish for wood floors.

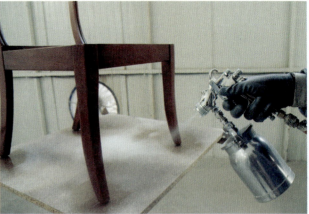

FIGURE 11-8 Lacquer is best applied with a sprayer because it dries so quickly.

A special slower-drying lacquer is available for brushing on the finish. The procedure for using the brush-type lacquer follows.

Procedure

1. Sand and clean the wood properly.
2. Use a clean, pure bristle brush.
3. Wipe the wood with a rag dampened with lacquer thinner.
4. Quickly apply a coat of lacquer. Move forward into new wood and do not brush over previously coated areas.
5. Replace the lid on the can promptly.
6. Clean the brush immediately in lacquer thinner followed with soap and water.
7. Allow the finish to dry thoroughly.
8. Apply a second coat.
9. After the lacquer has dried, rub paste wax into a pad of 000-grade steel wool. Rub the finish briskly with the wax and steel wool pad. Before the wax dries, polish the surface vigorously with terrycloth or a coarse towel.

Shellac

One of the very oldest of the clear finishes is shellac. This finish dates back to ancient Rome and was used extensively by the Middle Ages. It is derived from secretions of the shellac beetle that lives in Southeast Asia. It is easy to apply and creates a finish that shows the grain of the wood very well. In addition, it is perhaps the safest of all the clear finishes. In fact, shellac is approved by the U.S. Food and Drug Administration (FDA) as a food additive and is used in candies and as a coating for many types of pills. It appears to be making a comeback in recent years as a high-grade finish.

Shellac comes in a variety of grades ranging from very dark to "super blonde," which is the clearest of all the shellac grades. Orange shellac is the grade most commonly used in repairing and refinishing antique furniture. The darker grades can be used to darken the finish of various types of wood.

One of the disadvantages of shellac is that it has a short shelf life. To solve this problem, it is sometimes sold in packages of dry flakes that can be stored indefinitely. To prepare for use, the shellac flakes are dissolved in denatured alcohol. The thickness of the resulting mixture is usually expressed as either a 2-pound or 3-pound cut. This means that either 2 pounds or 3 pounds of shellac are dissolved into a gallon of alcohol.

Another disadvantage of shellac is that it does not resist water as well as some of the more modern clear finishes. If a wet glass is left on the finish, a white ring will form that greatly detracts from the finish. This problem can be solved by adding a couple of coats of a high-grade finishing wax to the shellac finish.

WOOD STAINS

Wood may be stained to many shades of black, brown, red, blonde, and white. In fact, stains can be used to stain wood to almost any color. Stains differ from paints in that stains allow the grain of the wood to show through. Often stains are used to make cheaper woods look like more expensive woods. They may also be used to enhance the grain and make furniture woods appear more attractive. The most common types of wood stains are those called penetrating stains, which penetrate the wood and enhance the grain (Figure 11-9). These stains are applied to the bare wood and top-coated with a clear finish. Among the woods that accept stains well are oak, mahogany, and ash. Certain types of wood such as cherry, pine, popular, and maple are difficult to stain because the wood structure prevents the stain from penetrating evenly. These woods can be stained using a wood dye or a gel stain. These products do not penetrate the wood and remain close to the surface of the wood. They are more difficult to use than penetrating stain. The procedure to use is as follows:

FIGURE 11-9 The most common types of stains penetrate the wood and enhance the grain.

Procedure

1. Sand the wood to a smooth surface, beginning with 120-grit sandpaper, then with a 150-grit, and finally a 220-grit sandpaper.

2. Wipe off all sanding dust using a tack cloth.

3. Using a lint-free cloth, apply a thin coat of prestain wood conditioner (Figure 11-10).

4. Let the wood conditioner set for about 10 minutes.

5. Using a lint-free cloth, apply a generous coat of gel stain (Figures 11-11 and 11-12).

6. Let the gel stain set for about 10 minutes.

7. Using a clean, lint-free cloth wipe off the excess stain. Make sure to wipe with the grain (Figure 11-13).

8. Dispose of the used cloths in an approved, metal disposal container.

9. Let the stain set overnight and apply two coats of top finish such as polyurethane.

FIGURE 11-10 Apply a thin coat of prestain wood conditioner.

FIGURE 11-11 Spread a generous amount of stain onto the conditioner.

FIGURE 11-12 Make sure that the stain covers the entire surface.

FIGURE 11-13 After the stain has set for 10 minutes, wipe it to an even finish.

Penetrating oil stains are popular and easy to use. To apply penetrating oil stains, use the following procedure.

Procedure

1. Sand and prepare the wood for finishing.
2. Stir the stain thoroughly.
3. Use a brush or soft, clean cloth to cover all parts of the wood with stain.
4. Allow about 5 minutes for the stain to penetrate the wood.
5. Wipe off the excess stain with clean, soft, absorbent rags.
6. Wipe and lightly polish the wood until no stain comes off when touched.
7. Set the project aside to dry for at least 24 hours.
8. Clean the brush in turpentine, Varsol, or paint thinner followed by soap and warm water.
9. Clean stain off your hands with the brush cleaner.
10. Apply a clear finish as described previously.

SOLVENTS AND THINNERS

Many problems in finishing wood can be traced to a lack of knowledge about solvents and thinners. The five solvents that are used extensively in the shop and home are denatured alcohol, turpentine, paint thinners (sometimes referred to as mineral spirits), Varsol, and lacquer thinner (Figure 11-14).

Denatured Alcohol

Denatured alcohol or wood alcohol is sold as shellac thinner. It will thin shellac and finishes that contain shellac. For the most part, alcohol is a single product solvent in the finishing room. However, it may be used elsewhere in the shop as a cleanup solvent when overhauling brake cylinders and other metal hydraulic parts.

Turpentine, Paint Thinners, and Varsol

These products are all suitable solvents or thinners for oil-based stains, paints, and varnishes. Although some cost more than others, most product labels indicate equally good results from all three. However, Varsol is especially desirable as a solvent in the shop. Varsol is safer than most solvents from the standpoint of fire

Thinner/Solvent*	Used with	Cleans Brushes Used In
Denatured alcohol	Shellac (both white and orange)	Shellac
	Sealers that contain shellac	Sealers that contain shellac
Turpentine and paint thinners	Oil-based stains and wood fillers	Oil-based stains, fillers, and paints
	Oil-based (alkyd) paints	Varnish
	Varnish	Polyurethane
	Polyurethane	
Lacquer thinner**	Brush lacquer	Lacquer and finishes containing lacquer
	Spray lacquer	Oil-based stains, fillers, and alkyd paints
	Finishes containing lacquer	(also softens hard lacquer, oil-based stains and fillers, and alkyd paints in brushes)
Water	Latex paints and other latex products	Latex paint and products (when water is warm and used with a detergent)
		Other thinners (when water is warm and used with soap)

*Caution: Protect eyes with safety devices and use all solvents in well-ventilated area.

**Caution: Lacquer thinner is especially hazardous because it evaporates quickly and dissolves the natural oils from skin. Avoid excessive contact with skin.

FIGURE 11-14 Thinners and solvents commonly used with finishing materials.

hazard. Secondly, it is a good, inexpensive grease solvent and can be rinsed away with water.

Lacquer Thinner

Lacquer thinner is a fairly universal solvent. Not only does it thin lacquer, but it softens hard paint as well. This makes it useful for cleaning paintbrushes used for lacquer as well as other materials.

Lacquer thinner evaporates quickly. This creates problems with fumes in poorly ventilated areas. Extra care is needed to avoid breathing lacquer thinner fumes. Similarly, special care must be taken to avoid sparks or fire when using lacquer or any other finishing material.

USE AND CARE OF BRUSHES

High-quality brushes make it possible to apply high-quality finishes. Without good brushes, high-quality finishes are impossible. Unfortunately, a good brush can be ruined the first time it is misused. Following is the procedure for correctly using a brush.

Procedure

1. Choose the right brush for the material you are using. Nylon and other man-made bristle brushes work best with latex paints. Natural bristle brushes are best for most other stain, varnish, and painting products.

2. Dip only about ½ inch of the brush into the material.

3. Brush back and forth to keep the bristles together. Do not bend or spread the bristles excessively.

4. Once a job is started, alternate dipping the brush into the material and applying it so the material does not dry in the brush.

5. When the job is finished, clean the brush promptly with the proper solvent and then wash it with soap and warm water until all evidence of the finishing material is gone.

After each use, brushes must be properly cleaned. The cleaning procedure should leave the brush perfectly clean. When dry, the bristles should be soft, compact, and ready for reuse. Like a good tool, a good brush, properly used and properly cleaned, can provide many years of excellent service. The procedure for cleaning a brush follows and is shown in Figure 11-15.

Procedure

1. Remove as much material from the brush as possible. Rags, newspapers, or paper towels may be used to wipe the brush.

(continued)

Copyright © 2015 Cengage Learning®.

A For oil-based paint, use mineral spirits or thinner to dissolve the paint in the brush; if the paint is latex, use water.

C Dry the clean brush prior to storing it.

B Use soap and warm water to remove the solvent and dissolved paint from the brush.

D Wrap the brush to keep the bristles compact and in shape while drying.

FIGURE 11-15 Brushes must be properly cleaned after each use.

Procedure, *continued*

2. Pour about ½ inch of solvent for the product being used into a container just a little bigger than the width of the brush. Using a small container reduces the amount of solvent needed.

3. Insert the brush and push it up and down bending the bristles slightly, but not flattening them.

4. Pour the used solvent into an empty solvent can marked "used."

5. Pour some more clean solvent into the container and again flush the brush.

6. Pour the solvent into the "used" solvent can. Replace the caps on all solvent cans. After several days the used solvent will become clear and usable again for cleaning brushes.

7. Wipe all excess solvent from the brush with a rag or paper towel.

8. Place a bar of soap in the palm of your hand, and run a small stream of warm water on the soap.

9. Rub the brush on the soap and in the water until all solvent and material are flushed away and the soap creates a good lather.

10. Rinse all soap from the brush. The bristles should feel like clean hair.

11. Lay a paper towel on a bench.

12. Position the brush on the left lower quarter of the towel.

13. Wrap the brush one or two turns, fold the paper down over the top of the brush, and continue the wrap until all of the towel is used. Twist the towel around the handle of the brush.

14. Lay the brush in a warm place to dry. It will take about 24 hours.

15. The properly cleaned brush will not show any evidence of previous use, as noted by color of material used or any bristle stiffness.

SUMMARY

A completed woodworking project is only as good as the finish applied to it. All wood must be preserved in some way. Whether it is a coat of paint or a finely polished oil finish, the preparation, procedures, and technique used will greatly enhance the quality and durability of the completed project.

Student Activities

1. Define the Terms to Know in this unit.

2. Take out the board that you used for activity 9 in Unit 10 and perform the following procedures.

   - Drill two $\frac{3}{16}$-inch holes about $\frac{1}{4}$ inch deep.
   - Fill the holes with natural-color plastic wood.
   - Use a file and sandpaper to smooth all edges of the board.
   - Sand all surfaces, remove all marks, and create clean, smooth surfaces on the board.
   - Apply stain to the back side (side without the dado).
   - Apply one or two coats of finish, such as polyurethane.
   - Apply wax.

3. Examine the finished board. What procedures did you do well? What errors did you make? How can the errors be avoided in the future?

Relevant Web Sites

DIY Doctor, Cleaning Paint Brushes
www.diydoctor.org.uk/projects/cleaningpaintbrushes.htm

ArtSparx.com, Solvents and Thinners
www.artsparx.com/undmatsolv.html

Self-Evaluation

A. Multiple Choice. Select the best answer.

   1. A dent may be removed from wood with
      a. a special tool
      b. a grinder
      c. sandpaper
      d. steam from a wet rag

   2. Holes in wood may be filled with
      a. glazing compound
      b. plastic wood
      c. putty
      d. all of these

3. Oil-based stains, varnish, and paints should be thinned with
   a. alcohol
   b. lacquer thinner
   c. turpentine
   d. water

4. Wood may be colored by applying
   a. polyurethane
   b. stain
   c. varnish
   d. wax

5. The material used to fill holes after wood has a final finish is
   a. glazing compound
   b. plastic wood
   c. putty
   d. putty stick

6. Plastic wood is recommended for
   a. filling nail holes
   b. hiding bad cuts
   c. removing sanding marks
   d. all of these

7. Final sanding should be done with
   a. coarse sandpaper
   b. the grain
   c. medium sandpaper
   d. a file

8. The correct sequence when applying finishes is
   a. wax, sealer, varnish, stain
   b. stain, sealer, varnish, wax
   c. stain, varnish, wax, sealer
   d. sealer, varnish, wax, stain

9. A clear, durable, water-resistant finish that is brushed on and requires no separate sealer is
   a. polyurethane
   b. shellac
   c. stain
   d. wax

10. The solvent that dissolves the greatest number of finishing products is
    a. alcohol
    b. lacquer thinner
    c. turpentine
    d. Varsol

B. **Matching.** Match the finishing materials in column I with the thinner/solvent they are used with in column II.

Column I

   1. oil-based stain and varnish
   2. latex paint
   3. polyurethane
   4. lacquer

Column II

   a. paint thinner
   b. lacquer thinner
   c. turpentine
   d. water

C. **Completion.** Fill in the blanks with the word or words that make each statement correct.

   1. Final sanding should always be done _____ the grain.

   2. _____ is popular because it is a clear finish that dries almost instantly if sprayed on.

   3. A _____ is a chemical layer that protects the surface of a material.

   4. After cleaning, one should _____ the brush to keep the bristles compact and in shape while drying.

   5. Varnish does not adhere very well to bare wood so a _____ is used.

D. **Brief Answer.** Briefly answer the following questions.

   1. What is a rag dampened with solvent used for?

   2. List the steps to follow when cleaning a varnish brush.

   3. Describe the process for applying an oil finish. What are the disadvantages of this technique?

   4. Why is Varsol especially desirable as a solvent in the shop?

   5. When applying stain or varnish, should one use natural or nylon bristle brushes?

UNIT 12

Identifying, Marking, Cutting, and Bending Metal

Objective

To identify, mark, cut, and bend cold metal.

Competencies to be developed

After studying this unit, you should be able to:

- Identify metals.
- Mark metal.
- Cut and file metal.
- Bend square, round, and flat steel.
- Form sheet metal.

Materials List

- Samples of:
  - Cast iron
  - Wrought iron
  - Mild steel (various shapes)
  - Tool steel
  - Stainless steel
  - Galvanized steel
  - Aluminum
  - Copper
  - Lead
- Scratch awl
- Scriber
- Assorted files

continued

Terms to Know

- ferrous
- nonferrous
- alloy
- malleable
- casting
- ductile
- malleable cast iron
- anneal
- tempering
- corrosion
- solder
- scratch awl
- scriber
- soapstone
- dividers
- center punch
- hacksaw
- single-cut file
- double-cut file
- draw filing
- stock
- snips
- shears
- cold chisel
- eye
- anvil

- Soapstone
- Chalk
- Dividers
- Center punch
- Hacksaw
- Metal snips
- Cold chisel
- 10-inch or 12-inch adjustable wrench
- Machinist's vise
- Anvil
- Magnet
- Combination square with scriber
- Stove bolt with nut ($\frac{1}{4}$" × 2")
- Flat steel ($\frac{1}{8}$" × 1"× 6")
- Blacksmith's or heavy ball peen hammer

Metals form much of the structures in the world around us. Skyscraper buildings became a reality only after it became possible to make structural steel (Figure 12-1). The transportation industry depends on metal for vehicles, rails, and reinforcement for concrete roads. Even the high technology of electronics and computers relies extensively on copper wire, silver connectors, and other metal components.

FIGURE 12-1 The use of structural steel has made the building of skyscrapers possible.

Most agricultural machines and equipment are made of steel and other metals. A successful career in many areas of agricultural mechanics depends upon a knowledge of metals and their uses in agriculture.

IDENTIFYING METALS

All metals can be classified as either ferrous or nonferrous. **Ferrous** metals come from iron ore. **Nonferrous** metals do not contain iron. The distinction is necessary because ferrous metals are used differently than nonferrous metals.

Most metals are not used in their pure metallic state. They are usually combined with one or more other metals. The combination of two or more metal elements is called an **alloy**. Alloys have characteristics that make them different from the original metals that were used to form the alloy. The alloy is made to improve the strength or some other quality of the original metals. For example, carbon is added to iron to make steel, which is tougher and more flexible than iron. Similarly, nickel and chromium may be mixed with iron to make non-rusting stainless steel. Figure 12-2 gives the characteristics and major uses of some common metals and alloys.

Ferrous Metals

Ferrous metals are composed primarily of iron. Iron is refined from iron ore. Iron ore is particles of iron mixed with other minerals found in abundance throughout the world (Figure 12-3). Because of the abundant supply, iron is relatively inexpensive and, therefore, used extensively. However, ferrous metals rust very easily and must be protected by coatings that resist corrosion. Ferrous metals are easy to identify because they are magnetic and give off sparks when ground on an emery wheel.

Iron is the most useful metal for making tools and machinery. Its usefulness is increased with the addition of carbon to make steel. Further, when combined with certain other metals, iron becomes a very corrosion-resistant material called stainless steel.

Cast Iron. Cast iron is a type of iron that is grainy and not easily shaped or bent. As the name implies, objects are made by pouring iron into molds. Cast iron is used in a variety of things ranging from pots and pans to automobile engine blocks (Figure 12-4). Therefore, it is not malleable. **Malleable** means workable. Parts with odd shapes may be made by pouring molten iron into a mold called a **casting**. When the metal cools and the form is removed, the item is said to be made of cast iron. Pure cast iron will break before it will bend.

Cast iron can be treated so it is somewhat workable. This is done by special heat treatments. The result is iron with a ductile outer layer. **Ductile** means that the metal can be bent slightly without breaking. The combination of the cast iron core and a

Metal	Origin	Characteristics	Major Uses
Cast iron	Iron ore	Forms into any shape; brittle	Machinery parts; engine blocks
Wrought iron	Iron ore	Malleable; tough; rust-resistant	Decorative fences; railings
Mild steel	Iron ore	Malleable; ductile; tough	Structural steel
Tool steel	Iron ore	High carbon; heat treatable; expensive	Tools; tool bits
Stainless steel	Iron ore, nickel, and chromium	Very corrosion resistant; bright appearance; hard; tough	Food handling equipment; milk tanks; restaurant equipment
Galvanized steel	Steel, zinc	Zinc-coated steel	Water tanks; towers; fencing; roofing; siding
Aluminum	Ore	Light; tough; relatively soft; good electrical conductor; silver-white color	Roofing; siding; truck bodies; automobiles; electrical wires and cables
Copper	Ore	Tough; malleable; corrosion-resistant; excellent heat and electrical conductor; reddish brown color	Pipe; electrical wire and cables; rain spouts and gutters; electrical equipment; bronze; brass
Brass	Copper and zinc	Soft; malleable; corrosion-resistant	Water valves; boat accessories; ornaments
Bronze	Copper, zinc, and tin	Soft; malleable; corrosion-resistant	Ornaments
Lead	Ore	Soft; very heavy; bluish gray	Batteries; cable coverings; shot; solder
Tin	Ore	Very malleable; corrosion-resistant; silver color	Plating; bronze; solder

FIGURE 12-2 Common metals and alloys.

Copyright © 2015 Cengage Learning®

FIGURE 12-3 Iron comes from ore such as this.

FIGURE 12-4 Cast iron is used to make pots and pans as well as a large variety of other things.

FIGURE 12-5 Wrought iron is used in fences and gates. It can be bent into almost any shape.

ductile metal outer layer is called **malleable cast iron**. Malleable cast iron is used extensively in farm and factory machinery.

Wrought Iron. Wrought iron is an almost pure form of iron. It is very malleable. It can be bent, shaped, welded, drilled, sawed, and filed. Wrought iron also resists rust. These qualities make it a popular material for fences and other ornamental uses (Figure 12-5).

Mild Steel. Mild steel is the workhorse of metals. It is made by adding small amounts of carbon to iron. Mild steel is tough, strong, ductile, and malleable. It is rolled into many shapes, such as flat bands, angles, channels, tees, I-beams, rods, and pipe (Figure 12-6). Mild steel is used extensively in the agricultural mechanics shop.

Tool Steel. Tool steel contains a specific amount of carbon, which permits it to be hardened. Tool steel can be annealed. **Anneal** means to heat a metal to the proper temperature and then slowly cool it. The process of annealing softens and toughens steel.

Tool steel can be hardened by heating to the proper temperature and then rapidly cooling the steel. The degree of hardness is determined by controlling both the temperature of the metal and the speed of

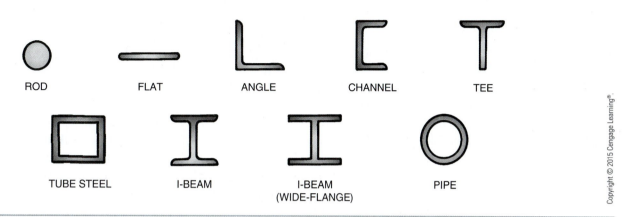

ROD FLAT ANGLE CHANNEL TEE

TUBE STEEL I-BEAM I-BEAM
(WIDE-FLANGE) PIPE

Copyright © 2015 Cengage Learning®.

FIGURE 12-6 Mild steel can be formed into many shapes for use in agriculture and industry.

cooling after heating. Only tool steel can be tempered. **Tempering** is the process of carefully controlled reheating and cooling of steel after it has been shaped. Tempering results in a specified degree of hardness, relieves stress, and prevents cracking in steel. A tempered product is hard, tough, and not easily bent.

Tool steel can generally be identified by the exploding, sparkling nature of the sparks given off when the metal is ground (Figure 12-7).

Stainless Steel. Stainless steel is made by adding nickel and chromium to steel. It is very tough, will not rust, and resists corrosion of all types. Stainless steel is used extensively for milk tanks, kitchen equipment, and factory equipment, where food is being processed and sanitation is necessary (Figure 12-8).

© iStockphoto/P_Wei

FIGURE 12-8 Stainless steel is used extensively in the food industry when sanitation is necessary.

Plated Steel. Steel may be made rustproof by applying a coating of metal such as tin or zinc. At one time tin was used on rolled steel roofing material. This gave rise to the term tin roof. When a relatively thick coating of tin was used, the roof stayed bright for many years without rusting. When rust started to appear, the roof was painted. Today, most steel for roofs and siding is coated with zinc. Zinc-coated steel is called galvanized steel. The rust resistance of galvanized steel depends upon the thickness of the zinc coating. The thickness is rated in ounces of zinc per square unit of metal surface.

Galvanized steel sheets, pipe, buckets, and tanks must be handled with care in the shop. When heated or welded, the zinc gives off a poisonous gas. Therefore, such operations must be done with great care in a well-ventilated area. Also, galvanized buckets and

© iStockphoto/Vitalina Rybakova

FIGURE 12-7 The carbon content of steel is indicated by the brightness and explosiveness of sparks produced when the metal is applied to a grinding wheel.

containers can be ruined if the zinc finish is damaged by cracking or chipping, by dropping, or other abuse.

Nonferrous Metals

Nonferrous metals and their alloys do not contain iron. They are more expensive than ferrous metals on a per weight basis and their supply is rather limited. However, nonferrous metals are needed and used extensively in industry and agricultural mechanics. Nonferrous metals include aluminum, copper, lead, tin, and zinc. Some nonferrous metals are light in weight.

Aluminum. Aluminum is identified by its toughness, light weight, and silver-white color. The combination of toughness and light weight makes aluminum the first choice for truck bodies, aircraft, ladders, equipment, containers, roofing, and siding (Figure 12-9).

Aluminum is a very good electrical conductor. Therefore, aluminum wire is used for high-voltage lines and often used when copper wire becomes expensive. Aluminum resists corrosion. **Corrosion** is the reaction of metals to liquids and gases that causes them to deteriorate or break down. Aluminum is often painted to provide choices in color when used in doors, siding, gutters, awnings, and toys.

Pure aluminum is very soft and malleable. Other materials, such as silicon, manganese, and magnesium, are added to give aluminum more strength and other qualities. Special techniques must be used to weld aluminum.

Copper. Copper is an excellent electrical conductor and does not corrode much in air or water. Therefore, copper is used extensively in electrical wiring, electric motors, and plumbing systems (Figure 12-10).

Copper makes up the bulk of the metal in the alloys brass and bronze. Brass is made of 60 to 90 percent copper; the rest is zinc. Bronze is copper and zinc plus about 10 percent tin. These alloys have been used for centuries. They are used extensively today for their decorative and corrosion-resistant qualities.

Lead. Lead is a very soft and heavy metal. It is used in the manufacture of batteries, lead pipe, wire coverings, and in combination with tin to make solder (Figure 12-11). However, lead solder used in drinking water systems may contribute to lead poisoning and is not permitted by building codes. **Solder** is a soft alloy used to join many kinds of metal. Lead is also used in washers on roofing nails to provide a watertight seal between the nail and roofing.

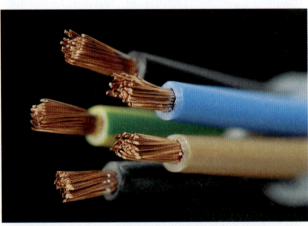

FIGURE 12-10 Copper is used extensively in electrical wiring. It is an excellent conductor of electricity.

FIGURE 12-9 Many ladders are made of aluminum because the metal is strong yet light weight.

FIGURE 12-11 Lead is a heavy, soft material that is used in car batteries.

Tin. Tin is used to coat steel for temporary rust protection. It is also a component of solder and bronze. Tin has long been used to coat steel used in cans, but the coating generally provides protection for only a short period of time. Steel, coated with tin, does not rust until it is discarded outdoors. The tin coat then wears away quickly and the steel rusts.

MARKING METAL

Steel is the most commonly used metal in agricultural mechanics. Therefore, most metalworking skills focus on steel. The student should realize, however, that most procedures used in working with steel apply to other metals as well.

Tools for marking metal and wood are discussed in Unit 8—the unit on layout tools and procedures. The main difference between marking wood and marking metal is the instrument used for marking. Metals are either very shiny or very dark in color. Therefore, a lead pencil mark does not show up well. The scratch awl, scriber, file, soapstone, dividers, and center punch are tools used to mark metal.

Scratch Awl. The scratch awl (or simply awl) is a tool with a wood or plastic handle and a pointed steel shank. It is held in the hand and used like a pencil. The awl is used to scratch the surface of metal to make a mark (Figure 12-12).

Scriber. The name scriber is generally given to very small, metal, sharp-tipped markers. Many combination squares have a scriber in the handle that can be removed and used for marking metal.

Files. Any file will work for marking metal. The sharper the edge or corner, the finer and more accurate the mark can be. To mark with a file, the sharp corner

FIGURE 12-13 Soapstone can be used to mark metal.

of the file is pushed over the flat side or the corner of the metal. The file works better on structural steel than it does on plate or sheet steel.

Soapstone and Chalk. Effective markers are made by cutting soapstone into thin pieces resembling pencils. Soapstone is a soft, gray rock that shows up well on most metals (Figure 12-13). It can be wiped off, but not easily. Soapstone can be sawed to a knifelike edge, if desired. This sharp edge makes a narrow line on metal.

Common chalk can also be used to mark metal. However, chalk will not leave a narrow mark and it brushes off easily.

Dividers. Dividers have two steel legs with sharp points that are used to make arcs and circles on metal. The tool is also useful for transferring measurements from one item to another. A third important use is to divide any measurement into two equal parts, hence the name dividers.

Center Punch. The center punch is a steel punch with a sharp point. The center punch is used to make a small indentation or hole in metal. This hole is used to help start a twist drill. The procedure for using the center punch is as follows.

FIGURE 12-12 Marking metal with a scratch awl.

Procedure

1. Place the point of the center punch exactly where the center of the hole should be.

2. Hold the punch straight.

(continued)

Procedure, *continued*

3. Give the punch a light blow with a steel hammer.

4. Check to see if the dent is exactly where the center of the hole should be.

5. Place the punch in the dent and give the punch one solid blow with the hammer.

CUTTING METAL

Metal can be cut or formed by sawing, shearing, filing, or grinding. Because metals are harder to work than wood, the choice of tools is somewhat more limited for working metal. The hacksaw, file, snips, and cold chisel are tools used to cut metal.

The Hacksaw

The tool used most often for cutting metal is the hacksaw (Figure 12-14). The **hacksaw** is a device that holds a blade designed for cutting metal. Such tools may be as simple as a handle that grips a common hacksaw blade. Another type of hacksaw is shaped like a miniature bow saw and holds a 6-inch blade with very fine

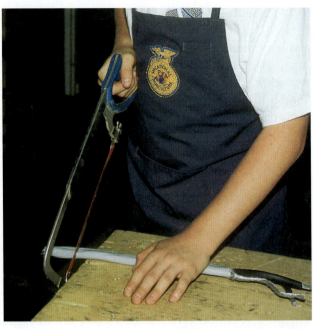

FIGURE 12-15 Electricians often use hacksaws to cut heavy wire.

teeth. Electricians favor such saws for cutting heavy copper and aluminum wire (Figure 12-15).

When using a hacksaw, a blade should be selected that is fine enough to allow at least three teeth at a time on the metal (Figure 12-16). Standard hacksaw blades come in 10-inch or 12-inch lengths. They may be bought with 14, 18, 24, or 32 teeth per inch. The blade with 32 teeth per inch has very small teeth and should be used to cut very thin metal. The thicker the metal, the coarser the teeth should be so the blade will cut faster.

Hacksaws are designed to cut on the forward stroke. Therefore, blades must be installed with the teeth pointing away from the handle. When using hacksaws, a small notch should be filed to help start the cut. The blade is pulled backwards and then pushed forward with slight pressure. After starting the cut, long full strokes with pressure on the forward stroke should be used. No pressure is used on the backward stroke.

Metal must be held tightly when sawing to prevent it from flexing or twisting and to prevent damage to the hacksaw blade. The best way to secure metal for sawing is to use a metal vise. The metal should be placed in the vise so the cut will be only ½ inch from the vise jaws (Figure 12-17). The metal should be positioned so that three or four teeth are cutting at one time.

A used hacksaw blade should not be replaced after the cut has been started. A new blade will have a wider set in its teeth. When a new blade is forced down into

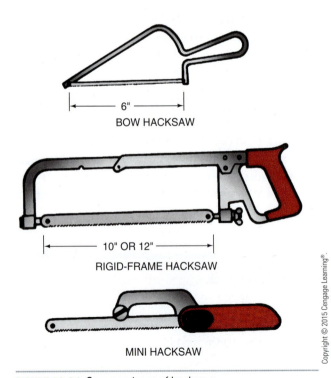

FIGURE 12-14 Common types of hacksaws.

BOW HACKSAW

6"

10" OR 12"

RIGID-FRAME HACKSAW

MINI HACKSAW

Copyright © 2015 Cengage Learning®.

Photo by Gary Farmer

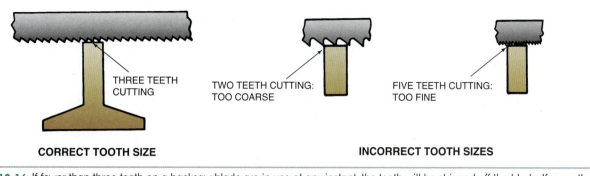

Copyright © 2015 Cengage Learning®

THREE TEETH CUTTING

TWO TEETH CUTTING: TOO COARSE

FIVE TEETH CUTTING: TOO FINE

CORRECT TOOTH SIZE

INCORRECT TOOTH SIZES

FIGURE 12-16 If fewer than three teeth on a hacksaw blade are in use at any instant, the teeth will be stripped off the blade. If more than three teeth are in contact, a lot of pressure will be needed to push the saw forward.

Copyright © 2015 Cengage Learning®

FIGURE 12-17 Position the metal in the vise so that the area to be sawed is ½ inch from the vise jaws.

the kerf left by an old blade, the teeth are damaged. To avoid such damage, a cut with the new blade should be started on the opposite side of the metal.

Band saws, power hacksaws, and saber saws all may be used to saw metal. However, the correct blades must be used. The use of power saws with hacksaw blades is discussed in another unit.

Files

Files are classified by length, shape, design, teeth, and coarseness (Figure 12-18). Files used in the agricultural mechanics shop are generally 8 inches to 12 inches long. The shop usually has one or more of each of the various file shapes (triangular, half-round, round, and flat) so any kind of filing job can be done.

If a file has straight teeth going in one direction, it is said to be a **single-cut file**. If a file has teeth going in two directions, it is a **double-cut file**. A double-cut file cuts faster than a single-cut file. Single-cut and double-cut files can be purchased in bastard (coarse), second-cut (medium), and smooth (fine) grades.

The rasp-cut file has raised, sharp, individual teeth. Rasps are used for shaping wood, horses' hooves for shoeing, and soft metal. They are not suitable for steel.

The curved-tooth file has teeth that follow a half-round pattern. The teeth are very coarse. The curved-tooth file is designed to cut soft metals such as aluminum and copper.

Filing Metal. When filing metal, every effort should be made to secure the metal firmly (Figure 12-19). Filing will cause metal to vibrate if it is not held tightly. A file cuts on the forward stroke only. Filing may be done across metal or along its length.

To file, the point of the file is grasped between the thumb and index finger of one hand. The handle is gripped with the other hand. The file is pushed forward while light pressure is applied. The file should be slightly lifted and brought back to the starting position.

Draw Filing. **Draw filing** is done by placing the file at a 90-degree angle to the metal and pushing or pulling the file in this 90-degree position. Pressure is applied with a push or pull movement of the file over the length of the stock. **Stock** refers to a piece of material such as wood or metal.

Cleaning Files. Files need to be cleaned because metal particles get stuck in the grooves of the file. They are cleaned by tapping the handle sharply on the bench top. Another cleaning method is to use a file card or wire brush when particles are difficult to remove.

Snips and Shears

The terms snips and shears are used for the same tool. **Snips** or **shears** are large scissorlike tools for cutting

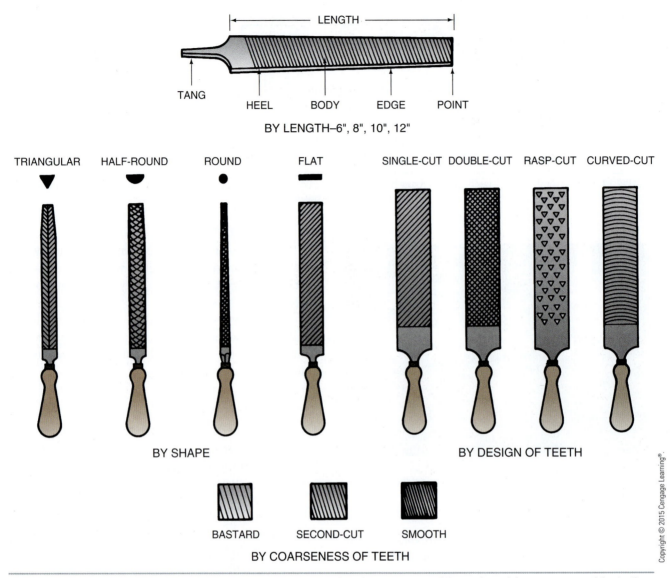

BY LENGTH–6", 8", 10", 12"

TRIANGULAR   HALF-ROUND   ROUND   FLAT   SINGLE-CUT DOUBLE-CUT RASP-CUT CURVED-CUT

BY SHAPE   **BY DESIGN OF TEETH**

BASTARD   SECOND-CUT   SMOOTH

BY COARSENESS OF TEETH

FIGURE 12-18 Files are classified according to the length and shape of the file, the pattern of the teeth, and the coarseness of the teeth.

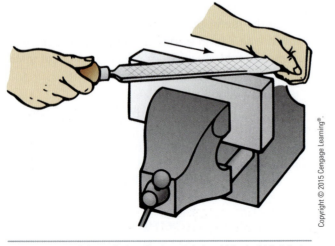

FIGURE 12-19 Prior to being filed, metal should be held tightly in a vise.

sheet metal and fabrics. They may be purchased for cutting straight, left-hand, or right-hand curves. Regular snips are designed like scissors. Snips for heavy cutting have compound handles and are also called aviation snips (Figure 12-20).

To cut sheet metal, a pair of snips that are sharp and free of nicks is selected. A piece of scrap should be cut first to be sure the snips are heavy enough. The snips are used like scissors. For snips to work well, the metal must curl or lift up and out of the snips as the cut progresses. Because curled metal is sharp and will cut, it is advisable to wear gloves when handling sheet metal. If cutting is difficult when using regular snips, let the lower handle rest on the bench top. Exert extra pressure by pushing down on the top handle (Figure 12-21).

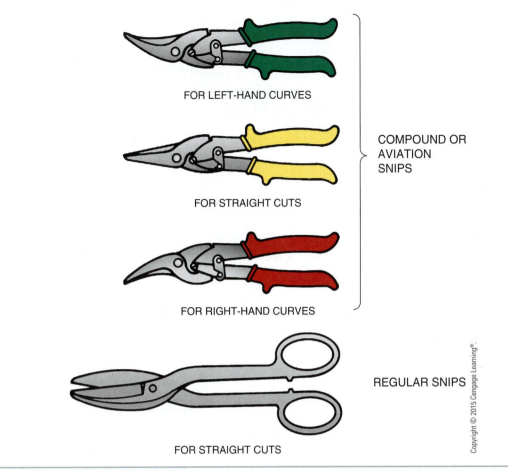

FOR LEFT-HAND CURVES

FOR STRAIGHT CUTS

COMPOUND OR
AVIATION
SNIPS

FOR RIGHT-HAND CURVES

REGULAR SNIPS

FOR STRAIGHT CUTS

FIGURE 12-20 Several types of snips are available. Compound or aviation snips provide extra leverage for heavy cutting.

FIGURE 12-21 Snips are used for cutting and shaping sheet metal.

Cold Chisels

A **cold chisel** is a piece of tool that is steel shaped, tempered, and sharpened to cut mild steel when driven with a hammer. A regular cold chisel has a wide cutting edge. Special chisels, such as the cape, round-nose, and diamond-point, have cutting edges designed to cut grooves (Figure 12-22).

CAUTION!

It is especially important to wear goggles when doing metalwork. A chisel or punch with a mushroomed head should not be used. Mushrooming can be corrected by reshaping the head using a grinder.

Cold chisels are used to cut mild steel rods and bands, and to cut rusted nuts and bolts. A cold chisel may also be used to loosen nuts that have rounded shoulders. The procedure for loosening such nuts with a cold chisel follows.

Procedure

1. Saturate the nut with a rust solvent.
2. Position the cutting edge of the chisel on a shoulder of the nut.

(continued)

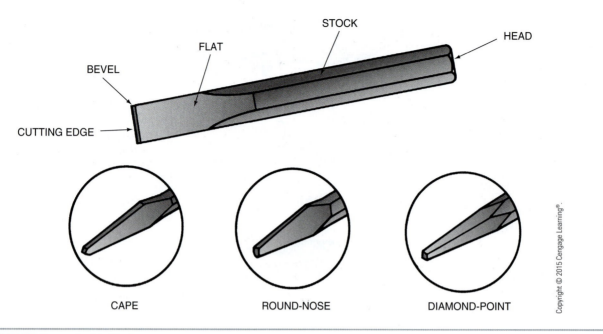

FIGURE 12-22 Types of cold chisels.

Procedure, *continued*

3. Angle the chisel so it will turn the nut counter-clockwise (backward) when struck.

4. Drive the chisel with a ball peen or blacksmith's hammer.

When using a cold chisel to cut nuts, the following procedure is recommended.

Procedure

1. Select a sharp chisel.

2. Examine the way in which the bolt and nut are mounted.

3. Place the chisel on the nut so it will be driven against a solid structure. The nut can be split by cutting downward, parallel to the bolt. The nut may also be cut into a right angle to the bolt.

4. Cut through the nut as far as possible.

5. Change the position of the chisel and drive it in such a way as to turn the nut off the bolt. The nut should either turn off or split off the bolt.

Sometimes it is easier to cut a bolt than to cut a nut. This may be true when the bolt is ⅜ inch or smaller. For best results, the bolt should be tight in its hole. The procedure for cutting a bolt follows.

Procedure

1. Place a sharp chisel at the base of the nut and at a right angle to the bolt.

2. Drive the chisel so it is forced under the nut. The lifting action of the chisel should cause the bolt to be exposed under the nut.

3. Continue driving until the bolt is cut.

The hacksaw is the best tool for cutting mild steel. However, there are times when the cold chisel may be preferred. When cutting rod or flat iron, the following procedure is recommended.

Procedure

1. Mark the metal where it is to be cut.

2. Place the steel in a heavy vise. The cutting mark should be even with the jaws of the vise.

3. Lay a heavy cold chisel on top of the movable jaw of the vise. The cutting edge of the chisel is to rest against the steel to be cut. If cutting flat iron, the cutting edge of the chisel should be placed on the corner of the iron.

4. Hold the chisel so it rests at about a 30-degree angle to the bench top. The flat side of the chisel should be parallel to the bench top.

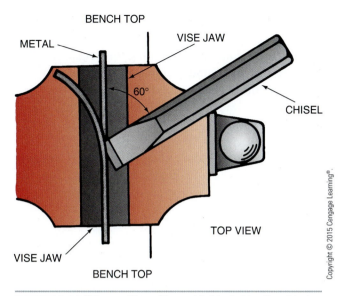

FIGURE 12-23 When cutting with a cold chisel, drive the chisel so that it acts as the second half of a pair of snips.

5. Drive the chisel until the metal is cut. The metal is cut by shearing action. The vise acts as one-half of the shear; the cold chisel is the other half (Figure 12-23).

Always use special face and hand protection when using driving tools. The occasional piece of metal that may fly from a chisel or hammer blow may become embedded in an eye or other parts of the face. Hands may be bruised or seriously injured by a chisel that slips or a hammer that misses its mark. Therefore, every precaution should be taken to protect the body.

BENDING METAL

Mild steel can be bent cold in sizes up to ½" square, ½" round, and ³⁄₁₆" × 1" flat. A vise anchored to a heavy bench makes the bending of cold metal possible. Metal can be twisted or bent for ornamental purposes such as for porch railings. Metal may also be bent at various angles to fit some object or to finish an assembly. Round metal may be bent to form an eye. An **eye** is a piece of metal bent into a small circle.

Bending Flat Metal

To bend a flat piece of metal up to 1" wide or a rod up to ½" in diameter, the following procedure is recommended.

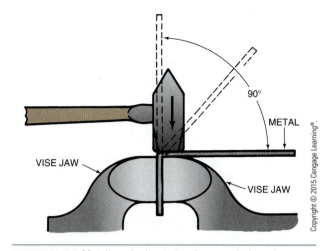

FIGURE 12-24 Metal can be bent at a sharp angle in a vise.

Procedure

1. Mark the metal where the bend is to occur.
2. Place the metal in a vise with the mark at the top of the vise jaws. It is best to have the longest part of the metal extending above the vise.
3. Tighten the vise securely.
4. Hold a heavy ball peen or blacksmith's hammer in one hand.
5. Grasp the metal near its end with the other hand.
6. Push the metal with one hand while hitting it with the hammer near the vise. Coordinate the movements so the metal bends sharply at the vise and not above it (Figure 12-24).

CAUTION!

Wear a glove or hold the chisel with a cloth to protect hands.

Rounding Metal

Metal can be rounded by using a vise and a piece of round stock or pipe. The procedure for rounding metal is as follows.

Procedure

1. Select a piece of round stock or pipe with a diameter equal to the desired bend.

(continued)

Procedure, *continued*

2. Place the end of the material to be bent tightly between the round stock and a vise jaw (Figure 12-25A).

3. Using a hand and a hammer, bend the metal around the round stock.

4. Open the vise and move the material around the round stock (Figure 12-25B).

5. Tighten the vise.

6. Bend and hammer the material to fit the round stock.

7. Repeat the above steps until the part is bent to the desired shape.

8. Reposition the metal as needed, and use the closing power of the vise to tighten and shape the bend.

Using an Anvil. An **anvil** is a heavy steel object that is used to help cut and shape metal (Figure 12-26). Metal may be rounded by holding the long part of the metal in one hand and bending the short end of the metal over the horn of an anvil (Figure 12-27).

Twisting Metal

Another useful procedure in agricultural mechanics is the twist. To make a 90-degree twist in flat iron, the following procedure is used.

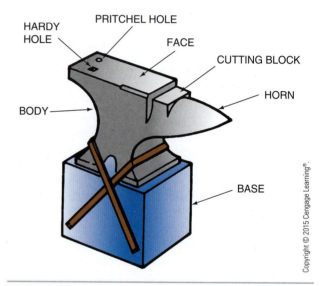

FIGURE 12-26 The parts of an anvil. The tapered square shank of an anvil chisel (or hardy) fits into the hardy hole. Metal placed on the cutting edge of the hardy is then struck with a hammer to cut it. The pritchel hole lets the smith punch holes in metal with a punch. The smith places red-hot metal over the hole and then drives the round, tapered part of the punch through the metal and into the hole.

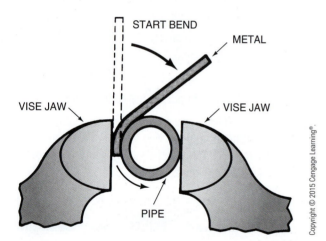

A Place the metal tightly between the pipe and the jaws of the vise.

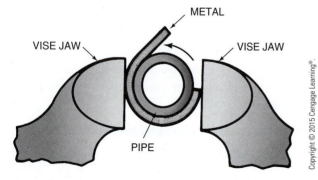

B Move the material around the cylindrical stock, and continue the bending process.

FIGURE 12-25 Using a pipe and a vise to round metal.

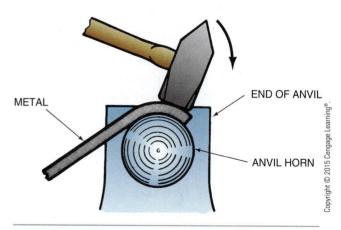

FIGURE 12-27 Using the horn of the anvil and a blacksmith's hammer to round metal.

Procedure

1. Mark the metal where the twist should start.

2. Measure up the metal a distance equal to 1½ times the width of the metal and make a second mark.

3. Place the metal in a vise with the twist mark at the top of the vise jaws.

4. Take a 10-inch or 12-inch adjustable wrench and adjust it to fit the thickness of the metal.

5. Place the wrench on the metal so the jaws extend the width of the metal. Move the wrench up the metal until the bottom of the wrench is on the upper line.

6. Hold the metal straight up in the air with one hand while turning the wrench with the other (Figure 12-28).

7. Stop when the upper section of the metal is at 90-degrees to the lower section.

> **NOTE**
>
> It may be necessary to move the wrench up or down the metal. Move it up if the metal starts to shear at the vise. Move it down if the bend is too long or gentle.

Bending Sheet Metal

Sheet metal is relatively easy to bend. Commercial benders are available; however, most agricultural shops use hand tools for this task. To bend sheet metal without special tools, the following procedure is used.

FIGURE 12-28 Using an adjustable wrench to twist flat iron.

Procedure

1. Mark the metal where the bend should occur.

2. Find a bench with a sharp 90-degree angle at its edge. If the edge is not sharp, place a 90-degree piece of angle iron over the edge and clamp the angle to the bench top.

3. Place the sheet metal so the metal to be bent extends over the bench. Be sure the bend line is directly over the edge of the bench.

4. Push the metal down along the edge to start the bend.

5. Complete the bend by tapping the metal with a hammer along the length of the bend line (Figure 12-29).

> **NOTE**
>
> For small projects, place an angle iron or heavy piece of flat steel of the suitable size in a vise. Form the metal using the steel to support the metal as it is hammered and shaped (Figure 12-30).

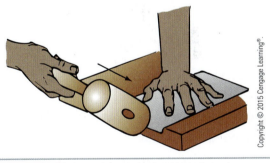

FIGURE 12-29 Sheet metal may be bent by using a hammer or mallet and the edge of a bench top.

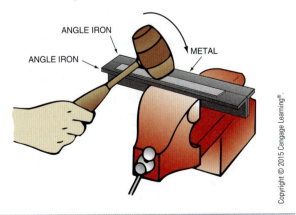

FIGURE 12-30 Two pieces of angle iron in a vise make a good device for bending sheet metal.

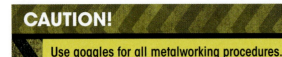

CAUTION!

Use goggles for all metalworking procedures.

CAUTION!

Warn others to stay away from the metalworking area while you are at work there, since the end of the bolt and the nut can become a projectile if driven hard.

SUMMARY

Identifying the specific type of metal is basic to any process using the metal. Different methods and techniques are used with different metals. Procedures that work with one metal may not be the best method to use with another. Properly identifying the metal, defining the task, and selecting the proper method is the proper sequence for most operations involving metal.

Student Activities

1. Define the Terms to Know in this unit.

2. With the use of a magnet, classify the following items as ferrous or nonferrous. Determine what metal or metals are in the items.
   - Penny
   - Quarter
   - Doorknob
   - Door hinge
   - Framing square
   - Automatic chalker
   - Piece of electric wire
   - Wrench
   - Steel water pipe

3. Break a piece of cast iron that has been discarded. Describe the appearance of the metal exposed by the break.

4. Examine a combination square. Does it have a scriber? If so, remove, examine, and replace it.

5. Examine five or six different files. Indicate the shape and type of teeth for each file.

6. Cut and bend flat steel.
   a. Obtain a piece of $\frac{1}{8}$" × 1" flat mild steel, 6 inches long.
   b. Measure 1 inch in from one end.
   c. Use a square and a scratch awl to make a line across the metal.
   d. Saw off the 1-inch section of metal, using a hacksaw. Check the work for squareness.
   e. Again, measure 1 inch from the end.
   f. Square a line across the metal with soapstone. (If soapstone is not available, use chalk.)
   g. Shear off the 1-inch section using a vise and a cold chisel.
   h. Again, measure 1 inch in from one end. Square across, using a scriber or scratch awl.
   i. Measure in $1\frac{1}{2}$ inches from the line, and square a line across.
   j. Place a 90-degree twist between the two lines, using a vise and an adjustable wrench.
   k. Secure 1 inch of the piece of metal in the vise and make a 90-degree bend in it.

7. Cut bolts.
   a. Obtain a $\frac{1}{4}$-inch stove bolt, 2 inches or more in length, and a $1\frac{1}{4}$-inch nut.
   b. Screw the nut onto the bolt about $\frac{1}{2}$ inch in from the tip.

   c. Grip the nut tightly in a machinist's vise, with the bolt in a horizontal position.

   d. Use a hacksaw to saw off the part of the bolt that extends beyond the nut.

   e. Smooth up the end of the bolt with a fine file.

   f. Unscrew the nut to straighten the threads.

   g. Screw the nut back on the bolt about $\frac{1}{2}$ inch in from the tip.

   h. Place the bolt vertically in the vise with the nut resting against the top of the vise. Tighten the vise.

   i. Use a cold chisel to shear off the nut. Drive the chisel slowly.

   j. Replace the bolt in the vise, leaving about $1\frac{1}{2}$ inch extended above the vise.

   k. Shear off the end with a cold chisel.

8. Shear and bend sheet metal.

   a. Lay out a piece of sheet metal 4 inches square.

   b. Cut out the 4-inch square sheet metal, with snips.

   c. Use a combination square and scratch awl to lay out a line 1 inch from one side of the sheet metal. Extend the line from edge to edge.

   d. Set a divider with the legs 1 inch apart. Place one leg of the divider on the line 1 inch from the end. Make an arc across the corner of the metal.

   e. Repeat step (d) on the other end of the line.

   f. Round the two corners with snips.

   g. Using a hammer and the edge of the bench, make a 90-degree bend on the line laid out in step (c).

Relevant Web Sites

Canadian Centre for Occupational Health and Safety, OSH Answers, Safety Hazards, Metal Working Machines—Metal Saws (Cold)

www.ccohs.ca/oshanswers/safety_haz/metalworking/metalsaws_cold.html

Self-Evaluation

A. Multiple Choice. Select the best answer.

1. When using a hacksaw, the number of teeth cutting at one time should be
   a. one
   b. two
   c. three
   d. five

2. Another name for snips is
   a. aviation
   b. combination
   c. scissors
   d. shears

3. Standard hacksaw blades come in lengths of
   a. 6 and 10 inches
   b. 7 and 10 inches
   c. 9 and 11 inches
   d. 10 and 12 inches

4. Choices of teeth per inch for hacksaw blades are
   a. 14, 18, 24, and 32
   b. 14, 20, 26, and 32
   c. 14, 24, and 36
   d. none of these

5. A hacksaw cuts on
   a. the backward stroke only
   b. the forward stroke only
   c. both the forward and backward strokes
   d. the stroke recommended by the manufacturer

6. Single-cut and double-cut refer to
   a. tooth patterns on files
   b. teeth on a hacksaw blade
   c. speed designations for blades
   d. widths of saw kerfs

7. Placing a file at a 90-degree angle to the metal and pushing or pulling is
   a. burnishing
   b. draw filing
   c. push filing
   d. none of these

8. It is very important to wear face protection when using a chisel because
   a. a chip of metal may hit you
   b. the hammer head is likely to come off
   c. chisels frequently break in half
   d. sight is improved by good face shields

B. Matching. Match the shape of the steel in column I with the correct name in column II.

Column I

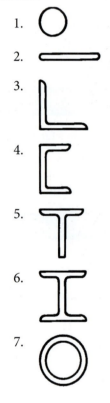

1.
2.
3.
4.
5.
6.
7.

Column II

a. channel
b. tee
c. pipe
d. rod
e. I-beam
f. flat
g. angle

C. Matching. Match the description in column I with the type of metal in column II.

Column I

1. malleable iron
2. flats, I-beams, pipe
3. zinc on steel
4. any shape; brittle
5. electrical wire; pipe
6. batteries
7. alloy of copper, zinc, and tin
8. plating for steel cans
9. alloy of copper and zinc
10. light, tough, roofing, and siding
11. food equipment, shiny
12. chisels, punches

Column II

a. cast iron
b. wrought iron
c. mild steel
d. tool steel
e. stainless steel
f. galvanized steel
g. aluminum
h. copper
i. brass
j. bronze
k. lead
l. tin

D. Completion. Fill in the blanks with the word or words that make the following statements correct.

1. A soft, gray rock that is used to mark metal is _____.

2. A tool with a handle and a pointed steel shank used to mark metal is the _____.

3. Ferrous metals are easily identified with a _____.

4. Thin metal may be bent into curves and circles by using a piece of _____ in a vise.

5. A 90-degree twist may be made in flat iron by using a vise and an _____ _____.

All images on this page are Copyright © 2015 Cengage Learning®.

UNIT 13
Fastening Metal

Objective
To fasten metals using procedures frequently used in agricultural mechanics.

Competencies to be developed
After studying this unit, you should be able to:
- Drill holes in metal.
- Tap threads in holes.
- Cut threads on bolts and pipe.
- Fasten metal with bolts and screws.
- Fasten metal with rivets.
- Solder sheet metal.
- Sweat copper pipe.

Materials List
- Center punch
- Set of twist drills
- Drill press or power hand drill
- Tap and die set or selected parts
- Oil for cutting
- Rags
- Pipe cutter
- Tubing cutter
- Pipe wrenches, adjustable wrench, and screwdriver
- Pipe reamer
- Pipe dies and stock

continued

Terms to Know
- soldering
- brazing
- welding
- high-speed drill
- jobbers-length drills
- tap
- tap wrench
- die
- die stock
- tap and die set
- screw plate
- taper tap
- plug tap
- bottoming tap
- bolt threads
- pipe fittings
- Teflon tape
- pipe joint compound
- pipe dope
- rivet
- pop rivet
- solder
- tinning
- flux
- 50–50 solder

continued

- hollow-core solder
- flux-core solder
- soldering copper
- sal ammoniac
- sweating

Materials list, *continued*

- A length of steel pipe—½", ¾", or 1" in diameter and 1" or more in length
- A pipe fitting for the length of steel pipe
- A length of copper tubing or pipe—⅜" or ½" in diameter and 6" or more in length
- Copper fitting for the length of copper tubing or pipe
- Metal countersink
- Sheet-metal screws
- Stove bolt, ⁵⁄₁₆" × ½" long
- Steel rivets, round-head and countersink-head
- 50–50 solder and flux
- Metal fruit or vegetable can
- One foot of #12, #14, or #16 stranded, covered copper wire
- Hot-rolled band iron—½" × 1" × 18" long
- Cold-rolled round stock, ⅜" diameter × 3" long
- Soldering copper and/or soldering gun
- Teflon tape or pipe compound

Metals are fastened with various devices such as screws, bolts, and rivets. Fasteners are needed to assemble machines, equipment, tractors, tools, and buildings. Screws are used for such purposes as holding the metal panels on appliances and stamped sheet-metal parts on machinery in place. Bolts are used to hold metal parts together on engines and transmissions. Rivets may be used to hold rain gutters and downspouts together on garages or barns. Rivets are also used in certain large bridges and steel buildings.

Metal pieces are also fastened together by melting a second metal between two pieces. This procedure is called **soldering** or **brazing**, depending on the material used. The copper pipes in houses, schools, and other buildings are held together with solder. Brazing and soldering differ from welding in that the metals being joined are not melted but are heated and a filler metal is melted to form the bond.

A third important way to hold metals together is to simply melt them together. This process is called **welding**. A filler rod may be used to add metal to strengthen the bond.

MAKING HOLES IN METAL

Holes are drilled in metal to permit screws to pass through a shank hole in the first piece. A pilot hole is needed in the second piece to accommodate the threaded part of the screw. In sheet metal, the pilot hole in the second piece is drilled so that the solid core of the screw will pass through. The threads of the screw hold by straddling the metal.

Holes are also drilled in metal to receive bolts and rivets. When threads are needed in a hole, the hole must be of a very specific size; otherwise, it cannot be threaded.

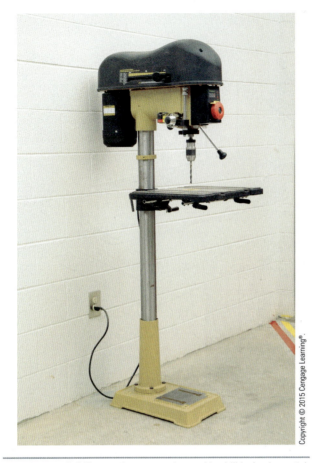

Copyright © 2015 Cengage Learning®.

FIGURE 13-1 A drill press is commonly used to drill holes in metal.

Tools for Drilling Holes

Holes are usually drilled in metal with a portable electric drill or power drill press (Figure 13-1). Some farm shops may still have a blacksmith's post drill. This is a hand-powered machine that is very effective for drilling metal.

Only high-speed twist drills are recommended for drilling metal. A **high-speed drill** is a drill that is hardened for use in metal. Such drills may be purchased in regular length, or longer length known as **jobbers-length drills**. It is especially important to use eye protection when drilling wood or metal. Thin metal should be backed with a piece of wood.

Drilling Holes in Metal

To drill a hole in metal, the following procedure is used.

CAUTION!

It is especially important to wear eye protection for all metalworking procedures.

Procedure

1. Use a center punch. Make a small dent to guide the center of the drill.

2. Secure the metal by clamping it to the drill-press table, holding it in a metal vise, or resting it against the post of the drill press (Figure 13-2).

3. Use a power drill that turns at the correct speed for the size of the drill being used (fast for small drills and slow for large drills) (Figure 13-3).

4. Select a high-speed twist drill of the desired size.

CAUTION!

Be sure to remove the key from the chuck after tightening.

5. Place the drill bit in the drill chuck. Tighten the chuck by using the key in at least two holes in the chuck; three holes are preferred (Figure 13-4).

NOTE

For holes larger than $\frac{3}{8}$" in diameter, first drill a small hole (pilot hole) about the size of the dead center of the twist drill. Proceed to drill with the full-size bit.

6. Apply cutting oil to carry heat away from the drill point. Cutting oil is a special oil that does not have additives used in oils designed for engines.

7. Apply firm, even pressure to the drill when starting and drilling.

8. Keep the drill perpendicular to the metal.

9. A uniform ribbon of metal is cut when the drill cuts steadily.

10. Use very little pressure as the drill breaks through the metal.

11. Be alert to the possibility of the bit seizing the metal as the drill breaks through. Hold portable drills very firmly at this point.

12. Avoid any side pressure on the drill bit. Small drill bits are brittle and break very easily.

13. After drilling the hole, remove any metal burrs with a file.

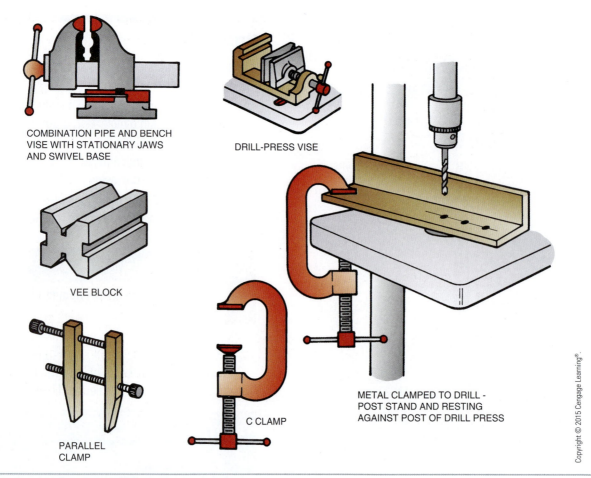

COMBINATION PIPE AND BENCH
VISE WITH STATIONARY JAWS
AND SWIVEL BASE

DRILL-PRESS VISE

VEE BLOCK

PARALLEL
CLAMP

C CLAMP

METAL CLAMPED TO DRILL -
POST STAND AND RESTING
AGAINST POST OF DRILL PRESS

FIGURE 13-2 Some devices for holding metal while it is being drilled.

Drill Speeds in rpm for Steel	
Drill Bit Size	**Recommended Drill Speed**
$1/8$	2100–2450
$3/16$	1400–1600
$1/4$	1000–1200
$5/16$	850–950
$3/8$	700–800
$7/16$	600–700
$1/2$	450–600
$9/16$	425–550
$5/8$	400–500
$11/16$	375–450
$3/4$	350–400
$13/16$	325–375
$7/8$	300–350
$15/16$	275–325
1	250–300

FIGURE 13-3 Selecting the correct drill speed for the size of drill bit being used will prolong the bit's life and ensure good results.

FIGURE 13-4 Tighten the drill chuck by turning the key in at least two holes. This will prevent the chuck from slipping and damaging the drill shank.

Punching Holes in Metal

Holes may be made in metal with punches. Soft, thin metal, like aluminum in storm windows, is punched rather than drilled. Punching is faster than drilling.

Steel may be punched if it is very thin. Steel up to $1/2$ inch thick can be punched with hand tools if the

Copyright © 2015 Cengage Learning®.

metal is red hot. Blacksmith's tools are needed to heat and punch steel. Heavy stationary punches driven by hand, hydraulics, or electricity may be used to punch cold steel. These machines are used when many holes of the same size are needed.

THREADING METAL

A common method of fastening metal is the use of bolts. Threads are a means in which the nut is turned onto the bolt. Sometimes a worker needs to cut threads in a hole or on a rod. This is accomplished by using a tap or die (Figure 13-5). Threads may be cut

into a hole in metal with a tap (Figure 13-6). A **tap** is a hardened, brittle, fluted tool that cuts threads into the hole. A tap is turned with a **tap wrench**. Threads are cut onto a rod or bolt with a **die** (Figure 13-7). A die is turned with a **die stock**. A set of tools for making threads is called a **tap and die set** or a **screw plate** (Figure 13-8). Common types of bolt threads are National Coarse (NC), National Fine (NF/SAE), and metric.

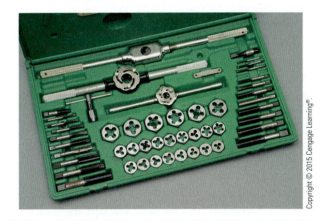

FIGURE 13-5 Taps and dies are used for cutting threads.

FIGURE 13-7 A die is designed to cut threads onto a rod.

FIGURE 13-6 A tap is designed to cut threads into a metal hole.

FIGURE 13-8 A set of taps and matching dies, together with the handles for turning them, is called a tap and die set (or a screw plate).

Tapping Threads

When tapping threads in a hole, the following procedure is used.

Procedure

1. Select the bolt that is to be threaded into the hole.

2. Determine the outside diameter of the bolt. Machine screw diameters are stated as Numbers 2, 4, 6, 8, 10, or 12. Other bolts are ¼ inch, ⁵⁄₁₆ inch, ⅜ inch, ⁷⁄₁₆ inch, ½ inch, etc. Metric bolts have diameters ranging from about 3 to 20 millimeters.

3. Determine the type of threads on the bolt by using a screw pitch gauge. Most bolt threads are National Coarse (standard), National Fine (NF/SAE), or standard metric. Metric bolts are also available with metric fine threads.

4. Select a taper tap of the correct size and thread type. The **taper tap** will start threads and complete them if the tap can extend through the material. If the hole does not go through the metal, a **plug tap** is used, followed by a **bottoming tap**. These taps will cut threads to the bottom of a blind hole (Figure 13-9).

5. Center-punch the location of the hole to be drilled.

6. Obtain a drill of the exact size specified for the tap selected, or use a tap-and-drill size chart (Figure 13-10). When selecting a drill size, the exact number or fractional size is needed when the hole is to be threaded.

7. Carefully drill the hole. The hole must be drilled at a 90-degree angle to the surface of the metal. Do not permit the drill bit to wobble.

8. Place the tap in a tap wrench.

9. Place the end of the tap in the hole. Be careful to keep the tap in line with the hole during the entire tapping process (Figure 13-11).

10. Apply enough cutting oil (if available) or motor oil to cover the threads of the tap and the hole in the metal.

11. Apply downward pressure on the tap, and turn the tap forward (clockwise) about one-half turn. Do not exert any sideways pressure or cause the tap to wobble. The tap will break very easily.

12. Turn the tap backward (counterclockwise) one-quarter turn to break the chip of metal that forms. Add oil to the threads if dry.

13. Turn the tap forward one or two turns.

14. Turn the tap backward to break the chip. Add oil.

15. Repeat steps 13 and 14 until the tap is through the hole or to the bottom of a blind hole. Keep the tap and metal covered with oil.

16. Remove the tap by turning it backward.

17. Wipe oil and metal chips from the parts and work area, using a cloth or paper towel.

18. Dispose of oily materials in the proper metal container.

> **CAUTION!**
>
> Do not brush metal chips away with bare hands, as the chips are sharp and will cause injury.

Threading A Rod or Bolt

The principles involved in threading a rod or bolt are similar to tapping threads. There is one major difference, however. Some dies are adjustable, but taps are not adjustable. With an adjustable die, the threads on a rod can be cut to a greater or lesser depth, as desired. This adjustment permits a rod to be threaded to fit a nut perfectly. The procedure for cutting threads on a rod follows.

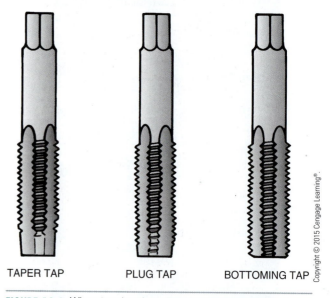

TAPER TAP   PLUG TAP   BOTTOMING TAP

Copyright © 2015 Cengage Learning®.

FIGURE 13-9 When tapping threads to the bottom of a blind hole, first use a taper tap, then use a plug tap, and finally use a bottoming tap.

National Coarse Threads (NC)

Size of Bolt or Screw and Tap	Size of Drill Bit to Use	Threads per Inch
#1	#53 or 1/16"	64
#2	#50	56
#3	#47 or 5/64"	48
#4	#43	40
#5	#38	40
#6	#36 or 7/64"	32
#8	#29	32
#10	#25	24
#12	#16	24
1/4"	#6 or 13/64"	20
5/16"	1/4"	18
3/8"	5/16"	16
7/16"	23/64"	14
1/2"	27/64"	13
9/16"	31/64"	12
5/8"	17/32"	11
3/4"	21/32"	10
7/8"	49/64"	9
1"	7/8"	8

National Fine Threads (NF)

Size of Bolt or Screw and Tap	Size of Drill Bit to Use	Threads per Inch
#1	1/16"	72
#2	#50	64
#3	#45	56
#4	#42 or 3/32"	48
#5	#37 or 7/64"	44
#6	#33	40
#8	#28 or 9/64"	36
#10	#21 or 5/32"	32
#12	#14	28
1/4"	#3 or 7/32"	28
5/16"	17/64"	24
3/8"	21/64"	24
7/16"	25/64"	20
1/2"	29/64"	20
9/16"	33/64"	18
5/8"	37/64"	18
3/4"	11/16"	16
7/8"	13/16"	14
1"	15/16"	14

Copyright © 2015 Cengage Learning®

FIGURE 13-10 Taps and appropriate drill bit sizes for producing coarse and fine threads.

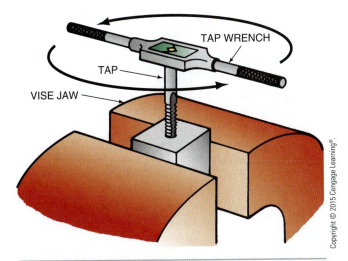

Copyright © 2015 Cengage Learning®

FIGURE 13-11 The tap must be held in line with the hole and turned carefully. Keep the tap and the area being threaded covered with oil. Turn the tap forward about one-half turn and then turn it backward one-quarter turn to break the chip. Proceed until the tap turns freely, indicating that the entire hole is threaded.

Procedure

1. Clamp a rod of cold-rolled steel in a vise (hot-rolled steel will not thread satisfactorily).

2. Bevel the end of the rod with a file (Figure 13-12).

3. Select a die of the correct size and type of thread.

4. Mount the die in a stock (handle).

5. Place the die over the rod, with its tapered side toward the rod.

6. Oil the die and rod.

7. Apply pressure on the die and turn it forward about one-half turn.

8. Back up one-quarter turn to break the chip.

(continued)

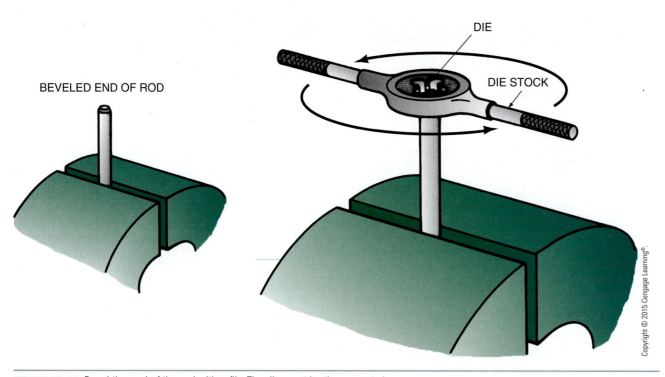

BEVELED END OF ROD

DIE

DIE STOCK

Copyright © 2015 Cengage Learning®.

FIGURE 13-12 Bevel the end of the rod with a file. The die must be the correct size.

Procedure, *continued*

9. Repeat steps 7 and 8 until a short length of the rod has been threaded. Add oil as needed to keep the die lubricated and cutting smoothly. If the die is adjustable, check to see if the nut threads on properly. If not, adjust the die to obtain a good fit before cutting threads over the entire area.

10. Continue cutting threads until the desired threaded length is obtained.

11. Remove the die and wipe all chips and oil from the rod, tools, and work area.

Cutting and Threading Pipe

Cutting Pipe. Pipe is cut with pipe cutters that consist of one steel cutting wheel plus two rollers. Some pipe cutters have three cutting wheels and no rollers. A small version of the pipe cutter is used for cutting copper and steel tubing. When using pipe or tubing cutters, the following procedure is used.

Procedure

1. Unscrew the cutter handle until the pipe fits between the rollers and wheels.

2. Slide the cutter onto the pipe until the cutter wheel(s) is on the mark.

3. Tighten the handle until the cutter wheel starts biting into the pipe.

4. Roll the pipe cutter around the pipe one or two times. The addition of oil will help the cutting process.

5. Repeat steps 3 and 4 until the pipe is cut. Do not tighten the handle more than one-half turn before rotating the cutter each time.

6. Insert a pipe reamer in the end of the pipe. Turn the reamer until the inside ridge is removed.

Threading Pipe. Pipe threads are not the same as bolt threads. **Bolt threads** are straight and can be cut the entire length of a rod. Pipe threads are tapered and can only be cut a short distance on the end of a piece of pipe (Figure 13-13). The die gets tight and binds if too many threads are attempted. The tapered threads permit pipe connections to be gas and liquid tight.

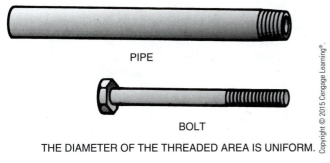

THE DIAMETER OF THE THREADED AREA IS LESS THAN THE DIAMETER OF THE PIPE. THE THREADS TAPER TO THE END OF THE PIPE.

PIPE

BOLT

THE DIAMETER OF THE THREADED AREA IS UNIFORM.

Copyright © 2015 Cengage Learning®

FIGURE 13-13 Pipe threads are tapered. No more than about 1 inch at the end of the pipe can be threaded. In contrast, bolt threads are straight and can be cut along the entire length of the bolt.

Pipe fittings screwed onto pipe with appropriate joint compound create joints that hold air, steam, and liquids under pressure without leaking. **Pipe fittings** are hollow connectors designed to attach to pipe. Threaded fittings for steel pipe require some material to seal the areas between threads if they are to be used for gases or liquids. Materials for this purpose are **Teflon tape** and **pipe joint compound** (also referred to as **pipe dope**). The procedure for threading pipe follows.

Procedure

1. Select the correct die size for the pipe to be threaded.
2. Place the die in the stock.
3. Tighten the pipe in a pipe vise with 8 to 10 inches extending beyond the pipe vise. A regular vise may crush the pipe.
4. Slide the tapered side of the die onto the end of the pipe.
5. Apply cutting oil to the die and pipe as needed.
6. Turn the die until the end of the pipe starts to extend through the die.
7. Remove the die. Check to see if a pipe fitting will thread on and become tight after 4 to 6 turns. If not, cut more threads on the pipe.
8. Use a cloth to wipe all oil and chips from the pipe, tools, and work area. Remove all chips and oil from the inside of the pipe. Such materials interfere with the use of the pipe if they are not properly removed.

FASTENING METAL USING BOLTS AND SCREWS

Flat pieces of metal are fastened quickly and easily with bolts. One of two methods is generally used. A nut is used with the bolt, or the bolt is screwed into the second piece of metal after threads are tapped.

Using a Bolt and Nut

When using a bolt and nut, the following procedure is recommended.

Procedure

1. Drill or punch a hole through both pieces of metal. The holes should be the same diameter as the bolt. Oversized holes may be used if there is a poor fit.
2. Select the desired type of bolt.
3. Countersink to accommodate the bolt head, if using a flat-head stove bolt.
4. Insert the bolt.
5. Add a lock washer.
6. Thread on the nut.
7. Line up all parts.
8. Tighten the nut until the lock washer is flattened.

Using Bolts without Nuts

Cap screws, machine bolts, and machine screws may be used without nuts. The procedure follows.

Procedure

1. Select the desired size and type of bolt.
2. Drill a hole in the first piece of metal the diameter of the bolt.
3. Drill the second piece of metal with the exact drill specified for the bolt size and thread type.
4. Tap threads into the second piece.
5. Place a lock washer on the bolt.
6. Push the bolt through the first piece.
7. Screw the bolt into the second piece.
8. Tighten the bolt until the lock washer is flattened.

Using Sheet-Metal and Drywall Screws

Sheet-metal screws are used to attach thin pieces quickly. The procedure follows.

Procedure

1. Select the desired type and size of sheet-metal screw.
2. Drill or punch a shank hole in the first piece of metal.
3. Drill or punch a pilot hole in the second piece.
4. Screw the two pieces together, but do not tighten the screw.
5. Align the parts.
6. Tighten the screw snugly. Be careful not to over-tighten. If the screw is overtightened, the pilot hole may be stripped and the screw will not hold.
7. If the material is soft or the metal is thin, self-tapping sheet-metal screws and drywall screws may cut their own threads.

FASTENING WITH RIVETS

A **rivet** is a fastening device held in place by spreading one or both ends. A special tool called a rivet set makes it easy to install rivets. Many shops do not have a rivet set and the peening method is used.

Most rivets are purchased with flat, round, or countersunk heads; they may be solid, tubular, or split (Figure 13-14). Tubular and split rivets have the advantage of being easy to rivet. This means they can be spread or shaped by hammering. They will not hold as well as solid rivets. Steel rivets, copper rivets, and pop rivets are commonly used.

The Steel Rivet

The round-head steel rivet is used to fasten structural and sheet steel in bridges, buildings, and truck bodies. An important use of steel rivets on the farm is to attach mower knives. For this application the flat-head rivet is used. To install a steel rivet, the following procedure is used.

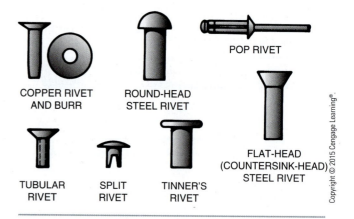

COPPER RIVET AND BURR ROUND-HEAD STEEL RIVET POP RIVET

TUBULAR RIVET SPLIT RIVET TINNER'S RIVET FLAT-HEAD (COUNTERSINK-HEAD) STEEL RIVET

Copyright © 2015 Cengage Learning®.

FIGURE 13-14 Rivets may be solid, tubular, or split, and they are available in different head styles. They are generally composed of steel, aluminum, or copper.

Procedure

1. Mark the spot, and center-punch the metal.
2. Drill a hole in both pieces of metal, the exact size of the rivet.
3. Determine the length of the rivet. The length should be the thickness of the two pieces of metal plus the diameter of the rivet. For example, when fastening two pieces of metal, each $\frac{1}{8}$ inch thick, with a rivet of $\frac{1}{8}$ inch diameter, the rivet length should be $\frac{3}{8}$ inch ($\frac{1}{8}$" + $\frac{1}{8}$" + $\frac{1}{8}$" = $\frac{3}{8}$").
4. Cut the rivet to length with a hacksaw.
5. Insert the rivet through the two pieces of metal.
6. Place the head of the rivet on a solid steel object such as an anvil or a metal vise.
7. Place a rivet set, or other object with a hole the size of the rivet, over the protruding rivet end.
8. Drive the metal that is to be riveted down tight against the rivet head.
9. Use a ball peen hammer to create a rivet head on the extended end. This is done by striking or peening the rivet stem on the edges until they are rounded down and mushroomed against the metal (Figure 13-15). A rivet set may also be used to form the riveted end (Figure 13-16).

The Copper Rivet

Copper rivets with a tight-fitting washer called a *burr* are used to fasten harnesses and other leather products.

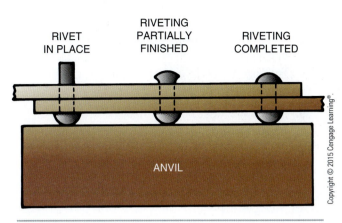

FIGURE 13-15 Using a ball peen hammer to install steel rivets.

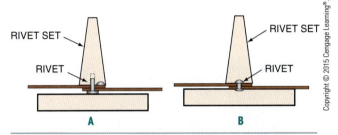

FIGURE 13-16 Using a rivet set to install steel rivets. First, place the rivet set over the rivet and tap the pieces of metal together (A). Then place the cupped part over the rivet and drive the rivet set to form the mushroom-shaped head (B).

A burr is as large as the head so soft materials are held without pulling the rivet through the material.

The copper rivet with burr is installed much like a steel rivet. The rivet is inserted in the material; the washerlike burr is then added to the rivet. The burr and material are hammered together. Finally, the end is riveted to keep the burr in place.

The Pop Rivet

The **pop rivet** is so named because it pops when installed. It is especially useful in situations in which it is difficult to hold one end of the rivet and hammer the other end. An example of this is repairing sheet metal on an automobile.

When purchased, the pop rivet appears to be mostly stem. However, during installation, the special pop rivet tool pulls the stem until its ball-like head forces the end of the rivet to expand (Figure 13-17). This enlargement holds the rivet in place. As the stem is pulled further by the tool, it breaks with a popping sound. The stem pulls out of the hollow rivet and is thrown away. The riveting process is then complete.

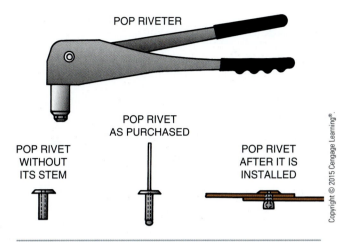

FIGURE 13-17 Pop rivets are installed from one side of the metal only. A built-in stem attached to a round ball is pulled into a hollow rivet, and the ball causes the rivet to spread on the opposite side of the metal.

FASTENING METALS WITH SOLDER

Solder is a mixture of tin and lead or other metals. The joining of two materials with solder is known as soldering. Solder is frequently used to join thin metal, to make electrical connections, and to join copper tubing and pipe. The following general procedure is used when soldering.

Procedure

1. Remove all dirt and corrosion from the metal with sandpaper or emery cloth. The metal should be shiny and bright.

2. Coat the tip of the soldering copper with solder. This process is called **tinning**. Tinning is done by cleaning the soldering copper with steel wool, sandpaper, or a file. The copper is heated and flux is applied. After the flux melts, solder is applied to the copper until it covers the tip. **Flux** is a material that removes tarnish or corrosion, prevents corrosion from developing, and acts as an agent to help the solder spread over the metal. Three fluxes that are commonly used are acid, rosin, and sal ammoniac.

3. Apply heat to the metal and apply flux.

4. Apply additional heat to the metal being soldered until that metal will melt solder.

(continued)

Procedure, *continued*

5. Tin the metal by applying solder from a wire spool or a solid bar.

6. Apply heat to both pieces of metal until the solder melts and flows between them.

7. Wipe off excess solder with a cloth or sponge dampened with water, while the solder is still molten.

8. Do not permit movement in the joint until the solder solidifies or hardens.

When soldering, several points should be kept in mind. It is important to select the appropriate flux and solder for the metal being soldered. Directions on the labels of commercial flux products as to the metals they will clean should be followed. Most soldering jobs in agricultural mechanics can be done with solder composed of 50 percent tin and 50 percent lead. This solder is called **50–50 solder**. Other solders are 30–70, 40–60, 60–40, and 80–20. Notice that each combination of numbers equals 100. Different solders are recommended for various special jobs.

Solder may be purchased with flux inside. This is referred to as **hollow-core solder** or **flux-core solder**. Many people find the flux in the core of the solder makes soldering more convenient and easy.

Tinning a Soldering Copper

A **soldering copper** is a tool used for soldering and consists of a wooden or plastic handle, steel shank, and copper tip (Figure 13-18). It may also have an electric heating element. The procedure for tinning a soldering copper is as follows.

Procedure

1. Remove burned residues with sandpaper, emery cloth, steel wool, or a file.

2. Heat the soldering copper until it will melt the flux or make smoke when rubbed in sal ammoniac. **Sal ammoniac** is a cube or block of special flux for cleaning soldering coppers (Figure 13-19).

3. Rub the heated copper in sal ammoniac, or apply paste flux with a small wooden stick. The copper should get shiny.

4. Unroll about 6 inches of solder from the spool and touch the end to the copper. The solder should melt and flow over all flat surfaces of the copper if the copper is clean, fluxed, and the correct temperature.

5. If the solder does not melt, heat the copper to a higher temperature. Then repeat steps 3 and 4.

6. Brush the copper lightly with a cloth dampened with water. The soldering copper should be a shiny silver color. The silver color indicates the copper is tinned.

Soldering Sheet Metal

To solder sheet metal, a large soldering copper is used. A soldering copper may be heated using a gas torch, a

FIGURE 13-18 A soldering copper consists of a wooden or plastic handle, a steel shank, and a copper tip.

FIGURE 13-19 The soldering copper must be kept clean. A small piece of steel wool can be used to clean and polish the tip.

gas furnace, or an electric element. Electric units are sometimes called soldering irons; however, the tip of the unit is made of copper, not iron. When soldering sheet metal, the following procedure is used.

Procedure

1. Clean the area to be soldered, using an abrasive or a file.
2. Heat each piece of metal with the flat part of a tinned soldering copper.
3. Add flux to the metal as it heats up.
4. As the metal gets hotter, keep the flat part of the soldering copper flat on the metal, near one edge.
5. Touch the solder wire to the metal, not the soldering copper.
6. Watch the point when the solder melts and begins to flow.
7. Move the soldering copper slowly across the metal (Figure 13-20).
8. Follow the soldering copper with the solder, as needed, to tin the area to be soldered.
9. Repeat steps 2 through 8 on the second piece of metal.
10. Lay the two tinned surfaces face-to-face on a piece of scrap wood, if support is needed.
11. Heat both sheets of metal with the soldering copper until the solder starts to flow.
12. Feed in additional solder to fill the joint as needed.
13. Wipe off excessive molten solder with a damp cloth.
14. Hold the two pieces together with a piece of scrap wood until the solder cools. The solder is cool when it changes from a shiny to a dull finish.

Some of the steps in this procedure must be done at the same time, or they must be done in close sequence. Therefore, exact instructions from an experienced teacher may be needed to learn the technique.

Wood is recommended to support or hold the metal because wood does not carry heat away from the metal. Metal tabletops, vises, and metal tools interfere with heat control. The control of heat is the heart of the soldering process.

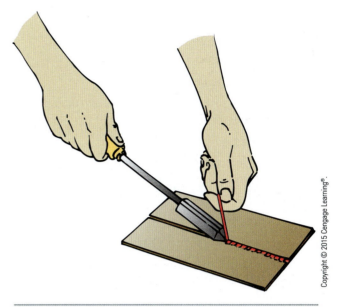

FIGURE 13-20 When soldering sheet metal, the soldering copper is moved slowly along the metal.

Copyright © 2015 Cengage Learning®.

Soldering Stranded Electric Wires

Copper wires for low-voltage wiring are soldered by first removing about 1 inch of insulation. The soldering procedure follows.

Procedure

1. Twist the two bare ends of the wire together.
2. Insert one of the wires in a vise and gently tighten the vise jaws against the insulation.
3. Hold a hot, tinned soldering copper under the twisted wires (Figure 13-21).
4. Hold rosin-core solder on the top of the wire.

NOTE

Acid flux is not recommended for electrical connections.

CAUTION!

Do not use an acid flux as this interferes with the function of the splice.

(continued)

Procedure, *continued*

5. Wait for the heated wire to cause the rosin to flow and clean the wire.
6. Watch the solder melt and run into and around the strands of wire.
7. Remove the heat.
8. Apply several thicknesses of vinyl electrical tape to the joint.

An electric soldering pencil or gun works well for soldering copper wires. These special soldering tools provide the correct amount of heat for electrical applications.

Soldering Copper Pipes and Fittings

The process of soldering a piece of copper pipe into a fitting is called **sweating** (Figure 13-22). The ability to sweat copper fittings is useful for repairing and installing water lines at home, on the job, or on the farm.

Common pipe fittings include elbows, tees, couplings, unions, and adaptors. Fittings are used to change the direction of a pipe line. They are also used to adapt one type of pipe to another, such as copper to plastic.

Fittings are used to connect pipe to faucets, valves, and other fixtures. To sweat a fitting onto a piece of copper pipe, the following procedure may be used.

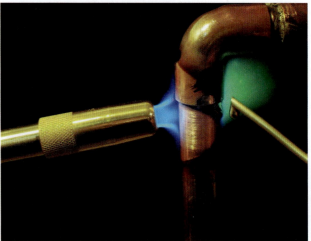

FIGURE 13-22 Soldering copper pipe into a fitting is called sweating.

© VladimirB/Shutterstock.com

Procedure

1. Use very fine sandpaper or steel wool to polish about ¾ inch at the end of the pipe.
2. Place the sandpaper or steel wool across the end of your index finger and polish the inside of the fitting.
3. Apply a paste flux to the polished surfaces of the pipe fitting.
4. Press the fitting onto the pipe.
5. Use a propane or other gas torch to heat the outside of the fitting near the opening where the pipe is inserted.
6. Heat the fitting ½ inch in from the opening. Use the tip of the inner flame of the torch to heat the underside of the fitting. Heat will rise to the upper areas.

> **NOTE**
>
> It is important to heat the fitting rather than the pipe to draw the solder into the joint.

7. Hold a length of lead-free solid core solder at the joint between the fitting and pipe. Be sure the solder is of the correct type for water lines.
8. Continue to heat the pipe until the solder starts flowing into the joint.

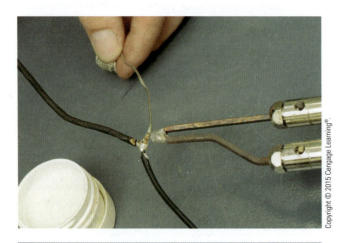

Copyright © 2015 Cengage Learning®.

FIGURE 13-21 An electric soldering pencil or soldering gun is used to solder stranded electrical wire.

9. Move the solder around the joint to permit solder to fill all areas between the pipe and fitting.

10. Remove the heat and solder when solder is seen all the way around the joint and the joint is full of solder.

11. Before the solder hardens, wipe around the joint with a damp cloth.

12. Cool the joint.

13. Turn on the water to pressurize the system.

14. Check for leaks.

When using a propane torch, several precautions are important. First, wear goggles and insulated gloves. Second, work only in areas that are free of paper, grease, fuel, and other materials that burn easily. Third, point the torch away from you when you turn it on and light it. Fourth, use approved fire safety procedures and materials.

Whenever any flame is used, be sure there is a fire extinguisher nearby. To reduce fire hazards, place sheet metal behind pipes where a torch is used. When finished with the torch, it is best to extinguish the flame. Certain propane torches have high-flame triggers. When the propane unit is set down, the flame becomes low and creates less of a fire hazard.

SUMMARY

Metal can be fastened using a variety of means. The correct method depends on the thickness of the metal, the appearance needed, the strength needed, and the cost. Choosing and applying the correct method is crucial to the success of any metalworking project.

Student Activities

1. Define the Terms to Know in this unit.

2. Make a bench stop using the following procedure.
   a. Cut a piece of band iron $\frac{1}{4}$" thick × 1" wide × 5" long.
   b. Make the ends square; slightly round all of the corners to remove their sharp edges.
   c. Center-punch the metal $\frac{1}{2}$ inch from each end and from each side.
   d. Drill a $\frac{5}{16}$-inch hole through the metal at the two punch marks.
   e. Tap National Coarse (NC) threads in each hole to receive a $\frac{3}{8}$-inch threaded rod.
   f. Cut two pieces of round cold-rolled steel $\frac{3}{8}$ inch in diameter and $1\frac{1}{2}$ inches long. Remove all sharp corners with a file.
   g. Thread each rod down $\frac{1}{4}$ inch.
   h. Screw the two threaded rods into the $\frac{1}{4}$" × 1" × 5" piece. The rods should thread in until tight. The end of the rods should be flush with the bottom of the metal strip.

   > **NOTE**
   >
   > The bench stop is useful to stop lumber when sanding, planing, or nailing it on a bench. To use the stop, drill two holes in the bench top $\frac{3}{8}$ inch or larger in diameter. The holes are drilled so the two rods drop into them. The bench stop is then ready for use.

3. Cut and thread a piece of steel pipe.
   a. Obtain a piece of steel pipe 1 foot or more in length. Pipe with an inside diameter of $\frac{1}{2}$ inch, $\frac{3}{4}$ inch, or 1 inch is recommended.
   b. Cut off any existing threads with a pipe cutter.

   c. Use a pipe reamer to remove any inside ridge.

   d. Cut threads on the pipe.

   e. Apply Teflon tape or pipe joint compound.

   f. Screw on a pipe fitting such as a tee or elbow.

   g. Have the teacher check your work.

   h. Remove the fitting, wipe off any pipe joint compound, and store the pipe.

4. Bolt and rivet two pieces of band iron.

   a. Cut two pieces of band iron $\frac{1}{4}$" × 1" × 6".

   b. Mark one piece A and the other B.

   c. Center-punch piece A at $1\frac{1}{2}$ inches, 3 inches, and $4\frac{1}{2}$ inches from one end and along the centerline of the metal.

   d. Lay A over B.

   e. Drill a hole through both pieces near one end with a $\frac{1}{8}$-inch drill bit.

   f. Redrill A with a $\frac{5}{16}$-inch drill.

   g. Redrill B with a $\frac{1}{4}$-inch drill.

   h. Tap $\frac{5}{16}$-inch NC threads into the hole in B.

   i. Screw the two pieces together with a $\frac{5}{16}$" × $\frac{1}{2}$" stove bolt, machine bolt, or cap screw.

   j. Drill a $\frac{1}{4}$-inch hole through both pieces at the center and end marks.

   k. Install a round-head rivet in the center hole.

   l. Countersink the third hole, and install a flat-head steel rivet.

5. Solder sheet metal.

   a. Obtain a clean steel fruit can and its lid.

   b. Cut the outer corrugations off the lid so that only the flat center remains.

   c. Turn the empty can upside down.

   d. Solder the round piece from the lid onto the flat round area on the bottom of the can.

   e. Punch several holes in the corrugations outside the patched areas, with a 4d nail.

   f. Close the holes by soldering.

6. Solder electric wires.

   a. Obtain two pieces of number 12, 14, or 16 stranded and insulated copper wire about 4 inches long.

   b. Remove 1 inch of insulation from one end of each wire.

   c. Cross the two bare wires and twist them around each other to form a smooth splice.

   d. Solder the joint.

   e. Tape the joint.

7. Sweat copper pipe.

   a. Obtain a short length of $\frac{3}{8}$-inch or $\frac{1}{2}$-inch copper pipe.

   b. Sweat a copper fitting onto the pipe.

   c. Have the instructor inspect the job.

   d. Remelt the solder and remove the fitting with pliers.

   e. Wipe off the hot solder with a damp cloth.

Relevant Web Sites

The Basic Electronics Soldering & Desoldering Guide, by Alan Winstanley
www.epemag.wimborne.co.uk/solderfaq.htm

Self-Evaluation

A. Multiple Choice. Select the best answer.

1. Metal may be fastened by
   a. bolts
   b. rivets
   c. screws
   d. all of these

2. The process of joining metal by melting a different metal between two pieces is known as
   a. gluing
   b. soldering
   c. washing
   d. welding

3. Holes are usually made in heavy metal by using
   a. an auger bit
   b. a forge
   c. a punch
   d. a high-speed twist drill

4. When drilling, the lightest pressure should be placed on the drill when
   a. breaking through
   b. midway drilling
   c. starting the hole
   d. none of these

5. Threads are cut onto a rod with a
   a. die
   b. ream
   c. stock
   d. tap

6. When tapping threads, always start with a
   a. bottoming tap
   b. plug tap
   c. taper tap
   d. die

7. The first tool to use in drilling a hole in metal is the
   a. center punch
   b. drill
   c. file
   d. ream

8. When cutting threads, oil is used to
   a. clean the tool
   b. harden the threads
   c. lubricate the tool
   d. soften the metal

9. Pipe threads differ from bolt threads in that
   a. pipe threads are tapered, bolt threads are not
   b. oil is needed to cut pipe threads but not bolt threads
   c. pipe is threaded with a die, a bolt is not
   d. all of these

10. If you cannot get to both sides of sheet metal, the rivet to use is the
    a. copper rivet and burr
    b. pop rivet
    c. split rivet
    d. solid steel rivet

B. Matching. Match the word or phrase in column I with the correct word or phrase in column II.

Column I

1. drill
2. pop
3. solder
4. countersink rivet
5. tin and lead
6. NC
7. NF or SAE
8. sal ammoniac
9. Teflon tape
10. sweating

Column II

a. rivet
b. solder
c. jobbers-length
d. National Fine
e. seal pipe threads
f. flux
g. 50–50
h. mower knives
i. National Coarse
j. soldering copper pipe

C. Completion. Fill in the blanks with the word or words that make the following statements correct.

1. The screw used to fasten thin metal is called a _____ _____ screw.

2. The metal used in the ends or tips of all soldering irons, guns, or pencils is _____.

3. The process of covering the tip of the soldering copper with solder is called _____.

4. When soldering electrical wire or connections, _____ core solder must be used.

5. After cleaning with an abrasive, metal must be treated with a _____ before soldering.

D. Brief Answer. Briefly answer the following questions.

1. What is the main difference between threading a rod or bolt and tapping threads? What is the benefit of this difference regarding threading rods?

2. When installing a steel rivet, how does one determine the length of the rivet?

3. What advantage and what disadvantage do tubular and split rivets have when compared with other rivets?

4. What are the purposes of applying flux in soldering?

5. Identify four precautions to take when using a propane torch.

6. Why is wood recommended to support or hold the metal when soldering sheet metal?

Section 4

POWER TOOLS IN THE AGRICULTURAL MECHANICS SHOP

UNIT 14
- Portable Power Tools

UNIT 15
- Woodworking with Power Machines

UNIT 16
- Adjusting and Maintaining Power Woodworking Equipment

UNIT 17
- Metalworking with Power Machines

UNIT 14

Portable Power Tools

Objective

To select and safely use major portable power tools in agricultural mechanics.

Competencies to be developed

After studying this unit, you should be able to:

- State recommended procedures for using portable power tools.
- Write a description of the uses of portable power tools.
- Name and properly spell the names of common portable power tools used in agricultural mechanics.
- Identify and spell the names of the major parts of portable power tools.
- Safely operate a portable power drill, belt sander, disc sander/grinder, finishing sander, saber saw, reciprocating saw, and circular saw.

Materials List

- Examples of portable electric and air-driven tools
- Unlabeled diagrams of portable power tools:
  - Drill
  - Belt sander
  - Disc sander
  - Grinder
  - Finishing sander
  - Saber saw
  - Reciprocating saw
  - Circular saw

Terms to Know

- sawhorse
- trestle
- ground-fault circuit interrupter (GFCI)
- double-insulated
- air tool
- compressed air
- chuck
- duty cycle
- continuous duty
- variable speed
- reversible
- hammer
- cordless
- pilot hole
- finishing sander
- orbital sander
- belt sander
- disc sander
- grinders
- power handsaw
- saber saw
- bayonet saw
- blind cut
- reciprocating saw
- portable circular saw
- bushing
- router

At one time almost all woodworking and metalworking was done using hand tools powered only by human muscles. Water wheels and, later on, steam-powered, larger machines came into use, but a large portion of the work was done by hand until electricity became accessible in the 20th century. Since that time, many types of electrically powered portable tools have been invented.

Portable power tools save labor and are relatively inexpensive to buy (Figure 14-1). Many people prefer to invest in portable power tools rather than stationary ones because of the expense and lack of space for stationary tools. Since work on the farm and in other agricultural settings requires that tools be taken to the job, portable tools are especially useful and efficient.

When using portable power tools, it is important that the work be well secured. There are several ways to do this. If possible, place the work in a vise or some other kind of holding device. If large pieces are being used, sawhorses, also called trestles, are recommended. A **sawhorse** or **trestle** is a wood or metal beam or bar with legs. It is used for the temporary support of materials. Large flexible panels require special handling to prevent the binding of saws and other equipment. If two or three 2 × 4s are placed on sawhorses, flexible panels can then be placed across them for support. Large panels can be cut safely when supported in this manner.

SAFETY PRECAUTIONS

Several safety problems are inherent with portable electric tools. When working with power tools, sharp blades, bits, and abrasives should be used to reduce the pressure needed to make the tool function. Reduced pressure decreases the likelihood of the tool binding or the hand or tool slipping. Even when using a sharp blade, never try to push the tool through material too rapidly. Give the blade or drill bit time to cut. Always make sure the blade or bit comes to a complete stop before laying the tool aside.

Wet areas are dangerous areas for using electrical power tools. Work areas should always be set up in dry locations. Wooden floors are the best, but concrete is safe if it is dry. Shoes with rubber soles or rubber boots reduce electrical hazards when power tools are being used. A **ground-fault circuit interrupter (GFCI)** must be used in circuits when the tool operator comes in contact with the ground or other damp surfaces (Figure 14-2). A GFCI breaks the electrical circuit when the operator is threatened by electrical shock. A newer type of safety breaker is called an arc-fault circuit interrupter (AFCI). This device is required on all bedroom circuits. The purpose is to break the circuit if an arc occurs anywhere along the circuit. When

FIGURE 14-1 Portable power tools save labor and are relatively inexpensive.

Copyright © 2015 Cengage Learning®.

© iStockphoto/SLRadcliffe

FIGURE 14-2 A ground-fault circuit interrupter must be used in damp areas.

electrical wires come loose, they can cause an arc that may result in a fire.

Another important safety consideration in using portable power tools is proper balance and footing. It is important to stay balanced at all times. Excessive reaching that may cause loss of balance or control should be avoided.

Eye and face protection are very important when using power tools. Under no circumstances should power tools be used without wearing safety glasses or goggles. Face shields are also recommended.

In addition to eye protection, coveralls and other protective clothing are recommended when using power tools. It is important to avoid all loose clothing. Leather shoes are recommended as a minimum. Steel-toed shoes are needed when heavy materials are being handled.

Figure 14-3 lists some basic safety rules to follow when using electric power tools.

SAFETY RULES

1. Plug into an outlet protected by a ground-fault interrupter when possible and at all times when wet conditions exist.
2. If the tool housing is metal, be sure to use a three-wire grounded power source.
3. Be certain the power cord, switch, and all electrical parts are in good condition.
4. Be certain that blades, bits, and other cutting units are clean and sharp.
5. Support work carefully to avoid any tendency for it to bend, buckle, or bind when power tools are used.
6. Keep power tools clean and free of dirt.
7. Keep vent holes free of dirt to permit ventilation and cooling of the motor.
8. Wear complete body protection.
9. Exercise care to avoid accidental electrocution caused by working in wet areas or by cutting power cords.
10. Do not force power tools to cut, drill, or otherwise work faster than designed to work.
11. Hold every power tool firmly and in control at all times.
12. In school, always obtain the instructor's permission before using a power tool.
13. If there is any question about the condition of a power tool, check with the instructor before proceeding.
14. Announce to others around you when you are ready to start a power tool.
15. Lay the power tool down safely so it will not damage cutting parts or injure others.
16. Report any faulty condition of the power tool and its cutting parts to the instructor as soon as it is noticed.

Copyright © 2015 Cengage Learning®.

FIGURE 14-3 Basic safety rules for using electric power tools.

POWER CORDS

The electrical cord of a power tool must be in good condition. The cord should be checked for broken insulation, broken plugs, bare wires, or other evidence of damage. Power cords often break, short out, or become electrical hazards due to breaks in the insulation where the cord leaves the power tool. It is advisable not to turn the first wind of the power cord too sharply if cords are wrapped around tools for storage.

All motors must have some method for protecting the operator against electrical shock in case of an internal electrical problem. Motors with metal housings should have a special ground wire and a plug with a ground prong (Figure 14-4). The cords of such tools contain three wires: two wires carry current for the motor, and the third wire is a safety ground wire to ground the housing of the tool. The ground wire connects to the longer prong on a 120-volt plug. This ground prong should never be broken off or damaged. If a tool in the agriculture shop is not grounded properly, the instructor should be notified immediately.

Motors with plastic housings generally are double insulated and do not need the special ground wire and ground prong. **Double-insulated** tools use two-wire, non-grounded cords. The electrical parts are insulated or separated from the user by special insulation inside the motor and by the plastic motor housing. Some electric motors have internal parts insulated in two different internal locations.

Extension Cords

In many cases an extension cord is needed with a portable power tool. Before a power tool is used, it should be determined if the tool is meant to be used with a three-wire extension cord. It is important that a

Copyright © 2015 Cengage Learning®/Ray Herren

FIGURE 14-4 A three-prong grounded plug.

three-wire grounded extension cord be used with tools that have metal bodies.

Extension cords should always be checked before use. Cords with frayed insulation or damaged ends should not be used. Extension cords longer than needed to reach the job should be used. If short cords are used, it is possible for tools to be pulled from benches if the cord is moved by a passerby. Workers must always be aware of the location of the power cord. Otherwise, electrical shock may result due to damage to the cord by sawing, drilling, sanding, binding, or crushing.

Extension cords should be of the proper size. This means that the diameter of the wires should be of sufficient size to safely carry the amount of electrical current required by the tool. Tools are rated by the amount of current or amperage they need. Always make sure the cord is large enough to handle the load. Remember that the longer the cord, the larger the wire size needed. A cord with 14-gauge wire should be sufficient for most portable power tool needs. However, heavy duty tools such as large saws and routers may require a 12-gauge wire, particularly if the cord is over 50 feet long.

AIR-DRIVEN TOOLS

Some agricultural mechanics shops have air tools. An **air tool** is a tool that is powered by compressed air. **Compressed air** is air pumped under high pressure and carried by special hoses. Compressed air provides pressure for spray guns; it also drives portable tools such as drills, grinders, sanders, cutters, and power nailers. Air tools have several advantages (Figure 14-5). One of the biggest advantages is that they do not require electricity at the tool. The only electrical current used is to power the compressor that can be many feet away from where the tool is used. This is particularly important when working on wet ground or in moist conditions, where the risk of electrical shock is greater with electrical tools. For example, to achieve a fine finish when spray-painting metal such as the hood of a tractor, it may be necessary to wet sand the surface, using water under the sanding pad. This would be dangerous using an electric sander, but the hazard is eliminated using an air-operated sander.

Power nailers are a fast, efficient method of driving nails into wood. They come in a variety of sizes that shoot nails from less than 1 inch in length to over 2 inches in length. Power nailers are used extensively in the construction industry. Extreme caution should be used with nail guns. These tools shoot nails out with a tremendous velocity. Although safeguards are built into the tools, they are still dangerous if not handled properly.

CAUTION!

Never try to shoot a nail from a nail gun with the safety guard disabled. Always make sure the gun is firmly pressed to the wood you intend to nail before you press the trigger.

Compressed air can be dangerous if not handled properly. It is particularly hazardous when air hoses are being coupled and uncoupled (Figure 14-6). High-pressure air lines must not be uncoupled when people are nearby. Streams of air must never be directed toward a person's face or body. Dust and dirt may be driven into the eyes, or the compressed air may damage ear drums and other organs of the body. Serious accidents of many types have been reported from the misuse of compressed air. Compressed air is not recommended for cleanup purposes unless the pressure is less than 30 pounds per square inch.

Types of Nail Guns

The most common types of nailers are air driven or pneumatic. Compressed air drives a piston inside a sleeve in the nailer to operate these tools. At the end of the piston is a pin that drives the nail into the wood. Most nailers require air compressors that can deliver about 60 pounds per square inch (psi) of compressed air. Obviously, the larger the nailer, the larger the compressor that is needed to operate the tool. Also, small compressors have to run longer to build up the amount of pressure and the quantity of air needed. Larger units can keep an adequate supply of compressed air in the tank.

Some small nailers are operated by electricity. A strong electromagnet inside the nailer drives a pin that in turn drives the nail in. These nailers work well in soft wood when small nails are required, but work less efficiently in hard wood. The advantages of these nailers are that they are relatively inexpensive and they do not require the use of an air compressor.

The newest types of nailers are driven by combustion. They use a fuel cell that is filled with compressed combustible gas. A small fan within the nailer forces the gas into a chamber, where it is ignited. The resulting combustion drives a piston down much like a piston in a gasoline engine. The advantages of these nailers are that they do not require the use of an air compressor and they can be used anywhere. They are generally more expensive than air nailers; however, that cost is offset by not needing to purchase an air compressor to run the nailer.

AIR ACCESSORIES

IMPACT CHISEL

AIR-POWERED HACKSAW

CUT-OFF SAW

DIE GRINDER

AIR-DRIVEN RATCHET

IMPACT WRENCH

ORBITAL SANDER

© 2008 The Stanley Works

FIGURE 14-5 Many types of tools use air power to perform various operations.

There are several types of nailers, which are classified according to use. The use is usually determined by the size of the nail driven or by the size of the lumber that is to be fastened. Following is a description of various types of nailers.

Brad Nailers

The smallest of the nailers is called a brad nailer (Figure 14-7). The name comes from the type of nail that is driven. A very small nail is sometimes referred to as a brad. These nailers shoot a small diameter (18 gauge) nail that can fasten thin strips of wood or molding without splitting the wood. An added benefit is that the brad nailer can be set so the head of the nail is driven to a depth just below the surface of the wood. The advantage of this adjustment is that the hole over the nail head can be filled with wood filler without using a nail set to countersink the nail head. The adjustment can be made for various degrees of wood hardness.

FIGURE 14-6 Air hoses must be coupled and uncoupled carefully. The short burst of air emitted during coupling or uncoupling can injure the operator or bystanders.

FIGURE 14-7 Brad nailers are used to fasten thin pieces of wood.

A variation of a brad nailer is one that shoots staples instead of nails. Staples are U-shaped wires that are driven in like a nail. The advantage of a staple over a nail is that a staple will hold very thin pieces of wood better than nails. Nails driven into thin strips of wood tend to pull through the wood. The disadvantage of staples is that they leave a larger hole in the wood, and the larger amount of filler needed to fill the hole is more noticeable in the finished project.

Finish Nailers

Finish nailers are generally used for putting up trim work such as baseboards, crown molding, and door facings (Figure 14-8). This is why they are called finish nailers. Typically these nailers use 15-gauge nails

FIGURE 14-8 Finish nailers are used to install trimming and molding. This is an angled finish nailer.

that are between 1 and 2½ inches in length. Finish nailers come in two types: straight and angled. The angled nailers allow the tool to be used at a steep angle, which allows easy nailing in tight areas. It also allows for toe nailing (driving a nail in at an angle).

Framing Nailers

Framing nailers drive nails that are similar to regular common nails that are up to 3½ inches long and about the size of a 16d common nail (Figure 14-9). These nailers are used during the construction of frame buildings. A building can be framed in a much shorter time using framing nailers rather than the conventional hammer.

FIGURE 14-9 Framing nailers, which can drive nails up to 3½ inches long, are widely used in construction.

The operator merely places the tip of the nailer at the correct place and pulls the trigger. This shoots a nail into the wood in a fraction of the time it would take to drive the nail with a hammer.

Safety Precautions

Always wear both eye and hearing protection when using power nailers. Make sure that the safety tip of the nailer is **ONLY** pressed against wood that you wish to nail. Severe injury or even death can result if a nail is driven into a person's body.

PORTABLE DRILLS

A portable drill is a small tool that can be easily moved to the work. The major parts of a portable drill are the:

- power cord
- handle
- motor housing
- gear chuck
- vents
- trigger switch
- trigger switch lock
- reversing switch
- chuck wrench or key

These parts are shown in Figure 14-10.

Portable drills may be classified by the chuck size. A **chuck** is the device used to hold a drill or tool bit in the machine. Common portable electric drills used in agricultural mechanics are $\frac{1}{4}$ inch, $\frac{3}{8}$ inch, and $\frac{1}{2}$ inch.

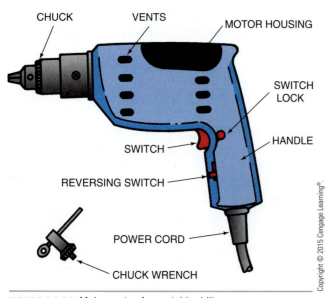

FIGURE 14-10 Major parts of a portable drill.

The $\frac{1}{4}$-inch and $\frac{3}{8}$-inch drills usually turn quite fast. The $\frac{1}{2}$-inch drills run more slowly. This type of drill has reduction gears that generate more torque than is provided by the drill motor itself.

Another way of classifying drills is by their power rating. Power drills typically draw from 2 to 5 amperes of electricity. A 115-volt motor using 5 amperes of electricity develops approximately $\frac{1}{2}$ horsepower. Such a drill is considered a powerful portable drill.

Portable drills may also be rated by duty cycle. **Duty cycle** refers to the amount of time a motor can run versus the time it needs to cool off. **Continuous duty** means a tool can be used all the time for a 6- or 8-hour day. Most drills, however, are not continuous duty cycle drills. Therefore, it is possible to overwork and overheat a power drill. When a drill gets too warm to hold comfortably, it is time to stop work and let it cool off. If overused, it is possible to burn out the motor. This same principle applies to all portable electric power tools.

Types of Drills

Drills may be single or variable speed. **Variable speed** means the speed of the motor can be controlled by the operator. Variable-speed drills are useful for special purposes such as running slowly enough to drive screwdriver bits. These bits are used to install and remove screws. A variable-speed drill will turn fast for small drill bits and more slowly, as needed, for larger ones.

All drills are reversible. **Reversible** means they will run backward as well as forward. A variable-speed, reversible drill is useful to back screws out as well as drive them in.

Another feature of some drills, called hammer drills, is their ability to hammer. The capacity to **hammer** means a drill will turn a bit and also provide a rapid striking action on internal gears that turn the bit, to provide more power when drilling in hard materials such as tempered steel or concrete (Figure 14-11).

Today, most of the drills in use are cordless (Figure 14-12). A **cordless** drill means it contains a rechargeable battery pack to drive the unit when it is not plugged into an electrical outlet. The tool uses a removable battery that can be easily recharged. The cordless drill is handy because it can be used in the field or any other location even though no electrical wiring exists. Modern cordless drills operate for many hours without recharging, and technology keeps improving so that the batteries can go even

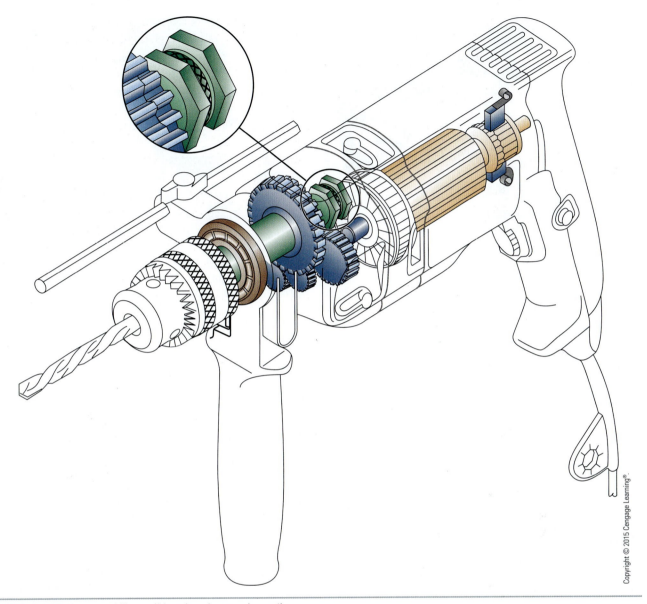

Copyright © 2015 Cengage Learning®.

FIGURE 14-11 Hammer drills are driven by a hammering action.

© Faith Kocyildir/Shutterstock.com

FIGURE 14-12 Today, most of the portable drills in use are cordless.

longer between charges. Also, the life of the batteries is now much longer than the cordless batteries of just a few years ago. A relatively new type of battery is the lithium-ion battery similar to the type of batteries used in cell phones. Of course the batteries used in drills are much larger than the ones used in cell phones but the technology is similar. These batteries are more powerful and weigh a lot less than the older nickel-cadmium batteries. The power of cordless drills is rated in volts. Common sizes are 9.6, 12, 14, 18, and 24 volts. The higher the voltage rating, the more powerful the drill.

Most other portable power tools are also now available in cordless models. Power handsaws, routers, reciprocating saws, and other types of power tools have adopted cordless technology (Figure 14-13).

FIGURE 14-13 Most portable power tools are cordless. The portable handsaw is a good example.

FIGURE 14-14 The portable power drill may be used to run nail-driving attachments, wood bits, hole-cutting saws, sanding discs, polishing bonnets, screwdriver bits, and other accessories.

Uses for Drills

Portable power drills have many uses (Figure 14-14). The most basic is the drilling of holes. Power drills can turn many different types of bits. Some power drills can turn screws in and out. Masonry drill bits permit the drilling of holes in brick, block, or stone walls. Hole saws are driven by portable electric power drills. Some people use sanding discs and polishing heads on portable electric drills. When using a portable power drill, the following procedure is recommended.

Procedure

1. Use only straight-shank bits.
2. When tightening a drill chuck, place the key into one hole and tighten the chuck securely. Place the key into a second hole and, again, tighten securely.

CAUTION!

Always remove the chuck key from the chuck after tightening a drill bit. Otherwise, the chuck key will be thrown when the drill is started.

3. Always center-punch metal to help start a bit.
4. Hold materials in a vise or other secure device.
5. Use slow-turning drills for large bits.
6. Use even pressure on the drill.

7. Ease off the pressure when the drill is breaking through the material.
8. Hold the drill so as to avoid binding the drill bit.
9. Keep operator positioned so that balance is always maintained.
10. Always remove the drill bit from the chuck when finished.
11. Store the portable power drill in its own case or in a special storage rack.

When drilling large holes, a pilot hole is used. A **pilot hole** is a small hole drilled in material to guide the center point of larger drills. By drilling a pilot hole, the bit stays exactly where planned and cuts with less power and pressure.

If a drill bit is not cutting, check the reversing switch to see if it is turned on. If the drill is turning counterclockwise, the bit will not cut. If the drill is turning clockwise and is not cutting, the drill bit is dull and must be sharpened.

PORTABLE SANDERS

Three types of portable sanders are used in agricultural mechanics. They are the portable belt sander, portable disc sander, and portable finishing sander. Belt sanders and disc sanders are used for coarse sanding; finishing sanders are used for the last operation before applying finishes.

Power sanders do the same work as hand sanding. However, power sanders remove wood or other materials faster and easier than hand sanding. It is the speed that makes power tools desirable for sanding or grinding. A person experienced in using sanders can create a very smooth finish on wood or metal.

Effective Sanding

When sanding, it is important to sand with the grain of the wood for the fine work. However, crossgrain sanding may be useful.

- if boards are uneven
- if extremely rough boards are encountered
- if very difficult finishes must be removed

Generally, coarse sandpaper is used first, then medium sandpaper, and finally, fine or very fine sandpaper is used to complete the job.

Sometimes, workers choose to use a belt or disc sander for the rapid removal of material. A **finishing sander** or hand sanding is then used to complete the job. The finishing sander moves randomly. Therefore, the sanding can always be with the grain of the wood or across the grain.

Almost all finishing sanders are **orbital sanders**. The orbital sander moves in a random circular pattern. This results in a faster cut, but still leaves a fine finish.

Safe Use of Portable Belt Sanders

The **belt sander** is a tool with a moving sanding belt (Figure 14-15). Major parts include the:

- motor housing
- handles
- belt
- belt rollers
- belt adjustment
- trigger switch
- switch lock
- power cord
- dust bag

Each part should be learned by the student.

The belt sander is often used in agricultural mechanics shops. It is a relatively safe tool to use, but care must be taken as with all power tools. When using the belt sander, the work should be secured well and both hands used on the machine to manage it very carefully (Figure 14-16).

When using belt sanders, the following general procedures are recommended for safe use and effective operation.

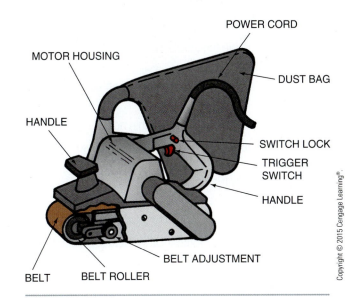

FIGURE 14-15 Major parts of a portable belt sander.

FIGURE 14-16 Both hands are needed to control a belt sander.

Copyright © 2015 Cengage Learning®

Procedure

1. Wear suitable face protection and protective clothing.

2. Check the power cord and extension cords for safety.

3. Install a sanding belt of suitable coarseness.

4. Lay the sander on its side when not in use. This prevents the sander from running off the bench if accidentally turned on.

5. Be sure the dust bag is empty or nearly so before starting to sand.

6. Always start the machine while holding it slightly above the material.

7. Keep the power cord out of the way of the belt.

8. After turning on the sander, touch the work with the front part of the belt first, then slowly settle the rest of the belt down onto the work.

9. Operate the machine with two hands at all times.

10. Sand with the grain. Move the machine from one end of the board to the other in a straight path; then move it slightly sideways and draw the machine back over new area. Gradually work across the board by slightly overlapping the forward and backward passes.

11. Keep the machine in motion. If permitted to sand in one spot, it will cut a depression in the wood.

12. The final movement is to lift the machine off the work while it is still running.

13. Examine the work carefully. If necessary, resand in order to create a perfectly level surface that is smooth.

14. Install a fine sanding belt and resand. This resanding leaves the work in its smoothest possible form using the belt sander.

15. Use a finishing sander or hand sand to obtain the degree of fineness desired.

Safe Use of Portable Disc Sanders and Grinders

Some portable tools may be used only as sanders or only as grinders. Others are designed to be used as either (Figure 14-17). By simply changing the sanding disc to a grinding wheel, many disc sanders become grinders. The manufacturer's instructions should be followed regarding any single tool. It is important to

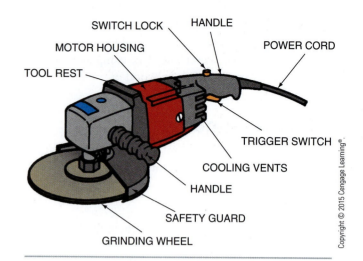

FIGURE 14-17 Major parts of a portable sander/grinder.

Copyright © 2015 Cengage Learning®.

use the proper guards and wear protection with either the sanding disc, grinding wheel, or wire brush.

Some important parts of portable sanders and/or grinders are the:

- motor housing
- handles
- power cord
- switch
- switch lock
- cooling vents
- wheel
- spindle
- safety guard
- tool rest

A **disc sander** is a tool with sanding materials, called grit, on a revolving plate. Disc sanders can be used for sanding wood or metal. However, only certain types of sanding discs are suitable for metal. Discs made from aluminum oxide may be used for sanding wood or metal. Discs made from flint paper are usable only for wood sanding.

Grinders have rigid grinding wheels instead of flexible discs. They cut metal only. Grinders may be used to shape metal, grind down welds, and remove metal, as needed. Portable grinders may also be used to turn wire brushes used for cleaning metal. In agricultural mechanics, wire brushes are often used to remove rust and scaling paint.

Some special safety precautions are in order when using a grinding wheel. Some of these are as follows.

- Wear a face shield.
- Always check the grinding wheel for cracks or damage before use. Do not use a wheel that shows any sign of damage.

- Be sure to use wheels that are designed for the machine.
- Tighten the wheel securely and carefully.
- Never use a grinding wheel that is less than one-half of its original diameter.
- When preparing to grind small pieces, secure them in a vise, if possible.
- Do not grind metal in areas where combustible gases or materials are present.
- Hold the machine with both hands at all times.
- Do not discharge sparks against persons, clothing, or other combustible materials.

The procedure for using a sander or grinder is as follows.

Procedure

1. Select the correct sanding disc, grinding wheel, or wire brush for the job.

CAUTION!

Be sure the wheel or disc is rated to turn at speeds higher than that of the machine.

2. Install the appropriate guard for the job being done.
3. Wear appropriate face protection and protective clothing.
4. Be sure the work is properly secured.
5. Keep the power cord out of the way of the machine.
6. Grip the machine firmly with both hands, and turn on the switch.
7. Settle the turning sanding disc, grinding wheel, or wire brush onto the work slowly.
8. Touch the work gently with the wheel. This is to avoid the wheel catching the work and throwing metal particles toward the operator, bystanders, or flammable materials.
9. Do not apply pressure to the machine. The weight of the machine is generally sufficient to sand, grind, or brush properly. Keep the wheel clean and sharp. Keep a fresh abrasive disc on the machine to grind or sand quickly and efficiently.

10. After turning the switch off, do not lay the machine down until it has completely stopped.
11. Most machines have a rest or a flat spot to rest on. Do not lay the machine down on its disc or wheel.
12. Remove the grinding wheel or sanding disc and store the machine properly after use.

Safe Use of Finishing Sanders

The finishing sander is a tool with a small sanding pad driven in a random forward-backward or circular pattern (Figure 14-18). An orbital sander is a finishing sander that moves in a random circular pattern. The finishing sander is generally the last power tool used on a project. It cuts slowly but gives the work a very smooth finish.

Important parts of the finishing sander are the:

- motor housing
- handles
- switch
- switch lock
- power cord
- paper clamps
- pad
- sandpaper

These parts are shown in Figure 14-19.

The recommended procedure for using a finishing sander is as follows.

FIGURE 14-18 The random orbital action of a finishing sander permits sanding across the grain.

Copyright © 2015 Cengage Learning®.

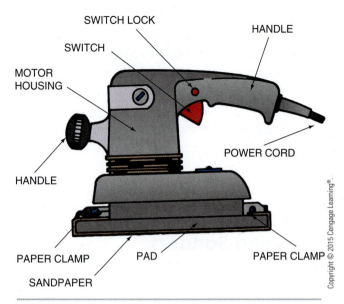

FIGURE 14-19 Major parts of a finishing sander.

Procedure

1. Use appropriate face and body protection.

2. Check the sander to determine if it is a straight-line or orbital type. Some sanders can be set to operate in either mode. Remember, straight sanding is the smoothest sanding.

3. When starting with rough work, use coarse sandpaper for the first sanding. Switch to medium sandpaper, and then to fine sandpaper for the final sanding.

4. To prepare sandpaper, cut the paper to the appropriate size so it fits the machine. Precut sandpaper can also be purchased.

5. To install sandpaper, loosen the clamps on the pad, insert the paper, and close the clamps. Some of the newer types of sandpaper use a system called "hook and loop," in which the paper sticks to the sanding pad.

6. Proceed to sand applying only slight pressure on the sander. The sander should be in constant movement over the work.

7. Remove the dust from the work frequently, to keep the material being removed from clogging the sandpaper.

8. Store the machine properly when finished.

PORTABLE SAWS

Portable saws are useful for carpentry projects and repair activities on the farm and in other agricultural settings. Their use will speed jobs and permit the operator to move tools to the job quickly and easily.

Portable saws include the saber saw, reciprocating saw, and circular saw. The action of saber and reciprocating saws is up and down or back and forth. They are compact, portable, and useful for cutting curves in plywood, paneling, drywall, and other sheet materials. The circular saw is generally known simply as a power handsaw. The **power handsaw** has a circular blade and is used extensively for cutoff work. All types of portable power saws are now available in cordless models. Another innovation is the use of a laser beam that is used to guide a power handsaw. The saw projects a red beam that shows where the saw will travel. By aligning the laser beam with a mark across the board, a straight cut can be made.

Safe Use of Saber Saws

The **saber saw** is also referred to as a **bayonet saw**. It is used primarily to cut curves or holes in wood, metal, cardboard, and similar materials (Figure 14-20). Saber

FIGURE 14-20 The saber saw is used to make curved and irregular cuts.

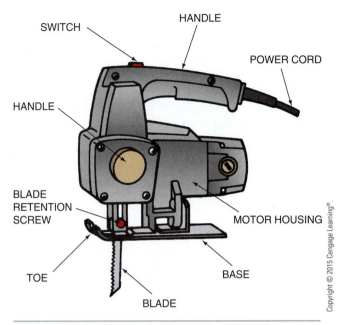

SWITCH
HANDLE
POWER CORD
HANDLE
BLADE RETENTION SCREW
MOTOR HOUSING
TOE
BASE
BLADE

Copyright © 2015 Cengage Learning®.

FIGURE 14-21 Major parts of a saber saw.

saws are compact. The saw blade's motion is up and down rather than circular and continuous. This feature makes saber saws less hazardous than circular saws; they are dangerous, nonetheless, if not handled properly. The major parts of the saber saw are the:

- motor housing
- base
- blade
- toe
- blade retention screw
- handles
- switch
- power cord

These parts are shown in Figure 14-21. A safe procedure for using the portable saber saw follows.

Procedure

1. Wear proper eye protection and protective clothing.
2. Select the correct blade for the job. Narrow blades permit shorter turns than wider blades but break more easily. Blades are available for cutting wood, metal, cardboard, and other material. Follow the manufacturer's instructions for the selection and use of blades.

3. Insert the blade and tighten the blade retention screw firmly. Do not apply so much pressure that the threads on the screw are stripped.
4. Adjust the base for the work being done. Some saws have bases that tilt to make angle cuts. Almost all saws have a depth control.
5. Carefully secure all material being cut.
6. Start the cut at the edge of the board. When making inside cuts, bore a hole inside the circle, insert the blade, and proceed.
7. A **blind cut** is a cut made by piercing a hole with the saw blade. To start blind cuts, place the toe of the base against the material, with the blade free to start without striking the material. Start the saw, and carefully roll the saw backward to permit the blade to saw a hole gradually into the material.
8. Use firm pressure and uniform forward movement when operating the saw.
9. Make turns slowly and carefully and give the saw an opportunity to cut as it is directed into the curve.
10. Near the end of the cut, take extra care to support the work. Also, reduce the rate of speed and pressure on the saw.
11. When finished, remove the blade and lightly retighten the blade retention screw to prevent accidental loss of the screw.
12. Store the tool properly.

Use of Reciprocating Saws

The **reciprocating saw** is held and operated much like a portable drill. The action is at the end rather than underneath the tool (Figure 14-22). It is used to make rough cuts where a circular saw cannot be used. The main parts of the reciprocating saw are the:

- motor housing
- blade retention screw
- blade
- shoe
- vents
- handles
- power cord
- trigger switch
- switch lock

FIGURE 14-22 The reciprocating saw is used for making rough cuts.

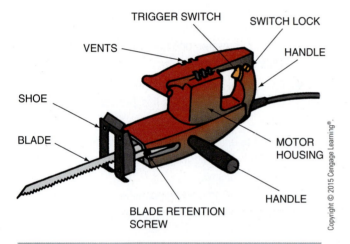

TRIGGER SWITCH
SWITCH LOCK
VENTS
HANDLE
SHOE
BLADE
MOTOR HOUSING
HANDLE
BLADE RETENTION SCREW

Copyright © 2015 Cengage Learning®

FIGURE 14-23 Major parts of a reciprocating saw.

These parts are shown in Figure 14-23. Safe operating procedures for using the reciprocal saw include the following.

Procedure

1. Wear eye protection and protective clothing.
2. Select the correct blade for the job. Blades are available for cutting wood, plastic, metal, and other materials. Follow the manufacturer's recommendations for selecting blades for the reciprocal saw.
3. Select the speed for the job. Some saws have high speeds for woodworking and low speeds for metalworking.

4. To make a blind or plunge cut, rest the saw on its shoe and gradually tilt the blade forward into the work.
5. Handle the saw as needed to make cuts in a fashion similar to the saber saw.
6. Be careful not to bind, pinch, or crowd the blade when sawing.
7. Operate the saw with the shoe against the work at all times.
8. Remove the blade for storage.
9. Keep extra blades on hand for a variety of jobs.
10. Be especially careful not to wear loose-fitting clothing. Observe other general safety precautions when using the reciprocating saw.
11. Always use the manufacturer's recommendations whenever there are questions of use and safety.

Safe Use of Circular Saws

The **portable circular saw**, also called a power handsaw, is a lightweight, motor-driven, round-bladed saw. It is perhaps the most popular saw used by people doing woodworking in agricultural mechanics. This saw is useful for many building and repair jobs on the farm and in other agricultural settings (Figure 14-24). The major parts of the portable circular saw are the:

- motor housing
- handles
- power cord
- trigger switch
- switch lock
- guard
- guard lift lever
- retractable guard
- blade
- spindle
- base
- angle adjustment lock
- angle scale

These parts are shown in Figure 14-25.

The portable circular saw can be a very useful tool. However, its high speed and tendency to kick back make it a very dangerous tool. The user is cautioned to pay attention to detail and hold the tool very carefully when using it. A safe procedure for using the power handsaw follows.

FIGURE 14-24 The circular saw is used extensively for cutoff work. Special blades are available for cutting concrete, brick, and stone.

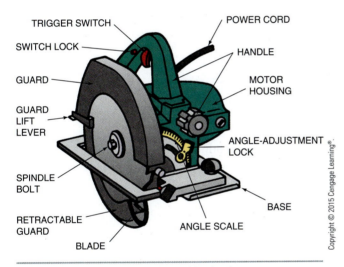

TRIGGER SWITCH
SWITCH LOCK
GUARD
GUARD LIFT LEVER
SPINDLE BOLT
RETRACTABLE GUARD
BLADE
POWER CORD
HANDLE
MOTOR HOUSING
ANGLE-ADJUSTMENT LOCK
BASE
ANGLE SCALE

FIGURE 14-25 Major parts of a portable circular saw.

Procedure

1. Always use a face shield and wear protective clothing.

2. Choose an appropriate blade for the work being done. Special blades include crosscut, rip, combination, hollow ground, and safety blades.

3. Install the blade with care. Use the securing washers in the appropriate sequence. Use the correct wrench to tighten the spindle bolt. If a **bushing** is used for the blade to make it fit the spindle, be sure the bushing is in place. A bushing is a sleeve that allows a blade with a larger mounting hole to fit on a smaller shaft. Install the blade so that the bottom saw teeth point toward the front of the saw.

4. Support the work with sawhorses or other solid materials to avoid pinching or binding of the blade when sawing.

CAUTION!

Do not attempt to saw a board between two sawhorses. The board should be placed so the waste piece can drop off without binding the saw.

5. Adjust the blade for depth so that only ¼ inch, or the length of a saw tooth, extends below the material.

6. Hold the saw securely with both hands at all times.

7. Start the saw while it is positioned near the work. Be careful that the blade is not touching another object or clothing.

8. Move the saw steadily into the work. If the saw stalls, back it up to clear its teeth from the material and correct the cause of the stalling.

9. When sawing, watch the line ahead of the saw and move the saw so that its reference mark stays on the line.

10. Near the end of the cut, reduce the pressure and release the switch as the cut is being finished.

11. When ripping, use the ripping guide if provided on the saw and if it is long enough for the cut being made.

12. For bevel cuts, adjust the angle of the base to the angle of the cut desired. Handle the saw the same as for other crosscuts or rip cuts.

13. When making a pocket cut, place the front of the saw base on the board, turn on the saw, and gradually lower the saw into the work.

14. When cuts are finished, be sure the saw blade stops completely before placing the saw at rest.

15. When the saw is not in use, unplug it to avoid accidental starting. Later model saws have a special switch or locking device that must be pushed before the switch will engage.

(continued)

Procedure, *continued*

16. If the blade is gummy, remove the gum with alcohol or another solvent. If the blade is becoming dull, sharpen or replace it. If the retractable guard does not move freely and cover the blade at all times except when in use, correct the problem.

17. Properly store the tool.

POWER ROUTERS

Power routers may be needed for specialized jobs in agricultural mechanics. A **router** is a power tool used to cut grooves and ornamental shapes on faces and edges of wood and other soft materials. It may be useful in agricultural mechanics for making rabbet, dado, and dovetail joints, installing butt hinges, creating decorative edges on wood, making signs, and other decorative work. Routers can range in size from ½ horsepower to 3 horsepower and can turn bits at very high speeds—up to 25,000 rpm (Figure 14-26). Bits for routers come in a wide variety of shapes and sizes. Figure 14-27 shows some of the types of bits used in routers. Bits usually come in ¼-inch or ½-inch shanks. The smaller, ¼-inch bits are usually used in smaller routers. The larger, ½-inch shank bits are used in routers with more horsepower and can give better cuts because they are

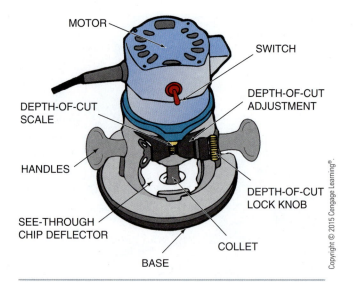

FIGURE 14-26 Major parts of a power router.

heavier and do not vibrate as much. The major parts of a router are the:

- base
- see-through chip deflector
- depth-of-cut lock
- depth indicator
- handles
- motor
- switch
- depth-of-cut adjuster
- collet

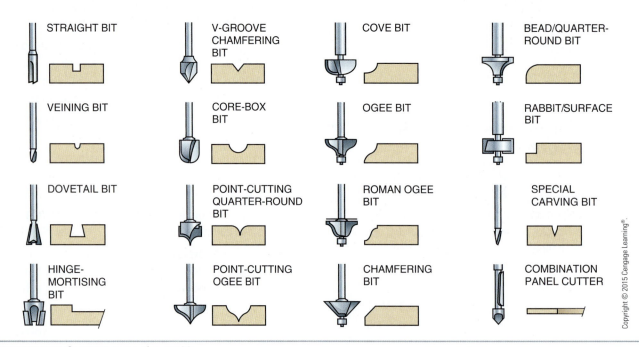

FIGURE 14-27 Common types of router bits.

Safe Use of Routers

The router has a smooth base that is moved across the material by the operator as the protruding bit turns and creates a groove or cut. The groove or cut is determined by the type, shape, and size of the bit. Bits are made of high-speed steel and may be carbide-tipped for extra resistance to dulling. Bits that are dull, rusted, or covered with gum from the wood are dangerous and should not be used. Clean, sharp bits will provide clean, even cuts if the router is held firmly to the wood. Carbide-tipped bits are sharper and last longer than regular bits. To create a straight cut or a cut along desired curved lines, some kind of guide or jig is generally needed. Some guides may come with the router (Figure 14-28). Others may be purchased from suppliers or devised by the craftsperson as needed. Guides are available to help do straight line, circular, or contour routing.

Some recommended procedures for the safe use of routers follow.

Procedure

1. Thoroughly read the operator's manual, and follow its instructions.
2. Clothe yourself with safety goggles and protective clothing.
3. Secure the work firmly before starting.
4. Select a sharp bit of the type needed for the job.
5. With the router unplugged, install the bit according to the manufacturer's instructions.
6. Install appropriate guides or jigs.
7. Set the depth carefully. Do not set deeper than the capacity of the bit.
8. Set up a piece of scrap wood for a trial cut.
9. Plug the router into a safe extension cord.
10. Set the router on its base with the bit clear of the wood, near the starting point, and turn it on.

(continued)

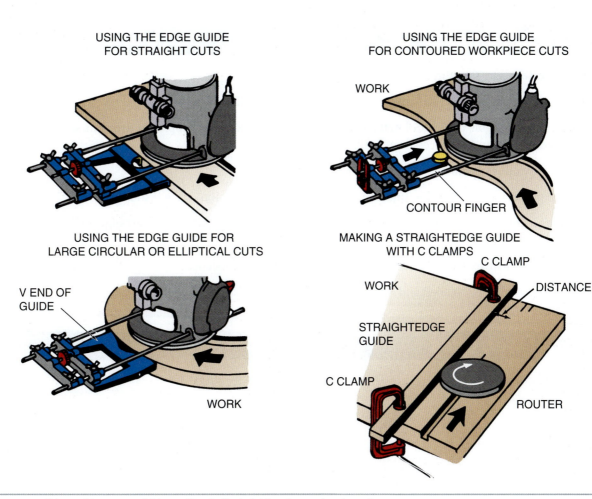

USING THE EDGE GUIDE FOR STRAIGHT CUTS

USING THE EDGE GUIDE FOR CONTOURED WORKPIECE CUTS

WORK

CONTOUR FINGER

USING THE EDGE GUIDE FOR LARGE CIRCULAR OR ELLIPTICAL CUTS

V END OF GUIDE

WORK

MAKING A STRAIGHTEDGE GUIDE WITH C CLAMPS

WORK

C CLAMP

DISTANCE

STRAIGHTEDGE GUIDE

C CLAMP

ROUTER

Copyright © 2015 Cengage Learning®.

FIGURE 14-28 Edge guides are useful for making both straight and curved router cuts.

Procedure, *continued*

11. Move the router slowly into the wood, and adjust the pace to produce tiny, cleanly severed chips. Move slowly enough into the wood that the motor maintains high speed, yet fast enough to keep cutting without heating the bit or glazing the wood.

12. Follow the guide or jig carefully.

13. Avoid putting side thrust on the bit. Small bits are easily broken and others easily ruined by forcing the bit rather than simply moving it into the wood and letting it cut its way as it progresses.

14. Finish the cut carefully, and turn off the router.

15. Place the router in an appropriate stand so it cannot roll and the bit cannot be touched. The bit turns at 20,000 rpm and cannot be totally guarded.

16. Examine the cut and adjust the machine or technique as needed.

17. Repeat the process while performing the work on the project.

18. Unplug, clean, and store the router and accessories properly.

SUMMARY

Every effort must be made to maintain a safe work environment when using power tools. A moment of carelessness or neglect can permit or cause an accident, which can lead to a lifetime of regret. Power tools are built to save labor and perform tasks quickly and accurately, and they are real assets to responsible users.

Student Activities

1. Define the Terms to Know in this unit.

2. Identify and correctly spell the names and parts of all power tools described in this unit.

3. Examine each portable power tool in the school agricultural mechanics shop. Determine its size classification, general condition, and the accessories available for use.

4. Study the operator's manual for each power tool in the shop.

5. Use each power tool described in this unit, under the close supervision of the instructor.

6. Examine all portable power tools you have at home. Work with your parent(s) to arrange for any repairs needed on any of these tools.

7. Obtain a three-wire extension cord for the power tools at home, if needed.

8. Arrange to have all bits, blades, and other accessories at home sharpened as needed.

Relevant Web Sites

Woodworkers' Tools Works
woodworkerstoolworks.com

Saw Dust Making 101: A Guide for the Beginning Woodworker
www.sawdustmaking.com/

U.S. Department of Energy, Office of Science, Advanced Photon Source, APS Guideline for Hand Tool and Portable Power Tool Usage
https://www.aps.anl.gov/Safety-and-Training/Training/For-Users-and-Employees/Required-Training-for-Users

Self-Evaluation

A. Multiple Choice. Select the best answer.

1. A power tool that is double insulated has parts internally insulated in two locations or has a
   a. plastic motor housing
   b. three-wire cord
   c. special carrying case
   d. continuous duty cycle

2. Turning the first wrap of a power cord tightly around a power tool can cause the
   a. housing to break
   b. tool to use more electricity
   c. warranty to be voided
   d. insulation in the cord to break

3. When using power tools, electrical shock hazards may be reduced by wearing
   a. gloves
   b. rubber-soled shoes
   c. safety glasses
   d. coveralls

4. Compressed air is especially dangerous when
   a. coupling and uncoupling hoses
   b. room temperatures are high
   c. portable tools are used
   d. air lines are long

5. Most portable power drills have chuck sizes of
   a. ⅛, ¼, or ½ inch
   b. ⅛, ³⁄₁₆, or ½ inch
   c. ¼, ⅜, or ½ inch
   d. ¼, ⅜, or ⁹⁄₁₆ inch

6. A variable-speed tool is one that will run
   a. a long time without damage
   a. backward as well as forward
   a. in hot locations
   a. at different speeds

7. A feature that makes some drills especially good for drilling masonry materials is
   a. reversible action
   b. variable speed
   c. a capacity to hammer
   d. a continuous duty cycle

8. When drilling metal with a power drill
   a. use a center punch
   b. make a pilot hole
   c. use a cordless drill if appropriate
   d. all of these

9. Routers are handy for
   a. crosscutting boards
   b. making holes
   c. setting screws
   d. creating grooves and decorative edges

10. A router bit rotates at
    a. 100 rpm
    b. 1,000 rpm
    c. 20,000 rpm
    d. 200,000 rpm

B. Matching. Select the word or phrase in column I that best matches the word or phrase in column II.

Column I

1. circular pattern
2. finishing sandpaper
3. flint paper
4. aluminum oxide
5. saber saw
6. reciprocating saw
7. power handsaw
8. metal housing
9. bushing
10. cordless
11. vent holes
12. crosscut

Column II

a. good for sanding only wood
b. good for sanding metal
c. three-wire, grounded cord
d. type of saw blade
e. tiger saw
f. fine or very fine
g. ventilation
h. battery pack
i. bayonet saw
j. circular saw
k. blade insert
l. orbital sander

C. Completion. Fill in the blanks with the word or words that make the following statements correct.

1. When using portable power tools, it is important to have the work well _____.

2. Another word for sawhorse is _____.

3. Under no circumstance should one use power hand tools without protecting the eyes with safety _____ or _____.

4. A three-wire power cord and plug with a ground prong is used on power tools with _____ bodies.

5. The type of saw with the blade extending from its end is called a _____ saw.

6. Sanding belts should run _____ the grain of the wood.

7. A good accessory to use on portable grinders to remove rust from metal is the _____.

8. Grinding wheels that are _____ should be discarded.

9. When sawing large panels, _____ are used on trestles to provide extra support to the material.

10. The portable power tool that has a retractable guard is the _____ _____.

D. Identification. Identify the parts of the following power tools.

1. Portable drill

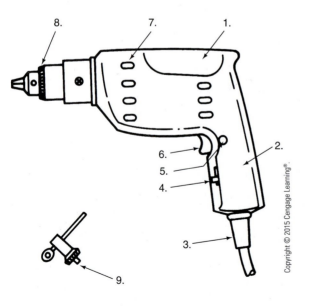

Copyright © 2015 Cengage Learning®.

2. Belt sander

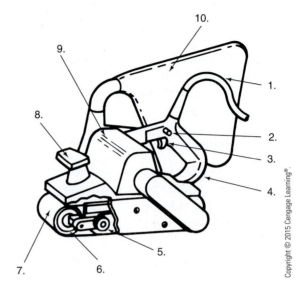

Copyright © 2015 Cengage Learning®.

3. Portable grinder

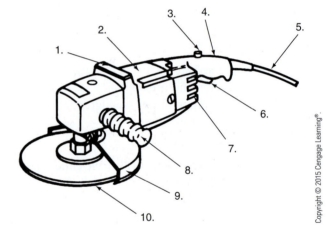

Copyright © 2015 Cengage Learning®.

4. Finishing sander

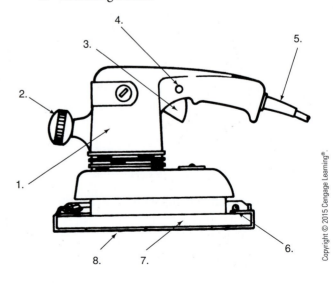

Copyright © 2015 Cengage Learning®.

5. Saber saw

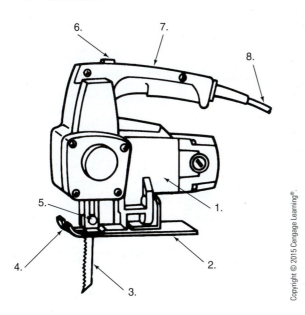

Copyright © 2015 Cengage Learning®.

7. Circular saw

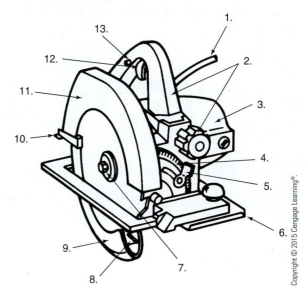

Copyright © 2015 Cengage Learning®.

6. Reciprocating saw

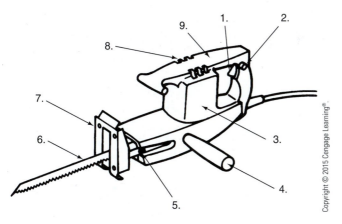

Copyright © 2015 Cengage Learning®.

8. Power router

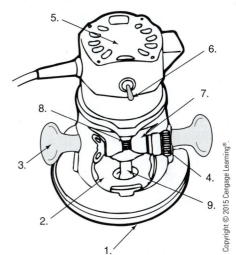

Copyright © 2015 Cengage Learning®.

UNIT 15

Woodworking with Power Machines

Objective

To use stationary power woodworking machines in a safe manner.

Competencies to be developed

After studying this unit, you should be able to:

- State basic procedures for using stationary power woodworking machines.
- Identify and properly spell major parts of specified machines.
- Operate a band saw.
- Operate a jigsaw.
- Operate a table saw.
- Operate a radial-arm saw.
- Operate a jointer.
- Operate a planer.
- Operate a sander.

Materials List

- Band saw
- Jigsaw
- Radial-arm saw
- Table saw
- Jointer
- Planer
- Bench brush and/or shop vacuum
- Roller stand or helper
- One piece full dimension (rough) lumber, 2" x 4" x 14"

Terms to Know

- power machine
- stationary
- band saw
- blade guide
- tilting table
- miter gauge
- rip fence
- jigsaw
- scroll saw
- reciprocal
- pulley
- table saw
- tilting arbor
- bench top saw
- contractor's saw
- cabinet saw
- parallel
- push stick
- dado heads
- radial-arm saw
- pivot
- miter saw
- jointer
- dress
- rough lumber

NOTE

Before proceeding with this unit, review the material in Unit 4 on personal safety.

The use of electricity has improved the quality of life for all who use it wisely. Electricity provides a way of moving energy from place to place. Electrical energy is generated from water power, wind power, or fuels and is then sent over long distances by power lines to be received in homes, shops, and industries, where it is put to use. In the agricultural mechanics shop, electrical energy is converted by relatively small motors to power in other forms used by shop machines. It is the power that makes machines hazardous to the careless operator. However, the careful operator uses the energy to get work done safely, quickly, and with ease.

SAFETY WITH POWER MACHINES

A **power machine** is a tool driven by an electric motor, hydraulics, air, gas engine, or some force other than, or in addition to, human power. Some tools such as metal shears may use levers and cams to increase human power. Such machines develop so much force that they, too, should be regarded as power machines.

Large power tools are stationary and should be placed in permanent locations in the shop. The word **stationary** means having a fixed position.

To reduce the likelihood of injury when working with stationary power machines and to ensure efficient use of the machines, the following are recommended.

- Plan the location of each machine carefully.
- Firmly anchor each machine to the floor.
- Have a licensed electrician provide electrical hookups if needed.
- Use a stripe or narrow line to mark the safety zone around each machine.
- Follow the manufacturer's recommendations for the installation, use, adjustment, and repair of each machine.
- Keep guards and shields in place on each machine at all times (Figure 15-1).
- Keep blades, knives, and bits sharp.

FIGURE 15-1 Operating a saw without the proper guards in place is extremely dangerous.

Safety Precautions

Safety rules are important and should be followed when using all shop machines. Some important precautions are:

1. Wear goggles and/or face shield. When using some larger machines, such as the planer, hearing protection is also a must (Figure 15-2).
2. Wear protective clothing that is not loose or baggy.
3. Walk—do not run around machines.
4. Only the operator is to be in the safety zone around a machine.
5. Do not use a machine without the instructor's permission.
6. Do not use a machine unless it is in good working order.
7. Perform only the procedures on a machine for which you have had instruction.
8. Do all operations slowly and cautiously.

FIGURE 15-2 In the shop, proper eye protection must be worn at all times. When running some machines such as a planer, hearing protection must be worn.

9. Use a push stick to help push or guide small pieces.
10. Get help with large pieces of stock.
11. If the machine is worked too hard, an overload protector should stop the motor. If this happens, notify the instructor, and the instructor will correct the problem.
12. Turn off the machine before leaving it.
13. Unplug the machine or switch off the circuit breaker when changing blades or doing repairs.
14. Clean woodworking machines with a brush or vacuum cleaner. Never clean the machines directly with the hand.

NOTE

Before continuing with this unit, learn these safety precautions.

BAND SAW

The **band saw** is a power tool with saw teeth on a continuous blade or band (Figure 15-3). The band saw will cut straight or curved lines and different kinds of materials. It is fast cutting and versatile when the proper blade and speed are used. Careful operation is absolutely necessary.

The blade of a band saw is flexible and is stretched over flat wheels. It is thin and narrow enough to allow the operator to cut curves (Figure 15-4). The face of each wheel is covered with flat rubber called a tire. The rubber protects the teeth of the saw blade. Generally, the

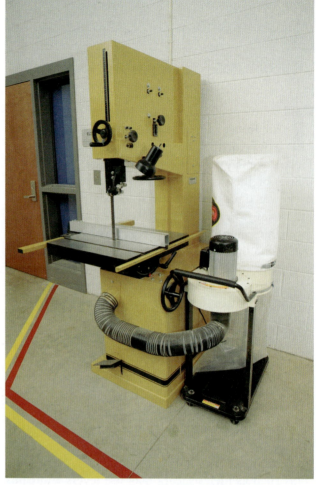

FIGURE 15-3 The band saw owes its name to a thin, continuous band of steel with teeth all along its edge that do the cutting.

wheel diameter determines the size of the saw. A band saw with a 16-inch wheel is classified as a 16-inch band saw. This means the saw can be used to cut to the center of stock up to nearly 32 inches wide.

FIGURE 15-4 The blade of the band saw is narrow enough to allow the operator to cut curves.

A **blade guide**, with carefully adjusted rollers, supports the blade. The blade guide and other parts of the guide and the table assembly of a band saw are shown in Figure 15-5. The rollers provide blade support as the operator pushes material into the teeth of the blade. In addition, material may be turned gradually to make curved cuts. The guides keep the blade from bending, breaking, or moving off the wheels during normal use. Blades are available in various types and widths. The more narrow the blade, the shorter the curve that may be cut. Blades used in school shops are generally ¼ inch to ½ inch wide.

There are several types of band saw blades (Figure 15-6). They can be classified by pitch and tooth type. The pitch refers to the number of teeth per inch of saw blade. For most 14- to 16-inch band saws, the pitch varies from 6 to 14 teeth per inch of blade. Generally,

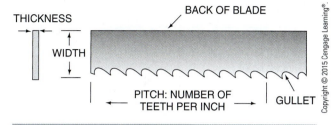

FIGURE 15-6 Band saw blades can be classified by the number of teeth per inch on the blade.

the more teeth per inch of blade, the smoother the cut; the fewer teeth per inch of blade, the rougher the cut. Thin wood (less than ¼ inch thick) should be cut with a blade having more teeth per inch. Blades with larger spaces between the teeth tend to hang on the thinner wood. Blades with fewer teeth per inch cut faster.

Tooth type refers to the shape of the tooth. Regular or standard tooth shape is used for cutting mild steel. The teeth are set straight and have a 0-degree rake angle, which is the angle the teeth are set from a straight line on the blade. Hook-tooth blades have teeth with a 10-degree rake angle. They are used to saw thicker wood where a faster cut is preferred; however, more saw marks are left on the edge of the wood than with other types of blades. Skip-tooth blades have a 0-degree rake angle but also have a deep gullet. The gullet refers to the distance from the tip of the tooth to the blade body. These blades are used to cut wood and soft metals such as aluminum (Figure 15-7).

Tooth set refers to the direction and pattern a band saw blade tilts. The two most common sets used for cutting wood are the raker set in which one tooth tilts left, the next tooth right, and the next tooth straight; and the alternate set in which one tooth tilts left and the next tilts right (Figure 15-8).

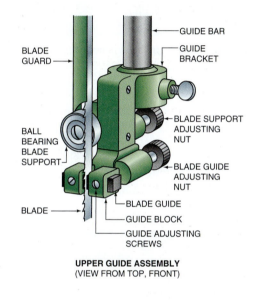

UPPER GUIDE ASSEMBLY
(VIEW FROM TOP, FRONT)

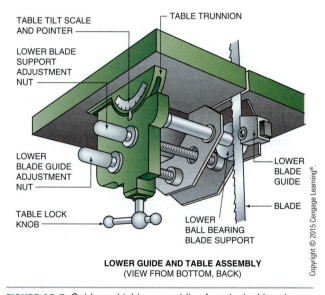

LOWER GUIDE AND TABLE ASSEMBLY
(VIEW FROM BOTTOM, BACK)

FIGURE 15-5 Guide and table assemblies for a typical band saw.

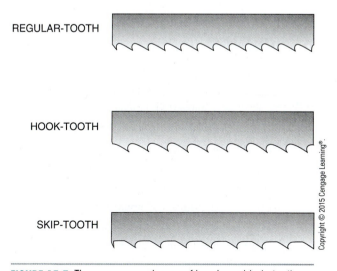

FIGURE 15-7 Three common shapes of band saw blade teeth.

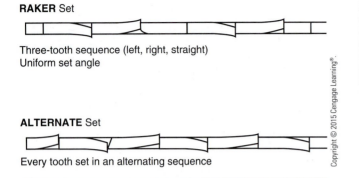

RAKER Set

Three-tooth sequence (left, right, straight)
Uniform set angle

ALTERNATE Set

Every tooth set in an alternating sequence

Copyright © 2015 Cengage Learning®

FIGURE 15-8 The term *set* refers to the direction and angle of the teeth in the blade.

Most band saws have a **tilting table**. The table can be set at various angles so that cuts of 45 to 90 degrees may be made.

Most band saw tables also are equipped with a miter gauge. A **miter gauge** is an adjustable, sliding device to guide stock into a saw at the desired angle. Some band saws have a rip fence. A **rip fence** is a guide that helps keep work in a straight line with the saw blade. It can be adjusted to a distance as close to, or as far from, the blade as the table size permits.

Band saws may have a speed control. This permits the saw to be used for cutting metal as well as wood. A slow-running saw with an appropriate blade is needed to cut metal.

Safe Operation of Band Saws

Band saws are adaptable to many cutting jobs. Therefore, specific instructions are needed for various operations. However, the following procedure is recommended for use of all band saws.

Procedure

1. Obtain the instructor's permission to use the band saw.
2. Put on a face shield and protective clothing.
3. Check to see if the saw is equipped with a suitable blade for the job.
4. Be sure the blade is tight and the guide is properly adjusted.
5. Move the upper guide assembly down until it is within ⅛ inch of the top of the stock.
6. Place the miter gauge or the rip fence in position, and adjust as needed.

CAUTION!

Never use both the miter gauge and the rip fence at one time, as the stock will bind and create a serious hazard.

7. Clear the table and blade of all materials. Keep hands out of the work area. Switch the machine on and off again.
8. If the blade seems to be tracking correctly, stand back from the blade and turn the machine on and let it run.
9. Slowly and carefully push the stock to be cut into the teeth of the blade.

CAUTION!

Keep your fingers and hands away from the front of the blade.

10. When straight cuts are made, move the stock in a straight line.
11. When curved cuts are made, rotate the stock slowly as the saw cuts.
12. Try to arrange all cuts so you can cut all the way into and out of the stock.

CAUTION!

If you must back out of a cut, turn the machine off to do so.

13. Make sure the blade has stopped before leaving the machine.

JIGSAW

The **jigsaw** is also known as a **scroll saw** and is designed for sawing curves (Figure 15-9). It is a poor tool for straight cutting. The saw cuts by means of the reciprocal action of the blade. **Reciprocal** means back and forth. The reciprocal action makes it a relatively safe tool if proper procedures are followed. However, the operator may receive serious injuries if the machine is not properly adjusted and if the work is not moved carefully into the blade. The major parts of a jigsaw are shown in Figure 15-10. Jigsaws may be mounted on a bench or on a stand.

FIGURE 15-9 The jigsaw or scroll saw can cut very short curves.

FIGURE 15-11 The narrower the jigsaw blade, the shorter the turn the blade can make.

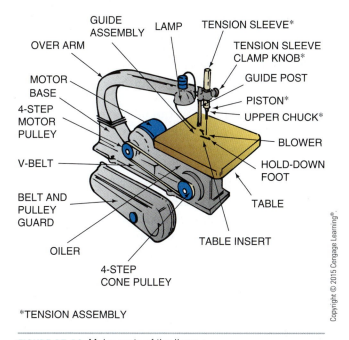

*TENSION ASSEMBLY

FIGURE 15-10 Major parts of the jigsaw.

Blades for the jigsaw are generally 6½ inches long. They resemble a coping saw blade except that they do not have pins in the ends. Blades range in thickness from 0.05 inch to 0.25 inch (¼ inch). When buying blades, specifications are stated by length, thickness, and width. For example, a 6½" × 0.020" × 0.110" 20T blade is 6½ inches long, 0.02 inch thick, and 0.11 inch wide, and has 20 teeth per inch. The number of teeth per inch indicates the size of the teeth and the speed at which the blade will cut. The narrower the blade, the shorter the turn the blade can make (Figure 15-11).

The tension assembly of a jigsaw consists of a sliding sleeve, a spring, and a pistonlike rod with a chuck. This upper chuck holds the upper end of the blade; a lower chuck holds the other end. The lower chuck is driven by a crank mechanism in the base. The crank pulls the blade down and a spring pulls it up. Blades must be installed with their teeth pointing down toward the table.

The jigsaw's hold-down foot puts spring tension on the material being cut. This tension prevents the material from being lifted from the table as the blade rises. Since the teeth point downward, the blade cuts on the downward stroke.

Many jigsaws have special pulleys. A **pulley** is a wheel-like device attached to a shaft and designed to fit and drive or be driven by a belt. Many jigsaws have pulleys with two or more steps. A motor with a large pulley drives a machine with a small pulley faster than the speed of the motor. Conversely, if the motor has a pulley smaller than the machine, the machine runs slower than the motor.

Safe Operation of Jigsaws

Jigsaws may be used to cut wood, plastic, or metal. However, the correct blade and machine speed must be used. The following procedure is recommended when installing a blade and using a jigsaw.

Procedure

1. Obtain the instructor's permission to use the machine.
2. Put on a face shield and protective clothing.
3. Unplug the motor.

(continued)

Copyright © 2015 Cengage Learning®

Procedure, *continued*

4. Select an appropriate blade and determine the appropriate speed to run the saw.

5. Place the belt on the correct pulley steps to achieve the desired speed.

6. Attach the blade in the lower chuck.

7. Rotate the motor pulley until the blade is at its highest point.

8. Attach the blade in the upper chuck.

9. Loosen the tension sleeve clamp knob.

10. Lift the tension sleeve until moderate spring tension is placed on the blade. Retighten the tension sleeve clamp knob.

11. Rotate the pulley by hand to be sure the blade will go all the way up without buckling. Readjust, if needed.

12. Adjust the hold-down foot so it puts slight pressure on the stock.

13. Plug in and turn on the machine.

14. Push the work slowly into the front of the blade. Keep your fingers to the side of the blade.

15. Make the saw kerf on the waste side of the line.

16. Rotate the stock as the saw cuts so it follows the curves.

17. Avoid backing out of cuts. Saw out across waste wood to the edge if necessary.

18. To make a cut inside a circle, first bore a hole in the stock inside the area to be cut. Remove the saw blade and reinstall it through the hole in the stock. Saw inside the circle.

19. Turn off the machine when finished.

TABLE SAW

The table saw is basic to any good woodworking shop. A **table saw** is a type of circular saw with either a tilting arbor or a tilting table for bevel cuts. A **tilting arbor** is a motor, belt, pulley, shaft, and blade assembly that moves as a unit (Figure 15-12). The operator can adjust the arbor for depth or angle of cut. There are three basic types of table saws. The smallest is called a **bench top saw**. It is small and designed to sit on a bench or tabletop. The motor typically will be around ½ horsepower, and uses an 8-inch blade. This type saw is designed for very light cutting and portability. Since

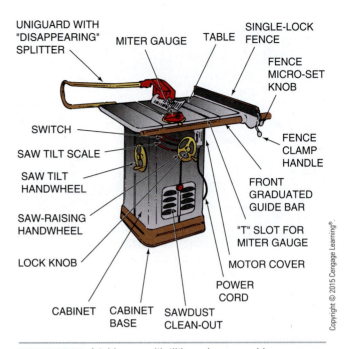

FIGURE 15-12 A table saw with tilting arbor assembly.

Copyright © 2015 Cengage Learning®.

it is compact and light, it can be moved around easily from place to place. The next larger size is called a **contractor's saw** (Figure 15-13). This saw usually sits on its own stand, which consists of legs made from angular shaped metal. It is a heavier duty saw that has a motor in the 1- to 2-horsepower range. As the name indicates, it is used by building contractors and is portable enough to be moved to a job site. These saws usually have a 10-inch blade. The largest of the table saws is the **cabinet saw**. This is a heavy duty machine that may weigh as much as 500 pounds. Cabinet saws are used by professional woodworkers because they are powerful, and very stable because of their weight, and make very accurate high-quality cuts. Motor sizes on these saws generally are 3 or 5 horsepower and drive a 10- or 12-inch blade.

Agricultural mechanics shops are most often equipped with cabinet saws with 10-inch or 12-inch blades. These are suitable for cutting large boards and rough-sawed lumber as well as small trim.

A relatively new type of table saw is equipped with a mechanism that instantly locks down the blade when it comes in contact with skin such as a finger or hand. The blade is connected to an electronic circuit that is activated when it detects moisture in the skin. A braking mechanism grasps the blade and prevents it from turning. Once this happens, the blade and brake have to be replaced. It should be noted that while this is a safety mechanism, all safety precautions taken with any other table saw should be strictly followed.

Copyright © 2015 Cengage Learning®

FIGURE 15-14 Setting the position of the rip fence on a table saw. To get an exact measurement, the operator checks the distance between the ripping fence and a tooth that is pointing toward the ripping fence. Notice the antikickback device on the splitter located behind the saw blade.

Saws should be equipped with antikickback devices for ripping Also table saws are provided with a blade guard to prevent the operator's hand from coming into contact with the blade (Figure 15-15). The guard shields the hands of the operator from the dangerous cutting blade. It can also provide protection from splinters, sawdust, sparks, or metal bits that can be thrown by the blade.

© flashgun. Image from BigStockPhoto.com

FIGURE 15-13 A contractor's saw is a midsize table saw.

Most saws have a hand wheel on the left side, which is used to adjust the tilt of the blade from 45 to 90 degrees. In addition, there are slots in the tabletop for a miter gauge. The miter gauge is also adjustable from 45 to 90 degrees. This permits crosscutting of boards at the desired angle.

A hand wheel on the front of the saw controls depth of cut. The wheel is near the operator and is easy to reach. The saw blade should be adjusted so it is never higher above the board than a distance equal to the depth of its teeth.

Table saws are equipped with a rip fence. The rip fence slides on the front and back edges of the table. It has a scale to indicate the width of cut that will be made at any given setting. However, the most accurate measurements are made with a ruler held on an outside blade tooth and the rip fence (Figure 15-14). Always make sure the saw is unplugged before measuring and setting the rip fence. When the fence clamp is tightened, the fence guides lumber through the saw for cuts with parallel edges. **Parallel** means two edges or lines are the same distance apart at all points along the length of the object.

Copyright © 2015 Cengage Learning®

FIGURE 15-15 Table saws are equipped with a blade guard and an antikickback device.

The blade guard should always be kept in place. The guard will be lifted by approaching stock, which glides under the guard. This provides maximum safety for the operator.

Sawing Large or Small Pieces

Two people are needed to safely manage pieces of material over 3 feet long or 2 feet wide. If a helper is not available, adjustable roller stands to help support the material should be used.

Pieces of wood being ripped to 3 inches or less should be pushed with a push stick. A **push stick** is a wooden device with a notch in the end to push or guide stock on the table of a power tool. Every effort should be made to keep hands several inches away from the sides of a saw blade. The operator should never place the fingers or hands in front of the blade. Also, the operator should never stand directly behind the board as it is being fed into the saw. If a kickback occurs, the operator can be injured as the board is thrown backwards.

When performing a crosscut on the saw, a miter gauge should always be used (Figure 15-16). It is extremely dangerous to attempt to "free hand" a cut. Never rely on the rip fence to do a crosscut or rip a board without a rip fence. Serious injury can occur.

Selecting the Correct Saw Blade

A wide variety of blades are used on table saws (Figure 15-17). These range from relatively inexpensive to very expensive. Obviously the more expensive blades give a better cut and last longer. Saw blades are classified generally by their diameter. Hence, a blade that has a diameter of 8 inches is called an 8-inch blade.

FIGURE 15-17 Many types of blades are available for table saws. Most are now tipped with carbide, which helps them stay sharper longer than ordinary steel blades.

Blades are also classified by the type of teeth. A blade may be a rip blade, crosscut blade, or combination blade. It may have large teeth for fast, aggressive cutting or small teeth for smooth cutting. All blades must be sharpened according to their design. Blades with carbide tips on their teeth stay sharp longer because of their extra hardness (Figure 15-18).

Blades are also designed and classified according to their unique function. For instance, **dado heads** are special blades that can be adjusted to cut kerfs from $\frac{1}{8}$ inch to $\frac{3}{4}$ inch wide in a pass. Dado heads may be a dial head that has staggered blades that can be dialed to set the width of the dado cut. Another, more common type is the stacked head. This type uses several blades and cutters that can be stacked to give different widths of cuts.

Care is needed when installing a blade on a saw. Always make sure the power is disconnected from the saw before attempting to remove the blade. Also, be careful of the blade. Position your hands so that the

FIGURE 15-16 When cross cutting, always use a miter gauge to support the board.

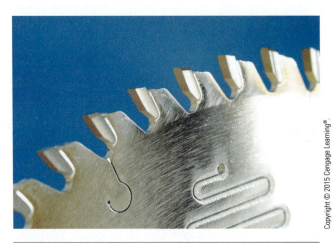

FIGURE 15-18 Blades with carbide tips stay sharper longer.

FIGURE 15-19 A thin block of wood can be used to hold the blade while the retaining nut is loosened. Note that the operator's hands are not in front of the blade.

blade won't cut you if a wrench slips (Figure 15-19). Most arbors (the shaft where the blade is mounted) have left-hand threads, which are backwards from standard bolt threads.

The hole in the blade must be the exact size of the shaft. A suitable bushing can be used if the hole is too large. Further, the slot in the throat plate in the saw table must be suitable for the blade being used. Be sure to place the blade on correctly. It is easy to get the blade on backward and this creates a dangerous situation. The teeth of the blade should be pointed downward toward the table when viewed from the operator's position (Figure 15-20). Tighten the blade until it is tight but do not overtighten the blade.

Safe Operation of Table Saws

Table saws are one of the most dangerous machines in the shop. Many people have lost fingers and even hands through careless use of the saw. Almost all of

FIGURE 15-20 The blade teeth closest to the front should be pointed toward the table when viewed from the operator's position.

FIGURE 15-21 This is a careless operator. How many safety violations can you see? Hint: There are at least five. See correct answers at the end of the chapter.

these accidents can be directly traced to carelessness on the part of the operator (Figure 15-21). If you carefully adhere to the procedures and cautions that follow, the likelihood of an accident is greatly lessened.

Procedure

1. Obtain the instructor's permission to use the saw.
2. Put on a face shield and protective clothing.
3. Unplug the motor.
4. Install the correct blade for the job, with the teeth pointing toward the direction of rotation.
5. Adjust the blade to the correct angle. Use the degree scale on the saw or use a sliding T bevel to establish the desired angle.
6. Adjust the height of the blade until only the teeth extend above the board to be sawed.
7. If ripping, move the miter gauge off the saw table. Always use either the miter gauge or the rip fence, but never both at the same time. Set the rip fence for the desired width of cut. This is checked by measuring from the fence to the tip of a tooth closest to the fence.
8. If crosscutting, move the rip fence to the right edge of the table or remove it. Adjust the miter gauge to the desired angle. The correctness of the angle is determined by using the degree scale on the gauge or by using a sliding T bevel.
9. Check to see that the guard assembly is in place and the saw is ready.

(continued)

Procedure, *continued*

10. Arrange for a helper or set up stands if a long or large piece of lumber is to be sawed. Make sure the helper knows not to pull the board but allow you to push the board into the blade.

11. Plug in the motor.

12. Stand to the side of the blade's path, and turn on the motor.

13. Avoid all distraction and keep your mind on operating the saw.

14. Push the work toward the blade with a slow and even movement.

15. Hold the work firmly so as to support both pieces on both sides of the blade.

16. Push the final work through with a push stick.

17. Turn off the machine.

RADIAL-ARM SAW

The radial-arm saw is a power circular saw that rolls along a horizontal arm. The saw can be raised or lowered. The arm will also pivot up to 45 degrees to the left and right.

The many movements of the radial-arm saw make it capable of numerous cutting operations. These same movements make it a very dangerous tool. Whole books are written on using the radial-arm saw. Only the simpler and more popular uses are described in this unit.

The radial-arm saw has many moving parts and major assemblies (Figure 15-22). To **pivot** means to turn or swing on. The arm pivots on the column, the yoke pivots under the arm, and the motor and blade assembly pivot at the bottom of the yoke. Each pivot point has a scale to indicate the position of the assembly. Each point also has a lock to hold the assembly once it is set. The major parts of the radial-arm saw are shown in Figure 15-23.

Special Safety Precautions

When using a radial-arm saw it is important to:

- Ask the instructor before attempting any operation.
- Always wear a face shield.
- Unplug or lock the switch in the "off" position when making adjustments.
- Be sure to wear a shirt that does not have loose sleeve cuffs. Make sure the cuff is securely buttoned.
- Always pull the saw toward the operator when sawing—never push the saw back into the wood.
- Always have all guards in place.

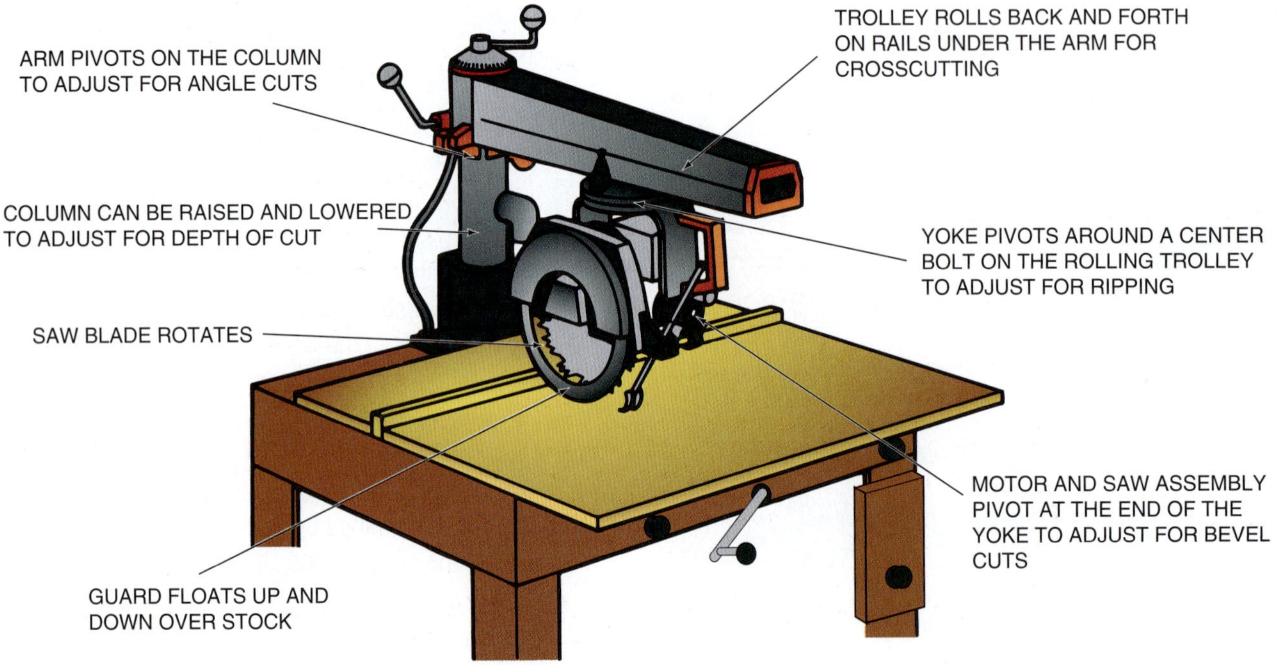

ARM PIVOTS ON THE COLUMN TO ADJUST FOR ANGLE CUTS

COLUMN CAN BE RAISED AND LOWERED TO ADJUST FOR DEPTH OF CUT

SAW BLADE ROTATES

GUARD FLOATS UP AND DOWN OVER STOCK

TROLLEY ROLLS BACK AND FORTH ON RAILS UNDER THE ARM FOR CROSSCUTTING

YOKE PIVOTS AROUND A CENTER BOLT ON THE ROLLING TROLLEY TO ADJUST FOR RIPPING

MOTOR AND SAW ASSEMBLY PIVOT AT THE END OF THE YOKE TO ADJUST FOR BEVEL CUTS

FIGURE 15-22 Moving assemblies on a radial-arm saw.

Copyright © 2015 Cengage Learning®.

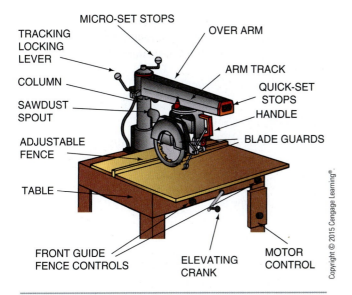

FIGURE 15-23 Major parts of a radial-arm saw.

Copyright © 2015 Cengage Learning®.

- Always have one hand on the saw handle when the saw is turning or is about to be turned on (it is easy for the saw to accidentally roll forward).
- Never turn the saw on when the blade is touching wood.
- Never stop a turning blade by pushing a piece of wood into it.

Operation of Radial-Arm Saws

The most popular use of the radial-arm saw is for cutoff work, including squaring boards, cutting them to length, cutting them at angles and bevels, cutting dados, and cutting rabbets.

Checking the Table for Levelness. The radial-arm saw cannot make accurate cuts unless the table is level and parallel to the arm. The following procedure may be followed to check the table.

Procedure

1. Obtain the instructor's permission to check the table.
2. Put on a face shield and protective clothing.
3. Unplug the machine, or lock the switch in the "off" position.
4. Loosen the column clamp handle.

5. Turn the elevating crank until the points of the teeth on the blade lightly touch the table.
6. Tighten the column clamp handle.
7. Loosen the trolley lock.
8. With the saw still turned off, grasp the saw handle and slowly pull the assembly across the table. The teeth should just touch the table at all points. If they do not, report the condition to the instructor. Do not use the saw until the problem is corrected.

Crosscutting Boards. The following procedure is recommended for crosscutting.

Procedure

1. Obtain the instructor's permission to use the saw.
2. Put on a face shield and protective clothing.
3. Mark the board where the saw will start the cut.
4. Elevate the saw until the teeth will just reach through the board and touch the tabletop.
5. Position the board on the table tightly against the fence so the saw will cut on the waste side of the mark.
6. Grasp the saw handle with one hand.
7. Turn on the saw with the free hand.
8. Hold the board with the free hand, being careful to position the hand so the hand and arm are not in line with the blade.
9. Stand out of line of the path of the blade.
10. Very slowly and firmly pull the saw forward until the cut is complete (Figure 15-24).
11. Push the saw back to the column.
12. Maintain the hold on the handle of the saw while you turn the saw off and lock the switch using the other hand.
13. Do not touch the lumber until the blade stops.

Crosscutting Bevels. The saw can be swung to the left or right to crosscut at angles. A beveled cut may also be made at any angle within the swing radius of the arm. When cutting a bevel, the following additional steps are required.

FIGURE 15-24 When using the radial-arm saw, hold the work securely across the fence and slowly pull the saw into the wood.

Procedure

1. Elevate the saw about 2 inches above the table.
2. Loosen the bevel latch.
3. Swing the saw assembly to the desired angle, as shown on the bevel scale.
4. Tighten the bevel latch.
5. Lower the saw to the table.
6. Proceed to crosscut.

Making Dados and Rabbets. Dados and rabbets are wide grooves cut only partway through the board. To cut dados and rabbets, the following procedure is used.

Procedure

1. Mark the board with a sharp pencil to show exactly where the grooves start and stop.
2. Mark the end of the board to indicate how deep the cut should be.
3. Install a dado blade, if available.
4. Place the board on the saw table, and pull the saw out so the blade is against the end of the board.
5. Elevate the blade until the points of the teeth touch the mark, which indicates the depth of the cut to be made on the board.

6. Push the saw back and position the board so the cut will remove the desired material.
7. Turn on the saw.
8. Make the pass.
9. Push the saw back.
10. If a dado blade is not being used, move the board ⅛ inch and make another pass.
11. Repeat until the groove is finished.
12. Turn off the saw.

Many other types of cuts can be made with the radial-arm saw. It is recommended that other books or bulletins on the subject be read if additional operations are desired.

MITER SAWS

A **miter saw** is a motor-driven circular blade that may be adjusted for angle cuts and is fed down into the material being cut (Figure 15-25). It can cut a bevel, angles, or a right angle. A miter saw for wood may be called a motorized miter box or compound miter box by some manufacturers. Compound means that it can cut a bevel on the edge of a board and at the same time cut an angle across the board. These are equipped with wood-cutting blades and are used for making quick, accurate cuts in moldings, floor boards, and framing. Miter saws may come in sizes that use either a 10- or 12-inch blade. The 12-inch blade will cut a wider and thicker board than the 10-inch saw. The saws are equipped with a circular gauge that indicates the

FIGURE 15-25 A miter saw can cut at a bevel, angle, or 90° cut.

FIGURE 15-26 A sliding compound miter saw can cut a wider board than a regular compound miter saw.

© John Lund/Blend Images/Getty Images

Procedure

1. Put on protective clothing and face shield.
2. Unplug the machine, and install the proper blade for the material being cut.
3. Set the saw for the desired angle of cut.
4. Stand to the side of the blade to avoid any thrown particles.
5. Grasp the handle with one hand, and turn on the saw with the other.
6. Slowly and carefully lower the saw into and through the stock.
7. Permit the saw to slowly rise to its parked position.
8. Turn off the saw and wait for the saw to stop turning.
9. Release the handle, and remove the stock.

Miter saws are stationary power tools that are light enough to move to the job. They are relatively low in cost and perform cutoff operations quickly and efficiently. Therefore, they are widely used throughout the construction industry.

JOINTER

The **jointer** is a machine with rotating knives used to straighten and smooth edges of boards and to cut bevels (Figure 15-27). The jointer is potentially a very dangerous tool. The knives can inflict severe cuts, and lumber may be thrown if not handled properly.

The size of a jointer is determined by the length of its knives. Most school shops have either 6-inch or 8-inch jointers. However, jointers are available up to 15¾ inches or more.

The main parts of the jointer are shown in Figure 15-28; these include the base, front infeed table, front table adjusting hand wheel, rear outfeed table, rear table adjusting hand wheel, rabbeting ledge, depth scale, knife assembly (or cutter head), guard, fence, fence clamp, and tilt scale.

All knives are installed so their cutting edges extend to the same height and leave the board smooth and even as the cutter head rotates. The rear outfeed table is adjusted so it is level with the cutting edges of the knives; the table is locked at this level. Any adjustment of the rear outfeed table should be made by the instructor. The fence may be moved across the table to

angle of the cut. Depending on the saw, a cut can be a single angle or a compound angle saw. Compound saws can cut angles in two different directions at once. One type of miter saw is the sliding compound miter saw (Figure 15-26). This saw is mounted on bars that allow the blade to be pulled across the wood in much the same manner as a radial-arm saw. The advantage is that wider boards can be cut than with a regular power miter saw.

Safe use of Power Miter Saws

Power miter saws have a spring-loaded mechanism to hold both the motor and the saw up and clear of the stock in its parked position. The blade is shielded by a guard, but care must be taken to stay clear of the blade, even when not in use. Material to be sawed must be held firmly or clamped in place to avoid any movement while the cut is being made. The saw may be set to cut any angle from 45 to 90 degrees.

It is important to follow good safety practices when using cutoff saws.

The height of the front infeed table determines the depth of cut. If the front infeed table is exactly level with the knives, the depth of cut will be 0 inch. The table is lowered to increase the depth of cut. The depth of cut is indicated by a scale on the side of the jointer.

Safe Operation of Jointers

The jointer is designed to remove a small amount of wood (1/16 inch or less) at a time. Generally, the operator should joint one edge smooth, and then use a table saw to cut the board to $\frac{1}{8}$ inch wider than needed. The board is then jointed down to the final width. This leaves the board with both edges smooth.

Strict attention must be given to the correct procedure to be used with the jointer.

A Using the jointer to dress the edge of a board.

B Using the jointer to bevel the edge of a board.

FIGURE 15-27 A jointer is used to straighten and bevel the edges of boards.

any point. Its position determines the maximum possible width of cut. In addition, the fence can be set at any angle from 45 to 90 degrees. This position determines the angle of the edge that is created.

Procedure

1. Obtain the instructor's permission to use the machine.
2. Put on a face shield and protective clothing.
3. Check to see that the guard is covering the knives.
4. Adjust the fence to the proper angle (usually 90 degrees).
5. Adjust the front infeed table for a $\frac{1}{16}$-inch cut. Only $\frac{1}{16}$ inch of material should be removed per cut. Never cut deeper than $\frac{1}{8}$ inch in one pass.
6. Be sure the board is 12 inches or more in length. Arrange for a helper if the board is longer than 4 feet.
7. Stand aside, and turn on the machine.
8. Place the board on its edge against the fence on the front infeed table.
9. Grip the board by placing the hands over the top edge.

CAUTION!

Keep your fingers high on the board and not over the cutter head.

10. Advance the board forward against the fence with moderate downward pressure.
11. When the leading hand is nearly over the knives, reposition it on the board over the rear outfeed table.

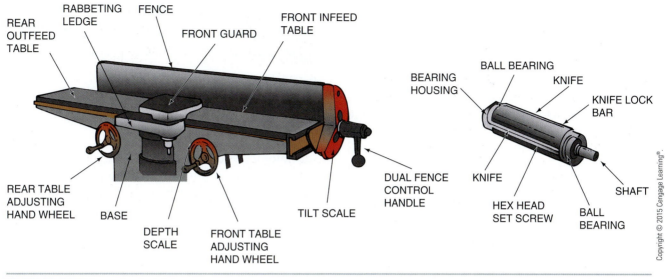

FIGURE 15-28 Major parts of a jointer.

> ## CAUTION!
> **Keep your hands away from the cutter knives.**

12. When most of the board has passed over the knives, transfer the second hand so it is also over the rear table.

13. Continue moving the board to finish the pass, with decreasing pressure near the end. Use a push stick to complete the pass for boards that are less than 4 inches wide. Do not use the jointer on boards less than 2 inches wide.

14. Reverse the board from end to end each time another pass is made. This decreases the tendency to create a wedge-shaped board after many passes.

15. Make the last pass with the grain, to get a smooth cut.

PLANERS

The planer (also known as a thickness planer or surface planer) is a machine with turning knives that dress the sides of boards to a uniform thickness. To **dress** means to remove material and leave a clean surface.

The planer is an excellent tool to convert low-cost, locally grown lumber into smooth, uniform materials. Planers are used to dress **rough lumber** to a desired thickness and smooth surface. Lumber is rough when it comes from a sawmill. Rough lumber may be thicker on one end than on the other. By making successive passes through the planer, a board can be dressed down to the desired thickness (Figure 15-29).

FIGURE 15-29 The planer dresses surfaces and leaves the board at the desired thickness.

The planer is also an excellent tool to level and smooth wide pieces made by gluing boards together. Agriculture students can make attractive projects such as gates, wagon bodies, and bench tops from rough lumber at a reasonable cost if a planer is available.

The planer is a massive machine with many parts. It includes an adjustable bed with smooth rollers to support lumber, corrugated rollers to draw lumber into the machine, and a rotating cutter head with knives that dress the lumber. Once a board is started in the machine, the machine is self-feeding. The parts of the planer are shown in Figure 15-30. However, most parts of the planer are shielded and normally hidden from view.

Safe Operation of Planers

Several critical adjustments on the planer must be made by the instructor or power tool specialist. These include installing knives, adjusting the feed rollers, and adjusting the smooth rollers.

CAUTION!

Students are not to make these adjustments.

If used carefully, the planer will safely dress large volumes of lumber. If used carelessly, the knives can be ruined by the first board planed after newly sharpened knives are installed. Careless use can also cause the operator to be seriously injured. The planer has a large motor capable of throwing massive pieces of lumber. Great care must be exercised when using the planer.

When using a planer, it is important that no metal or loose knots be in the lumber. The cutter heads may throw these items, or the planer may be damaged by them.

The planer generates large volumes of dust and wood chips. Therefore, a vacuum dust and chip removal system should be used to reduce the dust created by planing.

The following procedure is recommended for dressing rough lumber with a planer. This same procedure is

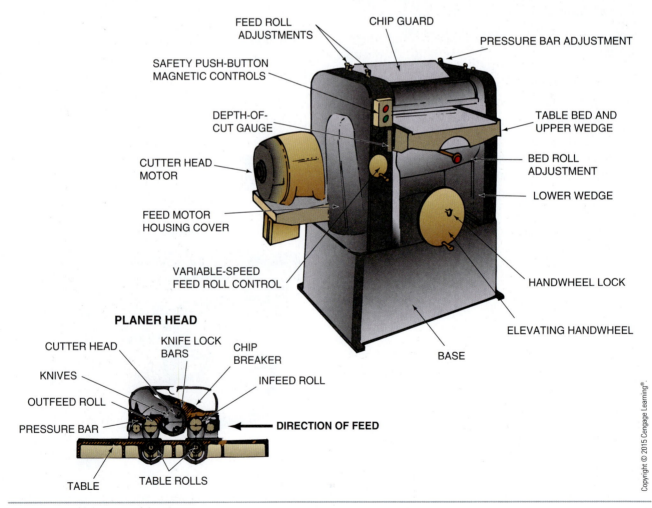

FIGURE 15-30 Major parts of the planer.

Copyright © 2015 Cengage Learning®.

used for edge-planing materials that are 2 inches thick or more. Lumber intended for edge planing should be reasonably straight.

Procedure

1. Obtain the instructor's permission to use the planer.
2. Put on a face shield and protective clothing.
3. Check the machine and be sure the chip guard and other shields are in place.
4. Sort the lumber according to its general thickness.

> **CAUTION!**
>
> Never plane lumber that has grit, nails, staples, or other metal in it.

5. Identify the thickness of the thickest end of the thickest piece of lumber.
6. Measure the thickest part of the thickest board.
7. Turn the thickness adjustment wheel until the thickness scale reads ⅛ inch less than the thickest piece to be planed. For instance, if the board is 2 inches thick, the planer should be set at 1⅞ inches.
8. Arrange for a helper.
9. Clear bystanders out of the path in front and back of the planer.
10. Set the feed control at its slowest rate.
11. Stand to the side, and turn on the planer motor. Never stand directly behind the board as it is fed into the planer. A kickback can cause severe injury (Figure 15-31).
12. Have the helper place the thinner end of the board on the front end of the bed.
13. Lift the other end of the board until it is slightly higher than the end on the planer bed.
14. While staying beside the board and keeping the end high, push the board slowly forward into the planer.

> **CAUTION!**
>
> When the thick part of the lumber is in the planer, the feed rollers will pull the lumber forward. Otherwise, the operator must push the lumber.

Copyright © 2015 Cengage Learning®

FIGURE 15-31 Always stand to the side of the board as it is fed into the planer. Never stand behind it.

15. When the lumber is halfway through the planer, the helper should move to the back of the machine to guide the lumber through and complete the first pass. The helper should take care not to stand in front of the lumber but to the side.
16. Run additional boards through the planer.
17. Raise the bed ⅛ inch with the thickness adjustment wheel.
18. Repeat steps 12 through 17 until both sides of all pieces are smooth and within ⅛ inch of the desired thickness.
19. Make one final pass on each side of the boards, going with the grain, and removing ¹⁄₁₆ inch per final pass.
20. Turn off the machine and wait for it to stop.

> **NOTE**
>
>
>
> The last pass on each side of the board should be made so it is planed with the grain and removes only ¹⁄₁₆ inch per pass. This technique results in a smooth finish. The depth of each cut may vary from as little as ¹⁄₃₂ inch to ⅛ inch. The final pass leaves the board at the desired thickness.

SANDERS

Some agricultural mechanics shops may have stationary sanders (Figure 15-32). For a list of procedures in sanding, refer to Portable Sanders in Unit 14. When

FIGURE 15-32 A stationary sander.

Sanders may be equipped with belts or discs suitable for use on wood or metal. The procedure for using a stationary sander is as follows.

Procedure

1. Obtain the instructor's permission to use the sander.
2. Wear safety glasses or other eye protection.
3. Wear a dust mask.
4. Use a belt or disc that is not clogged with particles.
5. Turn on the dust collector (if so equipped) and the sander motor.
6. Grip the material firmly, with fingers well away from the surface to be sanded.
7. On belt sanders, place the work on the belt lightly and move it against the stop.
8. On belts, move the material back and forth in a sideways motion. This helps prevent the belt from overheating and becoming clogged with particles.
9. On discs, use the half of the disc moving downward toward the table. Again, keep the stock moving on the table.
10. Use only moderate pressure to sand.
11. Reduce pressure before lifting or pulling the work from the belt or disc.
12. When doing freehand sanding of faces, ends, edges, and corners, use light pressure and keep the work moving at all times.
13. Turn off the machine when finished.

using portable sanders, the machine is moved to the material. To use stationary sanders, the material is moved to the machine.

SUMMARY

Power woodworking machines have been around for many years. They are efficient, powerful, and accurate. Cutting straight lines, such as crosscutting or ripping, can be accomplished using stationary or portable power tools. Similarly, cutting curves can be done by stationary or portable power tools as well. Choosing the proper tool is often determined by the cost and availability of the machine or tool. The inherent dangers of using these machines can be overcome by always observing all the safety rules and exercising caution.

Answers to Figure 15-21: (1) there is no saw guard; (2) the blade is adjusted too high; (3) he isn't using a push stick; (4) his fingers are too close to the blade; (5) he is talking on a cell phone.

Student Activities

1. Define the Terms to Know in this unit.

2. Identify and correctly spell the names of major parts of each machine in this unit.

3. Examine each stationary power woodworking machine in your shop, and make a list of items that need maintenance or repair. Submit the list to your instructor for review and appropriate action.

4. Properly clean a band saw, table saw, radial-arm saw, jointer, planer, and other woodworking machines in the shop.

5. Do the following Power Woodworking Machine Exercise.

Power Woodworking Machine Exercise

Purpose: The purpose of the Power Woodworking Machine Exercise is to obtain experience quickly. It requires the use of many power tools yet the material used is generally scrap material in an agricultural mechanics shop.

Note to the Teacher: Students should be instructed on the use of each machine before operations are performed.

Power Machine to Use	Procedures
Radial-Arm Saw or Portable Circular Saw	1. Obtain a rough 2 × 4 that is at least 2 inches thick and 4 inches wide. Cut it to a length of 14 inches (Figure 15-33).
Planer	2. Plane the 2 × 4 to a thickness of 1½ inches.
Jointer	3. Joint one edge of the 2 × 4 one time or until smooth. (The jointer should be set at a depth of 1/16 inch.)
Table Saw	4. Place the jointed edge of the 2 × 4 against the fence. Rip the 2 × 4 to a width of 3⅝ inches.
Jointer	5. Joint the rough side of the 2 × 4 to a width of 3½ inches.
Jointer	6. Joint the top edges of the 2 × 4 at 45-degree angles, ⅛ inch deep. This is done by setting the jointer for a 1/16-inch cut and making two passes.
Radial-Arm Saw	7. Cut the 2 × 4 to a length of 12 inches.
	8. On the bottom side of the 2 × 4, mark and label the angles shown in Figure 15-34, using a try square and pencil.
	9. Cut the pieces labeled "A" from the 2 × 4.
Band Saw	10. Cut the piece labeled "B" from the 2 × 4.
	11. On the bottom side of the 2 × 4, starting at the pointed end, make pencil marks at 3, 6, and 9 inches down the center. Locate the exact center so that cross marks can be drawn (Figure 15-35).
Portable Hand Drill	12. Drill a ½-inch hole through the 2 × 4 at the cross mark closest to the point.
Drill Press or Portable	13. Drill two 1-inch holes through the other two cross marks with a spade bit.
Power Drill	14. Connect the edges of the two 1-inch holes with a straightedge and pencil. Label the area inside as "C" (Figure 15-36).
Saber Saw or Jigsaw	15. Cut "C" out by sawing on the two pencil lines.
Router (optional)	16. Shape the inside edge of the top side of the 2 × 4, moving the router clockwise around the hole.
Disc Sander or Belt Sander	17. Sand the point of the 2 × 4 project lightly.
	18. Put your name on the bottom side of the 2 × 4 project in pencil.
	19. Turn your project in to your instructor to be graded (Figure 15-37).

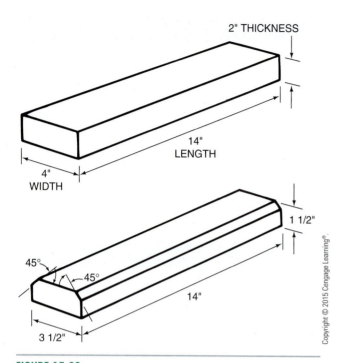

FIGURE 15-33

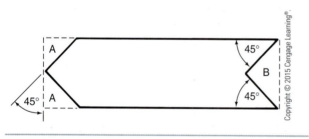

FIGURE 15-34

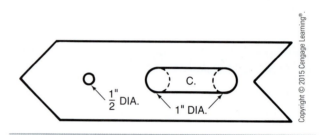

FIGURE 15-35

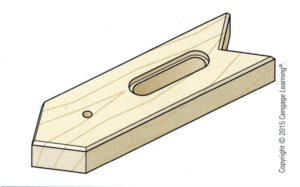

FIGURE 15-36

FIGURE 15-37 The completed project for this exercise should look like the piece shown in this illustration.

Relevant Web Sites

CDC, The National Agriculture Safety Database
https://www.cdc.gov/niosh/topics/aginjury/default.html

Canadian Centre for Occupational Health and Safety, OSH-Answers, Safety Hazards, Woodworking Machines—General Safety Tips
www.ccohs.ca/oshanswers/safety_haz/woodwork/gen_safe.html

www.BobsPlans.com, Free Woodworking Plans
www.bobsplans.com/FreeJigPlans/TableSawJig/TableSawJig.htm

Table Saw Techniques, by Rick Christopherson, Waterfront Woods
www.waterfront-woods.com/Articles/Tablesaw/tablesaw.htm

About.com, Woodworking, Tips for Using Your Band Saw Safely and Effectively
http://woodworking.about.com/od/safetyfirst/p/BandSawSafety.htm

Self-Evaluation

A. Multiple Choice. Select the best answer.

1. The number of people permitted in the safety zone around a machine is
   a. one
   b. two
   c. three
   d. any number

2. Woodworking machines should be cleaned with
   a. a brush
   b. the hand
   c. a rag
   d. an air gun

3. If a machine is worked too hard, the electric motor should stop because of
   a. burnout
   b. general fatigue
   c. overload protection
   d. voltage drop

4. When using a band saw, the operator should avoid
   a. backing out of cuts
   b. crosscutting
   c. cutting metal
   d. sawing curved lines

5. The band saw blade is held in position when cutting by
   a. wheels
   b. tires
   c. levers
   d. guides

6. Jigsaws are best for cutting
   a. very short curves
   b. straight lines
   c. rabbets
   d. dados

7. The blade on a table saw should extend how far above the work?
   a. 1 inch
   b. 2 inches
   c. 3 inches
   d. none of these

8. Small pieces of wood should be moved on a saw table by a
   a. bare hand
   b. gloved hand
   c. push stick
   d. hammer handle

9. Table saw blades are classified by the
   a. type of teeth
   b. diameter of the blade
   c. unique function
   d. all of these

10. The most popular use of the radial-arm saw is
    a. ripping
    b. cutoff work
    c. curve cutting
    d. dado cutting

11. Minimum protection when using power machines starts with
    a. steel-toed shoes
    b. a leather apron
    c. finger guards
    d. a face shield

12. When cutting with the radial-arm saw, the operator should
    a. move the wood into the saw
    b. pull the saw into the wood
    c. push the saw into the wood
    d. all of the above are safe

13. An adjustment that should be made on a jointer by the instructor is the
    a. rear outfeed table
    b. front infeed table
    c. fence
    d. miter gauge

14. The maximum safe depth per cut by a jointer is
    a. $\frac{1}{2}$ inch
    b. $\frac{1}{4}$ inch
    c. $\frac{1}{8}$ inch
    d. $\frac{1}{16}$ inch

15. The power machine that generates large volumes of wood chips is the
    a. band saw
    b. bench saw
    c. jointer
    d. planer

16. For good results when planing lumber, the operator should
    a. make each final pass while planing with the grain
    b. make each final pass with a shallow cut
    c. end up with the desired thickness
    d. all of these

B. Matching. Match the word or phrase in column I with the correct word or phrase in column II.

Column I

1. miter gauge
2. rip fence
3. radial-arm saw
4. many moving assemblies
5. tilting arbor
6. rotating cutter head
7. dado head
8. planer
9. sander
10. jigsaw

Column II

a. table saw
b. parallel to the blade
c. guides angle cuts
d. saw kerf ⅛ inch to ¾ inch wide
e. jointer
f. dress sides of lumber to a uniform thickness
g. also called a miter saw
h. radial-arm saw
i. special need for dust mask
j. reciprocal action

C. Completion. Fill in the blanks with the word or words that make the following statements correct.

1. Electrical wiring for shop machines should be installed by a(n) _____.
2. The safety zone around a machine should be marked with a _____.
3. Bits, knives, and blades should be kept _____.
4. Do only the procedures on a power machine for which you have had _____.
5. Three types of table saws are _____, _____, and _____.

D. Brief Answer. Briefly answer the following questions.

1. Describe the blades on a band saw, a jigsaw, and a table saw. For which types of cutting is each useful?
2. What makes the radial-arm saw particularly useful but at the same time particularly dangerous?
3. What is the first step in any procedure for using a power machine in the ag shop?
4. Planers are used to do what?
5. On belt sanders, what technique helps prevent the belt from overheating and becoming clogged with particles?

UNIT 16

Adjusting and Maintaining Power Woodworking Equipment

Objective

To properly adjust stationary woodworking machines to achieve the most efficiency and create the safest working conditions.

Competencies to be developed

After studying this unit, you should be able to:

- Explain the safety precautions one should take in adjusting stationary woodworking equipment.
- Discuss why it is important to maintain the proper adjustments on stationary machines.
- List five parts of the table saw that need to be properly adjusted.
- Properly align and adjust the parts of a table saw.
- List the components of a band saw that need to be adjusted.
- Adjust a band saw so that it will operate efficiently and safely.
- Make the proper adjustments to a jointer.
- Explain the working parts of a planer.
- Properly adjust a planer.

Materials List

- Table saw
- Band saw
- Jointer
- Planer

continued

Terms to Know

- adjust
- throat plate insert
- miter gauge
- trunnion
- blade guides
- cutter head
- snipe
- chip breaker

Materials list, *continued*

- Framing square
- Wrenches
- Straightedge
- Try square
- Rubber hammer
- Allen wrenches

Stationary power equipment in the shop represents a large investment. Even the best and most expensive equipment cannot operate efficiently if it is not properly adjusted. Most equipment has many moving parts, and these parts can come out of adjustment over time. In fact, even new equipment right out of the box can need tuning in order to achieve the most efficiency. Properly tuned machines last longer and achieve better results than machines that are not in the best working order.

Safety is an even more important factor in keeping machinery properly aligned and adjusted. Unless all the parts and systems of the machine are properly aligned, maximum safety can never be achieved. Woodworking machines need to cut cleanly, and wood should be able to move smoothly through the blade or cutters. Always keep in mind that machines have powerful electrical motors that can cause severe injury if improperly used. Never try to **adjust** any machine without disconnecting the power source. This means unplugging the power cord or turning off the current at the source. Never rely on the machine's switch alone to kill the power (Figure 16-1). The switch can accidentally be turned on while the machine is being adjusted, and severe injuries can result.

This unit discusses how to properly adjust four of the most widely used stationary machines—the table saw, the band saw, the jointer, and the planer. Keep in mind that while there are similarities in different brands and types of power equipment, there are also differences. With every machine there should be an instruction manual that accompanies the machine. Be sure to read the manual before making any adjustment.

FIGURE 16-1 The power source for any power machine must be disconnected before workers attempt to adjust the machine in any way.

ADJUSTING THE TABLE SAW

Remember from Unit 15 that the table saw is composed of several parts that allow the saw to be used for different operations. For example, the fence allows the operator to rip wood along its length, and the miter gauge allows wood to be cut across the grain. It is vitally important that all of the parts be properly adjusted for accurate cutting and for safety. A saw that is out of proper adjustment is neither accurate nor safe.

Leveling the Table

Begin the adjustments by making sure the table is level and the throat plate insert is properly aligned. The **throat plate insert** is the oval-shaped part that sets down in the table with a slot that the blade operates in. It is usually painted, and an indication that the insert is not fitting properly is paint that is rubbed off one corner (Figure 16-2). Use the small set screws at the corners of the insert to make all corners level with the table. This can be checked with a framing square or another straightedge.

Most table saws have tables made of cast iron. This material is heavy and stable enough to maintain its shape and prevent vibration when the saw is running. Usually the table consists of a central section with wings that are attached to each side. The wings should be perfectly level with the central part of the table. If the wings are not level with the top, wood that is moved on the table may not be level with the blade. This can result in cuts that are not square.

The following procedure describes how to perform the remaining steps involved in leveling the table of a table saw.

Procedure

1. Loosen the bolts underneath the table in order to move the wings level with the central part of the table (Figure 16-3).

2. Once the bolts are loose, lightly tap each wing with a rubber hammer to help move it. Cast iron can break, so tap lightly. If the wings don't move, loosen the bolts more.

3. Use a square or other solid straightedge to determine when the parts are level (Figure 16-4).

4. Retighten the bolts. Avoid overtightening the bolts as this may lead to stripping of the threads. Lock washers should be used under the bolts to prevent loosening when the saw vibrates during operation.

FIGURE 16-3 Adjusting the wings of a table saw involves loosening some bolts and then moving the wings up or down.

FIGURE 16-2 Paint rubbed off of the throat plate insert may be a sign that one side is higher than the other.

FIGURE 16-4 Proper alignment can be checked by using a square placed along the joint where the wing is attached to the table.

Copyright © 2015 Cengage Learning®.

Squaring the Blade

Once the table is level, check the saw blade to make sure it is perpendicular to the table. This means that the blade should be at a 90-degree angle to the table. If it is not perpendicular, the cut will not be at a proper 90-degree angle. The following procedure describes how to square the blade on a table saw.

Procedure

1. Make sure that the power is disconnected from the saw.

2. Use a framing square to check the blade's adjustment, by aligning the blade, the square, and the table as illustrated in Figure 16-5. Note that the tongue of the square should be between the teeth on the blade.

3. If there is any "daylight" between the tongue of the square and the saw blade, adjust the blade using the wheel that changes the angle of the blade to adjust the blade to the correct 90-degree angle.

4. Once the correction is made, be sure to lock the wheel in place by tightening the knob at the center of the wheel.

Squaring the Miter Gauge

When the table saw is used to make crosscuts, a **miter gauge** is always used to support the wood as it is fed into the saw blade. Never attempt to make a crosscut without using a miter gauge. In order to make a square cut, the miter gauge must be at a perfect 90-degree angle to the blade. Use a square to check the angle (Figure 16-6). If it is not at 90 degrees, loosen the knob at the back of the gauge and make the adjustment.

Truing the Trunnion

The next step is to make sure the blade is perfectly parallel with the table. This is accomplished by using a combination square to measure the distance between a tooth on the blade and the slot where the miter gauge operates (Figure 16-7). Use a combination square instead of a measuring tape because the square can give more accuracy.

The following procedure describes how to true the trunnion on a table saw.

Procedure

1. Measure the distance when the blade is toward the back of the saw, and again when the blade is all the way forward. Use the same tooth on the blade for each measurement. If the distance is the same, the blade is in proper alignment. If there is a difference, the trunnion will have to be adjusted. The **trunnion** is the heavy device under the table that mounts the blade.

2. There are several bolts that hold the trunnion to the table; these bolts are mounted in slots that allow the trunnion to be moved slightly (Figure 16-8). Loosen the bolts and move the trunnion until the blade is parallel to the slot on the table.

FIGURE 16-5 A square will indicate whether the saw's blade is at a 90-degree angle with the table. The blade shown here is not properly adjusted, as evidenced by the gap between the top of the square and the blade.

FIGURE 16-6 A square is used to check the squareness of the miter gauge to the blade.

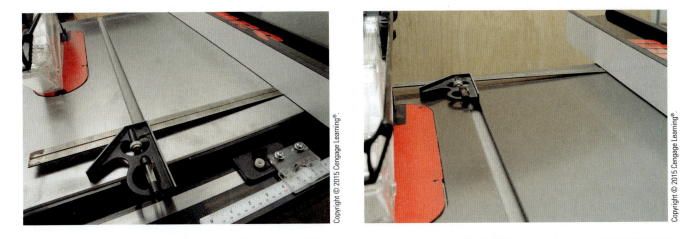

FIGURE 16-7 A combination square is used to measure whether the blade is parallel with the miter gauge slot.

3. While you are working with the trunnion, make sure the screw mechanisms that move the blade up and down as well as to an angle are clean of dust and rosin.

4. If the screw mechanisms need cleaning, they can be cleaned with mineral spirits.

5. Avoid using heavy lubricants on the screw mechanisms, because this causes sawdust and other particles to adhere to such lubricants, and to gum up the mechanisms. Instead, use a good-quality paste wax such as that used on cars. This wax should also be used on the table surface to prevent rust and to allow lumber to slide more freely across the table.

Aligning the Fence

The saw fence should be parallel with the table. Check this alignment by measuring the distance between a tooth on the front of the saw blade and the fence.

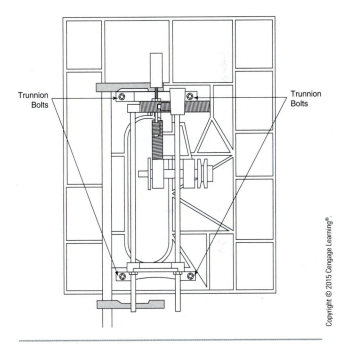

FIGURE 16-8 Loosen the bolts holding the trunnion and shift the table until the blade becomes parallel with the table.

CAUTION!

Make sure the power is disconnected from the saw before you begin.

Move the same tooth to the back of the blade by rotating the blade. Measure the distance between the tooth and the fence. If both the distances are the same, the blade and fence are parallel. If the fence is not properly aligned, make the adjustments at the end of the fence, as indicated in Figure 16-9.

ADJUSTING THE BAND SAW

Even though the main purpose of a band saw is not to make square cuts, it is important to make sure the machine is properly adjusted. Proper adjustment helps to ensure that the best cuts—whether curved or straight—are made. Also, as mentioned earlier, a well-adjusted machine is much safer than one that is out-of-alignment. Wood moves more freely through a band saw that has a sharp blade and that is in proper alignment.

A Determining whether the fence is parallel with the blade involves measuring the distance between a tooth on the front of the blade and the fence.

B Next, the operator rotates the same tooth to the back of the blade and performs the same measurement.

FIGURE 16-9 If the two measurements are the same, the fence is parallel to the blade.

Squaring the Blade and Table

The following procedure describes how to square the blade and the table of a band saw.

Procedure

1. Disconnect the saw from its power source.

2. Check the angle of the blade and the table. Just as in the table saw, the blade should be at a right angle, or perpendicular, to the table. To check this, use a square (Figure 16-10). The blade guard can be removed in order to use the square.

3. If there is a gap between the blade and square, the table must be adjusted. Adjust the trunnion (located under the table on the band saw) to change the angle of the table. Simply loosen the wheel that tightens the tilting portion of the trunnion, and move the table to the proper alignment with the blade (Figure 16-11).

4. Tighten the wheel and notice where the indicator is pointing. It should be pointing to zero. If it doesn't, loosen the screw in the center of the indicator and move it to zero, and retighten the screw.

Blade Tracking

Remove both the upper and lower blade guards and check the position of the blade. It should be running in the center of the tire. The tire is the rubber band that encircles the wheels where the blade runs (Figure 16-12).

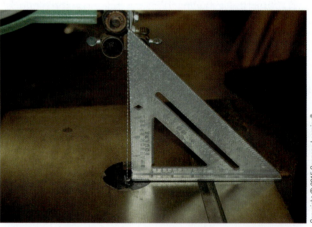

FIGURE 16-10 The blade's alignment can be checked by using a square as illustrated. The table and table blade shown here are not at the proper 90-degree angle.

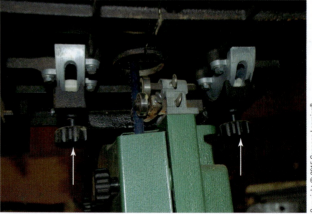

FIGURE 16-11 Squaring the band saw table with the blade involves loosening the knobs on the trunnion and rotating the table.

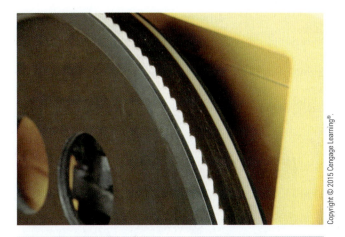

Copyright © 2015 Cengage Learning®

FIGURE 16-12 The band saw should run in the middle of the tire.

If the tire is worn or there is a groove worn in it, it may need replacing.

To move the blade to run in the center of the tire, adjust the knob on the back side of the upper wheel. Turning the knob moves the bottom of the upper wheel in or out. It only takes a slight turn to move the blade, so be careful with the adjustment.

To see if the blade is tracking properly in the center of the tire, turn the upper wheel by hand.

> **CAUTION!**
>
> Be careful of the blade, and do not get your fingers caught in the wheel. Never turn on the saw without the covers on the wheels.

While turning, the blade should track in the center of the tire without moving in or out.

Adjusting the Blade Guides

The band saw blade travels through a set of **blade guides** that prevent the blade from moving too far to either side or too far to the back. Adjustments to the blade guides are critically important. If out of adjustment, the teeth of the blade can come in contact with the guides and quickly damage both the blade and the guides. Also, it can present a dangerous situation.

The guides can be either aluminum blocks or wheels. The wheel or bearing type is generally considered to be superior because less friction is created by contact with the blade. The back guide is usually a bearing or wheel type of guide. When the blade is running free (not cutting wood), it should not touch the wheel. However, when the blade comes in contact with wood,

the guides serve to guide the blade. This means that the distance between the blade and guide must be correct. A simple way to gauge the correct distance is by using an advertising card that comes in many magazines (Figure 16-13). Fold the card around the blade and adjust the guides until they barely come into contact with the card. Set screws in front of the side guides and to the rear of the back guide can be loosened to move the guides back and forth. Position the side guides so that the teeth of the blade are not touching the guides or do not come in contact when cutting wood.

Setting the Blade Tension

In order to cut properly, the blade must have the proper amount of tension. If the blade is too loose, it will bend and proper cuts cannot be made. If the blade is too tight, it is more likely to break and the bearings on the guide wheels will wear prematurely.

Make sure the power is disconnected and remove the covers. Press on the blade midpoint between the blade guides and the place where the blade first touches the wheel.

> **CAUTION!**
>
> Make sure the power is disconnected from the band saw before removing the guards.

The blade should move about ¼ inch (Figure 16-14). The tension is adjusted by moving the tension knob at the back of the upper wheel guard. Turning the knob clockwise should tighten the blade, and turning the blade counterclockwise should loosen the blade.

ADJUSTING THE JOINTER

The main purpose of a jointer is to true the edge of a board. This means that the board will be straight from end to end and the edge will be at a perfect 90-degree angle with the face of the board. To work its best, the jointer must be properly adjusted.

Adjusting the Cutter Head

Perhaps the most critical adjustment is the set of the blades in the **cutter head**. This is a very precise and delicate setting. All of the blades must be set in the cutter head at exactly the same depth, and they must all be exactly parallel. If the blades are not set properly, the cut

FIGURE 16-13 An advertising card from a magazine can be used to gauge the clearance between the rollers and the band saw blade.

FIGURE 16-14 The blade should move about ¼ inch midway between the wheels.

on the edge of the board will have high places and resemble a washboard. Although there are gauges and jigs that can be used to set the blades in the cutter head, this job is best left to the professionals. Many stores and outlets that sell jointers and planers also sharpen blades. When the blades become dull, simply remove the entire cutter head and take it to the outlet for sharpening and resetting.

Aligning the Tables

The jointer has two tables: an infeed table and an outfeed table. Remember that to true the board, a small amount of wood must be taken off as the board is passed through the jointer. To accomplish this, the outfeed table must be slightly higher than the infeed table. Also, the outfeed table must be level with the cutting edge of the knives on the cutter head (Figure 16-15). To check this, lay a straightedge across the outfeed table and rotate the cutter head until the cutting edge of a knife is at its highest point.

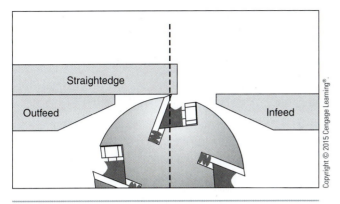

FIGURE 16-15 The infeed table should be slightly lower than the outfeed table. The outfeed table should be at the same height as the cutter blades.

CAUTION!

Make sure the power is disconnected from the jointer. Also, be careful of the cutting edges of the knives; they are very sharp and can cut your hand. Raise or lower the table until the outfeed table is level with the cutting edge. The infeed table should be about $\frac{1}{16}$ of an inch below the outfeed table. This allows about $\frac{1}{16}$ of an inch to be removed from the board's edge with each pass through the jointer.

Adjusting the Fence

In order to make a square edge on a board, the jointer fence must be perfectly perpendicular to the table. This can be checked using a try square. If the table is not square with the fence, move the fence by loosening the tilt lock on the rear of the fence and moving the fence until it is square with the fence. Lock the fence and check the indicator to make sure it indicates 0 to 90 degrees.

ADJUSTING THE PLANER

The planer is one of the most complicated pieces of equipment in the shop. Several different components have to be set exactly in order for the machine to do a proper job of surfacing boards. When a planer is out of adjustment, the results are usually obvious on the board being surfaced.

Setting the Cutter Head

Just as in the jointer, the main working part is the cutter head that holds the blades. If a board appears rippled

like a washboard, this means that the blades in the cutter head are out of adjustment. All of the blades have to be set to the exact same depth in the cutter head or some blades will be higher, and this results in the higher blades making a deeper cut on the surface of the board. This is what causes the ripple effect. To rectify the problem, the blades should be set at a shop that sharpens the blades. Here they have the proper gauges and equipment to not only sharpen the blades but also to set the blades the proper depth in the cutter head. This is usually a job too complicated to be done in the average agricultural mechanics lab or shop.

Setting the Tables

If a board comes out of the planer and it is thicker on one side than on the other, there is a problem with the setting of the cutter head in relation to the feed table. If the table is lower on one side than the other, the cut will be made deeper on one side of the board than on the other. Also, the cutter head may be lower on one end than on the other, resulting in the same problem. In either case the table and the cutter head must be perfectly parallel.

The following procedure describes how to adjust the tables on a planer.

Procedure

1. Prior to attempting to make any adjustments on a planer (or any other machine), make sure that the machine is disconnected from its power source.

2. To check whether the table and the cutter head are parallel, use a block of wood about 1 foot long, 3 to 4 inches wide, and 2 inches thick as a gauge (Figure 16-16).

3. Lower the table until you can insert the block between the table and the cutter head. As with all adjustments of any machine, make sure the power is disconnected before you begin.

4. Place the block on one side of the table, and gradually raise the table until it comes into contact with the cutter head.

5. Move the block to the other side of the table, and check for clearance. The block should have the same degree of contact on each side cutter head.

6. If there is more clearance on one side than on the other, adjust the table, the cutter head, or both to set the table and cutter head perfectly parallel.

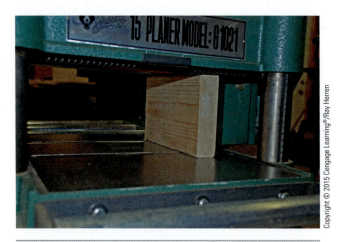

FIGURE 16-16 A wooden block can be used to gauge whether the cutter head is parallel to the feed table.

There are many different types and manufacturers of planers and they can vary widely in how the feed rollers are set. Refer to the owner's manual for instructions.

Setting the Feed Rollers

Planers have either a set of rollers or cast-iron tables that support the boards going into and coming out of the planer. If either the infeed rollers or the outfeed rollers are improperly positioned, the planed boards will have a condition called **snipe** where the boards have a dug-out or dished portion either at the beginning or at the end of the board (Figure 16-17). If both the infeed rollers and the outfeed rollers or tables are out of adjustment, snipe can occur at both ends of the boards. To correct the condition, the rollers must be set level with the table beneath the cutter head.

The following procedure describes how to adjust a planer to get rid of snipe.

FIGURE 16-17 Snipe is a condition caused when the planer's outfeed or infeed rollers are out of adjustment.

Procedure

1. Disconnect the planer's power source from the planer.
2. Lower the planer's feed tables or rollers to their lowest position.
3. Lay a long straightedge across the rollers that stretches from the beginning of the infeed roller to the outer edge of the outfeed roller (Figure 16-18).
4. Adjust both rollers so they are level with the center table.

Adjusting the Chip Breaker

As the cutters cut the surface of the board, the cut-off chips are thrown against a device called a **chip breaker** that breaks the removed pieces into smaller chips and deflects them away from the rollers and the table. If the chips fall on the rollers, the surface of the board may be dented and damaged (Figure 16-19). This aids in

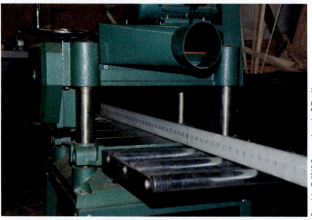

FIGURE 16-18 A long straightedge can be used in adjusting the infeed and outfeed tables.

FIGURE 16-19 An improperly adjusted chip breaker can put dents in a newly planed board.

removing the shavings by the dust collector. If the chip breaker is improperly set, the chips can clog against the chip breaker and clog the dust collector or can cause damage to the surface of the wood. If the board comes from the planer with dimple-like depressions in the surface, the most likely problem is an improperly adjusted chip breaker.

To correct this problem, remove the dust cover from the cutter head and measure the distance between the tip of the cutter and the chip breaker. If there is a dust collector on the planer, this distance should be about 1/16 inch. If the planer operates without a dust collector, the distance should be about 1/4 inch. Make the adjustment, and replace the cover.

SUMMARY

Machines cannot perform at their peak unless they are properly adjusted. Not only will the quality of the work be diminished, but the machines can be unsafe to operate as well. By closely observing the operation of the machine and the job it is doing, problems with machine adjustment can be determined and corrected by following a step-by-step process to properly adjust and align all the operating parts. Always remember to keep safety as a first priority. Never attempt to adjust any machine until the power is disconnected from the source. Never rely solely on the switch to disconnect the power; unplug the power cord or turn off the circuit breaker. Keep all machines in the shop properly adjusted and the quality of your work will be greatly enhanced as well as your safety.

Student Activities

1. Define each of the Terms to Know in this unit.
2. Create an inventory of all the stationary woodworking machines in the agricultural mechanics lab. Under the name of each machine, make a checklist of the functions that should be checked for proper adjustment.
3. Using the inventory and checklists created for activity 2, check the machines for proper adjustment. Remember to disconnect the power before beginning. Correct any machines that are out of adjustment.

Relevant Web Sites

American Furniture Design Co., Shop Tips, Band Saw Tune-Up & Resawing Secrets
www.americanfurnituredsgn.com/

DIY Network, How to Adjust a Table Saw
www.diynetwork.com/

About.com, Woodworking, Table Saw Tune-Up Tips
http://woodworking.about.com/

Self-Evaluation

A. Multiple Choice. Select the best answer.

1. Machines should be properly adjusted because such machines
   a. are more efficient
   b. make cleaner cuts
   c. are safer
   d. all of the above

2. Before adjusting any machine, the power should be
   a. left on for proper adjustment
   b. turned off at the machine's switch
   c. disconnected at the power source
   d. ignored as not important

3. The miter gauge on a table saw can be aligned with the blade using a
   a. level
   b. measuring tape
   c. framing square
   d. caliper

4. The fence on a table saw is checked for alignment by aligning the fence with the
   a. trunnion
   b. miter gauge
   c. miter gauge slot
   d. tabletop

5. The trunnion
   a. mounts the blade
   b. is mounted on the underside of the table
   c. is heavy
   d. is or does all of these

6. The blade on a band saw can be checked for squareness with the table by using a
   a. framing square
   b. try square
   c. sliding T bevel
   d. caliper

7. Clearance between the band saw's blade and the guides can be checked by using a
   a. square
   b. tape measure
   c. combination square
   d. advertising card

8. Snipe is a condition caused by an improperly adjusted
   a. planer
   b. jointer
   c. table saw
   d. band saw

9. If a newly planed board has dents in the planed surface, the likely cause is an improperly adjusted
   a. cutter head
   b. outfeed table
   c. infeed table
   d. chip breaker

10. Sharpening the blades on a planer is
    a. best left to a professional
    b. easy to do
    c. not important
    d. automatically done by the planer

B. Matching. Match the items in column I with those in column II.

Column I

1. tire
2. trunnion
3. chip breaker
4. blade guides
5. washboard
6. snipe
7. fence
8. safety
9. paste wax
10. framing square

Column II

a. aluminum blocks or wheels that guide a band saw blade
b. aids in removing chips from a planer
c. used to prevent corrosion on cast-iron tables
d. guides boards being ripped on the table saw
e. the most important factor in adjusting machines
f. a rubber band on the wheels of a band saw where the blade runs
g. a dug-out or dished-out portion of a board that has been planed
h. an effect caused by improperly adjusted jointer blades
i. used to properly align the miter gauge and the blade on a table saw
j. a heavy device that mounts the blade to the table on a table saw

UNIT 17
Metalworking with Power Machines

Objective
To safely use stationary power machines for metalworking in agricultural mechanics.

Competencies to be developed

After studying this unit, you should be able to:

- State basic procedures for using stationary machines for metalworking.
- Identify and properly spell major parts of specified machines.
- Operate a drill press.
- Operate a grinder.
- Operate power metal-cutting saws.
- Operate a power shear.
- Operate a metal bender.

Materials List

- Drill press
- Grinder and wire wheels
- Power hacksaw and/or metal-cutting band saw
- Metal shear
- Metal bender
- Drill bits, ⅛" and ½"
- Length of ⅛" × 1" flat steel (longer than 6 inches)

Terms to Know

- drill press
- grinding wheel
- grit
- face
- dress
- coolant
- horizontal
- horizontal band saw
- shear

NOTE

Before proceeding with this unit, review Unit 4, Personal Safety in Agricultural Mechanics, and the topic Safety with Power Machines in Unit 15.

Power is used in a number of ways for metalworking in agricultural mechanics. It is used to cut, make holes, shape, and sharpen metal. These functions will be discussed under the following headings of drill press, grinder, metal-cutting power saws, power shears, and metal benders.

THE DRILL PRESS

The **drill press** is a stationary tool used to make holes in metal and other materials. Its design and structure permit it to drive large drills and apply heavy pressure on bits. When material can be placed on a drill press, this tool is preferred over the portable power drill for drilling large holes. The drill press is also preferred for precision drilling or when many holes are needed.

Drill presses are available as floor models or bench models. The parts are similar for both. The drill press consists of a base, column, table, and head. The head is an assembly consisting of the motor, switch, drive belt, speed control, shaft, quill, chuck, and feed. Parts of the drill press are shown in Figure 17-1.

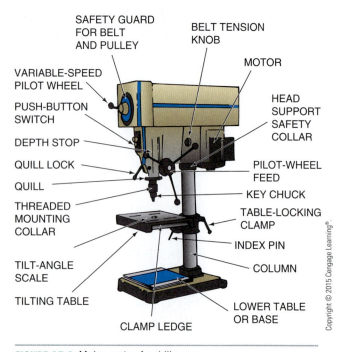

SAFETY GUARD FOR BELT AND PULLEY
BELT TENSION KNOB
MOTOR
VARIABLE-SPEED PILOT WHEEL
PUSH-BUTTON SWITCH
HEAD SUPPORT SAFETY COLLAR
DEPTH STOP
QUILL LOCK
PILOT-WHEEL FEED
QUILL
KEY CHUCK
THREADED MOUNTING COLLAR
TABLE-LOCKING CLAMP
INDEX PIN
TILT-ANGLE SCALE
COLUMN
TILTING TABLE
LOWER TABLE OR BASE
CLAMP LEDGE

Copyright © 2015 Cengage Learning®

FIGURE 17-1 *Major parts of a drill press.*

Safe Operation of the Drill Press

The motor of the drill press drives a shaft, which, in turn, drives the chuck. The speed of the chuck is controlled by the use of step pulleys, variable-speed pulleys, or a variable-speed motor. When the operator moves the feed handle, a gear moves the quill up and down. This movement raises and lowers the chuck to permit drilling (Figure 17-2). The movement of the quill may be limited by the depth stop, or the quill may be locked for special functions.

The table slides up or down on the column. This permits thick or thin material to fit on the machine. The table may be swung to one side so large objects can rest on the base in a drilling position. Some drill presses have tables that tilt. Tilting tables permit materials to be supported for angle drilling.

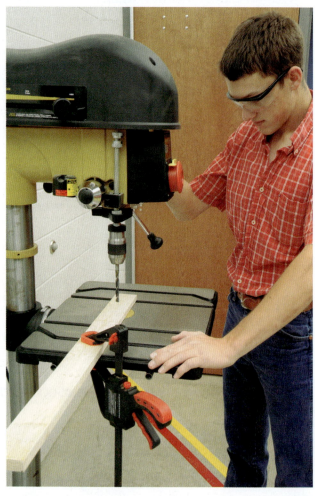

Copyright © 2015 Cengage Learning®

FIGURE 17-2 When the operator moves the feed handle, a gear moves the quill up and down, which raises and lowers the chuck to allow drilling to take place.

The following procedure is recommended for the use of the drill press.

Procedure

1. Obtain the instructor's permission to use the machine.

2. Wear a face shield and protective clothing. Leather gloves are recommended when drilling sharp or irregular-shaped metal.

3. When drilling wood, use a sharp pencil to mark where holes are to be drilled. A cross (+) or a caret (^) is useful to mark an exact location. Use a scratch awl to mark metal.

4. When drilling metal, use a center punch to aid in starting the drill.

5. Use only straight shank drills in gear chucks, and taper shank drills in taper chucks.

6. Select the correct drill by observing the size stamped on the shank, or use a drill gauge to determine drill size. Use a pilot hole for holes larger than 3/8 inch.

7. Tighten gear chucks securely by inserting the chuck wrench into at least two of the three holes, twisting the wrench as tightly as possible in each.

CAUTION!

Be sure to remove the chuck wrench after tightening.

8. Check table alignment to prevent damage to the table by the drill bit. To do this, move the feed handle until the drill passes down through the hole in the center of the table and then return it. Also, a flat piece of wood may be used to protect the table.

9. Clamp flat material to the table; clamp round material in a V block.

CAUTION!

The long end of the material should be on the operator's left and against the drill press column to prevent any rotation.

10. Hold metal or clamps with a gloved left hand.

11. Turn on the motor.

12. Grasp the feed handle and lower the drill to start the hole.

13. Apply steady pressure while drilling.

14. Add cutting oil to cool the drill when drilling steel.

15. Ease off and use very little pressure as the drill breaks through the metal.

16. Raise the drill.

17. Turn off the machine.

18. Use a file to remove burrs from the metal.

19. Use a bench brush to remove metal chips from the project and work area.

20. Use a cloth to clean oil from the project and work area.

NOTE

Refer to Unit 10, Fastening Wood, and Unit 13, Fastening Metal, for applications in which drilling is needed.

GRINDER

A grinder removes metal by abrasive action. Grinders are available in many types and sizes. They are used to sharpen tools, shape metal, prepare metal for welding, and remove undesirable metal.

Most shop grinders consist of a pedestal, a double-shafted motor, switch, wheels, guards, tool rests, and safety shields (Figure 17-3). Some grinders have a light to improve vision and a water pot for cooling metal. Small grinders have wheels that are 6 inches or 7 inches in diameter and 1 inch wide. Large grinders generally have wheels that are 12 inches in diameter and 2 inches wide.

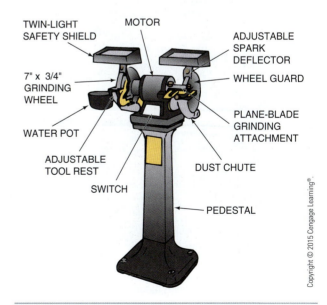

FIGURE 17-3 Major parts of a grinder.

Copyright © 2015 Cengage Learning®.

Grinders may be equipped with grinding wheels or wire brush wheels. A **grinding wheel** is made of abrasive cutting particles formed into a wheel by a bonding agent. The cutting particles are called **grit** (Figure 17-4).

When selecting a grinding wheel, it is important to use the correct wheel for the job. The wheel must have the correct speed rating for the motor. Failure to select the proper wheel can result in a serious accident. The wheel may fly into pieces, hurling rocklike fragments into the face and body of the operator and bystanders.

When replacing wheels, obtain specifications from the operator's manual or from the wheel on the machine. Bushings that fit the shaft must be used when the hole in the wheel is larger than the shaft.

A coarse or very coarse texture is recommended for wheels used to shape metal for sharpening, welding, or construction. A medium texture wheel is good for sharpening axes, mower blades, and other tools. Fine-textured wheels are not generally recommended for agricultural mechanics except for power oil stone or wet stone grinders. Fine wheels cut slowly and overheat the metal easily.

Grinding wheels are manufactured to be used on one surface only. The surface intended for use is called the **face**.

Wire wheels are useful for removing rust and dirt from small machinery parts such as bolts (Figure 17-5). Such cleaning is useful before welding or painting. Wheels with large, stiff wire are called coarse wire

FIGURE 17-5 Wire wheels are used for removing rust.

wheels. Coarse wheels cut aggressively and can be dangerous. Care must be exercised to use a tool rest to support metal being brushed. Otherwise, the wheel may drag the metal down between the wheel and the guard. Such accidents may result in injury to the operator, a damaged object, or a broken machine. Some instructors prefer to remove the tool rest when objects to be brushed are large. Fine wire wheels are good for moderate cleaning jobs. They are softer than coarse wheels and safer for the operator.

Handheld grinders, also called angle grinders, are used for grinding metal too large to use with a bench grinder (Figure 17-6). As with bench grinders, angle grinders have a variety of different types of grinding wheels. These tools are used to grind bevels on welding projects, smooth out welds, sharpen large mower blades, and a variety of other uses. A wire wheel or wire cup (Figure 17-7A) can be used to remove paint or rust from metal (Figure 17-7B).

FIGURE 17-4 A grinding wheel is made of abrasive cutting particles called grit.

FIGURE 17-6 Handheld grinders are used to grind metal too large for the bench grinder.

A

B

FIGURE 17-7 A wire cup can be used on a handheld grinder (A). A wire wheel or cup can be used to remove paint or rust (B).

Safe Operation of Grinders

In addition to the points previously mentioned, the following procedure is recommended for the safe operation of a grinder.

Procedure

1. Obtain the instructor's permission to use the grinder.
2. Wear a face shield.
3. Wear close-fitting leather gloves and a leather apron. Small work pieces should be held with lever lock pliers.
4. Check the wheels. Use the machine only if the wheel is clean and free of nicks, chips, and cracks.

CAUTION!

If any of these conditions are not met, have the instructor inspect the wheels and correct the problem(s).

5. Adjust the tool rest so its top surface is level with the center of the motor shaft and within $\frac{1}{16}$ inch of the wheel. For bevel grinding, set the tool rest at the appropriate angle.
6. Both bench grinders and handheld grinders can create a lot of sparks. Be sure the area is clean of any type of flammable materials (Figure 17-8).
7. Hold the metal to be ground firmly on the tool rest (Figure 17-9).
8. Move the metal back and forth or in a curved motion as needed. The movement will help keep the wheel clean and avoid overheating of the metal.

(continued)

FIGURE 17-8 Both bench grinders and handheld grinders can create a lot of sparks. Remove all flammable materials from the area.

FIGURE 17-9 The tool rest helps position and guide metal to be ground. The tool rest should be adjusted to 1/16 inch or less from the wheel.

Procedure, *continued*

> **CAUTION!**
>
> Do not use excessive pressure on the metal. The speed of grinding is determined by the grit and speed of the wheel, not by pressure. Never grind with the side of the wheel. Grinding creates heat; therefore, avoid handling the metal with bare hands.

9. Turn off the machine when finished, and wait for it to stop running.

Dressing the Grinding Wheel.

After proper instruction, the student may **dress** a grinding wheel. To dress a wheel means to remove material so the wheel is perfectly round with the face square to the sides and sharp abrasive particles exposed (Figure 17-10). Dressing a wheel removes clogged abrasives and slight bulges that cause the wheel to get out of balance. The procedure for dressing a wheel follows.

FIGURE 17-10 Dressing a grinding wheel removes material from it to leave the wheel perfectly round. Afterward the face is square with the sides of the wheel.

Procedure

1. Obtain the instructor's permission.
2. Wear a face shield. Leather gloves and leather apron are also recommended.
3. Wear a dust-type filter respirator.
4. Obtain a wheel-type grinding wheel dresser.
5. Turn on the grinder.
6. Place the dresser on the tool rest of the grinder.

7. Slowly rock the dresser wheels forward until they touch the grinding wheel.
8. Apply firm and even pressure on the dresser and move it back and forth across the wheel for about 30 seconds.

> **CAUTION!**
>
> Be prepared for abrasive particles to fly off the wheel. The wheel is cleaned, balanced, and squared by removing particles.

9. Remove the dresser and turn off the grinder.
10. Readjust the tool rest to within $\frac{1}{16}$ inch of the wheel. Then rotate the wheel by hand to check for roundness. Also, using a square, check the face of the stone for the proper angle (Figure 17-11).
11. Use a combination square or the tool rest to determine if the face of the wheel is square to the side.
12. Repeat steps 5 through 11 if additional dressing is needed.

Applications for grinders will be found in units on metalworking, welding, tool fitting, and others. The grinder is a basic tool and is found in most shops.

METAL-CUTTING POWER SAWS

Metal-cutting power saws are reciprocating hacksaws, band saws, or thin grinding-type wheels that cut metal. The hacksaw and band saw are used most often in agricultural mechanics.

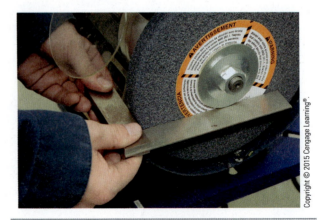

FIGURE 17-11 Checking the grinding wheel.

Hacksaws

The power hacksaw has a reciprocating movement that operates like a hand hacksaw (Figure 17-12). However, it is motor driven and cuts much faster. These fast-cutting saws have a system to pump a coolant onto the cutting area. A **coolant** is a liquid used to cool parts or assemblies.

The power hacksaw consists of a motor-driven frame mounted on a stand. The frame holds a rigid blade, which is ¾ to 1 inch wide and 12 to 18 inches long. Blades are available with fine to coarse teeth. Very hard blades are needed to cut hard steel. Such blades are brittle and break easily if the frame is dropped or if the saw binds. Extra care is needed to prevent blade breakage.

The power hacksaw stand is narrow. Therefore, metal must be supported to avoid tipping the machine. Sturdy models can support the weight of heavy steel. However, long stock must be supported by stands or a helper.

The power hacksaw is very useful in the agricultural mechanics shop. However, an expensive blade can be broken in an instant if the machine is not used properly. The following procedure is recommended for using a power hacksaw.

Photo provided by KATSO, Inc.

FIGURE 17-12 A power hacksaw works the same way a hand hacksaw does, with a reciprocating movement. The blade is hard and brittle, and can break easily. The metal must be clamped securely to prevent the blade from binding and breaking.

Procedure

1. Obtain the instructor's permission to use the saw.
2. Wear a face shield, leather gloves, and a leather apron.
3. Check the machine to be certain it has the proper blade and that the blade is tight.
4. Place the frame in the raised position.
5. Adjust the vise on the machine to hold the metal at the desired angle.

> **CAUTION!**
> If the vise does not hold the metal rigid, the blade will break.

6. Position the metal in the vise on the machine, and tighten securely.
7. Turn on the machine.
8. Lower the frame slowly and carefully until the blade is on the stock and starting to cut.
9. Turn on the coolant, if it does not turn on automatically.
10. Stay near and watch the machine while it is cutting.

> **CAUTION!**
> Do not put pressure on the blade or otherwise interfere with the machine while it is running.

11. Switch off the machine when the cut is finished if it does not turn off automatically.
12. Remove all scrap metal, and clean up all metal dust and coolant.

Horizontal Band Saw

The word **horizontal** means flat or level. The **horizontal band saw** has a blade that saws parallel to the ground. Generally, however, it is constructed like an upright band saw. It has a band-type blade that travels on wheels and moves through rollers and guides (Figure 17-13). Since blade movement is forward at all times, it cuts continuously. As a result, it cuts faster than a power hacksaw.

The procedure for cutting with a horizontal band saw is similar to the procedure described for a hacksaw.

FIGURE 17-13 The horizontal band saw is especially useful for cutting large pieces of structural metal such as round stock, flats, angles, I-beams, and channels.

However, the manufacturer's instructions should be followed. Machines vary in minor details of design and function.

POWER SHEARS

Sheet metal is easy to cut with hand shears. Therefore, power sheet metal shears may not be seen in many agricultural mechanics shops. To **shear** means to cut at an angle. Generally, shearing is done with two movable blades.

The cutting of flat, angle, and other structural steel is difficult by hand, and slow by power saw. The cutting of these materials is fast and clean when a shear is used (Figure 17-14). Some shears have features that

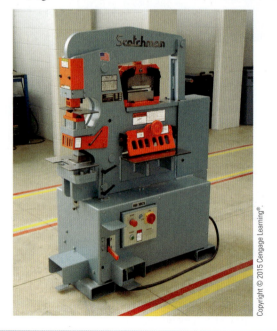

FIGURE 17-14 Power shears for cutting flat and round metal stock. Special knives for some shears enable them to cut angle iron.

permit them to bend, cut, and punch metal. To operate hand-operated power shears, the following procedure is recommended.

Procedure

1. Obtain the instructor's permission to use the machine.
2. Put on a face shield, leather gloves, and a leather apron.
3. Raise the handle of the shears.
4. Insert the metal to be cut or punched.
5. Carefully align the cut mark with the cutting edge of the stationary blade.
6. Support the metal so it is level.
7. Lower the handle until the cut is complete.

CAUTION!
Keep your hands away from the cutting shears or punch.

8. Store the handle so it cannot accidentally fall and operate the shears.

CAUTION!
The weight of the handle alone is sufficient to close the shears with enough power to cut off fingers.

The procedure described for use of hand power shears also applies to hydraulic shears. In the case of hydraulic shears and punches, power is delivered by cylinders driven by hydraulic fluid.

Structural steel (and other metals) can be cut using an abrasive cutoff saw (Figure 17-15). This saw cuts by using a thin, circular blade that is impregnated with abrasive material. As the blade turns, the metal is worn away quickly and a clean square cut can be made. When using all machinery, eye protection must be worn. A face shield that covers the entire face and a leather apron should be worn to protect the operator from sparks and metal particles.

FIGURE 17-15 An abrasive cutoff saw can cut structural steel up to 3 inches wide. Note the use of gloves and face shield.

FIGURE 17-16 The metal bender is useful for bending round, flat, or angle stock.

METAL BENDERS

Bending metal by hand is discussed in Unit 12. However, some shops have hand-operated benders called breaks. These are useful for bending small round and bar stock. By using cams, pins, and levers, such tools make fast and accurate bends (Figure 17-16).

Sheet-metal breaks are available for making rain gutters, heating and air-conditioning ducts, and other commercial uses (Figure 17-17). However, sheet-metal bending in agricultural mechanics is generally limited to repair work. Such occasional work may be done with hand tools.

FIGURE 17-17 The hand-operated sheet-metal break is useful for making angle bends and seams in sheet metal.

SUMMARY

Metal is a very tough, durable material. By its very nature it is difficult to cut, bend, or shape. Humans are not strong enough to bend and shape metal by hand, so power machines are used to give us an advantage in these operations. Caution must be used in the operation of these machines. Not only does the quality of the project depend on the cautious use of machinery, but the operator's safety as well. The proper, safe use of power equipment will give the results we desire.

Student Activities

1. Define the Terms to Know in this unit.

2. Examine the grinders in the shop to determine if the wheels need dressing. Ask the instructor to demonstrate how to dress a wheel.

3. Check the wheels on any grinders you have at home. Dress them if needed.

4. Do the following sawing, brushing, grinding, and drilling exercise.
   a. Use a metal-cutting power saw to cut a 6-inch piece of $\frac{1}{8}$" × 1" flat steel.
   b. Clean the $\frac{1}{8}$" × 1" × 6" piece with a power wire brush.
   c. Round the corners and edges of each end slightly with a power grinder.
   d. Measure 1 inch in from one end, and center-punch on the centerline.
   e. Drill the center-punched mark with a $\frac{1}{8}$-inch drill bit.
   f. Redrill the hole with a $\frac{1}{2}$-inch drill bit.
   g. Remove all burrs or rough spots with a power wire wheel.
   h. Measure $1\frac{1}{2}$ inches from the end without a hole.
   i. Make a 90-degree bend at the line with a metal bender.
   j. Submit your project to the instructor.

Relevant Web Sites

Metal Benders, Inc.
www.metalbenders.net

Thompson Rivers University, Safe Working Procedures, Angle and Pedestal Grinders
www.tru.ca/hsafety/workinglearningsafely/work/grinder.html

Self-Evaluation

A. Multiple Choice. Select the best answer.

1. After installing a drill in a gear chuck, the next important thing is to
   a. start the motor
   b. remove the chuck wrench
   c. place the table off center
   d. check the belt for tightness

2. Round stock is best held for drilling by a
   a. C clamp
   b. helper
   c. vise
   d. V block

3. A grinding wheel may fly apart when running if the wheel does not have
   a. a coarse texture
   b. a coolant device
   c. a clean surface
   d. an adequate speed rating for the motor

4. Grinding wheels are cleaned and restored to roundness with a
   a. grit cutter
   b. screwdriver
   c. wheel dresser
   d. any one of these

5. It is dangerous to grind
   a. with heavy pressure on the metal
   b. using the side of the wheel
   c. without wearing a face shield
   d. all of these

6. Blades for power hacksaws are generally
   a. less than 12 inches long
   b. 12 inches to 18 inches long
   c. $\frac{1}{2}$ inch to $\frac{3}{8}$ inch wide
   d. none of these

7. A critical step to prevent breaking of power hacksaw blades is
   a. buy brittle blades
   b. clamp the work securely
   c. cool the blade frequently
   d. provide some slack in the blade

8. Gloves that are worn while doing metalwork should be made of
   a. asbestos
   b. cotton
   c. leather
   d. all are recommended

9. Power metal shears can cut
    a. angle stock
    b. flat stock
    c. round stock
    d. all of these

B. Matching. Match the word or phrase in column I with the correct word or phrase in column II.

Column I

   1. chuck

   2. quill

   3. center punch

   4. taper shank

   5. shear

   6. grinder wheel

   7. used to shape metal

   8. used to sharpen small tools

   9. hydraulic

  10. abrasive cutoff saw

Column II

   a. type of drill bit

   b. bonded grit

   c. coarse grinding wheel

   d. medium grinding wheel

   e. power for some shears

   f. tighten in two holes

   g. helps start drill bit

   h. to cut at an angle

   i. controlled by the feed handle

   j. cuts with a circular blade

C. Completion. Fill in the blanks with the word or words that make the following statements correct.

   1. The _____ _____ of the drill press permits stock to be supported for angle drilling.

   2. Grinding wheels may be checked for squareness by using a _____ or a _____ _____.

   3. The liquid used to cool metal-cutting saw blades is called a _____.

   4. Removing material so the grinding wheel is perfectly round, with the face square to the sides and the abrasive particles exposed, is called _____ the wheel.

D. Brief Answer. Briefly answer the following questions.

   1. Before using the drill press, what should one do to prevent damage to the table?

   2. Why is it important to store the handle after using power shears?

   3. What determines the speed of grinding?

   4. When cutting metal, what protective items should be worn? Why?

   5. Which metal-cutting power saw cuts more rapidly—the hacksaw or the horizontal band saw? Why?

Section 5

PROJECT PLANNING

UNIT 18

- Sketching and Drawing Projects

UNIT 19

- Figuring a Bill of Materials

UNIT 20

- Selecting, Planning, and Building a Project

UNIT 18
Sketching and Drawing Projects

Objective

To use simple drawing techniques to create plans for personal projects.

Competencies to be developed

After studying this unit, you should be able to:

- Identify common drawing equipment.
- Match basic drawing symbols with their definitions.
- Distinguish between pictorial and three-view drawings.
- Use common drawing techniques to represent ideas.
- Read and interpret a drawing.
- Make a three-view drawing of a given object.

Materials List

- Unlined paper, 8½" × 11"
- Lined paper, 8½" × 11"
- Ruler, 12-inch
- Pencil
- Soft eraser
- Sample drawing blocks
- Optional:
  a. Drawing board
  b. T square
  c. Right triangle
  d. Tape

Terms to Know

- represent
- sketch
- drawing
- dimension
- pictorial drawing
- three-view drawing
- protractor
- border line
- object line
- hidden line
- dimension line
- extension line
- break line
- center line
- leader line
- border
- title block
- full scale
- scale
- show box

Many objects are represented by other things. To **represent** means to stand for or to be a sign or symbol. For instance, a map uses lines to represent roads in a geographic area, such as the community or state (Figure 18-1). Books and bulletins use words to describe and create images in the mind that stand for ideas or objects. A photograph represents a person or object.

REPRESENTING BY SKETCHING AND DRAWING

A **sketch** is a rough drawing of project (Figure 18-2). A **drawing** is a picture or likeness made with a pencil, pen, chalk, crayon, or other instrument.

Sketching and drawing are used in agricultural mechanics to put ideas on paper. For example, in a building project an accurate sketch records the details of construction and indicates how the finished product should look. A sketch is needed to determine the amount of lumber, nails, bolts, and other materials to be used. Any construction project will be more efficient if there is an accurate sketch or carefully drawn plan to follow.

The ability to make sketches and simple drawings is a valuable skill (Figure 18-3). It is a way to record ideas for present and future use. In this way, a project can be planned in every detail before investing time and materials in construction. As the plan develops, some ideas may prove unworkable. Since the discovery is made before lumber or metal is cut, other workable ideas can be substituted.

A sketch or plan can be simple; it does not have to be done with specialized drawing equipment. However, a sketch must be complete. It must include lines to represent edges and corners of sections or parts of an object. Dimensions must be included to indicate the size of each part of the object. A **dimension** is a measurement of length, width, or thickness (Figure 18-4).

Perspective

Figure 18-4 is a pictorial drawing. A **pictorial drawing** shows all three dimensions at once. That is, it shows the item turned so the front, side, and top are in view. Pictorial drawings are very useful, but it takes a good deal of training and experience to draw them accurately.

Almost anyone can make clear drawings by showing one view at a time. This method is called **three-view drawing**. It will be discussed in detail later in this unit.

Drawing Instruments

Sketches and simple drawings can be prepared with instruments found in most homes. These instruments include a sharp lead pencil, an eraser, a 12-inch ruler, a compass, and a protractor (Figure 18-5). A **protractor** is an instrument for drawing or measuring angles.

If much drawing is to be done, several additional items will help save time. These items include a small drawing board with a T square, a 30° × 60° × 90° plastic triangle, a plastic scale, and some masking tape (Figure 18-6).

To use these tools, place the T square across the bottom of the drawing board. Then place an 8½" × 11" piece of paper on the board above the T square. Adjust the paper so it rests on the T square, with its base parallel to the base of the board. Using masking tape, attach the paper to the board as shown. The T square can now be used to make horizontal lines. Place the base of the triangle on the T square. Using the triangle, lines can drawn at 30-degree, 60-degree, or 90-degree angles to the T square. Use of the scale will be discussed later.

Large-scale projects can be generated using a computer, using a type of software program called computer-assisted design (CAD). In only a few minutes these programs can create plans for large and complicated projects that would otherwise take days to draw by hand. However, it is better to complete most small projects by hand drawing.

Symbols Used in Drawing

Hundreds of figures have been adopted by architects and engineers for use as symbols in drawing. Only a few of these are needed for preparing sketches and simple drawings, however.

Several types of lines have specific meanings (Figure 18-7). They are as follows:

- **border line**: a heavy, solid line drawn parallel to the edges of the drawing paper
- **object line**: a solid line showing visible edges and form of an object
- **hidden line**: a series of dashes that indicates the presence of unseen edges
- **dimension line**: a solid line with arrowheads at the ends to indicate the length, width, or height of an object or part
- **extension line**: a solid line showing the exact area specified by a dimension
- **break line**: a solid, zigzag line used to show where the illustration stops but the object does not

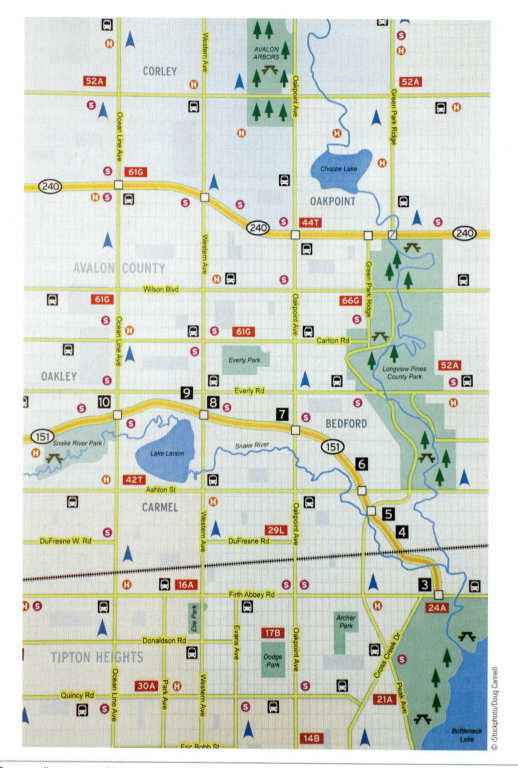

FIGURE 18-1 On maps, lines are used to represent roads and other geographical features.

- **center line:** a long-short-long line used to indicate the center of a round object
- **leader line:** a solid line with an arrow, used with an explanatory note to point to a specific feature of an object.

These lines and symbols help the person who is drawing to communicate with the person who will use the plan. Learning to recognize these lines and symbols is the first step in learning to read plans. Similarly, learning to use them is the first step in learning to draw plans.

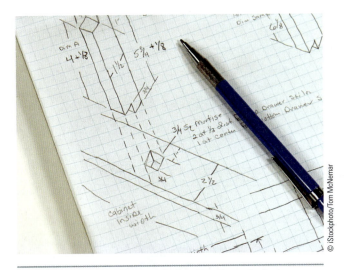

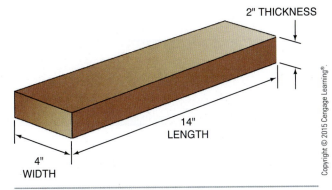

FIGURE 18-4 On project drawings, lines show the edges and corners of objects. Dimensions indicate length, width, and thickness.

FIGURE 18-2 A sketch is a rough drawing of a project.

FIGURE 18-3 The ability to make sketches and simple drawings is a valuable skill.

ELEMENTS OF A PLAN

Paper of any size may be used for drawing. However, it is advisable to start with 8½" × 11" paper. This is standard notebook size, is readily available, and easy to handle.

Border. When making a drawing, it is useful to draw a heavy line all around and close to the outer edges of the paper. This is called a **border**. The lines used to make a border are called *border lines*. It is suggested that borders be made ½ inch in from the edge of the paper.

Title Block. A few items of information about the total drawing are necessary. These include:

- the name of the person who prepared the drawing
- the date when the drawing was completed
- the name of the drawing
- the scale of the drawing

The section of a drawing reserved for information about the drawing in general is called the **title block**. For simple drawings, a ½-inch line drawn above the border line at the bottom of the paper works well. The information can be printed within these lines. The addition of very light guidelines drawn ⅛ inch from the top and bottom lines of the title block makes lettering easier.

Refer to Figure 18-8. Note that a border line has been drawn around the edges of the paper. The title block indicates that the drawing is a sample block, drawn to full scale by Bill Brown on January 1, 2014. **Full scale** means that the dimensions used in the drawing are the same size as those of the actual object it represents. In other words, 1 inch on the paper represents 1 inch on the block.

Views. The sample block is represented in this drawing by two views. The upper left-hand drawing is a top view of the block. In other words, the viewer is looking down on the block from above. The dimensions indicate that the block is 2 inches long and 2 inches wide. The inner circle indicates there is a hole in the center of the block. The dimension indicates it is 1 inch in diameter.

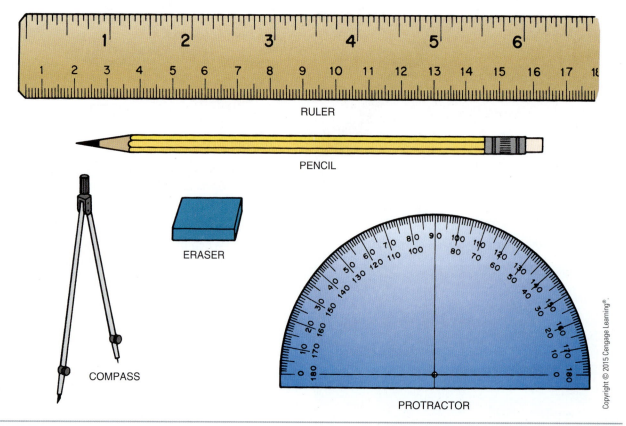

RULER

PENCIL

COMPASS

ERASER

PROTRACTOR

Copyright © 2015 Cengage Learning®.

FIGURE 18-5 Instruments for making sketches and simple drawings.

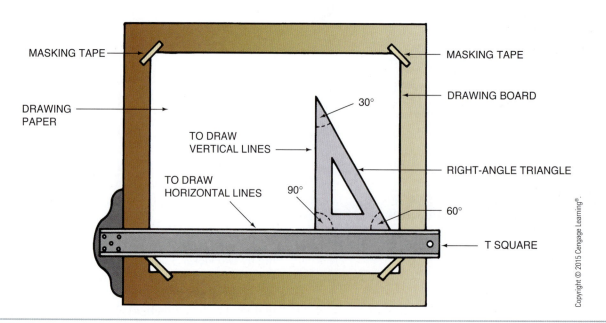

MASKING TAPE

DRAWING PAPER

TO DRAW VERTICAL LINES

TO DRAW HORIZONTAL LINES

30°

90°

60°

MASKING TAPE

DRAWING BOARD

RIGHT-ANGLE TRIANGLE

T SQUARE

Copyright © 2015 Cengage Learning®.

FIGURE 18-6 A drawing board with paper taped in place. The T square makes drawing horizontal lines easy. A right-angle triangle is useful for drawing vertical lines.

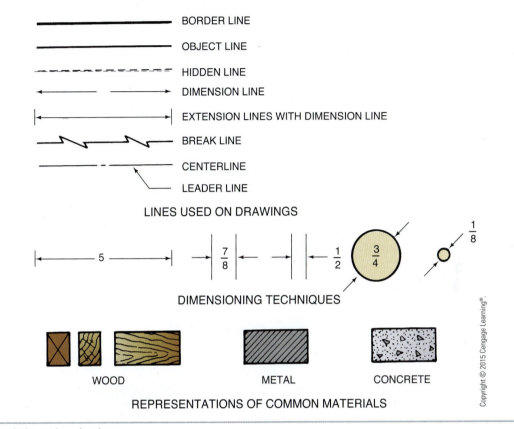

FIGURE 18-7 Some symbols used on drawings.

The outer circle shows that a larger hole goes partway through the block. The size of the outer hole is not shown. However, remember that the object is drawn to full scale. This means the drawing is the same size as the block. Measuring the outer circle on the drawing shows it to be 1¼ inches in diameter. Therefore, the hole in the block is the same size, or 1¼ inches.

The lower right-hand drawing in Figure 18-8 shows a front view of the block. This view shows that the height of the block is 1½ inches (¾ inch + ¾ inch = 1½ inches). It also shows that the larger hole is ¾ inch deep and the smaller hole goes through the remaining ¾ inch of the block. A note is provided just above the title block on the right to specify the material from which the block is to be made. Enough information is provided in Figure 18-8 to make the block correctly. Therefore, it is a good drawing.

Most objects require three views in a drawing to provide complete information. For example, consider the bookend shown in Figure 18-9. The top view is at the

upper left. The front view is at the lower left. However, the object is not recognized as a common bookend until the end view (provided at the lower right). In this case, the end view gives the best indication of the shape of the object. End views are generally placed to the right of the front view and in line with it.

The views are placed in the drawing with the front view in the lower left, the top view above it, and the end view to the right of the front view. This arrangement allows dimensions placed on the drawing to represent more than one view (Figure 18-10). By using this technique, dimensions are not repeated and the drawing is less cluttered (Figure 18-11).

SCALE DRAWING

The word **scale** as used here means the size of a plan or drawing as compared to that of the object it represents. In Figure 18-8, the drawing was made as large as the object.

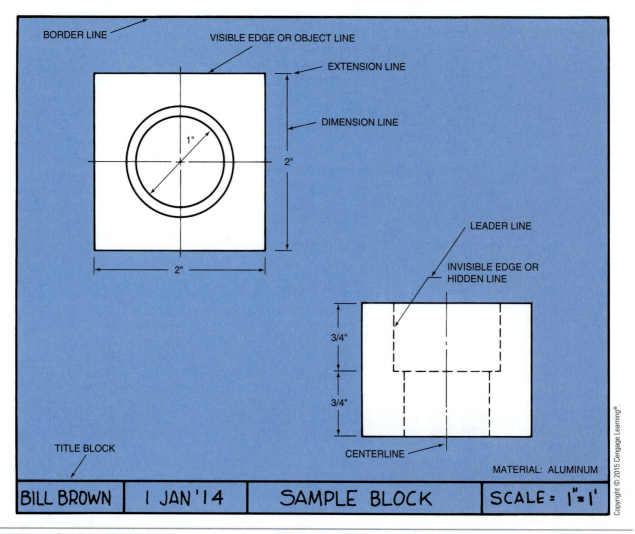

FIGURE 18-8 Drawing with a border, a title block, lines, and dimensions.

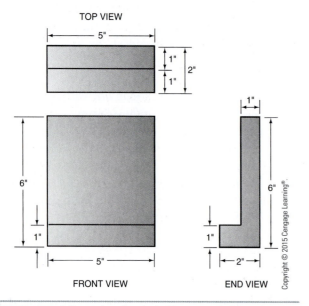

FIGURE 18-9 A three-view drawing with repeated dimensions. Each view is labeled here, but the labels would not appear on a final drawing.

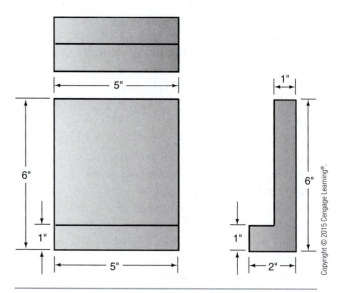

FIGURE 18-10 The preferred method of presenting three-view drawings requires each dimension to be shown only once.

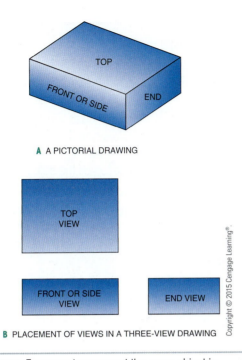

A A PICTORIAL DRAWING

B PLACEMENT OF VIEWS IN A THREE-VIEW DRAWING

Copyright © 2015 Cengage Learning®.

FIGURE 18-11 Two ways to represent the same object in drawings.

Since the size of the object was rather small, the drawing fit on the paper. However, most objects are much larger than a sheet of 8½" × 11" paper. Therefore, the scale to use in preparing the drawing must be determined.

Determining the Scale

A piece of 8½" × 11" paper, turned sideways, with borders and title block drawn, has 10 inches of horizontal drawing space and 7 inches of vertical drawing space. The three views must be planned to fit into this space without crowding.

A good place to start in deciding upon the scale is to assume that 1 inch equals 1 foot (1" = 1'). It is recommended that 1 inch of space for dimensions be allowed on every side of each view. Therefore, 1 inch to the left of the front view, 1 inch between views, and 1 inch to the right of the end view is reserved. The original 10 inches of horizontal space left after the borders are drawn, reduced by 3 inches for dimensions, leaves 7 inches of horizontal space for the two views.

A three-view drawing has a front view and a top view, one over the other. Three spaces of 1 inch each are for dimensions. This reduces the original 7 inches of vertical space by 3 inches, leaving 4 inches of vertical drawing space.

This procedure indicates that a scale of 1" = 1' is adequate to draw objects with front end horizontal dimensions that add up to 7 feet or less, and front and top

vertical dimensions that total no more than 4 feet. Larger objects require a larger scale, such as 1" = 2', 1" = 3', and so on. For example, consider a three-view drawing of a wagon body with hay racks. The dimensions are:

- Length (L) = 16'
- Width (W) = 8'
- Height (H) = 8'

Remember that there are 7 inches of horizontal drawing space and 4 inches of vertical drawing space. With a scale of 1" = 4', 1 inch of drawing represents 4 feet of wagon (divide each dimension by 4):

- L = 16' actual dimension, 4" scale value
- W = 8' actual dimension, 2" scale value
- H = 8' actual dimension, 2" scale value

For the horizontal views, the dimensions on the drawing would be:

4" (front view) + 2" (end view) = 6"

Since there are 7 inches of horizontal drawing space, the scale is usable for the horizontal views. The extra inch will simply be extra space.

For the vertical views, the dimensions would be:

2" (front view) + 2" (top view) = 4"

Since there are 4 inches of vertical drawing space, the scale will work.

Using a Scale

A scale is an instrument with all increments shortened according to proportion (Figure 18-12). On a 1/2 scale, 1-inch marks are set only ½ inch apart. It can be used to let ½ inch equal 1 inch, 1 foot, 1 yard, or 1 mile.

Some instruments have a different scale on each edge. A triangular scale has three sides but six scales, with each side showing two scales. Some instruments show different scales at each end for a total of four scales per side, or twelve scales per instrument.

An expensive scale should never be used as a straightedge for drawing because the edge can be damaged by running a pencil along the edge. However, low-cost plastic scales may be used for this purpose.

MAKING A THREE-VIEW DRAWING

To demonstrate the procedure for making a three-view drawing, a drawing of a show box will be made. A **show box** is used to carry livestock equipment and supplies

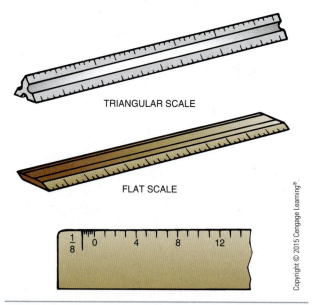

TRIANGULAR SCALE

FLAT SCALE

Copyright © 2015 Cengage Learning®.

FIGURE 18-12 Triangular and flat scales. On the bottom scale, 1/8 inch represents 1 inch.

used for preparing animals for showing. It can also be used to store and transport tools and materials for other purposes. The plans could be used to construct a trunk, chest, or toy box.

The show box will be 4 feet long, 2 feet wide, and 1½ feet high. The scale will be 1 inch equals 1 foot, or 1" = 1'. This can also be written as 1 inch equals 12 inches, or 1 = 12, or simply 1/12.

Blocking in the Views

The following procedure describes how to block in the views for a three-view plan.

Procedure

1. Select a plain, 8½" × 11" piece of heavy paper. Regular drawing paper is preferred, but copier paper will do. Turn the paper lengthwise in front of you (Figure 18-13). Draw a ½-inch border around the paper. Add a ½-inch title block.

2. In the title block, letter your name and the date. For the name of the project, print "Show Box"; for scale, Scale = 1" = 1'.

3. Make a light mark 1 inch in from the left margin, and draw a very light line through it from top to bottom, and parallel to the left border. This becomes the object line for the left side of the top and front views.

4. Make a light mark 1 inch up from the title block, and draw a very light line through it from border to border and parallel to the title block. This becomes the bottom object line.

5. The length of the box is 4 feet. The scale is 1" = 12", or 1" = 1'. Measure 4 inches from the left object line, and draw a very light vertical line from top to bottom. This becomes the object line on the right side of the top and front view.

6. Measure 1 inch to the right of the line you drew in step 5, and draw a very light vertical line from top to bottom. This becomes the right side of the end view.

7. Measure another 2 inches to the right, and draw another very light vertical line. This becomes the right side of the end view.

8. The show box is 1½ feet high. So, start at the bottom object line and measure up 1½ inches. Draw a very light line to form the top object lines of the front and end views.

NOTE

Do not extend the line into the spaces around the views.

9. Measure up 1 inch from the top object line and draw a light line to form the bottom object line of the top view.

NOTE

Do not extend the lines into the spaces around the views.

10. The show box is 2 feet wide. So, measure another 2 inches and draw a light line to form the upper object line of the top view.

11. Now all three views are blocked in and drawn to scale (Figure 18-14). Erase excess lines that extend beyond the views (Figure 18-15).

NOTE

Remember that you planned for 1 inch more of space around views. Therefore, when blocking in views, final lines generally stop before they touch the borders.

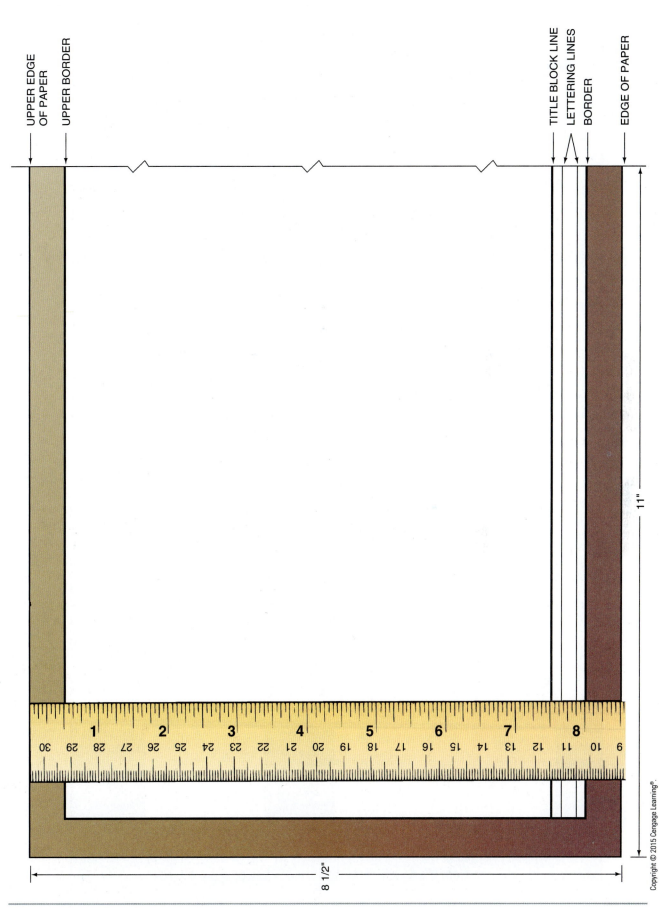

FIGURE 18-13 Proper placement of the ruler to make marks for drawing the lower border, the lettering guides, the title block line, and the upper border line.

Copyright © 2015 Cengage Learning®.

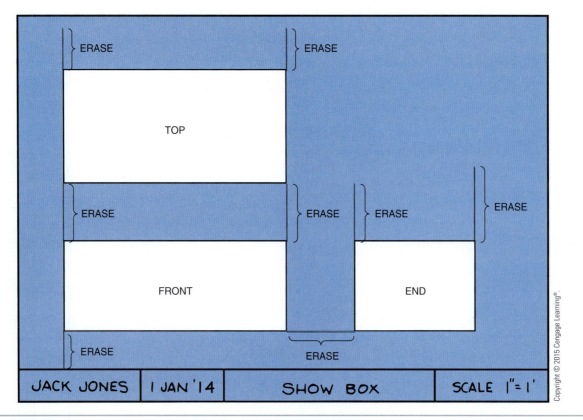

Copyright © 2015 Cengage Learning®.

FIGURE 18-14 Blocked-in, three-view drawing of the show box before erasure of lines.

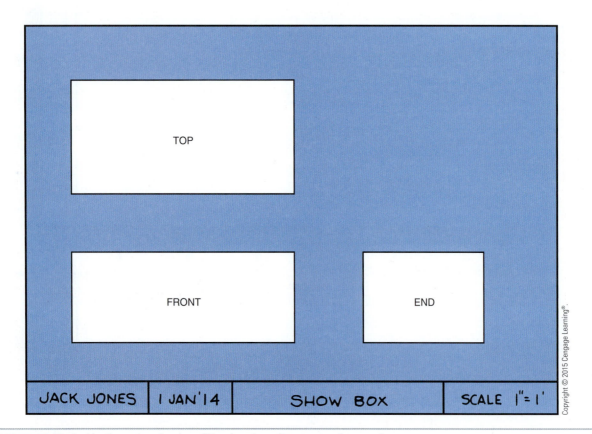

Copyright © 2015 Cengage Learning®.

FIGURE 18-15 Blocked-in, three-view drawing of the show box after erasure of excess lines.

All three views are now blocked in. This means the outside object lines are drawn. Next, the details of each view must be added.

Completing the Front View

The most prominent view is generally the front view. Therefore, the front view should be drawn first. It will probably tell the viewer more about the project than either the top or the end view.

Assume that the show box will be made of ¾-inch plywood. The sides will be rabbeted, and the ends set in. The bottom will be hidden by the sides and ends. The lid will fit on top of the sides and ends.

Before drawing, it must be decided how far apart two lines should be on paper to represent the thickness of ¾-inch plywood at 1/12 scale. For convenience, round off the ¾-inch thickness to 1 inch. Therefore, 1/12 inch on the drawing represents lumber that is ¾ inch thick. Rulers are laid out in eighths and sixteenths, not twelfths. Therefore, either $^1/_{16}$ or ⅛ inch must be chosen to represent an inch or a special 1/12 drawing scale must be obtained. For sketching purposes, ⅛ inch

will be used to represent the ¾-inch plywood to better show the detail of the joints.

To draw in the details of the front view, the steps are as follows.

Procedure

1. Measure ⅛ inch in from the left side of the front view, and draw a broken vertical line from top to bottom. This line is a hidden line because the edge of the end panel cannot be seen (Figure 18-16).

2. Measure ⅛ inch in from the right side of the front view, and draw a broken vertical line from top to bottom. This line represents the edge of the right-hand end panel.

3. Measure ⅛ inch up from the bottom ⅛ inch and draw a broken horizontal line between the two vertical broken lines. This line represents the hidden top edge of the bottom panel.

(continued)

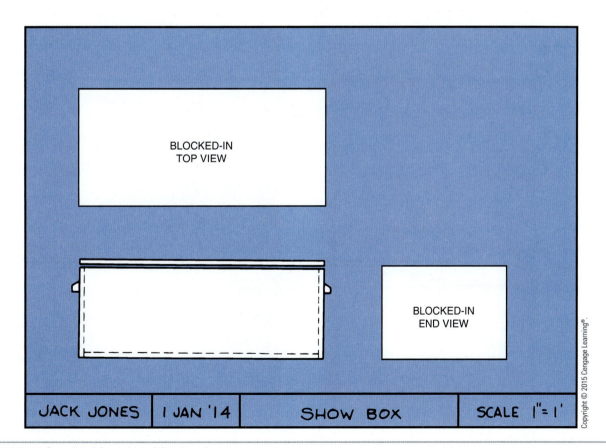

FIGURE 18-16 Completed front view of the show box.

Within the figure:

BLOCKED-IN
TOP VIEW

BLOCKED-IN
END VIEW

JACK JONES | 1 JAN '14 | SHOW BOX | SCALE 1"= 1'

Copyright © 2015 Cengage Learning®.

Procedure, *continued*

4. Measure ⅛ inch from the top of the front view, and draw a horizontal solid line from solid line to solid line to represent the thickness of the lid.

5. Erase any parts of the broken vertical lines that show inside the two lines representing the edge of the closed lid.

6. Show where handles will be attached, 12 inches up from the bottom of the chest.

This completes the front view, unless the exposed heads of nails or screws used to attach the front to the ends and bottom are to be shown. A hasp or other locking device could also be pictured.

Completing the End View

With experience in drawing, some of the parts of the top and end views can be drawn at the same time the front view is drawn. Many of the lines can be drawn by placing the ruler on the measurement and drawing the lines in two views without moving the ruler.

Refer to Figure 18-17. The steps in drawing the end view follow.

Procedure

1. Place the ruler on the bottom line of the lid in the front view. The ruler should extend across the end view also. Draw the bottom line of the lid in the end view. (Use a solid line.)

2. The end view of the sides needs special treatment to show the rabbet joints. That is, because ⅜ inch of the plywood end grain is seen, the other ⅜ inch is hidden. To show this, measure ¹⁄₁₆ inch in from the left-hand side of the end view and draw a solid vertical line. Measure another ¹⁄₁₆ inch farther to the right, and draw a vertical broken line. Repeat the procedure on the right-hand side of the end view, moving leftward this time.

3. Place the ruler on the hidden interior edge of the bottom panel in the front view. Draw in the hidden interior edge of the bottom panel in the end view. (Use a broken line.)

4. Draw in the handle. Notice that the handle is on the same level in both the front view and the end view. The handle consists of a 1" × 2" × 6" hollowed strip of wood. Since the scale is 1" = 12", the handle is ½" long on the drawing. Measure to the vertical center line of the end view, and then mark outward ¼ inch in each direction, using a solid line.

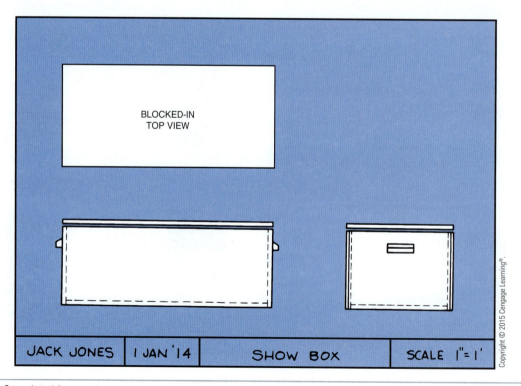

Copyright © 2015 Cengage Learning®.

FIGURE 18-17 Completed front and end views of the show box.

Completing the Top View

Refer to Figure 18-18. The steps in drawing the top view follow.

Procedure

1. Place the ruler on the hidden inner edge of the left end panel of the front view and extend it over the top view. (Use a broken line.)

2. Repeat step one on the right side of the view.

3. Measure ⅛ inch up from the bottom and ⅛ inch down from the top lines and draw the inner edges of the sides with broken lines.

4. Show the rabbet joints correctly in each corner. That is, show how the side panels are sawed and how the end panels fit into the rabbet grooves.

5. Add the handles on each end by locating the center line and measuring out ¼ inch in each direction. Assume a thickness of 1 inch wood, or ⅛ inch on paper.

Dimensioning the Drawing

The final step to complete the drawing is to add dimensions (Figure 18-19). The dimensions are the exact measurements, more exact than the drawing itself. The drawing focuses on details of the assembly. The dimensions tell the exact length, width, and height of the object. The dimensions also indicate the thickness of materials and other construction details as needed.

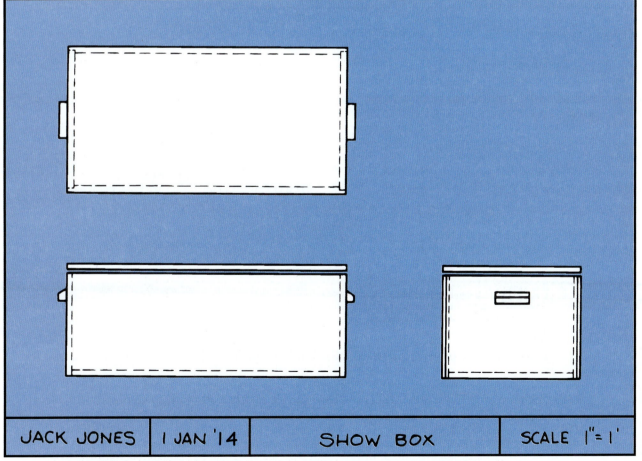

JACK JONES   1 JAN '14   SHOW BOX   SCALE 1"= 1'

Copyright © 2015 Cengage Learning®.

FIGURE 18-18 All three views of the show box completed, without dimensions.

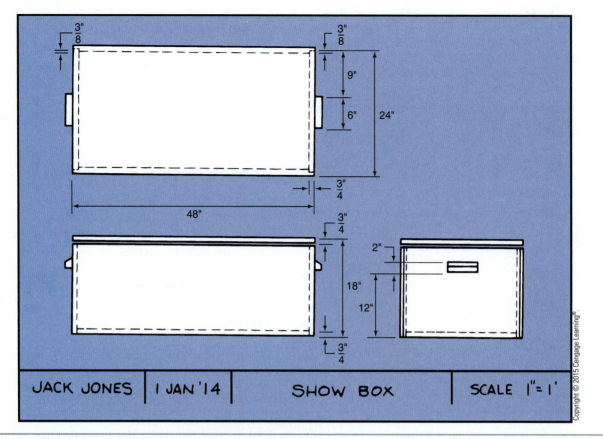

FIGURE 18-19 Completed drawing of the show box with dimensions.

SUMMARY

Quality projects begin at the planning stage. The proper dimensions and specifications have to be illustrated in a manner that is clear in detail and easy to follow. Mistakes are avoided by following a carefully drafted plan. Spending time in the planning and drawing of plans will result in a much higher-quality project.

Student Activities

1. Define the Terms to Know in this unit.

2. Study agricultural mechanics project plans to visualize how individual parts look and how they are fastened together.

3. Practice printing the alphabet in capital letters and in lowercase letters. Follow the form provided in Figure 18-7. Use lined paper.

4. Using a ruler, make twenty arrows following the form used in Figure 18-7 for dimension lines.

5. Practice making neat figures for dimensions by numbering from 0 through 9. Follow the form provided in Figure 18-7.

6. Practice measuring accurately by drawing lines on a sheet of paper the length of each figure stated below. Make some lines solid and some broken. Submit the paper to your instructor for evaluation.
    a. 1 inch
    b. 1½ inches
    c. 1¾ inches
    d. 3⅛ inches
    e. 3⅝ inches
    f. 1 1/16 inches
    g. 1 3/16 inches
    h. 1¼ inches
    i. 1 9/16 inches
    j. 1⅜ inches

7. Ask the instructor to provide blocks or other objects for you to practice drawing. Draw one or more of such objects.

8. Convert the following project dimensions to the measurements they would be on paper, first using a 1/12 scale (1" = 12") and then using a 1/2 scale (1" = 2").
    a. 12 inches = _____ _____
    b. 18 inches = _____ _____
    c. 36 inches = _____ _____
    d. 6 inches = _____ _____
    e. 3 inches = _____ _____

9. If a drawing instrument with 1/12 and 1/2 scales is available, draw lines on a sheet of paper representing the length of each item in activity 8.

Self-Evaluation

A. Multiple Choice. Select the best answer.

1. A simple plan must
    a. be done with specialized equipment
    b. be done by a draftsperson or engineer
    c. have dimensions
    d. all of these

2. The triangle recommended as an aid for simple drawing is
    a. 30° × 60° × 90°
    b. plastic
    c. three-sided
    d. all of these

3. Which scale is best for making a three-view drawing on a sheet of 8½" × 11" paper if the project is 4 feet long, 2 feet wide, and 1½ feet high?
    a. 1" = 12"
    b. 1" = 6"
    c. 1" = 3"
    d. 1" = 1"

4. Which scale is best for making a drawing of the front view only on an 8½" × 11" sheet of paper if the project is 4 feet long, 2 feet wide, and 1½ feet high?
    a. 1" = 12"
    b. 1" = 6"
    c. 1" = 3"
    d. 1" = 1"

5. The process of first drawing the outside object lines for all three views is called
    a. blocking in
    b. pictorial drawing
    c. scale drawing
    d. sketching

B. Matching. Match the terms in column I with those in column II.

Column I

1. dimension
2. pictorial
3. object line
4. border line
5. hidden line
6. extension line
7. leader line
8. center line
9. title block
10. Scale

Column II

a. ½" from edge of paper
b. use with dimensions
c. round objects
d. three views in one
e. targets special information
f. measurement
g. name, date, project, scale
h. proportion
i. broken line
j. visible edges

C. Completion. Fill in the blanks with the word or words that make the following statements correct.

1. The view placed in the upper left corner of a plan is the _____.
2. The view placed at the lower left corner of a plan is the _____.
3. The view placed in the lower right corner of the plan is the _____.
4. The _____ view is the most prominent and should be drawn first.
5. The _____ are more exact than the drawing itself.

D. Brief Answer. Briefly answer the following questions.

1. What is a drawing?
2. What is a sketch?
3. What is a protractor?
4. Using a scale of 1" = 12", convert the following measurements.
   a. 48 inches = _____
   b. 6 inches = _____
   c. 9 inches = _____
   d. 1 inch = _____

UNIT 19

Figuring a Bill of Materials

Objective

To state the use and format of a bill of materials and to make all calculations needed to develop a bill of materials.

Competencies to be developed

After studying this unit, you should be able to:
- Define terms associated with a bill of materials.
- State the components of a bill of materials.
- Record dimensions of structural metals and lumber.
- Calculate board feet.
- Calculate costs included in a bill of materials.
- Prepare a written bill of materials.

Materials List

- Paper
- Pencil

Terms to Know

- bill of materials
- item
- rounded up
- tongue-and-groove
- galvanize
- cadmium
- board foot

COMPONENTS OF A BILL OF MATERIALS

A **bill of materials** is a list and description of all the materials to be used in constructing a project. For each component, the bill of materials includes the following:

- item or part name
- number of pieces
- type of material
- size of pieces
- description of parts
- total feet
- unit cost
- total component cost

A bill of materials is used to determine the material that must be assembled for a project and its estimated cost. To determine the amount of lumber needed, the student must be able to compute board feet. The student must also be able to combine the costs of numerous pieces and types of materials to arrive at a total estimate of the cost of the project.

Several formats are used for bills of materials. These vary with the project or building being planned. One format for listing items is shown in Figure 19-1.

The **item** means a separate object. In a bill of materials, the term may refer to a part of a structure such as the side, end, bottom, or top. In a trailer body, an item may be "brackets" for attaching cross members to beams, "lag screws" for attaching pockets for side boards, "boards" for the bottom, "steel" for the sides, or "paint" for the whole project.

An item in a bill of materials may be paint, rafters, doors, nails, hinges, or 2 × 4s. The important thing to remember in making a bill of materials is that items of different types, materials, sizes, or costs must not be mixed and listed as one item.

The number of pieces is helpful since it can be used to buy materials efficiently. It is also needed in the calculation of total feet and total cost. For example,

consider a plan to build a small storage building. The door must be made from boards that are 1" × 6". Boards can be bought in 6-, 8-, 10-, 12-, 14-, and 16-foot lengths.

The plan indicates that the door is 6 feet high and 3 feet wide (Figure 19-2). This knowledge enables the worker to determine the number of pieces as follows: 36 inches (width of door) divided by 5½ inches (width of a finished 1" × 6") = 6.5 boards. This calculation indicates that 7 boards, each 6 feet long, are needed to make the 36-inch door. The number 6.5 is rounded up to 7 boards. When a number is **rounded up**, it means the next higher whole number is used.

The plan also shows cross boards at the top and bottom of the door to hold the vertical boards together. Each crosspiece is 36 inches (3 feet) long, equal to the distance across the 36-inch door. Finally, the plan calls for a 1" × 6" diagonal brace. Calculations based on the length and distance apart of the two cross boards would reveal that the diagonal brace needs to be 5 feet 2 inches long.

The bill of materials may be simplified by combining similar items. The total number of 1" × 6" pieces required to make the door can now be stated as follows:

7 pieces—6 feet long
2 pieces—3 feet long
1 piece—5 feet 2 inches long

Since the wooden pieces are all 1" × 6" boards, the items listed can be combined. To combine the three items into one item in the bill of materials, the two 3-foot pieces are cut from one 6-foot piece. The 5-foot-2-inch piece can also be cut from a 6-foot piece. Therefore, to make the door, nine pieces of 1" × 6" × 6' lumber are needed.

To make the door weathertight and strong, **tongue-and-groove** (T&G) lumber should be used. Tongue-and-groove boards have a tonguelike edge sticking out on one side and a groove cut into the other, which permits them to lock together.

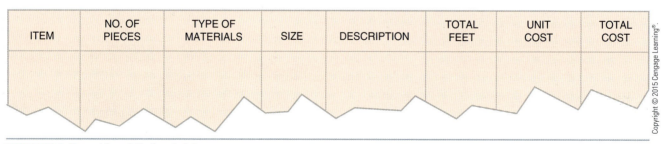

ITEM	NO. OF PIECES	TYPE OF MATERIALS	SIZE	DESCRIPTION	TOTAL FEET	UNIT COST	TOTAL COST

FIGURE 19-1 Format for a bill of materials.

Copyright © 2015 Cengage Learning®

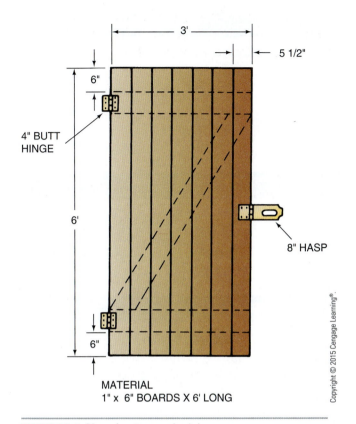

4" BUTT
HINGE

8" HASP

MATERIAL
1" x 6" BOARDS X 6' LONG

FIGURE 19-2 Plan of a storage shed door.

Assume the cost of the boards will be $0.30 per linear foot. The lumber requirement would be included in the bill of materials as shown for "door" in (Figure 19-3). The plan also calls for a pair of 4-inch butt hinges and an 8-inch hasp. These items of hardware, along with nails and screws, are also included on the bill of materials shown in Figure 19-3.

Some additional explanation is in order regarding the sample bill of materials. The 1" × 6" lumber

was priced by the linear foot since tongue-and-groove boards are used. This means that each running foot, or each foot of length in the board, costs $0.30. The amount of cost of a linear foot (or LF) is influenced by the width and thickness. The total number of linear feet is 54; thus, 54 × $0.30 = $16.20.

Rough lumber from the sawmill is sold by the board foot. Calculating board feet requires computation of the volume of the board. Calculating board feet will be discussed later. The *S4S* under "Description" is a symbol designating "surfaced 4 sides" or four surfaces of the board have been planed. The symbol *S4S* means the boards are planed or surfaced on four sides, while the symbol *S2S* would mean the boards were surfaced on two sides.

Since the hasp in the example is galvanized, the finish on the screws is specified as cadmium (Cd). To **galvanize** is to coat a metal with zinc. **Cadmium** is a similar metal used for rust-resistant plating of fasteners and other steel products. "No. 10 × ¾" size screws" means the screws are number 10 diameter and ¾" long. "N/A" means "not applicable" and is used when requested information does not apply to the situation.

If the door is a single project, the paint should be added to the bill of materials. However, the door could be part of a whole building under construction, in which case paint would appear as a single item for the entire building, unless the door was painted a different color.

UNITS OF MEASURE

A bill of materials provides much information in a small space. Therefore, standard abbreviations are helpful (Figure 19-4).

Item	No. of Pieces	Type of Materials	Size	Description	Total Feet	Unit Cost	Total Cost
Door	9	Boards	1″ × 6″ × 6′	S4S T&G Pine	54	$0.30	$16.20
Hinge	1 pr	Steel	4″	Butt	N/A	$4/pr	4.00
Hasp	1	Steel	8″	Galvanized	N/A	$3.50	3.50
Nails	1 lb	Steel	6d	Common	N/A	$0.75/lb	0.75
Screws	24	Steel	No. 10 × ¾″	Cd finish	N/A	$0.06 ea	1.44
Paint	1	Oil Base	Quart	Red	N/A	$6.50	6.50
							$32.39

FIGURE 19-3 Lumber, hardware, and paint requirements in a bill of materials.

Copyright © 2015 Cengage Learning®.

Standard Abbreviations

" or in = inch

' or ft = foot

yd = yard

mi = mile

ea = each

@ = at

N/A = not applicable

pt = pint

qt = quart

gal = gallon

LF = linear foot

BF = board foot

S1S = surface 1 side

S2S = surface 2 sides

S3S = surface 3 sides

S4S = surface 4 sides

no. or # = number

in² or sq in = square inch

ft² or sq ft = square foot

yd² or sq yd = square yard

square = 10′ × 10′ or 100 square feet

NC = national coarse threads

NF = national fine threads

NPT = national pipe threads

d = penny (nails)

lb = pound

oz = ounce

Cwt = hundredweight (100 pounds)

FIGURE 19-4 Useful abbreviations to include in a bill of materials.

Lumber

(All available in 6′, 8′, 10′, 12′, 14′, and 16′ lengths)

1″ × 6″	2″ × 6″
1″ × 8″	2″ × 8″
1″ × 10″	2″ × 10″
1″ × 12″	2″ × 12″
2″ × 4″	4″ × 6″

Plywood

(Interior and Exterior Grades)

$1/4$″ × 4′ × 8′	$5/8$″ × 4′ × 8′
$3/8$″ × 4′ × 8′	$3/4$″ × 4′ × 8′
$1/2$″ × 4′ × 8′	1″ × 4′ × 8′

Structural Steel

(Standard length = 20′)

Flat Band Iron—$1/8$″, $3/16$″, $1/4$″, $5/16$″, $3/8$″, and $1/2$″ thick

—$1/2$″, $3/4$″, 1″, $1\,1/2$″, 2″, 3″, and 4″ wide

Angle Iron—$1/8$″, $3/16$″, $1/4$″, $5/16$″, and $3/8$″ thick

—$1/2$″ × $1/2$″, $3/4$″ × $3/4$″, 1″ × 1″, $1\,1/2$″ × $1\,1/2$″, 2″ × 2″, and 3″ × 3″ (width of legs)

Round (Hot-rolled, Cold-rolled, or Tool Steel)

—$1/4$″, $5/16$″, $3/8$″, $1/2$″, $5/8$″, $3/4$″, 1″, $1\,1/2$″, and 2″ (diameter)

Black or Galvanized Steel Pipe

(Standard Length = 21′)

—$1/4$″, $3/8$″, $1/2$″, $3/4$″, 1″, $1\,1/4$″, $1\,1/2$″, 2″, $2\,1/2$″, and 3″ (inside diameter)

FIGURE 19-5 Common sizes and units of lumber and structural metals.

It is easier to plan and build projects with a knowledge of the sizes and weights of lumber and steel products. These are given in Figures 19-5 through 19-8.

CALCULATING BOARD FEET

Lumber in the form of trees, logs, or rough lumber is measured and sold by board feet. Sometimes dressed lumber is sold by the board foot as well.

A **board foot** is an amount of wood equal to a board 1 inch thick, 1 foot wide, and 1 foot long; or 1 board foot is 1″ × 12″ × 12″. A board foot is also 144 cubic inches (cu in. or in.³); that is, 1″ × 12″ × 12″ = 144 cu in. (Figure 19-9). Cubic measures are determined by multiplying thickness by width by length. To convert cubic inches back to board feet, divide by 144. Hence, 144 cubic inches divided by 144 = 1 board foot.

Two Methods for Calculating Board Feet

Two methods or formulas for calculating board feet are especially useful. Each method is useful for certain sizes of lumber.

Copyright © 2015 Cengage Learning®.

Angles—Bar Sizes

Size in Inches	Weight per Foot (Lb)	Size in Inches	Weight per Foot (Lb)
1 × 1 ×	$1/8$ .... .80	2 × 1$1/2$ ×	$1/8$ 1.44
	$3/16$ 1.16		$3/16$ 2.12
	$1/4$ 1.49		$1/4$ 2.77
1$1/8$ × 1$1/8$ × $1/8$ .90		2 × 2 ×	$1/8$ 1.65
1$1/4$ × 1$1/4$ × $1/8$ 1.01			$3/16$ 2.44
	$3/16$ 1.48		$1/4$ 3.19
	$1/4$ 1.92		$5/16$ 3.92
1$3/8$ × $7/8$ × $1/8$ .91			$3/8$ 4.70
	$3/16$ 1.32	2$1/4$ × 1$1/2$ × $3/16$ 2.28	
1$1/2$ × 1$1/4$ × $3/16$ 1.64		2$1/2$ × 1$1/2$ × $3/16$ 2.44	
1$1/2$ × 1$1/2$ × $1/8$ 1.23			$1/4$ 3.19
	$3/16$ 1.80		$5/16$ 3.92
	$1/4$ 2.34	2$1/2$ × 2 ×	$3/16$ 2.75
1$3/4$ × 1$1/4$ × $1/8$ 1.23			$1/4$ 3.62
1$3/4$ × 1$3/4$ × $1/8$ 1.44			$5/16$ 4.50
	$3/16$ 2.12		$3/8$ 5.30
	$1/4$ 2.77	2$1/2$ × 2$1/2$ × $3/16$ 3.07	
2 × 1$1/4$ ×	$3/16$ .. 1.96		$1/4$ 4.10
	$1/4$ 2.55		$5/16$ 5.00
			$3/8$ 5.90
			$1/2$ 7.70

Angles—Structural

Size in Inches	Weight per Foot (Lb)	Size in Inches	Weight per Foot (Lb)
3 × 2 ×	$3/16$ 3.07	3 × 3 ×	$3/16$ 3.71
	$1/4$ 4.10		$1/4$ 4.90
	$5/16$ 5.00		$5/16$ 6.10
	$3/8$ 5.90		$3/8$ 7.20
	$1/2$ 7.70		$7/16$ 8.30
3 × 2$1/2$ × $1/4$	4.50		$1/2$ 9.40
	$5/16$ 5.60	3$1/2$ × 2$1/2$ × $1/4$ 4.90	
	$3/8$ 6.60		$5/16$ 6.10
	$1/2$ 8.50		$3/8$ 7.20
3$1/2$ × 3 × $1/4$	5.40		$1/2$ 9.40
	$5/16$ 6.60	5 × 5 ×	$3/8$ 12.3
	$3/8$ 7.90		$1/2$ 16.2
	$1/2$ 10.20		$5/8$ 20.0
3$1/2$ × 3$1/2$ × $1/4$	5.80		$3/4$ 23.6
	$5/16$ 7.20	6 × 3$1/2$ × $5/16$ 9.8	
	$3/8$ 8.50		$3/8$ 11.7
	$7/16$ 9.80		$1/2$ 15.3
	$1/2$ 11.10	6 × 4 ×	$5/16$ 10.3
4 × 3 ×	$1/4$ 5.80		$3/8$ 2.3
	$5/16$ 7.20		$7/16$ 14.3
	$3/8$ 8.50		$1/2$ 16.2
	$7/16$ 9.80		$5/8$ 20.0
	$1/2$ 11.10		$3/4$ 23.6
	$5/8$ 13.60		$7/8$ 27.2
4 × 3$1/2$ × $1/4$	6.2	6 × 6 ×	$3/8$ 14.9
	$5/16$ 7.7		$7/16$ 17.2
	$3/8$ 9.1		$1/2$ 19.6
	$7/16$ 10.6		$5/8$ 24.2
	$1/2$ 11.9		$3/4$ 28.7
4 × 4 ×	$1/4$ 6.6		$7/8$ 33.1
	$5/16$ 8.2		1 .... 37.4
	$3/8$ 9.8	7 × 4 ×	$3/8$ 13.6
	$7/16$ 11.3		$1/2$ 17.9
	$1/2$ 12.8		$5/8$ 22.1
	$5/8$ 15.7	8 × 4 ×	$1/2$ 19.6
	$3/4$ 18.5	8 × 6 ×	$1/2$ 23.0
5 × 3 ×	$1/4$ 6.6		$5/8$ 28.5
	$5/16$ 8.2		$3/4$ 33.8
	$3/8$ 9.8		1 .... 44.2
	$7/16$ 11.3	8 × 8 ×	$1/2$ 26.4
	$1/2$ 12.8		$5/8$ 32.7
5 × 3$1/2$ × $3/16$	8.7		$3/4$ 38.9
	$3/8$ 10.4		$7/8$ 45.0
	$7/16$ 12.0		1 .... 51.0
	$1/2$ 13.6		
	$5/8$ 16.8		
	$3/4$ 19.8		

FIGURE 19-6 Weight per foot of angle steel. The cost per foot can be determined by multiplying the weight (in pounds) by the price per pound.

Copyright © 2015 Cengage Learning®

Concrete Reinforcing Bars

Bar Sizes Old (Inches)	New (Numbers)	Weight per Foot (Lb)	Diameter (Inches)
1/4	2	.167	.250
1/4	2	.167	.250
3/8	3	.376	.375
1/2	4	.668	.500
5/8	5	1.043	.625
3/4	6	1.502	.750
7/8	7	2.044	.875
1	8	2.670	1.000
1	9	3.400	1.128
1 1/8	10	4.303	1.270
1 1/4	11	5.313	1.410

Rounds

Size in Inches	Weight per Foot (Lb)	Size in Inches	Weight per Foot (Lb)
1/4	0.167	2 1/4	13.52
5/16	0.261	2 3/8	15.06
3/8	0.376	2 1/2	16.69
7/16	0.511	2 5/8	18.40
1/2	0.668	2 3/4	20.20
9/16	0.845	3	24.03
5/8	1.043	3 1/4	28.21
3/4	1.502	3 1/2	32.71
7/8	2.044	3 3/4	37.55
1	2.670	4	42.73
1 1/8	3.379	4 1/4	48.23
1 1/4	4.173	4 1/2	54.08
1 3/8	5.05	4 3/4	60.25
1 1/2	6.01	5	66.76
1 5/8	7.051	5 1/4	73.60
1 3/4	8.18	5 1/2	80.78
1 7/8	9.39	5 3/4	88.29
2	10.68	6	96.13
2 1/8	12.06		

Channels—Structural

Depth of Channel Inches	Weight per Foot (Lb)	Thickness of Web	Width of Flange (Inches)
3	4.1	0.170	1.410
	5.0	0.258	1.498
	6.0	0.356	1.596
4	5.4	0.180	1.580
	7.25	0.320	1.720
5	6.7	0.190	1.750
	9.0	0.325	1.885
6	8.2	0.200	1.920
	10.5	0.314	2.034
	13.0	0.437	2.157
7	9.8	0.210	2.090
	12.25	0.314	2.194
	14.75	0.419	2.299
8	11.5	0.220	2.260
	13.75	0.303	2.343
	18.75	0.487	2.527
9	13.4	0.230	2.430
	15.0	0.285	2.485
	20.0	0.448	2.648
10	15.3	0.240	2.600
	20.0	0.379	2.739
	25.0	0.526	2.886
	30.0	0.673	3.033
12	20.7	0.280	2.940
	25.0	0.387	3.047
	30.0	0.510	3.170
13	31.8	0.375	4.000
	50.0	0.787	4.412
15	33.9	0.400	3.400
	40.0	0.520	3.520
	50.0	0.716	3.716
18	45.8	0.500	4.000

FIGURE 19-7 Weight per foot of round and channel steel. The cost per foot can be determined by multiplying the weight (in pounds per foot) by the price per pound.

Copyright © 2015 Cengage Learning®.

Gauge	Thickness (Inches)	Weight/Sq Ft (Pounds)
No. 18	0.0478	2.00
No. 16	0.0598	2.50
No. 14	0.0747	3.125
No. 12	0.1046	4.375
No. 11	0.1196	5.000
No. 10	0.1345	5.625
No. 8	0.1644	6.875
No. 7	³⁄₁₆	7.500

Copyright © 2015 Cengage Learning®.

FIGURE 19-8 Weight per square foot of sheet metal. The cost per foot can be determined by multiplying the weight (in pounds per square foot) by the price per pound.

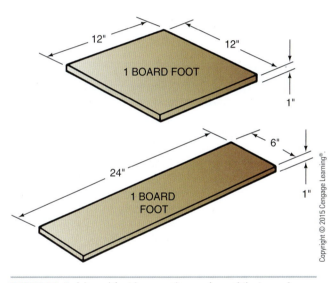

Copyright © 2015 Cengage Learning®.

FIGURE 19-9 A board foot is any volume of wood that equal 144 cubic inches.

Small Pieces of Lumber. Remember that 1 board foot equals 144 cubic inches. With this in mind, the board feet in small or odd-size pieces of lumber can be calculated. The following formula is:

$$BF = \frac{\text{Thickness (inches)} \times \text{Width (inches)} \times \text{Length (inches)}}{144}$$

Therefore,

$$BF = \frac{T'' \times W'' \times L'}{144}$$

What is the number of board feet in a piece of wood 1 inch thick, 2 inches wide, and 24 inches long? The answer is calculated as follows:

$$BF = \frac{T'' \times W'' \times L'}{144}$$

$$= \frac{1 \times 2 \times 24}{144}$$

$$= \frac{48}{144}$$

$$= 0.33$$

This number (0.33 board foot) is approximately equal to ⅓ board foot.

Consider a wooden nail box with two sides that are each 1" × 4" × 14", two ends that are each 1" × 4" × 9", and a bottom that is 1" × 9" × 12½". How many board feet are needed for the project? The answer is obtained by the steps shown in Figure 19-10.

Large Pieces of Lumber. There is a better formula for large pieces of lumber, such as those used in wagon bodies, feed bunks, or buildings. The formula is:

$$BF = \frac{T'' \times W'' \times L'}{12}$$

In this formula, the length is expressed in feet instead of in inches and the set is divided by 12 instead of by 144. The formula is derived by dividing both the top and the bottom by 12. An example would be a piece of lumber 1 inch thick, 6 inches wide, and 24 inches (2 feet) long. This board's size in board feet can be calculated equivalently as:

$$BF = \frac{1'' \times 6'' \times 24''}{144} = \frac{1'' \times 6'' \times 2'}{12} = 1$$

This formula can be used to determine the number of board feet in the door shown in Figure 19-11. The math is easier in the second formula because expressing the length in feet makes the figures smaller. Remember that lumber must be bought in lengths of even numbers of feet. Calculations must be based on the length of the lumber to be purchased, not just on the length to be used in the project. To ignore the waste lumber means the final cost of the project will be higher than calculated.

PRICING MATERIALS

Lumber is priced by the thousand board feet. The Roman numeral for 1,000 is a capital M. For example, lumber dealers may describe the price of lumber as

Item	No. of Pieces	Size	BF per Piece	Total BF
1. Sides	2	1" × 4" × 14"	0.39	0.78
2. Ends	2	1" × 4" × 9"	0.25	0.50
3. Bottom	1	1" × 9" × 12½"	0.78	0.78
			Grand Total	2.06

Step 1. Sides: $\dfrac{1" \times 4" \times 14"}{144} = \dfrac{56}{144} = 0.39 \times 2 \text{ pcs.} = 0.78$

Step 2. Ends: $\dfrac{1" \times 4" \times 9"}{144} = \dfrac{36}{144} = 0.25 \times 2 \text{ pcs.} = 0.50$

Step 3. Bottom: $\dfrac{1" \times 9" \times 12\frac{1}{2}"}{144} = \dfrac{112.5}{144} = 0.78$

FIGURE 19-10 Calculating the board feet needed for a project involving small pieces of wood.

Item	No. of Pieces	Size	BF per Piece	Total BF
1. Vertical	7	1" × 6" × 6'	3	21
2. Cross	2	1" × 6" × 3'	1.5	3
3. Diagonal	1	1" × 9" × 6'	3	3
			Grand Total	27

Step 1. Vertical: $\dfrac{1" \times 6" \times 6"}{12} = \dfrac{6 \times 6}{12} = \dfrac{36}{12} = 3 \times 7 \text{ pcs.} = 21 \text{ BF}$

Step 2. Cross: $\dfrac{1" \times 6" \times 3"}{12} = \dfrac{6 \times 3}{12} = \dfrac{18}{12} = 1.5 \times 2 \text{ pcs.} = 3 \text{ BF}$

Step 3. Diagonal: $\dfrac{1" \times 6" \times 6"}{12} = \dfrac{6 \times 6}{12} = \dfrac{36}{12} = 3 \times 2 \text{ pcs.} = 3 \text{ BF}$

Grand Total = 27 BF

FIGURE 19-11 Calculating the board feet needed for a project involving large pieces of wood.

$900/M. This means $900 for 1,000 board feet. In this example, to determine the price of 1 board foot, simply divide $900 by 1,000:

$$\frac{\$900}{1,000} = \$0.90$$

Wood products sold in sheets are sold by the square foot or panel. Examples of such products are plywood, pressed wood, chip board, wall paneling, and insulation board. These panels generally are sold in 4' × 8' sheets of various thicknesses. A ¾' × 4' × 8' piece of exterior-grade plywood may sell for $0.50 per square foot. However, an entire 4' × 8' sheet will probably have to be purchased for $16 (4' × 8' = 32 sq ft × $0.50/sq ft = $16).

Copyright © 2015 Cengage Learning®.

SUMMARY

Before any project can be successfully completed, all the needed materials have to be gathered and purchased. Basic mathematical skills are used to calculate the materials needed as well as the cost of the project. Costly errors can be avoided by devising an accurate bill of materials. Knowing the cost before starting will result in a much more successful and enjoyable project.

Student Activities

1. Define the Terms to Know in this unit.

2. List the eight components recommended for a bill of materials.

3. Calculate the number of board feet in each item listed below.
   a. 1 piece 1" × 6" × 12'
   b. 1 piece 1" × 1" × 12'
   c. 1 piece 2" × 6" × 8'
   d. 5 pieces 2" × 4" × 12'
   e. 1 piece 1" × 3" × 18'
   f. 3 pieces 1" × 12" × 14"

4. Calculate the cost of materials in each item listed below.
   a. 4 pieces 2" × 6" × 12' at a cost of $0.50/BF
   b. 20 pieces 2" × 4" × 16" at a cost of $0.40/BF
   c. 12 pieces 1" × 10" × 12' at a cost of $0.55 per LF
   d. 1 piece ¼" × 1" × 1" angle steel 20 feet long at a cost of $0.30 per pound. (Refer to Figure 19-6.)
   e. 1 piece ¾-inch black pipe 21 feet long at $0.45 per foot

Relevant Web Sites

WoodWeb, Inc., Calculating Board Footage
www.woodweb.com/knowledge_base/Calculating_board_footage.html

WoodBin Woodworking, Bill of Materials and Cut List
www.woodbin.com/ref/design/billmat.htm

NewWoodworker.com, Calculating Board Feet
http://www.newwoodworker.com/nwwref/index.html

Wisc-Online, Calculating Board Feet
www.wisc-online.com/

Self-Evaluation

A. Multiple Choice. Select the best answer.

1. A board with all four surfaces planed is specified as
   a. S2S
   b. S4S
   c. S2E
   d. S2+2

2. Wood products for building may be sold by the
   a. board foot
   b. linear foot
   c. square foot
   d. all of these

3. Boards that lock together are
   a. T&G
   b. BF
   c. LF
   d. S4S

4. Rough lumber is generally sold by the
   a. T&G
   b. BF
   c. LF
   d. piece

5. A board foot is based on the
   a. number of cubic inches
   b. length and width only
   c. length × width × thickness divided by 50
   d. none of these

6. Sheet steel is sold by the pound. To determine the cost per square foot,
   a. divide the cost per sheet by 12
   b. multiply length of the sheet by its width
   c. multiply price per pound by weight per square foot
   d. none of these

B. Matching. Match the terms in column I with those in column II.

Column I

1. steel, angle
2. dressed lumber
3. logs
4. pipe
5. $500/M
6. plywood

Column II

a. priced by the square foot
b. priced by BF and LF
c. priced by the pound
d. priced by board foot
e. priced by linear foot
f. price per 1,000 board feet

C. Completion. Fill in the blanks with the word or words that make the following statements correct.

1. A bill of materials is a _____.

2. The components of a bill of materials for agricultural mechanics projects are
   a. _____
   b. _____
   c. _____
   d. _____
   e. _____
   f. _____
   g. _____
   h. _____

3. In a board 1" × 4" × 12"
   a. the 1" refers to the _____
   b. the 4" refers to the _____
   c. the 12' refers to the _____

4. Five items that might be in a bill of materials for a trailer are
   a. _____
   b. _____
   c. _____
   d. _____
   e. _____

5. The two formulas used to calculate board feet are
   a. _____
   b. _____

D. Brief Answer. Briefly answer the following questions.

1. Which of the two formulas used to calculate board feet is best for small boards?

2. Calculate board feet for each of the following items.
   a. 1 piece 1" × 10" × 16' = _____ BF
   b. 1 piece 2" × 8" × 6' = _____ BF
   c. 1 piece 2" × 6" × 8' = _____ BF
   d. 2 pieces 2" × 4" × 8' = _____ BF

3. Calculate board feet and costs of a project
    with the following parts, if lumber costs
    $0.40 per board foot.

	Board Feet	Cost
2 sides, each 1" × 6" × 18" =	_____	_____
2 ends, each 1" × 6" × 10" =	_____	_____
1 bottom, 1" × 10" × 16½" =	_____	_____
8 screws, $0.07 each =		_____
1 pint of paint, $4.50 =		_____
Total =		_____

4. Calculate the cost of each item if the price of
    steel is $0.50 per pound.

    a. 1 piece ⅛" × 1" × 20" angle steel
        (pounds per foot = 0.80)
        Cost = _____

    b. 4 pieces ¼" × 2" × 1' (pounds per
        foot = 3.19)
        Cost = _____

    c. 2 pieces ½" × 10' concrete reinforcing bar
        (pounds per foot = 0.67)
        Cost = _____

UNIT 20

Selecting, Planning, and Building a Project

Objective

To select and plan projects that develop the woodworking and metalworking skills needed in agricultural jobs.

Competencies to be developed

After studying this unit, you should be able to:

- Determine the skills you now have in agricultural mechanics.
- Select projects that require the use of woodworking tools.
- Select projects that require the use of metalworking tools.
- Select projects that are appropriate for developing basic skills.
- Modify plans of projects to meet personal needs.

Materials List

- Lead pencil
- Eraser
- Ruler
- Graph paper, ¼"

Terms to Know

- project
- scope
- hands-on
- graph paper

This unit will show how to plan a project for construction in the agricultural mechanics shop. A **project** is a special activity planned and conducted to aid learning. At school, project work is supervised by an instructor. Shop construction projects are fun. They develop eye–hand coordination and skills for future work. In addition to personal development, there is a usable product when the project is completed.

PURPOSES OF PROJECTS

The project method of learning has been used in American schools for about a century. It stimulates students to think and learn. It helps students develop good and safe work habits. Any project a student builds for himself or herself stimulates interest and a desire to do a good job.

Students may design their own projects or they may modify plans from another source. Either way, the design phase requires thinking—to seek solutions and to solve problems. The project may stimulate the student to buy materials and get to know local suppliers. It can provide experience in shopping for the best prices and in looking for other ways to cut costs. These are good business skills.

School projects help to develop construction skills. These include measuring, cutting, drilling, assembling, and finishing. A good project will require the use of a wide variety of shop skills.

A good project will also help students judge their own achievement. When students are finished, they can decide what they did well. They can also see where they made mistakes, and decide how to avoid them the next time.

Finally, the project will leave a visible reminder of students' efforts (Figure 20-1). If they did well, the project will show it. If they were careless, lacked skill, or did not complete the project, this too will show. Students like to take home well-made projects. These are a source of pride for students and their families.

SCOPE OF PROJECTS

The **scope** of a project refers to its size and complexity. Students in agriculture should select and plan projects that help them learn. Learning is the most important reason for doing a shop project. Therefore, woodworking projects are appropriate when woodworking is being studied. Similarly, metalworking projects are

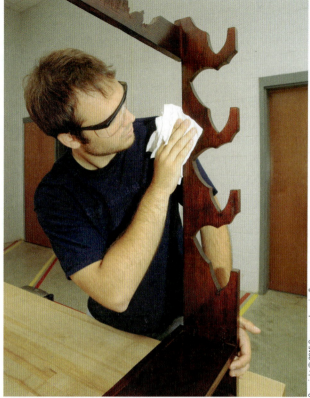

FIGURE 20-1 A properly constructed shop project can be rewarding.

appropriate when metalworking is being studied. The project provides an opportunity for hands-on experience. **Hands-on** means actually doing something, rather than reading or hearing others tell about it. It means practicing an activity to develop a skill.

Shop projects should be relatively simple during the first course in agriculture. At this time students are learning basic skills. Many tools and equipment will be used for the first time. Work may progress slowly. In addition, there may be competition for tools in the class. However, in completing these projects, the students will develop skills to help them move on to more complex projects in the future.

Projects such as nail boxes, toolboxes, feed scoops, birdhouses, stools, funnels, bench stops, foot scrapers, and gate staples have been favorites of first-year students and instructors (Figures 20-2 and 20-3). They have been popular because they require skills appropriate to the experience of beginning students.

For second-year students, more-complex projects are desired. Such projects typically are larger, require more skills, and take longer to complete. Examples are jack stands, car ramps, sawhorses, shop stands, storage units, yard gates, feed carts, pickup truck racks, and picnic tables (Figures 20-4 through 20-8).

FIGURE 20-2 Birdhouses make excellent projects for beginning agricultural mechanics students, providing appropriate woodworking skills and encouraging students to acquire and organize personal tools.

FIGURE 20-3 This hog panel is a good example of a student metalworking project.

Advanced students should choose more challenging projects. Advanced classes frequently are smaller. This may permit larger projects to be built. Such projects should be designed to provide additional skill development in carpentry, metal shaping and bending, welding, pipe cutting and bending, nailing, bolting, and painting (Figure 20-9). Good projects for developing such skills include loading chutes, farm and utility trailers, wagon bodies, truck bodies, horse trailers, cattle guards, cattle feeders, posthole diggers, portable buildings, hydroponics units, and aquaculture structures.

If farm tractor and machinery maintenance are being taught, farm machinery and tractors that need repair should be brought to the school shop from home or work. Appropriate repair and maintenance operations include engine tune-up, horsepower checks, minor engine repair, clutch and brake adjustments, servicing wheel bearings, replacing electrical wiring, repairing frames, replacing broken parts, and painting.

Whether for beginning or advanced students, projects must be matched with level of experience. Before proceeding too far with a project, students should discuss their ideas with their instructor.

FIGURE 20-4 Projects help students develop skills in metalworking.

FIGURE 20-5 Students built these deer-hunting stands as a school project.

FIGURE 20-7 A rack suitable for lumber storage makes a good woodworking project.

SELECTING A PROJECT

Projects should be selected carefully. Projects that are unsuitable may not be completed or may be poorly constructed. Either way, they do not achieve their intended purpose. The following steps are recommended for selecting a shop project.

FIGURE 20-6 This shelving project was completed in the agricultural mechanics lab.

FIGURE 20-8 Picnic tables are popular home-improvement projects. Several different plans are available for designing picnic tables.

FIGURE 20-9 Trailers can be relatively simple to build, or they can be a challenging project for advanced students, depending on the design used.

Procedure

1. Take a personal inventory of agricultural mechanics skills (competencies) (Figure 20-10).
2. Discuss the results of the inventory with your instructor.
3. Ask your instructor to suggest projects that will help you develop appropriate skills.
4. Search books, project bulletins, and departmental files for project ideas and plans.
5. Examine sample projects, and look for details of design.
6. Ask parents, guardians, friends, and employers for ideas and suggestions.
7. Make a tentative decision, and obtain approval from your instructor.

PREPARING A PLAN

Good plans for projects are often available in textbooks, special plan books, magazines, and agricultural mechanics shop files. Many project plans are presented in Appendix A of this text. A good plan with detailed drawings should be used. Such plans are generally well thought out and tested through actual construction.

If an exact plan is not available, a plan that is close to the need can be modified. If nothing is available, then an original design can be made.

Detailed information on how to make a simple project plan and a bill of materials is presented in previous units. However, project ideas can be tried out quickly on **graph paper**—paper laid out in squares of equal size. Paper with ¼-inch squares is recommended for preparing views of projects (Figure 20-11A). With graph paper, the blocks can be counted as a means of measuring. A ruler can then be used as a straight-edge to draw lines. On plain paper, however, every line requires two or more measurements to establish length and position (Figure 20-11B).

To design on paper, a tentative scale is first tried. If it doesn't work, another scale is tried. Graph paper with ¼-inch squares has 34 blocks one way and 44 blocks the other. To draw a project view on a sheet of this graph paper, follow these steps.

Procedure

1. Determine the longest dimension of the project. This may be the length or the height.

2. Turn the graph paper horizontally for long projects, vertically for high ones.
3. Divide 44 (maximum number of squares in the long direction) by the longest dimension of the project. The result tells how many feet or inches one square must represent.
4. Choose a scale that permits a one-view drawing on the paper with some room for dimensions and notes.
5. Count squares and draw lines to complete one view. Include dimensions.
6. Draw additional views on additional sheets of paper.
7. Modify the plan until it is what you want.
8. Review the plan with your instructor.
9. Draw better plans after obtaining the instructor's approval.

After the plan is drawn, a bill of materials can be prepared and the materials purchased. It is best not to change the design after construction has started. However, if errors in design become obvious during construction, design modifications must be made.

DESIGN

Design work is not easy. It generally requires a great deal of experience to be done well. When designing projects, the following principles and facts should be kept in mind:

- Base designs on engineered plans; that is, start with tested plans.
- Use materials of extra strength and size if safety is a factor. For example, car ramps and jack stands should be extra sturdy. Failure of these objects could cause injury or death.
- Structural steel generally holds up better than wood.
- Welding is stronger than riveting.
- Gluing is generally the most secure method of fastening wood.
- Bolts are more secure than screws, and screws are more secure than nails.
- Crimps and folds increase the strength of sheet-metal panels.
- Tongue and groove, dowels, and splines all increase the strength of joints in wood.
- Both wood and metal must be protected by preservative, paint, or other protective film if they are to be exposed to weather.

Examples of project plans are presented in Appendix A.

Agricultural Mechanics Competency Inventory

INSTRUCTION: Check the column that best describes your personal level of experience in each area of competency.

	Level of Experience			
	1 **Have** **Done**	**2** **Am** **Familiar** **with**	**3** **Will** **Learn** **This Year**	**4** **No Skill;** **Will Not** **Learn This Year**
1. Practice Safety and Shop Organization				
a. Work safely with hand tools	☐	☐	☐	☐
b. Work safely with power tools	☐	☐	☐	☐
c. Follow safety rules in shop	☐	☐	☐	☐
d. Maintain safe work areas	☐	☐	☐	☐
e. Eliminate fire hazards	☐	☐	☐	☐
f. Select and use fire extinguishers	☐	☐	☐	☐
g. Interpret product labels	☐	☐	☐	☐
h. Act correctly in an emergency	☐	☐	☐	☐
i. Store materials correctly	☐	☐	☐	☐
j. Clean the shop effectively	☐	☐	☐	☐
2. Identify and Fit Agricultural Tools				
a. Identify tools and equipment	☐	☐	☐	☐
b. Sharpen hand tools	☐	☐	☐	☐
c. Sharpen twist drills	☐	☐	☐	☐
d. Fit handles for hammers and axes	☐	☐	☐	☐
e. Fit handles for hoes and shovels	☐	☐	☐	☐
f. Dress grinding wheels	☐	☐	☐	☐
3. Maintain and Service a Home Shop				
a. Clean and maintain an orderly shop	☐	☐	☐	☐
b. Observe safety regulations	☐	☐	☐	☐
c. Place and use fire fighting equipment	☐	☐	☐	☐
d. Inventory tools and equipment	☐	☐	☐	☐
e. Plan shop layout	☐	☐	☐	☐
f. Select shop site	☐	☐	☐	☐
4. Apply Farm Carpentry Skills				
a. Select lumber for a job	☐	☐	☐	☐
b. Operate power saws	☐	☐	☐	☐
c. Operate a jointer	☐	☐	☐	☐
d. Operate a planer	☐	☐	☐	☐
e. Saw to dimension	☐	☐	☐	☐
f. Bore holes	☐	☐	☐	☐
g. Glue wood	☐	☐	☐	☐
h. Drive and remove nails	☐	☐	☐	☐
i. Set screws and install bolts	☐	☐	☐	☐

(continued)

Copyright © 2015 Cengage Learning®.

FIGURE 20-10 A detailed agricultural mechanics competency inventory.

	Level of Experience			
	1 **Have** **Done**	**2** **Am** **Familiar** **with**	**3** **Will** **Learn** **This Year**	**4** **No Skill;** **Will Not** **Learn This Year**
5. Properly Use Paint and Paint Equipment				
a. Select paints or preservatives	❏	❏	❏	❏
b. Compute area for painting	❏	❏	❏	❏
c. Prepare wood surfaces	❏	❏	❏	❏
d. Prepare metal surfaces	❏	❏	❏	❏
e. Apply paint with a brush or roller	❏	❏	❏	❏
f. Clean and store paint brushes	❏	❏	❏	❏
g. Mask areas prior to painting	❏	❏	❏	❏
h. Apply paint with a spray can	❏	❏	❏	❏
i. Clean and store spray gun	❏	❏	❏	❏
6. Operate an Arc Welder				
a. Practice safety in arc welding	❏	❏	❏	❏
b. Prepare metal for welding	❏	❏	❏	❏
c. Determine welder settings	❏	❏	❏	❏
d. Select electrodes	❏	❏	❏	❏
e. Operate AC and DC welders	❏	❏	❏	❏
f. Strike an arc	❏	❏	❏	❏
g. Run flat beads	❏	❏	❏	❏
h. Make weld joints	❏	❏	❏	❏
i. Weld horizontally	❏	❏	❏	❏
j. Weld vertically	❏	❏	❏	❏
k. Weld overhead	❏	❏	❏	❏
l. Cut with arc	❏	❏	❏	❏
m. Weld cast iron	❏	❏	❏	❏
n. Weld hard surface steel	❏	❏	❏	❏
7. Operate an Oxyacetylene Welder				
a. Check for leaks	❏	❏	❏	❏
b. Turn equipment on and off	❏	❏	❏	❏
c. Adjust equipment	❏	❏	❏	❏
d. Change cylinders	❏	❏	❏	❏
e. Cut with cutting torch	❏	❏	❏	❏
f. Choose proper tips	❏	❏	❏	❏
g. Braze thin metal	❏	❏	❏	❏
h. Run beads	❏	❏	❏	❏
i. Make butt welds	❏	❏	❏	❏
8. Perform Skills in Hot and Cold Metal Work				
a. Identify metals	❏	❏	❏	❏
b. Cut with hand hacksaw	❏	❏	❏	❏

FIGURE 20-10 (*continued*)

	Level of Experience			
	1 **Have** **Done**	**2** **Am** **Familiar** **with**	**3** **Will** **Learn** **This Year**	**4** **No Skill;** **Will Not** **Learn This Year**
c. Cut with cold chisel	❑	❑	❑	❑
d. Bend metal	❑	❑	❑	❑
e. Cut with tinsnips	❑	❑	❑	❑
f. Drill holes	❑	❑	❑	❑
g. Cut threads	❑	❑	❑	❑
h. Use files	❑	❑	❑	❑
i. Solder metal	❑	❑	❑	❑
j. Use power hacksaw	❑	❑	❑	❑
k. Heat metal with torch	❑	❑	❑	❑
l. Operate gas forge	❑	❑	❑	❑

9. Perform Skills in Concrete and Masonry Work

	1	2	3	4
a. Build and prepare forms	❑	❑	❑	❑
b. Treat forms	❑	❑	❑	❑
c. Reinforce concrete	❑	❑	❑	❑
d. Test aggregates for impurities	❑	❑	❑	❑
e. Mix concrete	❑	❑	❑	❑
f. Pour concrete	❑	❑	❑	❑
g. Embed bolts	❑	❑	❑	❑
h. Finish concrete	❑	❑	❑	❑
i. Protect concrete while curing	❑	❑	❑	❑
j. Trowel concrete	❑	❑	❑	❑
k. Remove forms	❑	❑	❑	❑
l. Lay concrete blocks	❑	❑	❑	❑
m. Drill holes in concrete	❑	❑	❑	❑

10. Operate and Maintain Farm Engines

	1	2	3	4
a. Read and follow operator's manual	❑	❑	❑	❑
b. Start and operate engines	❑	❑	❑	❑
c. Change oil and oil filters	❑	❑	❑	❑
d. Service air and fuel filters	❑	❑	❑	❑
e. Maintain battery water	❑	❑	❑	❑
f. Maintain and operate small gas engines	❑	❑	❑	❑
g. Identify engine components and systems	❑	❑	❑	❑
h. Remove and connect battery cables	❑	❑	❑	❑
i. Charge batteries	❑	❑	❑	❑
j. Read battery hydrometer	❑	❑	❑	❑
k. Test batteries	❑	❑	❑	❑
l. Troubleshoot fuel problems	❑	❑	❑	❑
m. Troubleshoot ignition problems	❑	❑	❑	❑

FIGURE 20-10 *(continued)*

	Level of Experience			
	1 **Have** **Done**	**2** **Am** **Familiar** **with**	**3** **Will** **Learn** **This Year**	**4** **No Skill;** **Will Not** **Learn This Year**
11. Perform Minor Tune-Up and Repair of Farm Engines				
a. Clean, gap, and replace spark plugs	❏	❏	❏	❏
b. Adjust carburetor mixture and speed crews	❏	❏	❏	❏
c. Install gap breaker points	❏	❏	❏	❏
d. Set dwell with meter	❏	❏	❏	❏
e. Install condensers	❏	❏	❏	❏
f. Time engines using timing light	❏	❏	❏	❏
g. Disassemble and reassemble distributors	❏	❏	❏	❏
h. Measure compression	❏	❏	❏	❏
i. Adjust valves	❏	❏	❏	❏
j. Set float levels in carburetors	❏	❏	❏	❏
k. Rebuild carburetors	❏	❏	❏	❏
l. Operate engine analyzers	❏	❏	❏	❏
m. Test and replace coils	❏	❏	❏	❏
n. Clean and install brushes	❏	❏	❏	❏
o. Clean commutator bars	❏	❏	❏	❏
12. Perform Electrification Skills				
a. Use safety measures in electrical wiring	❏	❏	❏	❏
b. Select correct fuse sizes	❏	❏	❏	❏
c. Replace fuses	❏	❏	❏	❏
d. Make splices	❏	❏	❏	❏
e. Repair electrical cords	❏	❏	❏	❏
f. Wire on-off switches	❏	❏	❏	❏
g. Select wire sizes	❏	❏	❏	❏
h. Know electrical terminology such as volts, amps, watts, ohms	❏	❏	❏	❏
i. Attach wires to terminals	❏	❏	❏	❏
j. Solder splices	❏	❏	❏	❏
k. Install wire nut connectors	❏	❏	❏	❏
l. Install light fixtures	❏	❏	❏	❏
m. Install electric motors	❏	❏	❏	❏
n. Wire buildings and structures	❏	❏	❏	❏
o. Wire three-way switches	❏	❏	❏	❏
p. Wire four-way switches	❏	❏	❏	❏
13. Learn Plumbing Skills				
a. Repair leaky faucets	❏	❏	❏	❏
b. Assemble pipe and pipe fittings	❏	❏	❏	❏
c. Thread pipe	❏	❏	❏	❏

FIGURE 20-10 (continued)

	Level of Experience			
	1 **Have** **Done**	**2** **Am** **Familiar** **with**	**3** **Will** **Learn** **This Year**	**4** **No Skill;** **Will Not** **Learn This Year**
d. Measure and cut pipe	❑	❑	❑	❑
e. Cut plastic tubing/pipe	❑	❑	❑	❑
f. Install plastic tubing/pipe	❑	❑	❑	❑
g. Flare copper tubing	❑	❑	❑	❑
h. Cut copper tubing	❑	❑	❑	❑
i. Ream pipe	❑	❑	❑	❑
j. Install fixtures on plastic pipe	❑	❑	❑	❑
k. Sweat joints on copper pipe	❑	❑	❑	❑
l. Adjust air control mechanism on pressure system	❑	❑	❑	❑
m. Install pressure pumps	❑	❑	❑	❑
n. Repair pumps	❑	❑	❑	❑
o. Install water heaters	❑	❑	❑	❑
p. Cut soil pipe	❑	❑	❑	❑
q. Caulk joints on soil pipe	❑	❑	❑	❑
r. Lay soil pipe	❑	❑	❑	❑

FIGURE 20-10 (*continued*)

A GRAPH PAPER WITH 1/4-INCH SQUARES IS EXCELLENT FOR PROJECT PLANNING. ON GRAPH PAPER, DISTANCES ARE DETERMINED BY COUNTING SQUARES AND USING A RULER TO CONNECT POINTS.

B A PROJECT DRAWN ON PLAIN PAPER REQUIRES MORE TIME AND EFFORT THAN ONE DRAWN ON GRAPH PAPER. ON PLAIN PAPER, EVERY LINE REQUIRES TWO OR MORE MEASUREMENTS TO ESTABLISH LENGTH AND POSITION.

FIGURE 20-11 Graph paper makes plan drawing easier.

Copyright © 2015 Cengage Learning®.

SUMMARY

Selecting an agricultural mechanics project can be a challenging task. Selection should be based on your skill level, the amount and types of material available, and access to the proper tools. Completing a quality project can be very rewarding. This can only be accomplished by developing the right skills, doing the proper planning, and following the correct procedures. Deviating from any of these areas will result in a poor project.

Student Activities

1. Define the Terms to Know in this unit.

2. Take a personal agricultural mechanics skills inventory.

3. Write down the skills that you need to develop through project construction.

4. Discuss project ideas with your instructor.

5. Search books, bulletins, files, and other sources for project ideas.

6. Select a project that fits your shop experience.

7. Make or modify plans as needed.

8. Plan a project from the conception to the completion. Use the principles laid out in Section 5 of the text. Make sure you include (1) a problem the project will solve and (2) how the completion of the project will solve the need or problem. Provide detailed step-by-step procedures and explain why every step is important.

9. Have your instructor evaluate your plan. Have you thought through the problem and solution? Are your steps workable? Will the project solve the need or problem?

Relevant Web Sites

Wood Work Zone
www.woodworkzone.com

Woodworking Plan Finder, Woodworking Links
www.woodworkingplanfinder.com/woodlinks.htm

Vintage Games Woodworking Plans
http://woodworkingclubs.tripod.com/

FurniturePlans.com, Woodworking Plans and Projects
www.furnitureplans.com

Self-Evaluation

A. **Multiple Choice.** Select the best answer.

   1. The project method of learning is used because it
     a. develops body coordination
     b. develops skills for jobs
     c. provides a usable product when finished
     d. all of these

   2. The project method has been used in U.S. public schools for about
     a. ten years
     b. two decades
     c. thirty years
     d. one century

3. The project method of learning usually
   a. causes students to think
   b. causes students to work carefully
   c. creates personal interest
   d. all of these

4. A suitable project must
   a. provide practice in skills taught in class
   b. earn the student money
   c. be completed in four weeks
   d. be paid for totally by the student's own funds

5. Projects that are not carefully chosen will
   a. result in student failure
   b. never be completed
   c. not achieve their intended educational purpose
   d. be too costly

6. Good project ideas may come from
   a. books
   b. teachers
   c. parents or guardians
   d. all of these

7. Good sample projects have the benefit of
   a. showing details of design
   b. showing mistakes to avoid
   c. eliminating the need for a bill of materials
   d. all of these

8. In general, preferred projects for first-year students are
   a. those already designed with published plans
   b. those designed by the student
   c. those modified by the student
   d. all of these

9. Graph paper for drawing permits
   a. easy drawing of angles
   b. quick measurement
   c. easy three-view drawing
   d. economy of cost

B. Matching. Match the terms in column I with the terms in column II.

Column I
   1. hands-on
   2. nail box
   3. feed cart
   4. farm trailer
   5. graph paper
   6. nailing
   7. gluing
   8. steel
   9. paint
   10. weld

Column II
   a. second- or third-year project
   b. third- or fourth-year project
   c. first-year project
   d. doing
   e. strongest wood joint
   f. weakest wood joint
   g. strongest metal joint
   h. stronger than wood
   i. ¼-inch squares
   j. protects from weather

C. Completion. Fill in the blanks with the word or words that make the following statements correct.

1. _____ is the most important reason for doing a shop project.

2. Shop skills developed by building suitable projects include _____, _____, _____, _____, and _____.

3. Students in agricultural mechanics should select and plan projects that _____.

4. It is best to base designs for projects on _____ plans.

5. It is preferable not to change the design after _____ has begun.

Section 6
TOOL FITTING

UNIT 21
- Repairing and Reconditioning Tools

UNIT 22
- Sharpening Tools

UNIT 21

Repairing and Reconditioning Tools

Objective

To restore worn, damaged, or abused tools to good working condition.

Competencies to be developed

After studying this unit, you should be able to:

- Remove rust from metal tools.
- Apply rust-inhibiting materials to metal surfaces.
- Repair split wooden handles.
- Replace broken wooden handles.
- Reshape screwdriver tips.
- Reshape the heads of driving and driven tools.

Materials List

- Tools that need repair or reconditioning
  a. Tools with rust showing
  b. Driving tool with a split wooden handle and a broken wooden handle
  c. Fork, rake, hoe, or shovel with a broken handle
  d. Tools with mushroomed heads
  e. Screwdriver that needs fitting
  f. Leather tools or protective clothing that needs reconditioning
- Replacement wooden handle for each handle in need of replacement

continued

Terms to Know

- abuse
- rust
- mushroomed
- tool fitting
- recondition
- saddle soap
- neat's-foot oil
- wedge
- eye
- ferrule
- tang
- crown
- torque

- Saddle soap or neat's-foot oil
- Stiff bristle scrub brush
- Soft cloths
- Paste finishing wax
- Varsol or kerosene
- Wire brush or wire wheel
- Waterproof silicon carbide paper, 220- and 400-grit
- Lightweight lubricating oil, No. 10 or lighter
- Medium-weight lubricating oil, No. 20 or No. 30
- Wood glue
- Vise, hand tools, and electric drill
- Assorted steel rivets
- Wood and metal handle wedges

Modern farms and many households make significant investments in tools (Figure 21-1). The most obvious use of tools is to build, repair, or adjust objects and machines. However, tools need repair, too.

This unit provides instruction in repairing tools. The student is advised to review Unit 14 and Unit 16 on the use of power drills and grinders. These tools are useful in performing the tasks described in this unit.

A high-quality tool will last for years if used and maintained properly. Normal use does cause tools to lose their shape, however. Also, tools are frequently abused or neglected. To **abuse** is to use wrongly, make bad use of, or misuse. Such practices necessitate tool repair. Even properly used tools require proper maintenance.

FIGURE 21-1 Many farms and many households make significant investments in tools.

Metal tools **rust** if they get wet and are not dried off quickly (Figure 21-2). Rust is a reddish brown or orange coating that results when steel reacts with air and moisture. It creates pits in the surface of steel, which lowers the usability of tools.

Wood handles may split or break, and metal handles may bend out of shape, leaving tools useless until repairs are made. Tools such as punches, chisels, wedges, and sledges get mushroomed faces. **Mushroomed** means a spread or pushed-over condition caused by being struck many times and occurs over a long period of use. A mushroomed head on a chisel or punch is dangerous because chips may fly off the tool when it is struck with a hammer (Figure 21-3). After much use, screwdrivers get rounded tips and cannot grip screws properly. Screw heads are then damaged by attempts to use such screwdrivers.

FIGURE 21-2 Most metal tools will rust if they are not kept dry.

FIGURE 21-3 A mushroomed head on a chisel or punch is dangerous.

The term **tool fitting** means to clean, reshape, repair, or resharpen a tool. To **recondition** means to do what is needed to put the tool into good condition. This unit will cover cleaning and protecting tool surfaces as well as reshaping and otherwise repairing tools. Unit 22 will cover tool sharpening. Many students in agricultural mechanics start or add to their tool holdings by restoring tools that others have discarded.

PROTECTING TOOL SURFACES

Most tool parts are made of wood, plastic, leather, or steel. Plastic is tough and is not affected by moisture. It is too soft to restore by grinding or buffing. Therefore, plastic parts must be handled carefully at all times. It is important to avoid cuts, nicks, or other damage to plastic.

Restoring Leather Parts

Leather is used in a few tools. It may be rolled very tightly to form a mallet head. Layers of leather may be stacked and glued on wooden chisel handles. This cushions the rest of the tool when it is struck with a mallet. Leather is also used in protective clothes such as gloves and aprons.

When leather becomes stiff or dry, it can be restored by rubbing with saddle soap. **Saddle soap**, when mixed with water, will clean, soften, and preserve leather. The following procedure describes how to use saddle soap for this purpose.

Procedure

1. Brush off all dirt with a stiff bristle brush.
2. Moisten a sponge or soft cloth with water.
3. Rub the cloth onto the surface of the saddle soap until suds develop into a lather.
4. Rub the sudsy mixture into all leather surfaces until the leather is clean and soft.
5. Remove the remaining suds with a damp cloth.
6. Let the leather dry.
7. Rub the leather with a soft, dry cloth or with a brush that has soft bristles.

Courtesy of Mary Herren

FIGURE 21-4 Leather may be restored with saddle soap, neat's-foot oil, or various commercial products that contain silicone and lanolin.

Courtesy of Mary Herren

FIGURE 21-5 Linseed oil or paste finishing wax will restore and protect wooden parts.

Another product used to soften, restore, and protect leather is neat's-foot oil. **Neat's-foot oil** is a light yellow oil obtained by boiling the feet and shinbones of cattle. Simply apply neat's-foot oil to clean leather surfaces with a lamb's wool pad or soft cloth.

Other products such as mink oil may be used to recondition leather. They contain various mixtures of silicone, lanolin, or other oils (Figure 21-4).

Restoring Wooden Surfaces

Wooden handles frequently become dried out, porous, and rough. Repeated use of such handles results in sore hands and sometimes splinters. Handles that dry out become loose and dangerous, too.

Wooden handles and other parts are easy to restore if treated as soon as drying is noticed. The procedure follows.

Procedure

1. Sand wooden parts with very fine sandpaper to remove all rough areas.
2. Rub all exposed wood with a soft cloth dipped in boiled linseed oil (Figure 21-5).
3. After the oil dries, rub the wood briskly with a soft, dry cloth.
4. Repeat steps 2 and 3 several times if the wood is very porous.

Another product that may be used to restore wooden tool handles is paste finishing wax. The procedure, which is similar to that for linseed oil, is as follows.

Procedure

1. Before applying the wax, sand the wood thoroughly.
2. Rub on a thin coat of wax, and let it dry.
3. Once the wax is dry, polish it with a soft cloth.
4. Repeat the process if needed to obtain a smooth surface that will prevent moisture from entering the wood.

Restoring Metal Surfaces

Metal surfaces may be quite difficult to restore if they have been badly neglected. However, a few standard procedures can stop rust and can restore a neglected tool to service.

Removing Grease and Oil. Grease and oil on tool handles are dangerous as well as messy. Tools are likely to slip out of the hand or be difficult to control when greasy.

To remove grease or oil, wipe the tool with a clean cloth dipped in kerosene, Varsol, or other commercial solvent for grease and oil. After cleaning with a solvent, dry the tool with a clean cloth or paper towel.

CAUTION!

Do not use gasoline to remove grease and oil from tools. Gasoline is very volatile. Its explosive fumes will travel to sparks, flames, or other sources of fire.

Removing Dirt. Tools such as shovels, hoes, and rakes may become caked with mud or concrete. When dry, these materials can be removed by tapping the tool with a metal object. The vibration generally causes the material to drop off. Any remaining material should be removed with a scraper, wire brush, or wire wheel.

Removing Rust. Light rust can be removed with Varsol, kerosene, or other solvents. However, once pitting starts, a wire brush, wire wheel, steel wool, emery cloth, or aluminum oxide paper must be used (Figure 21-6). These materials will dislodge deep rust, but may scratch polished surfaces. A smooth final surface may be obtained by using 400-grit silicon carbide paper. Use water to help keep the paper clean while sanding.

Immediately after sanding, dry the tool thoroughly and coat all smooth metal surfaces with a light oil such as No. 10. Rough surfaces should be coated with medium-weight oil such as No. 20 or 30. If the tools are not treated in this manner, the water will cause rusting to start again within a few hours.

FIGURE 21-7 Paste wax should be used on unpainted surfaces such as saw tables.

Protecting Steel from Rust. As noted before, rust occurs when steel is exposed to moisture and air. Therefore, rust is prevented by keeping air or moisture away from steel. Special rust-resisting primers and paints are the products most frequently used to protect steel from rust. However, many tool surfaces must remain smooth and unpainted in order to cut well. In addition, smooth, unpainted surfaces permit material to slide over the metal more easily. Tabletops on power machines are an example of this. Unpainted steel edges and surfaces are protected by rubbing with a good grade of paste wax (Figure 21-7). However, wax is useful only if the steel is free of rust when it is applied.

REPAIRING DAMAGED TOOLS

Tools may be repaired by welding or gluing of broken parts. Wooden handles on files, wood chisels, hammers, axes, shovels, and the like should be replaced when broken. Sometimes it is less expensive to replace the tool than it is to replace the handle. However, for most tools, replacing handles is the most economical step.

Repairing Wooden Handles

Split Handles. Wooden handles may split when exposed to excessive strain. If the split is clean and it does not go completely through the handle, it can be

FIGURE 21-6 Rust may be removed from steel implements with a wire brush, a wire wheel, steel wool, an emery cloth, sandpaper, or other abrasives.

repaired with wood glue. Be sure to use a wood glue that is water resistant and is meant for use outdoors. The procedure follows.

Procedure

1. Position the tool in a vise so the split can be forced open.
2. Force open the split and spread glue on both surfaces.
3. Close the split and clamp the handle in several places.
4. Wipe off all excess glue.
5. After the glue dries, thoroughly sand the handle to remove splinters and any roughness.
6. Treat the handle with boiled linseed oil, and polish with a clean cloth when dry.

Loose Handles. Loose handles on hammers, hatchets, sledges, and axes are very dangerous. Handles have a wooden wedge and one or two metal wedges driven into the end (Figure 21-8). A **wedge** is a piece of wood or metal that is thick on one end and tapers down to a thin edge on the other. Check frequently to ensure that wedges are securely in place. If they are not, drive them into place. Even when handles are properly attached, they can still loosen due to drying. To reduce this problem, soak the end of the handle in the head of the tool with linseed oil about once a year (Figure 21-9). An alternate method for treating tool heads with linseed oil follows.

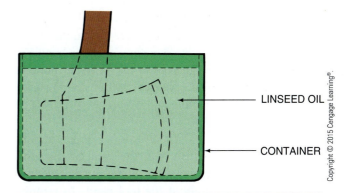

LINSEED OIL

CONTAINER

Copyright © 2015 Cengage Learning®.

FIGURE 21-9 Treating wooden handles with linseed oil for 12 to 24 hours prevents them from drying and loosening.

Procedure

1. Clamp the head of the tool in a vise with the end of the handle pointing upward. Brush enough boiled linseed oil over the end of the handle and wedges to saturate all parts. Keep them saturated for 12 to 24 hours if possible.
2. Permit the oil to dry while the head of the tool is still clamped in the vise.
3. Remove the tool from the vise, and wipe a coat of boiled linseed oil on the rest of the handle.
4. Polish the handle with a dry cloth after the oil dries.

Replacing Wooden Handles

To be safe, handles must be properly installed. Improperly installed handles are usually loose and can fly off in use (Figure 21-10). As mentioned before, most wooden handles are secured in the heads of tools by the use of

FIGURE 21-8 Wedges driven into the end of the handle help to secure the handle to the tool.

FIGURE 21-10 An improperly installed handle such as this one is dangerous.

wedges. The hole in the head of the tool is called the **eye**. This hole is smaller on the side where the handle enters than on the opposite side. Therefore, once the handle is inserted, it can be wedged out to fill the larger portion of the hole. As long as the wedges stay in place, the head remains tight on the handle.

Removing Broken Handles.

A wooden handle generally breaks off flush with the head of the tool. When this occurs, the wooden core and metal wedges are still lodged securely in the tool head. To remove the broken piece of handle, the following procedure is recommended.

Procedure

1. Place the head securely in a vise.
2. Use a ¼-inch or ⅜-inch metal-cutting drill bit and drill numerous holes into the wooden core.
3. Drive the remaining honeycomb of wood out of the head using a large punch or rod.

CAUTION!

Do not attempt to remove wooden pieces from metal tools by burning them out. The heat will cause the metal to lose its hardness.

Attaching Handles to Hammers and Axes.

When purchasing a replacement handle, it is important to buy the correct size. The end of the handle that is shaped to go into the head should be slightly longer than the head is deep and should have the same shape as the hole in the head. It should also be slightly larger in cross section than the smaller part of the eye (that is, the hole) into which it is to be inserted. Also, carefully inspect the wood grain of the new handle. Make sure the grain is straight, runs the entire length of the handle, and does not run to the side in any place. Such handles are much more prone to splitting (Figure 21-11).

Handles must be shaped to fit a head exactly. Start by placing the handle in a vise. The end to be worked on should be exposed so that it can be sawed, filed, and shaped. The procedure for attaching handles to hammers and axes follows. Figures 21-12 and 21-13 illustrate the procedure.

© iStockphoto/David Wiberg

FIGURE 21-11 The grain of the new handle should be straight and run the entire length of the handle. This grain ran to the side and split easily.

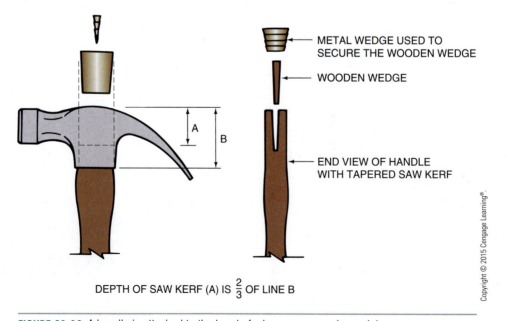

DEPTH OF SAW KERF (A) IS $\frac{2}{3}$ OF LINE B

METAL WEDGE USED TO SECURE THE WOODEN WEDGE

WOODEN WEDGE

END VIEW OF HANDLE WITH TAPERED SAW KERF

Copyright © 2015 Cengage Learning®.

FIGURE 21-12 A handle is attached to the head of a hammer or axe by wedging.

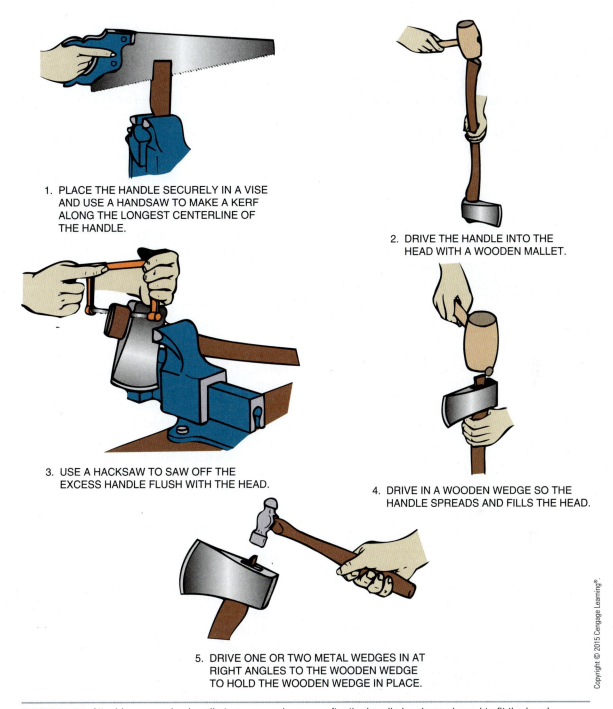

1. PLACE THE HANDLE SECURELY IN A VISE AND USE A HANDSAW TO MAKE A KERF ALONG THE LONGEST CENTERLINE OF THE HANDLE.

2. DRIVE THE HANDLE INTO THE HEAD WITH A WOODEN MALLET.

3. USE A HACKSAW TO SAW OFF THE EXCESS HANDLE FLUSH WITH THE HEAD.

4. DRIVE IN A WOODEN WEDGE SO THE HANDLE SPREADS AND FILLS THE HEAD.

5. DRIVE ONE OR TWO METAL WEDGES IN AT RIGHT ANGLES TO THE WOODEN WEDGE TO HOLD THE WOODEN WEDGE IN PLACE.

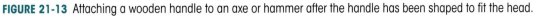

FIGURE 21-13 Attaching a wooden handle to an axe or hammer after the handle has been shaped to fit the head.

Copyright © 2015 Cengage Learning®.

Procedure

1. Place the head against the end of the handle to see if it will start on.

2. Observe where wood must be removed to shape the handle to enter the head.

3. Use a wood rasp or coarse file to shape the handle.

4. Try the head frequently as wood is removed to avoid a loose fit.

5. Work the handle down until the head slides on snugly. It should come to rest about ½ inch from the enlarged part of the handle.

(continued)

Procedure, *continued*

> ### CAUTION!
> Do not remove too much wood from the end. This will help get the head on, but it will also damage the wedged fit.

6. Mark the handle on both sides of the head.

7. Remove the head.

8. Reposition the handle vertically in the vise.

> ### CAUTION!
> Do not drive the head. It will bind and be very hard to remove.

9. Use a handsaw to make a kerf across the longest center line of the handle. The kerf should extend two-thirds of the distance between the two marks made in step 6.

10. Reposition the handle in the vise and squeeze the end until the saw kerf is completely closed.

11. Run the saw down through the kerf again. When released, the kerf will be wider at the end than further down in the handle.

12. Now make a wooden wedge as wide as the oval hole in the head and thick enough to spread the handle when driven in.

13. Slide the handle into the head and drive it in securely with a wooden mallet.

14. Grip the handle with the vise just below the head.

> ### CAUTION!
> Be sure to use a wooden mallet and not a steel hammer for this step. A steel hammer will ruin the end of the handle.

15. Use a hacksaw to saw off the excess handle flush with the head.

16. Drive the wooden wedge so the handle spreads and fills the head.

> ### CAUTION!
> Drive the wedge evenly, or it will split.

17. Use a hacksaw to saw off the excess wooden wedge.

18. Drive one or two metal wedges in at right angles to the wooden wedge. These will hold the wooden wedge in place.

19. Place the tool, head down, in a metal or plastic container about the size of the head.

20. Add several inches of boiled linseed oil to the container and brush oil around the handle and the head.

21. The handle and head should soak for several days to seal the wood.

22. Remove the tool, rub oil on all parts of the handle, allow it to dry, and polish it.

Attaching Handles to Rakes, Hoes, and Forks. Rakes, hoes, and forks are driven into their handles. The handles are fitted with metal collars to prevent splitting of the wood. This metal collar is called a **ferrule**. The tool has a metal finger called a **tang**, which is driven into the ferrule (Figure 21-14). In some handles, a hole is provided in the ferrule for a nail or rivet. The nail or rivet is driven through the hole into the handle and tang to help hold the parts together. Other handles rely on friction only to keep the parts together.

Attaching Handles to Shovels. Shovel handles are exposed to tremendous forces when digging. Therefore, a metal tube extends up the handle for extra support (Figure 21-15). Some shovel handles have a single bend, and others have a double bend. Shovel handles are made to fit specific shovels. Therefore, it

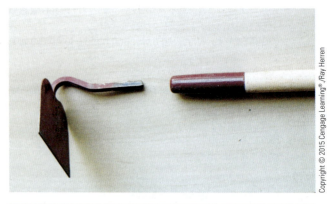

Copyright © 2015 Cengage Learning® /Ray Herren

FIGURE 21-14 Forks, hoes, and rakes have a tang that is held in the handle with a nail or rivet or by friction between the wood and the metal.

FIGURE 21-15 Shovels and spades have a split metal tube that is tightened around the handle for support. A rivet is installed through the metal and the handle to hold the handle in place.

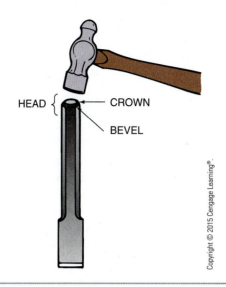

FIGURE 21-16 Driven tools such as punches, chisels, and wedges have a beveled or tapered edge around the crown to resist flattening when struck.

is best to take the shovel to the store when selecting a replacement handle. To attach a shovel handle, the following procedure is suggested.

Procedure

1. Grind off the head of the rivet that holds the handle in place.

2. Remove the rivet with a drift punch.

3. Spread the metal tube, and drive out the remains of the old handle.

4. Use a mallet to drive the new handle securely into place. If the handle stoutly resists driving, consider removing some wood to improve the fit.

5. Use a vise to close the metal around the handle and secure the assembly while drilling.

6. Insert a drill through the hole in the metal and drill through the wooden handle.

7. Install the replacement rivet securely, being careful to keep the metal tight to the wood.

Many other tools have replaceable handles. The procedures outlined for the tools in this unit will apply to other tools as well.

Reshaping Tools and Tool Heads

All metal tools that are driven will eventually flatten. If the tools are not reground, the surfaces will mushroom. A mushroomed tool head is dangerous.

When struck, fragments of steel are likely to break off and fly into the face and body of the operator or bystanders.

Driven tools, such as punches, chisels, and wedges, are designed with a slight bevel around the crown. The **crown** is the end of a tool that receives the blow of a hammer. The bevel and the crown together make up the head of the tool (Figure 21-16). The procedure to reshape the head of a driven tool follows.

Procedure

1. Examine the head of a similar tool that is in good condition.

2. With the good tool as a guide, use a medium grinding wheel to grind a taper from each flat surface to the crown.

CAUTION!

Use a face shield and leather gloves when grinding.

3. Finish the taper by twirling the tool slightly so that all corners are slightly rounded. Round tools will have an even taper in a circular pattern around the crown.

The heads of axes, sledges, and hammers can also be reshaped using the procedures just described. However, these tools do not have much of a taper around the crown. Instead, they have rounded edges. It is important to keep the edges slightly rounded to avoid mushrooming of the head.

Reshaping Screwdrivers. The tips of standard screwdrivers can be reshaped. Phillips and other cross-headed screwdrivers must be discarded when their tips become worn. Standard screwdriver tips must be reground often to maintain a shape that will grip screws securely. Properly shaped standard screwdriver tips have a flat end and parallel sides. The width and thickness of the tip are determined by the size of the slot in the screw to be used.

When reshaping a standard screwdriver tip, the following procedure is recommended.

Procedure

1. Obtain a screw of the size you wish to drive with the screwdriver.
2. Check to see that the screwdriver tip is as wide as the screw head. If it is not, obtain a wider screwdriver.
3. Check the face of the grinding wheel to make sure it is flat.

4. Check to see that the grinder's tool rest is properly adjusted.
5. Slowly push the flat end of the screwdriver tip across the tool rest and into the grinding wheel (Figure 21-17A).
6. Grind the tip until it is flat, square, and even.

CAUTION!

Avoid overheating the tip of the screwdriver. Cool the tip by placing it in a pail of water.

7. Place the tip into the screw slot. It should reach to the bottom of the slot and fit the sides snugly.
8. If the tip is too thick, grind the flat sides of the tip slightly to narrow the screwdriver blade (Figure 21-17B).
9. Dress the edges of the tip so they are equal to the width of the screw head.

CAUTION!

Be careful to maintain the flat end of the screwdriver that fits into the screw. The flat surfaces of the screw and the screwdriver must fit perfectly. Any taper in the tip will cause it to twist out of the screw slot when **torque** (a twisting force) is applied.

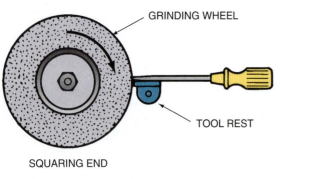

GRINDING WHEEL

TOOL REST

SQUARING END

A TO SQUARE THE END OF A SCREWDRIVER, SLOWLY PUSH THE TIP OF THE SCREWDRIVER ACROSS THE TOOL REST AND INTO THE GRINDING WHEEL.

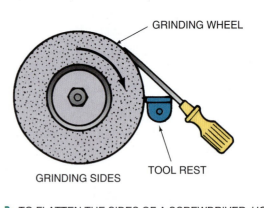

GRINDING WHEEL

GRINDING SIDES

TOOL REST

B TO FLATTEN THE SIDES OF A SCREWDRIVER, HOLD THE SCREWDRIVER DIAGONALLY ACROSS THE TOOL REST WITH THE SIDE FLAT AGAINST THE WHEEL WHILE GRINDING.

FIGURE 21-17 Reshaping the tip of a standard screwdriver.

Copyright © 2015 Cengage Learning®

SUMMARY

Quality tools are expensive. Proper maintenance and care can make them last much longer. Over time, even with the best of care, tools will need to be reconditioned and repaired. Using the correct procedures for maintenance and repair will greatly expand the life of the tools.

Student Activities

1. Define the Terms to Know in this unit.

2. Bring hand tools from home that need cleaning, rust protection, or repair.

3. Remove dirt and rust from tools.

4. Repair wooden handles on tools.

5. Replace handles on driving and digging tools.

6. Reshape the heads of driven tools such as chisels and punches.

7. Reshape screwdriver tips.

Relevant Web Sites

FurniturePlans.com, Woodworking Plans
www.furnitureplans.com

Tennessee Hickory Products
www.tennesseehickoryproducts.com

The Continental Line, Inc., Some Notes on Axes (article by John White, 1st Maryland Regiment), Newsletters, Fall 2006- Volume XIX, Number 2
http://www.continentalline.org

Reader's Digest, Improvement, section on home improvement
www.rd.com/home/improvement

Self-Evaluation

A. **Multiple Choice.** Select the best answer.

1. A common problem with wooden handles is
   a. rotting
   b. splitting
   c. rusting
   d. fatigue

2. Leather parts should be reconditioned with
   a. neat's-foot oil
   b. saddle soap
   c. lanolin products
   d. any of these

3. Hammer handles become loose if
   a. the handle dries out
   b. linseed oil is not applied occasionally
   c. the handle is not properly installed
   d. all of these

4. Wooden handles should be treated with
   a. linseed oil
   b. shellac
   c. paint
   d. all of these

5. A solvent recommended for removing grease and light rust is
   a. gasoline
   b. Varsol
   c. turpentine
   d. water

6. After water touches an unprotected steel surface, rusting starts within
   a. hours
   b. days
   c. weeks
   d. months

7. Rust may be prevented from forming on metal surfaces by applying
   a. oil
   b. water
   c. saddle soap
   d. any of these

8. Split wooden handles are best repaired with
   a. nails
   b. screws
   c. glue
   d. tape

9. Driving tool handles are held in place by
   a. bolts
   b. nails
   c. screws
   d. wedges

10. When standard screwdrivers have rounded tips, they
    a. work better
    b. need replacing
    c. slip out of screw slots
    d. should be heated and reshaped

B. Matching. Match the terms in column I with those in column II.

Column I

1. mushroomed
2. linseed oil
3. tool fitting
4. neat's-foot oil
5. mallet
6. ferrule
7. wedged handle

Column II

a. damaged head
b. hammer
c. reconditioning
d. treat handles
e. rakes and hoes
f. drive handles
g. leather

C. Completion. Fill in the blanks with the word or words that make the following statements correct.

1. When a tool head has mushroomed, it has _____.

2. The crown is the part of the tool that _____.

3. When reconditioning leather, a sponge may be used to apply _____.

4. Shovel handles are held in place with a _____.

5. A mushroomed condition is corrected by _____.

D. Brief Answer. Briefly answer the following questions.

1. How does one remove rust once pitting has begun?

2. Why is it important to keep wooden handles in good condition?

3. Why should gasoline never be used to remove oil and grease from tools?

4. How does one protect rust-free unpainted steel edges and surfaces?

5. What guidelines should one follow when selecting the correct size replacement for an axe or hammer handle?

UNIT 22
Sharpening Tools

Objective
To keep cutting tools sharp.

Competencies to be developed

After studying this unit, you should be able to:

- Examine tools and determine the design of the cutting edges.
- Select appropriate procedures for sharpening tools.
- Sharpen knives.
- Sharpen wood chisels and plane irons.
- Sharpen cold chisels and center punches.
- Sharpen axes and hatchets.
- Sharpen twist drills.
- Sharpen rotary mower blades.
- Sharpen digging tools.

Materials List

- Grinder with medium- and fine-grit wheels
- Eye or face protection and protective clothing
- Heavy-duty portable sander with fine aluminum oxide discs
- Large flat file with medium or fine teeth
- Machinist's vise
- Assorted hand tools
- Oil bench stone (one fine side and one coarse side)

continued

Terms to Know

- bench stone
- hand stone
- inclined plane
- convex
- concave
- hollow ground
- tool steel
- anneal
- temper
- draw the temper
- true
- whet
- serrated
- arbor
- ax stone
- field use
- high-speed drill
- clockwise
- balance

- Ax stone
- Tool fitting gauge
- Tools requiring repair and sharpening
  a. Pocket or pruning knife
  b. Wood chisel
  c. Cold chisel
  d. Center punch
  e. Axe or hatchet
  f. Twist drill
  g. Rotary mower blade
  h. Digging tools

A properly sharpened tool of good quality can make cutting, drilling, and digging jobs relatively easy. Such tools result in work of excellent quality when used by skilled craftspeople. Sharp tools are safer than dull ones because they require less pressure. A craftsperson should not attempt a job with a dull tool, because the resulting quality of work will be disappointing.

TOOLS USED FOR SHARPENING

Tools in the agricultural mechanics shop are generally sharpened with files, **bench stones**, **hand stones**, sanders, and grinders. The use of grinders is covered in Unit 14. A bench stone is a sharpening stone designed to rest on a bench. A hand stone is one designed to be held in the hand when in use (Figure 22-1). Care should be taken when using a hand stone.

© iStockphoto/Andy Olsen

FIGURE 22-1 A hand stone is held in the hand. Care should be taken to prevent cutting the hand.

FIGURE 22-2 Bench stones are mounted in a wooden box. Note the bottle of oil.

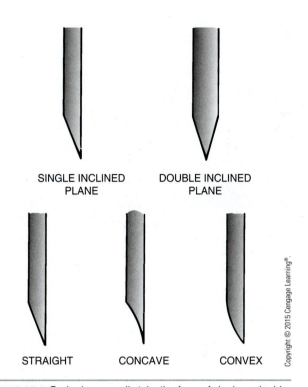

FIGURE 22-3 Tool edges usually take the form of single or double inclined planes. These planes may be straight, concave, or convex.

It is easy for the blade to slip and cut the hand. Sharpening stones include the common 6" × 2" × 1" bench stone, handheld slip stones, and power-driven wheels. Bench stones are generally mounted in a wooden box, which holds the stone during use and storage (Figure 22-2). The box also helps keep the stone clean and oil-filled. Many stones require either water or oil to keep them clean and to ensure that their abrasive materials cut during use.

Files needed for sharpening tools include flat, round, triangular, and special styles. Examples of special files are auger bit files and chainsaw files. Coarse files are useful for sharpening large objects such as blades for rotary mowers. Fine-toothed files are used for the majority of sharpening jobs.

CUTTING EDGES

Always observe the design and shape of the edges of cutting tools when they are new. The tool is designed to do a specific job. The nature of that job determines the shape of the cutting edge. For example, shaving should remove all evidence of hair. Hair is relatively soft. Therefore, a razor blade has a very thin, sharp edge. This edge does a good job in shaving but is too fragile for most other cutting jobs.

In contrast, steel is tough and hard. Therefore, tools used to cut steel must be tough and hard. Their edges must be thick and strong.

Wood is relatively soft, so wood chisels can be ground to fine, sharp edges. However, this means that they nick, break, or dull quickly when they are used on metal objects or come in contact with dirt or metal surfaces.

Tool edges are usually designed with either single or double inclined planes (Figure 22-3). An **inclined plane** is a surface that is at an angle to another surface. It is one of the six simple machines (along with the wedge, the screw, the lever, the pulley, and the wheel and axle). Wood chisels and plane irons have single inclined planes.

Other tools are designed with edges formed from two inclined planes. These tools must be sharpened from both sides. Some examples are knives, axes, hatchets, and cold chisels.

The term **convex** means curved out. An edge that is convex is strong because it has extra steel to back it up. However, the bulging nature of a convex edge means it requires more force to make it cut. Axes, and similar tools with convex edges, get their force from their speed and weight. Wedges get their force from a driving tool.

The term **concave** means hollow or curved in. Wood chisels and plane irons are ground to a concave edge. Their edges are very sharp and require little force to make them cut. Certain saws may be ground so the teeth are wider at the points than they are at their base. Such blades are said to be **hollow ground**. The teeth are wider at their points than the thickness of the blade behind the teeth. Therefore, the teeth have no set and the blade makes a very smooth cut.

FIGURE 22-4 Notice the blue color on the end of this chisel. The temper has been damaged.

Preserving Temper

Most tools are made from tool steel. **Tool steel** is steel with a specific carbon content that allows the tool to be annealed and tempered. To **anneal** is to heat and then cool the steel slowly so as to make it soft and malleable. To **temper** is to heat the steel and then cool it more quickly so as to control the degree of hardness. After tools are properly tempered, excessive heating will **draw the temper**, or modify it, to render the steel soft and useless until it is retempered. Tools need to be constantly cooled while being sharpened, particularly when using a grinder. If a blue color appears on the tool, it has overheated and the temper is damaged (Figure 22-4).

TESTING FOR SHARPNESS

Knives and chisels should be sharpened to a very fine edge. To check for sharpness, place the tool blade in a flat position on a narrow piece of soft wood. Raise the back slightly and push the cutting edge across the wood. The edge should catch the wood and shave off a fine sliver. If this does not happen, the tool is probably not very sharp.

> **CAUTION!**
>
> Any test of sharpness using the hands or body should be avoided.

SHARPENING TOOLS

When sharpening tools, first remove all metal burrs or rough spots. Burrs should be ground or filed away without changing the shape of the edge. If deep nicks are present, straighten the entire cutting edge. The edge should be ground to the proper shape and angle.

> **CAUTION!**
>
> Grind slowly to avoid drawing the temper. Figure 22-5 presents a guide for sharpening tools.

When using grinders, be sure to wear a face shield and protective clothing. Check the grinding wheel(s) for soundness. **True** and clean the wheel, if needed, and then proceed to sharpen tools.

Tools that require a very sharp edge must be whetted. To **whet** means to sharpen by rubbing on a stone designed for sharpening tools.

Knives

Knives with smooth edges may be sharpened in the shop. Those with serrated edges must be sharpened by the manufacturer or a business with special sharpening equipment. **Serrated** means notched, like the edge of a saw. Most schools have policies prohibiting students from carrying knives in school. Therefore, it is especially important that students obtain permission from their instructor before fitting and sharpening knives in the shop.

The following procedure describes how to sharpen a smooth-bladed knife.

Tool	Sides to Sharpen	Shape of the Edge	Recommended Angle
Ax/hatchet	Both	Convex	20°
Center punch	Point	Straight	60° to 75°
Cold chisel	Both	Straight	60° to 75°
Knife	Both	Straight	As original
Hoe/shovel	One	Straight	As original
Plane iron	One	Concave	29°
Rotary mower blade	One	Concave	As original or 45°
Scissors or snips*	One	Concave	80°
Twist drill	Both	Straight	118° to 120°
Wood chisel	One	Concave	25° to 29°

*Scissors and snips have two moving blades. Each blade is sharpened on one side only at an 80° angle.

FIGURE 22-5 A guide for sharpening tools.

Copyright © 2015 Cengage Learning®.

Procedure

1. Use a fine-grit oil or water stone grinder.

2. Remove the nicks and grind away some of the thickness of the blade along the edge. Grind slowly to avoid overheating.

CAUTION!

Hold the knife with the cutting edge away from the operator (Figure 22-6). Move the knife back and forth to create an even grinding mark behind the edge.

3. Use an oil stone to whet the blade to a sharp edge. Whetting is done by holding the knife flat to the stone, with the back of the blade slightly raised. Draw the knife, edge first, across the oiled stone; flip the knife over and push it in the opposite direction (Figure 22-7). Continue to alternate sides until the edge is even and sharp. Whetting is done by holding the knife flat to the stone, with the back of the blade slightly raised. Draw the knife, edge first, across the oiled stone; flip the knife over and push it in the opposite direction (Figure 22-7). Continue to alternate sides until the edge is even and sharp.

4. Some stones have one coarse and one fine side. Always make the final strokes on the fine side.

5. When finished, wipe the oil slurry from the stone. Cover the stone with an even coat of light oil, and store it in its wooden box or original carton.

Knives can also be sharpened by using a knife sharpener with abrasive wheels mounted at the proper angle on a frame. The knife blade is drawn through a slot in the sharpener and contacts the abrasive wheels (Figure 22-8). Another type of sharpener makes use of ceramic sticks mounted at the proper angle on a block of wood. The blade is drawn across the sticks until it is sharp (Figure 22-9).

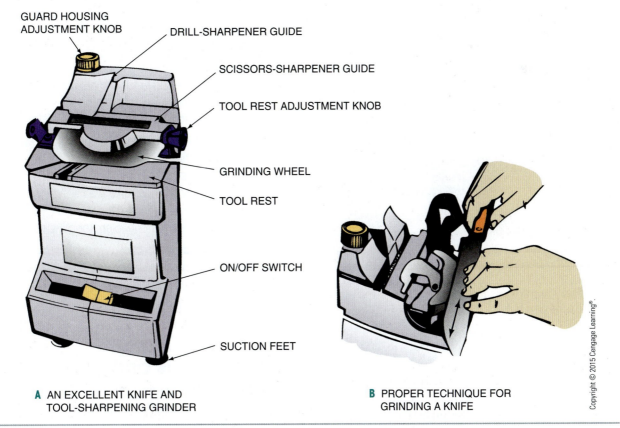

GUARD HOUSING
ADJUSTMENT KNOB

DRILL-SHARPENER GUIDE

SCISSORS-SHARPENER GUIDE

TOOL REST ADJUSTMENT KNOB

GRINDING WHEEL

TOOL REST

ON/OFF SWITCH

SUCTION FEET

A AN EXCELLENT KNIFE AND TOOL-SHARPENING GRINDER

B PROPER TECHNIQUE FOR GRINDING A KNIFE

Copyright © 2015 Cengage Learning®.

FIGURE 22-6 **A** An oil or water stone grinder should be used for grinding knives. **B** The knife should be held with the cutting edge away from the operator and moved back and forth across the wheel to give the blade an even taper.

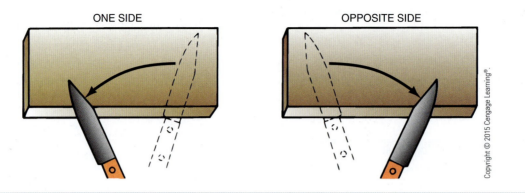

ONE SIDE   OPPOSITE SIDE

Copyright © 2015 Cengage Learning®.

FIGURE 22-7 Final sharpening of a knife is done by whetting it on an oil stone.

© Ilya Andriyanov/Shutterstock.com

FIGURE 22-8 This type of sharpener uses abrasive wheels mounted in a frame.

© Owl Mountain/Shutterstock.com

FIGURE 22-9 Ceramic sticks mounted on a block of wood can put a fine edge on a knife.

Wood Chisels

The edge of a wood chisel is sharpened on one side only to a 29-degree concave angle. A tool-sharpening gauge can be used to check the angle (Figure 22-10). To sharpen a wood chisel, grind the end until it is square and free of nicks. A special oil grinder is best, but

dry-wheel grinders may also be used, with special care. The following procedure describes how to sharpen a wood chisel.

Procedure

1. Loosen the tool rest on a grinder with a medium or fine wheel. Angle the tool rest so that the wood chisel lies on the curve of the wheel. Tighten the tool rest.

2. To grind the wood chisel, move it back and forth on the tool rest. After grinding, the cutting edge should be square to the side of the wood chisel.

3. Check the angle of the edge with a 29-degree gauge (Figure 22-11). It can also be compared to another wood chisel that is properly sharpened. Resharpen if necessary.

4. The next step is to put a fine edge on the wood chisel by whetting. To whet a wood chisel, place the ground section flat against the surface of an oiled stone, and then raise the heel off the ground area until it just lifts from the stone. Push the chisel across the stone to sharpen the edge.

5. Lay the blade on its back and push it across the stone to remove the wire edge (Figure 22-12). Alternate these two steps until the blade has a fine, sharp edge.

Cold Chisels and Center Punches

The end of a cold chisel is relatively thick, because it must withstand heavy blows from a hammer to cut steel. However, the thickness tapers at a second angle

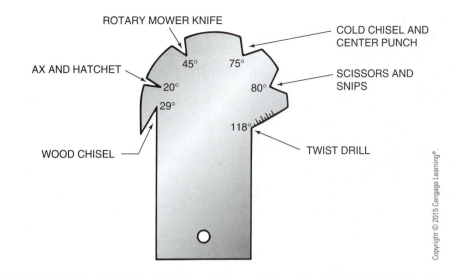

FIGURE 22-10 A tool-sharpening gauge greatly improves the accuracy of tool sharpening.

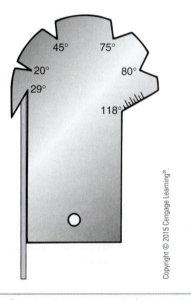

FIGURE 22-11 Gauging the correct angle for a wood chisel.

FIGURE 22-12 The final step in sharpening a chisel is to lay the blade on its back and remove the wire edge.

to a sharp edge. This secondary taper forms a 60- to 75-degree angle.

When sharpening a cold chisel, select a medium-grit grinding wheel. If the chisel edge has nicks, hold the chisel horizontally on the tool rest and ease the edge against the wheel. Move the chisel back and forth across the wheel, and grind the end until all nicks are removed.

To resharpen, use the following procedure.

Procedure

1. Hold the chisel on the tool rest. The chisel should point upward and lean toward the wheel. Form the cutting edge by making one pass across the wheel.

2. Then rotate the chisel half a turn and make one pass on the other side.

3. Alternate passes until a sharp edge is formed (Figure 22-13).

4. Check the edge for squareness.

5. Use a tool fitting gauge to check the edge for the correct angle.

6. Regrind the chisel until a perfect edge is formed.

Center punches are also sharpened at angles ranging from 60 degrees to 75 degrees. A center punch is sharpened by holding it at an angle to the wheel and twirling it as it is ground. The twirling action should create a round, even point.

FIGURE 22-13 Sharpening a cold chisel using a grinder. Here the operator is grinding the edge to a 60-degree angle.

Axes and Hatchets

Axes are large and hard to grind on standard grinding machines equipped with the proper guards. For this reason, some shops use grinding wheels mounted on a belt-driven **arbor**. Another method is to clamp the axe in a vise and use a portable disc sander. The edge of an axe is ground to a 20-degree angle.

The axe is slowly ground away by many sharpenings. This causes the angle to increase and the area behind the edge to get thicker. This problem is reduced by removing metal from as far back as an inch from the edge.

While cutting wood, the axe may be touched up or resharpened slightly with an ax stone. An **ax stone** is a small round stone held in the hand for field use. **Field use** means the area where an operation takes place. In this case field means the forest or woodpile.

Hatchets may be sharpened with a disc sander or grinder. The common shop grinder with a medium wheel may also be used to sharpen a hatchet. First remove any small nicks by pushing the blade horizontally across the tool rest and into the wheel. Make one pass across the wheel. Examine the edge to see if all small nicks are removed. Make another pass if needed.

In some cases, it may be better to leave large nicks in the blade than to grind away so much metal that the hatchet or axe becomes excessively thick. Generally, large nicks occur at the end of the blade. If so, the end of the blade may be rounded in order to grind the nick into a sharp edge. Needless to say, large nicks ruin an axe or hatchet. They should be avoided by never cutting wood that may contain metal. Another possible source of nicks is chopping through wood and into stones in the ground.

To sharpen a hatchet, use the following procedure.

Procedure

1. Draw a line on each side of the blade, parallel to and ½ to ⅝ inch in from the edge to serve as a guide.

2. Hold the head, edge up, against the wheel. Use a gloved hand to move the head quickly up and down ½ inch or more as the tool is moved slowly across the wheel (Figure 22-14). The up-and-down motion creates a wide grinding band and maintains the correct angle.

3. Reverse the position of the tool and make a similar pass on the other side.

4. Repeat the process until a good edge is obtained.

5. Remove grinding marks with a hand stone or fine flat file.

A large, sharp, medium-cutting, flat file provides an equally effective way to sharpen a hatchet or axe (Figure 22-15). The tool to be sharpened is first clamped securely in a machinist's vise, the end of the file is then placed on the flat of the tool, and metal is removed by filing.

CAUTION!

Use extreme care to prevent your fingers or hand from slipping into the cutting edge. Direct the file away from the cutting edge for safety reasons.

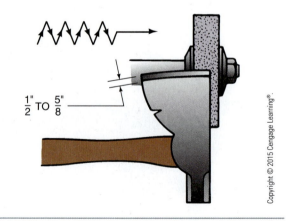

$\frac{1}{2}$ TO $\frac{5}{8}$"

FIGURE 22-14 When using the grinder to sharpen a hatchet, move the hatchet across the wheel with quick up-and-down movements (shown by the arrows) to produce a long taper on the edge.

FIGURE 22-15 When using a file to sharpen a hatchet, wear heavy leather gloves and direct the file away from the hatchet's cutting edge.

Twist Drills

Twist drills stay sharp for a long time when drilling wood or plastic. However, when drilling metal, they quickly become dull. Although aluminum is soft, it dulls drill bits rather quickly. Drilling steel has a moderate dulling effect on high-speed drills. A **high-speed drill** is a twist drill made and tempered specially to drill steel. It is easy to tell when a metal drilling bit is dull. When in use, a dull bit will turn out small bits of metal (Figure 22-16A), while a sharp bit will turn out curly ribbons of metal (Figure 22-16B).

Sharpening a twist drill requires familiarity with its design. The cutting tip of a drill consists of a dead center and two cutting lips (Figure 22-17).

The dead center must be exactly centered after the drill is sharpened. The dead center and two cutting lips must touch metal before the rest of the drill end. Otherwise, the drill will ride on its rounded end rather than cut into the metal.

The two cutting lips start at the two ends of the dead center. They are ground at 59 degrees to the center line of the drill shank, and together they form a 118-degree angle when new. When resharpening a twist drill, a 118-degree or 120-degree gauge is used to test the accuracy of the angle.

The metal recedes or falls back from the cutting tip to the heel. The resulting angle is 12 degrees on a new drill and should be kept as close to that as possible when resharpening. This provides clearance for the cutting lips and also gives adequate support to keep the lips from breaking.

Twist drills should be sharpened with a special jig to obtain the correct angles. However, with practice, they can be sharpened freehand by using the following procedure.

Procedure

1. Redress a fine grinding wheel, if needed to make it true.

2. Set the tool rest horizontal to the center of the wheel.

(continued)

A A dull drill bit.

B A sharp drill bit

FIGURE 22-16 When in use, a dull bit will turn out small bits of metal (A), while a sharp bit will turn out curly ribbons of metal (B).

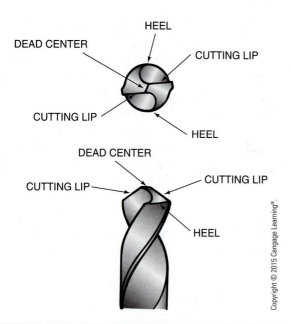

FIGURE 22-17 The cutting tip of a twist drill.

Procedure, *continued*

3. Study the shape of a new drill that is about the same size of the one to be sharpened.

4. Hold the drill between your thumb and index finger, with the tip of the drill exposed about 1 inch (Figure 22-18).

5. Place the back of your index finger on the tool rest with your thumbnail up and the drill at an angle to the wheel. The lip on the left should be visible and parallel to the stone.

6. Touch the lip against the grinding wheel and lower the opposite end of the drill as you give it a slight **clockwise** twist.

7. Rotate the drill half a turn so the other lip is visible.

8. Repeat step 6 to sharpen the second lip.

FIGURE 22-18 Sharpening a twist drill on a grinder.

Use a tool-sharpening gauge to check the angle of the lips, position of the dead center, and clearance of the heels (Figure 22-19). If the angle of the lips is less than 118 degrees, the bit will cut too fast. If the angle is greater than 118 to 120 degrees, it will cut too slowly. If correct, test the tool by drilling a piece of mild steel. If the drill does not cut well, it must be resharpened.

Rotary Mower Blades

Blades for both lawn and field rotary mowers need frequent sharpening. Sharp blades require less power, do a better looking job, and damage the plants less.

When sharpening blades, try to maintain the original angle. If this is not known, grind the blades to a 45-degree angle. One side of the blade is kept perfectly flat; the other side is ground to a cutting edge.

When grinding mower blades, remove the same amount of metal from both ends. This keeps the blade in balance. **Balance** means the weight is equally distributed on both sides of the center. Therefore, if one end is badly nicked, grind both ends equally until the nick is removed. Failure to do so will cause the blade to be out of balance and the mower to vibrate. This causes damage to the shaft, bearing, and mower body.

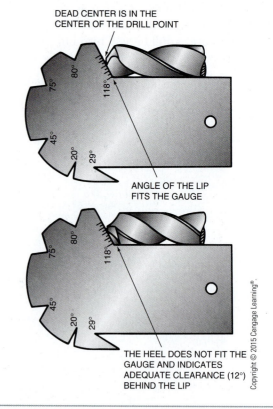

FIGURE 22-19 A tool-sharpening gauge is used to check the angle of the lips, position of the dead center, and clearance of the heels of a resharpened twist drill.

A First, remove the nicks using a grinder.

B Then, restore the edge using a flat file.

FIGURE 22-20 To sharpen a rotary mower blade, remove the nicks by grinding or filling the cutting edge (A). Clamp the blade in a vise and restore the flat side by using a flat file (B).

To sharpen a rotary mower blade, remove the blade from the mower, then remove the nicks by grinding or filing the cutting edge (Figure 22-20A), and then clamp the blade in a machinist's vise. Restore the flat side by using a large, medium-cutting, flat file (Figure 22-20B). Restore the cutting edge using a flat file or aluminum oxide disc on a heavy-duty portable sander.

Mower blades may be sharpened on a stationary grinder. Since the blades are hard to position properly on a grinder, this is a less desirable method.

CAUTION!

Whenever sanders or grinders are used, care must be taken to avoid overheating, which can cause the tool to lose its hardness.

Digging Tools

The edges of digging tools, such as hoes, shovels, spades, and spading forks, get ragged and dull with use. The same is true for scoop shovels. To reshape these tools, first observe the side on which they were ground

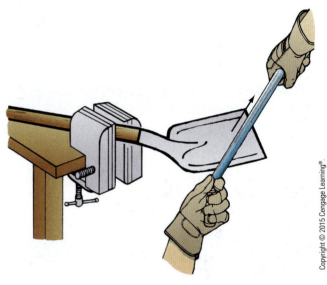

FIGURE 22-21 Restoring the cutting edge of a digging tool.

before. If metal is curled back, place the tool on an anvil and tap the bent metal back into place with a steel hammer. Restore the cutting lip with a file or grinder (Figure 22-21). All shovels are sharpened on the inside only. The outside or back remains straight.

SUMMARY

A cutting tool with a dull edge is inefficient and dangerous. The correct procedure for sharpening a tool depends on the tool and its use. If the proper procedure is followed, the correct edge will be achieved, and the tool will regain its efficiency. Use of improper techniques will result in a damaged cutting edge and possible injury to the operator.

Student Activities

1. Define the Terms to Know in this unit.
2. Copy the data in Figure 22-5 into your notes.
3. Use a grinder to sharpen a
   a. knife
   b. wood chisel
   c. cold chisel
   d. center punch
   e. hatchet
   f. twist drill
4. Use a bench stone to whet a knife and wood chisel.
5. Use a file to sharpen a hatchet, rotary mower blade, and shovel.
6. Use a portable power sander to sharpen an axe.
7. Create your own tool-sharpening gauge based on the plan shown in Figure 22-22.

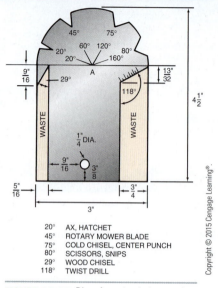

20°	AX, HATCHET
45°	ROTARY MOWER BLADE
75°	COLD CHISEL, CENTER PUNCH
80°	SCISSORS, SNIPS
29°	WOOD CHISEL
118°	TWIST DRILL

FIGURE 22-22 Plan for a tool-sharpening gauge.

Copyright © 2015 Cengage Learning®

Relevant Web Sites

Museum of Woodworking Tools, A Guide to Honing and Sharpening
www.antiquetools.com/sharp

Do It Yourself®, information on sharpening tools and lawn care equipment
www.doityourself.com

Self-Evaluation

A. Multiple Choice. Select the best answer.

1. The reason for learning tool-sharpening skills is that
   a. better work is possible with sharp tools
   b. sharp tools are easier to use
   c. sharp tools are safer
   d. all of these

2. A 6" × 2" × 1" oil stone is called
   a. an ax stone
   b. a bench stone
   c. a common stone
   d. a slip stone

3. The best tool to sharpen an axe is a
   a. grinder
   b. belt sander
   c. portable disc sander
   d. all of these

4. When sharpening tools with a grinder, the tool rest should be
   a. removed
   b. placed at a 90-degree angle
   c. placed to support the tool
   d. pushed down out of the way

5. A properly ground drill must have the
   a. proper angles
   b. proper clearance
   c. proper centering
   d. all of these

6. Hatchets may be sharpened with
   a. the common shop grinder with a medium wheel
   b. a disc sander
   c. a file
   d. all of the above

B. Matching. Match the terms in column I with those in column II.

Column I

1. concave
2. convex
3. serrated
4. anneal
5. temper
6. whet
7. dead center

Column II

a. make soft
b. curved out
c. curved in
d. sharpen
e. drill part
f. notched
g. make hard

C. Completion. Fill in the blanks with the word or words that make the following statements correct.

1. When grinding mower blades, remove the same amount of metal from both ends to keep the blade in _____.

2. For safety when filing, direct the file _____ from the cutting edge.

3. All shovels are sharpened on the _____.

4. After tools are properly tempered, excessive heating will _____ _____ _____.

5. Saw blades that are ground so the teeth are wider at the points than at the base are said to be _____ _____.

D. Brief Answer. Briefly answer the following questions.

1. List the number of sides to be sharpened and the recommended angle for grinding each of the following.
   a. axe
   b. cold chisel
   c. hatchet
   d. knife
   e. mower blade
   f. shovel
   g. wood chisel

GAS HEATING, CUTTING, BRAZING, AND WELDING

UNIT 23
- Using Gas Welding Equipment

UNIT 24
- Cutting with Oxyfuels and Other Gases

UNIT 25
- Brazing and Welding with Oxyacetylene

UNIT 23

Using Gas Welding Equipment

Objective
To use gas-powered heating, cutting, and welding equipment safely.

Competencies to be developed
After studying this unit, you should be able to:
- Identify major parts of propane and oxyacetylene welding equipment.
- Change oxygen and acetylene cylinders.
- Turn on and adjust oxyacetylene controls.
- Light and adjust oxyacetylene torches.
- Shut off and bleed oxyacetylene equipment.
- Check for leaks in gas equipment.

Materials List
- Oxyacetylene welding outfit
- Variety of welding tips
- Welding gloves and apron
- No. 5 shaded goggles
- Tip cleaners
- Spark lighter

Terms to Know
- weld
- fusion
- gas
- compress
- flammable
- apparatus
- manifold
- rig
- oxyacetylene
- torch
- cylinder
- valves
- regulators
- gauge
- hoses
- crack the cylinder
- seat
- purge the lines
- carbonizing flame
- neutral flame
- oxidizing flame
- tip cleaners
- bleeding the lines

Many gases will burn. This quality makes them both dangerous and beneficial. The burning qualities of gases are used in agricultural mechanics to heat, cut, braze, solder, and **weld** metals. Welding metals is the process of melting metals and allowing the melted metals to flow together in one piece. This process is called **fusion**. Brazing is a process where brass is melted and bonded with steel, cast iron, or other metals. Soldering is much the same process except it uses much softer metal than brass.

A **gas** is any fluid substance that can expand without limit when not contained by natural or artificial means. This expandable nature also means that a gas can be compressed. To **compress** means to apply pressure to reduce in volume. For example, an air compressor uses the force of a pump to compress great volumes of air into a small tank.

A compressed gas is dangerous simply because it is under pressure and is always trying to get free. It is like a spring in a compressed state. If a compressed gas is **flammable** (meaning that it burns easily), there is the additional danger that it may explode or burn out of control. Fortunately, suitable equipment and techniques are available for using compressed, combustible gases safely.

The most popular gases for heating, cutting, and welding metals are propane and acetylene. Before these gases will burn, they must be mixed with oxygen from the air or pure oxygen from tanks. The student is referred to Unit 5 ("Reducing Hazards in Agricultural Mechanics") for information on how to prevent and control fires.

Acetylene is particularly dangerous because of its chemical makeup. It is composed of 2 carbon atoms and 2 hydrogen atoms that combine to form molecules of acetylene. The particular way the atoms are bonded allows the storage of energy that can be released as heat when the compound is ignited. Unfortunately, this bond also makes the gas unstable under certain conditions. For example, the gas should never be compressed over 15 pounds per square inch (PSI). That is why the acetylene regulators should never be opened to more than 15 PSI. Obviously, the tanks have to be able to store a large amount of acetylene gas and this means compressing the gas. To prevent problems with too much pressure, the tanks contain liquid acetone. This material absorbs the acetylene gas and allows the gas to be compressed and stored with lower pressure. Because of the liquid, the acetylene tanks should never be laid on their side with the regulators mounted on the tanks. This would allow the acetone to flow into the regulator and cause damage.

GENERAL SAFETY WITH COMPRESSED GASES

Some general precautions in handling compressed gases are:

- Wear safety goggles or a face shield with properly tinted lens at all times.
- Obtain the instructor's permission before using compressed gases.
- Store fuel gas cylinders separately from oxygen cylinders.
- Keep gas cylinders upright and chained securely at all times (Figure 23-1). They should be stored outdoors or in well-ventilated, fire-safe areas.
- Do not bump or put pressure on pipes, connections, valves, gauges, or other equipment connected to compressed gas cylinders.
- When connections are opened or cylinders changed, check thoroughly for leaks before using the equipment.
- Never use equipment exposed to oil or grease. Spontaneous or instant fires may result.
- Follow specific procedures for turning systems on and off.
- Work only in areas that are free of materials that burn.
- Never cut or weld in an area that is not well ventilated.
- When cutting galvanized metal, take precautions not to breathe fumes from the process.

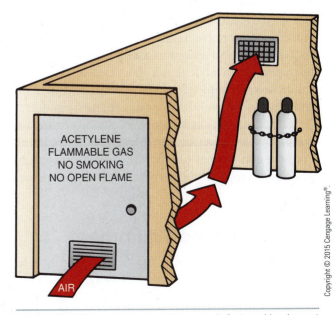

Copyright © 2015 Cengage Learning®.

FIGURE 23-1 Gas cylinders must be securely fastened in place at all times.

- Never use gas-burning equipment without approved fire extinguishers in the area.
- Always wear leather gloves and apron when using gas-burning equipment.
- Screw the caps on all cylinders that do not have regulators or other apparatus attached. An **apparatus** is any object necessary to carry out a function.
- All equipment or cylinders that may discharge gas should be pointed away from the operator, other people, and clothing. Fire will follow a gas stream.
- Never leave clothing where it can become saturated by oxygen or fuel gases.
- If gas equipment catches fire, immediately turn off the gas at the tanks. If this is not practical, or if this action does not extinguish the gas fire, evacuate the area and call for help.
- Learn to recognize the odors of combustible fuels.
- Protect gas cylinder storage areas with locked chain-link fences or concrete enclosures.

Gas-Burning Equipment for Agricultural Mechanics

Shops generally use propane and/or acetylene and oxygen for heating, cutting, and welding. Propane and acetylene are fuels. Oxygen is not a fuel, but it must be present for fuels to burn. Some shops have manifolds to which many cylinders and welding outfits may be attached. A **manifold** is a pipe with two or more outlets and, in this case, regulators are attached to the pipes in order to provide several stations for cutting and welding. Most agricultural mechanics shops have one or more portable oxyacetylene rigs. A **rig** is a self-contained piece of apparatus assembled to conduct an operation. **Oxyacetylene** is a shortened version of the words "oxygen" and "acetylene." It refers to equipment and processes where the two gases are used together.

Shops may also have propane and oxygen torches and propane furnaces for heating and soldering. Propane and oxygen torches are desirable as cutting units since propane gas is generally less expensive than acetylene.

Oxyacetylene Equipment

Major Parts. A portable oxyacetylene rig includes a cart, cylinders, valves, regulators, gauges, hoses, and torch assemblies (Figure 23-2). A **torch** is an assembly that mixes gases and discharges them to support a controllable flame (Figure 23-3). A gas **cylinder** is

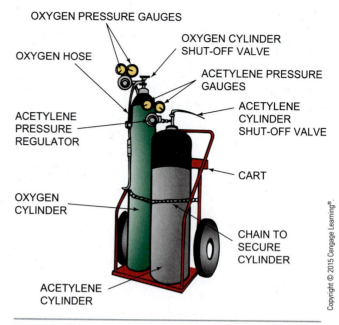

FIGURE 23-2 A portable oxyacetylene rig.

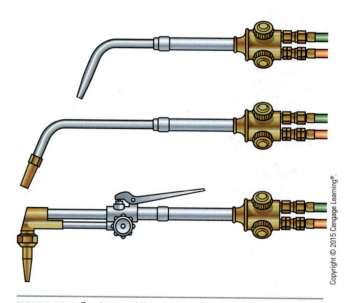

FIGURE 23-3 Torches and tips designed for brazing and welding (top), heating (center), and cutting (bottom).

a long round tank with extremely thick walls built to hold gases under great pressure. **Valves** and **regulators** are devices that control or regulate the flow of the gas. A valve is a device that allows the gas to flow from the cylinder; a regulator is a device that keeps the pressure of the flowing gas at a constant rate. A **gauge** is mounted on the regulator to measure and indicate the pressure in the hose, tank, or manifold (Figure 23-4). A check valve prevents a flame from reaching the gas in the cylinder. **Hoses** are flexible lines that carry the gases. They are rubber reinforced with nylon or other material to withstand high pressures and heavy use.

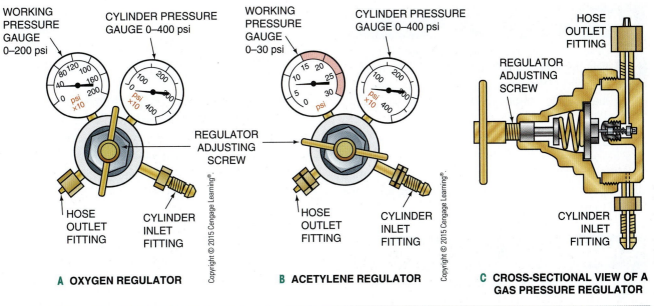

FIGURE 23-4 Oxygen and acetylene regulators and gauges.

The gas flow and control components of a portable oxyacetylene rig are as follows:

Acetylene Side
- acetylene cylinder
- cylinder valve
- red cylinder pressure gauge
- regulator
- hose pressure gauge
- red hose
- acetylene valve on torch

Oxygen Side
- oxygen cylinder
- cylinder valve
- green cylinder pressure gauge
- regulator
- hose pressure gauge
- green hose
- oxygen valve on torch
- torch mixing chamber
- torch tip

Setting Up. When setting up oxyacetylene equipment, one must be aware of certain design features. The valve on most acetylene cylinders is protected with a high collar. The valve is turned on and off with a handle. On cylinders with removable handles, care must be taken to leave the handle in place when turned on in case the gas must be turned off in an emergency. Acetylene equipment is color-coded red. Another important distinction is that all acetylene couplings have left-hand threads and notched fittings.

Oxygen cylinders generally have valves that are not protected during use. However, they have heavy caps that screw in place over the valves to protect them when not in use. Oxygen equipment is color-coded green. All oxygen couplings have right-hand threads.

Before attaching regulators to cylinders, **crack the cylinder** by turning the gas on and off quickly to blow any dust from the opening.

> **CAUTION!**
>
> When attaching regulators, hoses, gauges, or torch connections, the connectors must be threaded in the correct direction. The red color coding and opposite threads for acetylene are means of preventing the accidental mixing of acetylene and oxygen, except in the torch itself.

All connections are gastight when screwed on properly. Some parts such as regulators have metal-to-metal fittings. These draw together snugly and must be gastight. Some brass fittings used for connecting hoses are also metal-to-metal types. Gauge fittings have tapered threads. These are wrapped with Teflon tape to ensure a gastight fit when screwed into regulators or pipe fittings.

CAUTION!

Oxygen and acetylene valves on torches seal with light finger pressure. Overtightening will damage the seats of the valves. A seat is the point where the movable part of a valve seals off the gas.

Turning on the Acetylene and Oxygen. After all components are assembled, the following sequence is used to turn on the gases.

CAUTION!

The area must be properly ventilated. Wear protective goggles and clothing.

Activity

1. Close the acetylene valve on the torch.
2. Close the oxygen valve on the torch.
3. Turn the acetylene regulator handle counterclockwise until no spring tension is felt.
4. Turn the oxygen regulator handle counterclockwise until no spring tension is felt.
5. Open the oxygen cylinder valve slowly until the pressure gauge responds. Open the valve all the way.

CAUTION!

Do not stand in front of the gauges when gas is being turned on.

6. Open the acetylene cylinder valve slowly half a turn.

CAUTION!

The acetylene cylinder valve is never opened more than half a turn so it can be turned off quickly in an emergency.

7. Open the oxygen torch valve an eighth of a turn. Turn the oxygen regulator handle clockwise until the pressure gauge reads 10 psi (pounds per square inch). (The final desired pressure will vary with the specific torch equipment being used.) Close the oxygen valve on the torch.

8. Open the acetylene torch valve an eighth of a turn. Turn the acetylene regulator handle clockwise until the pressure gauge reads 5 psi. (The final desired pressure will vary with the equipment being used.) Close the acetylene valve on the torch.

Steps 7 and 8 purge the lines and set the regulators at safe starting pressures. To **purge the lines** is to remove undesirable gases. The unit is now pressurized and ready for use.

Testing for Leaks. A leak test should be performed when equipment is first set up, when cylinders are changed, or if the odor of acetylene is present when the unit is not in use. To test for leaks, put a small amount of water in a small jar or can. Add a sliver of non-detergent hand soap. Use a 1-inch paintbrush to produce a soapy lather; use the brush to apply the soap solution gently around each fitting and point where gas may escape. If a leak exists, it will cause bubbles in the solution.

If a leak is found, tighten the fitting. If this does not stop the leak, turn off both gases. Disassemble the joint and correct the problem. Usually the problem is a dried out or worn out O-ring that fits on the torch to prevent gas leaks. If the O-ring is bad, simply replace it with a new one. Turn on the gases again and recheck the entire system with the soap solution.

Lighting and Adjusting Torches. Once the system is pressurized, attention can be given to selecting torch parts. This is followed by lighting and adjusting the torch according to the work to be done.

Two types of torches are available: welding torches that are used for welding, brazing, or heating (Figure 23-5) and cutting torches (Figure 23-6). Both include a body or handle with hose connections and valves to control the oxygen and the fuel. Welding tips with mixing chambers are screwed onto the handle. Tips of different sizes are available. A cutting assembly with another set of valves may also be attached to the handle.

The following procedure describes how to light and adjust a welding or cutting torch.

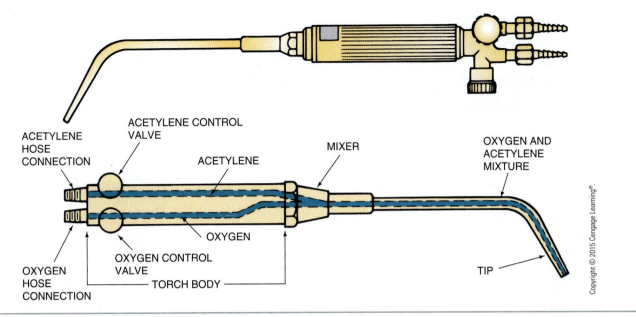

FIGURE 23-5 Exterior and interior views of a welding torch.

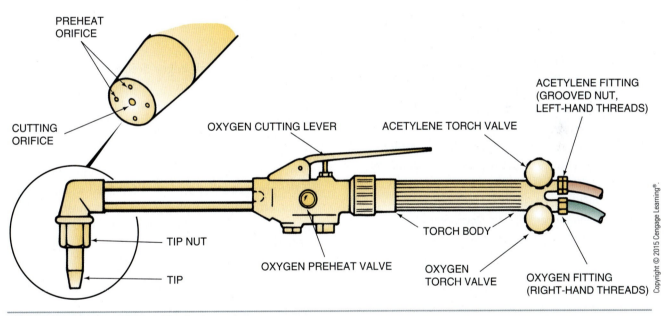

FIGURE 23-6 An oxyacetylene cutting torch.

Activity

1. Put on leather gloves and goggles with a No. 5 shaded lens (Figure 23-7).

2. Open the acetylene valve an eighth of a turn.

3. Use a spark lighter to ignite the torch (Figure 23-8).

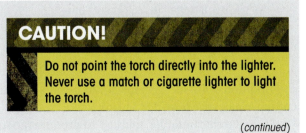

CAUTION!

Do not point the torch directly into the lighter. Never use a match or cigarette lighter to light the torch.

(continued)

WELDING
GOGGLES

LEATHER
GLOVES

Copyright © 2015 Cengage Learning®.

FIGURE 23-7 Before lighting a gas torch, the operator must put on the proper protective clothing.

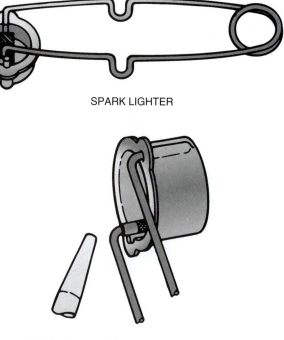

SPARK LIGHTER

CORRECT POSITION OF THE SPARK LIGHTER

Copyright © 2015 Cengage Learning®.

FIGURE 23-8 To light a gas torch safely, the operator should use a spark lighter.

Activity, *continued*

4. Open the acetylene valve slowly until the flame is ¼ inch off the tip of the torch. Increase or decrease the regulator pressure until the flame just touches the tip. At this point, the acetylene valve on the torch is open several turns. This will be the correct pressure for the tip being used. The flame is called a carbonizing flame. A **carbonizing flame** is one with an excess of acetylene (Figure 23-9). It is cooler than other types of flames.

5. Turn the oxygen valve on slowly and watch the inner flame shorten. Continue to add oxygen until the long inner flame just fits the cone. This is a **neutral flame**, or one with a correct balance of acetylene and oxygen. A neutral flame is correct for heating, cutting, and welding.

6. If additional oxygen is added, the cone becomes shorter and sharper and the flame more noisy. This is an **oxidizing flame**, or one with an excess of oxygen. It is the hottest type of flame, but it is not recommended except for special applications.

Additional instructions for lighting and adjusting cutting torches are provided in Unit 24.

If the torch cannot be adjusted to produce a neutral flame, clean the tip (Figure 23-10). **Tip cleaners** are rods with rough edges designed to remove soot, dirt, or metal residue from the hole in the tip. Use a tip cleaner that is equal to or smaller in size than the hole in the tip. If molten metal is fused to the tip, remove it with fine emery cloth.

Shutting Off Torches. Torches become a fire hazard if the proper procedure is not followed when shutting them off. Improper procedures may cause excessive popping noises of the torch, soot from unburned gases, or carbon deposits in the tip. To shut off a torch correctly, first close the acetylene valve on the torch, and then close the oxygen valve.

CAUTION!

Use gentle pressure on the torch valves as they have soft internal parts that may be damaged.

Bleeding Lines. When a torch is not in use, it is important to close every point where gas may escape. Gas should also be removed from all lines and equipment. This practice is called **bleeding the lines**. The following procedure describes how to bleed the lines.

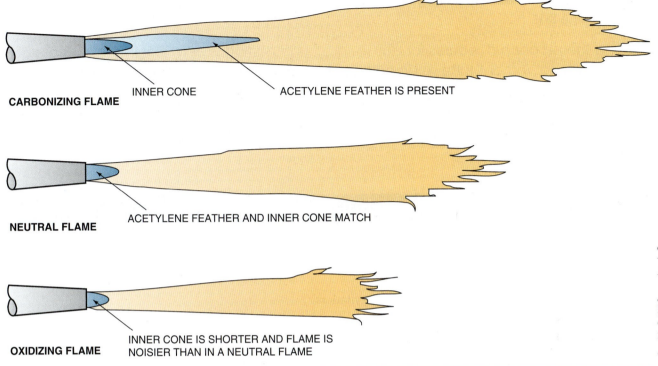

INNER CONE   ACETYLENE FEATHER IS PRESENT

CARBONIZING FLAME

NEUTRAL FLAME   ACETYLENE FEATHER AND INNER CONE MATCH

OXIDIZING FLAME   INNER CONE IS SHORTER AND FLAME IS NOISIER THAN IN A NEUTRAL FLAME

FIGURE 23-9 Carbonizing, neutral, and oxidizing flames. The neutral flame is correct for heating and cutting.

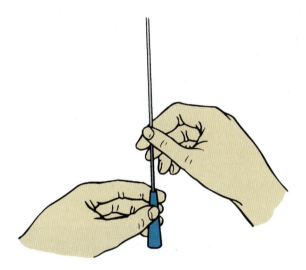

A. CLEANING A TIP WITH A STANDARD TIP CLEANER

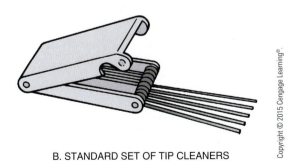

B. STANDARD SET OF TIP CLEANERS

FIGURE 23-10 The proper tip cleaner to use is the one whose diameter is equal to or smaller than the hole in the tip.

Activity

1. Turn off the acetylene at the cylinder.

2. Turn off the oxygen at the cylinder.

3. Open the acetylene valve at the torch until both regulators return to zero. Close the acetylene valve at the torch.

4. Open the oxygen valve at the torch until both regulators return to zero. Close the oxygen valve at the torch.

5. Turn all regulator handles counterclockwise until they are easy to turn (indicating that there is no pressure on the diaphragms).

6. Coil the hoses over the cart handles, special hose hangers, or the large part of the cylinders.

7. If portable, store the rig in a suitable place.

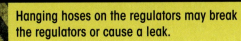

CAUTION!

Hanging hoses on the regulators may break the regulators or cause a leak.

Copyright © 2015 Cengage Learning®.

SUMMARY

Using the proper procedures for gas welding and cutting is essential to the successful use of the equipment. If these procedures are not followed, results will be poor. In addition, failure to observe the correct safety procedures can be catastrophic. The amount of power contained in the gas cylinders is awesome. The proper, safe release and use of that power requires careful adherence to the procedures described in this unit.

Student Activities

1. Define the Terms to Know for this unit.
2. Study the diagrams of all equipment pictured in this unit. Memorize all the names of all equipment components.
3. Attach regulators to oxygen and acetylene cylinders.
4. Turn on the oxygen and acetylene.
5. Test the apparatus for gas leaks.
6. Clean the tip of the torch.
7. Light the torch and adjust the gas flow to achieve a neutral flame.
8. Shut off the torch and bleed the lines.
9. Store the apparatus properly.

Relevant Web Sites

Great Britain's Health and Safety Executive, Welding Health and Safety
www.hse.gov.uk/welding

eFunda, Inc., Inert Gas Welding
www.efunda.com/processes/metal_processing/welding_inertgas.cfm

Self-Evaluation

A. Multiple Choice. Select the best answer.

1. Gas can be compressed with a
   a. cylinder
   b. lever
   c. pump
   d. valve

2. Acetylene may be dangerous because it is
   a. compressed
   b. flammable
   c. explosive
   d. all of these

3. Which gas is not a fuel used for torches?
   a. acetylene
   b. oxygen
   c. propane
   d. none of these

4. Oxygen and acetylene hoses can stand pressure because they are
   a. color coded
   b. extra thick
   c. made of steel
   d. reinforced

5. Oxygen hoses and related equipment are color-coded
    a. green
    b. ivory
    c. orange
    d. red

6. Acetylene hoses and related equipment are color-coded
    a. green
    b. ivory
    c. orange
    d. red

7. The acetylene pressure to light a torch should be
    a. 5 psi
    b. 15 psi
    c. 25 psi
    d. 50 psi

8. Gas leaks are checked with
    a. compressed air
    b. flame
    c. soapy water
    d. Teflon

B. Matching. Match the terms in column I with the terms in column II.

Column I
    1. weld
    2. gas
    3. compress
    4. oxyacetylene outfit
    5. tank
    6. neutral flame
    7. goggles
    8. Teflon

Column II
    a. pressure
    b. cylinder
    c. balanced
    d. tape

e. expands without limit
f. join by fusion
g. rig
h. No. 5

C. Completion. Fill in the blanks with the word or words that make the following statements correct.

1. Gas cylinders must be securely _____.

2. Never use gas-burning equipment without approved _____ _____ in the area.

3. When no apparatus is attached, a _____ should be screwed onto oxygen cylinders.

4. The presence of acetylene and propane is detected by _____.

5. The term *oxyacetylene* comes from _____ and _____.

6. The valve on an acetylene cylinder is protected by a _____.

7. Connectors for acetylene hoses have _____ threads.

8. Connectors for oxygen hoses have _____ threads.

9. Removing gas from oxyacetylene equipment is known as _____.

10. Hanging hoses on regulators may result in _____.

D. Brief Answer. Briefly answer the following questions.

1. Why should one never open the acetylene cylinder valve more than half a turn?

2. What is the purpose of color coding and using opposite threads for oxygen and acetylene?

3. Describe the differences among a carbonizing flame, a neutral flame, and an oxidizing flame. Which is correct for heating, cutting, and welding?

4. When should leak tests be performed?

5. How does one shut off a torch correctly?

UNIT 24

Cutting with Oxyfuels and Other Gases

Objective

To use oxygen and selected fuels to cut steel with a flame torch.

Competencies to be developed

After studying this unit, you should be able to:

- Write the names and characteristics of common fuels used for cutting.
- State and apply recommended safety practices for using oxyfuels.
- Select appropriate pressures for using oxygen and common fuel gases for cutting.
- Cut steel with oxyfuels.
- Pierce steel with oxyfuels.

Materials List

- Oxyacetylene cutting rig and welding area
- Welding safety equipment and protective clothing
- Two pieces of steel, 1/4" to 1/2" × 2" × 12"
- Plate steel, 1/4" to 1/2" × 6" × 12"
- Black pipe up to 3 inches in diameter, 1 to 2 feet long
- Soapstone

Terms to Know

- oxyfuel
- oxyfuel cutting
- kerf
- slag
- brazing
- brass
- base metal
- fusion welding
- backfire
- flashback
- slag box
- pierce
- plasma arc cutting
- plasma

Combustible gases combined with oxygen have been used since the 1800s for welding metals. Over the years, scientists have experimented to find the gases that work best. Unfortunately, no one gas is the safest and best for all jobs. A study of gas heating, cutting, and welding procedures is an important part of modern agricultural mechanics.

OXYFUEL PROCESSES

The term **oxyfuel** refers to the combination of pure oxygen and a combustible fuel gas to produce a flame. Oxyfuels are used for welding, brazing, cutting, and heating metals.

General Principles

Oxygen and Fuel Gases. These gases are stored under pressure in tanks or cylinders. They are released as individual gases through carefully designed valves, regulators, and hoses. The gases are mixed as they flow through torch assemblies. They burn as the mixture is discharged through carefully engineered tips.

The gas flame from an oxyfuel burns with intense heat. The temperature may range from 5,000°F to 6,000°F. These temperatures are hot enough to melt most metals, and permit cutting and fusion welding.

Cutting, Brazing, and Heating. Cutting, brazing, and heating differ from welding. **Oxyfuel cutting** is a process in which steel is heated to the point that it burns and is removed to leave a thin slit called a **kerf**.

To burn, hot steel must combine with oxygen. **Slag** is a product formed during the process. Slag is a good insulator. It hinders the cutting process because it forms a layer between the burning steel and the torch flame.

Cutting torches are designed to send a forceful stream of oxygen into preheated cherry red steel. This oxygen supports the combustion. The force of the oxygen stream drives the slag out of the area, and permits the heat from the torch to keep the steel burning. As a cutting torch moves forward, the oxygen stream pushes slag out to form the kerf. Torches can be adjusted to cut kerfs that are nearly as straight and clean as the cut produced by a saw (Figure 24-1).

Heating. In heating, the temperature may be raised enough to soften metal for bending or shaping. No melting takes place. The job may call for heat in a very narrow band to aid in making a sharp bend in a specific place. The heat may also be spread equally over a large area to reduce stress from welding. Different tips for oxyfuel torches are available for these purposes.

FIGURE 24-1 Torches can cut kerfs almost as straight and clean as a saw.

© iStockphoto/Glen Jones

Brazing. **Brazing** is the process of bonding with metals and alloys that melt at or above 840°F. **Brass** is a mixture of copper and zinc. It is an example of an alloy used for brazing. The process of brazing is similar to soldering: When brazing, the **base metal** (the main piece of metal) is heated until the brass melts, flows, and bonds to it (Figure 24-2). The base metal is not melted, and the metals do not mix during brazing.

Fusion Welding. Steel may be joined by fusion welding. **Fusion welding** is joining metal by melting it together. An oxyacetylene welding torch may be used to heat two pieces of steel until the metal from each runs together to form a joint. When properly done, such joints are as strong as the base metal itself.

CHARACTERISTICS OF OXYFUELS

Oxygen

Oxygen is not a fuel and it will not burn. However, it combines with other substances and causes them to burn. According to the fire triangle (see Unit 5), fuel

FIGURE 24-2 When brazing, the base metal is heated until the brass melts and bonds the base metal.

between 1,600°F and 1,800°F. This is about 800° below the melting temperature of iron.

Acetylene

The fuel most suitable for welding is acetylene. This gas produces a cleaner weld than most other fuel gases. It also produces a more controllable flame. However, acetylene gas is unstable and, therefore, it is hazardous. Acetylene must be handled very carefully.

Acetylene is more expensive than other oxyfuels. Therefore, it is used in welding only when its burning characteristics are needed. In such cases, the safety hazards must be tolerated. For high-volume cutting and heating, less expensive fuels are generally preferred (Figures 24-4 and 24-5). However, since acetylene performs most functions well, it is the choice of fuel for many agricultural mechanics shops.

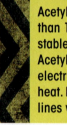

CAUTION!

Acetylene must not be used at pressures greater than 15 psi. The lower the pressure, the more stable the gas is and the safer it is to use. Acetylene containers must not be subjected to electrical shock, rough handling, or excessive heat. Do not use acetylene in copper lines or in lines with any grease or oil residues.

Propane and Natural Gas

Propane gas is available in cylinders and tanks. Natural gas is piped into buildings in urban and suburban areas through public utility lines. Both gases are used extensively for general heating, and may be used for

plus heat plus oxygen equals fire. Oxygen is necessary to use with the fuel gas to achieve an efficient, hot flame (Figure 24-3).

Oxygen must be 99.5 percent pure to support the combustion of iron. When iron is hot enough, it burns in oxygen the way wood or paper burns in air. This principle makes cutting with a torch possible. To burn in the presence of oxygen, iron must be heated to

Fuel (1 Cubic Foot)	Total O_2 Required (Cubic Feet)	O_2 Supplied Through Torch* (Cubic Feet)	% Total Through Torch	Cu Ft O_2 per Lb of Fuel
Acetylene—5,589°F/1470 Btu	2.5	1.3	50.0	18.9
MAPP gas—5,301°F/2406 Btu	4.0	2.5	62.5	22.1
Natural gas—4,600°F/1000 Btu	2.0	1.9	95.0	44.9
Propane—4,579°F/2498 Btu	5.0	4.3	85.0	37.2
Propylene—5,193°F/2371 Btu	4.5	3.5	77.0	31.0

*Balance of total oxygen demand is entrained in the fuel-gas flame from the atmosphere.

FIGURE 24-3 Cubic feet of oxygen (O_2) needed per cubic foot of fuel burned, for different fuel gases.

Source: Airco Welding Products.

Fuel	Neutral Flame Temp (°F)	Primary Flame (Btu/Ft³)	Secondary Flame (Btu/Ft³)	Total Heat (Btu/Ft³)
Acetylene	5,589	507	963	1,470
MAPP gas	5,301	517	1,889	2,406
Natural gas	4,600	11	989	1,000
Propane	4,579	255	2,243	2,498
Propylene	5,193	438	1,962	2,371

Key: MAPP gas—A multipurpose industrial fuel gas consisting of a mixture of methylacetylene and propadiene.
Btu/ft³—British thermal units per cubic foot of fuel. A Btu is a small unit of heat.

FIGURE 24-4 Heating values of major industrial fuel gases when burned with pure oxygen.
Source: Airco Welding Products.

Application	Acetylene	MAPP Gas	Propylene
Cutting			
Under ⅜ in thick	100	95	90
⅝ in to 5 in thick	95	100	95
Over 5 in thick	80	100	95
Cutting dirty or scaled surfaces	100	95	80
Cutting low-alloy specialty steels	100	90	80
Piercing	100	100	85
Welding	100	70	0
Braze welding	100	90	70
Brazing	100	100	90

FIGURE 24-5 Average performance ratings of three types of oxyfuel flames.
Source: Airco Welding Products.

torch heating and cutting, but not welding. For torch heating and cutting, however, both propane and natural gas consume large volumes of cylinder oxygen. This fact may offset other price advantages offered by these gases.

MAPP Gas

MAPP gas is a formulated mixture of methylacetylene and propadiene gases. (MAPP is a trade name of Dow Chemical.) The gas is reported to have many of the advantages of acetylene but is more stable and,

CAUTION!

Take every precaution in handling oxygen and fuel gases, setting up apparatus, and performing cutting and heating tasks.

therefore, safer. Its high-temperature flame is suitable for brazing, cutting, heating, and welding.

CUTTING STEEL WITH OXYFUELS

Before attempting any of the procedures discussed in the remainder of this unit, students should review Unit 23 ("Using Gas Welding Equipment").

The Flame

A description of a neutral flame and the procedure for obtaining it are given in Unit 22. However, it is more difficult to determine when a flame is neutral on a cutting torch than on a welding torch. A cutting torch must be adjusted to obtain a neutral flame both with and without the oxygen jet (Figure 24-6).

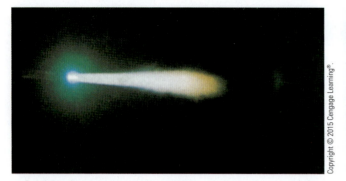

A Oxidizing flame—too much oxygen.

C Neutral flame—proper amount of oxygen and acetylene.

B Carbonizing flame—too much acetylene.

D Acetylene burning—note the black smoke and yellow-orange flame.

FIGURE 24-6 Oxyacetylene flame adjustments.

The following procedure describes how to obtain a neutral flame with a cutting torch.

Activity

1. Wear appropriate protective clothing and goggles with a No. 5 shaded lens.

2. Check the area and remove all fire hazards.

3. Set up the oxyacetylene equipment in a safe manner and check for leaks.

4. Set the pressure gauges at the recommended pressure for the tip being used. These will vary with the fuel and tip (Figure 24-7).

5. Light the torch using acetylene only.

6. Increase the acetylene pressure until the flame just starts to leave the tip when the acetylene torch valve is opened several turns.

7. Close the acetylene torch valve until the flame touches the tip.

8. With the oxygen preheat valve closed, open the oxygen (O_2) torch valve several turns.

9. Slowly open the O_2 preheat valve until the acetylene feathers just match the inner cones of the preheat flames.

10. Press the oxygen lever, and observe whether the flames stay neutral.

11. If the flame does not stay neutral, make slight adjustments in the oxygen preheat valve and/or the acetylene torch valve until the flame remains neutral both with and without the oxygen lever pressed.

12. If a neutral flame cannot be obtained, clean the tip. Use fine emery cloth on the outside and the appropriate tip cleaners for the inside of the holes.

Cutting Guide

Metal Thickness (in)	Airco Cutting Tip Size	Pressure (PSI)	
		Oxygen	Acetylene
1/8	00	30	1 1/2
1/4	0	30	3
3/8	1	30	3
1/2	1	40	3
3/4	2	40	3
1	2	50	3
1 1/2	3	45	3
2	4	50	3
3	5	45	4
4	5	60	4
5	6	50	5
6	6	55	5

Courtesy of Linde LLC

FIGURE 24-7 Oxygen and acetylene pressures vary according to the size of the cutting tip.

CAUTION!

Touching the tip against the work, overheating, incorrect torch adjustment, a loose tip, a dirty tip, or damaged valves may cause backfire. **Backfire** is a loud snap or popping noise that generally blows out the flame. The cause must be corrected before relighting.

CAUTION!

Backfire sometimes causes a flashback. A **flashback** is burning inside the torch that causes a squealing or hissing noise. When this occurs, quickly turn off the torch oxygen valve and then the torch acetylene valve. If fire is suspected in the hoses, rush to close the acetylene valve and then the oxygen valve at the tanks. After a flashback, only an experienced operator can determine if the torch is safe to relight.

Cutting Steel

Position the work for cutting, light the flame, and obtain a neutral torch flame. This is done by placing the metal to be cut over a slag box. A **slag box** is a metal container of water or sand placed to catch hot slag and metal from the cutting process. The work should be weighted or clamped so that it will not slip from the work area.

For general cutting, mark the line of the cut with soapstone. Place one gloved hand on the metal under the torch body. This permits control of the tip clearance. The other hand controls the handle and oxygen lever (Figure 24-8).

To start the cut, hold the flame over the corner and edge of the metal, at a slight angle away from the edge (Figure 24-9). The cones of the preheat flames should not quite touch the metal. Hold the torch steady until the edge of the metal turns cherry red. Press the oxygen lever and move the torch across the metal at a steady rate. Maintain the flame cones about 1/8 inch from the metal.

Different techniques may be used to control speed and clearance. When making short cuts, it is helpful to slide the torch over the gloved hand positioned near the torch head. In this case, one hand controls the clearance and the other controls the oxygen lever and movement across the metal.

When making cuts longer than about 2 inches, both hands grip the torch and slide over the metal. Special devices such as wheeled trolleys may be used to maintain the correct clearance (Figure 24-10). A rod with a sliding center point can be attached to a torch to aid in cutting perfect circles (Figure 24-11). A piece of steel angle or similar device can be used to guide the torch when making straight cuts (Figure 24-12).

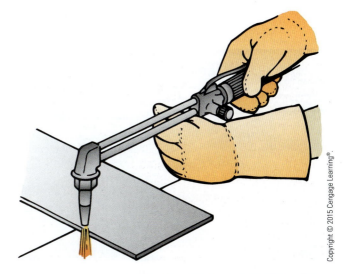

Copyright © 2015 Cengage Learning®

FIGURE 24-8 One gloved hand controls the tip clearance, and the other controls the oxygen lever.

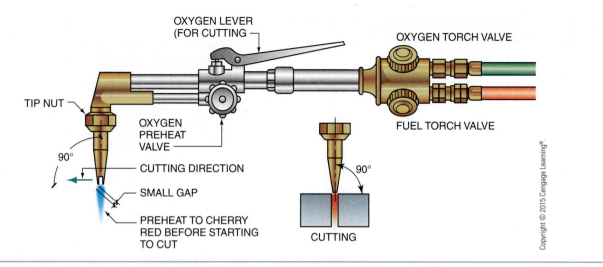

FIGURE 24-9 Torch positions for preheating and for cutting.

FIGURE 24-10 For long cuts, a wheeled trolley may be attached to a cutting tip to maintain the proper distance between the tip and the metal.

FIGURE 24-11 A rod with a sliding center pivot attached to a torch aids in cutting perfect circles.

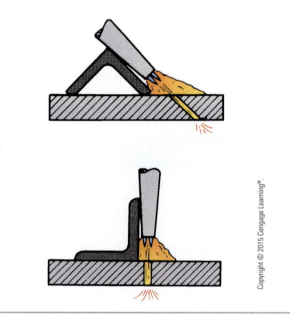

FIGURE 24-12 Angle steel can be used to guide the torch when cutting a straight line.

A smoother cut will be obtained by placing the guide metal on the waste piece since most of the slag will collect on the piece with the guide metal.

Improving the Cut

An examination of the metal along a cut reveals how the cut was made. Marks left by the flame provide clues to the preheat procedure, speed, and pressure.

Preheat. If the preheat flame is too hot, or the torch travels too slowly, the surface melts before the metal is heated through. This leaves a melted or rounded appearance along the top (Figure 24-13). The tip may be raised slightly to reduce the preheat. Increasing the speed slightly may also correct the problem.

Clearance. Clearance is the distance from the torch tip to the metal. Generally the clearance is correct when the tips of the primary flames are almost level with the surface metal.

Speed. Moving the torch too fast across the metal results in an incomplete cut and rough edges. Incomplete cuts generally occur at the bottom and end of the cut (Figure 24-14). On the other hand, moving too slowly results in a melted top edge and leaves gouges where the cutting stream has wandered (Figure 24-15).

Pressure. If the oxygen pressure is too high, the result may be a dish shape in the kerf near the top. On the other hand, if the pressure is too low, the cut may not be complete at the bottom.

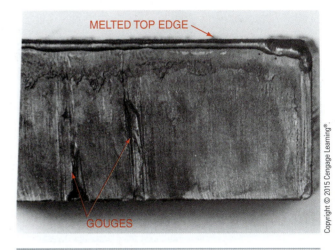

FIGURE 24-15 Moving the torch too slowly across the metal yields a melted top edge and gouges along the cut.

The Correct Cut. A correct cut is straight and square with a smooth face. Drag lines bend backward slightly at the bottom. If preheat, clearance, speed, and pressure are all correct, then cutting will be fast, clean, and accurate (Figure 24-16). Such a combination requires a steady hand and plenty of practice. While a good cut is being made, there is a smooth, even sound and a steady stream of sparks from the bottom of the kerf (Figure 24-17).

Piercing Steel

A flame cutting torch can also be used to pierce steel. To **pierce** means to make a hole. The operator must take special care to avoid being burned by molten slag and metal during the procedure.

FIGURE 24-13 Excessive preheating and/or too slow a pace across the metal results in a melted top edge.

FIGURE 24-14 When the torch travels too fast across the metal, rough edges and uncut metal result.

FIGURE 24-16 A correct cut is straight and square with a smooth face. Drag lines bend backward slightly at the bottom.

A

B

FIGURE 24-17 Note in the two different views (A) and (B) that the top edge of the plate has not been melted, and there are few if any sparks. This is a good cut.

CAUTION!

Use full face, head, shoulder, and body protection when piercing steel.

To pierce steel, the cutting torch is held above the mark at the normal preheat distance. When the spot becomes cherry red, raise the torch ½ inch or more above the surface (to reduce the hazard of molten metal) and slowly press the oxygen lever.

Move the tip sideways and into a circular motion until the hole breaks through (Figure 24-18). Enlarge the hole by cutting around the edges. If an inside cut is desired, proceed with the cut from inside the hole.

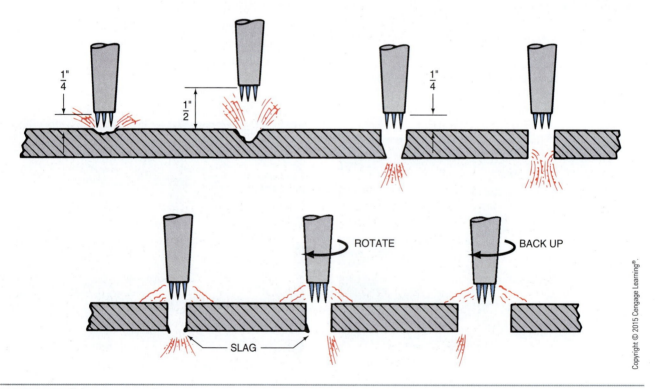

FIGURE 24-18 Using an oxyacetylene torch to pierce and enlarge a hole. After the flame breaks through the metal, the operator rotates the torch to enlarge the hole or backs it up slightly before proceeding with the cut.

CAUTION!

Do not cut galvanized pipe with oxyfuels. The melted zinc coating gives off poisonous fumes.

Cutting Pipe

Cutting pipe is much like piercing and cutting thin plate. To cut pipe up to 3 inches in diameter, first pierce a hole in the top of the pipe, then cut a kerf to the left side, followed by one to the right side (Figure 24-19). Rotate the pipe and repeat the process to cut the underside.

To cut large pipe, the torch is held at a right angle to the pipe. It is then moved around the pipe to make the cut. An alternate method is to rotate the pipe in steps or continuously (Figure 24-20).

PLASMA ARC CUTTING

One problem with using oxyfuels is that it is difficult to cut aluminum, copper, nickel alloys, and stainless steel. These metals do not react to oxygen in the same way as ferrous metals. A process called **plasma arc cutting** is used for cutting the nonferrous metals (Figure 24-21). This procedure involves the use of an electric arc and an inert gas called argon. **Plasma** is defined as a group of charged particles that conduct electrons across a gap. In this instance, argon gas is used to produce the plasma. A charge of electricity (approximately 20 amperes and 85 volts) is sent through a pressurized stream of argon to create heat. The cutting head of the torch gets so hot that water is circulated through it to cool the tip.

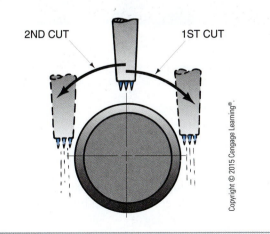

FIGURE 24-19 To cut pipe with a diameter of up to 3 inches, the operator should cut across the top, then rotate the pipe, and then cut the remaining section.

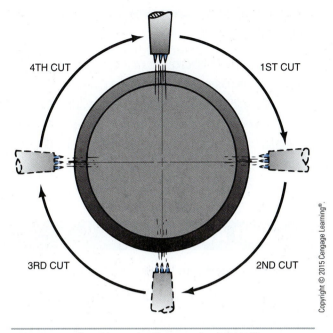

FIGURE 24-20 To cut large pipe, the operator should either move the torch around the pipe while maintaining a 90-degree angle to the pipe or hold the torch in a stationary position and rotate the pipe.

FIGURE 24-21 Plasma arc cutting is used to cut nonferrous metals and stainless steel.

The rapidly moving gas cuts the metal very quickly. In fact, one of the advantages of plasma arc cutting is that metal can be cut more rapidly than with the conventional oxyfuel torches. For this reason, plasma arc cutting is used in cutting plate steel under automated controls. This is true unless the metal is several inches thick. Plasma arc cutting is not as efficient in cutting the thicker metals.

SUMMARY

Ferrous metal can be cut quickly and efficiently using a mixture of oxygen and combustible gas. Using the proper procedure can result in a clean, smooth, and accurate cut. Use of incorrect techniques will not only result in a poor cut, but will also be extremely dangerous. Newer techniques involve the use of a plasma arc for cutting nonferrous metal. Following proper procedures is equally important when using a plasma arc torch as when using an oxyfuel torch.

Student Activities

1. Define the Terms to Know for this unit.

2. Light a cutting torch and adjust it to a neutral flame.

3. Obtain a piece of steel ¼" × 2" × 12" or longer. Square six lines across the metal at ½-inch intervals. Use soapstone to mark the lines. Practice cutting and inspect each cut. Discuss your progress with your instructor.

4. Obtain a piece of scrap plate steel. Draw a straight line about ½ inch in from an edge. Practice making a long straight cut.

5. Obtain a piece of scrap steel ¼" × 2" × 12". Practice piercing holes in the strip.

6. Practice cutting off rings about ½ inch wide from the end of black pipe.

Relevant Web Sites

Cutting Torch Fundamentals by the Welding Fanatic
YouTube

Victor Oxy-Fuel Safety Video by Victor Technologies
YouTube

Self-Evaluation

A. Multiple Choice. Select the best answer.

1. Combustible gases were first used for welding
   a. in the American space exploration program
   b. in the 1800s
   c. in the early 1900s
   d. during World War II

2. The temperatures from gas flames using oxyfuels are
   a. 150° to 200°F
   b. 400° to 500°F
   c. 1,000° to 2,000°F
   d. 5,000° to 6,000°F

3. The result of burning iron in the presence of pure oxygen is
   a. brass
   b. propane
   c. slag
   d. weld

4. When cutting steel, the oxygen stream
   a. aids in keeping the steel hot
   b. drives out slag
   c. supports combustion
   d. all of these

5. The fuel with the best qualities for welding and cutting is
   a. acetylene
   b. MAPP
   c. natural gas
   d. propane

6. A mixture of gases with excellent qualities for cutting is
   a. acetylene
   b. MAPP
   c. natural gas
   d. propane

7. A cutting torch must be adjusted so that it is neutral when
   a. cutting
   b. preheating
   c. the oxygen lever is down
   d. all of these

8. Correct oxygen and fuel pressure will vary with
   a. the tip
   b. the job
   c. the fuel
   d. all of these

9. When a correct torch cut is in progress, there will be a
   a. smooth, even sound
   b. spray of sparks
   c. slightly dished kerf
   d. all of these

10. When piercing, the clearance is increased after preheating to
    a. increase the force of the oxygen stream
    b. introduce more air into the process
    c. provide time for heat to move through the metal
    d. reduce the hazard from molten metal

11. Plasma arc cutting is used to cut
    a. aluminum
    b. stainless steel
    c. copper
    d. all of these

B. **Matching.** Match the terms in column I with those in column II.

Column I

1. fusion
2. brazing
3. propane
4. oxygen
5. fire triangle
6. acetylene
7. neutral flame
8. backfire
9. flashback
10. hot galvanized pipe

Column II

a. poisonous fumes
b. results in fire
c. maximum safe pressure is 15 psi
d. cone and feather are one
e. pop or snap at tip
f. bonding
g. oxyfuel not suitable for welding
h. supports combustion
i. fire inside the torch
j. melting metals together

C. **Completion.** Fill in the blanks with the word or words that make the following statements correct.

1. When cutting, slag should be caught in a _____.

2. When cutting, the edge of the metal should be preheated until it is _____.

3. When cutting, the flame cones should be about _____ from the metal.

4. When using proper equipment with a clean tip, most cutting problems can be traced to incorrect _____, _____, _____, or _____.

5. Cutting pipe is much like cutting _____.

D. **Brief Answer.** Briefly answer the following questions.

1. When piercing, why is the clearance increased after preheating?

2. Describe a correct cut.

3. Which actions and/or conditions can produce backfire?

4. What is plasma?

5. What does a squealing or hissing noise inside a torch indicate? What should be done?

UNIT 25

Brazing and Welding with Oxyacetylene

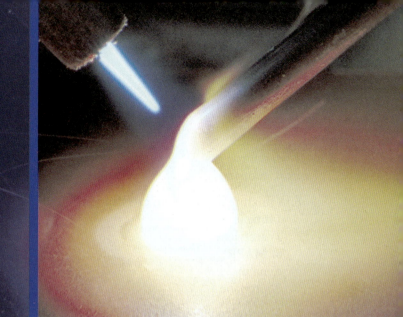

Objective

To braze and weld safely with oxyacetylene equipment.

Competencies to be developed

After studying this unit, you should be able to:

- Explain the nature and uses of braze welding.
- Prepare metal for welding.
- Select fluxes for welding.
- Identify joints commonly used in welding.
- Braze and braze-weld butt, lap, and fillet joints.
- Fuse-weld mild steel with and without filler rod.

Materials List

- Oxyacetylene rig, welding table, and protective clothing
- Welding torch with No. 1–4 tips
- Spool of 50-50 solder
- Rosin flux
- Brazing flux
- Brazing rod (about 1/16" in diameter)
- Welding rod (about 1/16" in diameter)
- Steel plate (not galvanized), 16-gauge × 6" × 16"
- Steel plate (not galvanized), 1/8" × 6" × 10"
- Copper pipe or tubing (3/8" or 1/2" in diameter), or sheet copper
- Power wire wheel or grinder
- Fine emery cloth

Terms to Know

- welding
- soldering
- brazing
- capillary action
- filler rod
- braze welding
- fusion welding
- oxide
- oxidation
- impurity
- flux
- play the flame
- tinning
- bead
- butt weld
- tacking
- fillet weld
- root
- puddle
- solidify
- burn-through
- puddling
- pushing the puddle
- back-stepping

BASICS OF SOLDERING, BRAZING, AND FUSION WELDING

The term **welding** may be defined as uniting metal parts by heating or compression. Early American blacksmiths welded wagon rims and other parts by heating and hammering. Parts were heated to a certain color, and a joint was formed by pounding the parts together using a hammer and anvil.

Soldering, brazing, and braze welding have much in common. All three processes are done without melting the base metal. All three use a metal or alloy as the bonding agent.

Soldering

The term **soldering** means bonding with metals and alloys that melt at temperatures below 804°F. The most common soldering material for non-plumbing use is 50-50 solder, consisting of 50 percent tin and 50 percent lead. Solder used for water lines cannot contain lead because of its toxic qualities. Other alloys are available for special applications (Figure 25-1). Tin-lead soldering is generally done with electric irons, soldering guns, and propane torches. Common applications of soldering are to join electrical wires, join copper gutters and spouts, sweat copper pipes, and fasten thin tin-plated steel. The basic procedure for soldering is covered in Unit 13 ("Fastening Metal").

Brazing

The term **brazing** means bonding with metals and alloys that melt at or above 840°F when capillary action occurs (Figure 25-2). The term also refers to joining parts that are fitted extremely well. Only a very thin layer (0.025 inch or less) of alloy is needed to fill the void between parts (Figure 25-3). This small spacing allows the alloy to be drawn into the joint by capillary action. **Capillary action** is the rising of the surface of

Base Metal	Brazing Filler Metal
Aluminum	BAlSi, aluminum silicon
Carbon Steel	BCuZn, brass (copper-zinc) BCu, copper alloy BAg, silver alloy
Alloy Steel	BAg, silver alloy BNi, nickel alloy
Stainless Steel	BAg, silver alloy BAu, gold base alloy BNi, nickel alloy
Cast Iron	BCuZn, brass (copper-zinc)
Galvanized Iron	BCuZn, brass (copper-zinc)
Nickel	BAu, gold base alloy BAg, silver alloy BNi, nickel alloy
Nickel-Copper Alloy	BNi, nickel alloy BAg, silver alloy BCuZn, brass (copper-zinc)
Copper	BCuZn, brass (copper-zinc) BAg, silver alloy BCuP, copper-phosphorus
Silicon Bronze	BCuZn, brass (copper-zinc) BAg, silver alloy BCuP, copper-phosphorus
Tungsten	BCuP, copper-phosphorus

KEY: B = Brazing   Ag = Silver
      Al = Aluminum   Ni = Nickel
      Si = Silicon   Au = Gold
      Cu = Copper   P = Phosphorus
      Zn = Zinc

FIGURE 25-2 Primary (or base) metals and the common brazing filler metals used for each.

Alloy	Use On
Tin-Lead	Copper and copper alloys Mild steel Galvanized metal
Tin-Antimony	Copper and copper alloys Mild steel
Cadmium-Silver	Copper and copper alloys needing high strength Mild steel Stainless steel
Cadmium-Zinc	Aluminum and aluminum alloys

FIGURE 25-1 Alloys used for soldering.

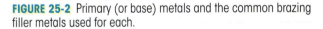

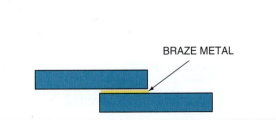

BRAZE METAL

FIGURE 25-3 A very thin layer of brazing alloy joins two pieces of primary (base) metal.

Copyright © 2015 Cengage Learning®

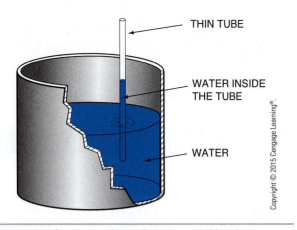

FIGURE 25-4 Capillary action pulls water up a thin tube.

a liquid that is in contact with a solid. Capillary action may be observed by placing a thin tube in a container of water (Figure 25-4).

The temperature of properly prepared metal must first be raised to the melting point of the brazing alloy to be used. The brazing alloy is then added by means of a long, thin metal rod called a **filler rod**. The most popular filler rods for brazing and braze welding are copper-zinc alloys. The alloy is drawn between the parts and spreads throughout the spaces that are narrow enough for capillary action to work. This principle makes brazing an attractive method for repairing broken castings.

Braze Welding

Braze welding refers to bonding with alloys that melt at or above 840°F when capillary action does not occur. Here the alloy is bonded to each part, and the void between or around the part is filled with the melted alloy (Figure 25-5). Since many applications require brazing as well as braze welding, the term brazing is frequently used for both operations.

Advantages of Soldering and Brazing

The advantages of soldering and brazing are that they (1) are low-temperature processes, (2) permit easy disassembly, (3) allow different metals to be joined,

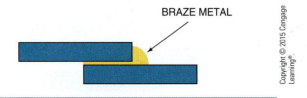

FIGURE 25-5 Braze welding bonds pieces of metal together and fills the space between the pieces with the brazing alloy.

(4) can be done at high speed, (5) do little damage to parts, (6) permit easy realignment of parts, and (7) permit parts of various thicknesses to be joined.

Fusion Welding

The term **fusion welding** means joining parts by melting them together. The base metals are melted and mixed together to form a continuous piece of the same kind of metal. Fusion welding can be done with or without a filler rod. A sawed cross section of a skillfully made fusion weld will show such complete fusion that it looks like a single piece of metal.

Fusion welding is done extensively on steel. The strength of steel is such that only fusion welding is acceptable for most work. Oxyfuels are used only for welding on very thin steel. Most other fusion welding of steel is done using an electric arc welder as the source of heat.

Preparing Metals for Welding

Metals must be clean before soldering, brazing, or welding can take place. This means that oxides and other impurities must be removed from the metal. An **oxide** is the product resulting from oxidation of metal. **Oxidation** means combining with oxygen. Rust on iron and steel is an oxide. The dull coatings that form on brass and copper are also oxides.

Metal may be cleaned for soldering or welding by using a wire brush, wire wheel, grinder, sander, steel wool, file, or other mechanical process.

After cleaning metal by mechanical means, chemicals are used to remove any impurities. An **impurity** is any product other than the base metal.

The term **flux** is given to any chemical used to clean metal. Fluxes remove tarnish and corrosion caused by oxidation. They also prevent corrosion from developing. Additionally, they promote wetting or movement of molten alloys on metal. Finally, fluxes aid in capillary action, which moves molten alloys into very small cracks. Most fluxes work best when hot. If the metal gets too hot, however, the flux will burn and must be cleaned off before proceeding. Fluxes are available as solids, pastes, powders, liquids, sheets, rings, and washers (Figure 25-6). They may also be in the hollow core or on the surface of filler rods.

Using Flux. In order to control the heat needed to manage the fluxing action, the following procedure is recommended.

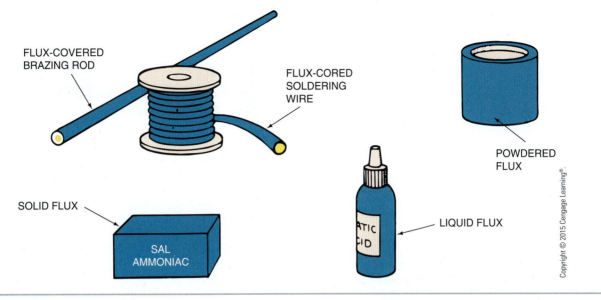

FLUX-COVERED BRAZING ROD

FLUX-CORED SOLDERING WIRE

POWDERED FLUX

SOLID FLUX

SAL AMMONIAC

LIQUID FLUX

Copyright © 2015 Cengage Learning®.

FIGURE 25-6 Fluxes come in many forms.

Activity

1. Obtain a piece of copper pipe, tubing, or sheet. The copper should be clean but not shiny.

2. Set up, light, and adjust a small torch.

3. Put a pea-sized spot of rosin flux on the copper.

4. Heat the flux and the surrounding area by playing the flame on the rosin. To **play the flame** means to alternately move a flame into and out of an area to achieve temperature control.

5. Observe the point when the copper suddenly becomes shiny under the flux. Remove the heat. The shiny surface indicates that the flux has reacted with the oxide and removed it (Figure 25-7).

6. Heat an area near the shiny spot. The flux should flow to the newly heated area and clean it, too.

7. Observe the effect of too much heat, by continuing to heat the clean areas. Notice that too much heat will burn the flux and the area will turn dark or black.

8. Remove the burnt flux with steel wool or fine sandpaper.

9. Reapply the flux and reheat. Note that the fluxing action can be repeated only after mechanical removal of the burnt flux. It is important to recognize fluxing action. It is necessary to learn how to control the fluxing process. Otherwise, soldering or brazing fillers will not bond to the metal.

Tinning. Tinning is the key to success in soldering and brazing. The term **tinning** means to bond filler material to a base metal. The term probably originated from the tinlike appearance of solder when it is applied to a base metal. Before two pieces of metal can be joined by soldering, brazing, or braze welding, both pieces must be tinned. Most unsuccessful attempts at soldering fail at this point. Learn to tin before attempting to solder, braze, or braze weld.

CAUTION!

Never braze or weld galvanized steel without special training and special precautions. Galvanized coatings give off toxic fumes when melted or heated above soldering temperatures.

AREA PROTECTED BY THE FLAME AND FLUX

AREA OF LIGHT TO HEAVY OXIDATION

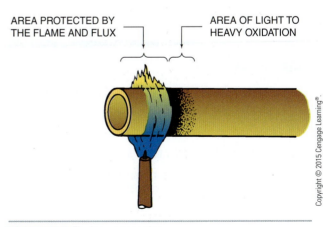

Copyright © 2015 Cengage Learning®.

FIGURE 25-7 Fluxing copper pipe.

The following procedure is suggested for tinning.

Activity

1. Obtain a piece of copper pipe or sheet.
2. Clean the metal with fine emery cloth.
3. Obtain a spool of tin-lead solder, paste flux, and a damp cloth.
4. Light and adjust an oxyfuel torch with a small tip and neutral flame.

NOTE

The operator controls the torch with one hand. The other hand holds the filler rod like a pencil, about 12 inches from the end.

5. Heat the copper until the fluxing action starts.
6. Touch the end of the solder to the hot metal. Feed the solder down to the metal as it flows across the fluxed area.
7. Use the torch to heat an enlarged area and attract flux and solder across the metal.
8. Play the torch over the tinned area to get a thin, smooth, and shiny surface.
9. To improve the appearance and to remove excess solder, wipe the solder-covered area quickly with a damp cloth. Note that the damp cloth removes excess solder and leaves a thin coating of solder bonded to the copper.

The following procedure describes how to tin with brazing filler.

Activity

1. Obtain a piece of nonrusted steel plate, approximately ⅛" × 2" × 6".
2. Clean the plate carefully with fine emery cloth.
3. Obtain a brazing filler rod and can of brazing flux.
4. Place the plate on firebricks.
5. Light and adjust an oxyfuel torch using a small tip and neutral flame.

6. Heat the center of the plate until it is dull red.

CAUTION!

Do not permit the steel to melt.

7. Heat the end of a braze filler rod by placing it in the flame as the base metal is heating.

NOTE

A propane torch without oxygen may also be used.

8. With the tip of the filler rod heated, dip it into the can of flux (Figure 25-8).
9. Place the flux-covered end of the rod into the flame and touch the hot base metal.
10. When the flux reaches the correct temperature, fluxing action occurs (the chemical flows across the hot metal and removes impurities).
11. Once fluxing occurs, watch for the end of the filler rod to melt. Then play the flame on the filler metal, flux, and plate, causing the filler to flow as you create the correct temperature on the base metal.
12. Play the heat over the entire area until the filler material flows into a thin layer of bonded metal.

CAUTION!

Do not overheat the filler material and base metal or the metals will turn dark from oxidation and will require mechanical cleaning again.

If difficulty is encountered with the process, clean the metal completely and start again. The critical task is to keep the filler rod heated very nearly to its melting point. Then, when it is needed for fluxing or filling, the operator can easily control these processes. Quick melting of the filler rod is achieved by touching the base metal with the rod while playing the torch on the rod.

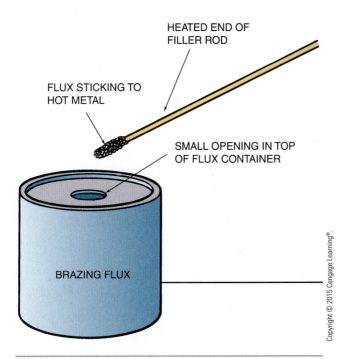

HEATED END OF
FILLER ROD

FLUX STICKING TO
HOT METAL

SMALL OPENING IN TOP
OF FLUX CONTAINER

BRAZING FLUX

Copyright © 2015 Cengage Learning®

FIGURE 25-8 The heated end of a brazing filler rod can be used to transfer flux from the container to the piece of metal to be tinned.

NOTE

The flame must be directed to the thick piece more than to the thin one. Practice with pieces of various sizes until you can make both pieces turn dull red at the same time. This skill is essential for soldering, brazing, and welding.

BRAZING AND BRAZE-WELDING PROCEDURES

Brazing and braze welding are generally done in agricultural mechanics using an oxyfuel torch. Acetylene is perhaps the most widely used fuel. However, acetylene has the disadvantage of producing a very hot inner flame and a relatively cold outer flame. For the brazing and braze welding, a hot outer flame is useful for heating large areas uniformly. Such fuel gases as MAPP, propane, butane, and natural gas have the advantage of providing more uniform heating.

Heating of large areas is necessary when entire assemblies, such as pipe fittings or large castings, are being bonded. A good way to control heat when brazing large assemblies is to first heat the basic assembly until it is near the melting point of the braze filler. It is fairly easy then to add the additional heat required to melt the braze metal and fill the voids in specific locations.

Controlling Heat. Controlling heat is the key to successful soldering and brazing. Metals are excellent heat conductors. Therefore, heat applied in one place soon moves to surrounding areas. The larger the piece of metal, the faster the heat gets away from the heated area. This must be kept in mind when joining two pieces of metal. If the torch directs an equal amount of heat onto two pieces of different size, the smaller piece will get hot faster. When this occurs, the flux will burn on the small piece before the large piece is up to fluxing temperature.

The operator must manipulate the torch to heat pieces of different size equally. The following procedure is helpful.

Torch Tips. Tips used for oxyfuel welding have a single hole. They are identified by a number that designates the size of the hole. Tip numbers range from 000 (the smallest) to 10 or larger. A small tip is used for thin metal. The tip size must be larger as the thickness and mass of the metal increase. The torch manual should provide proper guidelines for tip selection.

Torch Adjustment. A good general rule for torch adjustment is that most soldering, brazing, and welding is done with a neutral flame. Only specialized applications required slightly carbonizing or slightly oxidizing flames.

Brazing

Since brazing works by capillary action, it requires that parts fit together very closely. Some common types of joints used to fasten metal are shown in Figure 25-9.

Activity

1. Obtain one piece of steel $\frac{1}{8}$" × 2" × 2" and another piece of steel 16-gauge × 2" × 2".

2. Place the two pieces end-to-end on a firebrick.

3. Light and adjust the torch.

4. Play heat on the joint and in a circular area covering both pieces until both are dull red.

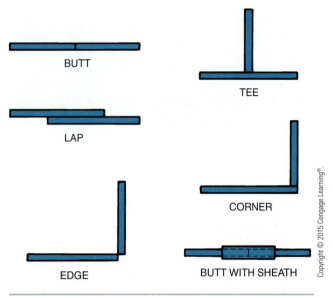

FIGURE 25-9 Common types of joints used to fasten metal.

The following procedure is suggested for brazing flat material in a lap joint. A lap joint is simply a welding joint made when one piece of metal overlaps another.

Activity

1. Obtain three perfectly flat, clean pieces of nongalvanized steel plate $\frac{1}{8}$" × 2" × 2".

2. Clean a 1-inch strip along one edge of each of two pieces.

3. Place the two pieces on firebricks so that clean, $\frac{1}{2}$-inch strips of metal overlap one another. Use the third piece to support the top pieces (Figure 25-10). If possible, clamp the assembly or hold the top piece in place with a brick.

4. Obtain a brazing rod and flux.

5. Light and adjust the oxyfuel torch.

6. Apply heat to the clean joint until both pieces are dull red.

7. Add flux to the joint by using the heated filler rod.

8. Continue to raise the temperature of the assembly and move the filler rod in and out of the flame.

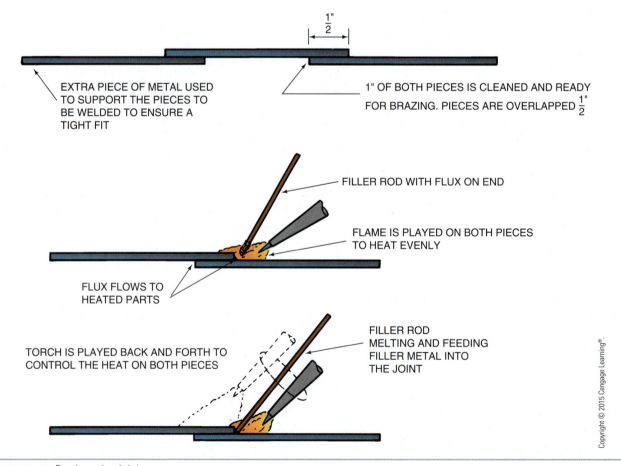

FIGURE 25-10 Brazing a lap joint.

9. When both pieces are dull red, touch the metal with the rod and feed filler into the joint.

> **NOTE**
>
> Bend a loop into the cold end of the brazing rod before using it.

10. Play heat more on the top piece and back from the edge, to draw the filler into the joint.
11. Let the assembly cool.
12. Turn the assembly over and examine it for completeness of brazing. Filler metal should be present on the back edge of the joint.

Cast iron assemblies of irregular shapes may be brazed. Appropriate fluxes must be used to correct the presence of free graphite in some iron. Additionally, a large torch is needed to heat large assemblies. Brazing may be preferred over fusion welding of cast iron since there is less likelihood of heat distortion and breaking of the casting.

Braze Welding

When braze welding, many of the principles of brazing apply. The major difference is that a perfect fit is not needed. Still, good fits generally result in stronger joints.

Braze welding is useful on metal that is 16-gauge or thinner. Thicker metal can be braze-welded, but electric arc welding is faster and easier and generally the preferred process for thicker metals.

Running a Bead. Most welding requires the ability to run a bead. A **bead** is a continuous and uniform line of filler metal. To run a bead with a braze filler rod, the following procedure is suggested.

Activity

1. Obtain a clean piece of ⅛" × 2" × 6" steel.
2. Clean both sides with fine emery cloth.
3. Position the metal on a flat firebrick. The bead will be run across the metal, ½ inch in from one end.
4. Light and adjust the torch.

5. Heat the end of a filler rod and a spot on one edge of the plate, about ½ inch in from the end.
6. Dip the hot rod into flux, and touch the fluxed rod to the heated spot.
7. When the rod starts to melt, move it in and out of the flame to add filler metal as needed.
8. The bead is formed by depositing filler material along a fluxed path across the metal.

> **NOTE**
>
> Run additional beads across the metal at 1-inch intervals until you can run complete, straight, and even beads.

The filler rod must be dipped into the flux frequently. This provides flux to cover the new bead as it forms.

Braze Welding Butt Joints. A **butt weld** is essentially a bead laid between two pieces of metal set edge-to-edge. This may be end-to-end or side-to-side. When braze welding a butt joint, it is important that the filler metal is down between the pieces and bonds the edges at all points. To achieve this, the two pieces are placed with a slight gap between them. To braze-weld a butt joint, the following procedure is suggested.

Activity

1. Cut two pieces of nongalvanized 16-gauge × 2" × 2" steel.
2. Clean the edges and faces of the steel pieces with fine emery cloth.
3. Place them edge-to-edge on a flat firebrick with about a 1/32-inch gap between them.
4. Weight the two pieces with firebricks to help hold them in position.
5. Heat a spot at one end of the joint. Flux the area and deposit a spot of filler material.
6. Repeat step 5 at the other end of the joint. These small welds are called tacks. The process of making a small weld to hold two pieces together temporarily is called **tacking**.
7. Lay a bead over the entire length of the joint (Figure 25-11).

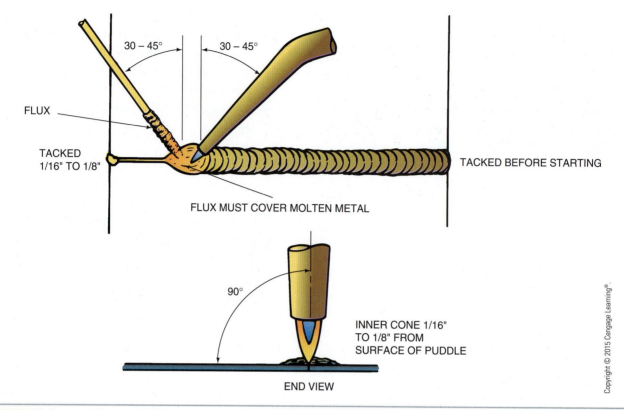

FIGURE 25-11 Braze welding a butt joint.

Braze Welding Fillets. A **fillet weld** is a weld placed in a joint created by a 90-degree angle. The joint may be in the form of an L or a T.

When making a fillet weld, the torch must be played on the flat and vertical pieces with care. If both pieces are equal in thickness and mass, the heat is applied equally. However, if one piece is thicker, that piece must receive more heat (Figures 25-12 and 25-13).

Special care must be taken to ensure that sufficient heat reaches the root of the weld. The **root** is the deepest point in a weld. In this case, it is the place where the two pieces melt to form an angle. A braze-welded fillet should have the filler material under the vertical piece and a bead along the 90-degree corner (Figure 25-14).

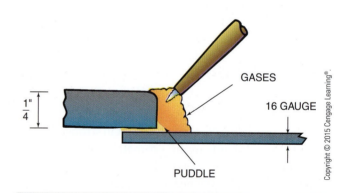

FIGURE 25-13 Torch position for directing heat onto a vertical surface that has more mass than the horizontal piece does.

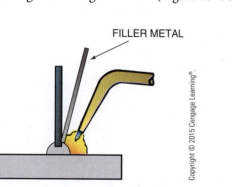

FIGURE 25-12 Torch and rod positions required to balance heat when the horizontal piece is thicker than the vertical piece.

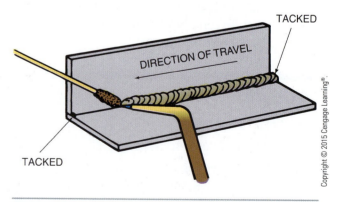

FIGURE 25-14 Making a fillet weld on pieces of steel of equal thickness.

Copyright © 2015 Cengage Learning®.

FUSION WELDING WITH OXYACETYLENE

Fusion welding of steel with oxyacetylene is generally limited to thin metal or small jobs where portability of equipment is a factor. Electric arc welders are now so versatile that they account for most fusion welding done in agricultural mechanics. Still, it is useful to be able to do small jobs and weld thin metal with oxyacetylene.

Factors Affecting the Weld

The Tip. Selection of the torch tip will depend upon the width and penetration of the bead desired. The larger the tip, the more heat the flame produces. The heat from a given tip is controlled by the distance the cone is held from the **puddle** and the angle of the tip to the work. The greatest melting ability is obtained when the angle is 90 degrees. The ideal angle is 45 degrees. The ideal cone distance is $\frac{1}{8}$ to $\frac{1}{4}$ inch from the puddle.

The Rod. The filler rod for fusion welding should be made of the same material as that being welded. Remember, fusion welding is a melting and mixing of base metals. The filler rod adds volume so the joint will be filled and slightly reinforced, or strengthened. The larger the rod, the greater is its tendency to cool the molten puddle, increase the size of the bead, and reduce penetration.

> **CAUTION!**
>
> Always bend one end of the filler rod to form a loop. The loop identifies the unheated end and decreases chances of those in the area being hit by the sharp end or burned by the hot end.

Controlling the Process

The gases in the air will contaminate, or pollute, an unprotected welding puddle. Fortunately, the presence of the torch flame prevents air from getting to the molten puddle. However, this makes it necessary to keep the torch on or nearly on the puddle at all times. To complete a weld properly, the torch is lifted slowly so the puddle can cool and solidify without contamination. To **solidify** is to turn from a liquid to a solid.

During the welding process, some components of the base metal burn and produce sparks. Changes in welding temperature can be detected by observing the amount of sparks given off. Hence, the alert welder can detect when a burn-through is about to happen and reduce the heat to prevent it. **Burn-through** is the process whereby the cool side of metal becomes molten and a hole opens in the surface due to excessive heat from the opposite side. A burn-through is generally preceded by an increase in spark activity.

Pushing the Puddle

The first step in oxyacetylene welding is to control the puddle. This is called **puddling**. To create a bead by puddling without a filler rod is called **pushing the puddle**. The following procedure describes how to push the puddle.

> ### Activity
>
> 1. Obtain a clean piece of 16-gauge × 2" × 6" steel.
> 2. Light and adjust a torch with a number 2 or number 3 tip, using 3-psi acetylene pressure and a neutral flame.
> 3. Place the strip of steel on a firebrick, and weight it with another firebrick.
> 4. Start at one end of the metal, with the torch tip turned toward the plate and with a gap of $\frac{1}{8}$ inch from cone to metal.
> 5. When the metal starts to melt, move the torch in a small circular pattern down the plate to create a bead.
> 6. Lift the tip slowly at the end of the bead. Make additional practice beads until the metal is covered.

The puddle should move ahead as the torch moves down the plate. A bead will form as the flame passes. If the size of the pool changes, speed up or slow down (Figure 25-15). The size of the pool should be kept constant.

The ideal torch angle is 45 degrees. As the angle of the torch increases to 90 degrees, the bead will become wider and wider. If the angle is too flat—such as 30 degrees—the bead will be too narrow (Figure 25-16).

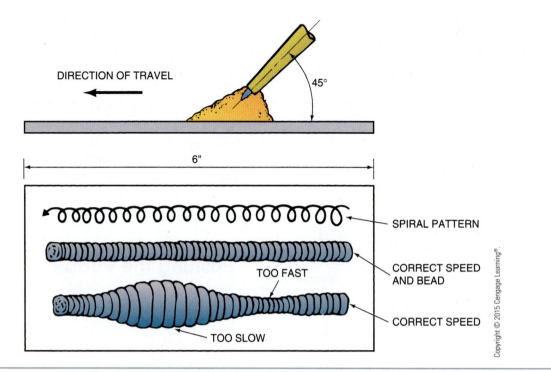

FIGURE 25-15 Torch speed affects the size and shape of the bead.

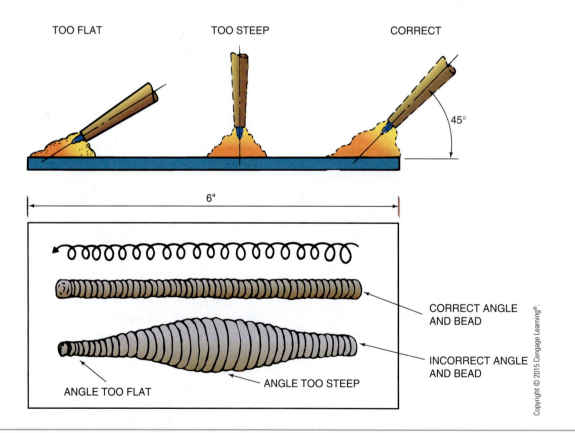

FIGURE 25-16 Torch angle affects the size and shape of the bead.

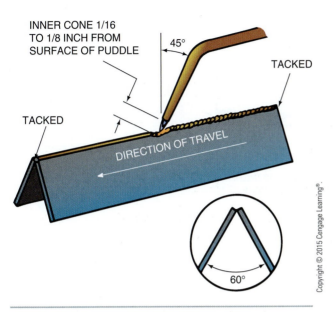

INNER CONE 1/16 TO 1/8 INCH FROM SURFACE OF PUDDLE

45°

TACKED

TACKED

DIRECTION OF TRAVEL

60°

Copyright © 2015 Cengage Learning®

FIGURE 25-17 Making a corner weld without a filler rod.

Making a Corner Weld without Filler Rod

A corner weld is one made on edges laid together to form a corner (Figure 25-17). The following procedure describes how to make a corner weld.

Activity

1. Obtain two pieces of clean, 16-gauge × 1½" × 6" steel.

2. Place the pieces on their edges on a firebrick to form a tentlike shape with a 60-degree angle.

3. Tack weld the joint at both ends and two places in between.

4. Make a weld bead along the entire corner.

5. Evaluate your speed and torch angle by comparing your results with those shown in Figure 25-16.

6. Open the joint, examine the penetration, and discuss the results with your instructor. Grind off the rough edges and save the metal for the next exercise.

Laying a Bead with Filler Rod

The movements for laying a bead with a filler rod in fusion welding are similar to those used in brazing. Practice laying a bead with a filler rod in a butt weld as follows.

Activity

1. Use the two pieces of 16-gauge × 1½" × 6" metal left from the previous exercise.

2. Place the two pieces edge-to-edge on a firebrick to form a 6-inch-long butt joint.

3. Adjust the pieces until a ⅟₁₆-inch gap between them. Weight them with firebricks.

4. Tack weld at both ends of the joint and at two places in between.

5. Point the torch toward one end of the joint at an angle of 45°.

6. Establish a molten puddle with a ⅟₁₆- to ⅛-inch cone clearance.

7. Add filler metal to the front edge of the puddle.

8. Rotate the flame and move the rod in and out to create an even bead (Figure 25-18).

9. Complete the bead by slowly raising the torch while filling the puddle as you withdraw the torch.

10. Discuss the resulting weld with your instructor.

The Finished Weld

The completed weld should be thoroughly fused to the base metal throughout the joint (Figure 25-19). The weld metal should penetrate to the root of the joint, with a small amount extending below the surface to assure a full section weld. At the face of the weld, there is usually a buildup known as "reinforcement." This reinforcement should be slight, blending smoothly with the base metal surfaces. It is especially important to avoid undercutting or overlapping at the juncture of the weld and base metal.

Bending or distortion of the metal can be reduced by back-stepping. **Back-stepping** is making short welds in the backward direction as the process progresses in the forward direction (Figure 25-20). Repeat the previous exercise using the back-step method.

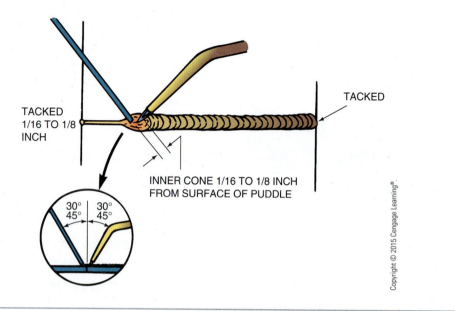

TACKED
1/16 TO 1/8
INCH

TACKED

INNER CONE 1/16 TO 1/8 INCH
FROM SURFACE OF PUDDLE

30°
45°
30°
45°

FIGURE 25-18 Laying a bead to form a butt weld.

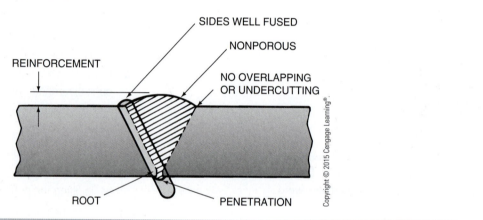

REINFORCEMENT

SIDES WELL FUSED

NONPOROUS

NO OVERLAPPING
OR UNDERCUTTING

ROOT

PENETRATION

FIGURE 25-19 Characteristics of a proper fusion weld.

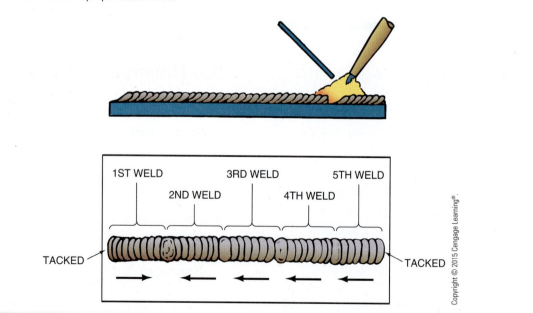

1ST WELD   3RD WELD   5TH WELD

2ND WELD   4TH WELD

TACKED   TACKED

FIGURE 25-20 Bending or distortion can be controlled by back-stepping. The first weld is made from left to right, but all subsequent welds are made from right to left.

Making a Fillet Weld

A fillet weld is made by laying a bead in a 90-degree angle formed by two pieces of metal. To make a fillet weld, the following procedure is recommended.

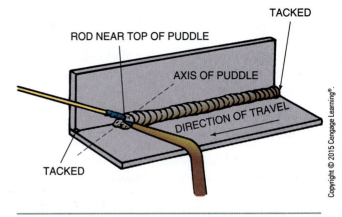

FIGURE 25-21 Making a fillet weld.

Activity

1. Obtain two pieces of clean, 16-gauge × 1½" × 6" steel.
2. Place one piece flat on firebricks. Set the other piece on its edge on the first piece, to form a 90-degree angle.
3. Prop the assembly with firebricks.
4. Tack weld both ends and the middle.
5. Lay a bead along the entire joint (Figure 25-21).

Other joints can be welded using the procedures outlined in this unit. The student is encouraged to learn well all basic procedures presented.

SUMMARY

One of the most important tools for use in working metal is that of oxyacetylene. Using these gases, metal can not only be cut, but can also be joined by using such procedures as brazing, soldering, and fusion welding. In order to achieve the best results with these procedures, the proper techniques must be followed. The use of oxyacetylene can be safe if the proper procedures are followed. Many safety features are built into the use of torches, regulators, and tanks. If the proper safety rules are not followed, the use of oxyacetylene can be quite dangerous.

Student Activities

1. Define the Terms to Know for this unit.

2. Using a torch and flux, heat a piece of copper and practice fluxing copper without overheating the flux.

3. Using a torch, flux, 50-50 solder, and copper, tin the copper surface.

4. Using an oxyfuel torch, brazing filler rod, flux, and ⅛" × 2" × 6" steel plate, tin the steel plate without burning the flux.

5. Prepare an oxyfuel torch, a piece of ⅛" × 2" × 2" steel, and a piece of 16-gauge × 2" × 2" steel. Place the pieces side-by-side, and heat both at the same time to an even red color.

6. Prepare a torch, flux, brazing filler rod, and ⅛" × 2" × 6" steel plate. Practice laying braze-weld beads across the plate at ½-inch intervals.

7. Prepare a torch, flux, brazing filler rod, and three pieces of ⅛" × 2" × 2" steel plate. Use one plate for support and braze a ½-inch lap joint between the other two.

8. Prepare a torch, flux, brazing filler rod, and two pieces of 16-gauge × 2" × 2" steel. Butt-weld the two pieces together.

9. Prepare a torch, flux, brazing filler rod, and two pieces of 16-gauge × 1" × 2" steel. Lay one piece flat on a firebrick. Bend the other piece slightly, and place it on its side on top of the first piece. Braze-weld the two pieces to form a fillet weld.

10. Prepare a torch, welding flux, and one piece of 16-gauge × 1½" × 6" steel. Push the puddle to form three beads across the steel without using a filler rod.

11. Prepare a torch, welding flux, and two pieces of 16-gauge × 1½" × 6" steel. Prop the two pieces edge-to-edge to form a 60-degree angle. Make a corner weld over the ridge of the assembly without using a filler rod.

12. Prepare a torch, flux, ¹⁄₁₆-inch welding filler rod, and two pieces of 16-gauge × 1½" × 6" metal. Use the filler rod to butt-weld the two pieces together.

13. Prepare a torch, flux, ¹⁄₁₆-inch welding filler rod, and two pieces of 16-gauge × 1½" × 6" steel. Join the two pieces to form a T, using a fillet weld.

14. Prepare a torch, flux, ¹⁄₁₆-inch welding filler rod, and two pieces of 16-gauge × 1½" × 6" steel. Weld a lap joint.

Relevant Web Sites

Oxfordplasticsinc.com, Blog Focus on Business, Agriculture Category
http://oxfordplasticsinc.com/category/agriculture

OxyAcetylene Welding by ChuckE2009
YouTube

Welding Safety—Oxygen Acetylene by freddytk421
YouTube

Self-Evaluation

A. Multiple Choice. Select the best answer.

1. Welding is
   a. uniting
   b. heating
   c. fusion
   d. all of these

2. Brazing is much like
   a. soldering
   b. painting
   c. fusion welding
   d. arc welding

3. The most popular soldering alloy is
   a. aluminum silicon
   b. copper-zinc
   c. silver
   d. tin-lead

4. The most popular brazing alloy is
   a. aluminum
   b. copper-zinc
   c. silver
   d. tin-lead

5. Rust is a form of
   a. dirt
   b. galvanize
   c. oxidation
   d. weathered paint

6. Metal may be cleaned before brazing with a
   a. brush
   b. flux
   c. emery cloth
   d. all of these

7. Two pieces lying flat and end-to-end may be joined by a
   a. butt weld
   b. corner weld
   c. fillet weld
   d. lap weld

8. Bonding solder or braze material to a piece of metal is called
   a. brazing
   b. fusion
   c. soldering
   d. tinning

9. The best distance between cone and puddle when fusion welding is
   a. ½ inch
   b. ¼ inch
   c. ⅛ inch
   d. 1/32 inch

10. The best torch angle for flat welding is
    a. 30°
    b. 45°
    c. 70°
    d. 90°

B. **Matching.** Match the terms in column I with those in column II.

Column I
1. soldering
2. brazing process
3. fusion welding
4. bead
5. tacking
6. flux
7. small tip
8. large tip
9. emery cloth
10. galvanize

Column II
a. joins electrical wires
b. fastens temporarily
c. used before fluxing
d. capillary action
e. size 000
f. size 5
g. mixes metals
h. line of filler metal
i. toxic when heated
j. removes oxides

C. **Completion.** Fill in the blanks with the word or words that make the following statements correct.

1. The first step in oxyacetylene welding is to control the _____.

2. A weld placed in a joint created by a 90-degree angle is a _____ weld.

3. When pushing the puddle, the size of the pool should be kept _____.

4. The deepest point in a weld is the_____.

5. A burn-through is generally preceded by an _____ _____.

D. **Brief Answer.** Briefly answer the following questions.

1. What are three advantages of brazing over fusion welding?

2. What is the difference between brazing and braze welding?

3. Why is back-stepping used?

4. When running a bead, how should one move the torch?

5. Describe the characteristics of a properly completed weld.

UNIT 26
* Selecting and Using Arc Welding Equipment

UNIT 27
* Arc Welding Mild Steel and GMAW/GTAW Welding

UNIT 26

Selecting and Using Arc Welding Equipment

Objective

To select electric arc welders, equipment, and materials needed for welding in agricultural mechanics.

Competencies to be developed

After studying this unit, you should be able to:

- Describe the shielded metal arc welding process.
- Distinguish types of electric welding machines.
- Select suitable supplies and equipment for shielded metal arc welding.
- Recognize color and numerical code markings on electrodes.
- Select electrodes for use in agricultural mechanics.

Materials List

- A variety of welding machines
- A safe and fully equipped welding area
- Electrodes for identification

Terms to Know

- arc welder
- arc
- welder
- electrodes
- shielded metal arc welding (SMAW)
- arc welding
- stick welding
- slag
- amperage
- duty cycle
- ampere (A)
- amp
- conductor
- volt (V)
- voltage
- watt (W)
- wattage
- transformers
- alternating current (AC)
- 60-cycle current
- generator
- direct current (DC)
- polarity
- straight (negative) polarity (SP)

continued

Heat for arc welding is obtained by using electricity. In this process, electric current flows from a transformer connected to lines from a power plant. Another source of electricity for welding is electric motors or gas engines driving special generators. Regardless of the source of the electrical energy, a machine that produces current for welding is known as an **arc welder**. An **arc** is the discharge of electricity through an air space. A person who welds is known as a **welder**.

ARC WELDING PROCESS AND PRINCIPLES

Shielded Metal Arc Welding

This unit focuses on a welding process that uses flux-coated metal welding rods called **electrodes**. The process is called **shielded metal arc welding (SMAW)**. Some welders call the process **arc welding**, or **stick welding**. The term *shielded* refers to the gaseous cloud formed around the weld by the burning flux. The gases help the metal electrode burn evenly as it mixes with the base metal (Figure 26-1). The flux also removes impurities from the base metal, preventing oxidation. The flux and impurities float to the top of the weld to form a layer called **slag**.

Advantages of Arc Welding. Arc welding is used extensively in agricultural mechanics. Some reasons for its popularity are as follows:

- Electricity is relatively inexpensive as a source of heat for welding.
- Electric welders suitable for farm welding are relatively inexpensive.
- Welders are available that work on ordinary 230-volt household or farmstead wiring.
- Engine-driven portable welders are available.
- Arc welding is fast and reliable.
- Agricultural students and workers can become good arc welders quickly.
- An arc welder can be used for heating, brazing, and hard-surfacing as well as welding.

Temperature. The arc from a welder has a temperature of about 9,000°F. The exact temperature varies with the length of the arc, size of the electrode, and **amperage** setting. Typical amperage settings for welders used in

Terms to know, *continued*

- reverse (positive) polarity (RP)
- electrode holder
- ground clamp
- chipping hammer
- National Electrical Manufacturers Association (NEMA)
- end marking
- spot marking
- group markings
- American Welding Society (AWS)
- tensile strength
- carbon arc torch

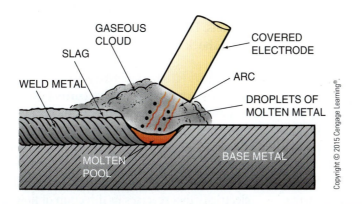

FIGURE 26-1 Components of the shielded metal arc welding process.

383

Machine Model	Type (60 Hertz)	Nema Rating Current (Amperes)	Arc Volts	Duty Cycle	Output Current Range (Amperes)	Type	Required Power System	Bulletin Number
AC-225-S	K-1170	225	25	20%	40-225	AC	1 phase	E320
AC-250	K-1051	250	30	30%	35-300	AC	1 phase	E330
AC/DC-250	K-1053	250	30	30%	40-250#	AC & DC	1 phase	E330
TM-300	K-1103	300	32	60%	30-450	AC	1 phase	E340
TM-400	K-1105	400	36	60%	40-600	AC	1 phase	E340
TM-500	K-1108	500	40	60%	50-750	AC	1 phase	E340
TM-300/300	K-1104	300	32	60%	45-375#	AC & DC	1 phase	E340
TM-400/400	K-1107	400	36	60%	60-500#	AC & DC	1 phase	E340
TM-500/500	K-1110	500	40	60%	75-625#	AC & DC	1 phase	E340
TM-650/650	K-1126	650	44	60%	75-750#	AC & DC	1 phase	E340

Source: Lincoln Electric Company

FIGURE 26-2 Different welding machines have different current ranges, enabling the operator to select one whose performance characteristics match the tasks it is needed for.

agricultural mechanics range from about 20 amps to 225 amps. Upper limits for industrial and commercial welders are higher (Figure 26-2).

Duty Cycle. Welding machines get hot during use. The design of the machine determines how long it can operate (Figure 26-3). The **duty cycle** is the percentage of time that a welder can operate without overheating. Stated another way, it is the number of minutes out of 60 that a welder can operate at full capacity. A 20-percent duty cycle welder should weld only 12 minutes out of every hour, or 20 percent of 60 minutes. Welders are available with duty cycles ranging from 20 percent to 100 percent.

Duty cycle ratings are based on a midrange setting. The duty cycle is shorter if the welder is used at higher settings. Similarly, it is longer when the welder operates at lower settings.

Electricity for Welding

Three terms are basic to understanding how electricity is used for welding. They are amperes, volts, and watts. An **ampere (A)** or **amp** is a measure of the rate of flow of current in a conductor. A **conductor** is any material that permits current to move through it. Amperes of electricity flowing through a wire can be compared to gallons per minute of water flowing through a pipe.

A **volt (V)** or **voltage** is a measure of electrical pressure. Volts in a wire can be compared to pounds per square inch of water pressure in a pipe.

A **watt (W)** or **wattage** is a measure of energy available or of work that can be done. For example, a 100-watt bulb gives out a specific amount of light. To figure the number of watts consumed, multiply volts by amperes: $W = V \times A$ or $W = VA$. This is called the West Virginia (W.VA.) formula to make it easier to remember. If any two values are known, the third can be calculated (Figure 26-4).

Welding machines put out a high amperage and relatively low voltage. It is the low voltage that permits the welder to be a relatively safe machine while putting out so much energy. It requires about 60 volts to push current through the human body. At low to moderate settings, some welders put out less than 60 volts. However, reasonable caution to prevent electrical shock must be exercised at all times.

Alternating and Direct Current. Some welders are transformers that receive current directly from utility power lines. **Transformers** convert high voltage and low amperage to low voltage and high amperage. These are called alternating current welders. **Alternating current (AC)** is current that reverses its direction of flow frequently. In the United States, power plants produce electricity that reverses its direction of flow 60 times per second. This electricity is referred to as **60-cycle current**.

FIGURE 26-3 Welders have a duty cycle that determines how long the machine can continuously operate.

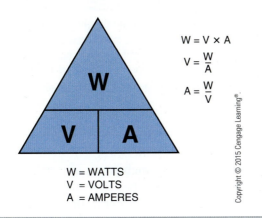

$$W = V \times A$$
$$V = \frac{W}{A}$$
$$A = \frac{W}{V}$$

W = WATTS
V = VOLTS
A = AMPERES

FIGURE 26-4 When two of the three electrical values W, V, and A are known, the third value can be calculated.

Other welders are really generators, driven by an electric motor or a gas or diesel engine. A **generator** is a machine that produces direct current. **Direct current (DC)** flows in one direction only, in accordance with how the welder is set. DC welders can be set for straight or reverse polarity. **Polarity** refers to the direction of flow of electricity in the welding circuit. Certain electrodes and certain jobs work better with current flowing a certain way. **Straight (negative) polarity (SP)** is the term given to DC current flowing in one direction, while **reverse (positive) polarity (RP)** is direct current flowing in the opposite direction. Most DC welders have a switch to operate at either straight or reverse polarity as needed for any given electrode or job. Otherwise, polarity on DC welders can be switched by reversing the welding cables at the machine.

WELDING EQUIPMENT

AC Welders

AC welders rated at 180 and 225 amperes are popular for farm use (Figure 26-5). Such welders can be purchased for approximately $200 and are adequate for most farm welding jobs. They utilize standard 220 volt current. They do not draw excessive current, so electricity bills are usually not greatly affected by them. Such machines are also available as AC/DC welders. Large farming operations may benefit from larger and more versatile welders for specialized welding.

DC Welders

Some DC welders are driven by electric motors. Such welders must be located near a suitable electrical outlet. Other welders are driven by gasoline or diesel

FIGURE 26-5 Welders rated at 180 or 225 amperes are popular for farm use.

Courtesy of The Lincoln Electric Co., Cleveland, OH

FIGURE 26-6 Diesel engine-driven AC/DC welder and generator.

engines. These units may be transported to the job by truck or trailer. Some machines are both DC and AC (Figure 26-6). These machines can be used as welders and as portable generators to operate lights and equipment (Figure 26-7). The control panel on such machines includes controls to operate the engine as well as the welder (Figure 26-8).

Courtesy of The Lincoln Electric Co., Cleveland, OH

FIGURE 26-7 Some welders are portable and can act as generators to produce electricity for lights and power tools.

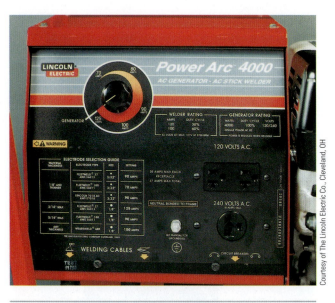

Courtesy of The Lincoln Electric Co., Cleveland, OH

FIGURE 26-8 Control panel of an engine-driven welder/generator.

Welding Cables

The diameter of electrical wire is stated in terms of gauge. The lower the gauge number, the larger the size of the wire. Wire sizes range from 18 for lighting circuits in automobiles to 0 or lower for large service cables. The gauge used for electrical wiring in homes is generally 14 for lights and 12 for heavier loads.

CAUTION!

Careful attention should be given to cable care. Always protect cables from forceful impact and contact with sharp objects that may damage their coverings. These coverings insulate the conductors and prevent dangerous shocks and electrical shorts.

A welder puts out high amperage, so it needs large electrical cables to safely carry the current. Cables carry current from the welder to the electrode, across the arc, and back to the welder. The cable that comes with a welder as original equipment is the correct diameter for its length. If longer cables are used, they may also need to be of larger diameter (Figure 26-9). If the cables are too small, they will not deliver as much current as the welder is designed to provide. Under these conditions, the force of the arc is not as intense as the machine is capable of delivering.

Amperes		Copper Welding Lead Sizes								
		100	150	200	250	300	350	400	450	500
ft	m									
50	15	2	2	2	2	1	1/0	1/0	2/0	2/0
75	23	2	2	1	1/0	2/0	2/0	3/0	3/0	4/0
100	30	2	1	1/0	2/0	3/0	4/0	4/0		
125	38	2	1/0	2/0	3/0	4/0				
150	46	1	2/0	3/0	4/0					
200	61	1/0	3/0	4/0						
250	76	2/0	4/0							
300	91	3/0								
350	107	3/0								
400	122	4/0								

Amperes		Aluminum Welding Lead Sizes								
		100	150	200	250	300	350	400	450	500
ft	m									
50	15	2	2	1/0	2/0	2/0	3/0	4/0		
75	23	2	1/0	2/0	3/0	4/0				
100	30	1/0	2/0	4/0						
125	38	2/0	3/0							
150	46	2/0	3/0							
175	53	3/0								
200	61	4/0								
225	69	4/0								

Length of Cable

FIGURE 26-9 Copper and aluminum welding lead sizes (gauges) for various current loads and cable lengths.

Electrode Holders and Ground Clamps

A welder has two cables: one ends in an electrode holder, and the other ends in a ground clamp. An **electrode holder** is a spring-loaded device with insulated handles used to grip a welding electrode (Figure 26-10). Welding cables have a copper or aluminum core. This core is clamped or soldered to the metal part of the electrode holder. An insulated sleeve slides over the connection to protect the operator from electric shock. All other parts of the holder are insulated except the jaws.

A **ground clamp** is a spring-loaded clamp attached to an electrical cable (Figure 26-11). It is not insulated and is not a shock hazard. It carries current between the welding table, project, and to the welder.

FIGURE 26-10 Electrode holder.

Copyright © 2015 Cengage Learning.

FIGURE 26-11 Ground clamp.

Welding Table and Booth

A booth and metal table for welding are convenient and efficient (Figure 26-12). The proper table permits the operator to stand or sit comfortably.

FIGURE 26-12 A welding booth equipped with a welding table is convenient and efficient.

FIGURE 26-13 Fire-resistant curtains must cover the opening of a welding booth to protect workers in the area from light flashes.

The booth must have curtains to protect other workers from blinding light flashes (Figure 26-13). The booth should also have panels for the storage of welding tools and equipment. Exhaust equipment must be provided to remove the fumes resulting from welding. The entire area must be fire-resistant and free of flammable materials.

CAUTION!

It is essential to have a fire blanket and appropriate fire extinguishers in the welding area.

Welding is often done on the floor of a shop or outdoors. In such situations, portable curtains must be set up to protect others in the area from welding flashes (Figure 26-14). It is important that all combustible materials be removed before welding in such areas.

Equipment for Cleaning Welds

Welding areas should be equipped with a large grinder with a wire wheel. The grinder is used extensively in preparing metal for welding, and the wire wheel is useful for cleaning beads after welding.

Tongs or pliers of various types are useful for handling hot metal. Hand wire brushes and chipping hammers are recommended for removing slag. A **chipping hammer** is a steel hammer with a sharp edge and/or point.

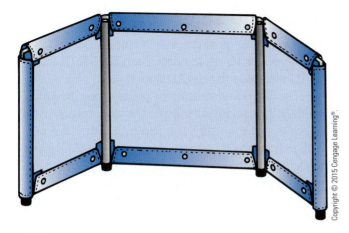

FIGURE 26-14 Portable curtains for use around welding done on the shop floor or outdoors.

ELECTRODES

Electrodes are specially coated metal rods that conduct electricity from the welder to the metal being welded (Figure 26-15). Arc welding technology has developed to a high level. A wide variety of electrodes is available to enable a welder to do many different jobs, including welding in all positions. Much of the difference in electrodes is in the different types of coating on the metal rods.

Welding positions include flat, horizontal, vertical up, vertical down, and overhead. (These are described in Unit 27.) Welding in agricultural mechanics may call for any of these positions. Some electrodes are suitable for all-position welding. Others are suitable for only one or two positions. The parts of an electrode are shown in Figure 26-16.

FIGURE 26-15 Electrodes are specially coated metal rods.

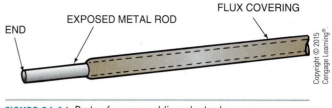

FIGURE 26-16 Parts of an arc welding electrode.

NEMA Color Coding

The **National Electrical Manufacturers Association (NEMA)** developed a system to permit manufacturers to mark their electrodes by color codes. When used, the markings are placed on three areas of an electrode: (1) the exposed end of the metal rod, (2) the exposed surface of the metal rod, and (3) the flux near the exposed rod. The color markings are given specific names. Color on the end is called an **end marking**; color on the bare surface is called a **spot marking**; and color on the flux is called a group marking (Figure 26-17). Thus electrodes are identified by the colors of their end, spot, and **group markings**. Most manufacturers simply stamp the AWS classification number on each electrode, rather than use the color-coding system.

AWS Numerical Coding

The **American Welding Society (AWS)** is an organization of individuals and agencies that support education in welding processes. The society has developed a numerical system for coding electrodes. The code condenses a great deal of information into a four- or five-digit number for mild steel electrodes. The number is preceded by the letter "E" to indicate that the number describes an electrode.

The first two digits (or the first three if it is a five-digit number) refer to the tensile strength of 1 square inch of weld. **Tensile strength** refers to the amount of tension or pull the weld can withstand.

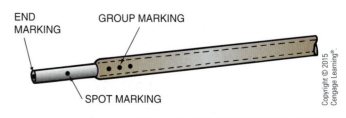

FIGURE 26-17 Areas on an electrode where the color-coded markings of the NEMA system appear.

The number represents thousands of pounds of tensile strength per square inch. For example, E60 stands for 60,000 pounds per square inch (psi) of tensile strength.

The two right-hand digits in the AWS number refer to the type of welding the electrode is capable of doing. The third digit (or fourth if it is a five-digit number) refers to the welding positions for which the electrode is suited. The fourth digit (or fifth if a five-digit number) refers to the depth of penetration and/or welding current (Figures 26-18 and 26-19). For complete identification, it is more useful to read the last two digits together, as defined in tables available from the American Welding Society (www.aws.org; see Figure 26-20 for some examples).

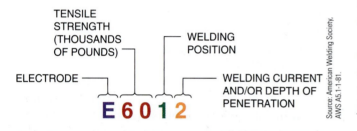

Source: American Welding Society, AWS A5.1-1-81.

FIGURE 26-18 Components of a typical AWS classification number for electrodes. A five-digit AWS number is used if three digits are needed to express tensile strength.

The Second Digit from the Right Indicates Welding Position	
E _ _ 1 _	Usable in all directions
E _ _ 2 _	Usable in flat and horizontal positions only
E _ _ 4 _	Usable for vertical down only

The Right-Hand Digit Indicates Type of Welding	
E _ _ _ 0	DC reverse polarity only
E _ _ _ 1	AC and DC reverse polarity
E _ _ _ 2	AC and DC straight polarity
E _ _ _ 3	AC and DC
E _ _ _ 4	AC and DC
E _ _ _ 5	DC reverse polarity
E _ _ _ 6	AC and DC reverse polarity
E _ _ _ 8	AC and DC reverse polarity

Copyright © 2015 Cengage Learning®.

FIGURE 26-19 The two farthest-right digits in the AWS electrode numbering system specify the suitable welding positions and the type of welding current.

Electrode Selection

All-purpose, all-position AC/DC mild steel electrodes are the choice for most shops. Both the E6013 and the E6011 electrode fit this description.

The E6013 electrode is popular for beginning welders. It is relatively inexpensive and easy to use. It deposits metal quickly and leaves an attractive bead. The slag is easy to remove. However, the E6013 electrode has shallow penetration. Therefore, it is recommended primarily for thin metal, ⅛ inch or less.

The E6011 electrode is perhaps the most commonly used in agricultural mechanics. It is not easy for the beginner to make an attractive weld using the E6011. However, the deep penetration it provides generally results in a stronger weld than can be obtained with the E6013. Commercial welding shops can choose from many types and suppliers of electrodes.

The experienced welder needs at least a pound of each additional electrode in the shop. These include special electrodes for welding cast iron and stainless steel. Such electrodes are handy for specialized repair jobs.

Other useful electrodes include hard-surfacing types. One type deposits material that resists soil abrasion. Layers deposited on parts such as plowshares, landslides, cultivator shovels, and bulldozer blades extend the life of the part. Another type of hard-surfacing electrode leaves a deposit on metal chains, drags, or bars to resist metal-to-metal wear.

Some shops have arc welders but do not have gas cutting and welding torches. Such shops may benefit from special arc welding electrodes designed for cutting and/or brazing. Additionally, the carbon arc torch may be a useful addition to the arc welder for heating. The **carbon arc torch** is a device that holds two carbon sticks and produces a flame from the energy of an electric welder.

Storing Electrodes

If not stored properly, electrodes can corrode and absorb moisture (Figure 26-21). This is particularly true in humid climates. Moist electrodes will not arc properly and will not produce a quality weld. The best method of storing electrodes is in a storage bin specially designed for electrodes (Figurer 26-22). The bin has shelves so different types of electrodes can be stored. The unit is air tight and has a drying unit run by electricity that removes and keeps out moisture.

Electrode Classification

AWS Classification	Type of Covering	Capable of Producing Satisfactory Welds in Position Shown[a]	Type of Current[b]
E60 Series Electrodes			
E6010	High cellulose sodium	F, V, OH, H	DCEP
E6011	High cellulose potassium	F, V, OH, H	AC or DCEP
E6012	High titania sodium	F, V, OH, H	AC or DCEN
E6013	High titania potassium	F, V, OH, H	AC or DC, either polarity
E6020	High iron oxide	H-fillets	AC or DCEN
E6022[c]	High iron oxide	F	AC or DC, either polarity
E6027	High iron oxide, iron powder	H-fillets, F	AC or DCEN
E70 Series Electrodes			
E7014	Iron powder, titania	F, V, OH, H	AC or DC, either polarity
E7015	Low hydrogen sodium	F, V, OH, H	DCEP
E7016	Low hydrogen potassium	F, V, OH, H	AC or DCEP
E7018	Low hydrogen potassium, iron powder	F, V, OH, H	AC or DCEP
E7024	Iron powder, titania	H-fillets, F	AC or DC, either polarity
E7027	High iron oxide, iron powder	H-fillets, F	AC or DCEN
E7028	Low hydrogen potassium, iron powder	H-fillets, F	AC or DCEP
E7048	Low hydrogen potassium, iron powder	F, OH, H, V-down	AC or DCEP

a. The abbreviations, F, V, V-down, OH, H, and H-fillets indicate the welding positions as follows:

F = Flat
H = Horizontal
H-fillets = Horizontal fillets
V-down = Vertical down
V = Vertical   For electrodes 3/16 in. (4.8 mm) and under, except 5/32 in. (4.0 mm) and under for classifications
O H = Overhead   E7014, E7015, E7016, and E7018.

b. The term DCEP refers to direct current, electrode positive (DC reverse polarity). The term DCEN refers to direct current, electrode negative (DC straight polarity).

c. Electrodes of the E6022 classification are for single-pass welds.

Source: Reproduced from AWS A5.1-81, American Welding Society, with permission.

Source: American Welding Society. (2004.) Specification for carbon steel electrodes for shielded metal arc welding. AWS A5.1/A5.1M: 2004.

FIGURE 26-20 Specifications for covered carbon-steel arc welding electrodes.

© spaxiax/Shutterstock.com

FIGURE 26-21 These electrodes were not properly stored. Note the rust and crumbling coating on the electrodes.

Copyright © 2015 Cengage Learning®.

FIGURE 26-22 The best way to store electrodes is in a bin specially designed for electrodes.

SUMMARY

The use of arc welding for joining metal is almost 100 years old. Electric arc welding is done on almost all major construction projects where metal is joined together. The proper selection of equipment, metal, and electrodes is essential to achieving a quality weld. When the proper equipment and materials are selected, the proper technique must be used in order to achieve a strong weld. Following the proper procedures also results in a safer working environment.

Student Activities

1. Define the Terms to Know for this unit.

2. Examine the welders in your school agricultural mechanics shop. Classify them according to AC or DC type, and determine if the DC welders are motor-driven or engine-driven.

3. Examine the electrodes used in your agricultural mechanics shop. Is an AWS code stamped on the flux? Are the electrodes color-coded? If so, determine what the code(s) mean.

4. Identify the tools and equipment in the welding area of your shop.

Relevant Web Sites

•Lincoln Electric, Education Center
www.lincolnelectric.com

•American Welding Society
www.aws.org

•Miller Welds, See Resources section for information on Improving Your Skills and Welding Safety
www.millerwelds.com

Self-Evaluation

A. **Multiple Choice.** Select the best answer.

1. Voltage is a measure of
   a. rate of current
   b. electrical pressure
   c. available energy
   d. all of these

2. A welder that gets its energy directly from a utility power plant is
   a. an alternator
   b. a generator
   c. a rectifier
   d. a transformer

3. When welding, the flux on an electrode
   a. forms a gas
   b. creates slag
   c. shields the weld
   d. all of these

4. Shielded metal arc welding is also called
   a. brazing
   b. electrode welding
   c. oxyacetylene welding
   d. stick welding

5. Suitable arc welders for farm use are
   a. fairly easy to use
   b. relatively inexpensive
   c. reliable
   d. all of these

6. The output of a welder is relatively
   a. low voltage and high amperage
   b. high voltage and low amperage
   c. high voltage and high amperage
   d. low voltage and low amperage

7. Types of polarity on DC welders include
   a. reversed
   b. straight
   c. both of these
   d. none of these

8. The proportion of time that a welder can operate without overheating is known as its
   a. AC/DC
   b. AWS classification
   c. duty cycle
   d. voltage drop

9. A chipping hammer is used to
   a. prepare edges for welding
   b. remove scale from steel
   c. remove slag
   d. temper beads

10. An organization of people and agencies interested in promoting welding is the
   a. AWS
   b. EPA
   c. SAE
   d. WPA

B. **Matching.** Match the items in column I with those in column II.

Column I
   1. welder
   2. voltage
   3. transformers
   4. amperage
   5. tensile strength
   6. polarity
   7. wattage
   8. generator

Column II
   a. alternating current welders
   b. a measure of the rate of flow of current
   c. amount of pull a weld can withstand
   d. one who welds
   e. produces direct current
   f. the direction of flow of electricity
   g. a measure of energy available or work that can be done
   h. a measure of electrical pressure

C. **Completion.** Fill in the blanks with the word or words that make the following statements correct.

   1. The color markings on an electrode will be _____, _____, or _____ markings.

   2. The E in electrode code stands for _____.

   3. In the American Welding Society's electrode code, going from left to right, the first two (or three) digits refer to _____; the next digit refers to _____; and the final digit refers to _____.

   4. Three types of welding electrodes other than those used for welding mild steel are those used for _____, _____, and _____.

D. **Brief Answer.** Briefly answer the following questions.

   1. How does one determine the number of watts consumed?

   2. A welder has two cables, each ending with a different device. Identify each and state its purpose. How do they differ?

   3. Welding areas need to be equipped with a number of items to aid in cleaning welds. List these and tell what their purpose is.

   4. Why are the curtains on the welding booth so important?

   5. Which electrode is the most popular for beginners? What is a drawback of this electrode?

UNIT 27

Arc Welding Mild Steel and GMAW/GTAW Welding

Objective

To use arc welding equipment and procedures in cutting and welding.

Competencies to be developed

After studying this unit, you should be able to:

- Use safety equipment and protective clothing for arc welding.
- Strike an arc and run beads.
- Make butt and fillet welds.
- Make flat, horizontal, and vertical welds.
- Weld pipe.
- Pierce holes and cut with electrodes.
- Identify major parts of GMAW and GTAW welding outfits.
- Explain and perform GMAW and GTAW welding.

Materials List

- Welding booth
- Electric welder, 180 amperes or larger
- Protective clothing: welding helmet, clear goggles, leather gloves, welding jacket, leather apron, and leather leggings
- Chipping hammer
- Pliers
- Wire brush
- Bucket with water

continued

Terms to Know

- pad
- crater
- stringer bead
- weaving
- pass
- root pass
- horizontal welds
- vertical down welds
- vertical up welds
- gas metal arc welding (GMAW)
- gas tungsten arc welding (GTAW)

Materials list, *continued*

- Fire extinguisher and fire blanket
- Welding pads, ¼" × 2" × 6", ¼" × 4" × 6", and ⅛" × 1" × 6"
- Grinder with coarse abrasive wheel and wire wheel
- Soapstone markers
- Electrodes, ⅛-inch E6011 and E6013

The term *arc welding* as used in this unit refers to shielded metal arc welding (SMAW). In this type of welding, an electric welder and flux-covered electrodes are used. Before starting to weld, the operator must be thoroughly familiar with shop safety and fire prevention.

SAFETY PROCEDURES

Fire Protection

The temperature of the electric arc (about 9,000°F) creates a very real danger of burns as well as fires. Nevertheless, with reasonable care and use of the proper equipment, the welding area is no more dangerous than many other areas of the shop.

The welding area should be equipped with metal benches. These serve both for fire protection and to electrically ground the work. Welding booths should be constructed of fireproof or fire-resistant materials, such as metal sheets or concrete block. Other materials such as pressed wood panels, plywood, and special fire-resistant canvas are also used.

Fire extinguishers suitable for Class A, B, and C fires, safety equipment, and a first-aid kit should be within easy reach at all times (Figure 27-1). A wool fire blanket is another important piece of fire control equipment (Figure 27-2). If human hair or clothing catches on fire, the fire blanket is used to wrap the victim and smother the fire.

© oriontrail/Shutterstock.com

FIGURE 27-1 Because welding can create a lot of sparks, it is extremely important to keep a fire extinguisher nearby.

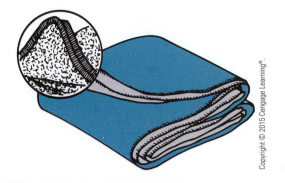

FIGURE 27-2 A wool fire blanket is the first line of defense if clothing catches on fire; the victim can be wrapped in the blanket to smother the fire quickly.

Buckets of water are frequently used to receive sparks and cool metal in the welding areas. Cooling of metal used for welding practice reduces the chances of personal burns. The water may also be useful in extinguishing accidental fires when electric shock is not a hazard. However, any water spilled on the floor of the work area must be removed because it will create an electric shock hazard.

All grease, oil, sawdust, paper, rags, and other flammable materials must be removed from areas where welding is done. Good housekeeping is a major factor in reducing fire hazards.

Personal Protection

The No. 10 shade lens must always be worn when welding, but this lens is too dark to see through except when the welding arc is burning. Therefore, the operator must either wear safety glasses under the helmet or use a helmet with a flip-up lens for processes such as chipping (Figure 27-3). Newer helmets automatically darken as the light intensifies (Figure 27-4). When the flip-up lens helmet is not used, clear safety glasses **must** be worn under the helmet.

When not welding, the operator may exchange the helmet for another type of clear face and eye protective

FIGURE 27-3 Some welding helmets have a flip-up shaded lens. When no welding is being done and the lens is up, the operator can see through the clear lens, which protects the eyes from accidents.

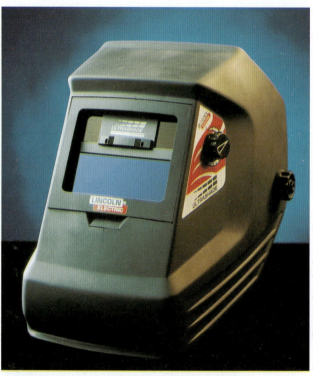

FIGURE 27-4 Some newer helmets have lenses that automatically darken as the light intensifies.

CAUTION!

The human eye must not be exposed to direct light from a welding arc. The welding arc contains light rays that burn, even from a distance. Fortunately, it is easy to protect the eyes from these rays. Any material that blocks the light also stops the damaging rays. The welder's face and eyes are protected by using a face shield with a dark viewing glass. The glass must be classified as a No. 10 shade or higher.

CAUTION!

Attempting to chip hot slag from welds without eye protection is likely to result in eye injury. Pants with cuffs should be avoided. Sparks may fall into the cuffs.

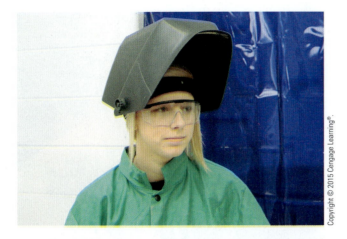

FIGURE 27-5 When using a welding helmet that lacks a flip-up shaded lens, the operator should wear standard safety glasses or goggles under the helmet.

covering (Figure 27-5). Prescription eyeglasses are not approved safety glasses unless ordered as such.

Fire-resistant coveralls and high leather shoes are recommended as standard clothing when welding. In addition, leather gloves are needed to protect hands and wrists. All skin areas must be covered or "sunburning" will occur from the rays of the arc. For many welding operations, protective clothing such as a leather apron, sleeves, jacket, or pants may be needed to protect the body from sparks and hot metal (Figure 27-6). Further, suitable safety glasses or goggles are essential when chipping, cleaning, and examining welds.

SETTING UP

When preparing to weld, the appropriate electrode must be selected. For the beginning welder, this is likely to be an E6011 or E6013 electrode. As described in Unit 26, the best single choice is probably the E6011. The E6013 is easier to use, but is recommended primarily for thin metal.

The next step is to check the welding area to eliminate fire hazards. Gather all the necessary materials, including a piece of ¼" × 4" × 6" plate

CAPE SLEEVES/
DETACHABLE BIB   FULL-SIZED JACKET   18-INCH SLEEVE   23-INCH SLEEVE

WAIST APRON   BIB-APRON   SPLIT LEG APRON   LEATHER GLOVES

FIGURE 27-6 Certain welding jobs require the operator to wear leather gloves and other special protective clothing.

steel for practice, wire brush, chipping hammer, pliers to hold hot metal, and bucket of water to cool the metal. The practice metal is called a **pad**. It may be thicker and larger than specified, if desired. To improve handling, an electrode may be bent into a U shape and the ends welded to the pad to provide an insulated handle.

The welding machine should have suitable welding cables. One cable will end in an electrode holder and the other in a ground clamp. The metal pad to be used for practice is clamped to the welding table using the ground clamp, a vise grip, or a spring clamp. When an extra clamp is used, the ground clamp is attached to either the practice metal or the table.

The operator must wear a welding helmet or use a face shield with a No. 10 lens. It is recommended that tight-fitting coveralls, leather shoes, and leather gloves be worn. Pants must not have cuffs as these will catch burning steel in the form of sparks. Figure 27-7 shows a welder properly dressed for safety.

Once dressed, the operator is ready to set the welder. Most welders have a rotary switch or lever that moves a pointer to the desired ampere setting. On other welders, there is a selection of sockets into which the cables can be plugged. The desired amperage setting is obtained by plugging the cables into the correct pair of sockets. For practice welding, electrodes ⅛ inch or ⁵⁄₃₂ inch in diameter are to be used. Appropriate welder settings range between 70 and 220 amperes, depending on the welder and the electrode selected (Figure 27-8). A recommended starting setting is 90 amperes for ⅛ inch E6011 or E6013 electrodes. The amperage must then be adjusted according to the individual machine.

STRIKING THE ARC

Before striking an arc, the metal should be placed on the welding table. It should be located so the operator can reach across the metal and weld in a comfortable

FIGURE 27-7 This welding arc operator is properly dressed for safety.

Courtesy of The Lincoln Electric Co., Cleveland, OH

position. The metal should be clean and free of grease, oil, or rust. The following procedure is suggested for striking an arc.

Procedure

1. Set the proper amperage, and place a ⅛-inch E6011 electrode in a 90-degree position in the electrode holder.

2. Close all curtains in the welding booth.

Electrode Size	Classification					
	E6010	E6011	E6012	E6013	E7016	E7018
3/32 in	40–80	50–70	40–90	40–85	75–105	70–110
1/8 in	70–130	85–125	75–130	70–125	100–150	90–165
5/32 in	110–165	130–160	120–200	130–160	140–190	125–220

FIGURE 27-8 Welding amperage ranges for various electrode classifications.

Copyright © 2015 Cengage Learning®.

3. Warn others that you are about to start welding by saying "Cover up!" This means they should place shielding in front of their eyes if they can see your work.

4. Turn on the welder.

5. Position the tip of the electrode about ¼ inch above the metal and lean it slightly in the direction you plan to move. Right-handed welders move from left to right.

6. Drop the face shield or cover lens over your eyes.

7. Strike the arc by quickly lowering, touching, and lifting the electrode (Figure 27-9). The action is similar to striking a match. The lift should only be about ⅛ inch.

   a. If you do not lift the electrode in time, it will stick to the metal. If it does, use a quick whipping action to free it, or release the electrode holder.

CAUTION!

Do not turn the welder off with the electrode frozen in the circuit as the welder would be damaged. First, release the electrode holder from the electrode.

   b. If the flux becomes cracked or broken on the end of the rod, lay the electrode aside for a more experienced welder to use. Obtain another electrode for your next attempt to strike an arc.

   c. If the arc goes out, try again; keep the electrode closer to the metal after lifting it.

8. Feed the electrode very slowly as the weld metal burns away, keeping it about ⅛ inch from the metal. As you feed the electrode, move very slowly and evenly across the plate (generally from left to right for right-handed welders).

9. Lift the electrode slowly to break the arc after traveling about 1 inch.

10. Continue practicing steps 5 through 9 until you can start and stop the arc at will. When you stop the bead, lift the electrode slowly to permit the crater to fill. The **crater** is a low spot in metal where the force of a flame has pushed out molten metal.

NOTE

As you practice this technique, strike the arc and hold a slightly longer arc for a second or two before lowering the rod and moving across the metal. This practice preheats the metal and results in a better bead.

NOTE

For practice, make 16 strikes with 1-inch beads. Compare your results with those diagrammed in Figure 27-10. Discuss your welds with your instructor.

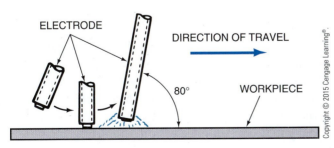

FIGURE 27-9 Striking an arc.

ELECTRODE

DIRECTION OF TRAVEL

80°

WORKPIECE

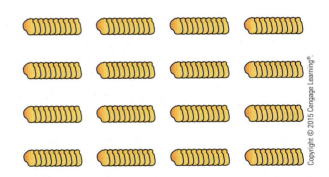

FIGURE 27-10 Students should practice striking and maintaining an arc until they can make a short (½- to 1-inch-long) bead. The practice beads should be made in parallel lines across the metal.

Copyright © 2015 Cengage Learning®.

The method described is called the scratch method of striking an arc. Some people find it easier simply to touch the metal and raise the electrode. This is known as the tapping method. Both methods are acceptable. Striking an arc and keeping it going takes a lot of practice. However, it is well worth the time and patience required, since it is the starting point for arc welding.

RUNNING BEADS

A bead is produced by handling the electrode in a way that results in a proper mix of filler and base metal. A bead is called a **stringer bead** if it is made without weaving. **Weaving** means moving the electrode back and forth sideways to create a bead that is wider than the electrode would normally make. After learning to strike and maintain an arc, the student should practice making stringer beads.

Making stringer beads helps the student learn to:

- Start a bead at the desired location.
- Hold the electrode at about an 80-degree angle.
- Feed the electrode at an even rate.
- Move across the metal in a straight line.
- Move across the metal at an even rate of speed.
- Stop or terminate a weld at the desired spot and in an acceptable manner.

Most of these skills must be developed at the same time. This means that attention to technique is important. It is helpful if someone who can make suggestions for correcting the student's technique watches the student practice.

Angle of the Electrode

Experienced welders vary the angle of the electrode according to the electrode being used and the job. However, the beginner should lean the electrode slightly in the direction of travel. A 75- to 80-degree angle, 10 to 15 degrees from the vertical (straight up) position, is suggested (Figure 27-11).

Arc Length

The arc length indicates the skill with which the electrode is fed as it burns away. The sound of the arc is a good guide. An arc that is the correct length sounds like bacon frying. If the arc is too long, a booing sound is heard. If the arc is too short, the electrode sticks to the metal.

Copyright © 2015 Cengage Learning®.

FIGURE 27-11 The electrode should be positioned at a 75° to 80° angle, and 10 to 15 degrees from the vertical position.

Movement across the Metal

The ability to weld in the intended direction is not easy. Looking through a dark lens, it is difficult to see a soapstone mark, a groove in the metal, another welded bead, or the molten puddle. However, the arc lights the surrounding area. With experience, it becomes possible to observe the welding process with ease.

Speed

Achieving a uniform rate of travel across the metal requires practice in arm movement. It requires the welder to be in a comfortable position that permits such movement. The best way to control the movement is to watch the welded metal solidifying behind the puddle. When the electrode is moving at an even rate of speed, the weld material forms evenly spaced semicircles behind the puddle.

Amperage Setting

The correct amperage can be obtained by observing the welding process and the weld that results. If the amperage is too low, it will be difficult to strike the arc and keep it running. Low amperage results in a narrow, stringy bead. If the arc seems to struggle or start and stop, it is advisable to try a higher amperage. Increase the setting by 10 to 15 amperes and try again.

Learn to observe the shape of the puddle as it is forming. Long, narrow semicircles indicate the heat is too low. Semicircles that are wider than they are long indicate that the correct heat is being used. Gourd-shaped marks indicate that the heat is too high (Figure 27-12).

Other symptoms or indicators of excessive heat include the following:

- The electrode covering turns brown.
- The bead does not have clear markings.
- The puddle burns through the plate.
- Managing the puddle is difficult.
- The arc is very noisy.

When one or more of these conditions exist, decrease the current setting by 10 to 15 amperes and try again.

Making a Practice Bead

To practice running beads, it is best to make short beads about ¾ inch apart on the practice pad. This is far enough apart to evaluate the individual beads and also permit the most efficient use of the metal.

It is advisable to use both hands to hold the electrode holder, or use the other hand to steady the wrist of the hand holding the electrode holder. It is important to wear proper protective clothing so hot slag cannot reach the skin and a comfortable position can be maintained (Figure 27-13). To practice running beads, the following procedure is suggested.

FIGURE 27-13 This person is wearing proper protective clothing and is ready to weld in any position.

Procedure

1. Clamp a ¼" × 4" × 6" welding pad in a suitable position on the table.
2. Set the welder at 110 amperes, and turn on the welder.
3. Insert a ⅛-inch E6011 electrode in the holder at an angle of 90 degrees.
4. Cover up, and remind others nearby to do so as well.
5. Strike the arc, and quickly move the electrode to the starting point near the edge of the pad.
6. Watch the puddle, feed the electrode, move from left to right across the pad, observe the angle of the electrode, and concentrate on coordinating your movements.
7. At the end of the bead, near the edge of the plate, lift the electrode gradually until the crater fills and the arc goes out.
8. Put on clear goggles or a clear face shield (if you are not wearing safety glasses or goggles under your helmet), and remove the slag by striking the edges of the bead with a chipping hammer.
9. Compare your results with those shown on a welding chart or with sample beads provided by the teacher (Figure 27-14).

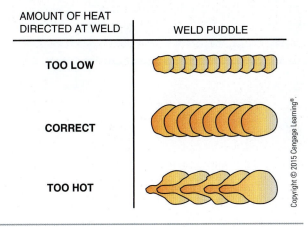

AMOUNT OF HEAT DIRECTED AT WELD	WELD PUDDLE
TOO LOW	
CORRECT	
TOO HOT	

FIGURE 27-12 The amount of heat directed onto the welding puddle affects the puddle's shape.

(continued)

Procedure, *continued*

10. Repeat the process until you can strike the arc and run a good bead over and over again. Readjust the amperage setting if necessary. As the pad accumulates heat from welding, the amperage may need to be lowered to prevent overheating of the electrode and metal.

Weaving

After learning to run stringer beads, it is useful to practice running wider beads by weaving. Many patterns of weaving are used by experienced welders to help control the heat in the puddle. These patterns are generally practiced as welding in different positions is learned. The beginning welder should use a circular pattern, which provides good control of the bead width (Figure 27-15).

WELDING JOINTS

The objective of welding is generally to fasten metal pieces together. There are many types of joints, but most are variations of the butt joint and fillet joint. In a butt joint, the pieces are placed end-to-end or edge-to-edge. In a fillet joint, the two parts come together to form a 90-degree angle.

Butt Welds

Butt welds require the proper preparation. There must be a gap between the pieces to be welded of about the thickness of the electrode core. If the metal is thicker than the electrode core, it must be ground down. Grinding may be done on one or both sides of the metal. Grinding on both sides will produce stronger welds. The thickness of the unground metal should be no more than the diameter of the electrode core. For example, when welding with a ⅛-inch electrode, the top and/or bottom of the metal is ground at a 30-degree angle to leave a ⅛-inch thickness (Figures 27-16 and 27-17).

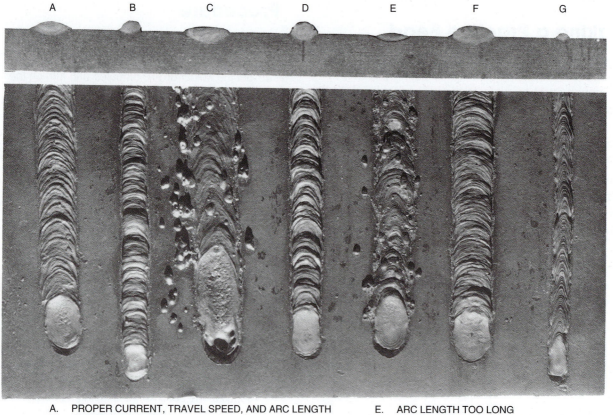

A. PROPER CURRENT, TRAVEL SPEED, AND ARC LENGTH
B. CURRENT TOO LOW
C. CURRENT TOO HIGH
D. ARC LENGTH TOO SHORT
E. ARC LENGTH TOO LONG
F. TRAVEL SPEED TOO SLOW
G. TRAVEL SPEED TOO FAST

Courtesy of The Lincoln Electric Co., Cleveland, OH

FIGURE 27-14 A careful study of welding beads will reveal the errors made during their welding. In this way, the operator can correct faulty techniques and develop proficiency.

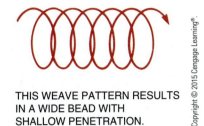

THIS WEAVE PATTERN RESULTS
IN A NARROW BEAD WITH
DEEP PENETRATION.

THIS WEAVE PATTERN RESULTS
IN A WIDE BEAD WITH
SHALLOW PENETRATION.

FIGURE 27-15 Weaving patterns are used to create wider beads and to control the heat of the puddle during welding.

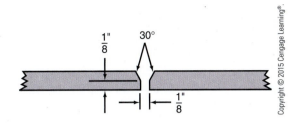

FIGURE 27-16 Before a butt joint can be welded properly, the metal pieces must be ground and correctly positioned. When welding is done with a ⅛-inch electrode, the top and/or bottom of the metal should be ground to a 30-degree angle to leave a thickness of ⅛ inch at the root of the wheel.

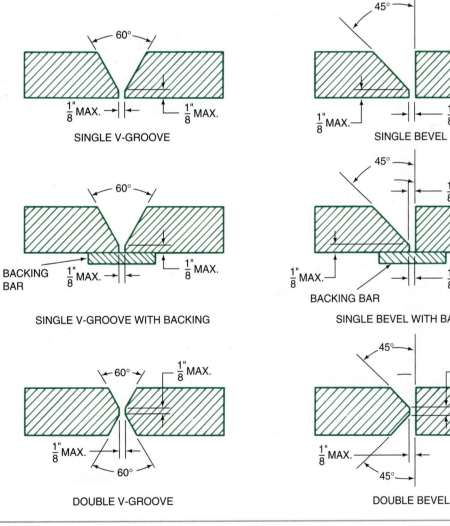

SINGLE V-GROOVE

SINGLE BEVEL

SINGLE V-GROOVE WITH BACKING

SINGLE BEVEL WITH BACKING

DOUBLE V-GROOVE

DOUBLE BEVEL

FIGURE 27-17 Typical butt joint positions.

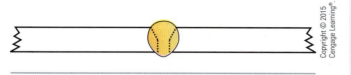

Copyright © 2015 Cengage Learning®.

FIGURE 27-18 Butt weld completed by using one heavy pass.

When welding, one bead or layer of filler metal is called a **pass**. The first pass made in a joint is called the **root pass**. The root pass is the most important pass in a weld. In some instances, one heavy pass will produce a complete butt weld (Figure 27-18).

When making a butt weld, the following procedure is suggested.

Procedure

1. Obtain two pieces of ¼" × 2" × 6" flat steel.
2. Grind one edge of each piece of metal so the matching surfaces are ⅛ inch thick.
3. Position the two pieces ⅛ inch apart.
4. Set the welder at 110 amperes, and obtain ⅛-inch E6011 electrodes.
5. Wear protective clothing.
6. Tack-weld each end of the joint with about ¼ inch of weld.
7. Straighten the joint so the two pieces are flat.
8. Strike the arc, and keep a long arc as you move to the edge of the metal. Shorten the arc to the correct length and proceed to weld. Rotate the electrode in a circular pattern to create an even bead about ¼ inch wide.
9. Chip the weld to remove all slag. Use a chipping hammer blade or point and a stiff wire brush to remove the slag.
10. Examine the weld for evenness and penetration. The answer should be "yes" to the following questions.
    a. Do the sides blend in evenly with the base metal?
    b. Are all the semicircles of the bead evenly spaced?
    c. Does the weld go all the way through to the bottom?
    d. Does the weld start at the starting edge?
    e. Does the weld fill the groove at the starting edge?
    f. Does the weld go all the way to the finishing edge?
    g. Does the weld fill the groove at the finishing edge?
11. If the bead does not fill the gap, run one additional wide bead over the first bead (Figure 27-19). If thicker metal is used, two or more beads may be needed to complete the weld (Figure 27-20).

CAUTION!

All slag must be removed between beads to ensure a solid weld with no voids. Voids result when slag is trapped inside the weld.

NOTE

For further practice, prepare another butt weld, correcting any problems observed in the first weld.

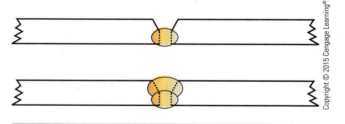

FIGURE 27-19 Butt weld completed by using two passes.

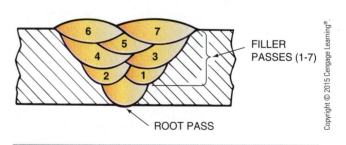

FILLER PASSES (1-7)

ROOT PASS

FIGURE 27-20 Butt weld completed by using many passes.

Copyright © 2015 Cengage Learning®.

The following procedure describes how to make an internal check of a practice weld.

Procedure

1. Saw 1 inch off the end of the welded piece.

2. Examine the cross section of the weld where it was cut in step 1. Voids that are open or filled with slag indicate weak parts of the weld (Figure 27-21).

3. Saw off another inch-long section of the weld.

4. Clamp one side of the 1-inch section in a heavy vise with the welded section exposed.

5. Bend the welded section by driving the exposed piece with a heavy hammer. The welded section should be as tough and strong as the rest of the metal. If the welded section breaks, determine what steps are needed to improve the weld.

Fillet Welds

Most of the principles and procedures that apply to butt welds also apply to fillet welds. To practice making fillet welds, prepare the piece that will be vertical so the weld metal will fuse both pieces completely. It is not necessary to grind ⅛ inch-thick metal. However, thicker pieces need to have one or both edges beveled to permit the weld to penetrate their entire thickness. A ⅛-inch electrode should penetrate about ⅛ inch deep. Therefore, metal that is thicker than ⅛ inch needs to be ground on one or both sides.

The recommended procedure for making a fillet weld follows.

Procedure

1. Obtain one piece of ⅛" × 1" × 6" flat steel.

2. Lay that piece on its edge on a piece of ⅛" × 2" × 6" or ¼" × 2" × 6" metal.

3. Clamp the assembly to the table, or steady the parts with firebricks.

4. Select a ⅛-inch E6011 electrode, and set the welder at 110 amperes.

5. Tack-weld both ends of the fillet joint.

6. Square the vertical piece so it is at a 90-degree angle to the bottom piece.

7. Hold the electrode at a 45-degree angle to the tabletop, and lean it in the direction of travel.

8. Run a bead from end-to-end along the joint (Figure 27-22). Watch the puddle to be sure it penetrates both pieces equally. To create more penetration in one piece or the other, direct the end of the electrode more to that piece.

(continued)

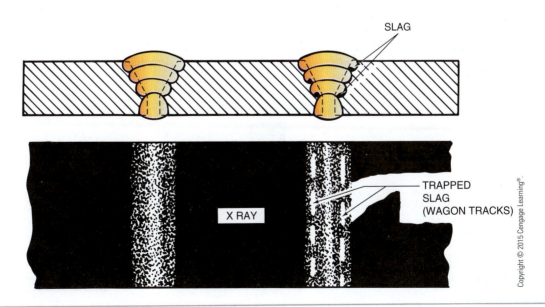

SLAG

X RAY

TRAPPED SLAG (WAGON TRACKS)

Copyright © 2015 Cengage Learning®

FIGURE 27-21 Slag trapped between passes is visible in a cross section X-ray examination.

Procedure, *continued*

Adjust the angle of the electrode as needed to obtain equal penetration. Adjust the amperes, if needed. Check for undercut or overlap in the sides of bead (Figure 27-23).

9. Run a bead on the other side of the vertical piece. Alternating passes from one side to the other reduces the problem of distortion.

10. Chip and examine the bead.

11. Make additional passes, as illustrated in Figure 27-24.

It is important to remember that heat rises in metal. The welder must make allowances for this fact in certain types of welds. For example, when making a fillet weld to form a lap joint, the "J" weave is useful to help control heat (Figure 27-25). When welding a thin piece of metal on top of a thicker piece, the J pattern concentrates more heat on the thicker bottom piece. This permits proper weld penetration without overheating the vertical piece.

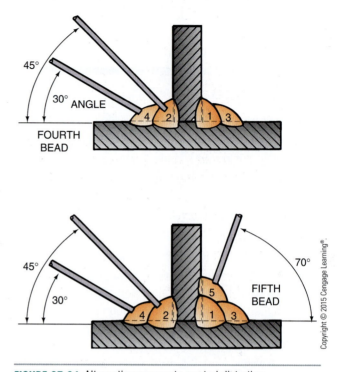

FIGURE 27-24 Alternating passes to control distortion.

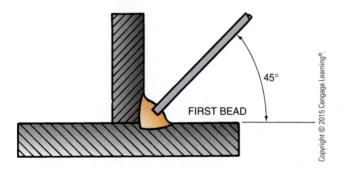

FIGURE 27-22 Making the first pass of a fillet weld.

WELDING IN VARIOUS POSITIONS

Many welding jobs in agricultural mechanics require welding in a variety of positions. Fortunately, modern welding electrodes are available that make welding in all positions possible. The welder should have reasonable skill in flat or down-hand welding before attempting to weld in other positions. Even the experienced welder will try to position materials so they can be welded in a down-hand manner.

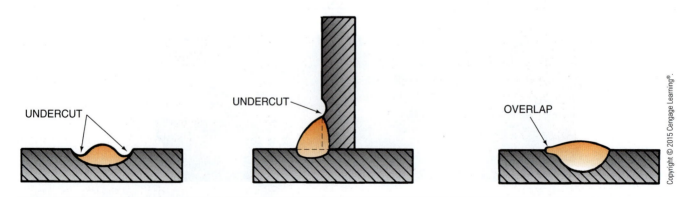

FIGURE 27-23 Undercut and overlap should be avoided in welding beads.

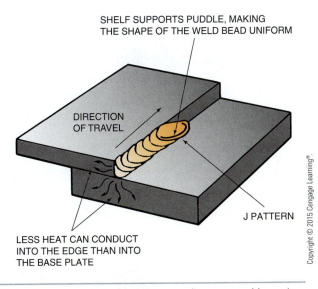

SHELF SUPPORTS PUDDLE, MAKING
THE SHAPE OF THE WELD BEAD UNIFORM

DIRECTION
OF TRAVEL

J PATTERN

LESS HEAT CAN CONDUCT
INTO THE EDGE THAN INTO
THE BASE PLATE

Copyright © 2015 Cengage Learning®.

FIGURE 27-25 The J-weave pattern permits proper weld penetration without overheating the upper piece.

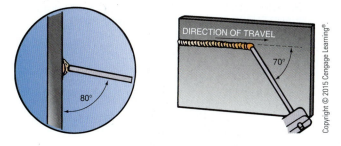

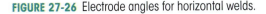

DIRECTION OF TRAVEL

80°

70°

Copyright © 2015 Cengage Learning®.

FIGURE 27-26 Electrode angles for horizontal welds.

> **NOTE**
>
> Adjust the amperage if needed, and run additional beads to develop your skill.

Horizontal Welds

Welds made by moving horizontally across a vertical piece of metal are called **horizontal welds**. When making horizontal welds, the metal is placed so that it is comfortable to run a bead across the surface. The metal should be at eye level so welding from a sitting position is recommended. To make a horizontal bead, the following procedure is suggested.

Procedure

1. Obtain a ⅛" or 14" × 4" × 6" welding pad.

2. Weld two small pieces of metal on the edge of the pad so that the pad can be set on edge and clamped securely to the table.

3. Put on helmet, gloves, arm and shoulder protection, and a leather apron.

4. Sit in front of the table.

5. Select a ⅛-inch E6011 electrode, and set the welder at 110 amperes.

6. Hold the tip of the electrode near the left side of the pad. Lean it in the direction of travel at an angle of about 70 degrees to the plate. Tilt the electrode so it points slightly upward at an 80-degree angle to the plate (Figure 27-26).

7. Run a bead across the plate at a rate of speed that creates a proper bead, without sagging, undercutting, or overlapping.

Vertical Welds

Welds made by moving downward across a vertical piece of metal are called **vertical down welds**. Welds made by moving upward across the metal are called **vertical up welds**. Vertical down welds are easy to make. However, they are shallow in penetration and so should be made only on materials that are ⅛ inch or less in thickness.

To make a vertical down weld, set up a practice pad as described in the section on horizontal welds. A sitting position is recommended for welding. The electrode is held with the tip pointing upward at an 80-degree angle to the plate. Strike the arc at the top of the plate, and let the force of the arc push the electrode downward across the plate at an even rate. Weaving is not recommended for this procedure. Adjust the amperes as needed to obtain an even bead.

Vertical up welding is rather difficult to manage, because heat builds up in the metal as the weld progresses up the metal. Special movements of the electrode are used to control the heat and permit individual parts of the puddle to cool. Unless this is done carefully, the weld metal will sag or drop out of the puddle. The shape of the puddle indicates the temperature of the surrounding base metal (Figure 27-27).

A shorter arc, lower amperage, a steeper electrode angle, and more rapid movement of the electrode may all be helpful in controlling the puddle. To make a vertical up weld on a flat area, the following procedure is recommended.

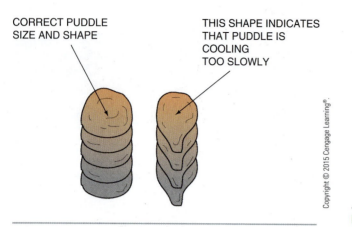

FIGURE 27-27 The shape of the puddle indicates the temperature of the surrounding main piece of metal.

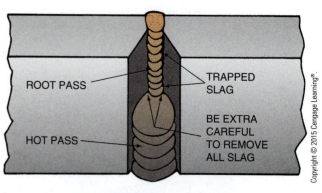

FIGURE 27-28 Making a vertical weld on a flat area.

Procedure

1. Set up a practice pad as outlined under horizontal welding.

2. Start at the bottom of the pad and work upward.

3. Use a wide weaving pattern, moving the tip of the electrode from side to side. Some welders create the bead by making a series of very short horizontal beads as they progress upward. Other welders use a figure-eight pattern. Others use a wide, crescent-shaped, back-and-forth motion.

4. If the bead gets too hot and is not controllable, stop. Chip and clean the weld. Then start again. Make corrections by reducing the amperage and using more sideways motion to permit parts of the puddle to cool more quickly. Make additional passes, if needed, after the root pass is cool and all slag chipped (Figure 27-28).

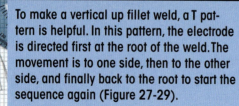

NOTE

To make a vertical up fillet weld, a T pattern is helpful. In this pattern, the electrode is directed first at the root of the weld. The movement is to one side, then to the other side, and finally back to the root to start the sequence again (Figure 27-29).

Overhead Welding

Overhead welding is not a difficult procedure. However, it can be dangerous without protective clothing. The metal is positioned above the welder and the

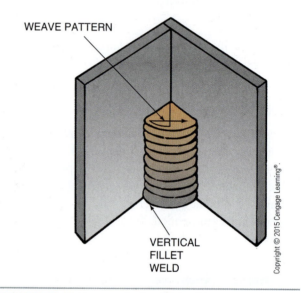

FIGURE 27-29 Making a vertical up fillet weld.

welder assumes a sitting position, if possible. The electrode is held nearly straight up and down. Therefore, some welders prefer to place the electrode so it extends out the end of the holder. The holder is then gripped like a pencil.

When welding overhead, use an E6011 electrode, a normal amperage setting, electrode angle of 10 to 15 degrees from vertical and normal speed. Welds may be single pass or multiple pass according to the need (Figure 27-30).

WELDING PIPE

Welding of pipe is done extensively in agricultural mechanics. Pipe may be used to make gates, trailers, wagon sides, athletic equipment, and other projects. Used pipe is a cheap source of construction material in many communities.

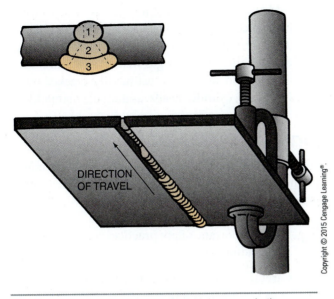

FIGURE 27-30 A weld pad showing multiple passes in the overhead position.

Pipe is made of thin metal. Therefore, pipe welding requires the use of electrodes that produce shallow penetration. The E6013 electrode is a good choice. It is a fast-filling electrode that makes it suitable when fits are not perfect.

Pipe is generally welded without grinding. As the electrode proceeds around a pipe, the rate of travel must be faster at the electrode holder than at the electrode tip. It is important to maintain the same relationship of electrode-to-weld-surface as used for straight areas (Figure 27-31). Welding pipe ends is easier if the ends are first flattened. This permits the welder to make straight passes.

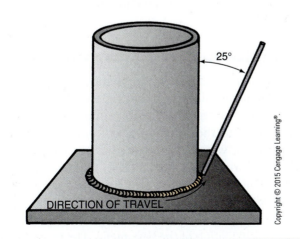

FIGURE 27-31 Making a fillet weld on pipe.

CAUTION!

Do not weld galvanized pipe due to the hazard of poisonous fumes except under the supervision of an expert.

PIERCING, CUTTING, AND SPOT WELDING

Piercing and cutting with arc welders are not as fast and clean as with an oxyacetylene torch. However, it is a useful procedure for those who have arc welders but not an oxyacetylene torch.

Piercing

To pierce a hole with an arc welder, the following procedure is recommended.

Procedure

1. Clamp a piece of ¼" × 2" × 6" steel so it projects 4 inches from the tabletop.

2. Place a bucket half filled with water or a sandbox under the metal.

3. Put on protective clothing, including leather shoes and a leather apron and/or leggings.

4. Determine the correct amperage for welding, and increase this value by 50 percent. Set the amperage at the increased value.

5. Select a ⅛-inch E6011 electrode or a special cutting electrode.

6. Strike the arc above the metal and hold a long arc (about ¼ to ⅛ inch) until the metal is molten.

7. Rotate the tip of the rod to create an enlarged molten puddle.

8. Quickly push the electrode into the molten puddle to create a hole. If the rod does not go through, quickly withdraw the electrode or whip it loose if it sticks. Start again, but make the molten puddle deeper before thrusting the electrode through the metal.

9. Keeping the arc burning, rotate the tip of the electrode around the hole to enlarge it as needed.

(continued)

Procedure, *continued*

10. If a smooth hole is required, quickly place the hole in the metal over the hole in an anvil. Then drive a blacksmith's punch into the hole while the metal is still very hot.

11. Smooth the underside of the piece with a grinder, as needed.

NOTE

If additional amperage gives better results, use a higher setting. However, excessive amperage burns and overheats the rod and creates a heavier load on the machine.

Cutting

To cut with an arc welder, the following procedure is recommended.

Procedure

1. Obtain a piece of ¼" × 2" × 6" steel.

2. Use an E6011 electrode or a special cutting electrode.

3. Set the amperage 50 percent higher than for welding.

4. Strike the arc and direct a long arc at the edge of the metal until it becomes molten.

5. Use a quick up-and-down chipping movement to gouge out the molten metal and create a kerf across the metal.

6. Smooth the cut with a grinder, if needed.

GMAW WELDING

Gas metal arc welding (GMAW) became popular when manufacturers began using thin-gauge, high-strength, low-alloy (HSLA) steels. Manufacturers insisted that the only correct way to weld HSLA and other thin-gauge steel was with a GMAW system. Once the GMAW welder was in place, it was easy to see that it provided clean, fast welds for all applications.

GMAW welding is ideal for exhaust system work, repairing mechanical supports, installing trailer hitches and truck bumpers, and any other welds that could be done with either an arc or gas welder. In addition, it is possible to weld aluminum castings like cracked transmission cases, cylinder heads, and intake manifolds.

Safety in GMAW Welding

GMAW welding involves the use of electrical equipment as well as gas welding equipment. Therefore, a study of both gas and electrical arc welding should be completed before doing GMAW and GTAW welding. Some important safety practices to observe when GMAW welding are:

1. Carefully follow the manufacturer's recommendations provided with the welder.

2. Check all welding cables to be sure they are in good repair and properly connected. Be sure the equipment is properly grounded.

3. Wear leather gloves, body jackets, and chaps for protection against burns.

4. Use an approved helmet with a minimum #11 shaded lens for nonferrous and #12 for ferrous metals.

5. When the electrical switch is on, never touch electrical connections or the welding wire.

6. Never weld in wet locations or with wet hands, feet, or clothing.

7. Be sure there are no matches or other flammable materials in your pockets. They could ignite.

8. Handle hot metal with pliers or tongs.

9. Weld only in well-ventilated places.

10. Have only competent professionals do repair work on welding equipment.

11. When finished welding, be certain the welding equipment is turned off and safely stored.

Principles and Characteristics

The GMAW welding method uses a welding wire that is fed automatically at a constant speed as an electrode. A short arc is generated between the base metal and the wire, and the resulting heat from the arc melts the welding wire and joins the base metals together. Since the wire is fed automatically at a constant rate, this method is also called semiautomatic arc welding. During the welding process (Figure 27-32), either inert or active gas shields the weld from the atmosphere and prevents oxidation of the base metal. The type of inert or active gas used depends on the base material to be

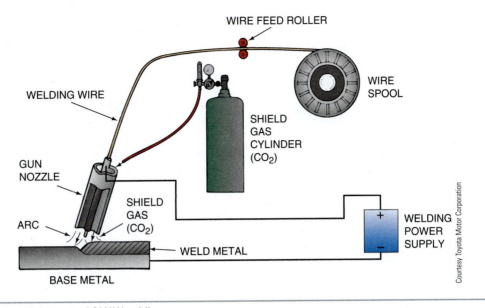

WIRE FEED ROLLER

WELDING WIRE

WIRE SPOOL

SHIELD GAS CYLINDER (CO_2)

GUN NOZZLE

SHIELD GAS (CO_2)

ARC

WELD METAL

WELDING POWER SUPPLY

BASE METAL

Courtesy Toyota Motor Corporation

FIGURE 27-32 The principle features of GMAW welding.

welded. For most steel welds, carbon dioxide (CO_2) is used (Figure 27-33). With aluminum, either pure argon gas or a mixture of argon and helium is used, depending on the alloy and the thickness of the material. It is even possible to weld stainless steel by using argon gas with a little oxygen (between 4 and 5 percent) added.

The use of carbon dioxide or oxygen in arc welding is sometimes called metal inert gas (MIG) welding. Actually, MIG welding uses a fully inert gas such as argon or helium as a shield gas, so the term MIG is incorrect. Since carbon dioxide gas is not a completely inert gas, it is more accurately called GMAW welding. The term GMAW is used to describe all gas metal arc welding processes. In fact, many welders on the market can use

carbon dioxide (a semi-active gas) or argon (inert gas) by simply changing the gas cylinder and the regulator.

GMAW uses the short circuit arc method, which is a unique method of depositing molten drops of metal onto the base metal. Welding of thin sheet metal can cause welding strain, blow holes, and warped panels. To prevent these problems, it is necessary to limit the amount of heat near the weld. The short circuit arc method uses very thin welding rods, a low current, and low voltage. By using this technique the amount of heat introduced into the panels is kept to a minimum and penetration of the base metal is quite shallow.

As shown in Figure 27-34, the end of the wire is melted by the heat of the arc and forms into a drop, which then comes in contact with the base metal and creates a short circuit. When this happens, a large current flows through the metal and the shorted portion is torn away by the pinch force or burnback, which reestablishes the arc. That is, the bare wire electrode is fed continuously into the weld puddle at a controlled, constant rate, where it short-circuits, and the arc goes out. While the arc is out, the puddle flattens and cools, but the wire continues to feed, shorting to the workpiece again. This heating and cooling happens on an average of 100 times a second. The metal is transferred to the workpiece with each of these short circuits. Generally, if current is flowing through a cylindrically shaped fluid (in this case molten metal) or current is flowing through an arc, the current is pulled toward the weld. This works as a constricting force in the direction of the center of the cylinder. This action is known as the pinch effect, and the size of the force is called the pinch force (Figure 27-35).

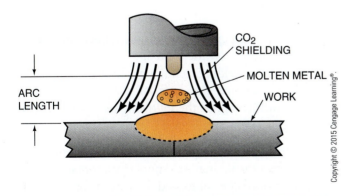

CO₂ SHIELDING

MOLTEN METAL

ARC LENGTH

WORK

Copyright © 2015 Cengage Learning®.

FIGURE 27-33 During the GMAW welding process, carbon dioxide (CO_2) protects the molten metal from contamination by the atmosphere.

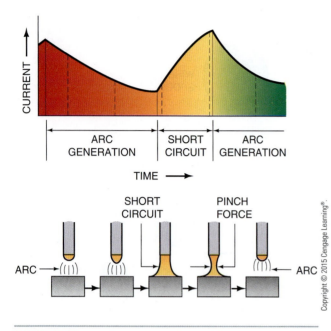

FIGURE 27-34 How the short circuit arc method works.

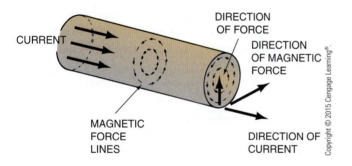

FIGURE 27-35 Typical pinch force and how it is formed.

In summary, the GMAW process works like this:

- At the weld site, the wire undergoes a split-second sequence of short-circuiting, burnback, and arcing (Figure 27-36).
- Each sequence produces a short arc transfer of a tiny drop of electrode metal from the tip of the wire to the weld puddle.
- A gas curtain or shield surrounds the wire electrode. This gas shield prevents contamination from the atmosphere and helps to stabilize the arc.
- The continuously fed electrode wire contacts the work and sets up a short circuit. Resistance heats the wire and the weld site.
- As the heating continues, the wire begins to melt and thin out or neck down.

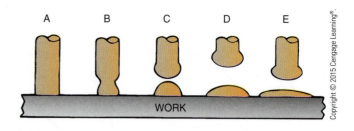

FIGURE 27-36 Typical action of the welding wire as it burns back from the work during the GMAW welding process.

- Increasing resistance in the neck accelerates the heating in this area.
- The molten neck burns through, depositing a puddle on the workpiece and starting the arc.
- The arc tends to flatten the puddle and burn back the electrode.
- With the arc gap at its widest, it cools, allowing the wire feed to move the electrode closer to the work.
- The short end starts to heat up again, enough to further flatten the puddle but not enough to keep the electrode from recontacting the workpiece. This extinguishes the arc, reestablishes the short circuit, and restarts the process.
- This complete cycle occurs automatically at a frequency ranging from 50 to 200 cycles per second.

GMAW Welding Equipment

Most GMAW welding equipment is semiautomatic. This means the machine's operation is automatic, but the gun is hand-controlled. Before starting to weld, the operator sets:

- Voltage for the arc
- Wire speed
- Shielding gas flow rate and presses the power button

Then the operator has complete freedom to concentrate entirely on the weld site, the molten puddle, and whatever welding technique is used.

Regardless of the type of GMAW equipment used, it will comprise the following basic components (Figure 27-37):

- Supply of shielding gas with a flow regulator to protect the molten weld pool from contamination
- Wire/feed control to feed the wire at the required speed
- Spool of electrode wire of a specified type and diameter

FIGURE 27-37 A modern GMAW welder.

• GMAW type of welder machine connected to an electrical power supply
• Work cable and clamp assembly
• Welding gun and cable assembly that the welder holds to direct the wire to the weld area

GMAW spot welding is termed consumable spot welding because the welding wire is consumed in the weld puddle. Consumable spot welds can be made in a variety of methods and in all positions using various nozzles supplied with this option.

In spot welding different thicknesses of materials, the lighter gauge material should always be spotted to the heavy material.

Spot welding usually requires greater heat to the weld than continuous or pulse welding. It is best to use sample materials when setting the controls for spot welding. To check a spot weld, pull the two pieces apart. A good weld will tear a small hole out of the bottom piece. If the weld pulls apart easily, increase the weld time or heat. After each spot is complete, the trigger must be released and then pulled for the next spot.

GMAW spot has the advantage of an easily grindable crown. The procedure does not leave any depression requiring a fill.

The pulse control allows continuous seam welding on the material, with less chance of burn-through or distortion. This is accomplished by starting and stopping the wire for preset times without releasing the trigger. The weld "on" and "off" time can be set for the operator's preference and metal thickness.

The burnback control on most GMAW equipment gives an adjustable burnback of the electrode to prevent it from sticking in the puddle at the end of the weld.

In GMAW welding, the polarity of the power source is important in determining the penetration to the workpiece. DC power sources used for GMAW welding typically use DC reverse polarity. This means the wire (electrode) is positive and the workpiece is negative. Weld penetration is greatest using this connection.

Weld penetration is also greatest using CO_2 gas. However, CO_2 gives a harsher, more unstable arc, which leads with increased spatter. So when welding on thin materials, it is preferable to use argon/CO_2 (Figure 27-38). If the material is extremely thin, weld in straight polarity. In straight polarity, the wire (electrode) would be negative and the workpiece would be positive. This would put more heat in the wire, providing less penetration. The disadvantage of using straight polarity would be a high rope-like bead, requiring more grinding.

Voltage adjustment and wire-feed speed must be set according to the diameter of the wire being used. It should be noted that when setting these parameters, manufacturers' recommendations should be followed to reach approximate settings. When rough parameters

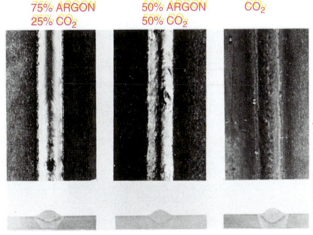

FIGURE 27-38 Typical GMAW welding bands produced using different shielding gases.

Courtesy of The Lincoln Electric Co., Cleveland, OH

Copyright © 2015 Cengage Learning®.

are selected, change only one variable at a time until the machine is fine-tuned for an optimum welding condition. GMAW welders can be tuned in using both visual and audio signals.

GMAW Weld Defects

Defects in GMAW welds and their causes are summarized in Figure 27-39. Proper welding techniques ensure good welding results. If welding defects should occur, think of ways to change their method of operation to correct the defect.

When making any GMAW repairs, the materials and panels must be similar enough to allow mixing when they are welded together. The melting and flowing of metals can be accomplished by many methods, depending on the materials being joined. The combinations of cleanliness of the welded area, the mixing of proper metals, and the right heat application will result in a good GMAW weld.

Defective Weld Conditions and Their Main Causes			
Welding Defect	**Defect Condition**	**Remarks**	**Main Causes**
Pores/Pits		A hole is made when gas is trapped in the weld metal.	1. There is rust or dirt on the base metal. 2. Rust or moisture is adhering to the wire. 3. Shielding action is improper (the nozzle is blocked, or wind or the gas flow volume is low). 4. The weld is cooling off too fast. 5. The arc length is too long. 6. The wrong wire is being used. 7. Gas is sealed improperly. 8. The weld joint surface is not clean.
Undercut		The overmelted main metal has made grooves or an indentation; its section is made smaller, and therefore the weld zone's strength is severely lowered.	1. The arc length is too long. 2. The gun angle is improper. 3. The welding speed is too fast. 4. The current is too large. 5. The torch feed is too fast. 6. The torch angle is tilted.
Improper Fusion		This is an unfused condition between the weld metal and base metal or between deposited metals.	1. The torch feed operation is faulty. 2. The voltage is too low. 3. The weld area is not clean.
Overlap		More apt to occur in a fillet weld than in a butt weld, this condition causes stress concentration and leads to premature corrosion.	1. The welding speed is too slow. 2. The arc length is too short. 3. The torch feed is too slow. 4. The current is too low.
Insufficient Penetration		An insufficient deposition of weld is being made under the panel.	1. The welding current is too low. 2. The arc length is too long. 3. The end of the wire is not aligned with the butted portion of the panels. 4. The groove face is too small.

FIGURE 27-39 Defective weld conditions and their likeliest causes in the GMAW system.

Copyright © 2015 Cengage Learning®.

Defective Weld Conditions and Their Main Causes

Welding Defect	Defect Condition	Remarks	Main Causes
Excess Weld Spatter		Weld spatter occurs as speckles and bumps along either side of the weld bead.	1. The arc length is too long. 2. Rust is on the base metal. 3. The gun angle is too severe.
Spatter (short throat)		Spatter is prone to occur in fillet welds.	1. The current is too great. 2. The wrong wire is selected.
Vertical Crack		Cracks usually occur on top surface only.	There are stains on welded surface (paint, oil, rust).
Bead Is Not Uniform		The weld bead is misshapen and uneven rather than streamlined and even.	1. The contact tip hole is worn or deformed and the wire is oscillating as it comes out of the tip. 2. The gun is not steady during welding.
Burnthrough		Holes are burned into the weld bead.	1. The welding current is too high. 2. The gap between the metal is too wide. 3. The speed of the gun is too slow. 4. The gun-to-base metal distance is too short.

FIGURE 27-39 (continued)

Gas Tungsten Arc Welding (GTAW)

The welding process known as **gas tungsten arc welding (GTAW)** is sometimes referred to as Tungsten Inert Gas welding or Heli-Arc welding. It was developed by the aircraft industry to weld very light, thin metals such as magnesium, which forms the frame and skin of some airplanes. The first GTAW welders used a direct current (DC), but later technology developed a type of alternating current (AC) that produced better welds than those made with DC currents. GMAW welders lay down weld beads at the average of 25 inches per minute. GTAW welding is much slower, with weld speeds ranging between 5 to 10 inches per minute. However, this slower speed gives much more control, and the end result is the best-looking weld obtainable. A GTAW unit can be used to repair cracks in aluminum cylinder beads and reconstruct combustion chambers and other automotive components that need to be welded.

Like GMAW welding, GTAW welders use an inert gas such as argon or helium to surround the weld area and prevent oxygen and nitrogen in the atmosphere from contaminating the weld (Figure 27-40). Instead of having a wire-feed welding electrode, like GMAW units, GTAW machines use a tungsten electrode with a very high melting point (about 6,900°F) to strike an arc between the welding gun and the work.

Since the tungsten electrode has such a high melting point, it is not consumed during the welding process, so a filler rod must be used when welding thicker materials. Because the torch is held in one hand and the filler rod is held in the other (Figure 27-41), GTAW welding is similar in some ways to oxyacetylene welding.

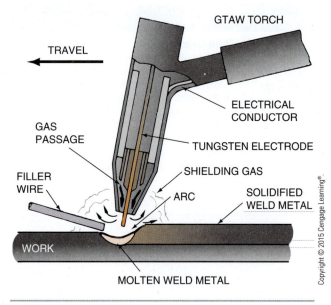

FIGURE 27-40 Typical action of welding wire as it burns back from the work during the GMAW welding process.

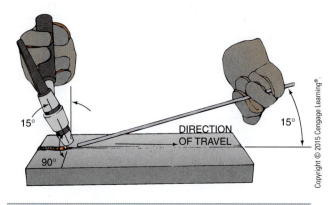

FIGURE 27-41 Proper position of the torch and of the filler rod for manual GTAW welding.

Safety in GTAW Welding

When doing GTAW welding, all safety practices listed earlier under "Safety in GMAW Welding" should be applied. Some major safety procedures to apply when GTAW welding are:

1. Observe all safety procedures that apply for GMAW welding.
2. Wear hearing protection with pulsed power and high-current settings to protect against sound waves from the arc during high-current pulses.
3. Never touch the tungsten electrode with the filler rod or body parts, since the tungsten electrode carries electrical current.

4. Be careful to keep the GTAW high-frequency unit setting within the limits prescribed by the manufacturer.

GTAW Equipment

Three major components make up a GTAW welding machine: the power supply, the welding gun, and the gas cylinder with flowmeter (Figure 27-42). Most GTAW units have a sophisticated power supply system (Figure 27-43) that can supply current to the electrode

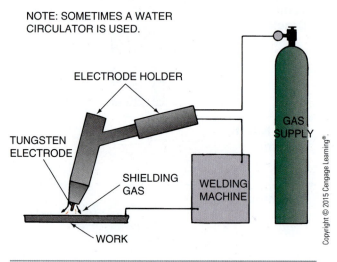

FIGURE 27-42 Components of a manual GTAW welding outfit.

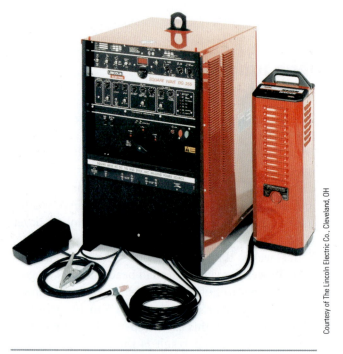

FIGURE 27-43 Typical GTAW welding machine.

as alternating current (AC), direct current straight polarity (DCSP), or direct current reverse polarity (DCRP). The type of current used depends on the type of material to be welded as well as the desired shape and penetration of the weld bead.

Welding with AC is primarily used for nonferrous metals like aluminum. The two DC currents are used primarily for welding ferrous metals like steel and cast iron. With DCSP, electron flow at the arc is from the electrode to the workpiece. The electrons strike the weld at a high rate of speed, which builds heat at the weld, resulting in a narrow weld with a high degree of penetration.

With DCRP, the electron flow is reversed, and heat builds up in the electrode instead of the weld. To handle the extra heat, a larger diameter electrode is needed for DCRP welding. The resulting weld bead tends to be wide, with minimum penetration.

The welding gun holds the tungsten electrode with a collet that screws into the body of the gun (Figure 27-44). The gas nozzle consists of a ceramic cup that blows the shielding gas around the weld area. The cups are available in various sizes and flow rates to provide optimum shielding for different types of welds. Some guns also have a screened gas nozzle to eliminate troublesome turbulence, which can interfere with the shielding gases' ability to surround the weld.

Some GTAW welders have a trigger on the gun to control current and gas flow. Other models have a foot control. When using filler rod, GTAW welding becomes a two-handed procedure, and a foot-operated trigger is a handy thing to have. Electrodes are available in diameters of $\frac{1}{16}$, $\frac{3}{32}$, and $\frac{1}{8}$ inch. The correct electrode diameter is based on the thickness of the material and the amount of current used. When DCRP or DCSP welding is used, the electrode must be ground to a point. For AC welding, the tip is rounded off.

Plasma Arc Welding

Plasma arc welding was developed in the 1960s for use in specialty and precision welding applications. The process works by creating an ionized gas called plasma. Many common materials, such as water and most gases, have three states: solid, liquid, or gas. The state the material is in depends on how much heat is applied to the material. For example, water is in a solid state (ice) when the temperature is below 32°F at atmospheric pressure. When heat is applied and the temperature rises above the freezing point, it turns to a liquid. If more heat is applied, it turns to a gas (steam) when the temperature rises above 212°F. If enough heat is applied, the steam will become super heated and convert to a gas that will conduct electricity. This gas is called plasma and may be thought of as a sort of fourth state.

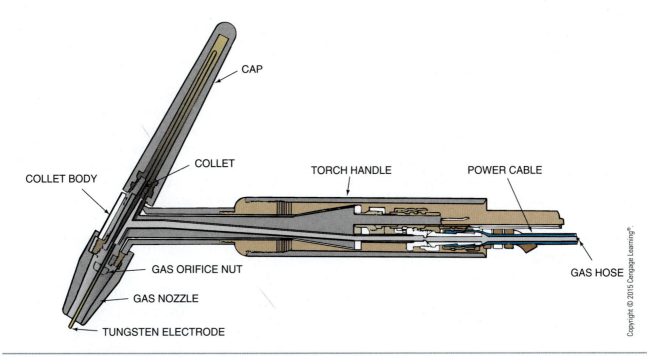

CAP

COLLET

TORCH HANDLE

POWER CABLE

COLLET BODY

GAS ORIFICE NUT

GAS NOZZLE

TUNGSTEN ELECTRODE

GAS HOSE

Copyright © 2015 Cengage Learning®

FIGURE 27-44 *Cross section and parts of a GTAW welding torch.*

FIGURE 27-45 Plasma arc welding works by creating an ionized gas called plasma.

In the case of plasma arc welding, an inert gas called argon is heated until it converts into plasma, becomes ionized, and conducts electricity (Figure 27-45). In this process, a tungsten electrode begins an electric arc when electric current is passed through it. The plasma gas created by heating the argon is forced through a small opening that surrounds the tungsten electrode tip. The plasma conducts electricity that in turn produces a very intense heat in a small, very concentrated area. Unshielded arc welding creates a temperature of about 11,000°F, but a plasma arc creates a temperature of about 43,000°F. The metal to be welded is melted and fused together very quickly. A secondary shield of argon or helium gas protects the molten puddle of metal.

The advantage of plasma arc welding is that it provides a very stable arc using a small amp current. This allows the welding of fine wire and instruments smaller than a needle. Also, there is very little metal distortion with the plasma arc welding process.

AUTOMATED AND ROBOTIC WELDING

The newest processes in welding involve the use of computer controls and robots. These methods are usually used on assembly lines in shops and factories

FIGURE 27-46 Automated or robotic welding uses a computer program to guide the movement of a welding arm.

where many of the same type of welds are used. In automated welding, a computer is programmed to guide the movement of a welding arm (Figure 27-46). The weld is done in exactly the same manner each time until the computer controls are reprogrammed. This process is very precise, efficient, and produces high-quality welds.

Robotic welding uses several steps in the manufacturing process. Only one of the steps includes welding. For instance, the robot grasps the components to be welded, positions them, and welds the pieces together. The robot then places the welded part on a conveyor. This is a very expensive process that is reserved for high-output productions.

SUMMARY

There are several types of arc welding. All involve the use of an electric current that melts metal. The correct type depends on the type of metal being welded and the use of the project. Each method has a specific set of procedures for making the proper weld. The correct type of welder, electrode, and technique should result in strong, attractive welds.

Student Activities

1. Define the Terms to Know for this unit.

2. In the welding area of the shop, examine all the protective clothing that is available.

3. Examine all fire protective equipment in the welding area. Report any irregularities to your instructor.

4. Ask your instructor to provide samples of weld showing good beads and poor beads (due to improper welding).

5. Using a ¼" × 4" × 6" welding pad, make 16 beads, 1 inch long each (four rows of four beads each).

6. Fill a ¼" × 4" × 6" welding pad with practice beads, ¾ inch apart.

7. Using two pieces of ¼" × 2" × 6" steel, make a butt weld 6 inches long with an even bead.

8. Make an internal check of the bead from activity 6 by examining a cross section of a 1-inch piece sawed from the assembly. Bend open the 1-inch weld and analyze the bead.

9. Weld a ⅛" × 1" × 6" piece of steel to a ¼" × 2" × 6" piece of steel, with a fillet weld on each side.

10. Run a straight, 6-inch horizontal weld across a ⅛" or ¼" × 4" × 6" welding pad.

11. Make a 2-inch vertical down butt weld.

12. Make a 2-inch vertical up butt weld.

13. Make a 2-inch vertical up fillet weld.

14. Pierce a 1-inch hole in a ¼" × 2" × 6" pad.

15. Cut three 1-inch strips from the pad used in activity 14, using an electric welder and electrode.

16. Observe GMAW and GTAW welding demonstrations.

17. Perform GMAW and GTAW welding.

Relevant Web Sites

TheFabricator.com, a publication of the Fabricators & Manufacturers Association, Int.®
www.thefabricator.com

American Metallurgical Consultants, website on Welding Procedures & Techniques
www.weldingengineer.com

Self-Evaluation

A. Multiple Choice. Select the best answer.

1. The temperature of an electric welding arc is about
   a. 400°F
   b. 840°F
   c. 1,800°F
   d. 9,000°F

2. Welding tables should be made of
   a. concrete
   b. Masonite
   c. metal
   d. wood

3. Fire extinguishers for welding areas should be suitable for
   a. Class A fires
   b. Class B fires
   c. Class C fires
   d. all of these

4. Burning clothes on a human should be extinguished with
   a. a fire blanket
   b. a fire extinguisher
   c. sand
   d. any of these

5. Water in a welding area is useful for
   a. receiving sparks from piercing
   b. extinguishing fires
   c. cooling metal
   d. all of these

6. Injury to eyes can result from
   a. chipping without goggles
   b. viewing welding without shielding
   c. welding with less than a No. 10 lens
   d. all of these

7. If only one kind of electrode for all arc welding is to be purchased, the best choice is an
   a. E6010
   b. E6011
   c. E6013
   d. E7018

8. For most welding in agricultural mechanics, the best electrode size is
   a. $\frac{1}{16}$ inch
   b. $\frac{1}{8}$ inch
   c. $\frac{3}{16}$ inch
   d. $\frac{1}{4}$ inch

9. Correct arc length is approximately
   a. $\frac{1}{8}$ inch
   b. $\frac{1}{4}$ inch
   c. $\frac{3}{8}$ inch
   d. $\frac{1}{2}$ inch

10. When welding, the operator sees by
    a. daylight
    b. fluorescent light
    c. light from the arc
    d. all of these

11. The appearance and strength of a bead are influenced by
    a. amperage
    b. angle
    c. speed
    d. all of these

12. The recommended position of the welder for horizontal and vertical welding is
    a. standing
    b. sitting
    c. the most comfortable one
    d. lying flat

13. The recommended weave pattern for the beginning welder doing down hand welding is
    a. circular
    b. figure eight
    c. J
    d. T

14. A second pass should never be done if
    a. the first was a poor weld
    b. the joint was ground
    c. the slag has been removed
    d. the slag has not been removed

15. In metal, the most rapid movement of heat is
    a. down
    b. equal in all directions
    c. horizontal
    d. up

16. GMAW welding is especially useful for welding
    a. aluminum castings
    b. thin steel
    c. exhaust system parts
    d. all of these

17. The welding process that uses an automatic wire feed is
    a. GMAW
    b. oxyacetylene
    c. stick
    d. GTAW

B. Matching. Match the items in column I with those in column II.

Column I

1. lens for welding
2. lens for chipping
3. leather
4. E6013
5. E6011
6. root pass
7. vertical up
8. vertical down
9. J weave
10. electrode for pipe

Column II

a. protective clothing for welders
b. electrode for thin metal
c. electrode for most welding
d. first pass
e. procedure results in a weld with shallow penetration
f. clear
g. relatively difficult
h. useful technique when welding metal of different thickness
i. E6013
j. No. 10

C. Completion. Fill in the blanks with the word or words that make the following statements correct.

1. The _____ weld is the most difficult weld to make.

2. Making a hole with an electrode is called _____.

3. The best amperage setting for cutting is _____ percent higher than for welding.

4. The correct motion for cutting is _____.

5. GMAW stands for _____.

6. GTAW stands for _____.

D. Brief Answer. Briefly answer the following questions.

1. Describe the standard protective clothing, headgear, and hand protection to be used when arc welding.

2. What is the range of appropriate welder settings? What is the recommended starting setting for $\frac{1}{8}$ inch E6011 or E6013 electrodes?

3. Identify six indicators of excessive heat. If these conditions exist, what should one do?

4. What is a crater? What techniques allow it to fill when one stops the bead?

5. When vertical up welding, what adjustments in technique and amperage may be helpful in controlling the puddle?

6. GMAW welding safety procedures are also applied when GTAW welding, but three additional safety procedures should be observed. What are these?

Section 9

PAINTING

Objective

To prepare wood and metal for painting.

Competencies to be developed

After studying this unit, you should be able to:

- Prepare unpainted wood for painting.
- Prepare previously painted wood surfaces for repainting.
- Steam clean machinery.
- Prepare unpainted metal for painting.
- Remove rust and scale from metal surfaces.
- Feather chipped paint on metal surfaces.
- Mask tractors and other machinery for spray painting.
- Estimate materials for paint jobs.
- Paint buildings and machinery.

Materials List

- Paint can label
- Mineral spirits
- Rags
- Putty
- Scraper and wire brush
- Assortment of emery cloth and sandpaper
- Silicon carbide paper, 200- and 400-grit
- Window sash in need of repainting

continued

Terms to Know

- waterproof
- paint
- pigment
- vehicle
- paint film
- wood preservatives
- new wood
- new work
- seal
- sealer
- rot
- warp
- air-dried lumber
- old work
- sandblasting
- primer
- caulk
- steam cleaner
- feather

Materials list, *continued*

- Wooden project to be painted
- Wooden project to be repainted
- Piece of machinery that needs repainting
- Portable power wire wheel
- Steam cleaner

Wood, steel, and concrete are the primary materials used to construct agricultural buildings and equipment. Wood may rot and steel will rust if not protected from moisture. Concrete is not damaged by moisture in the air, but may need special treatment to make it waterproof. **Waterproof** means that water cannot enter the material.

Most wood exposed to the weather must be painted or treated with a preservative. **Paint** is a substance consisting of pigment suspended in a liquid known as a vehicle. A **pigment** is a solid coloring substance. The liquid is called a **vehicle** because it carries the pigment.

Paint must be applied in thin layers so some of the vehicle can evaporate. As the vehicle evaporates, the paint is said to be drying. After paint dries, the material that is left is called a **paint film**. Wood and other materials must be properly prepared or the paint film will peel off rather than wear off. In such cases, the paint film first dries out excessively. Hairline cracks then start to appear. The film then breaks and starts to curl away from the wood. Finally, the paint film separates from the wood, and leaves the surface exposed (Figure 28-1).

EXPOSED WOOD ⌐   PAINT CURLING
                                AWAY ⌐

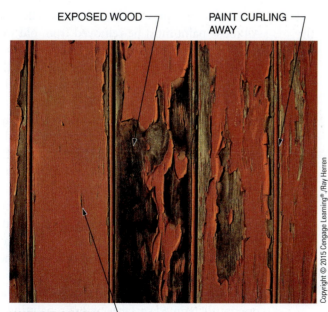

HAIRLINE CRACKS ⌐

Copyright © 2015 Cengage Learning®./Ray Herren

FIGURE 28-1 Three stages of paint peeling.

PRESERVATIVES

Wood may be covered with a liquid that kills wood-rotting fungus and wood-eating insects such as termites. Such liquids are called **wood preservatives**. These chemicals are not usually applied by hand. Wood is preserved by treating it with chemicals in large pressurized vats and is done commercially. Most home improvement centers sell preserved wood. Most often, southern yellow pine is the preferred wood for treatment. This lumber is usually green or yellow in color.

PREPARING WOOD FOR PAINTING

Wood that has never been painted or sealed is called **new wood**. Wood used in original construction is frequently referred to as **new work**. To **seal** wood means to apply a coating that fills or blocks the pores so no material can pass through the surface. The coating is called a **sealer**.

A sealer prevents moisture from entering the wood and causing it to rot or warp. To **rot** means to decay or break down into other substances. To **warp** means to bend or twist out of shape. A sealer also keeps moisture and natural liquids inside the wood. This prevents the wood from drying out and cracking.

Preparing New Wood

New wood is easy to prepare for painting. See Unit 11 ("Finishing Wood") for procedures to prepare wood for clear or stained finishes. The procedures are the same for preparing wood for painting.

FIGURE 28-2 Wood preservatives can be brushed on to help protect wood.

New wood must be dry when it is painted; that is, there should be no indication of moisture on the surface, and it should have been air dried. **Air-dried lumber** means that the boards are separated by wooden strips stacked and are then protected from rain and snow for six months or more. Air drying permits the excessive natural moisture to leave the wood. If the wood contains excess moisture, that moisture will tend to lift the paint as it tries to escape to the drier surrounding air.

The wood must be free of grease, wax, dirt, and other substances or, again, the paint will not adhere to the wood. Grease and oil are removed by wiping the wood with a cloth soaked in mineral spirits. Wax is removed by washing the wood with a solution of household ammonia. Dirt from shoes and other sources is removed by using a stiff bristle brush.

Planing of rough lumber greatly reduces the amount of paint needed to cover the surfaces. No sanding is needed except to remove objectionable roughness or to remove grease, wax, and other residues.

Bark must be removed before painting or planting posts and poles. Bark holds moisture and also serves as a home for disease organisms and insects that destroy wood. Posts may be dried by standing them on end around a tree trunk or other vertical support so air can flow around each one. Bark can then be removed with a draw knife or axe. After the bark is removed, the post must be left until all surfaces are thoroughly dry before painting is attempted.

Preparing Old Work

Wood that was previously painted is called **old work**. All loose paint must be removed from old work. This is done with a scraper, wire brush, and sandpaper (Figure 28-3). The job is made easier with the use of abrasive flaps, wire wheels, and discs attached to a portable power drill (Figure 28-4). If a smooth paint job is needed, the paint must be removed by a chemical paint remover or a torch and a scraper, or by complete sanding. When removing old paint, caution must be taken not to breathe any of the dust from the old surface. Not only is all dust harmful to your lungs, but a lot of the older paints also contain lead. This substance is toxic and care should be taken not to inhale any dust or to come in contact with the residue. Always wear a respirator and a disposable set of coveralls when there is any possibility that the old paint contains lead.

Paint manufacturers recommend the washing of previously painted surfaces before repainting. This can be done with a cloth and household detergent in water. The detergent removes grease, dirt, and weathered

FIGURE 28-3 A wire brush is useful for removing loose paint.

FIGURE 28-5 Bare wood must be coated with primer before paint is applied.

paint. A uniform, clean surface is left to which the new paint will adhere.

High-pressure water cleaners are frequently used to clean floors, walls, siding, and concrete surfaces. Care must be used to avoid injury to bystanders or stripping of paint if repainting is not intended. **Sandblasting** where fine sand particles are thrown by compressed air is especially useful for cleaning metal and masonry.

FIGURE 28-4 A power drill can be equipped with a wire wheel, abrasive flaps, or a sanding disc to remove peeling paint quickly.

A final step in preparing old work is to fill all holes, caulk all cracks, and seal any problem spots. If the holes and cracks are not sealed before filling, the liquids in the filling material will be absorbed by the dry wood. This means that the filler will shrink and loosen, causing the paint to crack and peel around the filler.

A **primer** is a special paint used to seal bare wood or metal (Figure 28-5). Paint will not adhere to bare wood or metal, so a primer must be applied to the bare surface. Paint will then adhere to the primer. Some of the newer paints contain a primer with the paint so the base coat of primer is not necessary. Primer should be applied to the wood with a brush so the material can be worked into holes, cracks, and crevices. A thin coat is recommended because the job of the primer is to penetrate the wood as it is applied. When the primer dries, the pores are sealed and only a thin film is left on the surface. Knots or other problem areas that seem to absorb too much of the primer should be sealed with a primer containing shellac.

After priming, all cracks must be filled and the joints between materials must be caulked. **Caulk** is a material that stretches, compresses, and rebounds as materials expand and contract. This characteristic permits it to bridge the air gap between two materials such as brick and wood (Figure 28-6).

Holes must be filled and window glass secured with putty or glazing compound on outdoor work. Plaster-like fillers may be used for indoor surfaces. Putty and glazing compound are soft materials containing oils that help keep them pliable over a long period of time (Figure 28-7). Pliable means it will give if pushed or pulled. Putty is workable and may be pushed into holes and formed into special shapes. An example is the

FIGURE 28-6 Caulking between two different materials or different pieces of wood permits them to expand and contract without breaking the paint film. A rag is used to smooth the caulking bead.

triangular bead of putty used to seal glass in a window (Figure 28-8). Priming followed by applying new putty or glazing compound are important steps in repainting window sashes.

Scraping, brushing, washing, sanding, priming, caulking, and filling should prepare any old work for painting.

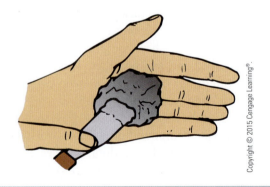

FIGURE 28-7 Putty and glazing compound are soft materials that contain oils to keep them pliable over long periods of time. Working the material into the hand warms it and makes it more pliable prior to application.

FIGURE 28-8 Window glass is sealed to wood or metal frames with a bead of putty that is smoothed into a triangular shape. The putty on this window should be replaced before the window is painted.

STEAM CLEANING MACHINERY

Tractors, engines, and machinery may be cleaned quickly and thoroughly with a **steam cleaner** or high-pressure washer. A steam cleaner is a portable machine that uses water, a pump, and a burner to produce steam (Figure 28-9). Special chemicals in solution are added to the steam to dissolve grease and remove paint, dirt, manure, and other materials from metal, wood, and concrete.

A steam cleaner is powerful enough to strip paint from the metal if the appropriate chemical solution and pressure are used. This combination of heat and pressure and chemicals used can make the steam cleaner hazardous in the hands of careless operators.

Courtesy of Landa Water Cleaning Systems

FIGURE 28-9 A high-pressure washer can remove grease and dirt from machinery. A clean machine can be painted or repaired more easily.

High-pressure washers have almost replaced steam cleaners for use in general cleaning for paint preparation. In conditions where there is not very heavy grease, pressure washers are preferred over steam cleaners. A high-pressure washer uses a pump to put water under high pressure (around 150 psi to around 3,500 psi). The stream of pressurized water cleans away oil, dirt, and grease.

CAUTION!

Steam cleaners can cause serious burns to the operator or bystanders if not used carefully. The controls of the power washer will vary with the manufacturer. Read the operator's manual before using a steam cleaner.

A general procedure for operating a steam cleaner or pressure washer follows.

Procedure

1. Cover all electrical parts on the machinery to be cleaned with plastic. Food bags, garbage bags, or vinyl film can be used. Wrap the parts with string or rubber bands to hold the plastic in place.
2. Place the machine at least 15 feet from any building.

3. Plug the motor into a three-wire grounded electrical extension cord attached to a proper outlet.

CAUTION!

Do not allow the electrical plug to get wet. Place the plug up on the steam cleaner or other dry surface to keep it out of water that may accumulate in the area.

4. Attach a garden hose to the water inlet of the cleaner and to a cold water faucet. Make all connections watertight.
5. Fill the fuel tank of the cleaner with recommended fuel.
6. Mix the cleaning chemical(s) in the solution tank according to the manufacturer's instructions.

CAUTION!

Cleaning materials are powerful chemicals that may burn eyes and skin. Use the proper goggles and protective clothing.

7. Position the steam cleaner so you can observe all gauges and controls while operating the machine.
8. Unwrap and lay out the cleaner's outlet hose.

CAUTION!

Have a helper to watch the gauges and manage the controls of the machine as you perform the power wash. The helper should wear the same protective clothing as the operator.

9. Hold the handle with the nozzle pointed toward concrete or free air.

CAUTION!

Never point the nozzle toward a person. Do not allow water or steam to touch electrical cords, wires, motors, outlets, or lights.

10. Open the water faucet.

(continued)

Procedure, *continued*

11. Turn on the switch to start the steamer motor.

12. Turn on the fuel valve.

13. Adjust the fuel valve to create the pressure recommended by the manufacturer.

14. When water comes from the nozzle, turn on and adjust the solution valve to obtain the desired cleaning power.

15. Direct the water to the high parts of the machine and move downward. The nozzle is held 8 to 12 inches from surfaces needing gentle cleaning, and 2 to 4 inches from surfaces needing harsh cleaning.

> **CAUTION!**
>
> The nozzle should be positioned so water and dirt are not forced into bearings or electrical components by the steam and water.

16. Reclean the entire machine lightly to wash off all loose dirt.

Shutting down and storing the steam cleaner properly is very important. If stored incorrectly, the machine may freeze up, which will cause extensive damage. Some general recommendations for a safe shutdown are as follows.

Procedure

1. Close the solution valve.

2. Turn off the fuel.

3. Hold the nozzle until cold water comes out.

4. Turn off the motor.

5. Unplug the extension cord at the building, and then unplug the steam cleaner.

> **CAUTION!**
>
> Since the surrounding area is wet, electric shock is a real hazard, so make sure you plug the cord into a properly grounded outlet.

6. Disconnect the garden hose from the machine and hose down the work area.

7. If the machine will be exposed to freezing temperatures, straighten the hose. Use compressed air to push water from the inlet side through the coils and out through the steam nozzle at the end of the hose.

8. Store the steam cleaner with all cords and hoses protected.

9. Remove all plastic coverings from the clean machine.

10. If the machine has an engine, start the engine and let it warm up to help speed the drying process.

11. Dry the machine quickly and thoroughly.

Steam cleaning may produce moisture in the engine systems that will prevent starting. When this occurs, the trouble can usually be corrected by drying the spark plug wires and the inside of the distributor cap. A common hair dryer is effective for drying out electrical systems.

> **CAUTION!**
>
> Since steam cleaning may dissolve grease in bearings, these should be checked and repacked if necessary.

PREPARING MACHINES FOR PAINTING

Machines used in shops, buildings, and fields all need to be painted to prevent rusting. The paint film on a machine is damaged by mechanical contact, normal wear, and weathering. Such damage results in chipping, cracking, peeling, flaking, and fading of paint.

New Metal

Painting must follow new construction or repairs, such as welding, that leave metal exposed. To prepare new metal, the following steps are suggested.

Procedure

1. Chip all welds and wire-brush them thoroughly.

2. Wire-brush or sand (using emery cloth) to remove all dirt or rust from the metal.

3. Steam-clean the metal or wipe it down with mineral spirits to remove grease.

4. Clean the metal with a commercial preparatory solvent.

5. Coat the metal with a primer immediately.

Previously Painted Metal

Previously painted metal surfaces include castings, parts, assemblies, and smooth surfaces (such as guards, hoods, and fenders) where an automobile-type finish is used.

All surfaces must first be cleaned of dirt, mud, manure, and grease. Steam cleaning is the fastest and easiest method. If steaming is not possible, high-pressure water cleaning machines may be used. Or materials may be removed by hand with a solvent, a hose, water, scrapers, and brushes.

Heavy grease deposits are removed by scraping, followed by treatment with Varsol or another commercial grease solvent. Generally, high-pressure water must then be used to remove the grease and solvent from the metal. Any grease or oil film left will prevent paint from adhering to the metal.

Scaling paint and rust can be removed and rough or pitted areas can be smoothed by wire brushes, wire wheels, disc sanders, or emery paper. A combination of hand and power methods is commonly used (Figure 28-10).

Smooth surfaces such as vehicle hoods and fenders need to be stripped of all paint if they are badly

FIGURE 28-11 Chipped areas on a painted surface should be sanded smooth or feathered so that no roughness can be felt between the painted and sanded areas.

chipped. Otherwise, every chip will show in the new paint. Metal can be stripped using a paint remover or by sanding. If only a few chips exist, wet-type sandpaper can be used to **feather** the chipped areas. To feather means to sand so the chipped edge is tapered and no roughness can be felt between the painted and unpainted areas (Figure 28-11). If a chip in a painted surface is feathered properly, its edge cannot be felt.

The following procedure describes how to perform feathering.

Procedure

1. Tear a piece of 200-grit silicon carbide waterproof paper into quarters.

2. Pour about 1 quart of water into a bucket or pan.

3. Dip the paper into the water.

4. Grip the paper between your thumb and index finger so the balls of your fingers serve as a backing pad for the paper.

5. Sand across the chipped area in all directions until a feathered edge is obtained. Dip the paper into water frequently to rinse the pigment from the paper.

6. Finish with 400- or 600-grit paper to remove sanding marks and leave a smooth surface.

7. Wash the part thoroughly with water to remove all pigment.

FIGURE 28-10 A power sander can be used to remove rust from metal and to smooth rough or pitted surfaces prior to painting.

(continued)

Procedure, *continued*

CAUTION!

Paint pigment remains suspended in water, so the part or the area must be thoroughly washed and rinsed with clean water to avoid leaving a paint residue after the water dries.

8. Dry the part thoroughly or rusting will start immediately.

MASKING FOR SPRAY PAINTING

A good-looking spray job requires careful masking of all glass, chrome, and other areas that should not be painted. To mask means to cover so paint will not touch. Masking may be done with tape or paper (Figure 28-12).

Small items such as a light lens, distributor cap, and chrome trim are generally covered with masking tape. High-quality masking tape is required for good results. A sharp pocket knife is used to trim off the tape that extends beyond the part. For larger parts, paper is cut so that it is about ¼ inch short of covering the area. Then ½-inch or ¾-inch masking tape is used to finish covering the exposed area and tape the paper in place. Newspaper is frequently used for masking because of its availability and low cost. Very small areas such as grease fittings may be coated with a heavy film of general-purpose grease. When the paint job is dry, the grease and paint are wiped from the fitting.

Copyright © 2015 Cengage Learning®./Ray Herren

FIGURE 28-12 Masking tape can be used to cover areas that are not supposed to be painted.

ESTIMATING PAINT JOBS
Flat Surfaces

For most types of paint, 1 gallon will cover 400 to 500 square feet or about 100 to 125 square feet per quart. On flat surfaces such as doors, it is easy to calculate the square footage. For example, if two shed doors cover an opening 12 feet by 20 feet, the square footage is determined by multiplying the length (20 feet) by the height (12 feet). The amount is therefore 240 square feet. A quart of paint will cover at least 100 square feet, so 240 is divided by 100 to get 2.4, which is rounded to 3. Thus, three quarts of paint will do the job. However, paint generally is cheaper by the gallon. For example, a quart may cost $7 while a gallon costs $21. In that case, 3 quarts of paint in quart cans will cost as much as 4 quarts in a 1-gallon can. A good rule to use is: When the estimate indicates that 2 quarts will not be enough for the job, buy a gallon.

If the job called for two coats of paint, then the decision to buy a full gallon would be an easy one. If the gallon did not cover the entire area with two coats, an additional quart of paint could be purchased.

When estimating paint for a whole building, the square footage for all four sides plus the ceiling should be added up. The estimate for a shed 15 feet wide by 30 feet long with an average height of 11 feet is as follows:

Front = 11' × 30' = 330 square feet

Back = 11' × 30' = 330 square feet

End = 11' × 15' = 165 square feet

End = 11' × 15' = 165 square feet

Total = 990 square feet for the walls

The ceiling is 15' × 30' so 30' × 15' = 450 square feet. This added to the 990 square feet for the walls equals 1,440 square feet for the entire surface.

A gallon of paint is expected to cover 400 to 500 square feet. Dividing 1,440 square feet by 400 equals 3.6. If the building were very smooth, 3 gallons might cover the entire surface. However, an extra quart or two may be needed to finish the job. If it is a long distance to the paint store, it would be wise to buy 4 gallons. This would eliminate the chance of running out of paint and avoid having to interrupt the painting once the job is started.

For buildings with shed-type roofs, the average height of the four sides is satisfactory for estimating purposes. Buildings with gable ends require an additional amount of paint to cover the areas in the triangles of the roof. The area (A) of a triangle is determined by multiplying the length of the base (b) by the height (h) of the triangle and dividing by 2. Thus, A = ½ bh.

Spray-Painting Machinery

Machinery has many irregular surfaces. When spraying, a lot of paint is wasted. For instance, the cone of spray coming from the gun may be 12 inches wide, and the part being sprayed may be only 3 inches wide. Therefore, if the paint is distributed evenly across the 12-inch cone, only one-fourth of the paint is being used. The rest escapes into the air and settles on other surfaces.

Experience shows that 1 gallon of machinery enamel will paint a small farm tractor. Generally, 1 quart of paint is sufficient to paint the trim or to apply a second color to a tractor. One quart of enamel thinner may be sufficient to thin 5 quarts of enamel paint, but it may be wise to purchase a gallon of thinner and have some available for cleanup purposes. A cheaper solvent such as mineral spirits should be used first to clean the spray gun and equipment. Therefore, the estimate for a tractor paint job would be as follows:

> Main color 1 gallon
>
> Trim color 1 quart
>
> Enamel thinner 1 gallon
>
> Cleanup solvent 1 gallon

Estimates for other jobs can be made by comparing them with the example given. Some useful guidelines for purchasing paint are as follows:

- One quart of spray enamel will be sufficient for one piece of lawn or garden equipment.
- One gallon of enamel is needed for a small farm tractor.
- If the paint is custom-mixed, be sure to buy enough paint for the job. A second batch may not match the original batch perfectly.
- If 2 quarts are not likely to do the job, buy 1 gallon.
- Estimate your paint requirements on the high side (overestimate).
- Running out of paint during a spray paint job results in loss of time and money. Cleaning the spray equipment twice is time-consuming and requires extra solvent. A second trip to the paint store costs time and money.
- When buying paint, purchase an extra amount with the understanding that unopened cans may be returned.

SUMMARY

Without the proper finish, most wood and metal will deteriorate rapidly. To obtain the proper finish, material must be correctly prepared to accept the particular type finish to be used. Without proper cleaning and preparation, paint will not adhere to either wood or metal. Learn and follow the recommended preparation procedures before applying a finish to your projects.

Student Activities

1. Define the Terms to Know for this unit.

2. Examine some buildings and machinery for signs of chipped paint and paint that has scaled or blistered.

3. Examine the label on a can of paint. Write down the ingredients listed as pigments. Write down the ingredients listed as the vehicle. Record the percentage of the total volume made up by each ingredient.

4. Examine a window with loose and missing putty. Note the cross-sectional view of the putty. Clean, prime, and apply new putty to the window. Repaint the window when the putty hardens.

5. Prepare a wooden project for painting.

6. Prepare a wooden project or building for repainting.

7. Steam clean a piece of machinery.

8. Wire brush and clean a piece of machinery for spray painting.

9. Sand a piece of metal with chipped paint. Feather the chipped area so that its presence cannot be felt with bare fingers.

10. Mask a machine for spray painting.

11. Estimate the amount of paint needed to apply one coat to the sides and ends of a building with the following dimensions: L = 40 feet, W = 24 feet, H = 12 feet. Assume that there are no windows.

Relevant Web Sites

HGTV's Painting Guide
www.hgtv.com

Lowe's, How-To Library topic on Paint
www.lowes.com

Self-Evaluation

A. **Multiple Choice.** Select the best answer.

1. The solid substance that gives color to paint is
   a. film
   b. pigment
   c. thinner
   d. vehicle

2. The vehicle in paint is a
   a. film
   b. liquid
   c. pigment
   d. solid

3. The material that helps steam to dissolve grease is
   a. electricity
   b. fuel
   c. pressure
   d. a chemical solution

4. The first step in starting a power washer is to connect the
   a. electricity
   b. fuel
   c. solution
   d. water

5. The first step in shutting down a steam cleaner is to
   a. stop the motor
   b. shut off the fuel
   c. close the solution valve
   d. turn off the water

6. A hazard generally not associated with the steam cleaner is
   a. chemical damage
   b. electrical shock
   c. falling objects
   d. steam burns

7. Steam cleaning will not
   a. cause hard starting
   b. dry out bearings
   c. increase paint preparation time
   d. strip paint

8. Scaling paint is removed with
   a. brushes
   b. scrapers
   c. wire wheels
   d. all of these

9. A 1-gallon can of paint generally costs about the same as
   a. two 1-quart cans
   b. three 1-quart cans
   c. five 1-quart cans
   d. none of these

10. One gallon of paint is generally expected to cover
    a. 400 to 500 square feet
    b. 300 to 400 square feet
    c. 200 to 300 square feet
    d. 100 to 200 square feet

B. Matching. Match the terms in column I with those in column II.

Column I

1. new work
2. old work
3. film
4. seal
5. pressure treated
6. pliable
7. chipped paint
8. waterproof paper
9. tape and paper
10. square feet

Column II

a. dry paint
b. method for preserving wood
c. purpose of a primer
d. correct by feathering
e. previously painted
f. silicon carbide
g. length × height
h. masking
i. putty
j. never before painted

C. Completion. Fill in the blanks with the word or words that will make the following statements correct.

1. New wood must be _____ when it is painted.

2. To _____ is to sand so the chipped edge is tapered and no roughness can be felt between the painted and unpainted areas.

3. In _____, fine sand particles are thrown by compressed air to clean metal or masonry.

4. Wood and other materials must be properly prepared or the _____ _____ will peel off rather than wear off.

5. _____ with tape or paper should be done to protect any areas that should not be painted.

D. Brief Answer. Briefly answer the following questions.

1. What is a reasonable estimate of the amount of paint thinner and cleanup solvent needed to paint a small farm tractor?

2. How many quarts of paint would be needed to give a 10-foot by 10-foot door one coat?

3. The area of a triangular area to be painted is determined by doing what?

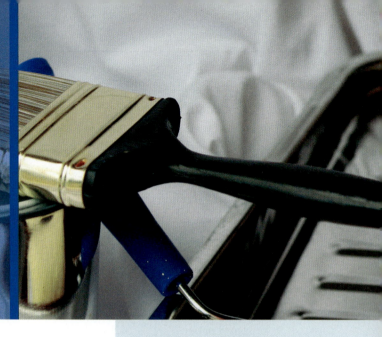

UNIT 29

Selecting and Applying Painting Materials

Objective

To select and apply paint in agricultural settings.

Competencies to be developed

After studying this unit, you should be able to:

- Select paint for wood and metal.
- Apply paint with brushes.
- Spray-paint with aerosol containers.
- Select and use spray-painting equipment.
- Clean painting equipment.

Materials List

- Labels from oil-based and latex paints
- Paintbrushes
- Paint thinner
- Latex paint
- Oil-based machinery enamel
- Spray aerosols with enamel
- Spray gun and equipment
- Paint room or outdoor area for painting
- Goggles, respirators, fire extinguishers, and other safety materials
- Cloths, waste cans, paint storage cabinet, drop cloths, and other painting equipment
- Projects to paint

Terms to Know

- enamel
- gloss
- semigloss
- flat finish
- water-based paints
- oil-based paints
- latex
- alkyd
- epoxy
- exterior
- hiding power
- lead
- titanium dioxide
- iron
- zinc
- aluminum
- calcium
- magnesium
- silicon
- asphalt
- undercoaters
- compatible
- drop cloth
- fingering
- loading the brush
- paint roller

continued

Painting may be regarded as both a luxury and a necessity. Changing the colors of the surroundings just for fun is a luxury, adding enjoyment to life. However, most objects made of steel must be painted to prevent them from rusting. Similarly, wood that is exposed to high moisture levels must be painted or otherwise protected or it may rot. In these cases, painting is a necessity. Thus, there are two major reasons for painting: to improve appearance and to preserve.

Unit 11 ("Finishing Wood") discusses thinners and solvents, procedures for painting with brushes, and brush cleaning and storage. Unit 28 ("Preparing Wood and Metal for Painting") discusses the stages of prep work that are necessary before successful painting can occur.

SELECTING PAINT

The chemistry of paint is complex. Therefore, the recommendations of the manufacturer must be relied upon heavily when choosing paints. However, some basic materials used in paints must be recognized to choose the right paint for the job and to buy paints that are good value for the money.

Types of Finishes

Paint with a gloss or semigloss finish is referred to as **enamel**. **Gloss** means shiny. Paint with a slight shine is called **semigloss**. Most enamel is formulated to be used either inside or outside. Enamels are resistant to wear. There are many different types of enamel designed for a wide variety of uses. For example interior enamels are useful on the floors of porches, patios, and other heavy-wear areas. Exterior enamel is recommended for outside doors, windows, and trim. Other types of enamel are used on machinery that is exposed to the elements. Some of the newer enamels such as those used on cars and trucks are especially durable and are good choices for painting tractors and other equipment (Figure 29-1). Unlike the older enamels, these new paints withstand sunlight and weather without fading.

© iStockphoto/David Arthur

FIGURE 29-1 Newer types of enamel, such as the type used on cars, withstand the elements much better than the older, less expensive paints.

Some paints dry with a flat finish. **Flat finish** means dull or without shine. A flat-finish paint is preferred for walls that are not generally touched or exposed to friction or moisture.

Vehicle

As discussed in Unit 28, the liquid portion of paint is called the vehicle. The vehicle in most paints is either water or oil. Paints containing water are called **water-based paints**. Those containing oil are called **oil-based paints**.

The word **latex** on a paint label generally indicates a water-based paint. Most latex paints can be thinned with water, and brushes used to apply latex paint can be cleaned with water and detergent. The term **alkyd** indicates an oil-based paint. Such paint must be thinned with an oil, various petroleum products, or turpentine.

In recent years, epoxy paints have been added to the traditional line of paint materials. **Epoxy** is a synthetic material with special adhesive and wear-resistant qualities. Epoxy paints may be more difficult to apply, but their excellent durability makes them desirable for use in barns, milking parlors, public restrooms, commercial kitchens, and other hard-use areas.

Interior Versus Exterior

The word interior on a label means the paint will not hold up if exposed to weather. The term **exterior** designates a paint that will withstand moisture and outside weather conditions. Both water-based and oil-based paints may be formulated for interior or exterior use. Oil-based paints are becoming obsolete as the quality of water-based paints improve. The paint label will indicate whether it is intended for interior or exterior use.

Pigments

Pigments give color and hiding power to paint. **Hiding power** is the ability of a material to create color and mask the presence of colors over which it is spread. Many materials are used for paint pigments. In the past, lead was used extensively. However, **lead** is a metal that has a cumulative toxic effect and stays in the body once it is ingested. Lead has now been banned as a paint ingredient.

A high-quality pigment used in many paints is **titanium dioxide**. In addition, compounds of **iron**, **zinc**, and **aluminum** are associated with high-quality paints. Therefore, as a rule of thumb, the higher the proportion of these pigments in a paint, the better its

© iStockphoto/Kyoungil Jeon

FIGURE 29-2 Using modern techniques, an almost endless variety of colors and shades can be mixed.

quality. At one time it was very difficult to match the color on paint. Once a paint color was applied, finding new paint that was an exact match was a problem. However, today, paint color can be matched almost perfectly through the use of a computer. A sample of the old paint is scanned, and pigment colors and the correct ratio are determined by the computer. An almost endless variety of colors and shades can be mixed using modern techniques (Figure 29-2).

> **CAUTION!**
>
> Lead paint has been found to act as a slow poison to livestock that lick it. The same is true of children who put objects painted with lead paint in their mouths. In the United States, laws have been passed to stop the use of lead paint, but old paint containing lead remains a hazard.

Low-quality pigments provide little or no hiding power. Some low-quality pigments are **calcium**, **magnesium**, and **silicon** compounds. **Asphalt**—a tarry substance derived from various chemical refining processes—provides excellent black hiding power, but it wears away faster than the high-quality pigments.

Primers and Undercoaters

Primers and **undercoaters** are used to prepare surfaces for high-quality top coats. They stick to and seal surfaces. They must be compatible with the material being painted as well as with the top coats that follow. **Compatible** means they go together with no undesirable reactions. Primers and undercoaters will not stand weather, wear, or exposure. They must be covered with a proper top coat. It is generally wise to use the manufacturer's recommended primer or undercoat with any paint or finishing material.

Rust Resistance

Some paints consist of a vehicle and pigments that enable them to adhere well to rusted surfaces. Such paints work well if all loose rust is removed before the paint is applied. The primer and top coat must be compatible for best results.

USING BRUSHES

Brushes are useful for small paint jobs. They are especially adapted to use on irregular areas such as window sashes and building trim. The cost of equipment is low for both brush and roller painting. Therefore, both types of applicators are used extensively for home and farm projects. There are several types of paintbrushes for a wide variety of uses (Figure 29-3). Latex paint is generally applied with a paintbrush with bristles made of synthetic materials such as polyester or nylon. Oil-based paints or varnishes should be applied with natural bristle brushes. These are made of animal hair. Brushes also come in a variety of sizes—usually from ½ to 4 inches in width. Generally, the longer the bristle, the more paint the brush will hold. The bristles may be either straight across or tapered. Tapered brushes are used for trim work while straight brushes are used for larger, flat surfaces.

A new paint job can do almost as much to lift the spirits as a new purchase. However, a poorly done job may be ugly and messy, and may interfere with the proper operation of the object that is painted. For example, improperly painted windows may not open due to sticking. The recommended procedure for applying paint with brushes follows.

Procedure

1. Prepare the object or area as outlined in Unit 28.

2. Mask all parts that are not to be painted. Cover areas under surfaces to be painted with drop cloths. A **drop cloth** is any material used to protect floors, furniture, shrubbery, and other objects from paint spatter or droppings (Figure 29-4).

3. Use masking tape to make a straight line where paint should stop (Figure 29-5). When the tape is removed, a straight paint line will be left, giving the job a professional appearance.

4. Mix paint thoroughly. Unopened oil-based paints should be shaken by machine. Once opened, the paint should be divided between two containers and stirred thoroughly (Figure 29-6). Pouring the paint back and forth between containers is generally the final step in mixing.

5. Use pure bristle brushes for oil paints, and nylon or other artificial brushes for latex paints. Select a brush that has extralong, thick bristles and one that matches the job in size and shape—round, beveled, tapered, or flat.

6. Paint with the flat side of a flat brush. Painting with the edge causes **fingering**, or dividing, of the bristles into finger-like clumps of bristles (Figure 29-7).

7. Dip about one-third the length of the bristles into the paint. Touch the bristles to the side of the paint container to remove the excess paint and prevent dripping. This step is called **loading the brush**.

8. Touch the loaded brush to the work in several places in a small area to deposit paint. Then smooth the paint by brushing back and forth to fill in the area. End all strokes in the paint rather than on unpainted surface (Figure 29-8).

9. Paint adjacent sections before the paint dries. This reduces the marks that result from uneven blending of paint layers.

(continued)

Copyright © 2015 Cengage Learning®.

FIGURE 29-3 Paintbrushes come in a variety of types.

Procedure, *continued*

CAUTION!

Latex paints dry quickly and should not be brushed more than necessary to spread them.

10. Stir the paint frequently.

11. If it is necessary to stop painting briefly, wrap the bristles in a plastic food bag to prevent the paint from drying in the brush.

12. At the end of the work day, clean latex paintbrushes with warm water and detergent. Wrap them in a paper towel and place them in a warm place to dry. This is necessary to maintain brush shape and compactness. Oil-based

FIGURE 29-6 Oil-based paints should be divided between two containers, stirred thoroughly, and then poured back and forth between the containers to ensure proper mixing.

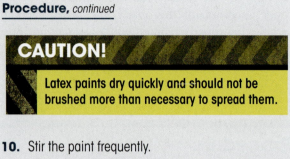

FIGURE 29-4 Most paint jobs require the use of drop cloths to protect shrubbery, walks, blacktop, floors, or furniture.

FIGURE 29-7 Use the flat side of the brush when painting. Painting with the edge will cause the bristles to divide into finger-like clusters.

FIGURE 29-5 Masking tape can be used to paint a perfectly straight line.

FIGURE 29-8 End all brushstrokes in the freshly painted area.

paintbrushes may be suspended in linseed oil or thinner until the next day.

13. Before restarting the oil-based paint job, wipe excess thinner from the bristles with rags or paper towels. Proceed with painting.

14. When the painting is finished, clean oil paint from brushes using thinner. Wash the brush thoroughly with soap and water to remove all paint and thinner residue. Clean bristles will squeak like clean hair. Wrap the brush in a paper towel to form the bristles in their proper shape. The bristles will then dry and provide a brush that is like new.

USING ROLLERS

Paint can be poured into a wide, shallow pan and applied with rollers more quickly and easily than with brushes (Figure 29-9). A **paint roller** is a hollow cloth- or fabric-covered cylinder that turns on a handle. The type of covering used depends on the type of surface that is being painted. A smooth, flat surface should be painted using a short-nap (fiber) cover. Surfaces that are rough or textured should be painted using a long-nap cover. The **nap** refers to the soft, woolly, thread-like surface of the roller cover. A newer type of roller is the *power roller*, which makes use of an electric-powered pump that brings the paint from the bucket to the roller. This eliminates the need for a pan or having to roll paint onto the roller. This makes painting much faster and easier (Figure 29-10).

However, rollers tend to throw off tiny droplets of paint that spatter the surfaces below. Therefore, using drop cloths is especially important when applying paint with a roller. Some suggestions for using rollers to apply paint follow:

- Use a 3- or 4-inch roller for narrow or uneven surfaces such as fence boards and siding. A 9-inch roller can be used for large, level surfaces such as plywood, wall sections, and ceilings.
- Use a roller with a short nap for smooth surfaces, and one with a longer nap for rougher surfaces (Figure 29-11).
- Line the paint pan with aluminum foil. The foil is discarded after painting, leaving the pan clean for the next job (Figure 29-12).
- Use small rollers or sponges on bevels or grooves in siding or on rounded surfaces such as rails and posts (Figure 29-13).

FIGURE 29-9 Paint can be applied more quickly and easily with a roller than with brushes.

FIGURE 29-10 Power rollers, which run on electricity, make painting much faster and easier.

FIGURE 29-11 A roller with a very long nap should be used to paint rough, irregular surfaces.

FIGURE 29-12 Cleanup time can be reduced if the paint tray is lined with aluminum foil. After the painting project is done, the foil can be removed and discarded.

- Paint corners and edges first; then roll the larger areas. When painting large areas, apply paint in a diagonal or cross pattern (Figure 29-14). Then, roll it again in the direction of movement across the work. Finish the surface by rolling toward the painted area—not toward the unpainted area.
- Stir the paint frequently, especially when the pan is refilled.
- When work is stopped, temporarily—or even for several days—store the roller in an airtight plastic or foil bag or wrapper (Figure 29-15).
- When the job is finished, discard low-cost roller covers. High-quality covers are generally made with a washable core and may be cleaned. However, the process requires large amounts

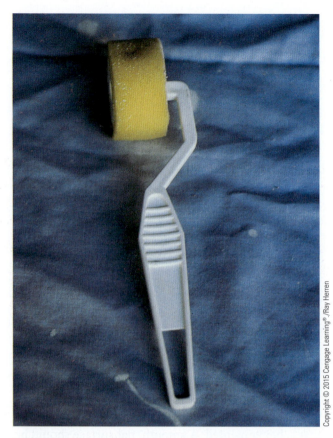

FIGURE 29-13 Small rollers or sponges should be used on rounded surfaces such as rails and posts.

FIGURE 29-14 The corners and edges of the surface should be painted first. Then the large areas should be painted after first applying paint with the roller in a cross pattern.

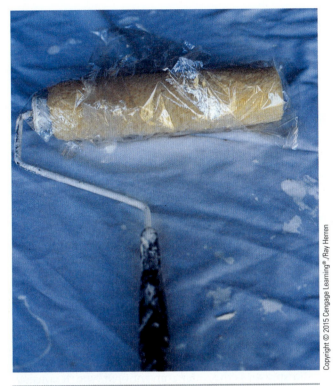

Copyright © 2015 Cengage Learning®./Ray Herren

FIGURE 29-15 A used roller may be stored for several days without being cleaned if it is wrapped in an airtight material such as plastic.

Copyright © 2015 Cengage Learning®./Ray Herren

FIGURE 29-16 Aerosol spray paint is especially useful on small jobs and on surfaces that are hard to paint with brushes.

of solvent. Therefore, cleaning may be practical only when using latex paints, which are removed with warm water and detergent. If all paint is not removed, the **roller cover** will not be usable again.

USING AEROSOLS

Small paint jobs can be done quickly at a low cost with aerosol spray cans. An **aerosol** can is a high-pressure container with a valve and spray nozzle (Figure 29-16). The container is partially filled with paint, which is put under pressure with compressed gas. The gas is called a **propellant**, since it pushes or propels the paint toward the object being painted.

When painting with aerosols, the following procedure is suggested.

Procedure

1. Degrease and sand the object to be painted.
2. Wash all dust and residue from the object.
3. Choose a clean, well-ventilated area to paint in.
4. Cover or mask all adjacent areas.

5. Wipe areas to be painted with a tack rag and solvent.
6. Shake the aerosol can for several minutes.
7. Practice spraying on a piece of cardboard before starting on the project.
8. Hold the can 10 to 12 inches from the object. Start the spray by pushing the button on top of the can, and then move it across the surface to be painted. Spray with rapid, uniform strokes.
9. Spray on a mist coat, and let it become tacky or sticky. Gradually build a full cover by successive overlapped strokes to create an even paint film.

CAUTION!

Do not apply too much paint at one time on vertical surfaces or it will develop unsightly runs.

10. When finished, invert the can and press the nozzle until there is no more paint in the spray. This leaves the nozzle free of paint so that it can be used another time.

(continued)

Procedure, *continued*

11. If additional coats are needed for good coverage, follow the directions on the can regarding sanding and drying time between coats.

12. Allow to dry thoroughly before heavy use. Paint may take 30 days to achieve maximum hardness.

A great selection of paints, varnishes, and other coatings is available in aerosol cans. Therefore, their use is encouraged for touch-up work on machinery and full paint jobs on small tools and equipment. Paint in aerosols is generally more expensive than an equal amount of paint in a regular can. However, the time saved and the attractive appearance obtained make the aerosol spray can the preferred method for many small jobs. Paint should be purchased in aerosols that use environmentally safe propellants.

SPRAY-PAINTING EQUIPMENT

Spray Guns

A **spray gun** is a device that releases paint in the form of a fine spray. Spray guns may be the air atomization type or they may be airless. **Air atomization** means

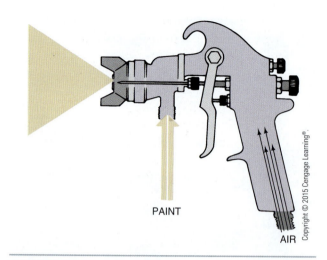

FIGURE 29-17 A spray gun uses compressed air to split paint into tiny droplets.

that the paint is split into tiny droplets using compressed air (Figure 29-17). The most familiar type of paint sprayer is the older conventional type. It consists of a gun body, a fluid control knob, a pattern control knob, an air valve, a paint needle valve, packing nuts, a fluid nozzle, an air nozzle, and a trigger (Figure 29-18). Air pressure is kept constant by a regulator. When the trigger is pulled a short distance, air flows through the passageways of the gun and out the air nozzle. The air flow creates a vacuum in the nozzle. When the trigger is pulled further, the fluid needle valve opens and

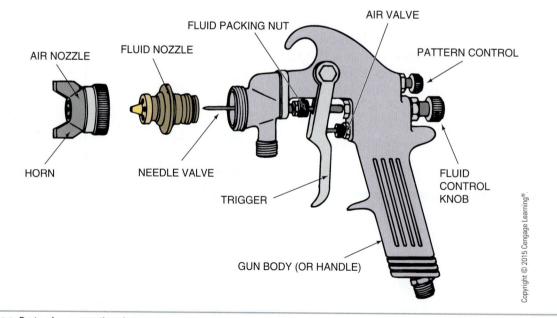

FIGURE 29-18 Parts of a conventional spray gun.

paint rushes out the fluid nozzle. Holes are positioned in the air nozzle to discharge air at angles that force the paint into a fan or cone pattern, depending on the adjustment.

A newer type of spray gun is the high-volume, low-pressure (HVLP) type. These spray guns differ from conventional spray guns in that they operate at much lower air pressure. This means that more paint goes onto objects that are intended to be painted. The older type of high-pressure spray gun tends to spray paint beyond the object being painted. With less pressure driving the atomized paint, more paint can be directed onto the intended surface. An HVLP sprayer can reduce overspray by as much as 80 percent over the older high-pressure sprayer (Figure 29-19). This means that the job can be done more neatly and with less cleanup. Also, there is less paint in the air, and that lowers the chance of paint spray getting inside the mask that covers the operator's mouth and nose. Another advantage is that less paint is used. Since there is less wastage from **overspray**, the savings can be substantial. A third advantage is that the HVLP sprayers do a better painting job. The flow of the paint can be controlled better and a finer finish can be achieved— atomized as it is released into a spray pattern.

Airless spray guns are similar in design to air spray guns. However, no air is used in the gun. Instead, the paint is pumped through hoses under high pressure. The pressurized paint then rushes through the nozzle.

FIGURE 29-20 Airless sprayers are used to paint large, flat surfaces such as walls and floors.

These types of sprayers range from the inexpensive diaphragm-type pumping systems to the more expensive types that use a piston-style pump to pressurize the paint. Airless sprayers are generally used when spraying large, flat surfaces such as walls or floors (Figure 29-20).

Fluid Systems

Air spray guns may be connected to paint supplies in one of three ways: with a siphon feed cup, with a pressure feed cup, or with a pressure feed tank. All three systems must include a regulator and extractor in the air line. The **regulator** keeps the air pressure at a set level. The **extractor** removes moisture and oil from the compressed air. Moisture and oil in the air interfere with the spray pattern and ruin the paint job.

> ### CAUTION!
> Never point an airless gun at any part of your body or at anyone else's. If the trigger is pulled, the paint is discharged at sufficient pressure to penetrate the skin. If this occurs, get medical attention immediately.

The **siphon feed cup system** (Figure 29-21) has a small vent hole in the cap of the cup. This hole must be kept open at all times to permit atmospheric pressure on the paint. Paint may cover the hole if the cup is tipped too far. As a result, there will be a vacuum in the cup and the gun will not work. The hole may be reopened by using a small piece of wire or a stiff broom straw. The siphon-type system works by creating a

FIGURE 29-19 HVLP spray guns operate at much lower air pressure than conventional air guns do.

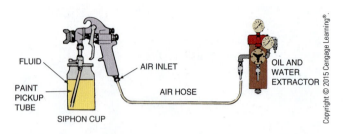

FLUID

PAINT
PICKUP
TUBE

SIPHON CUP

AIR INLET

AIR HOSE

OIL AND
WATER
EXTRACTOR

Copyright © 2015 Cengage Learning®.

FIGURE 29-21 A spray gun that uses the siphon feed cup system.

low-pressure area in the nozzle where paint can move in by force of atmospheric pressure.

The **pressure feed cup system** and **pressure feed tank systems** operate with the paint under air pressure at all times. The pressure pushes the paint through the system. These systems do not use a vent hole in the cap. The pressure feed cup holds only 1 quart of paint; the pressure feed tank may hold 5 gallons or more. The pressure tank system is good for large jobs such as barns. The paint tank can remain on the ground while the operator is painting the sides or roof. The operator needs to carry only the gun and hoses.

Air Supply

Good spray-painting systems permit fast and efficient painting. However, the initial cost is relatively high. For shop and farm use, a fairly large compressor is needed to operate a good spray gun. It is a good idea to buy the spray gun, compressor, air hose, regulator, extractor, and related equipment from one source

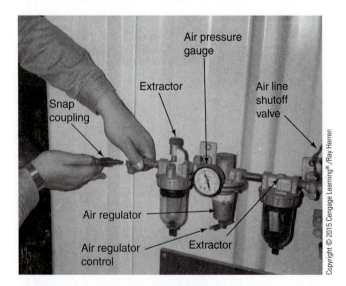

Snap
coupling

Air pressure
gauge

Extractor

Air line
shutoff
valve

Air regulator

Air regulator
control

Extractor

Copyright © 2015 Cengage Learning®./Ray Herren

FIGURE 29-22 To prepare for spray painting, an air hose is attached between the air gun and other parts of the spray system. This system includes a snap coupling, an oil and water extractor, a pressure regulator with gauge and control, a shutoff valve, an air hose, a tank, and a compressor.

(Figure 29-22). The supplier should provide assurance that the components work well together to provide an efficient system.

It is especially important that the compressor provide as many **cubic feet per minute (CFM)** of air as the gun requires. The air must also be delivered to the gun at the required pressure. To achieve this requires adequate sizing of the motor, compressor, electric wiring, air pipe, and hoses. Another important component is a water trap. When air is compressed, water vapor is turned to water. If water is allowed to escape into the spray gun, it will ruin the finish. Water traps are installed in the line between the regulator and the paint sprayer.

SPRAY-PAINTING PROCEDURES
Preparing the Surface

Work should be prepared for painting as outlined in Unit 28. Expensive equipment and a skillful painter cannot make paint attractive and permanent if the surface has not been prepared properly. All surfaces must be clean of blistered paint, rust, dirt, oil, and wax. Glossy surfaces must be sanded to create a dull finish to which paint can stick.

Choosing a Location

Gun spraying is not recommended indoors unless a commercial spray booth is available. Paint spray is a fire and explosion hazard. In addition, the spray will settle on surfaces and ruin ventilation equipment. Spraying may be done outdoors if the temperature is 50°F or above and there is no wind. Wherever the spraying is done, an approved respirator must be used to prevent the inhalation of the paint fumes and paint particulates. Breathing these materials can cause severe lung damage.

Following Safety Rules

Spray painting can be hazardous activity. Some safety rules for spray painting are as follows:

- Mix paints and solvents in well-ventilated areas.
- Spray paint outdoors or in special booths with recommended electrical fixtures and ventilation.
- Never smoke or have any other source of fire in the vicinity while working with a paint sprayer.

CAUTION!

All automobiles and other objects with good paint jobs should be at least 100 feet from the painting area. Otherwise, particles of paint will drift, settle on them, and ruin their appearance.

- Wear a respirator or paint mask while painting.
- Use ladders that are in good condition. Do not use ladders with rungs or sides cracked or damaged.
- To avoid electrocution, be especially careful to keep aluminum ladders away from all electrical wires.

Preparing to Paint

Nothing is more frustrating than to have the project ready, spray equipment set up, and paint in the gun, only to find that the gun will not spray paint. Such problems occur from the improper preparation of paint and equipment. Paint is prepared for spraying as follows.

Procedure

1. Thoroughly mix and stir the paint until no streaks can be seen. If more than one container of paint is to be used, mix them together to assure a uniform color.

2. Thin the paint at room temperature according to the manufacturer's instructions. Since all thinners and paints are not compatible, mix a small amount of paint and thinner to see if they form a smooth paint with uniform color.

3. To be safe, use a **viscosimeter** to confirm that the paint is thin enough. A viscosimeter is an instrument used to measure the rate of flow of a liquid. (A fluid's resistance to flowing is called its **viscosity**.)

4. Using a commercial paint strainer, strain the paint into a clean container. Strainers are available from paint and automotive parts stores.

5. Pour the thinned paint into a clean sprayer cup or tank.

Adjusting the Spray Gun

Fluid and air pressures should be set as low as possible, yet do the job. Pressure requirements vary with the design of the gun, the viscosity of the paint, and the temperature of the paint. All guns should be adjusted according to the manufacturer's instructions. However, some general recommendations for adjusting spray guns follow.

Procedure

1. Set up all equipment.

2. Put on safety glasses and a paint respirator.

3. Set up a sheet of cardboard to use as a practice surface.

4. Turn on the air supply, and set the regulator at the level specified for the gun.

5. Open the pattern control and fluid control knobs.

6. Prepare to make a pass by pointing the gun off to the left side of the panel. Pull the trigger slowly until air is heard coming from the nozzle (if you are using an air-type gun). Pull the trigger further, until a faint paint fan is seen.

7. Position the gun with the nozzle 6 to 10 inches from the work and perpendicular to it. Slowly move the gun across the surface, maintaining the same distance from it, to create an even spray band (Figure 29-23).

8. Readjust the pattern control knob to obtain a fan-shaped pattern.

9. Adjust the fluid control knob to permit the desired amount of paint to flow when the trigger is pulled all the way.

10. Adjust the distance from the object and the rate of travel until a good, even pattern is obtained. Each stroke should overlap the previous one by 50 percent.

Operating the Spray Gun

Spray guns must be held the correct distance from the work (Figure 29-24). The gun stroke is made by moving the gun parallel to the work.

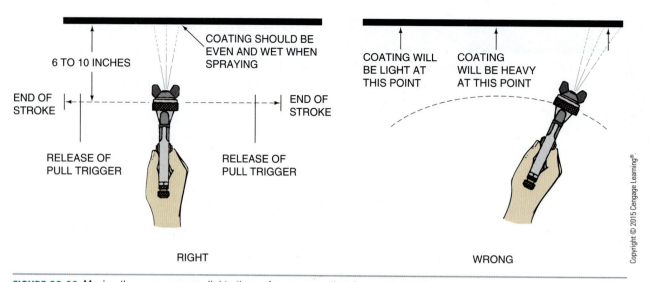

Copyright © 2015 Cengage Learning®.

RIGHT

WRONG

FIGURE 29-23 Moving the spray gun parallel to the surface ensures that the coating will be even.

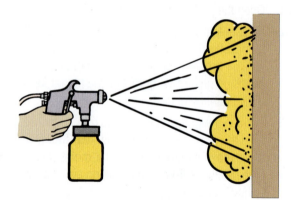

A NOZZLE IS TOO FAR FROM THE SURFACE. DUSTING RESULTS IN A ROUGH PAINT FILM WITH A DULL FINISH.

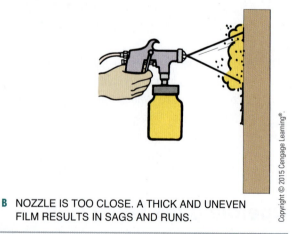

Copyright © 2015 Cengage Learning®.

B NOZZLE IS TOO CLOSE. A THICK AND UNEVEN FILM RESULTS IN SAGS AND RUNS.

FIGURE 29-24 Holding a spray gun's nozzle too close to or too far from the object that is being painted produces an undesirable effect.

The closer the gun is held to the work, the faster the gun must be moved to prevent sags and runs. A **sag** occurs when a large area of paint shifts downward because it has been applied too thickly. A **run** is a narrow stream of paint flowing downward due to excessive buildup. On the other hand, holding the gun too far from the work causes dry spray and excessive spray dust. These result in a flat, dull, and rough finish.

When painting tractors and machinery, many irregular surfaces must be sprayed. Under these conditions, the spray pattern should be narrow. This permits paint to reach parts that are farther than the ideal distance from the nozzle. Large, smooth areas, such as hoods and fenders, require a broad fan pattern and should be removed from the machine being painted. The shape of the fan is controlled by adjusting the pattern control knob.

Modern machinery enamels are designed to be sprayed in two or more layers at one time. To do this, spray on a thin layer and let it become **tacky** (this means sticky). By the time a machine such as a tractor or baler has been given one coat, the paint applied at the beginning of the job will be tacky. Go back to the starting point and apply a heavy top coat. The paint layer should shine but not run or sag. When a run or sag occurs, the entire part or panel must be wiped clean with solvent and then repainted. Another method of correcting sags or runs is to allow the part to dry thoroughly, and then sand the area and repaint.

CAUSE
Dried material in side port "A" restricts passage of air through port on one side. Results in full pressure of air from clean side of port in a fan pattern in direction of clogged side.

REMEDY
Dissolve material in side port with thinner. Do not use metal devices to probe into air nozzle openings.

CAUSE
Dried material around the outside of the fluid tip at position "B" restricts the passage of atomizing air at one point through the center ring opening of the air nozzle. This faulty pattern can also be caused by loose air nozzle, or a bent fluid nozzle or needle tip.

REMEDY
If dried material is causing the trouble, remove air nozzle and wipe off fluid tip, using rag wet with thinner. Tighten air nozzle. Replace fluid nozzle or needle if bent.

CAUSE
A split spray pattern (heavy on each end of a fan pattern and weak in the middle) is usually caused by: (1) atomizing air pressure too high, (2) attempting to get too wide a spray with thin material, (3) not enough material available.

REMEDY
(1) Reduce air pressure. (2) Open material control "D" to full position by turning to left. At the same time turn spray width adjustment "C" to right. This reduces width of spray but will correct split spray pattern.

SPITTING

CAUSE
• Air entering the fluid supply.
• Dried packing or missing packing around the material needle valve that permits air to get into fluid passageway.
• Dirt between the fluid nozzle seat and body or a loosely installed fluid nozzle.
• A loose or defective swivel nut, siphon cup, or material hose.

REMEDY
• Be sure all fittings and connections are tight.
• Back up knurled nut "E," place two drops of machine oil on packing, replace nut, and finger tighten. In aggravated cases, replace packing.
• Remove air and fluid nozzles "F" and clean back of fluid nozzle and nozzle seat in the gun body, using a rag wet with thinner. Replace and tighten fluid nozzle using wrench supplied with the gun. Replace air nozzle.
• Tighten or replace swivel nut "G."

CAUSE
A fan spray pattern that is heavy in the middle, or a pattern that has an unatomized "salt-and-pepper" effect indicates that the atomizing air pressure is not sufficiently high, or there is too much material being fed to the gun.

REMEDY
Increase pressure from air supply. Correct air pressures as discussed elsewhere in this manual.

Copyright © 2015 Cengage Learning®

FIGURE 29-25 Faulty spray patterns and possible ways to correct them.

Correcting Faulty Spray Patterns

Faulty spray patterns can be caused by incorrect pressure, clogged air passages, loose packing nuts, dried packing, thick paint, or unstrained paint.

After each use, a drop of light machine oil should be placed on each needle in a spray gun where it enters the packing nut. This procedure helps eliminate some problems with faulty patterns. Other recommendations for correcting faulty patterns are given in Figure 29-25.

Cleaning Spray Equipment

Cleaning spray guns and equipment requires a great deal of solvent. Thinner is recommended for oil paints because it is low in cost. Another advantage of thinner is that the fire hazard is not as great as it is with other solvents.

To clean the gun, cup, and other equipment, the following procedure is suggested.

Procedure

1. Pour all unused paint back into the container and reseal it.

2. Wipe most of the paint from the interior of the feed cup and spray gun parts with paper towels or clean rags.

3. Pour about 1 inch of thinner into the cup.

4. Attach the cup to the spray gun, and shake the gun so the thinner can dissolve the paint residue.

5. Spray a piece of cardboard until the cup is empty. Cardboard absorbs the thinner.

6. Wipe all visible surfaces, both inside and outside of the gun and cup, with clean thinner on a cloth.

7. Repeat steps 3 through 6 one or two times using paint thinner until all traces of paint are absent from the thinner.

(continued)

Procedure, *continued*

8. Remove the air nozzle and the paint nozzle.

9. Use a cloth with thinner to wipe away any remaining evidence of paint.

10. Dry all parts and reassemble the gun.

11. Clean all other equipment using a cloth and solvent.

CAUTION!

The spray gun will not work well the next time unless every trace of paint is removed. A good rule to follow is that you should not be able to tell the color of the paint used previously when examining the gun.

12. Discard all paper, cloths, used strainers, and paint containers in an approved metal trash container.

13. Store all equipment properly.

14. Turn off the air pressure and/or compressor.

Spray painting can be very satisfying. However, care must be taken to prepare the work and paint carefully. Absolute cleanliness is required to avoid spray gun problems. Since paint and thinners are very flammable, there is always a danger of fire. Every precaution must be exercised to keep fire and electrical sparks away from the spray-painting area.

SUMMARY

Paint should be applied to most wood and metal surfaces. There are several different types of paints and finishes that require techniques particular to the type of paint or finish. Unless the proper procedures and techniques are used, the results will be less than desirable. Correctly applied paint or finish not only makes a project look better, but the project will last longer as well.

Student Activities

1. Define the Terms to Know in this unit.

2. Examine the paint on a wooden building, a wooden fence, and a piece of used farm machinery or garden equipment. What evidence is there to indicate a need for repainting?

3. Examine the labels on a can of oil-based paint and a can of latex paint. List the ingredients and the percentage of each. What is recommended as a thinner for each?

4. Prepare a wooden project for painting. Brush-paint the project.

5. Prepare a metal project for painting. Spray-paint the project using an aerosol spray container.

6. Prepare a piece of machinery for spray painting. Use a spray gun to paint the machine.

7. Use a pan and roller to paint a fence, a room, or a building.

Relevant Web Sites

Al's Home Improvement Center, do it yourself and how-to tips
http://alsnetbiz.com/homeimprovement/index.html

This Old House®
www.thisoldhouse.com

Self-Evaluation

A. Multiple Choice. Select the best answer.

1. Painting is not a necessity when it
   a. is done for personal preference
   b. preserves wood
   c. prevents rot
   d. prevents rust

2. An example of a low-quality pigment is
   a. aluminum
   b. calcium
   c. titanium
   d. zinc

3. Lumpy particles in paint should be
   a. ignored
   b. mashed up
   c. reduced by thinning
   d. removed by straining

4. Perfectly straight lines are obtained when painting by use of
   a. a steady hand
   b. masking tape
   c. a ruler and pencil
   d. a yardstick

5. Oil paints should not be applied with
   a. sponges
   b. nylon bristle brushes
   c. pure bristle brushes
   d. rollers

6. Latex paints should not be
   a. brushed excessively
   b. applied with rollers
   c. used outdoors
   d. used on wood

7. An aerosol for painting is
   a. not pressurized
   b. relatively expensive
   c. difficult to use
   d. too expensive for small jobs

8. All spray guns are
   a. airless
   b. pressurized
   c. siphon-type
   d. none of these

9. Painting may be hazardous because of
   a. explosive materials
   b. flammable materials
   c. compressed air
   d. all of these

10. Painting hazards are greatly increased by
    a. excessive ventilation
    b. smoking
    c. spraying outdoors
    d. using a respirator

B. Matching. Match the terms in column I with the correct definition in column II.

Column I

1. formulate
2. chalking
3. gloss
4. flat
5. latex
6. titanium dioxide
7. primer
8. lead
9. extractor
10. regulator

Column II

a. dull
b. thin with water
c. first coat
d. poisonous
e. removes water
f. controls pressure
g. put together
h. safe, high-quality pigment
i. shiny
j. wash off

C. Completion. Fill in the blanks with the word or words that make the following statements correct.

1. Gun spray painting should not be done indoors unless a _____ is available.

2. Aluminum ladders are a special hazard around _____.

3. Dry paint in a spray gun will _____.

4. Spray guns should be lubricated with thin machine oil on each _____ where it enters the _____.

5. The tendency of a liquid to resist flowing is known as its _____.

D. Brief Answer. Briefly answer the following questions.

1. How does one correctly load a brush?

2. What is epoxy? Although more difficult to apply, what is the advantage of epoxy paints?

3. Spray guns can be connected to paint supplies in three ways. What are they? Which is best for large jobs? Why?

4. What procedure helps eliminate some faulty spray patterns by a spray gun?

5. What are two techniques for correcting runs or sags?

6. What is "nap"?

Section 10
POWER MECHANICS

UNIT 30

Fundamentals of Small Engines

Objective

To identify and explain the principles of operation and the systems of small engines.

Competencies to be developed

After studying this unit, you should be able to:

- Discuss the uses of small engines.
- Observe all safety precautions when working with small engines.
- Distinguish between two- and four-cycle engines.
- Explain how a four-cycle engine operates.
- Explain how a two-cycle engine operates.
- Identify all the systems in a small engine.
- Explain how the different systems of a small engine work.

Materials List

- two- and four-cycle engines
- spark plug
- carburetor
- armature
- valve
- piston
- camshaft

Terms to Know

- internal combustion engine
- external combustion engines
- four-cycle engine
- piston
- cylinder
- crankshaft
- block
- valves
- intake stroke
- compression stroke
- power stroke
- exhaust stroke
- flywheel
- two-cycle engine
- piston rings
- carburetor
- venturi
- camshaft
- governor
- air vane
- magneto
- primary circuit
- secondary circuit

Much of the advancement made in American agriculture is a result of the efficiency of machines, which has increased the amount of work that can be done by the producer. A very large portion of this efficiency is because of the development of the **internal combustion engine**, which provides power for machines. However, the first engines developed were **external combustion engines**. This type of engine powered everything from locomotives to cars to farm machinery such as tractors. Power was derived from fuel consumed outside the engine. Usually, the heat from the combustion turned water into steam that in turn powered pistons inside the engine. The drawbacks were that these types of engines were large, cumbersome, and heavy; thus, they saw limited use on farms (Figure 30-1).

In the late 1800s, engines referred to as *internal combustion engines* were developed to replace the steam engines. These engines burned fuel inside the engine and provided more efficient power at a much reduced size and weight. As in the steam engine, the power was derived as a result of a piston moving inside a cylinder. Almost all machinery used on modern farms is powered by these engines. Large engines run tractors, combines, and other harvesting equipment that requires a lot of power. Generally, these engines are expensive and require regular maintenance. People who do repairs on these large engines must be specially trained to work in repair shops that have equipment capable of handling the heavy engines.

Small single-cylinder engines play an indispensable role in any agricultural operation. They power everything from weed trimmers to conveyor belts. Small gasoline engines are used on boats, in recreational vehicles, in snowmobiles, and on bikes (Figure 30-2). Small engines run chainsaws, portable grass cutters, leaf blowers, and lawn mowers, which are common in most neighborhoods. Small engines power garden tractors, tillers, bale throwers, and elevators on farms. They are lightweight, powerful, and relatively inexpensive. Generally they are simple to operate, maintain, and repair. Almost anyone with a minimal amount of tools and know-how can do the maintenance and repairs required to use these machines.

SAFETY PRECAUTIONS

There is always some inherent danger when using machines, and small engines are no exception. There are moving parts, which are dangerous, and gasoline is an added hazard in small engines. Suggestions for avoiding fire and injury when operating or working on small engines follow:

- Wear safety glasses, hearing protection, and leather shoes (Figure 30-3).
- Keep your hands and feet a safe distance from all moving parts.
- Do not allow burning material near oil, solvents, or gasoline. Do not smoke near the openings of gasoline tanks or containers.
- Stop the engine and allow it to cool before refueling.
- Whenever possible, handle gasoline outdoors.
- Disconnect the spark plug cable prior to making any adjustments on machines, to prevent the possibility of an accidental engine start-up.
- Never disengage any of the safety equipment or devices on the engine or machine.
- Keep all shields in place.
- Do not operate engines above the specified speed.
- Do not overload engines or force equipment beyond the designed capacity.
- Never start and run an engine inside a building where the ventilation is inadequate.
- Never use gasoline as a solvent or to clean engine parts.
- Dispose of gasoline properly, by pouring it into an approved container. Never pour it onto the ground or down a drain.

TYPES OF COMBUSTION ENGINES

Two types of internal combustion engines have been developed: the four-cycle engine and the two-cycle engine. With the exception of small, lightweight engines that run implements such as chainsaws and

FIGURE 30-1 The first engines were powered by steam. They were heavy and cumbersome.

© Alan Egginton/Shutterstock.com

FIGURE 30-2 Small gasoline engines are commonly used on machines that require a portable power source.

FIGURE 30-3 When operating small engine equipment, always wear eye protection, hearing protection, and leather shoes.

string trimmers, almost all engines are four-cycle. In fact, many of the modern very small engines are now four-cycle. Today's engines are tremendously more complicated than the first gasoline engines, although the basic operation principles are the same.

The Four-Cycle Engine

The **four-cycle engine**, as the name implies, operates on a series of four strokes, or piston movements, per cycle (Figure 30-4). The engine consists of a **piston** that operates within a **cylinder** that is only slightly larger in diameter than that of the piston. The piston is connected to a **crankshaft** that converts the piston's up-and-down motion into rotary motion, which in turn rotates a gear, a wheel, or some other implement. The cylinder is capped with a thick plate called a *head*, which seals the cylinder. Within the **block** (the mass of metal that contains the cylinder) or within the head are two **valves**: the *intake valve* lets in the fuel mixture, and the *exhaust valve* lets the exhaust fumes out. Some modern large engines may have as many as four valves to every cylinder.

FIGURE 30-4 A typical four-cycle small engine.

A four-stroke or four-cycle engine must complete four movements of the piston before the cycle is complete (Figure 30-5). The movement of the piston is all the way to the top (called *top dead center*) and all the way to the bottom (called *bottom dead center*) of the cylinder. Each of these movements is called a *stroke*, and each stroke serves a particular purpose.

Intake Stroke. The four-stroke cycle begins with the piston moving down. When this occurs, the intake valve opens, and a mixture of fuel and air enters the cylinder. This is called the **intake stroke**. At the completion of this stroke, the piston is at the bottom of the cylinder and both valves are closed.

Compression Stroke. When the piston reaches the bottom of the cylinder on the intake stroke, it starts upward on the **compression stroke**. Both valves are closed. At the top of this stroke, the fuel mixture is compressed tightly into the space between the top of the piston and the cylinder head. At the top of the compression stroke, the air-fuel mixture has been compressed into the area in the head called the combustion chamber. The relationship between the volume of the cylinder and the volume of the combustion chamber at the beginning and end of the compression stroke is known as the *compression ratio*. Compression ratios for small gas engines are about 6 to 1, which is written as 6:1.

Power Stroke. As the piston nears the top of the cylinder on the compression stroke, a spark from the spark plug ignites the mixture (only in gasoline-powered engines; diesel engines are different), and the resulting rapid expansion of the burning mixture pushes the piston downward on the **power stroke**. During the

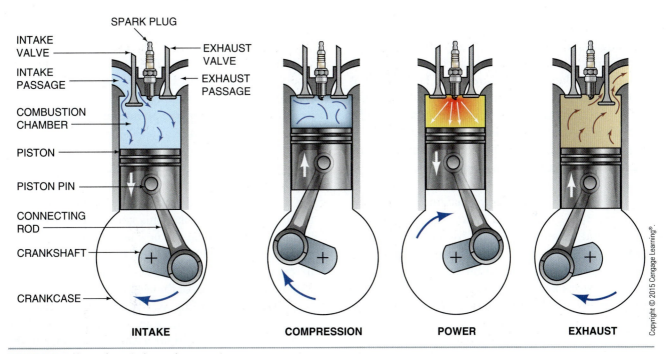

FIGURE 30-5 How a four-stroke engine operates.

power stroke, both valves are tightly closed. Usually, the larger the cylinder and the piston are, the more power the engine can produce. This is because there is more room for the air-fuel mixture; the larger the room, the greater the amount of energy that is released on ignition of the gases.

Exhaust Stroke. When the piston reaches the bottom, it starts upward on the **exhaust stroke**. As the piston moves up, the exhaust valve opens and the exhaust fumes are pushed out of the cylinder. The exhaust stroke is completed with the piston at top dead center, the exhaust valve opens, and the exhaust gases are removed. As the piston starts downward, the exhaust valve closes, the intake valve opens, the intake stroke begins, and the cycle is repeated.

Notice that the crankshaft makes two complete turns or revolutions in the cycle in response to the four strokes of the piston. The momentum or turning force of the weights on the **flywheel** and other moving parts carries the engine through the three nonpower strokes. In engines with more than one cylinder, one cylinder will be on the power stroke while other cylinders are on nonpower strokes.

The Two-Cycle Engine

The **two-cycle engine** completes the intake, compression, power, and exhaust stages in two strokes. The crankcase of the engine does not contain oil. Instead, it is airtight and contains a reed valve to admit the air-fuel mixture from the carburetor when the piston moves away from the crankshaft. In the newer two-cycle engines, there are no reeds and the engine relies on open ports that are covered and uncovered by the movement of the piston (Figure 30-6). The mixture is compressed when the piston returns toward the crankshaft. The crankcase then holds the mixture under pressure until the cylinder is ready to receive it. Oil is mixed with the gasoline to lubricate the moving parts of a two-cycle engine. Hence, there is no need for oil in the crankcase.

The operation of a two-cycle engine is illustrated in Figure 30-7. To begin a cycle, the piston moves upward and creates a vacuum in the crankcase of the engine. The air-fuel mixture from the carburetor rushes into the crankcase to fill the vacuum. When the piston stops its upward movement, the reed valve closes by its own spring action. The piston then moves downward and puts pressure on the mixture in the crankcase. Now the engine is ready to start a normal cycle.

EXHAUST PORT

INTAKE PORT

© Piotr Krzeslak/Shutterstock.com

FIGURE 30-6 A two-cycle engine uses ports for valves. This cutaway shows the intake and exhaust ports. This type of two-cycle engine uses no reeds.

Intake and Exhaust Strokes. As the piston nears the bottom of its stroke, it uncovers the intake and exhaust port(s) (Figure 30-7C). Because the air-fuel mixture in the crankcase is under pressure, the mixture rushes through a passage to the intake port and enters the cylinder. This incoming gaseous mixture pushes air or exhaust out of the cylinder. Therefore, intake and exhaust functions occur with very little movement of the piston.

Compression Stroke. The cylinder is now filled with an air-fuel mixture. The piston moves upward, closes the intake and exhaust ports, and compresses the air-fuel mixture trapped in the cylinder. At the same time, a new supply of air and fuel rushes into the crankcase, as shown in Figure 30-7A.

Power Stroke. At or near top dead center (TDC), the spark plug fires to ignite the mixture. The burning and expanding gases drive the piston downward through the power stroke. This same downward movement puts pressure on the new air-fuel mixture in the crankcase (Figure 30-7B). Thus, the engine completes its cycle of intake, compression, power, and exhaust with only two strokes of the piston.

The advantage of the two-cycle engine is that it is lighter weight because there are no mechanical valves. This eliminates the need for a camshaft to operate the valves. Also, the lubricating oil is mixed with the fuel

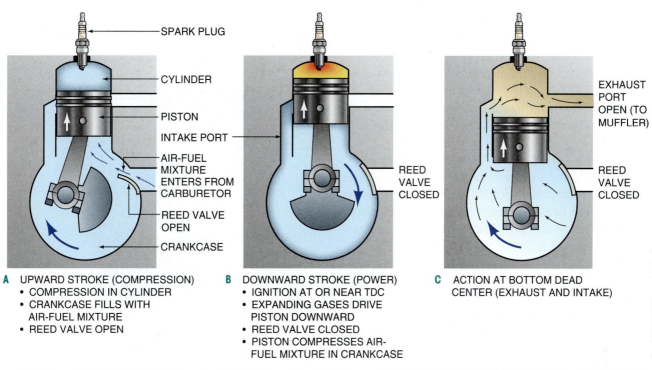

SPARK PLUG

CYLINDER

PISTON

INTAKE PORT

AIR-FUEL MIXTURE ENTERS FROM CARBURETOR

REED VALVE OPEN

CRANKCASE

REED VALVE CLOSED

EXHAUST PORT OPEN (TO MUFFLER)

REED VALVE CLOSED

A UPWARD STROKE (COMPRESSION)
- COMPRESSION IN CYLINDER
- CRANKCASE FILLS WITH AIR-FUEL MIXTURE
- REED VALVE OPEN

B DOWNWARD STROKE (POWER)
- IGNITION AT OR NEAR TDC
- EXPANDING GASES DRIVE PISTON DOWNWARD
- REED VALVE CLOSED
- PISTON COMPRESSES AIR-FUEL MIXTURE IN CRANKCASE

C ACTION AT BOTTOM DEAD CENTER (EXHAUST AND INTAKE)

FIGURE 30-7 How a two-stroke or two-cycle engine operates.

and there is no sump to hold engine oil. This allows the engine to be operated in any position, even upside down. Chainsaws, string (or line) trimmers, leaf blowers, and other small power tools are examples of implements that use two-cycle engines (Figure 30-8). Some large engines are two-cycle engines, as well.

ENGINE SYSTEMS

An engine operates through the coordination of several different systems that must function properly for the engine to operate. These systems are much like the systems of the body that must work together. Our digestive, skeletal, circulatory, and nervous systems must all function properly before the body can operate efficiently. All systems of the body seldom get sick at the same time. Likewise, when an engine malfunctions, it is usually only a part of one system that malfunctions. That part, however, can cause the entire engine to malfunction.

The Compression System

The upward movement of the piston in the cylinder (the compression stroke) compresses the gases into the combustion chamber at the top of the piston.

FIGURE 30-8 A string (or line) trimmer is powered by a two-cycle engine.

Copyright © 2015 Cengage Learning®

Copyright © 2015 Cengage Learning®./Ray Herren

Downward movement of the piston (power stroke) turns the crankshaft and creates power for the engine. It is essential that a tightly closed cylinder be maintained during both of these strokes. If gases leak from the cylinder, either the fuel mixture will not be compressed tightly enough for proper ignition during the intake stroke or power will be lost during the power stroke.

Compression is generally lost in one of three ways: the fit of the piston in the cylinder, a bad head gasket, or by valves that close improperly. Obviously, the piston cannot fit too tightly in the cylinder because it has to move up and down freely. Also, as combustion occurs the piston gets hot and needs some room to expand. The problem is solved by fitting rings made of cast iron and/or steel into grooves near the top of the piston. These **piston rings** have a spring action that creates a seal between the piston and the cylinder wall (Figure 30-9).

Both the intake and the exhaust valves must seat tightly when closed (Figure 30-10). Any amount of open space between the valve and the valve seat will cause gases to escape and compression will be lost. Also, if the exhaust valve is left open, the heat from the combustion may destroy the edge of the valve (called a "burned" valve), causing a gap to open in the valve. When the valve is properly seated, the heat is dispensed into the cylinder head or cylinder block. Valve springs must be strong and tight, and the proper space between valve lifters and valves must be maintained to allow for proper valve seating.

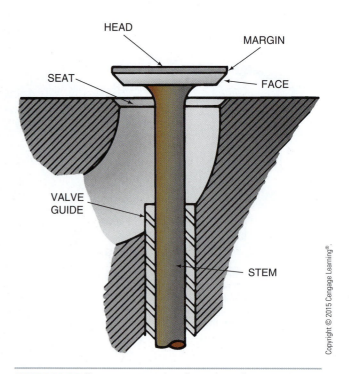

FIGURE 30-10 Proper compression depends on the correct seating of valves.

Copyright © 2015 Cengage Learning®.

The Fuel System

The fuel system consists of the fuel tank, fuel lines, the carburetor, and the intake system. The **carburetor** supplies fuel and air to the engine in appropriate proportions and volume. Liquid gasoline must be vaporized in the proper ratio of air if the mixture is to ignite and burn efficiently. As the piston moves down the cylinder on intake stroke, a vacuum is created that pulls through the intake valve and through the carburetor. This action causes the intake of air from outside the engine. An air filter mounted at the opening of the carburetor cleans the air before it enters the engine.

As air is drawn in, it passes through restriction in the throat of the carburetor called a **venturi**. The venturi is narrower in the middle than elsewhere along its length, creating a restriction that increases the flow of air and fuel by creating a low-pressure area that makes the air come through more rapidly. This is the same effect that provides lift for the wings of an aircraft or makes a breeze pull through an open barn hallway. As the air rushes through the venturi, gasoline is released into the flow by a regulated valve (Figures 30-11 and 30-12). The correct mixture passes through the carburetor, along the intake tube, through the intake valve, and into the cylinder at the correct time. Valves are opened and closed by a lobed rod called a **camshaft**. The camshaft is timed to open and close the valves

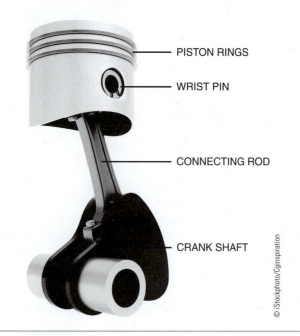

FIGURE 30-9 Piston rings provide a tight seal between the piston and cylinder wall.

© iStockphoto/Cginspiration

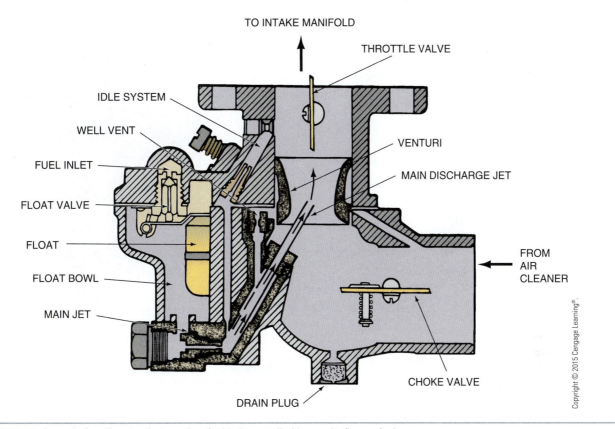

FIGURE 30-11 An updraft carburetor that requires fuel to be supplied by gravity flow or fuel pump.

Copyright © 2015 Cengage Learning®.

at precisely the correct time. Larger, more modern engines no longer use a carburetor. Instead, fuel is supplied to the cylinders by an injector pump that forces the fuel through injectors mounted to the combustion chamber.

Engine speed, power, and acceleration are controlled by the amount of gasoline that flows through the carburetor. To prevent the engine from running too fast, a **governor** attached to the carburetor slows the engine when a certain speed is reached. Spring tension in the linkage is countered by the action of the governor. The engine's speed is controlled by a balance of forces exerted by the cable in one direction and the governor in the other. As the cable pulls harder on the spring, the governor will permit the engine to run faster before leveling off to a constant speed. The governor controls engine performance through linkage to the throttle valve (Figure 30-13).

One type of governor consists of weights on a shaft turned by a gear that is driven by the engine. Centrifugal force causes the weights on the shaft to move outward and move a lever attached to the throttle shaft as the engine speed increases. This spreading force counteracts the action of the throttle cable on the spring.

Another type of governor is the **air vane**, which is used on small engines. It is activated by the flow of air created by flywheel fins under the engine shroud. As a rule, the only service or repair needed on governors is to keep the parts clean (Figure 30-14).

The Ignition System

When the fuel mixture has entered the combustion chamber and has been compressed, a spark must occur for the fuel to ignite. Most small engines have an ignition system known as a **magneto** in which the spark is created by an electric current generated by a process called *electrical induction*. In this process, a low-voltage current is generated when a strong magnet located in the flywheel passes very close to an armature coil. The armature coil is comprised of many layers of thin iron strips that have been riveted together. These strips are known as a *laminated core* (Figure 30-15). Around the laminated core are two separate coils of wire. One coil that is made of several hundred turns of heavy-gauge wire is called the **primary circuit**. Another coil of wire around the laminated core consists of several thousand turns of a fine-gauge wire and is referred to as the **secondary circuit**.

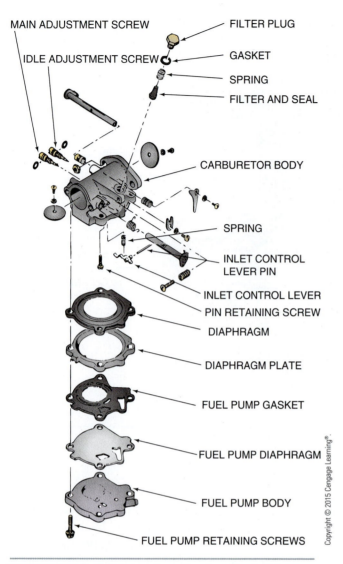

MAIN ADJUSTMENT SCREW
IDLE ADJUSTMENT SCREW
FILTER PLUG
GASKET
SPRING
FILTER AND SEAL
CARBURETOR BODY
SPRING
INLET CONTROL LEVER PIN
INLET CONTROL LEVER
PIN RETAINING SCREW
DIAPHRAGM
DIAPHRAGM PLATE
FUEL PUMP GASKET
FUEL PUMP DIAPHRAGM
FUEL PUMP BODY
FUEL PUMP RETAINING SCREWS

Copyright © 2015 Cengage Learning®.

FIGURE 30-12 Exploded view of a carburetor with a built-in fuel pump.

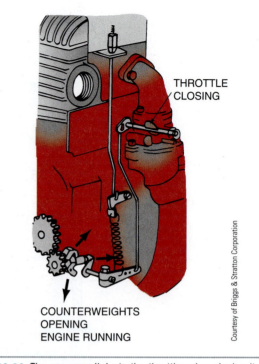

THROTTLE CLOSING

COUNTERWEIGHTS OPENING ENGINE RUNNING

Courtesy of Briggs & Stratton Corporation

FIGURE 30-13 The governor links to the throttle valve, closing it to stabilize engine speed.

When the magnet on the flywheel passes by the armature coil, a low-voltage electrical current is sent through the primary circuit. If the primary circuit is suddenly opened, the decaying magnetic field around the primary circuit sends a high-voltage charge through the secondary circuit. This action can produce up to 40,000 volts of electricity. This charge travels along a wire to the spark plug.

Basically, the spark plug consists of two parts: the center electrode, which runs the length of the middle of the plug; and the ground electrode, which is grounded to the block of the engine (Figure 30-16). The end of the spark plug extends into the combustion chamber. When the surge of electricity runs through the center electrode with sufficient voltage, a spark will jump from the end of the electrode as the current attempts to cross the air gap between the center electrode and the ground electrode.

The spark must occur at exactly the right time to ignite the fuel mixture. This means that the surge of electricity must be sent through the secondary circuit at the correct time. For this to happen, the primary circuit must be opened at precisely the right time. Until about 1982, most small-engine primary circuits were opened and closed through the use of breaker points that operated off a rod pushed up and down by the camshaft. When the points opened, the primary circuit died and the surge of electricity was sent through the secondary circuit. The current in the primary circuit was deleted rapidly by sending the current into a small device called a condenser. The gap in the breaker points had to have the correct space or the system would not operate properly. The gap spacing was set using a feeler gauge (Figure 30-17).

Modern ignition systems use transistors and integrated chips in roughly the same way as a computer. Instead of a set of mechanical points opening and closing the primary circuit, a solid-state system opens and closes the circuit electronically. The solid-state systems are very dependable and seldom cause trouble. They are much more reliable than the old mechanical breaker points.

FIGURE 30-14 One type of governor uses a vane that is activated by the flow of air created by the fins on the flywheel.

FIGURE 30-15 An armature is composed of many stacks of thin iron strips that have been riveted together.

The Cooling System

A small engine can fire and ignite the fuel mixture over 2,000 times per minute. The heat generated by the combustion can be as much as 3,000°F. If the heat is allowed to build, the moving parts in the engine will expand to the point where the engine will seize up. This means that the heat must be dissipated to keep the engine operating at the proper temperature.

Many large multi-cylinder engines are cooled by a system of liquid coolant that circulates around and through the engine. However, most small engines are cooled by air that circulates around the engine. The cooling system on air-cooled engines generates a velocity of moving air by the use of fins on the flywheel. These fins serve as a fan to move air across and around the engine. Stationary fins on the engine head

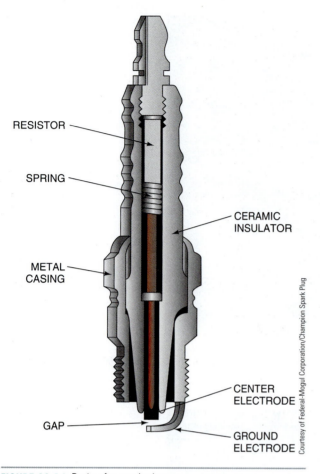

Courtesy of Federal-Mogul Corporation/Champion Spark Plug

FIGURE 30-16 Parts of a spark plug.

and block create more surface for the rapidly moving air to circulate around (Figure 30-18). A sheet metal shroud encircles the engine to route air over these fins. If all the fins are kept clean and the shroud is in place, the engine should run at the proper operating temperature.

The Lubrication System

Throughout the inside of the engine, metal parts move against each other. This movement builds up friction that leads to wear and heat generation. It is essential that oil be circulated to all internal parts within the crankcase. In most four-cycle engines, the oil is held in a reservoir called an oil sump (Figure 30-19). When the engine is not running, the oil settles in a pool at the bottom of the crankcase. When the engine starts, a device on the end of the rod cap, called an *oil dipper*, slings oil as the crankshaft turns. An alternative is a gear-like device that is turned by the crankshaft and agitates the pool of oil to coat all the interior surfaces. As mentioned earlier, two-cycle engines are lubricated by the oil that is mixed in the fuel.

The oil in an engine serves four purposes:

1. It lubricates moving parts.
2. It cleans off carbon deposits that build up from the combustion process.

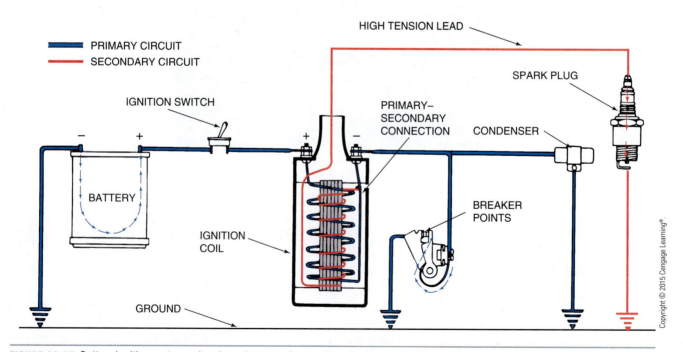

FIGURE 30-17 Battery ignition systems showing primary and secondary circuits.

Copyright © 2015 Cengage Learning®

FIGURE 30-18 Cooling fins on the engine block and cylinder head help to dissipate heat.

3. As it circulates and splashes around the inside of the engine, it cools the engine's interior and dissipates heat.
4. It helps seal the gap between the rings and cylinder wall.

A thorough discussion of the different types and grades of oil is included in Unit 31 ("Small Engine Maintenance and Repair").

FIGURE 30-19 Oil is held inside a compartment called an oil sump.

SUMMARY

Small engines are an important part of agricultural industry. Although they have existed for a long time with a wide variety of types, they all operate in basically the same way. The fundamentals of operation use the elementary principles of physics to make the engine run. Although the engines seem very complicated, they are actually composed of several less-complicated systems. A thorough understanding of the basic principles of operation and how the systems work together is essential to the operation, maintenance, and repair of the engines.

Student Activities

1. Disassemble a small engine, and label the parts. Reassemble the engine.

2. Make a list of the different machines at your home and school that use small engines. Tell whether each engine is a two- or a four-cycle engine.

3. Collect different types of carburetors from discarded engines. Label them according to their proper names, and tell how they differ in operation.

Relevant Web Sites

Two-Stroke Engine by Matthew Tomaini
YouTube

Four-Stroke Engine by Matthew Tomaini
YouTube

Self-Evaluation

A. Multiple Choice. Select the best answer.

1. The thick plate that covers the cylinder is called a
   a. cap
   b. head
   c. seal
   d. valve

2. The four-stroke cycle begins with the piston moving
   a. right
   b. left
   c. up
   d. down

3. The compression ratio for most small engines is
   a. 6:1
   b. 7:2
   c. 16:1
   d. 2:1

4. Piston rings are needed to
   a. keep oil from getting in the combustion chamber
   b. provide better compression
   c. provide more power on the power stroke
   d. all of these

5. The valves are opened and closed by the action of
   a. the throttle
   b. the wrist pin
   c. the camshaft
   d. the bearings

6. When the exhaust valve is open and the intake valve is closed, the engine is on the
   a. power stroke
   b. intake stroke
   c. exhaust stroke
   d. compression stroke

7. When the engine is on the compression stroke
   a. the intake valve is open
   b. the exhaust valve and the intake valve are open
   c. both valves are closed
   d. only the exhaust valve is closed

8. The part of a coil that is made of several hundred turns of heavy-gauge wire is called
   a. the main circuit
   b. the secondary circuit
   c. the primary circuit
   d. the complete circuit

9. The purpose of oil in an engine is to
   a. lubricate
   b. cool
   c. clean
   d. all of these

10. Heat in an engine is dissipated by means of
    a. cooling fins
    b. a radiator
    c. a water jacket
    d. none of these

B. **Matching.** Match the terms in column I with the correct definition in column II.

Column I

1. piston
2. governor
3. crankshaft
4. primary circuit
5. magneto
6. power stroke
7. valve
8. intake stroke
9. cylinder
10. secondary circuit

Column II

a. high-voltage circuit
b. speed-control device
c. controls the flow of gas into and exhaust out of the cylinder
d. engine process of moving fuel and air into the cylinder
e. low-voltage circuit of an ignition system
f. shaft with an offset projection
g. capped, sliding cylinder within a cylinder
h. engine process in which burning fuel expands
i. tube in which the piston works
j. produces electricity by magnetism

C. **Completion.** Fill in the blanks with the word or words that make the following statements correct.

1. In the late 1800s, engines referred to as _____ _____ _____ were developed to replace steam engines.

2. An _____ _____ mounted at the opening of the carburetor cleans the air before it enters the engine.

3. To prevent the engine from running too fast, a _____ is attached to the carburetor that slows the engine when a certain speed is reached.

4. The _____ _____ connects the piston to the crankshaft.

5. The pin in the center of the piston is called a _____ pin.

6. Breaker points have been replaced by _____-_____ systems.

7. The fuel system consists of the _____, _____, _____, and _____ _____.

8. Most small engines have an ignition system known as a _____ in which the spark is created by an electric current generated by a process called _____ _____.

9. The heat generated by the combustion of the fuel mixture can be as much as _____.

10. The electrical system has a device called an _____ that is comprised of many layers of thin metal.

D. **Brief Answer.** Briefly answer the following questions.

1. Until about 1982, breaker points were used to open and close most small engine primary circuits. How is this done today? What is the advantage of the modern systems?

2. What are the components of the fuel system?

3. What is the ignition system that most small engines use called? How is the spark created in this system?

4. Why is oil not needed in the crankcase of a two-cycle engine?

5. What are the three most common causes of compression loss?

UNIT 31

Small Engine Maintenance and Repair

Objective

To maintain and perform repairs on small engines.

Competencies to be developed

After studying this unit, you should be able to:

- Practice safety precautions.
- Identify tools needed for engine maintenance and repair.
- Perform the basic maintenance procedures required for small engines.
- Explain the procedures for repairing the systems of a small engine.
- Troubleshoot using a sequential procedure.

Materials List

- Spark plug
- Flywheel wrenches
- Flywheel puller
- Torque wrench
- Micrometer
- Telescoping gauge
- Ring expander
- Ring compressor
- Valve grinding tool

Terms to Know

- air filter
- oil-foam filter
- dry-element filters
- dual-element filter
- octane rating
- air-fuel mixture
- overhaul
- service classification
- viscosity
- Society of Automotive Engineers (SAE)
- micrometer
- telescoping gauge
- Plastigage
- torque

SAFETY PRECAUTIONS

When working with small engines, it is essential to observe all safety precautions to avoid injury. Safety precautions include the following:

- Store gasoline and other fuels in approved containers.
- Wipe up all oil and fuel spills immediately. Place rags that have fuel or oil on them in an approved fireproof container.
- Keep paper and other flammable materials away from the work area.
- Read and observe all safety precautions found in the operator's manual and the repair manual.
- Work only in well-ventilated areas.
- Keep an approved fire extinguisher in the work area.
- Do not use an open flame or electrical equipment in the work area.
- Wear protective goggles at all times.
- Use the proper tool for the proper job.

TOOLS FOR ENGINE REPAIR

There is an endless list of tools used for mechanical work on engines and drive systems. However, a limited set of basic tools is adequate for small engine work. Always use the proper tool for the job to be done. Use of the wrong tool can result in damage to the engine and/or injury to the person using the tool.

One set each of socket, open-end, and box wrenches ranging in size from $\frac{1}{4}$ inch to 1 inch will generally be sufficient for small engine work. Sockets in this size range may have $\frac{1}{4}$-inch, $\frac{3}{8}$-inch, or $\frac{1}{2}$-inch drive handles. Similar wrench sets in metric sizes will enable work to be done on a wider range of engines.

Other general tools needed for small engine repair are a set of hex wrenches, 6-inch and 10-inch adjustable wrenches, slip-joint pliers, long-nose pliers, an 8- to 12-ounce ball peen hammer, chisels, and punches. Several sizes of standard and Phillips-head screwdrivers round out the basic tools. These and other tools are pictured in Unit 7 ("Hand Tools, Fasteners, and Hardware"). An individual will encounter the need for additional tools as the range of repair work increases.

Small engines have parts such as valves, pistons, rings, and bearings that require specialized tools for repair. Gear pullers and wrenches used for removing the flywheel are also needed. These are pictured in Figure 31-1.

Copyright © 2015 Cengage Learning®

FIGURE 31-1 Some of the special tools needed to repair small engines.

All machinery needs periodic maintenance, and small engines are no exception. If the engines are properly maintained, they can give years of satisfactory service. Usually the service is not complicated and does not take time. As with the engines themselves, maintenance and repairs are usually very simple. Once you understand the principles of operation discussed in Unit 30 ("Fundamentals of Small Engines"), both maintenance and repairs are logical and relatively easy to do (Figure 31-2). Remember that each of the engine systems needs some care and maintenance and will occasionally need repair.

MAINTENANCE OF SMALL ENGINE SYSTEMS

Fuel System

The fuel system needs the most maintenance. Dirt or water in the fuel will make the engine misfire or stall, and dust in the air can enter the cylinder and scratch the cylinder walls. The engine will then lose compression and power. Care of the air filter is essential. There are a few basic steps to follow in keeping the air filter clean.

An **air filter** is a cleaner attached to the carburetor that is designed to remove dirt and dust from the air before the air mixes with fuel. An engine can be ruined in just a few hours of operation if the air cleaner is not working properly. This is especially true under dusty conditions. Most small engine air filters should be cleaned after every 25 hours of operation, or more frequently under dusty conditions. There are essentially three types of air filters used on modern small engines. These include the oil-foam, dry element, and

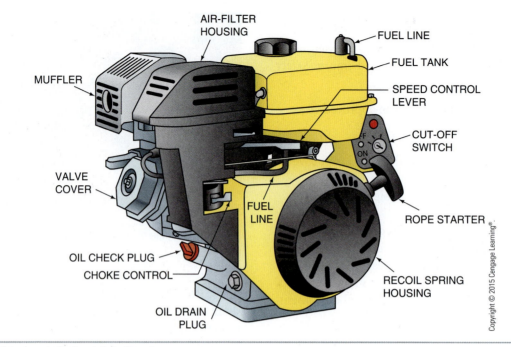

FIGURE 31-2 A four-cycle horizontal shaft engine.

dual-element types. Some of the older models may have another type known as an oil-bath filter.

Oil-Foam Type. To service an **oil-foam filter**, the following general procedure is suggested. This procedure is in Figure 31-3.

Procedure

1. Remove the cover to the air filter.
2. Carefully remove the foam filter.

3. Wash the foam in kerosene or water with detergent.
4. Squeeze out the excess liquid.
5. Dry the foam by squeezing it with a dry cloth.
6. If the foam is torn or crumbles, replace it.
7. Saturate the foam with the recommended motor oil for the engine.
8. Squeeze the foam to remove the excess oil.
9. Carefully reassemble the air filter.

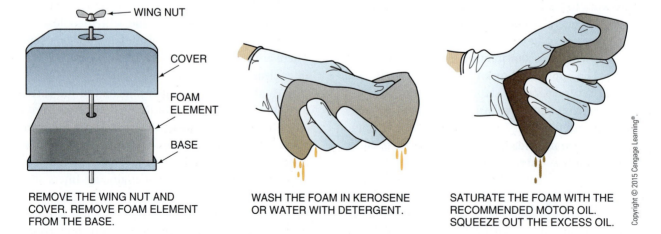

REMOVE THE WING NUT AND COVER. REMOVE FOAM ELEMENT FROM THE BASE.

WASH THE FOAM IN KEROSENE OR WATER WITH DETERGENT.

SATURATE THE FOAM WITH THE RECOMMENDED MOTOR OIL. SQUEEZE OUT THE EXCESS OIL.

FIGURE 31-3 Servicing an oil-foam air filter.

FIGURE 31-4 A dry-element filter in need of cleaning.

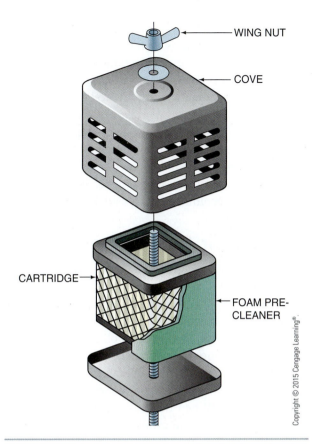

FIGURE 31-5 A duel-element filter has both a dry element and an oil-foam cover.

Dry-Element Type. **Dry-element filters** are efficient and easy to maintain. To clean a dry element, simply remove the metal cover and element (Figure 31-4). Tap the bottom of the element on a flat surface to dislodge large particles of dirt and chaff and then reinstall the element. Make sure the filter is not completely clogged. If it is clogged, replace it and be sure to replace the element at the time interval recommended by the engine manufacturer. Replacement is recommended whenever the element becomes damaged or oily, or appears to be restricting airflow. Any element that has even a small hole or slight tear must be replaced.

Dual-Element Type. A **dual-element filter** has both a dry element and an oil-foam cover (Figure 31-5). The same procedures described for the oil-foam and dry-element cleaners are used here. The oil-foam cover keeps most dirt from entering the cleaner and serves as a precleaner. The foam should be serviced every 25 hours. The dry element needs servicing less frequently. Some manufacturers recommend changing the dry element after about 100 hours of service. If any part of the filter element is torn or has a hole in it, replace the element.

Oil-Bath Type. Oil-bath cleaners are found on some older-model engines. They consist of a metal container with a metal mesh core. The following procedure is used to service this type of cleaner.

Procedure

1. Remove the nut and top cover.
2. Lift out the wire mesh core.
3. Lift off the metal container.

4. Pour the oil and dirt out of the container into a proper disposal container.
5. Wash the container in kerosene, and wipe it dry.
6. Add motor oil to the fill mark on the container.
7. Flush the wire mesh core in kerosene or other appropriate solvent.
8. Remove the excess solvent from the core by wiping it with a clean cloth.
9. Reassemble the unit.

Another key to proper maintenance is to use only clean, fresh gasoline of the appropriate type. Gasoline becomes stale if it sits for a few weeks. This is because part of the ingredients of gasoline evaporates and leaves a fuel that is more difficult to ignite. Old gasoline is said to be less volatile. Also, gasoline that sits for a long period of time has a tendency to have water in it as a result of moisture condensation within the tank. Additives can be added to the gasoline that greatly extend the volatility. They can be found at any auto parts store. Simply add the product to gasoline in the ration recommended on the label. Gasoline can be extended for a year by using the additive. Gasoline is usually designated by the **octane rating**. This rating refers to the volatility of the gasoline and how easily it ignites. Small engines are designed to run on regular gasoline that is rated about 87 octane. Using higher octane gasoline may cause the engine to knock. Knocking is usually caused by the mixture of gasoline and air igniting inside the engine's combustion chamber at the wrong time. Using a higher rating than the manufacturer recommends is not advisable.

Carburetor Adjustment

Small engine carburetors are adjusted correctly at the factory, and they seldom get out of adjustment on their own. In fact, some states do not allow the sale of engines with adjustable carburetors. A poorly adjusted carburetor may cause more air pollution than a correctly adjusted one. Unfortunately, carburetor settings are frequently changed by operators when engines do not start or when they run incorrectly because of other problems. When the real cause of the problem is corrected, the engine cannot perform well because the carburetor is out of adjustment. Typical carburetor adjustment points are shown in Figure 31-6.

Adjusting the Idle-Speed Adjustment. Most
carburetors have an idle-speed adjusting screw. The engine's idle speed is increased when the screw is turned clockwise and decreased when the screw is turned counterclockwise. The engine idle speed should be set with a speed indicator. A speed indicator is a device used to measure the speed at which the crankshaft or other parts turn, in revolutions per minute (rpm or r/min). It is important to keep the idle speed within the range specified by the manufacturer.

Adjusting the Air-Fuel Mixture. The **air-fuel
mixture** must be carefully controlled if small engines are to run correctly. Most engines have one slotted-head needle valve to adjust the air-fuel mixture at idle, and one for medium- and high-speed operation. These valves are found on the carburetor and are called the idle-mixture adjustment and the main mixture adjustment. On most engines, these adjustment screws are turned clockwise to reduce the proportion of fuel and make the mixture leaner. A leaner mixture has less fuel and more air. Turning the screw counterclockwise increases the proportion of fuel to air, creating a richer mixture.

Adjusting the Idle Mixture. Idle mixture is the
air-fuel mixture needed when the engine is running without a load on the engine. The operator's manual should be consulted before any attempt is made to adjust a small engine. However, in the absence of a manual, the following steps are generally followed to make an idle-mixture adjustment. Note that most modern engines have the idle mixture set at the factory and cannot be changed.

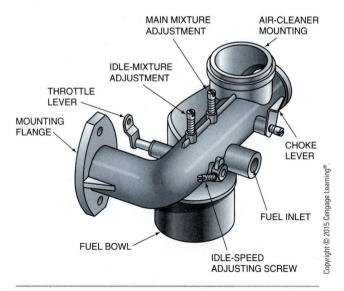

FIGURE 31-6 Typical carburetor adjustment points.

MAIN MIXTURE ADJUSTMENT

AIR-CLEANER MOUNTING

IDLE-MIXTURE ADJUSTMENT

THROTTLE LEVER

MOUNTING FLANGE

CHOKE LEVER

FUEL INLET

FUEL BOWL

IDLE-SPEED ADJUSTING SCREW

Copyright © 2015 Cengage Learning®.

Procedure

1. Start the engine and let it warm up to the recommended operating temperature.

2. Set the idle-speed adjustment so that the engine idles at the recommended speed.

3. Slowly turn the idle-mixture-adjustment screw clockwise until the engine slows down and starts to labor. To labor means to struggle or work hard to keep running.

4. Slowly turn the screw counterclockwise until the engine runs smoothly, and continue slowly turning it counterclockwise until the engine decreases in speed.

5. Slowly turn the screw in the opposite direction to the point where the engine runs smoothly.

6. Reset the idle-speed adjustment so that the engine idles at the specified speed.

Adjusting the Main Mixture. When making a main mixture air-fuel adjustment, the following steps are commonly used.

Procedure

1. Warm up the engine to its operating temperature.
2. Set the idle speed to the manufacturer's specification.
3. Be certain the idle mixture is correctly adjusted.
4. Set the throttle on the engine at its high-speed operating position.
5. Slowly turn the main mixture adjustment screw clockwise until the engine labors.
6. Slowly turn the screw counterclockwise until the engine runs fast and then begins to slow down.
7. Slowly turn the screw in the opposite direction until the engine runs smoothly.
8. Put the engine under load to check its performance.
9. If the engine does not accelerate or pull well, turn the screw counterclockwise slightly to make the mixture richer.

Carburetor Overhaul

Carburetor overhaul is needed occasionally. When **overhaul** is necessary, an overhaul kit consisting mostly of gaskets and needle valves with matching jets, or seats, should be used. A needle valve is a long, tapered shaft and a jet, or seat, is a hole shaped to receive the needle and control the flow of fuel. Some types of carburetors have a diaphragm that may need to be replaced. This part is responsible for pumping gasoline from the tank to the carburetor. Follow the instructions provided in the kit. A commercial carburetor solvent may be necessary if fuel deposits have accumulated in hidden passages.

Carburetor disassembly and reassembly must be done with careful attention to details. Use appropriate-size screwdrivers and wrenches to avoid damage to screw slots, needle valves, and seats. Clean all parts in kerosene or special carburetor solvent (never use gasoline). Reassemble with new gaskets, seats, and needle valves. Tighten screws with moderate torque; do not turn needle valves tightly against their seats, or the needle will be damaged and make proper carburetor adjustment impossible. Use the information provided with the overhaul kit and the manufacturer's materials.

Ignition System

Ignition systems in modern small engines seldom need servicing. The armature components are sealed and well protected. Other parts such as the flywheel magnet and wiring last a long time with reasonable care. The ignition part that needs the most care is the spark plug. After 100 hours of operation, the spark plug should be replaced. At one time, plugs were removed and cleaned. Now, spark plugs are replaced because they are so inexpensive and cleaning them can sometimes do more harm than good. If the plug continues to foul, there is a problem with the fuel system or oil is getting into the combustion chamber (Figure 31-7).

(LEFT)   (RIGHT)

FIGURE 31-7 The spark plug on the right has the correct color and appearance. The plug on the left needs to be replaced.

A A wire gauge should be used to check the gap.

B A tool on the gauge is used to properly set the gap.

FIGURE 31-8 Setting the proper gap is essential when changing the spark plug.

When replacing spark plugs, be sure to buy the exact type recommended by the engine manufacturer. Plugs that do not fit properly can do irreparable damage to the engine. Before the old plug is removed, all dirt and debris should be cleaned from around the plug. This prevents any foreign material from getting into the combustion chamber through the spark plug hole. The new plug will have to be gapped to the proper distance. Usually the gap is about 0.030 inch, but be sure to look up the specification recommended by the manufacturer. The gap is set using a spark plug gauge. Do not use a blade-type feeler gauge because the gap in the plug is usually concave in shape. The wire spark plug gauge provides the correct thickness and shape to properly set the gap. The gap is opened and closed using a tool on the gauge (Figure 31-8). Install the new plug and tighten until it is snug. Be sure not to over-tighten because the threads can be stripped out if too much torque is applied.

Lubrication System

It is important to keep the crankcase of a four-cycle engine filled to the correct level with the recommended type of oil. Also, some of the newer two-cycle engines, such as outboard boat engines, use an oil reservoir that automatically mixes the oil and gasoline at the correct ratio. It is important to keep the oil level checked on these engines also. The correct oil level for most engines is indicated by a mark on a dipstick or by a filler plug located so that oil will fill to the level of the plug. Before the engine is started each time, the oil level must be checked (Figure 31-9). If the engine does not have the proper amount of oil, undue wear will occur in the engine.

FIGURE 31-9 The correct oil level can be determined with a dipstick or by using a filler plug that overflows when the crankcase has received the proper amount of oil.

Remember that engine oil has four major functions: to lubricate, to cool, to seal, and to clean by carrying contaminants away from the engine. Oils for internal combustion engines are classified according to their performance under various operating conditions. Performance ratings are then designated by code letters in the classification. Most manufacturers of small gas engines recommend the use of oil with a **service classification** of SE, SF, SD, or SC. The service classification is clearly marked on every container of motor oil.

Oil is also rated according to the **viscosity**, or thickness of the oil. It is essential that the appropriate oil viscosity be used. Oil that is too thin will not function properly, and an oil that is too thick may not be able to circulate through all the parts. Engines need lighter weight oil as temperatures drop. Oil thickens

FIGURE 31-10 An oil's weight and service rating are clearly marked on every container.

as it gets cold, so that it cannot flow between tight-fitting moving parts. Therefore, in many areas, the oil must be changed as the seasons change. Multiviscosity oil can satisfy the viscosity requirements of both hot and cold temperatures. For example, oil rated 10W-30 will have the viscosity of 10-weight oil when the weather is cold and the viscosity of a 30-weight oil when the weather is hot. It is important to use oil of the correct viscosity grade, as rated by the **Society of Automotive Engineers (SAE)**. Always refer to the engine manufacturer's recommendation for proper oil viscosity (Figure 31-10).

After each 25 hours of engine use, the crankcase oil should be drained while the engine is warm. Hot oil flows easier and carries the contaminants with it. The crankcase should then be refilled to the proper level using the recommended oil. Additional attention may be needed if the engine has a reduction gear box. In this case, there is generally one hole for filling, one hole for checking the oil level, and one hole for draining the oil. Each hole has a plug (Figure 31-11). Again, refer to the manufacturer's recommendations for the oil type and change interval.

Other Maintenance

The cooling system and the compression system need little maintenance other than making sure the proper amount of oil is in the crankcase. Also the cooling fins on the engine need to be kept clean and free from debris. Engines that have been under heavy load should never be stopped when they are hot. Always let the engine run at idle speed for a few minutes before shutting it down. To prevent any possible fire from a hot engine, let the engine completely cool before storing the equipment away. If the engine has a series of belts and pulleys, make sure the belts are in good operating condition with no frays or breaks. Adjust the tension on the belts to the manufacturer's specifications. Linkages such as those on the throttle and governors should be periodically lubricated with a light oil (Figure 31-12).

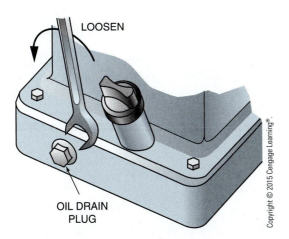

A To change engine oil, remove the oil drain plug and drain the oil in the crankcase while the engine is warm. Replace and tighten the drain plug, and then refill the crankcase with new oil through the oil filler plug.

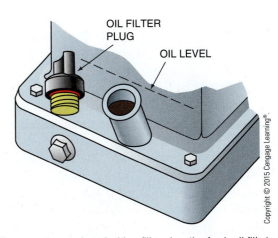

B On an engine equipped with a filler plug, the fresh oil fills to the level of the filler plug.

FIGURE 31-11 Changing oil on an engine that has a filler plug.

FIGURE 31-12 Linkages should be lubricated with a light oil.

TROUBLESHOOTING AND REPAIR

Even with the best maintenance, engines will occasionally malfunction and need repair. Locating the problem area is the key to efficient and effective repairs of the engine. Almost always, the problem can be isolated to one of three areas: ignition, fuel, or compression. If all of these systems are functioning properly, the engine should run. However, if even one of these three systems is not functioning properly, the engine will not run or will operate very poorly. In order to locate and repair the problem, a systematic check of the systems is essential.

Ignition System

Because several problems can occur with the ignition system, a logical sequence of checks is needed. First,

make sure the spark plug wire is correctly in place. Next, check the spark output by using a spark tester connected between the spark plug and spark plug wire (Figure 31-13). If there is no spark or a weak spark, the problem is probably either a bad armature, a grounded wire, or a bad ignition module. Check wires running to all kill switches to make sure there are no bare spots and that the wire is not grounding improperly.

If a good strong spark is occurring, the armature is producing electricity, and it is getting to the plug. However, the timing of the spark may be off. As described in Unit 30, the electricity for the system is generated by a magnet on the flywheel passing by the armature. If the magnet is not passing by at the correct time, the timing will be off.

The flywheel is generally held in correct position with the crankshaft by a soft metal key in the keyway. This ensures correct engine timing. The key will shear off if the engine is subjected to an abrupt stop such as a mower blade hitting an object. The shearing of a key decreases the chances of crankshaft breakage. If the key is only partially sheared, the timing will still be off (Figure 31-14).

To check the key, first remove the shroud from the engine. This usually can be accomplished by removing several small bolts. When the shroud is removed, take off the screen and remove the flywheel nut, using the correct tools (some of the flywheel nuts have left-hand threads). You should then be able to see the key

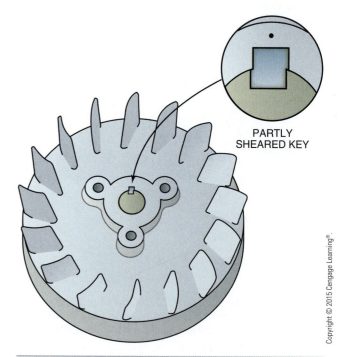

PARTLY SHEARED KEY

FIGURE 31-14 The flywheel key helps keep the engine in proper time and helps protect the engine.

FIGURE 31-13 A spark tester indicates whether the ignition system is delivering a strong enough spark.

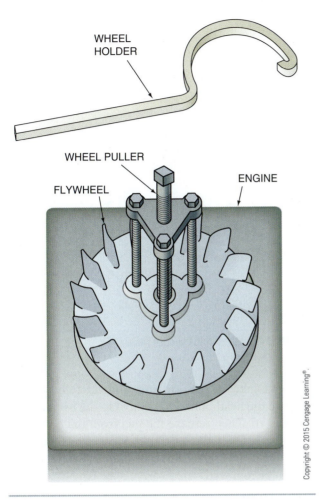

WHEEL HOLDER

WHEEL PULLER

FLYWHEEL

ENGINE

FIGURE 31-15 Pulling a flywheel.

FIGURE 31-16 In troubleshooting fuel system problems, remove the spark plug and check for an odor of gasoline. If gasoline is not present, the fuel system is probably not functioning properly.

in the keyway, and a sheared or partially sheared key should be apparent. To replace the key, it is necessary to remove the flywheel. Appropriate pullers are needed to remove flywheels without breakage. To pull the flywheel, remove the crankshaft nut, install the puller, tighten the puller to exert pulling force, and tap the center bolt of the puller with a light blow using a steel hammer (Figure 31-15). The impact should cause the wheel to pop off of the tapered shaft. Be sure to replace the key with an exact replacement. Any variation will result in poor performance or damage to the engine.

Fuel System

Once the electrical system is ruled out, begin a check of the fuel system. Start by making sure there is gas in the tank. Next, remove the spark plug after attempting to start the engine several times. Check the end of the plug for gasoline (Figure 31-16). If the odor of gasoline is present on the plug or in the cylinder, the fuel system is probably working correctly. If the odor of gasoline is not present, trace the flow of fuel from the tank to the

carburetor and look for fuel line blockages or restrictions. Also, there may be a tear in the diaphragm in the carburetor. This prevents the gasoline from being lifted from the tank to the carburetor. Replace the diaphragm or the entire carburetor.

Another problem might be flooding. Flooding refers to an excessive amount of gasoline in the cylinder. Such a condition can be caused by applying the choke too long. A more difficult problem to correct involves the float needle valve sticking in the carburetor. If this is the case, the carburetor must be disassembled and corrections made.

Compression System

The most difficult system to repair is the compression system. It contains many parts, and most are on the inside of the engine. Repairs require a disassembly of the engine to reach the parts. Compression is checked by the use of a compression tester. If the system is not building sufficient compression, the engine will malfunction. If the compression system is weak, a decision must be made whether to overhaul the engine or to replace it. Most of the less-expensive small engines probably should be replaced rather than overhauling the entire engine. Because the block is made of aluminum, the cylinders may eventually wear to the point where repairs are ineffective. However, if the engine is expensive and the cylinders do not have too much wear, then an overhaul may be feasible.

To begin the overhaul, make sure you have a clean, orderly place to work (Figure 31-17). Have several containers on hand to store parts as they are removed.

FIGURE 31-17 A clean, orderly place to work is necessary for performing an engine overhaul.

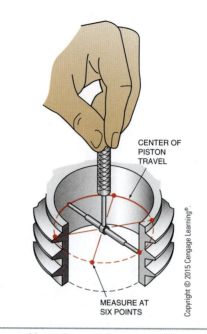

CENTER OF PISTON TRAVEL

MEASURE AT SIX POINTS

FIGURE 31-19 Measuring cylinder wear.

Reference books are essential so bring a repair manual with the specifications for your particular engine to the work area. Schematic drawings of how the engine is assembled will be helpful.

Begin by removing the gasoline tank and pouring the gasoline into an approved container. Never pour it on the ground or down a sink. Carefully remove the linkages. Fuel system linkage parts may be damaged if forced. Care is needed when disconnecting or reconnecting fuel system parts. During disassembly, observe which hole each link goes in, to ensure proper reassembly. Disconnect links by loosening components and rotating them until the link separates on one end. It is best not to disconnect any more linkage than is necessary.

Remove the shroud and flywheel using the procedure previously discussed. Be sure to place all bolts and parts in containers where you can find them during reassembly (Figure 31-18). Once the flywheel is

removed, take off the armature and carefully store it. Remove the head bolts and gently tap the head until it is loosened. Then remove the pan from the bottom of the engine. Once the head and pan are removed, the rod cap, piston, crankshaft, camshaft, and valve assembly can be removed.

After removing the piston, rings, and connecting rod assembly, the cylinder must be analyzed for wear, using an inside **micrometer** or **telescoping gauge** to determine the nature and extent of cylinder wear (Figure 31-19). Cylinders tend to wear to an elliptical out-of-round shape; if the wear is too great, new piston rings will not seal properly. Heavy cast-iron blocks may be rebored to a larger diameter and oversized pistons installed. However, most inexpensive small engines have aluminum blocks and cannot be rebored. Engines with aluminum blocks must be replaced if wear exceeds specification limits.

Piston and Rings. Piston rings are removed and installed with a ring expander (Figure 31-20). The ring grooves and piston are then cleaned with a soft wire brush to remove all carbon deposits. New piston rings are then installed and ring-gap and ring-groove clearances checked. Pistons are designed with various ring configurations and designs. Pistons typically have two compression rings designed to prevent compression blowby or leakage (Figure 31-21). The oil ring (or bottom ring) is designed to prevent crankcase oil from getting past the piston. Oil getting past the piston rings is burned in the cylinder, causing smoking, oil consumption, and

FIGURE 31-18 All parts of the disassembled engine should be neatly laid out for easy reassembly.

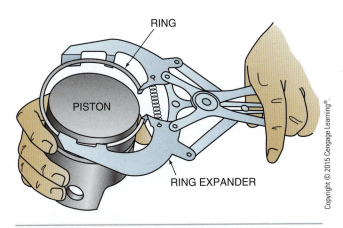

FIGURE 31-20 Using a ring expander to remove rings.

carbon deposits. A piston ring compressor is used to force the rings into their grooves so that the piston can be pushed down into the cylinder (Figure 31-22).

Bearings. Bearings are located where the crankshaft connects to the block and where the rod and rod cap connect to the crankshaft. On small engines, they are generally polished aluminum surfaces. A coat of oil serves to reduce friction. These bearings may wear to the point where they are no longer functional. To

determine the degree of wear, a threadlike material called **Plastigage** is used to measure connecting rod bearing clearance. Plastigage is a carefully designed material that flattens out uniformly when pressed. The material is laid across the bearing surface, and the bearing cap is attached and **torqued**. The bearing cap is removed again, and the width of the Plastigage measured with a special gauge. The gauge converts the measurement to thousandths of an inch of bearing clearance. If the bearing clearance is not within factory specifications, the rod and bearing cap should be replaced (Figure 31-23).

The micrometer is used to measure outside surfaces of shafts to determine the nature and extent of wear on the shaft (Figure 31-24). Generally, if the crankshaft and camshaft are too worn, the engine is replaced because the expense of replacing the shafts is not justified when compared to the cost of a new engine. In larger, more expensive engines, costs may justify replacing the parts.

Valves. Cylinder valves control the entry of air and fuel into and the exit of exhaust gases out of the cylinder. Valves are either intake or exhaust type. Valves, stems, and valve heads must fit the valve guides and valve seats according to specifications. Valves are

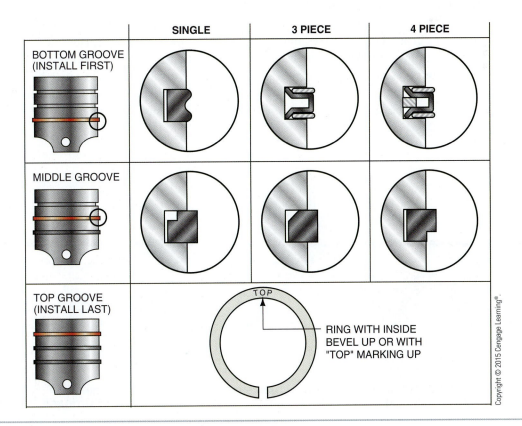

FIGURE 31-21 Types of piston rings for a three-ring (and three-groove) design.

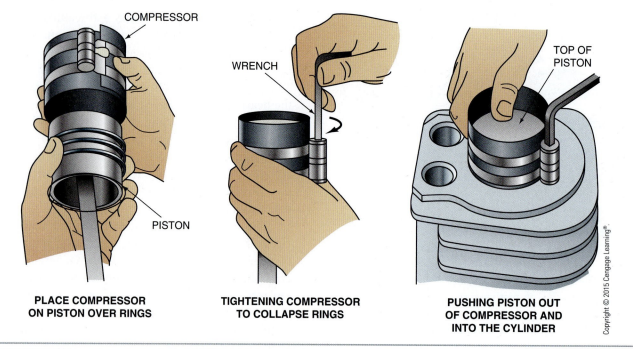

PLACE COMPRESSOR
ON PISTON OVER RINGS

TIGHTENING COMPRESSOR
TO COLLAPSE RINGS

PUSHING PISTON OUT
OF COMPRESSOR AND
INTO THE CYLINDER

COMPRESSOR

PISTON

WRENCH

TOP OF
PISTON

Copyright © 2015 Cengage Learning®.

FIGURE 31-22 Using a ring compressor to install a piston in a cylinder.

removed from the engine using a valve spring compressor to remove the valve keepers or valve pin.

The flat part of a valve is called the head and the long, round section is the stem. The outer edge of the valve head is the margin and the tapered section is the face. The valve face may be ground or re-dressed until the margin is one-half its original thickness. The face and seat must each be ground at matching 30-degree or 45-degree angles, according to specifications. New or reground valves must be lapped in to fit the

seat for a perfect seal. This is done by putting a gritty valve-lapping compound on the valve seat and rotating the valve back and forth with a valve grinding tool (Figure 31-25).

The valve stem is held in alignment by a valve guide. Valve guides should be replaced and reamed to correct specifications when new valves are installed. Valves must have a valve stem clearance of about 0.010 inch ($^1/_{100}$ inch) or as specified by the manufacturer. The correct gap is achieved by grinding the end of the valve

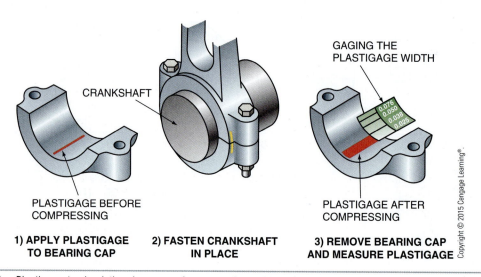

1) APPLY PLASTIGAGE
TO BEARING CAP

2) FASTEN CRANKSHAFT
IN PLACE

3) REMOVE BEARING CAP
AND MEASURE PLASTIGAGE

CRANKSHAFT

GAGING THE
PLASTIGAGE WIDTH

PLASTIGAGE BEFORE
COMPRESSING

PLASTIGAGE AFTER
COMPRESSING

Copyright © 2015 Cengage Learning®.

FIGURE 31-23 Using Plastigage to check the clearance of a connecting rod bearing.

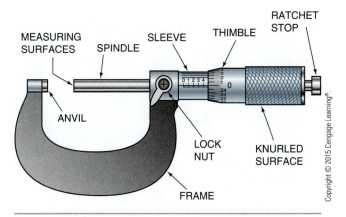

FIGURE 31-24 An outside 1-inch micrometer. Micrometers are used to measure the thickness of round and other shaped objects very accurately.

FIGURE 31-26 The letters "OHC" stand for "overhead cam."

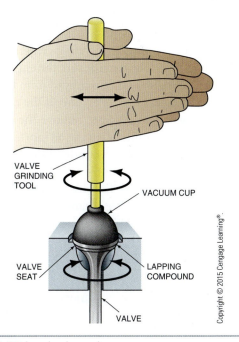

FIGURE 31-25 Lapping in a valve.

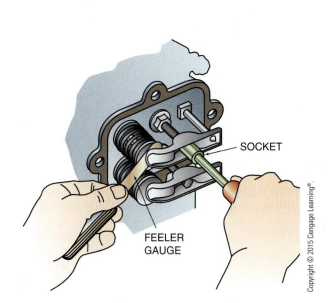

FIGURE 31-27 Checking valve stem clearance with a feeler gauge.

stem or by adjusting valve tappets. In some small engines (and most large engines) the valve tappets are rods that are moved by the camshaft to open the valves. These engines have the valves and the camshaft in the head instead of the block. These are designated by the letters OHC (for overhead cam) (Figure 31-26). Valve clearance is measured using a feeler gauge, with the valve closed and the camshaft removed (Figure 31-27).

Reassembly. Reassembly is done in reverse order of disassembly. All parts should be clean, and all matching surfaces must be clear of grease, dirt, or carbon

deposits. All gaskets should be replaced. Before assembling moving parts, coat them with oil to prevent them from being dry when the engine is first started. Nuts and bolts must be tightened to a specific torque specification. A torque wrench is used to give the exact torque. If bolts are too loose, they can fall out while the engine is running. If they are tightened too much, parts can be distorted or the bolts can break. Special care is needed when tightening a cylinder head. The head bolts must be torqued gradually and in the correct sequence. Failure to do so can result in a warped head and leaking head gasket (Figure 31-28).

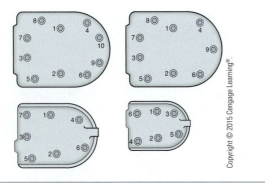

FIGURE 31-28 Recommended sequence for tightening head bolts on various cylinder heads. Check engine repair manual for torque specifications.

FIGURE 31-29 The starter rope uses the pull of a rope wrapped around a pulley for turning power needed to start the engine.

Rope and Wind-up Starters

Small engines are frequently started by hand. The rope starter utilizes the pull of a rope wrapped around a pulley for turning power to start the engine. The rope and pulley have a spring mechanism that rewinds the rope onto the pulley when the starter rope is pulled (Figure 31-29). Starter ropes need replacing and starter assemblies need to be cleaned and lubricated periodically. Follow the instructions in the operator's manual for servicing starter assemblies (Figure 31-30).

Storing Small Engines

Whenever a gas engine needs to be stored, it is important to leave it in good condition. A little care when storing can save time when the engine is to be restarted. To properly store an engine, drain all the fuel from the tank into an approved disposal container. Used oil and gasoline can be recycled at recycling centers located in most communities. Next, make sure the engine is clean and dry. All debris should be removed from the cooling fins and other places on the engine. If the engine is to be stored longer than 60 days, remove the spark plug and pour a few spoonfuls of engine oil into the opening. Pull the starter rope slowly until the piston is at the top of the cylinder. This spreads oil on the cylinder, piston, and combustion chamber. Replace the spark plug. Then make sure all linkages are cleaned, oiled, and move freely. Store the engine and implement in a dry location.

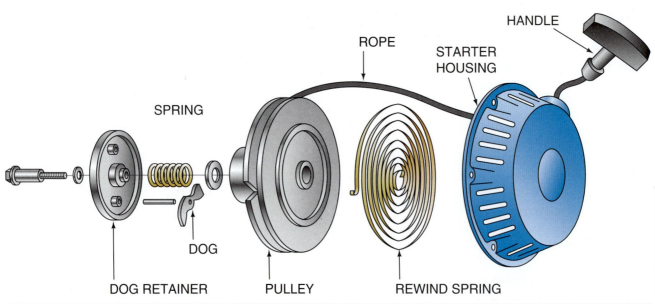

FIGURE 31-30 Rope starter assemblies.

SUMMARY

Anyone can learn to properly service and maintain small engines. The most important tool is the service manual that belongs with each engine. By following the manufacturers' recommendations, you will be sure to keep the engine properly maintained. Problems with malfunctioning engines can be located by following a logical sequence of troubleshooting steps (Figure 31-31). Remember that all problems can be traced to ignition, compression, or fuel malfunctions. Servicing and repairing small engines can be enjoyable and productive. It can keep engines running longer and more efficiently. In addition, engine servicing skills can lead to lifelong savings and excellent career opportunities in mechanics, sales, and service.

Common Troubles and Remedies

The following charts list the most common troubles experienced with gasoline engines. Possible causes of trouble are given along with probable remedy. Paragraph references direct the reader to the correction portion of the handbook.

4-Cycle Engine Troubleshooting Chart

Cause	Remedy and Reference
Engine Fails to Start or Starts with Difficulty	
No fuel in tank.	Fill tank with clean, fresh fuel.
Shut-off valve closed.	Open valve.
Obstructed fuel line.	Clean fuel screen and line. If necessary, remove and clean carburetor.
Tank cap vent obstructed.	Open vent in fuel tank cap.
Water in fuel.	Drain tank. Clean carburetor and fuel lines. Dry spark plug points. Fill tank with clean, fresh fuel.
Engine overchoked.	Close fuel shut-off and pull starter until engine starts. Reopen fuel shut-off for normal fuel flow.
Improper carburetor adjustment.	Adjust carburetor.
Loose or defective magneto wiring.	Check magneto wiring for shorts or grounds; repair if necessary.
Faulty magneto.	Check timing, point gap, and if necessary, overhaul magneto.
Spark plug fouled.	Clean and regap spark plug.
Spark plug porcelain cracked.	Replace spark plug.
Poor compression.	Overhaul engine.
No spark at plug.	Disconnect ignition cut-off wire at the engine. Crank engine. If there is a spark at spark plug, ignition switch, safety switch or interlock switch is inoperative. If no spark, check magneto. Check wires for poor connections, cuts or breaks.
Electric starter does not crank engine.	See 12-volt starter troubleshooting chart.
Engine Knocks	
Carbon in combustion chamber.	Remove cylinder head and clean carbon from head and piston.
Loose or worn connecting rod.	Replacing connecting rod.
Loose flywheel.	Check flywheel key and keyway; replace parts if necessary. Tighten flywheel nut to proper torque.

FIGURE 31-31 Troubleshooting guide.

Source: Tecumseh Corporation; Products Company, Engine and Gear Service Division.

4-Cycle Engine Troubleshooting Chart (Cont.)

Cause	Remedy and Reference
Worn cylinder.	Replace cylinder.
Improper magneto timing.	Time magneto.

Engine Misses Under Load

Cause	Remedy and Reference
Spark plug fouled.	Clean and regap spark plug.
Spark plug porcelain cracked.	Replace spark plug.
Improper spark plug gap.	Regap spark plug.
Pitted magneto breaker points.	Replace pitted breaker points.
Magneto breaker arm sluggish.	Clean and lubricate breaker point arm.
Faulty condenser (except on Tecumseh Magneto).	Check condenser on a tester; replace if defective.
Improper carburetor adjustment.	Adjust carburetor.
Improper valve clearance.	Adjust valve clearance to 0.010 cold.
Weak valve spring.	Replace valve spring.

Engine Lacks Power

Cause	Remedy and Reference
Choke partially closed.	Open choke.
Improper carburetor adjustment.	Adjust carburetor.
Magneto improperly timed.	Time magneto.
Worn rings.	Replace rings.
Lack of lubrication.	Fill crankcase to the proper level.
Air cleaner fouled.	Clean air cleaner.
Valves leaking.	Grind valves and set to 0.010 cold.

Engine Overheats

Cause	Remedy and Reference
Engine improperly timed.	Time engine.
Carburetor improperly adjusted.	Adjust carburetor.
Air flow obstructed.	Remove any obstructions from air passages in shrouds.
Cooling fins clogged.	Clean cooling fins.
Excessive load on engine.	Check operation of associated equipment. Reduce excessive load.
Carbon in combustion chamber.	Remove cylinder head and clean carbon from head and piston.
Lack of lubrication.	Fill crankcase to proper level.

Engine Surges or Runs Unevenly

Cause	Remedy and Reference
Fuel tank cap vent hole clogged.	Open vent hole.
Governor parts sticking or binding.	Clean and, if necessary, repair governor parts.
Carburetor throttle linkage or throttle shaft and/or butterfly binding or sticking.	Clean, lubricate, or adjust linkage and deburr throttle shaft or butterfly.
Intermittent spark at spark plug.	Disconnect ignition cut-off wire at the engine. If spark, check ignition switch, safety switch, and interlock switch. If no spark, check magneto. Check wires for poor connections, cuts or breaks.
Improper carburetor adjustment.	Adjust carburetor.
Dirty carburetor.	Clean carburetor.

FIGURE 31-31 (*continued*)

4-Cycle Engine Troubleshooting Chart (Cont.)

Cause	Remedy and Reference
Engine Vibrates Excessively	
Engine not securely mounted.	Tighten loose mounting bolts.
Bent crankshaft.	Replace crankshaft.
Associated equipment out of balance.	Check associated equipment.
Engine Uses Excessive Amount of Oil	
Engine speed too fast.	Using tachometer adjust engine rpm to spec.
Oil level too high.	To check level turn dipstick cap tightly into receptacle for accurate level reading.
Oil filler cap loose or gasket damaged causing spillage out of breather.	Replace ring gasket under cap and tighten cap securely.
Breather mechanism damaged or dirty causing leakage.	Replace breather assembly.
Drain hole in breather box clogged causing oil to spill out of breather.	Clean hole with wire to allow oil to return to crankcase.
Gaskets damaged or gasket surfaces nicked causing oil to leak out.	Clean and smooth gasket surfaces. Always use new gaskets.
Valve guides worn excessively thus passing oil into combustion chamber.	Ream valve guide oversize and install $1/32$" oversize valve.
Cylinder wall worn or glazed, allowing oil to bypass rings into combustion chamber.	Bore hole, or deglaze cylinder as necessary.
Piston rings and grooves worn excessively.	Reinstall new rings and check land clearance and correct as necessary.
Piston fit undersized.	Measure and replace as necessary.
Piston oil control ring return holes clogged.	Remove oil control ring and clean return holes.
Oil passages obstructed.	Clean out all oil passages.
Oil Seal Leaks	
Crankcase breather.	Clean or replace breather.
Old seal hardened and worn.	Replace seal.
Crankshaft seal contact surface is worn undersize causing seal to leak.	Check crankshaft size and replace if worn excessively.
Crankshaft bearing under seal is worn excessively, causing crankshaft to wobble in oil seal.	Check crankshaft bearings for wear and replace as necessary.
Seal outside seat in cylinder or side cover is damaged, allowing oil to seep around outer edge of seal.	Visually check seal receptacle for nicks and damage. Replace P.T.O. cylinder cover, or small cylinder cover on the magneto end if necessary.
New seal installed without correct seal driver and not seating squarely in cavity.	Replace with new seal, using proper tools and methods.
New seal damaged upon installation.	Use proper seal protector tools and methods for installing another new seal.
Bent crankshaft causing seal to leak.	Check crankshaft for straightness and replace if necessary.
Oil seal driven too far into cavity.	Remove seal and replace with new seal, using the correct driver tool and procedures.
Breather Passing Oil	
Engine speed too fast.	Use tachometer to adjust correct rpm.
Loose oil fill cap or gasket damaged or missing.	Install new ring gasket under cap and tighten securely.

FIGURE 31-31 (*continued*)

4-Cycle Engine Troubleshooting Chart (Cont.)

Cause	Remedy and Reference
Oil level too high.	Check oil level. Turn dipstick cap tightly into receptacle for accurate level reading. DO NOT fill above full mark.
Breather mechanism damaged.	Replace reed plate assembly.
Breather mechanism dirty.	Clean thoroughly in solvent. Use new gaskets when reinstalling unit.
Drain hole in breather box clogged.	Clean hole with wire to allow oil to return to crankcase.
Piston ring end gaps aligned.	Rotate end gaps so as to be staggered 90 degrees apart.
Breather mechanism installed upside down.	Small oil drain holes must be down to drain oil from mechanism.
Breather mechanism loose or gaskets leaking.	Install new gaskets and tighten securely.
Damaged or worn oil seals on end of crankshaft.	Replace seals.
Rings not properly seated.	Check for worn, or out of round cylinder. Replace rings. Break in new rings with engine working under a varying load. Rings must be seated under high compression, or in other words, under varied load conditions.
Breather assembly not assembled correctly.	See section on Breather Assembly.
Cylinder cover gasket leaking.	Replace cover gasket.

Troubleshooting Carburetion

Points to Check for Carburetor Malfunction

Trouble	Corrections
Carburetor out of adjustment.	3-11-12-13-15-19.
Engine will not start.	1-2-3-4-5-6-8-11-12-14-15-22.
Engine will not accelerate.	2-3-11-12.
Engine hunts (at idle or high speed).	3-4-8-9-10-11-12-14-20-21-24.
Engine will not idle.	4-8-9-11-12-13-14-19-20-21-22.
Engine lacks power at high speed.	2-3-6-8-11-12-19-20-22-23.
Carburetor floods.	4-7-17-20-21-22-23.
Carburetor leaks.	6-7-10-18.
Engine overspeeds.	8-9-11-14-15-18-19.
Idle speed is excessive.	8-9-13-14-15-18-19-22-23.
Choke does not open fully.	8-9-15.
Engine starves for fuel at high speed (leans out).	1-3-4-6-11-15-17-20-23.
Carburetor runs rich with main adjustment needle shut off.	7-11-17-18-20-22-23.
Performance unsatisfactory after being serviced.	1-2-3-4-5-6-7-8-9-10-11-15-16-17-18-19-20-22-23.

1. Open fuel shut-off valve at fuel tank—Fill tank with fuel.
2. Check ignition, spark plug, and compression.
3. Clean air cleaner—Service as required.
4. Dirt or restriction in fuel system—Clean tank and fuel strainers, check for kinks or sharp bends.
5. Check for stale fuel or water in fuel—Fill with fresh fuel.
6. Examine fuel line and pick-up for sealing at fittings.
7. Check and clean atmospheric vent holes.
8. Examine throttle and choke shafts for binding or excessive play—Remove all dirt or paint, replace shaft.
9. Examine throttle and choke return springs for operation.
10. Examine idle and main mixture adjustment screws and "O" rings for cracks and damage.
11. Adjust main mixture adjustment screw; some models require finger-tight adjustment. Check to see that it is the correct screw.

FIGURE 31-31 (*continued*)

Troubleshooting Carburetion

Points to Check for Carburetor Malfunction

12. Adjust idle-mixture-adjustment screw. Check to see that it is the correct screw.
13. Adjust idle speed screw.
14. Check for bent choke and throttle plates.
15. Adjust control cable or linkage, to assure full choke and carburetor control.
16. Clean carburetor after removing all non-metallic parts that are serviceable. Trace all passages.
17. Check inlet needle and seat for condition and proper installation.
18. Check sealing of welch plugs, cups, plugs, and gaskets.
19. Adjust governor linkage.

Specific Carburetor Checks for Float

20. Adjust float setting, if float type carburetor.
21. Check float shaft for wear and float for leaks or dents.

Specific Carburetor Checks for Diaphragm

22. Check diaphragm for cracks or distortion and check nylon check ball for function.
23. Check sequence of gasket and diaphragm for the particular carburetor being repaired.

Troubleshooting 12-Volt Starters

Problem	Probable Cause	Fix
Does not function.	Weak or dead battery.	Check charge and/or replace battery.
	Corroded battery terminals and/or electrical connections.	Clean terminals and/or connections.
	Brushes sticking.	Free brushes. Replace worn brushes and those which have come in contact with grease and oil.
	Dirty or oily commutator.	Clean and dress commutator.
	Armature binding or bent.	Free armature and adjust end play, replace armature, or replace starter.
	Open or shorted armature.	Replace armature.
	Shorted, open or grounded field coil.	Repair or replace housing.
	Loose or faulty electrical connections.	Correct.
	Load on engine.	Disengage all drive apparatus and relieve all belt and chain tension.
	Electric starter cranks, but no spark at spark plug.	Disconnect ignition cut-off wire at the engine. Crank engine. If there is a spark at spark plug, ignition switch, or safety switch is inoperative. If no spark, check magneto. Check wires for poor connections, cuts or breaks.
Does not function.	Electric starter does not crank engine.	Remove wire from starter. Use a jumper battery and cables and attach directly to starter. If starter cranks engine the starter is okay; check solenoid, starter switches, safety switches, and interlock switch. Check wires for poor connections, cuts or breaks.
Low rpm.	Unit controls engaged.	Insure all unit controls are in neutral or disengaged.
	Worn bearings in cap assemblies.	Clean bearings or replace cap assemblies.

FIGURE 31-31 (continued)

Troubleshooting 12-Volt Starters (Cont.)

Problem	Probable Cause	Fix
	Bent armature.	Replace armature.
	Binding armature.	Free up armature. Adjust armature end play.
	Brushes not seated properly.	Correct.
	Weak or annealed brush springs.	Replace springs.
	Incorrect engine oil.	Ensure the correct weight of oil is being used.
	Dirty armature commutator.	Clean commutator.
	Shortened or open armature.	Replace armature.
	Loose or faulty electrical connections in motor.	Correct.
Motor stalls under load.	Shorted or open armature.	Replace armature.
	Shorted field coil.	Correct, or replace housing assembly.
Intermittent operation.	Brushes binding in holders.	Free up brushes. Replace worn brushes and those that have come in contact with grease and oil.
	Dirty or oily commutator.	Clean and dress commutator.
	Loose or faulty electrical connections.	Correct.
	Open armature.	Replace armature and interlock switch.
	Break in electrical circuit.	Disconnect ignition cut-off wire at the engine. Crank engine; if spark, check ignition switch, safety switch and interlock switch. Check wires for poor connections, cuts or breaks.
Sluggish disengagement of the drive assembly pinion gear.	Dirt and oil on assembly and armature shaft.	Clean drive assembly and armature shaft and relubricate shaft splines.
	Bent armature.	Replace armature.

FIGURE 31-31 (*continued*)

Student Activities

1. Define the Terms to Know in this unit.

2. Locate a small engine that is used at your home. A lawnmower, garden tiller, snowblower, chainsaw, boat, or go-cart are examples. Develop a service schedule for the engine that will cover an entire year. Keep records on the maintenance you perform.

3. Find a small engine that is not running. Go through all the procedures outlined in this unit and locate the problem. Find the cause of the malfunction, and repair the engine based on your diagnosis.

4. Visit a small engine repair shop, and ask the owner to explain how repairs are done in the shop. Ask him or her what the most common repairs are.

Relevant Web Sites

Self Help & More, Home Improvement Tips and Advice, Small Engines
www.selfhelpandmore.com/small-engines

Self-Evaluation

A. Multiple Choice. Select the best answer.

1. Most small engine air filters should be cleaned after how many hours of service?
   a. 45 hours
   b. 35 hours
   c. 25 hours
   d. 15 hours

2. Any element in an air filter that is torn should be
   a. replaced
   b. repaired
   c. left alone
   d. oiled and dried

3. Some states do not allow small engine carburetors to be adjusted because improper adjustments cause
   a. engines to overheat
   b. too much power to the engine
   c. more air pollution
   d. too much noise

4. The device in a carburetor that controls the fuel flow is called a
   a. seat
   b. port
   c. linkage
   d. needle valve

5. If a flywheel key is partially sheared, the engine will not run because it is
   a. too hot
   b. out of time
   c. broken
   d. without fuel

6. An oil rated 10W-30 will have the viscosity of
   a. a 10-weight oil when the weather is cold and the viscosity of a 30-weight oil when the weather is hot
   b. a 30-weight oil when the weather is cold and the viscosity of a 10-weight oil when the weather is hot
   c. a 40-weight oil
   d. 10W-30 oil does not have viscosity

7. Oil should be changed when the engine is
   a. cold
   b. cool
   c. hot
   d. warm

8. Excessive fuel in the combustion chamber is called
   a. flooding
   b. swamping
   c. floating
   d. drowning

9. The most difficult system to repair is the
   a. fuel system
   b. compression system
   c. ignition system
   d. lubrication system

10. Loss of compression can be caused by
    a. worn piston rings
    b. stuck piston rings
    c. worn valves
    d. all of the above

B. Matching. Match the terms in column I with those in column II.

Column I

1. micrometer
2. torque
3. octane
4. viscosity
5. Plastigage
6. dual element
7. volatile
8. 0.030 inch
9. SF, SE, SD, SC
10. feeler gauge

Column II

a. service classification of oil
b. threadlike material used to gauge tolerances
c. used to set valves to the proper clearance
d. twisting power
e. proper gap for most small engine spark plugs
f. used to gauge thicknesses
g. rating for gasoline
h. thickness of an oil
i. air filter type
j. easy to burn

C. **Completion.** Fill in the blanks with the word or words that make the following statements correct.

1. Place rags that have fuel or oil on them in an approved _____ container.

2. Keep an approved _____ _____ in the work area.

3. Never use _____ as a solvent.

4. Wear protective _____ at all times.

5. Use the proper _____ for the proper _____.

6. When replacing spark plugs, buy the exact type recommended by the _____.

7. Almost always, an engine problem can be narrowed down to one of three areas: _____, _____, or _____.

8. Compression can be checked by using a _____ _____.

9. When a block is made of _____, the cylinders may eventually wear to the point where repairs are ineffective.

10. The head bolts must be torqued gradually and in the correct _____.

D. **Brief Answer.** Briefly answer the following questions.

1. What is the purpose of cylinder valves? Identify the parts of a valve.

2. An engine problem can almost always be isolated to one of which three areas?

3. Why may cylinders in most inexpensive small engines eventually wear to the point where repairs are ineffective?

4. How should head bolts be torqued?

5. How does one store an engine properly, if it is to be stored more than 60 days?

UNIT 32

Diesel Engines and Tractor Maintenance

Objective

To explain how diesel engines work and explain how tractors are maintained.

Competencies to be developed

After studying this unit, you should be able to:

- Explain why most modern farm machinery is powered by diesel engines.
- Explain how diesel engines work.
- List the systems of a modern tractor.
- Discuss the principles of tractor maintenance.

Materials List

- Bottles of various types of oil
- Battery hydrometer
- Tire gauge
- Tractor maintenance manuals
- Grease gun

Terms to Know

- internal combustion engine
- compression ratio
- stroke
- torque
- viscosity
- pour-point depressant
- antifoam agent
- oxidation inhibitors
- detergents
- quick coupling

HOW A DIESEL ENGINE OPERATES

Most of the power for operating agricultural machinery is supplied by **internal combustion engines**. As discussed in Unit 30, these engines may be classified as two-cycle or four-cycle engines, depending on the engine design. Engines are also classified as gasoline or diesel powered (Figure 32-1). The diesel engine is named after its German inventor, Rudolf Diesel (Figure 32-2). Diesel obtained a patent for the design in 1892, and over the next five years worked at perfecting the engine. The engine, which operated on peanut oil, was demonstrated at the Exhibition Fair in Paris in 1898.

FIGURE 32-1 Diesel engines are used to power most modern farm machines.

FIGURE 32-2 The diesel engine was invented by a German inventor named Rudolf Diesel.

The use of the vegetable oil was highlighted as a cheap means of operating an engine through the use of a renewable oil crop. In fact, diesels ran on vegetable oil up until the 1920s, when petroleum-based diesel fuel was refined on a large scale. Petroleum oil was less viscous than vegetable oil, and was relatively cheap. It is ironic that today, a lot of research is conducted on the use of biodiesel fuel. Biodiesel is a fuel derived from vegetable oil such as peanut or soybean oil.

In the early years (the first half of the twentieth century), diesel engines were used mostly for powering trains, ships, stationary machinery, and large earth-moving equipment (Figure 32-3). Most farm tractors used gasoline engines because of the initial cost and the ease of farm home repair of the tractors. In the 1960s, diesels became commonplace in farm machinery. Today, practically all of the new tractors and equipment have diesel engines. Diesels are popular for agricultural and industrial machinery because they tend to have more power, operate more efficiently, and last longer than gasoline engines. The drawbacks are that diesel engines are larger, heavier, and noisier; they smell bad; and they are more expensive to buy. They also must be more closely maintained than gasoline engines.

Gasoline differs from diesel fuel in that it is much more volatile, and it ignites more rapidly. Diesel fuel is heavier and thicker, tends to ignite more slowly, and is generally safer to handle and store. Although the operating procedures of the two engine types are much the same, there are basic differences in the fuel and ignition systems. The strokes of a four-cycle

FIGURE 32-3 In the first half of the twentieth century, diesels were used mostly to power trains, ships, stationary engines, and large earth-moving equipment.

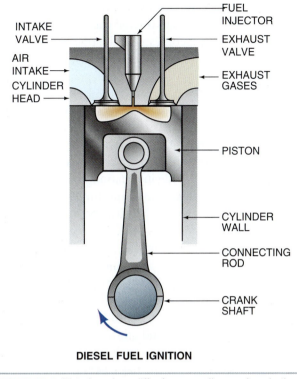

DIESEL FUEL IGNITION

FIGURE 32-4 Diesel engines differ from gasoline engines in the way fuel is introduced and in the way ignition occurs.

Source: U.S. Dept. of Energy, Office of Energy Efficiency and Renewable Energy, Freedom CAR & Vehicle Technologies Program, Just the Basics Document.

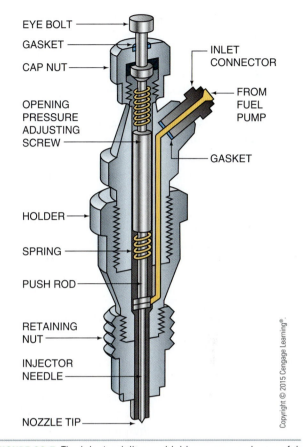

FIGURE 32-5 The injector delivers a highly compressed, powerful spray of fuel.

Copyright © 2015 Cengage Learning®.

diesel engine are the same as in the four-cycle gasoline engine. The difference is in the delivery of the fuel and the system of igniting the fuel (Figure 32-4). In the gasoline engine, the fuel is introduced through the intake valve and comes in as a mixture of gasoline and air. In older gasoline engines, a carburetor supplies air and turns the gasoline into a vapor-like consistency. The fuel mixture is sucked into the cylinder by the vacuum-like action of the piston moving down in the cylinder. More-modern gasoline engines use injectors to deliver the fuel mixture into the combustion chamber. The mixture is compressed at the top of the cylinder and ignited by a properly timed spark from the spark plug.

In a diesel engine, only air is brought into the cylinder through the intake valve. The air is tightly compressed at the top of the cylinder, where a precisely timed spray of diesel fuel is injected into the heated air. An injector pump that delivers a small amount of diesel fuel to the injectors at the proper time supplies the fuel. The injectors then deliver a very powerful blast of vaporized fuel into the compressed air at the top of the cylinder (Figure 32-5). When air is tightly compressed, heat is generated; in the diesel engine, the heat of compression is enough to ignite the fuel.

The temperature of the compressed air in the cylinder of a diesel engine is more than 800°F.

The **compression ratio** is the ratio of the combined volume of the cylinder and the combustion chamber at the beginning of the compression stroke (when the piston is all the way down) to the volume of the cylinder and combustion chamber at the end of the compression stroke (when the piston is all the way up). The compression ration for a typical gasoline engine might be about 8 to 1 (written as 8:1), while for a typical diesel engine it might be 18:1 or more. Because the compression ratio is so much greater in a diesel engine, the components of the engine must be heavier and more durable.

As mentioned earlier, diesel engines are generally more powerful than gasoline engines. They are also slower running and weigh considerably more. The increased power comes primarily as a result of a longer stroke. The **stroke** refers to the distance the piston must travel from the bottom of the cylinder to the top of the cylinder. This longer stroke gives more leverage to the crankshaft and provides more torque. **Torque**, which refers to the turning power of an engine or other machine, is measured in foot-pounds or horsepower.

Another characteristic of the diesel engine is a rattle or consistent "knock" in the engine. This is in contrast to the steady sound of a properly tuned gasoline engine. The noise from a diesel engine is due to the timing of the ignition. In a gasoline engine, the spark plug sets off the combustion at precisely the same time on each power stroke. In a diesel engine, there can be slight variations in the timing of the combustion because the ignition is created by the less-precise heat of combustion. This process is not nearly as precise as ignition set off by carefully timed spark plugs. The slight inconsistency in diesel ignition causes the characteristic sound of a diesel engine. This is another reason why diesel engines have to be made of heavier components than gasoline engines.

MAINTAINING DIESEL ENGINES AND FARM TRACTORS

The useful lifespan of any equipment is dependent on the maintenance given at the proper intervals. This is especially true of power equipment such as tractors. Modern tractors have several different systems that must be maintained (Figure 32-6). If any of the systems are neglected, the tractor does not function properly and will not last as long as it should. Most manufacturers provide maintenance schedules for the tractors they produce. This is the very best source of information about the correct service and maintenance needed to keep the tractor functioning properly. Following are some of the basic maintenance requirements, which are basically the same for all tractors. The specifics of maintaining a particular tractor can be found in the owner's manual that comes with the tractor.

The Engine

One of the most vital components of an engine is the lubricating system. If oil is not properly circulated to all the moving parts of the engine, the life of the engine will be short. Remember from a previous chapter that not only does oil provide lubrication between moving parts, but it also cools, cleans, and provides a cushion between the parts. Not only must the oil be properly circulated, but it must also be of the proper **viscosity** (thickness) and grade for the particular engine. In addition, the oil must be clean and fresh. Oil that has been in the engine too long will be dirty and worn out. This means that it must be changed at regular intervals. The owner's manual will tell you the proper interval, and the proper viscosity and grade of oil.

The Society of Automotive Engineers (SAE) rates oil. Code letters in a particular classification indicate performance ratings. The service classification is clearly marked on every container of motor oil (Figure 32-7).

These ratings denote how well the oil will perform under severe or adverse conditions. An "S" means that

FIGURE 32-6 Modern diesel tractors will provide many years of service if they are properly maintained.

FIGURE 32-7 The service classification assigned to a particular motor oil by the Society of Automotive Engineers (SAE) is clearly marked on every container.

the oil is intended for use in engines with a spark ignition, such as gasoline engines. For example, oil with a service classification of SE, SD, or SC is meant to be used in a gasoline engine. Oil grades such as CA to CF are intended for use with compression-ignited engines such as diesels. Always read and follow the manufacturer's recommendations regarding which classification of oil to use in an engine.

The SAE has also rated oil according to the viscosity (also discussed in Unit 31) or thickness of the oil. It is essential that oil with the proper viscosity be used. Oil that is too thin will not function properly, and oil that is too thick may not be able to properly circulate through all the parts. Engines need lighter-weight oil as temperatures drop because the cold temperature causes the oil to become more viscous. In very cold climates, oil can become so thick that it cannot get between tight-fitting moving parts. Therefore, in many areas, the oil must be changed as the seasons change. Oil can be rated as having different viscosity as the temperature changes. For example, oil rated as 10W30 will have the viscosity of 10-weight oil when the weather is cold and that of 30-weight oil when the weather is hot. Always refer to the engine manufacturer's recommendation for the proper oil viscosity.

Oil is more than refined petroleum. It contains several additives to improve the oil's function within the engine. The following are typical additives:

- **Pour-point depressants** allow oil to flow in cold weather.
- **Antifoam agents** counteract the natural tendency of oil to foam as it circulates under high pressure to all moving engine parts.
- **Oxidation inhibitors** prevent the oil from breaking down because of reaction with oxygen.
- **Detergents** clean engine parts and suspend tiny contaminants in the oil.

An integral part of the lubrication system is the oil filter. All the oil in the system flows through the filter, where particles of carbon, dirt, and other contaminants are trapped (Figure 32-8). While good-quality oil filters trap most of these particles, even the best filters must be replaced before they become clogged with contaminants. When the oil is changed, it is essential that the oil filter be replaced also.

When the oil filter is changed, the air filter should also be serviced or changed. In some older tractors, the air filter is a type known as an oil-bath filter. These should be disassembled, thoroughly cleaned, and filled to the proper level with the correct weight of clean oil. Newer tractors have a disposable air filter that should

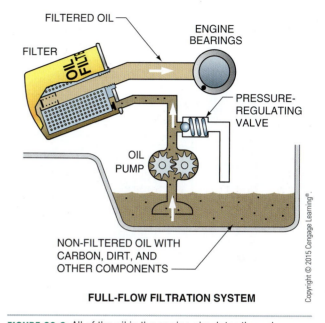

FULL-FLOW FILTRATION SYSTEM

Copyright © 2015 Cengage Learning®.

FIGURE 32-8 All of the oil in the engine circulates through the filter, which traps particles of carbon, dirt, and other contaminants.

be removed and replaced when it gets dirty. Remember to check the oil level in the engine. Too much or too little oil can lead to engine failure (Figure 32-9).

The Cooling System

As combustion occurs, a large amount of heat is built up in the engine. Unless this heat is dissipated, the engine parts will expand to the point where the engine will seize up. Most tractors use a circulating liquid coolant that absorbs the heat from the engine block. The coolant is circulated to a radiator where it is cooled. The radiator is composed of a series of small tubes that run the length of the radiator. The coolant

Copyright © 2015 Cengage Learning®./Ray Herren

FIGURE 32-9 Remember to keep checking the oil level in the engine. Too much or too little oil can lead to engine failure.

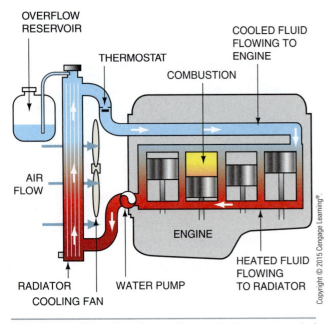

FIGURE 32-10 The radiator is a cooling mechanism composed of a series of small tubes that run along its entire length.

is forced into these small tubes where a fan driven by the engine rapidly cools the liquid (Figure 32-10). The water pump mounted on the front of the engine recirculates the coolant back through the engine where it again absorbs heat and then returns to the radiator for cooling. A device called a *thermostat* opens when the coolant reaches a certain temperature and allows the coolant to circulate. When the coolant falls below a certain temperature, the thermostat closes until it reaches the proper operating temperature. This helps the engine run at the proper temperature.

Coolant, often referred to as antifreeze, is mixed with water before it is put in the cooling system. The manufacturer will recommend the proper mixture for a particular climate. The colder the climate, the higher the concentration of antifreeze is needed. If the concentration is too great, the cooling system will not operate properly, so be sure to follow manufacturer's recommendations. During routine maintenance checks, remove the radiator cap and make sure the coolant is at the proper level in the cooling system. Examine the radiator cap to make sure that the gasket on the cap is in good shape. If the rubber gasket is cracked, the system will lose power and will not cool the engine effectively. Always use high-quality antifreeze because the coolant not only protects from freezing, but also prevents corrosion in the system.

Coolant must be changed periodically to maintain the ability to cool the engine and also to protect the engine from freezing. Remember that the major ingredient in the coolant is water. If water freezes

CAUTION!

Never open the radiator cap on a hot engine. Pressure is built up in the system and hot coolant can severely burn you. Also, be careful when using coolant. Most antifreezes are highly toxic.

inside an engine, the ice will expand enough to burst the engine block and head. Freezing has ruined many engines. A simple device called a *hydrometer* can be used to determine if the coolant should be changed (Figure 32-11). If the coolant is changed, make sure that the old coolant is properly disposed of.

The cooling system must also be kept clean. If the tractor is used for tasks such as mowing where it comes in contact with weeds, brush, or crop residue, the fins in the radiator may become clogged. If this happens, air cannot circulate through the radiator and cooling is hampered. A garden hose can be used to wash the

FIGURE 32-11 A simple device called a hydrometer is used to check the coolant to determine if it should be changed.

FIGURE 32-12 The grill and radiator should be washed with a garden hose. This tractor has residue caught in the grill and radiator.

radiator to get rid of the debris (Figure 32-12). Never use a pressure washer because the high pressure can damage the radiator fins. Be sure that the tractor is cool before washing the radiator. Spraying water on a hot tractor can result in damage.

The Fuel System

Proper maintenance of the fuel system is absolutely essential for diesel engines. This system consists of a fuel tank, an injector pump, injectors, and a filtering system. The injector pump and the injectors are machined to very precise tolerances, and any foreign materials can cause major damage. One of the worst problems is water. Water combined with the diesel fuel can cause enough corrosion to ruin the injector pump. Water filtering mechanisms must be closely monitored and serviced at regular intervals (Figure 32-13). Most fuel filter housings have valves in the bottom that can be opened to drain water. If the tractor is not used for an extended period, the fuel tank should be kept full to prevent water condensation within the tank. Fuel filters must also be changed regularly. Periodically the injectors must be cleaned and calibrated. Black smoke coming from the exhaust of the tractor is an indication that the injectors and/or the injector pump need service. Blue smoke coming from the exhaust indicates that oil is burning and the piston rings are worn. A professional in a shop equipped for diesel repair must do the servicing of injectors and injector pumps. Calibration must be done on a machine that is designed for that purpose.

Another important component of the fuel system is the air filtering system. Most engines have a series of filters that clean the air before it enters the injector pump. A very tiny foreign particle can cause a lot of damage to the pump. At the top of the intake is the first filter. This removable cap can be lifted off and cleaned. Some older tractors have oil-bath filters that have oil in the bottom of the filter that traps foreign material such as dust and other residue. Most modern tractors have dry filters that can be removed and cleaned. This can be done by shaking the filter or using low-pressure air to clean the filter. When the filter gets too dirty, it can be replaced (Figure 32-14 A–C). Always follow the tractor manufacturer's maintenance instructions.

The Hydraulic System

A section in Unit 39 ("Hydraulic, Pneumatic, and Robotic Power") explains how a hydraulic system works. A thorough understanding of the operating principles will help you appreciate the importance of maintaining the hydraulic system. Most modern tractors have several hydraulic systems, such as the lift, remote cylinders, power steering, and power brakes. Following are some of the principles of maintaining the hydraulic system.

Hydraulic System Maintenance

Hydraulic systems must have regular maintenance service to operate trouble-free. Otherwise, unnecessary and expensive repairs will result. Fluid of the proper type and grade must be kept clean and free of dirt, moisture, and other contaminants.

Fluid Level. Fluid levels must be checked frequently, and care must be taken not to introduce dirt in the process. Most tractors and other machines with hydraulic systems use a dipstick to monitor fluid levels (Figure 32-15).

Fluid Filters. Filters must be serviced at regular intervals as prescribed by the manufacturer. This involves simply unscrewing a disposable filter canister, wiping the surfaces clean, and screwing on a new filter canister (Figure 32-16).

Quick Couplings. Some hydraulic lines go to equipment that can be separated from the machine. In such cases a quick coupling is generally used in the hoses. A **quick coupling** is a device with two connected parts attached to two hose ends. The parts can be connected by retracting a spring-loaded collar on one part,

Copyright © 2015 Cengage Learning®./Ray Herren

FIGURE 32-13 Water filtering systems should be regularly serviced. The filters have a drain on the bottom.

A The top of the air intake can be removed and cleaned.

B The air filter is removed for cleaning.

C The filter can be shaken to remove dust and residue.

FIGURE 32-14 Air filters must be cleaned regularly.

inserting the second part, and releasing the collar to lock the parts together. When quick couplings are disconnected, the exposed ends should be capped to prevent dirt from entering the hose; when reconnected, the parts must be clean (Figure 32-17).

Fluid Change. Fluid systems must be drained and clean fluid and filters installed, as prescribed by the manufacturer. Such service must be done with care to ensure that dirt is not introduced into the system, that proper fluid is put back into the system, and that the old

FIGURE 32-15 Most hydraulic systems have a dipstick to check the fluid level.

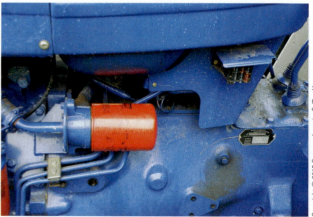

FIGURE 32-16 Most hydraulic systems have spin-on, disposable filters.

FIGURE 32-17 When quick couplings are disconnected, the exposed ends should be capped to prevent dirt from entering the hoses.

fluid is discarded properly. The latter means it should be recycled and not left to pollute the environment.

Tire Maintenance

For a tractor to operate at peak efficiency, the tires must be properly maintained (Figure 32-18). This includes keeping the correct air pressure as well as the recommended amount of liquid weight. Most tractor tires contain a solution of water and calcium chloride or other material to prevent the water from freezing in the tires. This solution is generally added to about three-fourths of the volume of the tire. This adds a great amount of weight to the rear tires that, in turn, aids in traction. Even though the tires contain a large amount of the solution, proper inflation pressure is important. Tires should be checked periodically

to ensure that they are properly inflated. This should be done with the valve stem at the top to prevent the liquid from getting in the tire gauge.

Belts and Hoses

Belts are used to drive devices such as the fan, alternator, water pump, power steering, and air conditioner. All belts should be checked periodically for cracks. As belts age, they dry out and lose much of their flexibility. When this happens, the belt can break, and the tractor will be out of service until the belt is replaced. It is much better to prevent breakage by replacing belts as they dry and crack. The tension should be properly adjusted on all belts. As a general rule, you should be able to move the belt downward about 12 inch per foot of belt span by pressing a finger on the midpoint of the belt (Figure 32-19). A belt that is too loose will not function properly, and a belt that is too tight will cause excessive wear on the pulley bearings.

Hoses are used to circulate coolant from the radiator to the engine block. Hydraulic hoses connect pumps and cylinders as well as controls in the system. A break in these hoses can cause the coolant to leak out and the engine to overheat. These, too, can become old, brittle, and cracked. If the hoses break, the system cannot function and the hydraulic fluid is lost. Both hoses and belts should be inspected periodically and replaced when they show signs of aging (Figure 32-20).

FIGURE 32-18 A tire gauge is used to check the pressure in the tires for proper inflation.

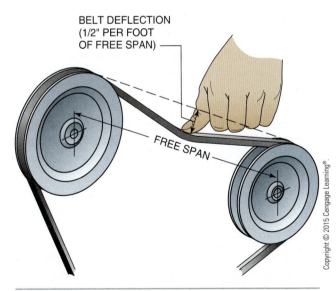

BELT DEFLECTION (1/2" PER FOOT OF FREE SPAN)

FREE SPAN

FIGURE 32-19 As a general rule, a person should be able to move a drive belt about ¼ inch downward per foot of belt span by pressing on the midpoint of the belt with a finger.

Copyright © 2015 Cengage Learning®./Ray Herren

FIGURE 32-20 Hoses used to circulate coolant from the radiator to the engine block should be checked regularly for cracks or breaks.

Copyright © 2015 Cengage Learning®./Ray Herren

FIGURE 32-21 Grease fittings should be thoroughly cleaned before the grease gun is attached.

Chassis Maintenance

The chassis of the tractor includes the axles and wheels, as well as the steering, clutch, lift, and brake mechanisms. All of the wheels have bearings that must be lubricated. Some tractors have rear-axle bearings that are lubricated from the fluid in the transmission case. If the tractor is a four-wheel drive, the front wheel bearings might also be lubricated through the drive train. Others must be lubricated with a grease gun. The front wheels on two-wheel-drive tractors have bearings that must be periodically packed with axle grease. The steering mechanism also has grease fittings for lubricating the moving parts. Before the grease gun is attached, the fittings should be cleaned (Figure 32-21). If dirt or grit is on the outside of the fitting, it will be forced into the bearings or moving parts. This will cause excessive wear. All linkages such as those on the brakes and clutch should be inspected and oiled. If the clutch is manual, the linkage should be adjusted so that there is about 1 inch of play before the clutch is engaged. If the clutch is hydraulic, be sure all fluids are at the proper level.

The Electrical System

Although most modern tractors do not have electric ignition like a lot of the older models did, they do have an electrical system. This system is responsible for charging the battery, operating the lights, powering the instrument panel, and more. Like all of the other systems, the electrical system needs periodic maintenance. In this system, the battery is the most commonly maintained component. Batteries contain acid that can cause corrosion on the terminals and connections (Figure 32-22). This corrosion must be removed or the system may not charge properly. A solution of baking soda and water can be used with a stiff brush to clean the terminals and connections. All of the solution must be washed off when cleaning is completed. A light coating of oil on the terminals can prevent corrosive buildup. Make sure that all the insulation on the wiring is intact and that the wires are securely fastened to the terminals. A loose connection can cause electrical failure.

Copyright © 2015 Cengage Learning®./Ray Herren

FIGURE 32-22 Batteries should be kept clean to prevent corrosive buildup. This battery needs to be cleaned.

SUMMARY

Almost all modern farm machinery is powered by diesel engines. They are relatively more powerful and efficient, and they last longer than gasoline engines. Proper maintenance is essential for the efficient operation and longevity of farm tractors. Systems that must be maintained include the engine, the cooling system, the fuel system, the hydraulic system, and the chassis. Understanding the basic principles on which these systems operate will help you understand why maintenance needs to be done.

Student Activities

1. Define the Terms to Know in this unit.

2. Develop a service and maintenance schedule for a tractor. If your family owns one, you may use that tractor or you may use one owned by your school.

3. Bring a basketball or football pump to class. Tightly compress the air in the pump and feel the barrel. What causes the heat? What does this have to do with a diesel engine?

4. Conduct research and write a paper on the use of biodiesel. What are the advantages? What are the disadvantages?

Relevant Web Sites

How Diesel Engines Work
www.howstuffworks.com/diesel1.htm

Squeeze Play—How a Diesel Engine Works
www.dieselpowermag.com/tech/general/0610dp_how_diesel_engine_works/viewall.html

Self-Evaluation

A. **Multiple Choice.** Select the best answer.

1. Until the 1920s, diesels ran on
   a. petroleum-based fuel
   b. steam
   c. vegetable oil
   d. alcohol

2. The compression ratio for a typical diesel engine is
   a. 8:1
   b. 18:1
   c. 6:1
   d. 16:1

3. Oil must be
   a. of the proper viscosity and grade
   b. properly circulated
   c. clean and fresh
   d. all of the above

4. A hydrometer is used to
   a. flush the radiator
   b. verify levels of pressure
   c. measure water temperature
   d. determine if coolant needs changing

5. Compression-ignited engines may require an oil grade of
   a. SC
   b. CA
   c. SD
   d. CS

6. Water combined with fuel can cause enough corrosion to ruin the
   a. injector pump
   b. filtering system
   c. fuel tank
   d. radiator

7. Blue smoke indicates that
   a. the injectors need service
   b. the piston rings are worn
   c. the filter needs replacement
   d. the injector pump is corroded

8. To check for appropriate belt tension, press at the midpoint of the belt. It should move
   a. 1 to 2 inches
   b. ½ to 1 inch
   c. ⅛ to ¾ inch
   d. ⅝ to ¾ inch

9. The lubrication of front-wheel bearings on a two-wheel-drive tractor is done
   a. through the drive train
   b. from the fluid in the transmission case
   c. by packing with axle grease
   d. with a solution of water and calcium chloride

10. If the clutch is manual, the linkage should be adjusted so that there is about _____ of play before the clutch is engaged.
    a. 1 inch
    b. 2 inches
    c. ¾ inch
    d. ⅝ inch

B. Matching. Match the items in column I with those in column II.

Column I

1. stroke

2. Rudolf Diesel

3. pour-point depressants

4. oxidation inhibitors

5. biodiesel

6. antifreeze

7. antifoam agents

8. calcium chloride

9. detergents

10. compression ratio

Column II

a. distance the piston travels from the bottom to the top of the cylinder

b. allows oil to flow in cold weather

c. comparison of the volume of the space at the top of the cylinder when piston is down, to the volume of the cylinder when piston is up

d. keeps oil fluid in spite of high pressure

e. inventor of diesel engine

f. prevents oil breakdown

g. suspends contaminants

h. derived from vegetable oil

i. prevents freezing of water in tires

j. coolant

C. Completion. Fill in the blanks with the word or words that make the following statements correct.

1. Most of the power for operating agricultural machinery is supplied by _____ _____ _____.

2. In the 1960s, _____ became commonplace in farm machinery.

3. The temperature of the compressed air in the cylinder of a diesel engine is more than _____.

4. Because the compression ratio is so much greater in a diesel engine than in a gasoline engine, the engine's components must be _____ and _____ _____.

5. The increased power of diesel engines comes primarily as a result of a _____ _____.

6. The colder the climate, the _____ the concentration of antifreeze that is needed.

D. Brief Answer. Briefly answer the following questions.

1. Why are diesel engines popular for agricultural and industrial machinery?

2. How does diesel fuel compare with gasoline?

3. Explain how fuel is delivered and ignited in a diesel engine.

4. What causes the characteristic "knock" of a diesel engine?

5. Describe the cooling system used by most tractors.

ELECTRICITY AND ELECTRONICS

UNIT 33

Electrical Principles and Wiring Materials

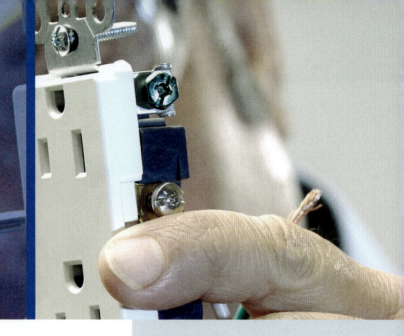

Objective

To use principles of electricity and safety for planning simple wiring systems.

Competencies to be developed

After studying this unit, you should be able to:

- Describe some basic principles of electricity and magnetism.
- Use safety practices with electricity.
- Steam-clean machinery.
- Describe the relationship among volts, amperes, and watts.
- Select materials for electrical wiring.
- Design simple wiring systems.

Materials List

- Horseshoe magnet
- Samples of wire, nonmetallic sheathed cable, armored cable, and conduit
- Electric meter dial

Terms to Know

- electricity
- filament
- fluorescent light
- resistance
- conductor
- insulator
- amperes
- volts
- watts
- ohm
- Ohm's Law
- volt-ohm-milliampere meter (VOM meter)
- milliampere
- magnetism
- permanent magnet
- poles
- north pole
- attraction
- repulsion
- south pole
- magnetic flux
- magnetic field
- reversing the polarity
- electromagnet
- commutator
- armature

continued

Electricity is the major power source for stationary equipment in houses, farm and ranch buildings, and agribusinesses. It is the energy source commonly used for driving machinery, and for lighting, heating, and cooling. Some knowledge of electricity is essential for the safe use of electrical equipment. Understanding how to wire simple circuits and make minor electrical repairs is also useful. It is important to maintain electrical circuits and equipment properly to ensure their long life and safe operation.

PRINCIPLES OF ELECTRICITY

Heat and Light

Electricity is a form of energy that can produce light, heat, magnetism, and chemical changes. Light can be produced by heating a special metal element or **filament** in a vacuum tube called a bulb. The flow of electricity into the bulb must be carefully controlled so that the filament glows without burning out. In addition, electricity flowing through certain gases causes them to glow. A **fluorescent light** glows as a result of electricity flowing through a gas (Figure 33-1). The tubes contain a small amount of mercury and an inert gas, usually argon. When a current is passed through electrodes and across the mercury in the tube, mercury gas is created. Electricity interacts with the atoms in the light to create light.

Heat is produced when electricity flows through metals with some difficulty. Any tendency of a material to prevent electrical flow is called **resistance**. If electricity flows easily, the metal is said to be a good **conductor**. Silver is an excellent conductor, copper is a very good conductor, and aluminum is a good conductor. Because silver is so expensive, copper is generally used in wiring systems; aluminum may be used when the price of copper is high. Aluminum is also used extensively in outside lines where heat buildup is not a problem. Applications include the high-voltage power lines that cross the countryside, as well as the overhead wires that stretch from building to building in residential areas or on farms. Aluminum is no longer used in branch circuits with wires smaller than 8 gauge.

© Yufeng Wang/Shutterstock.com

FIGURE 33-1 Fluorescent lights glow as the flow of electricity through gases creates light.

Heat in wires, switches, outlets, motors, and lights is not desirable. It wastes electrical energy, causes materials to deteriorate, heats up the surrounding areas, and may cause fires. Therefore, the proper design of wiring systems is important. In properly designed electrical systems, heat is kept within acceptable limits. In heating systems, when heat is generated by design, the heat is produced by using elements with just the right amount of resistance. Heat from electricity is clean and easy to control.

A material that provides great resistance to the flow of electricity is called an **insulator**. Examples of good insulators include rubber, glass, vinyl, and air. Anyone working with electricity must be aware that insulation is relative. For example, a very thin layer of rubber or vinyl will prevent electrical flow if the voltage is low, such as 12 volts for automobile lights. However, to prevent electricity at 30,000 volts in a spark plug wire from jumping to ground requires a much thicker layer of rubber.

Amperes, Volts, and Watts

Amperes, volts, and watts must be understood to design electrical circuits or install electrical materials safely. **Amperes** are a measure of the rate of flow of electricity in a conductor. **Volts** are a measure of electrical pressure. **Watts** are a measure of the amount of energy or work that can be done by amperes and volts.

The following equations define how amperes, volts, and watts are related:

$$\text{Watts} = \text{Volts} \times \text{Amperes}$$

$$\text{Volts} = \frac{\text{Watts}}{\text{Amperes}}$$

$$\text{Amperes} = \frac{\text{Watts}}{\text{Volts}}$$

For example, if a 200-watt lightbulb operates at 120 volts, it will draw 1.67 amperes of electricity:

$$A = \frac{W}{V} = \frac{200}{120} = 1.67 \text{ amperes}$$

A 5-ampere motor running on 120 volts will consume 600 watts of electricity:

$$W = V \times A = 120 \times 5 = 600 \text{ watts}$$

A refrigerator motor rated at 3 amperes that consumes 360 watts of electricity should be plugged into a 120-volt circuit:

$$V = \frac{W}{A} = \frac{360}{3} = 120 \text{ volts}$$

These relationships are known as the West Virginia formula: W = VA. By remembering the basic formula, any one of the three quantities can be determined if the other two are known. Notice that the formulas used in these illustrations are all from the basic W = VA formula.

Ohm's Law

A physicist named George Simon Ohm made a number of important discoveries about electric current. As a result, the unit used to measure a material's resistance to the flow of electrical current is known as the **ohm**. Ohm discovered that the flow of electricity through a conductor is directly proportional to the electrical or electromotive force that produces it. The relationship he discovered between electric current, electromotive force, and resistance is called **Ohm's Law**.

Electromotive force is measured in volts and is represented by a capital E in Ohm's Law. The rate of flow of electricity through a conductor is measured in amperes, represented by a capital I in Ohm's Law. The tendency of a material to prevent electrical flow, or resistance, is represented by a capital R.

$$E = \text{Volts}$$

$$I = \text{Amperes}$$

$$R = \text{Resistance}$$

Ohm's Law states that electromotive force is equal to amperes times resistance, or E = IR. From this basic formula, amperes (I) and resistance (R) can also be computed: I = E/R and R = E/I. In any situation, one of the values can be computed if the other two are known. This knowledge is useful when designing, testing, or using electrical equipment.

Voltage, amperage, and resistance can be measured using various types of meters. Voltage is measured with a voltmeter; amperage is measured with an ammeter; and resistance is measured by an ohmmeter. A popular combination meter used in electronics is a **volt-ohm-milliampere meter (VOM meter)**. A **milliampere** is a thousandth of an ampere.

Magnetism and Electricity

Electricity flowing through a conductor results in magnetism. **Magnetism** is a force that attracts or repels iron or steel (Figure 33-2). Magnetism can also be created by exposing iron or steel to magnetic forces. If iron or steel holds its magnetism, it is said to be a **permanent magnet**. Permanent magnets may be used to create or generate electricity.

FIGURE 33-2 Magnetism is a force that attracts or repels iron or steel.

Magnets have two ends, called **poles**. These are designated as the north pole and the south pole. The **north pole** of one magnet **attracts** the south pole of another magnet but **repels**, or pushes away, the north pole of another magnet. In a similar manner, the **south pole** of one magnet will repel the south pole of another but attract the north pole. In summary, like poles repel and unlike or opposite poles attract (Figure 33-3).

Lines of magnetic force, often called **magnetic flux**, occur in patterns in the air between the two poles. The pattern is referred to as the **magnetic field** (Figure 33-4).

Electric Motors

The principle of magnetism is the basis upon which electric motors operate. A strong magnet can be made

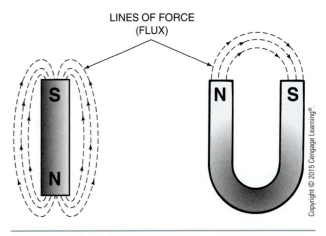

LINES OF FORCE (FLUX)

FIGURE 33-4 Lines of magnetic force occur in a pattern called the magnetic field.

by wrapping many coils of insulated wire around an iron core and passing electrical current through the wire (Figure 33-5). When the current stops flowing, the magnetic field ceases to exist (Figure 33-6). If the wires are switched to opposite battery terminals, the

FIGURE 33-5 A strong magnet can be made by wrapping many coils of insulated wire around an iron core and passing electrical current through the wire.

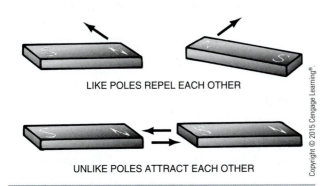

LIKE POLES REPEL EACH OTHER

UNLIKE POLES ATTRACT EACH OTHER

FIGURE 33-3 Like poles repel and unlike poles attract.

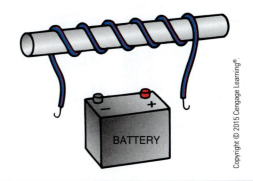

BATTERY

FIGURE 33-6 When the current stops flowing, the magnetic field ceases to exist.

poles of the magnet will reverse, but the magnetic field will remain (Figure 33-7). However, reversing the direction of current flow **reverses the polarity** of the magnet. The unit is called an **electromagnet**.

A motor may be made by bending the electromagnet into the shape of a horseshoe and suspending a permanent magnet on a bearing between the poles. When current flows, the unlike poles of the electromagnet and the permanent magnet will attract each other, causing the permanent magnet to make a half turn (Figure 33-8). If the wires are reversed at the battery, the direction of the current and polarity of the magnet will be reversed, and the permanent magnet will make another half turn (Figure 33-9). In a motor, the current is reversed by a part called a **commutator**. The commutator causes the motor to run continuously in the same direction. The rotating (turning) magnet is called the **armature**, and the magnetic forces around the armature are called the **field** (Figure 33-10).

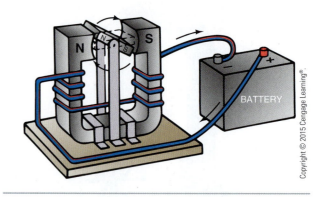

FIGURE 33-9 If the wires are reversed at the battery, the polarity of the electromagnet reverses, too, and the permanent magnet makes another half turn.

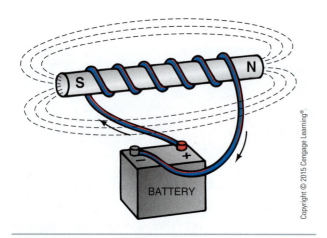

FIGURE 33-7 If the wires of the electrical circuit are switched to opposing battery terminals, the poles of the magnet reverse, but the magnetic field remains.

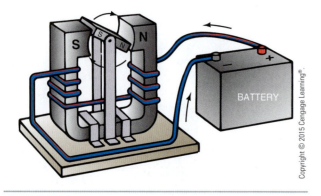

FIGURE 33-8 The attraction between unlike poles causes the permanent magnet to make a half turn.

Devices that produce electricity by magnetism are called generators and alternators. A **generator** produces direct current, which means the electricity flows in one direction. An **alternator** produces alternating current, which reverses in direction, or alternates, each time the armature turns. Generators and alternators must be powered by human or animal power, waterwheel, gas or diesel engine, turbine, or an electric motor getting electricity from a separate source. A **turbine** is a high-speed rotary engine driven by water, steam, or other gases.

Electric Motor Maintenance. Modern electric motors are designed for long life with little maintenance (Figure 33-11). Many have permanently lubricated bearings.

Some older motors have two oil cups or tubes with a wick to hold oil for the two shaft bearings. The operator's manual supplied by the manufacturer typically calls for about 10 drops of oil per year in these tubes. More oil than specified will cause dirty oil accumulations on interior and exterior parts. On the other hand, failing to use oil as specified by the manufacturer shortens the life of the motor due to excessive bearing wear. Therefore, it is important to follow the instructions provided, and to adjust oil volumes and intervals according to operating conditions.

Circuits

A source of electricity plus two or more wires connected to a load such as a light, heater, or motor are known collectively as a **circuit**. A circuit may be likened to a circle. If the circle is broken so that current cannot flow through it, it is said to be an **open circuit**.

A The commutator reverses the current and causes the motor to run continuously in one direction.

B The armature is a rotating magnet that fits inside the commutator.

FIGURE 33-10 The main parts of an electric motor are the commutator and the armature.

FIGURE 33-11 Most modern electric motors are sealed and need very little maintenance.

All circuits must include an object with resistance that falls within a certain range. Otherwise, the electricity will flow through the circuit and back to its source too rapidly and blow fuses, burn wires, and drain batteries. Such a condition is called a **short circuit**.

Electricity can travel back to its source through the earth as well as through a wire. A circuit always includes a wire to carry the current back to its source. However, safety requires that an additional wire or conductor be provided to all metal parts in the system in case the current gets out of its circuit. Making this additional connection between a piece of equipment and the earth is called **grounding**. It is accomplished by driving a solid copper rod or a 1-inch-diameter or larger galvanized steel pipe into soil that is always moist. The rod or pipe, called a ground **rod**, is connected to the service panel where electricity enters a building. The electrical equipment is connected to the grounding bar at the service panel by a conductor

called a ground **wire**. Grounding equipment is a standard safety practice. In this way electricity that accidentally gets out of its circuit is channeled to the earth through a ground wire rather than through the body of a human or animal.

ELECTRICAL SAFETY

There are two deadly hazards associated with electric current: shock and fire. **Shock** refers to the body's reaction to an electric current. Shock can interfere with a normal heartbeat and result in the injury or death of the victim. Fire may result when electrical conductors overheat or when a spark is produced by an electrical current jumping an air space and igniting flammable materials nearby.

All people who work around electric current should observe safety practices at all times to avoid shock and fire. Some precautions advised by authorities are:

- Never disconnect or damage any safety device that is provided by the manufacturer or specified by electrical codes.
- Do not touch electrical appliances, boxes, or wiring with wet hands or wet feet.
- Do not remove the long ground prong from a three-prong 120-volt plug.
- Use ground fault interrupters in kitchen, bathroom, laundry, and outdoor circuits, or wherever moisture may increase shock hazard. A **ground-fault circuit interrupter (GFCI)** is a device that cuts off the electricity if even very tiny amounts of current leave the normal circuit. This device reduces the likelihood of human shock injury. GFCIs must be installed in accordance with the instructions provided by the manufacturer.

- Immediately discontinue the use of any extension cord that feels warm or smells like burning rubber.
- Do not place extension cords under carpeting.
- Install all electrical wiring according to the specifications of the National Electrical Code.
- Use only double-insulated portable tools or tools with three-wire grounded cords.
- If a fuse is blown or circuit breaker is tripped, determine and correct the problem before inserting a new fuse or resetting the breaker.
- Fuses and circuit breakers are designed to prevent the circuit from being overloaded. Do not create a fire hazard by installing higher-capacity fuses or breakers than the system is designed to handle.
- Do not leave heat-producing appliances, such as irons, hair dryers, and soldering irons, unattended.
- Place all heaters and lamps away from combustible materials.
- Keep the metal cases or cabinets of electrical appliances grounded at all times.
- Do not remove the back of a television set. There is danger of causing a 20,000- or 30,000-volt shock even when the unit is unplugged.
- Keep electrical motors lubricated and free of grease and dirt.
- Keep appliances dry to reduce shock hazard and prevent rust.
- Do not use any switches, outlets, fixtures, or extension cords that are cracked or damaged in any way.
- Follow manufacturer's instructions for installation and use of all electrical equipment.

ELECTRICAL WIRING

An electrical system must meet several conditions to be satisfactory. It must:

- be safe
- be convenient
- be expandable
- look neat
- provide sufficient current

Service Entrance

Electrical power comes to the home or farm by overhead or underground wires. The power company provides a transformer, service drop, and appropriate wiring to an entrance head. An **entrance head** is a waterproof device used to attach exterior wires to interior wires of a building. Most modern systems have entrance cables that are buried underground. This eliminates the need for an entrance head. The **transformer** converts high voltage from the power lines to 240 volts for home and farm installations. The **service drop** is an assembly of electrical wires, connectors, and fasteners used to transmit electricity from the transformer to the entrance head and on to the service entrance panel. The **service entrance panel** is a box with fuses or circuit breakers where electricity enters a building. Wires feed from the service drop through the insulated entrance head or underground cables to a meter box. The electric meter plugs into the meter box. The power continues on to the service entrance panel for distribution to branch circuits (Figure 33-12).

Meter

All electricity that passes through the system is measured by the **meter**, in kilowatt-hours (Figure 33-13). **Kilo-** means 1,000. A **watt-hour** is the use of 1 watt for 1 hour. (A 100-watt bulb that burns for 1 hour will consume 100 watt-hours of electricity.) A **kilowatt-hour** is the use of 1,000 watts for 1 hour.

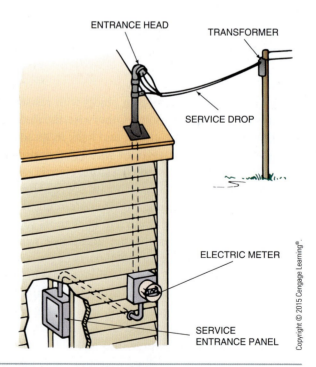

FIGURE 33-12 Traditionally, electrical power moved from high-voltage power lines to local homes and farms through a transformer, a service drop, an entrance head, an electric meter, and a service entrance panel.

Copyright © 2015 Cengage Learning®.

FIGURE 33-13   Electric meters gauge the amount of electricity used in a month, in kilowatt-hours.

The meter will convert any amount of electricity into kilowatt-hours. A power-company representative reads the meter at regular intervals, and the customer is billed according to the number of kilowatt-hours used.

Branch Circuits

Most circuits begin from the service entrance panel. They are called **branch circuits** since they branch out into a variety of places and for a variety of purposes (Figure 33-14). A branch circuit generally includes only one motor or a series of outlets or a series of lighting fixtures (Figure 33-15). It is important to provide the correct size of wire and fuse or circuit breaker for the load on each circuit.

A **fuse** is a plug or cartridge containing a strip of metal that melts when more than a specified amount of current passes through it. A **circuit breaker** is a switch that trips and breaks the circuit when more than a specified amount of current passes through it. A circuit breaker can be reset after it trips. A fuse must be replaced if it blows.

> ## CAUTION!
>
> Fuses and circuit breakers are used to protect circuits from damage and fire. Never destroy this protection by installing fuses or breakers with a larger ampere rating than is recommended for the size of wire being protected.

When planning electrical wiring systems or reading blueprints, it is useful to know standard electrical symbols (Figure 33-16). The symbols are used on drawings to indicate the location of outlets, receptacles, switches, and special appliances.

Types of Cable

Three types of wiring are widely used in home, farm, and commercial installations: nonmetallic sheathed cable, armored cable, and conduit.

Nonmetallic sheathed cable consists of copper or aluminum wires covered with paper and vinyl for insulation and protection (Figure 33-17). Certain types of nonmetallic sheathed cable are waterproof and are suitable for burial in soil.

Armored cable consists of a flexible metal sheath with individual wires inside (Figure 33-18). The wires are insulated with vinyl and wrapped in paper for protection. Armored cable provides good protection against mechanical damage. It is fairly easy to install in narrow and irregular building sections. However, its tendency to rust and develop short circuits makes it unsuitable for use in damp areas.

Conduit is tubing that contains individual insulated wires (Figure 33-19). Such tubes are available in ½-inch, ¾-inch, 1-inch, or larger diameters. Conduit may be rigid or bendable. It is available in metal or plastic. The bendable type of metal conduit is known as **electrical metallic tubing (EMT)**. Special benders permit shaping EMT in bends of up to 90 degrees. Couplings and connectors are used to assemble conduit

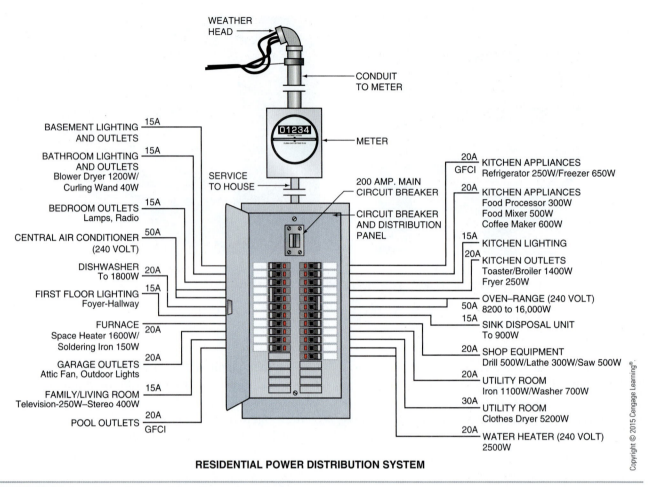

WEATHER HEAD

CONDUIT TO METER

METER

SERVICE TO HOUSE

200 AMP. MAIN CIRCUIT BREAKER

CIRCUIT BREAKER AND DISTRIBUTION PANEL

BASEMENT LIGHTING AND OUTLETS — 15A

BATHROOM LIGHTING AND OUTLETS — 15A
Blower Dryer 1200W/
Curling Wand 40W

BEDROOM OUTLETS — 15A
Lamps, Radio

CENTRAL AIR CONDITIONER — 50A
(240 VOLT)

DISHWASHER — 20A
To 1800W

FIRST FLOOR LIGHTING — 15A
Foyer-Hallway

FURNACE — 20A
Space Heater 1600W/
Soldering Iron 150W

GARAGE OUTLETS — 20A
Attic Fan, Outdoor Lights

FAMILY/LIVING ROOM — 15A
Television-250W–Stereo 400W

POOL OUTLETS — 20A
GFCI

20A GFCI KITCHEN APPLIANCES
Refrigerator 250W/Freezer 650W

20A KITCHEN APPLIANCES
Food Processor 300W
Food Mixer 500W
Coffee Maker 600W

15A KITCHEN LIGHTING

20A KITCHEN OUTLETS
Toaster/Broiler 1400W
Fryer 250W

OVEN–RANGE (240 VOLT)
50A 8200 to 16,000W

15A SINK DISPOSAL UNIT
To 900W

20A SHOP EQUIPMENT
Drill 500W/Lathe 300W/Saw 500W

20A UTILITY ROOM
Iron 1100W/Washer 700W

30A UTILITY ROOM
Clothes Dryer 5200W

20A WATER HEATER (240 VOLT)
2500W

RESIDENTIAL POWER DISTRIBUTION SYSTEM

Copyright © 2015 Cengage Learning®.

FIGURE 33-14 Branch circuits fan out from the service entrance panel.

and attach it to boxes. Waterproof plastic fittings provide waterproof seals where needed for metal conduit. Conduit systems are used to provide the most protection for wiring and are required for most commercial jobs.

Wire Type and Size

Individual wires within cabling or conduit may be made of aluminum or copper. Wire size is designated by gauge or AWG (American Wire Gauge) number.

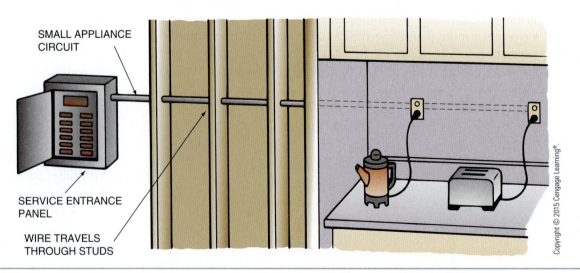

SMALL APPLIANCE CIRCUIT

SERVICE ENTRANCE PANEL

WIRE TRAVELS THROUGH STUDS

Copyright © 2015 Cengage Learning®.

FIGURE 33-15 A branch circuit generally includes a series of outlets, a series of lights, or a single electric motor.

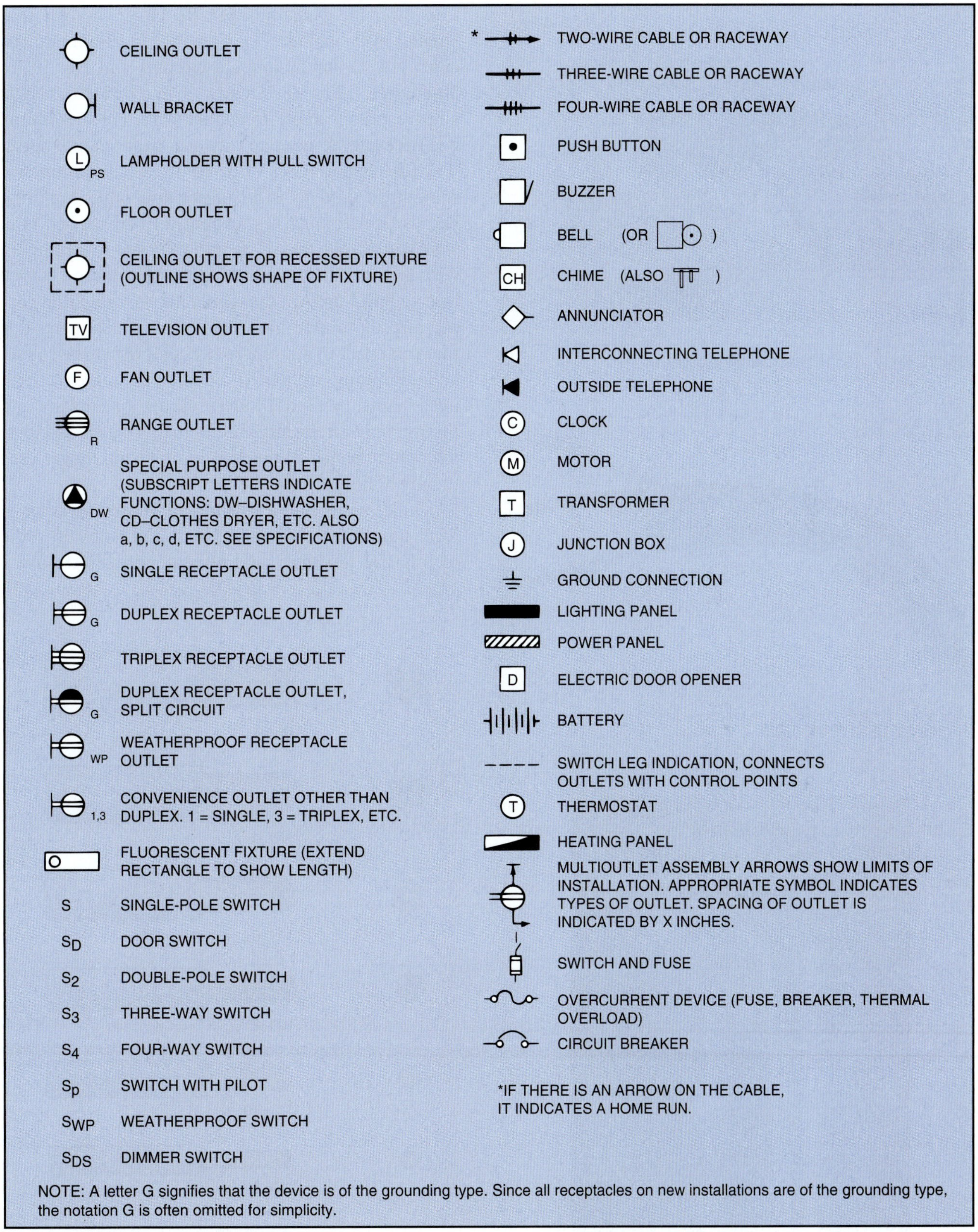

Symbol	Description
⊕	CEILING OUTLET
◯⊢	WALL BRACKET
(L)$_{PS}$	LAMPHOLDER WITH PULL SWITCH
⊙	FLOOR OUTLET
⟦⊕⟧	CEILING OUTLET FOR RECESSED FIXTURE (OUTLINE SHOWS SHAPE OF FIXTURE)
TV	TELEVISION OUTLET
(F)	FAN OUTLET
⊕$_R$	RANGE OUTLET
▲$_{DW}$	SPECIAL PURPOSE OUTLET (SUBSCRIPT LETTERS INDICATE FUNCTIONS: DW–DISHWASHER, CD–CLOTHES DRYER, ETC. ALSO a, b, c, d, ETC. SEE SPECIFICATIONS)
⊢⊖$_G$	SINGLE RECEPTACLE OUTLET
⊢⊖$_G$	DUPLEX RECEPTACLE OUTLET
⊢⊖	TRIPLEX RECEPTACLE OUTLET
⊢⊖$_G$	DUPLEX RECEPTACLE OUTLET, SPLIT CIRCUIT
⊢⊖$_{WP}$	WEATHERPROOF RECEPTACLE OUTLET
⊢⊖$_{1,3}$	CONVENIENCE OUTLET OTHER THAN DUPLEX. 1 = SINGLE, 3 = TRIPLEX, ETC.
▭	FLUORESCENT FIXTURE (EXTEND RECTANGLE TO SHOW LENGTH)
S	SINGLE-POLE SWITCH
S$_D$	DOOR SWITCH
S$_2$	DOUBLE-POLE SWITCH
S$_3$	THREE-WAY SWITCH
S$_4$	FOUR-WAY SWITCH
S$_P$	SWITCH WITH PILOT
S$_{WP}$	WEATHERPROOF SWITCH
S$_{DS}$	DIMMER SWITCH

Symbol	Description
* ⊢⊦⊦→	TWO-WIRE CABLE OR RACEWAY
⊢⊦⊦⊦	THREE-WIRE CABLE OR RACEWAY
⊢⊦⊦⊦⊦	FOUR-WIRE CABLE OR RACEWAY
▣	PUSH BUTTON
▢	BUZZER
◻	BELL (OR ▭⊙)
CH	CHIME (ALSO ⊤⊤)
◇	ANNUNCIATOR
◁	INTERCONNECTING TELEPHONE
◀	OUTSIDE TELEPHONE
(C)	CLOCK
(M)	MOTOR
T	TRANSFORMER
(J)	JUNCTION BOX
⏚	GROUND CONNECTION
▬	LIGHTING PANEL
▨	POWER PANEL
D	ELECTRIC DOOR OPENER
⊣⊢⊣⊢	BATTERY
- - - -	SWITCH LEG INDICATION, CONNECTS OUTLETS WITH CONTROL POINTS
(T)	THERMOSTAT
◣	HEATING PANEL
⊖	MULTIOUTLET ASSEMBLY ARROWS SHOW LIMITS OF INSTALLATION. APPROPRIATE SYMBOL INDICATES TYPES OF OUTLET. SPACING OF OUTLET IS INDICATED BY X INCHES.
⊟	SWITCH AND FUSE
⌒⌒	OVERCURRENT DEVICE (FUSE, BREAKER, THERMAL OVERLOAD)
⌒	CIRCUIT BREAKER

*IF THERE IS AN ARROW ON THE CABLE, IT INDICATES A HOME RUN.

NOTE: A letter G signifies that the device is of the grounding type. Since all receptacles on new installations are of the grounding type, the notation G is often omitted for simplicity.

FIGURE 33-16 Standard electrical symbols identify the locations of outlets, switches, and equipment.

Copyright © 2015 Cengage Learning®.

FIGURE 33-17 Nonmetallic sheathed cable.

FIGURE 33-18 Armored cable.

FIGURE 33-19 Rigid and EMT conduit.

The lower the gauge number, the larger the wire size. Smaller sizes include No. 14 wire for 15-ampere circuits, No. 12 for 20-ampere circuits, and No. 10 for 30-ampere 120-volt circuits (Figure 33-20). These ampere ratings apply to copper wire. For aluminum, it is necessary to use one wire size larger than is specified for copper wire. Prior to 1972, aluminum was sometimes used to wire houses and other buildings because it was much less expensive than copper. Fires were sometimes caused by improper use of aluminum wiring. Too-small wire overheated circuits and caused heat to build up. Another problem was attaching copper directly to aluminum wire. This causes corrosion that can result in an improper circuit. Anytime copper and aluminum are joined together, an approved connector must be used. Modern wiring systems use aluminum only for the large cables that bring electricity to the service panel. Aluminum wiring is no longer used to wire branch circuits.

Small wires called **strands** are placed together to form bundles for wire size No. 8 and larger in order to improve flexibility and conductivity. Electricity is

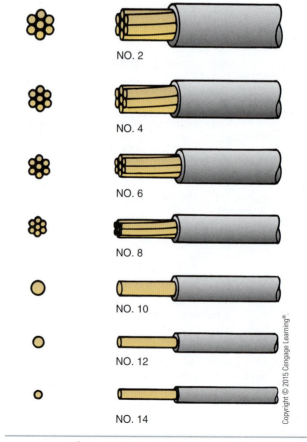

NO. 2

NO. 4

NO. 6

NO. 8

NO. 10

NO. 12

NO. 14

FIGURE 33-20 Common sizes of wire used in cabling and conduit.

carried on the outer surfaces of wire, and stranded wire has more total surface area than solid wire of equal diameter, and carries more current. Either solid wire or stranded wire may be placed in cables or conduit.

Voltage Drop

A common problem encountered in homes and on farms is voltage drop. **Voltage drop** refers to a loss of voltage as it travels along a wire. It is caused by electrical resistance. Although copper is a very good conductor of electricity, there is still some resistance to the flow of electricity. The longer the conductor, the more resistance is encountered.

Voltage drop causes lights to dim, heaters to put out less heat, and motors to put out less power and to overheat. The larger the wire, the less problem there is with voltage drop for a given amount of current. The longer the wire, however, the greater the problem.

Allowance for long wires must be made by using either lighter loads or larger wires. For example, to obtain 20 amperes of current with standard voltage at the end of a circuit, the wire size must be increased according to the distance (Figure 33-21). This is especially true when installing special circuits for motors.

WIRE IDENTIFICATION

The type of outer covering, individual wire covering, cable construction, and number of wires all help determine where a cable can be used. The wire type

Copper Up to 200 Amperes, 115-120 Volts, Single Phase, Based on 2% Voltage Drop

Minimum Allowable Size of Conductor

In Cable, Conduit, Earth			Overhead In Air	Length of Run in Feet																		
Load in Amps	Types R,T, TW	Types RH, RHW, THW	Bare & Covered Conductors	Compare size shown below with size shown to left of double line. Use the larger size.																		
				30	40	50	60	75	100	125	150	175	200	225	250	275	300	350	400	450	500	550	600	650	700
5	12	12	10	12	12	12	12	12	12	12	10	10	10	10	8	8	8	8	6	6	6	6	4	4	4
7	12	12	10	12	12	12	12	12	12	10	10	8	8	8	8	5	6	6	6		4	4	4	4	3
10	12	12	10	12	12	12	12	10	10	8	8	8	6	6	6	5	4	4	4	4	4		4	4	3
15	12	12	10	12	12	10	10	10	8	6	6	6	4	4	4	4	4	3	2	2	1	1	1	0	0
20	12	12	10	12	10	10	8	8	6	6	4	4	4	4	3	3	2	2	1	1	0	0	00	00	00
25	10	10	10	10	10	8	8	6	6	4	4	4	3	3	2	2	1	1	0	0	00	00	000	000	000
30	10	10	10	10	8	8	8	6	4	4	4	3	2	2	1	1	1	0	00	00	000	000	000	4/0	4/0
35	8	8	10	10	8	8	6	6	4	4	3	2	2	1	1	0	0	00	00	000	000	4/0	4/0	4/0	250
40	8	8	10	8	8	6	6	4	4	3	2	2	1	1	0	0	00	00	000	000	4/0	4/0	250	250	300
45	6	8	10	8	8	6	6	4	4	3	2	1	1	0	0	00	00	000	000	4/0	4/0	250	250	300	300
50	6	6	10	8	6	6	4	4	3	2	1	1	0	0	00	00	000	000	4/0	4/0	250	250	300	300	350
60	4	6	8	8	6	4	4	4	2	1	1	0	00	00	000	000	000	4/0	250	250	300	300	350	400	400
70	4	4	8	6	6	4	4	3	2	1	0	00	00	000	000	4/0	4/0	250	300	300	350	400	400	500	500
80	2	4	6	6	4	4	3	2	1	0	00	00	000	000	4/0	4/0	250	300	300	350	400	400	500	500	600
90	2	3	6	6	4	4	3	2	1	0	00	000	000	4/0	100	250	250	300	350	400	500	500	500	600	600
100	1	3	6	4	4	3	2	1	0	00	000	000	4/0	4/0	250	250	300	350	400	500	500	500	600	600	700
115	0	2	4	4	4	3	2	1	0	00	000	4/0	4/0	250	300	300	350	400	500	500	600	600	700	700	750
130	00	1	4	4	3	2	1	0	00	000	4/0	4/0	250	300	300	350	400	500	500	600	600	700	750	800	900
150	000	0	2	4	2	1	1	0	000	4/0	4/0	250	300	350	350	400	500	500	600	700	700	800	900	900	1M
175	4/0	00	2	3	2	1	0	00	000	4/0	250	300	350	400	400	500	500	600	700	750	800	900	1M		
200	250	000	1	2	1	0	00	000	4/0	250	300	350	400	500	500	500	600	700	750	900	1M				

FIGURE 33-21 Sizes of copper and aluminum wire needed to carry given loads with 2 percent or less voltage drop.

Source: National Food and Energy Council, Inc., Columbia, Missouri.

Aluminum Up to 200 Amperes, 115-120 Volts, Single Phase, Based on 2% Voltage Drop (Cont.)

Minimum Allowable Size of Conductor

Length of Run in Feet — Compare size shown below with size shown to left of double line. Use the larger size

Load in Amps	Types R, T, TW	Types RH, RHW, THW	Bare & Covered Conductors Single Triplex	30	40	50	60	75	100	125	150	175	200	225	250	275	300	350	400	450	500	550	600	650	700
5	12	12	10	12	12	12	12	12	10	10	8	8	8	8	6	6	6	6	4	4	4	4	3	3	3
7	12	12	10	12	12	12	12	10	10	8	8	6	6	6	6	4	4	4	4	3	3	2	2	2	1
10	12	12	10	12	12	10	10	8	8	6	6	6	4	4	4	4	3	3	2	2	1	1	0	0	0
15	12	12	10	12	10	8	8	8	6	4	4	4	3	3	2	2	2	1	0	0	00	00	00	000	000
20	10	10	10	10	8	8	6	6	4	4	3	3	2	2	1	1	0	0	00	00	000	000	4/0	4/0	4/0
25	10	10	10	8	8	6	6	4	4	3	2	2	1	1	0	0	00	00	000	000	4/0	4/0	250	250	300
30	8	8	10	8	6	6	6	4	3	2	2	1	0	0	00	00	00	000	4/0	4/0	250	250	300	300	350
35	6	8	10	8	6	6	4	4	3	2	1	0	0	00	00	000	000	4/0	4/0	250	300	300	350	350	400
40	6	8	10	6	6	4	4	3	2	1	0	0	00	00	000	000	4/0	4/0	250	300	300	350	350	400	500
45	4	6	10	6	6	4	4	3	2	1	0	00	00	000	000	4/0	4/0	250	300	300	350	400	400	500	500
50	4	6	8	6	4	4	3	2	1	0	00	00	000	000	4/0	4/0	250	300	300	350	400	400	500	500	600
60	2	4	6	6	4	3	3	2	0	00	00	000	4/0	4/0	250	250	300	350	350	400	500	500	600	600	700
70	2	2	6	4	4	3	2	1	0	00	000	4/0	4/0	250	300	300	350	400	500	500	600	600	700	700	750
80	1	2	6	4	3	2	1	0	00	000	4/0	4/0	250	300	300	350	350	500	500	600	600	700	750	800	900
90	0	2	4	4	3	2	1	0	00	000	4/0	250	300	300	350	400	400	500	600	600	700	750	800	900	1M
100	0	1	4	3	2	1	0	00	000	4/0	250	300	300	350	400	400	500	600	600	700	750	900	900	1M	
115	00	0	2	3	1	1	0	00	000	4/0	300	300	350	400	500	500	600	600	700	800	900	1M			
130	000	00	2	2	1	0	00	000	4/0	250	300	350	400	500	500	600	600	700	800	900	1M				
150	4/0	000	1	2	0	00	00	000	250	300	350	400	500	500	600	600	700	800	900	1M					
175	300	4/0	0	1	0	00	000	4/0	300	350	400	500	600	600	700	750	800	900							
200	350	250	00	0	00	000	4/0	250	300	400	500	600	600	700	750	900	900								

Copper Up to 400 Amperes, 230-240 Volts, Single Phase, Based on 2% Voltage Drop

Minimum Allowable Size of Conductor

Length of Run in Feet — Compare size shown below with size shown to left of double line. Use the larger size.

Load in Amps	Types R,T TW	Types RH, RHW, THW	Bare & Covered Conductors	50	60	75	100	125	150	175	200	225	250	275	300	350	400	450	500	550	600	650	700	750	800
5	12	12	10	12	12	12	12	12	12	12	12	12	12	12	10	10	10	10	8	8	8	8	8	6	6
7	12	12	10	12	12	12	12	12	12	12	12	10	10	10	10	8	8	8	8	6	6	6	6	6	6
10	12	12	10	12	12	12	12	12	10	10	10	10	8	8	8	8	6	6	6	6	4	4	4	4	4
15	12	12	10	12	12	12	10	10	10	8	8	8	6	6	6	6	4	4	4	4	3	3	3	2	
20	12	12	10	12	12	10	10	8	8	8	6	6	6	6	4	4	4	4	3	3	2	2	2	1	1
25	10	10	10	12	10	10	8	8	6	6	6	6	4	4	4	4	3	3	2	2	1	1	1	0	0
30	10	10	10	10	10	10	8	8	6	6	4	4	4	4	3	2	2	1	1	1	0	0	0	0	00
35	8	8	10	10	10	8	8	6	6	4	4	4	4	3	3	2	2	1	1	0	0	0	00	00	00
40	8	8	10	10	8	8	6	6	4	4	4	4	3	3	2	2	1	1	0	0	00	00	00	000	000
45	6	8	10	10	8	8	6	6	4	4	4	3	3	2	2	1	1	0	0	00	00	00	000	000	000
50	6	6	10	8	8	6	6	4	4	4	3	3	2	2	1	1	0	0	00	00	000	000	000	4/0	4/0

FIGURE 31-21 (continued)

Copper Up to 400 Amperes, 230-240 Volts, Single Phase, Based on 2% Voltage Drop (Cont.)

Minimum Allowable Size of Conductor

Length of Run in Feet — Compare size shown below with size shown to left of double line. Use the larger size.

Load in Amps	Types R,T,TW	Types RH, RHW, THW	Bare & Covered Conductors (Overhead In Air)	50	60	75	100	125	150	175	200	225	250	275	300	350	400	450	500	550	600	650	700	750	800
60	4	6	8	8	8	6	4	4	4	3	2	2	1	1	1	0	00	00	000	000	000	4/0	4/0	4/0	250
70	4	4	8	8	6	6	4	4	3	2	2	1	1	0	0	00	00	000	000	4/0	4/0	4/0	250	250	300
80	2	4	6	6	6	4	4	3	2	2	1	1	0	0	00	00	000	000	4/0	4/0	250	250	300	300	300
90	2	3	6	6	6	4	4	3	2	1	1	0	0	00	00	000	000	4/0	4/0	250	250	300	300	350	350
100	1	3	6	6	4	4	3	2	1	1	0	0	00	00	000	000	4/0	4/0	250	250	300	300	350	350	400
115	0	2	4	6	4	4	3	2	1	0	0	00	00	000	000	4/0	4/0	250	300	300	350	350	400	400	500
130	00	1	4	4	4	3	2	1	0	0	00	00	000	000	4/0	4/0	250	300	300	350	400	400	500	500	500
150	000	0	2	4	4	3	1	0	0	00	000	000	4/0	4/0	4/0	250	300	350	350	400	500	500	500	600	600
175	4/0	00	2	4	3	2	1	0	00	000	000	4/0	4/0	250	250	300	350	400	400	500	500	600	600	600	700
200	250	000	1	3	2	1	0	00	000	000	4/0	4/0	250	250	300	350	400	500	500	500	600	600	700	700	750
225	300	4/0	0	3	2	1	0	00	000	4/0	4/0	250	300	300	350	400	500	500	600	600	700	700	750	800	900
250	350	250	00	2	1	0	00	000	4/0	4/0	250	300	300	350	350	400	500	600	600	700	700	750	800	900	1M
275	400	300	00	2	1	0	00	000	4/0	250	250	300	350	350	400	500	500	600	700	700	800	900	900	1M	
300	500	350	000	1	1	0	000	4/0	4/0	250	300	350	350	400	500	500	600	700	700	800	900	900	1M		
325	600	400	4/0	1	0	00	000	4/0	250	300	300	350	400	500	500	600	600	700	750	900	900	1M			
350	600	500	4/0	1	0	00	000	4/0	250	300	350	400	400	500	500	600	700	750	800	900	1M				
375	700	500	250	0	0	00	4/0	250	300	300	350	400	500	500	600	600	700	800	900	1M					
400	750	600	250	0	00	000	4/0	250	300	350	400	500	500	500	600	700	750	900	1M						

Conductors in overhead spans must be at least No. 10 for spans up to 50 feet and No. 8 for longer spans. See NEC, Sec. 225-6(a).
See Note 3 of NEC "Notes to Tables 310-16 through 310-19."

Aluminum Up to 400 Amperes, 230-240 Volts, Single Phase, Based on 2% Voltage Drop

Minimum Allowable Size of Conductor

Length of Run in Feet — Compare size shown below with size shown to left of double line. Use the larger size

Load in Amps	Types R,T,TW	Types RH, RHW, THW	Bare & Covered Conductors Single Triplex (Overhead In Air)	50	60	75	100	125	150	175	200	225	250	275	300	350	400	450	500	550	600	650	700	750	800
5	12	12	10	12	12	12	12	12	12	12	10	10	10	10	8	8	8	8	8	6	6	6	6	4	4
7	12	12	10	12	12	12	12	12	10	10	10	8	8	8	8	6	6	6	6	4	4	4	4	4	4
10	12	12	10	12	12	12	10	10	8	8	8	8	6	6	6	6	4	4	4	3	3	3	2	2	
15	12	12	10	12	12	10	8	8	8	6	6	6	4	4	4	3	3	2	2	2	1	1	1	0	
20	10	10	10	10	10	8	8	6	6	6	4	4	4	4	3	3	2	2	1	1	0	0	0	00	00
25	10	10	10	10	8	8	6	6	4	4	4	4	3	3	2	2	1	1	0	0	00	00	00	000	000
30	8	8	10	8	8	8	6	4	4	4	3	3	2	2	2	1	0	0	00	00	00	000	000	000	4/0
35	6	8	10	8	8	6	6	4	4	3	3	2	2	1	1	0	0	00	00	000	000	000	4/0	4/0	4/0

FIGURE 31-21 (*continued*)

Aluminum Up to 400 Amperes, 230-240 Volts, Single Phase, Based on 2% Voltage Drop (Cont.)

Minimum Allowable Size of Conductor

In Cable, Conduit, Earth		Overhead In Air		Length of Run in Feet — Compare size shown below with size shown to left of double line. Use the larger size																					
Load in Amps	Types R,T, TW	Types RH, RHW, THW	Bare & Covered Conductors Single	Triplex	50	60	75	100	125	150	175	200	225	250	275	300	350	400	450	500	550	600	650	700	750	800
40	6	8	10		8	6	6	4	4	3	3	2	2	1	1	0	0	00	00	000	000	4/0	4/0	4/0	250	250
45	4	6	10		8	6	6	4	4	3	2	2	1	1	0	0	00	00	000	000	4/0	4/0	250	250	250	300
50	4	6	8		6	6	4	4	3	2	2	1	1	0	0	00	00	000	000	4/0	4/0	250	250	300	300	300
60	2	4	6	6	6	6	4	3	2	2	1	0	0	00	00	00	000	4/0	4/0	250	250	300	300	350	350	350
70	2	2(a)	6	4	6	4	4	3	2	1	0	0	00	00	000	000	4/0	4/0	250	300	300	350	350	400	400	500
80	1	2(a)	6	4	4	4	3	2	1	0	0	00	00	000	000	4/0	4/0	250	300	300	350	350	400	500	500	500
90	0	2(a)	4	2	4	4	3	2	1	0	00	00	000	000	4/0	4/0	250	300	300	350	400	400	500	500	500	600
100	0	1(a)	4	2	4	3	2	1	0	00	00	000	000	4/0	4/0	250	300	300	350	400	400	500	500	600	600	600
115	00	0(a)	2	1	4	3	2	1	0	00	000	000	4/0	4/0	250	300	300	350	400	500	500	600	600	600	700	700
130	000	00(a)	2	0	3	2	1	0	00	000	000	4/0	250	250	300	300	350	400	500	500	600	600	700	700	750	800
150	4/0	000(a)	1	00	2	2	1	00	000	000	4/0	250	250	300	300	350	400	500	500	600	600	700	750	800	900	900
175	300	4/0(a)	0	000	2	1	0	00	000	4/0	250	300	300	350	400	400	500	600	600	700	750	800	900	900	1M	
200	350	250	00	4/0	1	0	00	000	4/0	250	300	300	350	400	400	500	600	600	700	750	900	900	1M			
225	400	300	000			1	0	00	000	4/0	250	300	350	400	500	500	500	600	700	750	900	1M	1M			
250	500	350	000			0	00	000	4/0	250	300	350	400	500	500	500	600	700	750	900	1M					
275	600	500	4/0			0	00	000	4/0	250	300	400	400	500	500	600	600	750	900	1M						
300	700	500	250		00	00	000	250	300	350	400	500	500	600	600	700	800	900	1M							
325	800	600	300			00	000	4/0	250	300	400	500	500	600	600	700	750	900	1M							
350	900	700	300			00	000	4/0	300	350	400	500	600	600	700	750	800	900								
375	1M	700	350		000	000	4/0	300	350	500	500	600	700	700	800	900	1M									
400		900	350		000	4/0	250	300	400	500	600	600	700	750	900											

Conductors in overhead spans must be at least No. 10 for spans up to 50 feet and No. 8 for longer spans. See NEC, Sec. 225-6(a). See Note 3 of NEC "Notes to Tables 310-16 through 310-19."

FIGURE 31-21 (continued)

is stamped on the outer surface of the wire or cable or cable covering. Some common types of wire and cable are as follows:

- Type T—for use in dry locations
- Type TW—for use in dry or wet locations
- Type THHN—for use in dry locations with high temperature
- Type THW and THWN—for use in wet locations with high temperature
- Type XHHW—for high moisture and heat resistance
- Type UF—for direct burial in soil but not concrete

The National Electrical Code® and local building codes must be consulted for specific guidance in wire selection. All instructions and diagrams in this chapter conform to the National Electric Code.

Individual wires are color-coded to help identify their function in the circuit. Black wires, red wires, and blue wires are **positive (hot) wires**, which carry current to the appliances. White wires are **neutral wires** that carry current back to the source. Green or bare wires are used to ground all metal boxes and appliances in the circuit.

Nonmetallic sheathed cable is generally stamped at regular intervals with a mark that identifies its type, wire size, and number of conductors. Typical markings and their meanings are as follows:

- 12-2—two strands of No. 12 wire, one black and one white
- 12-2 w/g—two strands of No. 12 wire plus a ground wire (w/g means "with ground"), one black, one white, and one green or bare

- 12-3—three strands of No. 12 wire, probably one black, one red or blue, and one white
- 12-3 w/g—same as 12-3 cable with the addition of a green or bare wire for grounding the circuit

The same system is used for all wire sizes. Letters such as T, TW, or THHN on the cable indicate the type of insulation on the cable. It is very important to install wire and cable with the correct type designations to heat, moisture, fungus, and other insulation-destructive factors.

SUMMARY

Electricity is perhaps our most important source of power. In order to achieve the most efficient use, the natural laws governing electricity must be observed. Circuits have to be properly installed, and grounded efficiently and safely. Unit 34 ("Installing Branch Circuits") discusses how to install circuits correctly.

Student Activities

1. Define the Terms to Know in this unit.

2. Collect samples of cable and label them with the name, type of conductor, and type of insulation.

3. Use the formula W = VA to calculate the unknown value in the following problems.
   a. A 100-watt bulb in a 120-volt circuit draws _____ amperes.
   b. A 120-volt hair dryer is rated at 1,200 watts. It will draw _____ amperes.
   c. A toaster draws 13 amperes at 120 volts. How many watts does it use?
   d. A circuit has three 100-watt lamp bulbs and a 1,600-watt toaster. How many watts are used when all four appliances are in operation?
   e. Would a 15-ampere fuse carry the load described in item (d)? Explain.
   f. A grain elevator motor has a 1-horsepower motor. A general rule of thumb is that 10 amperes at 120 volts equals 1 horsepower. How many amperes does this motor draw?
   g. A microwave oven draws 1,400 watts at 12 volts on a 20-ampere circuit. It is 40 feet between the outlet and the service entrance panel. What size of copper wire is needed for the circuit?

4. Cover a horseshoe magnet with a piece of paper. Sprinkle iron filings over the paper. Observe the magnetic lines of force exhibited by the filings.

5. Read the electric meter at your home. Read it again two days later. How many kilowatt-hours of electricity did your family use in two days?

6. Clean an electric motor. Oil it according to instructions in the operator's manual.

Relevant Web Sites

Don Vandervort's Home Tips, Electrical Wiring, Types of Wires and Cables
www.hometips.com
http://hyperphysics.phy-astr.gsu.edu/hbase/magnetic/mothow.html
http://edisontechcenter.org/electricmotors.html

Self-Evaluation

A. Multiple Choice. Select the best answer.

1. The major power source for stationary equipment in houses, farm buildings, and agribusinesses is
   a. diesel fuel
   b. electricity
   c. gasoline
   d. steam

2. Electricity produces
   a. chemical changes
   b. heat and light
   c. magnetism
   d. all of these

3. A device that produces direct current by means of magnetism is
   a. a generator
   b. an alternator
   c. a turbine
   d. a pole

4. A device used to protect circuits that can be reset is a
   a. three-way switch
   b. ground-fault interrupter
   c. fuse
   d. circuit breaker

5. A device in circuits to protect against human shock is a
   a. three-way switch
   b. ground-fault interrupter
   c. fuse
   d. circuit breaker

6. Electricity is distributed to branch circuits by
   a. an electric meter
   b. an entrance head
   c. a service drop
   d. a service entrance panel

7. Tubes used to carry wires are called
   a. armored cable
   b. conduit
   c. nonmetallic sheathed cable
   d. pipe

8. A suitable wire for high temperature, high moisture locations is
   a. Type T
   b. Type THHN
   c. Type THW
   d. WVA

9. A cable consisting of No. 12 wire, one black, one red, one white, and a ground wire will be stamped
   a. 12-2
   b. 12-3
   c. 12-3 w/g
   d. 12-3 BRW

10. In order to run copper wire in a building 100 feet to a 10-ampere, 120-volt motor, and hold the voltage drop to 2 percent, the minimum size of wire must be
    a. No. 12
    b. No. 10
    c. No. 8
    d. No. 6

B. Matching. Match the terms in column I with those in column II.

Column I

1. conductor
2. fluorescent
3. filament
4. insulator
5. amperes
6. volts
7. watts
8. magnet
9. armature
10. circuit

Column II

    a. great resistance

    b. electrical pressure

    c. electrical energy or work

    d. north and south poles

    e. circle

    f. rotating part of motor

    g. rate of electrical flow

    h. glowing element

    i. glowing gas

    j. carries electricity

C. **Completion.** Fill in the blanks with the word or words that make the following statements correct.

    1. Two deadly hazards of electricity are _____ and _____.

    2. Electric meters measure the amount of electricity used and express it in _____.

    3. Electrical circuits in a building are called _____ circuits.

    4. Three types of cable used for wiring are _____, _____, and _____.

    5. Strands of wire are gathered together in bundles to increase the _____ of the wire and, therefore, its conductivity.

D. **Brief Answer.** Briefly answer the following questions.

    1. Give good examples of (a) good conductors and (b) good insulators.

    2. What is Ohm's Law? When is this law useful to know?

    3. Explain how grounding is accomplished.

    4. What does the commutator in the motor do?

    5. In older motors that require lubrication, failure to use oil as specified by the manufacturer results in what?

UNIT 34

Installing Branch Circuits

Objective

To run wires and safely install boxes, switches, outlets, and fixtures.

Competencies to be developed

After studying this unit, you should be able to:

- Select electrical boxes, outlets, and switches.
- Install and replace switches, outlets, and fixtures.
- Install, extend, and modify branch circuits.
- Test electric circuits.

Materials List

- Hand tools suitable for electrical wiring
- Cable ripper
- Wire stripper
- Test light
- 14-2 cable with ground
- 14-3 cable with ground
- Two switch boxes
- One octagon box
- One single-pole switch
- Two three-way switches
- One porcelain lamp holder
- One duplex receptacle

continued

Terms to Know

- new work
- fixture
- wire nuts
- receptacle
- switch
- duplex (double) receptacle
- splitting the receptacle
- single-pole switch
- knockout
- ground (gee) clip
- continuity tester
- continuity
- three-way switch
- common terminal
- switching (traveler) wires
- four-way switch

Materials list, *continued*

- Two switch covers
- One outlet cover
- Wire nuts
- Ground (gee) clip
- Electrical tape

Knowledge of electrical principles and wiring materials is important. The rules for the proper and safe installation of electrical devices and circuits given in the current National Electrical Code must be strictly observed at all times. All the instruction and diagrams in this chapter conform to the National Electric Code.

Cable and conduit are run through the floors, ceilings, and partitions of new buildings during construction. Such installations are called **new work** (Figure 34-1). Electrical codes may permit wiring to be placed in view in old work installations. This is due to the difficulty of concealing wires after a building is completed. In surface installations, however, the wiring must be protected from mechanical damage.

Utility buildings and other farm structures generally do not have finished interior walls. Such structures are relatively easy to wire at any time. However, the National Electrical Code has established special requirements for such wiring.

© Christina Richards/photos.com/Getty Images

FIGURE 34-1 Cable is run through the framework of new construction.

THE ENTRANCE PANEL

Electricity enters a building by way of an entrance panel (also known as a load center) (Figure 34-2). As discussed in Unit 33, large cables bring electricity from the transformer to the entrance panel. If the cables come in from underground, they are attached to the meter base, which is then connected by cables to the entrance panel. Overhead cables attach to cables that run through the entrance head.

Entrance panels come in different sizes, depending on the amount of current that will be distributed from the panel. The panel consists of large main breakers (where all electricity to the building can be cut off), metal bars for attaching the breakers, and grounded strips. The large entrance cables from the meter usually attach at the top of the panel but may also attach at the bottom. Two hot wires bring in 120 volts each. Combined they can create a 240-volt circuit. In most circuits these wires have black insulation to distinguish them from the neutral and grounding wires. The grounding bars are attached to a metal rod that is driven deep into the ground to ground the circuit.

As each new circuit is attached to the entrance panel, a circuit breaker is attached to the bars that are connected to the entrance hot wires. A black and/or a red wire is attached to the circuit breaker. The neutral wire (white) is attached to the neutral grounding bar, and the bare or green wire is attached to the equipment grounding bar. These bars are then attached to the metal grounding rod that is driven into the ground. This provides power for various branch circuits throughout the building.

WIRING BOXES

All wiring systems require the use of fuses or circuit breakers, and protected wires and boxes. Boxes must be used for all connections, and fixtures such as outlets and switches. The boxes are metal or plastic (Figure 34-3). Boxes have several important functions:

- They hold the cable or conduit so stress cannot be placed on the wire connections.
- They are nailed, screwed, or clamped to the building to support switches, outlets, or fixtures. A **fixture** is a base or housing for lightbulbs, fan motors, and other electrical devices.
- They contain all electrical connections that are not made inside fixtures.
- They prevent mice and other vermin from nesting around electrical connections.

Electrical boxes are available in rectangular or octagonal shapes and in various depths. Some switch and outlet boxes have removable sides so that the boxes can be installed in gangs (series). Such boxes can hold multiple switches or outlets.

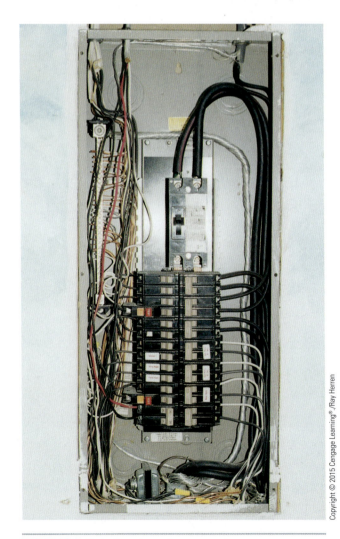

Copyright © 2015 Cengage Learning®./Ray Herren

FIGURE 34-2 Typical service entrance panel.

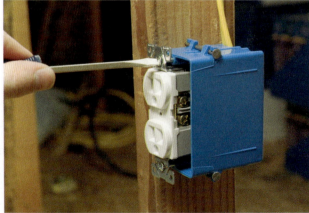

© iStockphoto./Jason Gayman

FIGURE 34-3 All connections, switches, and outlets must be installed in a box. The box is made of plastic.

Large steel boxes are used for service entrances. They contain the main fuses or circuit breakers. They also contain fuses or circuit breakers for one or more branch circuits.

Certain basic rules apply to wiring all electrical boxes. The National Electrical Code and local building codes must be checked for specific regulations. Some basic requirements follow:

- The box must be fastened securely to the building (Figure 34-4).
- The cable or conduit must be clamped securely to the box.
- Cables running from box to box must run through the interior of the building's walls, floors, and ceilings, be secured by staples or clamps near each box, and be secured as needed to prevent the cable from being accidentally caught and pulled.
- The box must be grounded if it is metal. The ground wire from the cable is attached to the box by a screw or by a grounding clip (Figure 34-5).
- Wires in boxes must be connected to each other by insulated solderless connectors called **wire nuts** (Figure 34-6). The proper size of wire nut must be used. The sizes are distinguished by color (Figure 34-7). Bare areas are not permitted on wires except ground wires.
- Ground wires must be held together by a special metal clamp or a solderless connector.
- Wires must be attached to terminals of switches and receptacles by tightening no more than one wire under one screw or spring clamp. A **receptacle** is a device for receiving electric plugs. A **switch** is a device used to stop the flow of electricity. If the receptacle or switch is equipped with special clamps, one wire may be inserted into each clamp provided.

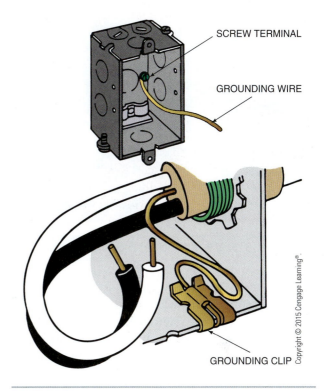

FIGURE 34-5 A screw or grounding clip is used to attach a ground wire to a metal box.

- Positive or hot wires (black, red, or blue) must always be attached to brass-colored screws. Neutral wires (white) must always be attached to aluminum-colored screws. Ground wires (bare or green) are attached to green screws.
- When a white wire must be used as a positive wire (as in a light switch), the insulation visible in the box should be painted black or marked with black tape.

FIGURE 34-4 Boxes must be fastened securely to the building.

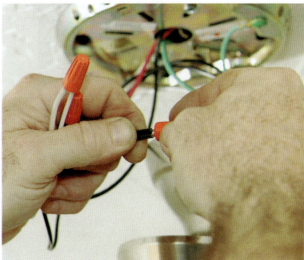

FIGURE 34-6 Connections are made with wire nuts.

FIGURE 34-7 Sizes of wire nuts are distinguished by color.

The most common type of receptacle is the **duplex (double) receptacle**, which is wired so that both outlets are on the same circuit (Figure 34-8). The current comes in on a black wire, flows from one screw through a metal strap to the other screw, and, when so wired, continues on to the next electrical box. The two receptacles in the duplex may be wired to two different circuits if the metal strap between the two screws is removed. This is called **splitting the receptacle**.

A standard switch box will only hold four wires safely. To determine the number of wires in a box, only the positive and neutral wires are counted, not the ground wires. In other words, one incoming cable with a black wire, a white wire, and a ground wire is counted as two wires. With an identical cable taking the current on to the next box, the total of four wires has been reached.

WIRING A SWITCH AND LIGHT

A switch is a device used to stop the flow of electricity. A **single-pole switch** is a switch designed to be the only switch in a circuit. The cable carrying current from the service entrance panel may come to an outlet box where a light is mounted or to one where a switch is mounted. For example, a power cable coming to a box where a light is mounted and controlled by a single-pole switch is shown in Figure 34-9. To wire this circuit in accordance with NEC regulations, a cable with a white wire, a black wire, and a ground wire is used.

The cable is prepared for insertion into the box by slitting 6 to 8 inches of the outside cable covering with a cable ripper or splitter (Figure 34-10).

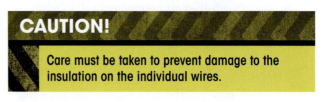

CAUTION!

Care must be taken to prevent damage to the insulation on the individual wires.

The wires are then separated from the jacket and the excess jacket material is cut off.

Electrical boxes are provided with a **knockout**, or partially punched impression, that can be punched out and removed. The cable is pushed through this hole. Some boxes have cable clamps provided with the box; others require the addition of a cable clamp, as illustrated in Figure 34-9. The cable is inserted until $\frac{1}{16}$ inch of the cable jacket extends beyond the clamp. The clamp is then tightened, and cable is run between the switch box and the light box. If there is more than one light in the circuit, a cable may be installed to carry current from one light box to the next. All cables must be clamped securely in the boxes and stapled or clamped to the structure within 12 inches of each box, at intervals of $4\frac{1}{2}$ feet or less.

To install the switch, a wire stripper is used to remove about 34 inches of insulation from the ends of individual wires (Figure 34-11).

CAUTION!

Care must be taken not to nick the wire during this procedure.

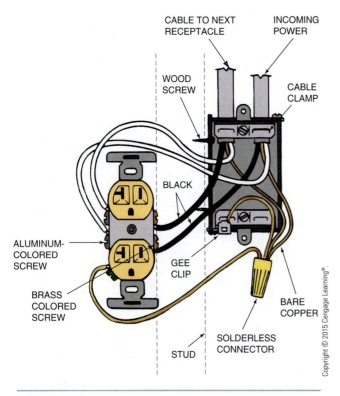

FIGURE 34-8 Typical duplex receptacle.

CABLE TO NEXT RECEPTACLE

INCOMING POWER

WOOD SCREW

CABLE CLAMP

BLACK

ALUMINUM-COLORED SCREW

GEE CLIP

BRASS COLORED SCREW

BARE COPPER

SOLDERLESS CONNECTOR

STUD

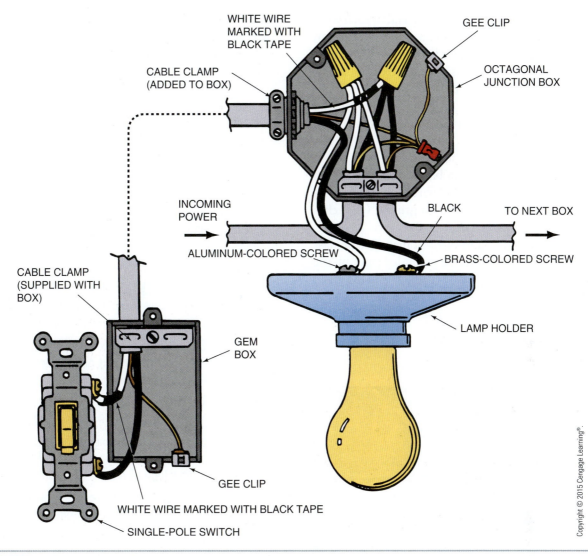

WHITE WIRE MARKED WITH BLACK TAPE

GEE CLIP

CABLE CLAMP (ADDED TO BOX)

OCTAGONAL JUNCTION BOX

INCOMING POWER

BLACK

TO NEXT BOX

ALUMINUM-COLORED SCREW

BRASS-COLORED SCREW

CABLE CLAMP (SUPPLIED WITH BOX)

GEM BOX

LAMP HOLDER

SINGLE-POLE SWITCH

WHITE WIRE MARKED WITH BLACK TAPE

GEE CLIP

FIGURE 34-9 A light fixture controlled by a single-pole switch.

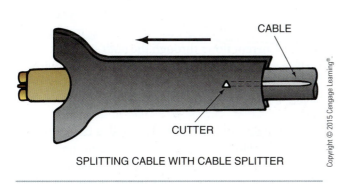

CABLE

CUTTER

SPLITTING CABLE WITH CABLE SPLITTER

FIGURE 34-10 A cable splitter is used to cut a slit 7 inches long in the outer covering of nonmetallic sheathed cable. The individual wires are then pulled out of the jacket, and the loose jacket is cut off and discarded.

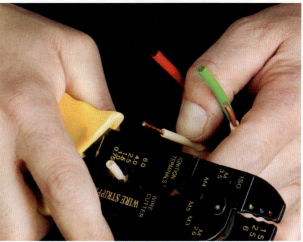

FIGURE 34-11 Using a wire stripper.

Copyright © 2015 Cengage Learning®

Copyright © 2015 Cengage Learning®

© iStockphoto/boiling

A round loop is then made in the end of each wire. Each loop is wrapped around a brass-colored screw in the direction (clockwise) the screw turns when being tightened. For all connections, whether fixtures, lights, switches, or outlets, the wire must be looped in a clockwise direction to ensure the wire is securely installed behind the screw (Figure 34-12). The screws are then tightened securely.

Switches are placed in hot wires only—never in neutral wires. Thus, a piece of black tape is placed on the white wire to mark it as a hot wire. The ground wire is attached to the metal box with a **ground (gee) clip**. The switch is then screwed in place and a switch cover is installed. Some modern installations make use of plastic boxes that do not need to be grounded since plastic is an insulator. The ground wire is attached to the green grounding screw and is not connected to the box.

Wiring the lamp holder box is more difficult. However, the work progresses in a logical sequence according to the following procedure.

FIGURE 34-12 The wire must always be looped in a clockwise direction.

© iStockphoto/GarysFRP

Procedure

1. Strip the ends of all insulated wires, leaving ⅝ inch of wire exposed. The wires are now ready for connection with wire nuts.

2. Strip ⅝ inch from both ends of an 8-inch length of bare wire, and ground it to the box with a ground clip or screw.

3. Hold the four ground (bare) wires so they form a bundle. Twist a wire nut onto the bundle and tighten it as much as possible with the hands. For large bundles, use a wire nut handle or pliers to tighten the wire nut.

4. Mark the white wire coming from the switch with black tape. This is now regarded as a black wire.

5. Attach the black wire from the switch to the brass terminal of the fixture or lamp holder.

6. Cut an 8-inch piece of white wire and strip ⅝ inch of insulation from each end. Attach one end to the aluminum-colored screw to the lamp holder.

7. Select and use a wire nut of the correct size to connect the loose ends of the three white wires.

8. Use another wire nut to connect the ends of the three remaining black wires. Treat the white wire with the black tape as a black (hot) wire.

9. Check each connection for tightness by holding the wire nut and pulling hard on each of the wires. If one or more wires is loose, the entire bundle must be retightened.

The flow of electricity in this circuit proceeds from the service entrance box to the light box through the black wire of the incoming cable. It flows to the switch by way of the white wire marked with black tape. If the switch is in the "on" position, the current flows through the switch and through the black wire to the lamp holder. Current flows through the bulb and back to the service entrance box by way of white (neutral) wires.

TESTING A CIRCUIT

After all lights and switches are wired, a circuit should be tested before turning on the circuit breaker or inserting the fuse. Use a continuity tester to be sure the circuit is not open. A **continuity tester** is a device used to determine if electricity can flow between two points. **Continuity** means connectedness. A circuit is open if there is a break or poor connection anywhere in it. The circuit should also be checked for shorts, or places where current can get from the black wires to the ground wires or boxes, or where it can bypass the bulb and flow directly to the neutral wires. The final step is to test all boxes to ensure that they are properly grounded.

The following procedure describes how to test a circuit.

Procedure

1. Place all switches in the "on" position.

2. Remove all bulbs.

3. Test the circuit by connecting to the wires at the fuse box or circuit breaker box before the cable is wired into the fuse block or circuit breaker. Using an ohmmeter, touch one lead to the black wire and the other to the white wire of the cable. There should be no reading. This indicates no continuity. That is, there are no points where electricity can flow from the black wires to the white wires.

4. Install a good lightbulb in the last fixture of the circuit. Repeat the test in step 3. There should be continuity through the filament of the bulb, as indicated by a meter response. This indicates that current can flow through all black wires, through the bulb, and back through the white wires.

5. Test to see if each metal box in the circuit is grounded. Continuity is checked with an ohmmeter by touching one test lead to the box and the other to the ground wire. There should be a reading.

When all cover plates and fixtures have been installed, the cable can be wired into the fuse block or circuit breaker.

CAUTION!

Do not add electricity to a circuit until a qualified electrician has checked it and all required inspections have been completed.

WIRING LIGHTS WITH THREE-WAY SWITCHES

A **three-way switch** is a switch that, when used in a pair, permits a light or receptacle to be controlled from two different locations (Figure 34-13). The wiring of a circuit with three-way switches will vary according to the location of the incoming power cable, the light, and the two switches. However, certain basic rules apply regardless of the location of components.

A three-way switch has three terminals. The current feeds into a **common terminal**, which is identified by its dark screw. The current flows out through either one of the remaining two terminals identified by light-colored screws. The light-colored screw that carries the current is determined by the position of the switch at any particular time. The two wires attached to the light-colored screws are called the **switching (traveler) wires**. A wire running from the common terminal or dark screw of the second switch is attached to the brass screw on the light fixture.

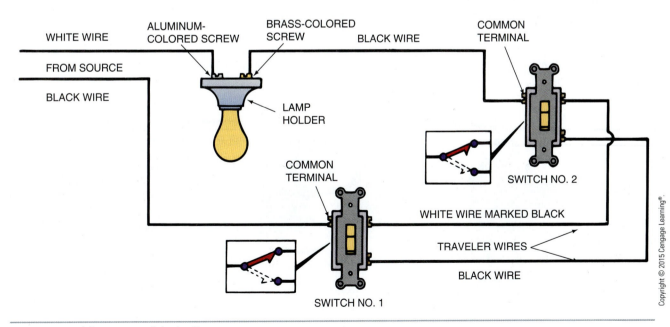

FIGURE 34-13 A three-way switch circuit.

Copyright © 2015 Cengage Learning®

The following procedure describes how to wire three-way switches.

Procedure

1. Connect the white wire of the incoming power cable to the silver terminal of the light fixture.

2. Connect the black wire of the incoming power cable to the dark screw (common terminal) of the switch.

3. Connect a black wire from the dark screw of the other switch to the brass terminal of the light fixture.

4. Run a pair of traveler wires between the light-colored terminals of the two switches.

Three-wire cables are useful when wiring circuits controlled by three-way switches (Figure 34-14 to Figure 34-16). Such cables typically have black, red,

white, and ground wires. The white wire may be used as a hot wire if it is connected to a switch or to a black wire, and if it is clearly marked with black tape.

Black and red wires are generally used as the traveler wires in a three-way switch circuit. If metal boxes are used, the ground wire grounds the boxes.

Wiring Lights Using Four-Way Switches

A **four-way switch** is a switch connected in the pair of traveler wires between two three-way switches (Figure 34-17). Any number of four-way switches can be added to a three-way switch circuit to accommodate the total number of switch locations needed. A four-way switch permits the electricity to be changed from one traveler to the other, depending on the position of the switch lever (Figure 34-18).

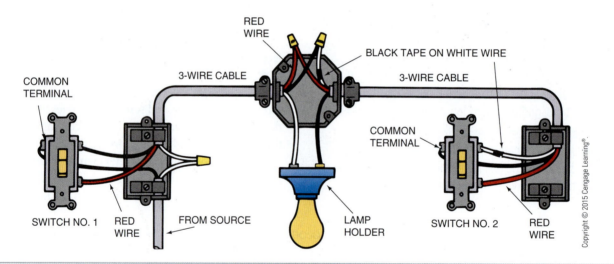

FIGURE 34-14 A three-way circuit with current coming to one switch and with a light located between two three-way switches.

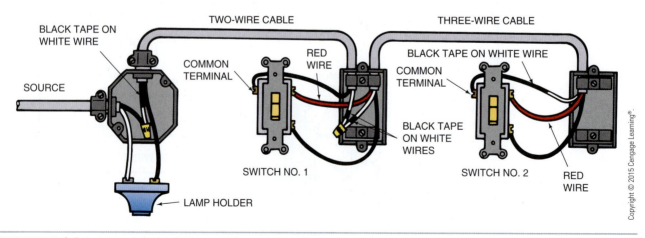

FIGURE 34-15 A three-way switch circuit with current coming in at the light.

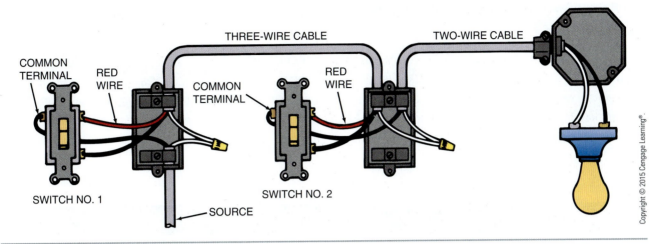

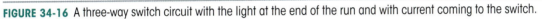

FIGURE 34-16 A three-way switch circuit with the light at the end of the run and with current coming to the switch.

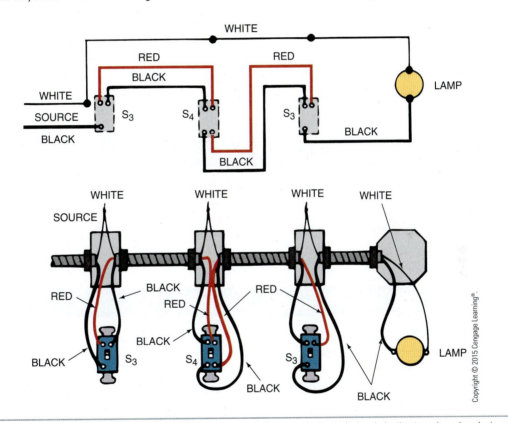

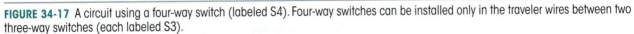

FIGURE 34-17 A circuit using a four-way switch (labeled S4). Four-way switches can be installed only in the traveler wires between two three-way switches (each labeled S3).

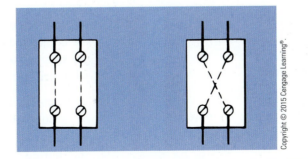

FIGURE 34-18 Two possible positions of a four-way switch.

SUMMARY

Branch circuits distribute electricity throughout a building. The type of circuit, size of the wire, and the type of receptacle used all depend on the appliances that will be operated when plugged into the circuit. All the proper materials, procedures, and specifications for circuits are outlined in the National Electrical Code. By following this code, electricians can be sure the circuits they install are done properly.

Student Activities

1. Define the Terms to Know in this unit.

2. Wire a circuit with an outlet in compliance with NEC regulations.

3. Wire a circuit with several outlets in compliance with NEC regulations.

4. Wire a circuit with a light and a single-pole switch in compliance with NEC regulations.

5. Wire a circuit with a light and two three-way switches in compliance with NEC regulations.

Relevant Web Sites

Hammerzone.com, Installing a Circuit Breaker
www.hammerzone.com

Ace Hardware, Installing or Replacing Electrical Switches
www.acehardware.com

Self-Evaluation

A. **Multiple Choice.** Select the answer that best represents NEC-compliant installation procedure:

1. All electrical connections in a circuit are made
   a. in boxes or fixtures
   b. by screws
   c. with solder
   d. with tape

2. All metal electrical boxes must
   a. be grounded
   b. be securely fastened
   c. secure the cable or conduit
   d. all of these

3. Neutral wires are attached to screws colored
   a. white
   b. green
   c. silver
   d. yellow

4. The device that receives electrical plugs is a
   a. box
   b. cap
   c. circuit breaker
   d. receptacle

5. White wires used as positive wires must be
   a. connected to black wires
   b. connected to fixtures
   c. stripped of all insulation
   d. taped or painted black

6. A properly wired circuit will be
   a. grounded
   b. open
   c. shorted
   d. all of these

7. In three-way switch circuits, electricity passes from one switch to the other through
   a. traveler wires
   b. neutral wires
   c. common terminals
   d. none of these

8. Three-way switch circuits usually include
   a. two-wire cables
   b. three-wire cables
   c. grounded boxes
   d. all of these

B. Matching. Match the terms in column I with those in column II.

Column I

1. old work
2. duplex
3. solderless connector
4. switch
5. positive or hot
6. red wire
7. neutral wire
8. circuit protector
9. electrical contact
10. switching wire

Column II

a. used in place of black wire
b. always white
c. traveler
d. fuse
e. extensions to existing systems
f. receptacle
g. continuity
h. stops current
i. black wire
j. wire nut

C. Completion. Fill in the blanks with the word or words that make the following statements correct.

1. Rules for the proper and safe installation of electrical devices and circuits are given in the _____ _____ _____.

2. All wiring systems require the use of _____ or _____ _____ and protected wires and boxes.

3. A standard switch box will hold _____ wires only, not counting _____ wires.

4. Switches are placed in _____ wires only—never in _____ wires.

5. Black and red wires are generally used as the _____ wires in a three-way switch circuit.

D. Brief Answer. Briefly answer the following questions.

1. What does the entrance panel consist of?

2. What is "splitting the receptacle"? What does the operation require?

3. All wiring systems require the use of boxes. What are their four functions?

4. What is a continuity tester used for? When should it be used?

UNIT 35

Electronics in Agriculture

Objective

To utilize electronic principles and applications in agricultural settings.

Competencies to be developed

After studying this unit, you should be able to:

- Cite major historical landmarks in electronics development.
- State basic principles of electronics.
- Name typical applications of electronics in agriculture.
- Discuss strategies for maintaining electronics equipment.

Materials List

- Publications with photos of electronics applications
- Poster materials
- Bulletin-board materials
- Reference materials on electronics

Terms to Know

- lodestone
- conductance
- resistivity
- insulators
- ohm
- siemen
- reciprocal
- resistor
- potentiometer (pot)
- rheostat
- series circuit
- parallel circuit
- series-parallel circuit
- electrical meter
- analog meter
- digital meter
- ammeter
- voltmeter
- multimeter
- inductance
- henry (H)
- inductor
- capacitance
- capacitor
- plates
- dielectric
- cycle
- hertz (Hz)
- sine wave

continued

Electronics is certainly the most pervasive and influential technology of the twentieth century. Set your digital watch; use your computer; click on your television; listen to your stereo; start your car, truck, or tractor; ride a motorcycle or run a boat—you have used electronics (Figure 35-1).

Radio, television, audio and video recorders, computers, satellites, factory automation, space flight, aircraft of all types, and traffic control devices all rely on electronics (Figure 35-2). Homes, schools, supermarkets, department stores, gas stations, banks, and offices are more efficient and productive because of electronic equipment.

FIGURE 35-1 Electronic devices are a constant presence in modern life.

FIGURE 35-2 Electronic systems serve as the core of our society's communications.

© bloomua. Image from BigStockPhoto.com

© nruboc. Image from BigStockPhoto.com

535

DEVELOPMENT OF ELECTRONICS

Computers and telephones in homes, scanners and monitors in hospitals, audiovisual equipment in homes and classrooms, electron microscopes in laboratories, electronic feeding and milking systems on farms, and communications satellites in space are electronic realities of today. However, it took many years of investigation and discovery to reach our current level of electronics technology.

The Ancients

As early as 600 B.C. Thales of Miletus, Greece, discovered that when a piece of amber was rubbed with a cloth, the amber would attract bits of feathers and pith from plants. Thales believed that there was a connection between electricity and magnetism. He was one of the seven wise men of Greece.

Some people of early civilizations tried to use electric shocks from electric eels in efforts to cure stiffening diseases of the body. This suggests a realization that somehow electronic impulses are used in the body to send messages from the brain to muscles. We know this today as the body's nervous system.

Archaeologists have discovered strange pots in ancient Arabic ruins, which seem to indicate someone was electroplating jewelry thousands of years ago!

The first compass is believed to have been used on a cart in the high-Asian plains. The cart utilized a piece of lodestone in a wood spindle to keep an arm pointing north. It also had a counter that ticked off units of distance traveled. **Lodestone** is a magnetized piece of black iron oxide.

Pioneers from 1600 to 1900

From 1600 to 1900, many discoveries were made that make today's electronics possible. Some of these discoveries and the pioneers include:

- Numerous discoveries about static electricity and magnetism: William Gilbert
- The discovery that lodestones attract iron: William Gilbert
- The discovery that only certain substances conduct electricity: Stephen Gray
- The observation that substances with like charges repel and substances with opposite charges attract: Charles du Fay
- The laws of attraction and repulsion: Charles Augustin de Coulomb

Copyright © 2015 Cengage Learning®.

FIGURE 35-3 The first battery was built by Volta, who demonstrated that certain metal and chemical combinations cause the flow of electricity.

- The development of the first battery: Allesandro Volta (Figure 35-3)
- The demonstration that lightning is electricity: Benjamin Franklin
- The recognition of different charges as negative and positive: Benjamin Franklin
- The use of a battery to power the first electric light: Humphry Davy
- The first machine to make electricity from mechanical energy: Michael Faraday
- The first electric motor: Joseph Henry
- The first telegraph: Samuel F. B. Morse
- The telephone: Alexander Graham Bell
- The incandescent lightbulb: Thomas Edison

After the 1800s, research and development became more characterized as group activities and company projects. Whom to credit with new discoveries became more difficult as science entered the twentieth century.

Twentieth-Century Discoveries

The twentieth century was ushered in with Guglielmo Marconi's first transatlantic wireless message in 1901. From then on, discoveries and developments mushroomed. A few highlights are the following:

- The deForrest Audion (vacuum radio tube): led to radios in homes
- Superheterodyne receiver: gave the Allies a great advantage in World War I
- Discovery of radio waves from outer space: the beginning of radio astronomy

- Television: developed in the 1930s; became a household item in the 1950s
- Radar: a method of detecting distant objects through analysis of high-frequency radio waves received from their surfaces
- Sonar: a system using transmitted and reflected acoustic waves to detect and locate submerged objects
- Transistor: a three-terminal semiconductor device used for amplification, switching, and detection
- Printed circuit: electrical conducting material placed on a flat board by means of printing techniques
- Direct dialing: the ability to dial from a home telephone to any other telephone
- Molecular electronics: placement of material only a millionth of an inch thick in electrical circuits
- Radio/TV satellites: orbiting instruments capable of receiving and forwarding electronic signals between points thousands of miles apart on Earth
- Videocassette recording: technology for recording visual and audio signals on magnetic tape for playback on television
- Laser: light energy focused in a very small and powerful beam
- Computers: machines that perform high-speed mathematical or logical calculations or processes and print information
- Electron microscope: instrument that views areas as small as one-billionth of an inch
- Radiotherapy: treatment of diseases with radiation

PRINCIPLES OF ELECTRONICS

Magnetism and Electricity

Before proceeding with the remainder of this unit, students should review the material in Unit 33 ("Electrical Principles and Wiring Materials"). There, some basic principles of electricity and magnetism are presented. Further, safety practices with electricity and volts, amperes, and watts are discussed. Elements of electrical wiring are covered in Units 33 and 34 (Electricity and Electronics "Installing Branch Circuits"). Additional information on electricity and electronics is also given in Unit 26 ("Selecting and Using Arc Welding Equipment"), Unit 27 ("Arc Welding Mild Steel and GMAW/GTAW Welding"), Unit 30 ("Fundamentals of Small Engines"), and Unit 31 ("Small Engine Maintenance and Repair").

Resistance

Resistance, as defined earlier, is the tendency of a material to prevent electricity flow. **Conductance** is the opposite. It is a measure of how easily a material allows electricity to pass through it.

The **resistivity** of a material is defined as the resistance in a wire made from a material that is 1 foot long and 1 mil (0.001 inch) in diameter at 20°C. Materials that do not allow the flow of electricity are called **insulators**. The symbol for resistance is R.

Resistance (R) is measured in ohms. An **ohm** is the amount of resistance that allows 1 ampere of current to flow when 1 volt of electrical pressure is applied. The symbol for ohm is the Greek letter omega (Ω).

The ability of a material to carry electrons or electricity is called *conductance*. The symbol for conductance is (G). The unit of measure for conductance is the **siemen**; its symbol is S.

Resistance is the **reciprocal**, or opposite, of conductance. The relationship is illustrated by the following formulas:

$$R = 1/G \text{ and } G = 1/R$$

Example #1
Problem: What is the conductance of a 250-ohm resistor?

> Formula: $G = 1/R$
> Solution: $G = 1/250 = 0.004$ siemen

Example #2
Problem: What is the resistance of a conductor with a conductance of 600 siemens?

> Formula: $R = 1/G$
> Solution: $R = 1/600 = 0.0017$ ohm

Resistors

A **resistor** is a device that offers a specific amount of resistance to a current. Resistors are the most commonly used components in a circuit. They come in various sizes and shapes to meet various conditions (Figure 35-4). They may be designed with fixed resistance values, or they may permit resistance to be varied.

Fixed Resistors. Fixed resistors are manufactured to deliver their rated resistance within certain tolerances above or below that rating. Lower-tolerance resistors tend to cost more. Resistors are generally available with tolerance ratings of plus or minus 20, 10, 5, 2, or 1 percent.

FIGURE 35-4 Resistors are designed to work in various types of circuits.

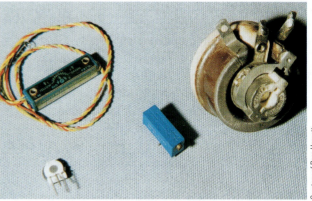

FIGURE 35-6 A rheostat is a variable resistor used to control current.

Variable Resistors. Variable resistors have three terminals (Figure 35-5). Two of the terminals form the ends of a resistor strip, and the third terminal attaches to a conductor on the end of a rotating shaft. When the shaft is rotated, the conductor moves to various points along the resistor to provide the desired level of resistance between the center terminal and each of two side terminals. A variable resistor used to control voltage is called a **potentiometer (pot)**. A variable resistor used to control current is called a **rheostat** (Figure 35-6).

Resistor Identification. The small size of some fixed resistors prevents printing the resistance value and tolerance on its case. Therefore, they are color-coded. The industry uses a standard system of color coding established by the Electronics Industries Association (EIA).

The color code is represented by a system of colored bands. The first band is the band closest to an end of the resistor. The first band represents the first number of the resistance value, the second band represents the second number, and the third band represents the number of zeros to add to the first two digits. The fourth band represents the tolerance of the resistor (Figure 35-7). There are certain exceptions to these color interpretations that the technician should also learn.

Use the top table in Figure 35-7 to see why a resistor with bands that are brown, green, red, and silver would have a resistance value of 1,500 ohms. Its tolerance level is 10 percent (Figure 35-8). There is also a number system for use on larger resistors.

Types of Resistive Circuits. Resistive circuits may be of three types (Figure 35-9). A **series circuit** provides a single path for current flow. A **parallel circuit** provides two or more paths. A **series-parallel circuit** combines the two types.

Checking Resistors. A resistor may be checked with an ohmmeter. However, to check a resistor, it must be disconnected at both ends from the circuit.

Electrical Meters

An **electrical meter** is a device that measures the activities of electrons and provides a visual interpretation of these activities. Meters may be analog or digital. The **analog meter** has a graduated scale and a pointer (Figure 35-10). The **digital meter** provides a direct numerical readout (Figure 35-11).

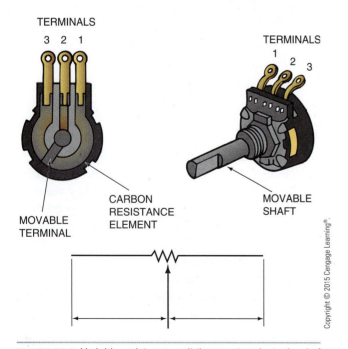

FIGURE 35-5 Variable resistors permit the user to select a level of resistance suitable for the job.

TERMINALS
3 2 1

TERMINALS
1 2 3

MOVABLE
TERMINAL

CARBON
RESISTANCE
ELEMENT

MOVABLE
SHAFT

Resistor Color Codes

Two-Significant-Figure Color Code

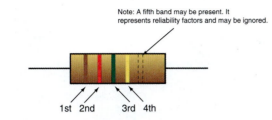

Note: A fifth band may be present. It represents reliability factors and may be ignored.

1st 2nd 3rd 4th

COLOR BANDS			
	1st	**2nd**	**Number of Zeros**	**Tolerance 4th**
Black	0	0	–	
Brown	1	1	0	
Red	2	2	00	
Orange	3	3	000	
Yellow	4	4	0,000	
Green	5	5	00,000	
Blue	6	6	000,000	
Violet	7	7		
Gray	8	8		
White	9	9		
Gold			0.1	5%
Silver			0.01	10%
No Color				20%

Three-Significant-Figure Color Code*

COLOR BANDS			
	1st	**2nd**	**3rd**	**Multiplier 4**
Black	0	0	0	–
Brown	1	1	1	0
Red	2	2	2	00
Orange	3	3	3	000
Yellow	4	4	4	0,000
Green	5	5	5	00,000
Blue	6	6	6	000,000
Violet	7	7	7	
Gray	8	8	8	
White	9	9	9	
Gold				0.1
Silver				0.01

*All one-percent tolerance resistors.

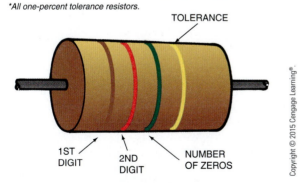

TOLERANCE

1ST DIGIT   2ND DIGIT   NUMBER OF ZEROS

FIGURE 35-7 The color bands on a resistor indicate its resistance and tolerance.

Some suggestions for effective use of meters are:

1. Properly connect meters observing correct polarity. The red terminal should be positive, and the black terminal should be negative.

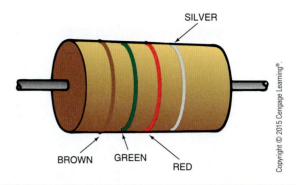

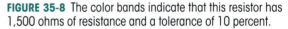

SILVER

BROWN   GREEN   RED

FIGURE 35-8 The color bands indicate that this resistor has 1,500 ohms of resistance and a tolerance of 10 percent.

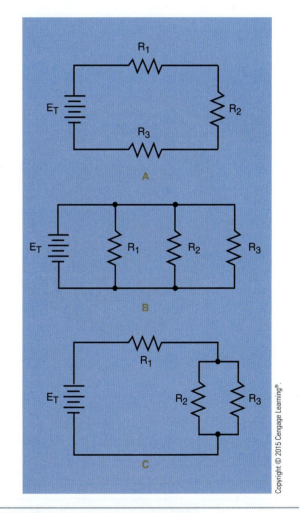

FIGURE 35-9 (A) Series, (B) parallel, and (C) series-parallel circuits.

2. Before using an analog meter, adjust the pointer to indicate zero on the scale.
3. To measure current (amperes) with an **ammeter**, open the circuit and place the meter into the circuit in series.
4. To measure voltage with a **voltmeter**, place the meter's leads in the circuit in parallel.

Copyright © 2015 Cengage Learning®.

FIGURE 35-10 An analog meter.

FIGURE 35-12 A volt-ohm-milliammeter (VOM) is also known as a multimeter.

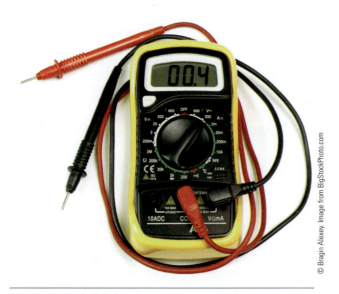

FIGURE 35-11 A digital meter.

7. Study the scale on the meter to be certain you understand what each mark and space represents (Figure 35-13).

8. Amperes in an alternating-current circuit may be measured without opening the circuit utilizing a clamp-on meter that measures magnetism and then converts its readings out into amperes (Figure 35-14).

5. To measure resistance, take the resistor, bulb, motor, or conductor out of the circuit. Zero the ohmmeter, and then connect the ohmmeter's prods or clips.

CAUTION!

An error in connecting an ammeter or voltmeter will damage the instrument!

6. When using a **multimeter**, which is a combination volt-ohm-milliammeter (or VOM), make sure the instrument is set for the appropriate function and scale before attaching the leads (Figure 35-12).

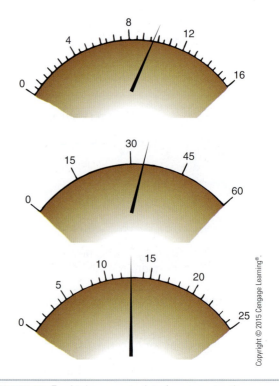

FIGURE 35-13 To obtain accurate data from a meter, the user must read the scale carefully.

FIGURE 35-14 A clamp-on meter senses magnetic intensity and converts that measurement into amperes of current.

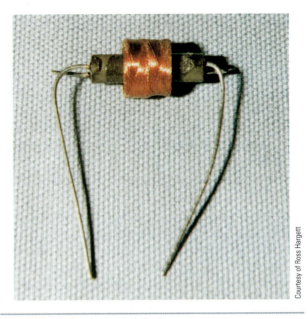

FIGURE 35-15 An electronic inductor.

Inductance

The ability to draw energy from a source and store it in a magnetic field is called **inductance** (L). For example, if current increases in a coil, the magnetic field expands. If the current decreases, the magnetic field collapses. However, the collapsing magnetic field induces a voltage back into the coil, which maintains the current flow. The symbol for inductance is L.

The Henry. The unit of measure of inductance is called the **henry (H)**. A henry is the amount of inductance required to induce an electromotive force (EMF) of 1 volt when the current in a conductor changes at the rate of 1 ampere per second.

Inductors. An **inductor** is a device that provides inductance. Both fixed and variable inductors are available (Figure 35-15).

Capacitance

The ability of a device to store electrical energy in an electrostatic field is **capacitance**. A **capacitor** is a device that possesses a specific amount of capacitance. An electrostatic field is created when two objects with different charges come close together.

A capacitor is made of two conductors separated by an insulator. The conductors are called **plates** and the insulator is called **dielectric** (Figure 35-16). Capacitors are generally made by rolling the plates and the dielectric into a cylinder (Figure 35-17).

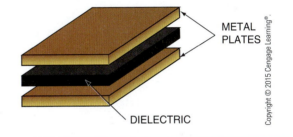

FIGURE 35-16 A capacitor consists of plates (conductors) separated by a dielectric (insulator).

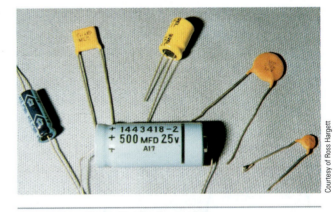

FIGURE 35-17 Various types of capacitors.

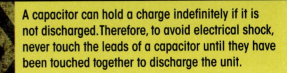

CAUTION!

A capacitor can hold a charge indefinitely if it is not discharged. Therefore, to avoid electrical shock, never touch the leads of a capacitor until they have been touched together to discharge the unit.

Direct and Alternating Current

Current from batteries flows in one direction and is called direct current (DC). Direct current can also be produced with DC generators. On the other hand, alternating current (AC) reverses its direction of flow at rapid intervals caused by rotations of the armature of the generator or alternator creating the current.

Cycle. The flow of current through one positive thrust followed by flow of current through one negative thrust is called a **cycle**. One cycle is thus created by one complete turn of the generator's armature in the generator (Figure 35-18). One cycle per second is defined as a **hertz (Hz)**. The wave created by the flow of current through one cycle is called a **sine wave**. AC current provided by power suppliers in the United States is generated at 60 cycles per second. Therefore, appliances operated on such current must be rated 60 Hz.

Oscilloscopes. An **oscilloscope** is an important piece of test equipment for working on electronic equipment and circuits. It provides a visual display of the frequency and duration of a signal, the phase relationships, the wave form of the signal, and the amplitude (strength) of the signal (Figure 35-19).

Semiconductors

A **semiconductor** is a material with characteristics that fall between those of insulators and conductors. Three pure semiconductor elements are carbon (C), germanium (Ge), and silicon (Si). Silicon and germanium are the most commonly used semiconductors. **Silicon** is the most used and is abundant in sand, quartz, agate, and flint.

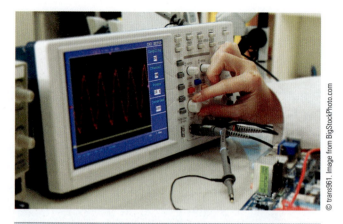

FIGURE 35-19 An oscilloscope shows visual displays of electronic signals.

Common uses of semiconductors include diodes (which rectify alternating current by converting it into direct current), transistors (which amplify or strengthen electrical signals), thyristors (which act as switches), and integrated circuits (which may be used to switch or amplify).

Semiconductors must undergo highly technical processes to modify their structure to achieve their performance in modern technology. The primary use of semiconductor devices is to control voltage or current for a desired result. Their advantages are small size and weight, low power consumption, great reliability, instant operation, and economical mass production.

Diodes. **Diodes** are the simplest type of semiconductor (Figure 35-20). They allow current to flow in one direction only. A popular use of diodes is on automobile alternators to change AC current to behave like DC current, since it permits current to flow in one direction only.

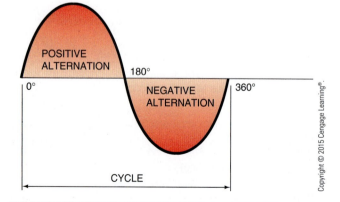

FIGURE 35-18 One complete (360-degree) turn of the generator's armature produces one cycle.

FIGURE 35-20 Common diode packages.

Transistors. A **transistor** is a three-element, two-junction device used to control electron flow. By varying the voltage applied to the three elements, the amount of current can be controlled for amplification, oscillation, and switching (Figure 35-21).

Thyristors. A **thyristor** is a semiconductor device used for electronically controlled switches. Thyristors are used to apply power to a load or remove power from a load. They can also be used to regulate power or adjust power to a load. Examples include dimmer controls for lights and motor speed controls. *Silicon-controlled rectifiers (SCRs)* are the best known of the thyristors (Figure 35-22).

Integrated Circuits. An **integrated circuit** is a miniature electronic circuit. The integrated circuit includes diodes, transistors, resistors, and capacitors. Their main advantage is smallness. An integrated circuit may be contained in a chip of semiconductor material only $\frac{1}{8}$" × $\frac{1}{8}$" in size.

Computer systems, once the size of rooms, are now desktop or laptop accessories because of integrated circuits. Integrated circuits cannot be repaired, so faulty or damaged circuits are replaced in their entirety (Figure 35-23).

Optoelectric Devices. An **optoelectric device** is a circuit component designed to interact with light energy in the visible, infrared, and ultraviolet ranges. They include light-detecting, light-converting, and light-emitting devices. The manufacturing method used for the semiconductor diode will determine the light wavelength of a particular device.

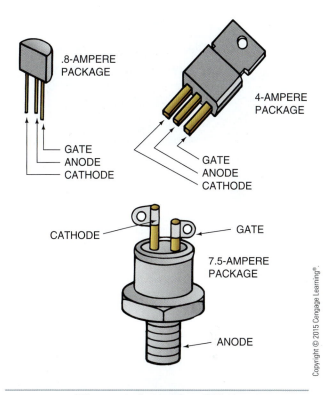

FIGURE 35-22 Silicon-controlled rectifiers (SCRs).

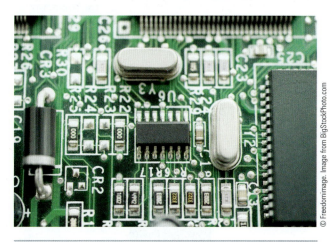

FIGURE 35-23 Examples of integrated circuit packages.

A **photoconductive cell** is a light-sensitive device. It is made from cadmium sulfide (CdS) or cadmium selenide (CdSe). It is useful for low-light and relatively high-voltage applications (Figure 35-24). Photoconductive cells are popular automatic switches for all-night lights.

A **photovoltaic cell (solar cell)** converts light energy into electrical energy (Figure 35-25). Solar cells are popular for powering pocket calculators, light meters for photography, sound-track decoders for motion picture projectors, and battery chargers on satellites. A **photodiode** is similar to a photovoltaic cell except that it is used to control current flow, not generate it (Figure 35-26).

FIGURE 35-21 Typical transistor packages.

FIGURE 35-24 A photoconductive cell.

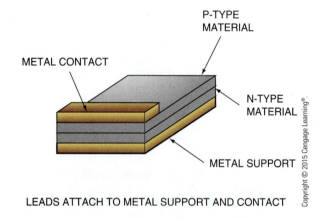

LEADS ATTACH TO METAL SUPPORT AND CONTACT

FIGURE 35-25 Composition of a solar cell. Solar cells are used to convert energy from the sun to electrical energy.

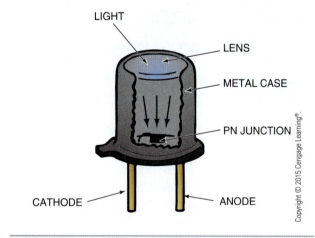

FIGURE 35-26 A photodiode package.

A **phototransistor** is packaged like a photodiode except it has three leads. The third lead permits adjustment of the turn-on point. Phototransistors can produce higher output than photodiodes. They are used in phototachometers (a device that uses a light beam

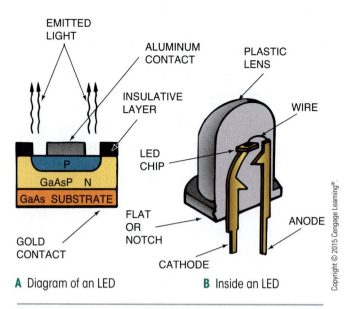

A Diagram of an LED **B** Inside an LED

FIGURE 35-27 Composition of a light-emitting diode (LED).

to measure the speed of an object such as a wheel or gear), photographic exposure controls, flame detectors, object counters, and mechanical positioners.

A **light-emitting diode (LED)** produces light when subjected to a current (Figure 35-27). These devices provide digital readouts for many types of electronic equipment.

Power Supplies. A **power supply** is a device that provides electricity of the proper type and amount to circuits. The main functions are to increase or decrease the incoming AC voltage by means of a transformer, convert AC to DC current by using a **rectifier**, and control voltage output with a **voltage regulator**. Further, a **filter** converts pulsating DC voltage into smooth voltage. DC voltage may be stepped up, if desired, with a **voltage multiplier**. Additionally, circuit protection devices, such as **overvoltage protection circuits, fuses**, and **circuit breakers**, may be used.

Amplifiers. An **amplifier** is an electronic circuit used to increase the amplitude (extent of) of an electronic signal. A circuit may be designed to function as a voltage amplifier or a current amplifier. A **video amplifier** is used to amplify pictorial information.

Oscillators. An **oscillator** is a nonrotating device for producing alternating current. Oscillators are used extensively in radios, televisions, communications systems, computers, industrial controls, and timekeeping devices. Most electronics equipment cannot function without oscillators.

Computers

Computers have become the backbone of the communications revolution. As early as the 1930s, progress was being made in the development of the modern digital computer. The **digital computer** is a device that automatically processes data using digital techniques. **Digital** means using digits. **Digits** are the numbers 0 through 9. Computers record, store, and manipulate data using the **binary system**. Known simply as the binary system, it contains only two digits, 0 and 1. **Binary** means two parts or components.

Computer Chips. Data are stored, processed, displayed, or printed in or from tiny electronic circuits in computers called **chips**. One chip can hold millions of bits of information. The term **bit** comes from the *b* in binary and the *it* in digit. One bit is equal to one digit, one number, or a yes/no answer to a question. A computer can have many chips.

Computer Data Storage. A computer can store vast amounts of information on its **hard drive**, which consists of multiple rotating platters that have magnetic surfaces. Data can also be stored on removable media such as compact discs (CDs), digital video discs (DVDs), and flash drives.

Types of Computers. The early computers occupied entire rooms. They operated on relatively large vacuum tubes that burned out quickly. Therefore, the early computers were large, heavy, generated a lot of heat, and failed frequently. All the computer capability of an institution was placed in a large unit called a **mainframe** computer.

With the development of computer chips and integrated circuits, the microcomputer has revolutionized business, government, education, and the home. The **microcomputer** is a self-contained small computer suitable for desktop, portable, or on-board use (Figure 35-28). It may work independently or in a system (or network) with other computers. Microcomputers are found in kitchen appliances, automobiles, trucks, tractors, boats, aircraft, offices, factories, and elsewhere.

While computers (commonly referred to simply as *computers*, *personal computers*, or *PCs*) are very popular, industry, universities, government, and other large systems still use mainframe computers and supercomputers. **Supercomputers** have enormous capacity.

FIGURE 35-28 Microcomputers such as the digital tablet are a common business tool.

AGRICULTURAL APPLICATIONS

Electronics provide the mechanisms for high-technology applications in every conceivable area of agriculture/agribusiness and renewable natural resources.

In the Field

In land leveling, drainage, irrigation, and construction, laser equipment is essential for determining accurate elevations and sightings. In field planting and harvesting equipment, electronic sensors, lights, sound alarms, and digital readouts inform operators of proper equipment operation (Figure 35-29).

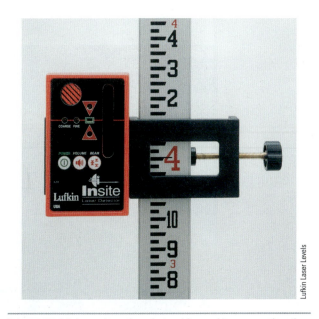

FIGURE 35-29 Laser-based equipment is used to gauge elevation. The part shown here reads a laser beam and gives the elevation.

Planting equipment utilizes electronics to monitor seed drop, seed spacing, and planting rate per acre. Fertilizer application rate and placement, acreage covered, ground speed, and rate per acre can be monitored similarly.

A grain combine has devices to monitor and/or control speed of the machine over the ground, reel speed, cylinder speed, engine speed, and **throughput** or rate of flow of grain and straw through the combine. Warning devices alert the operator to header overload, cylinder clearance and load, walker area overload, excessively dirty or unthreshed grain, and grain tank fullness. Management data provided by electronic devices include bushels per acre of yield, bushels per minute being harvested, and percent of moisture in the grain (Figure 35-30).

Tractor performance is also monitored by electronic devices. Digital readouts may provide engine rpm, percentage of wheel slippage, forward ground speed, axle load, drawbar load, **power takeoff (PTO)** speed, and PTO load. **Infrared light sensors** detect heat or motion.

The latest system in monitoring technology is the **Global Positioning System (GPS)**. This technology was developed by the U.S. Department of Defense and is now widely available for civilian use. The system uses a series of satellites to pinpoint any location on Earth (Figure 35-31). When used in conjunction with harvest monitoring equipment, GPS permits a producer to determine which areas of a field are producing the most or the least amount of grain.

FIGURE 35-30 This device monitors crop yields on harvesting equipment.

This helps the producer locate areas that are not producing well. He or she can then take measures to remedy the problem.

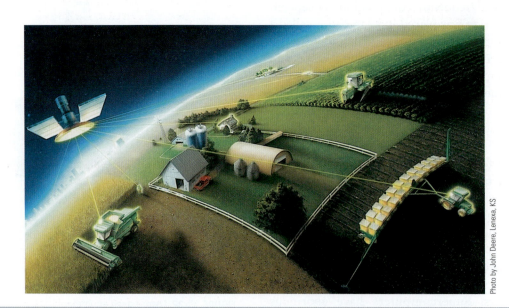

FIGURE 35-31 The Global Positioning System (GPS) can be used to pinpoint locations in a field where production is especially bad or good.

Livestock and Dairy Operations

Feedlots and milking areas are becoming more high-tech every year. **Biosensors**—devices that measure processes of the body—are used on cattle and other animals to monitor their body temperature, breeding status, feed requirements, and disease problems, so the producer can supply additional feed or other supplements if appropriate (Figure 35-32).

Heat sensors in the **claw** or cup of milking machines can signal the possibility of mastitis and other udder infections. Similarly, **viscosimeters** or flow meters can indicate thickening of milk or presence of blood clots, which indicate mastitis. Various temperature-sensing devices can monitor milk-cooling operations to ensure proper cooldown for maximum milk quality.

Robots for putting on milkers are making milking a totally automated process. Cameras, visual imagery, computers, and various sensors all guide the activities of robots.

Humidistats gauge moisture buildup from livestock, thermostats sense temperature changes, and gas analyzers determine when ammonia is being produced from improper conditions in poultry and livestock manure. Such equipment typically controls heating, cooling, and ventilation in buildings.

Grain and Feed Processing

Computers monitor and control the drying, grinding, and handling of grains for human and livestock consumption. Computers also determine proportions for rations and control the machinery that mix and transport the ingredients and final products.

Laboratories

Soil analysis, fertilizer ingredient proportioning, milk quality control, chemical analysis, pesticide monitoring, and other laboratory functions are now automated. Entire feed-processing plants are typically fully automated today.

Aquaculture

Aquaculture uses electronics for depth finding, fish finding, and navigational aids. Automated water analysis equipment for fish culture, pollution monitoring, and water media management will be increasingly important in the future (Figure 35-33).

FOOD DISTRIBUTION

Food distribution centers for modern supermarkets are highly computerized and automated. Such centers have huge buildings laid out in cells for pallet storage. A **pallet** is a portable wooden platform for storing or moving cargo.

Food and household items needed by a given store are ordered by way of a computer. The computer directs the automatic forklift unit to the appropriate item in the warehouse, the item is picked up, and taken to and loaded on the appropriate truck at the dock. The process is repeated until all items are loaded. The computer automatically adjusts the inventory in the warehouse and the store to reflect the change.

FIGURE 35-32 Biosensors on livestock provide data about the animal to producers.

FIGURE 35-33 Water analysis equipment is used for agriculture and water management.

Smart House

Homes are now being equipped to accommodate the physically handicapped. Verbal commands and other automated mechanisms can turn water on and off, open doors, lock windows and doors, and adjust bed positions. Such equipment can be designed to do almost any task that people with handicaps find difficult to do for themselves. The use of electronic devices can alert homeowners or the police of intruders and have become valuable home security devices.

STRATEGIES FOR MAINTAINING ELECTRONICS EQUIPMENT

Electronics components contain very tiny, complex, and complicated circuitry. Many components are difficult or impossible to repair once they are damaged. Therefore, service of electronic devices should only be done by specially trained personnel. Such individuals have analytical and testing equipment to isolate the problem and indicate the faulty component. The faulty component or module can generally be replaced with ease.

General maintenance procedures for all electronic equipment should include:

- Keep the item free of dust, dirt, metal objects, and liquid spills.
- Wipe off routinely with a dry, clean cloth.
- Lubricate equipment as specified by the manufacturer.
- Maintain a cool, dry environment for computers, copiers, sound systems, and most electronic devices.
- Read the operator's manual and follow the instructions carefully.
- Call the supplier, factory representative, or qualified repair technician when needed; do not tamper with electronic devices.

SUMMARY

Electronics have become a major part of our everyday lives. Electronics permit us to communicate, automate many processes, and be more productive. They increase health and safety, and improve quality of life. The career outlook in electronics and its applications in agriculture is excellent.

Student Activities

1. Define the Terms to Know in this unit.

2. Ask your teacher to divide the class into groups of five. Each group should select a recorder and a chairperson. Have each group spend 5 minutes listing activities or processes in which electronics equipment is used. Have the recorders report the lists to the class.

3. Develop a collage depicting electronic devices.

4. Prepare and present a class report on one major discovery that contributed to today's electronics technology.

5. Develop an Electronics Hall of Fame poster. Include persons, discoveries, and dates for major landmark activities in the discovery and development of electricity and electronics.

6. Research the manufacturing and service procedures for a major type of electronic component, such as semiconductors, power supplies, or computer chips.

7. Design a security system for your home utilizing modern electronic devices.

8. Construct a chart or bulletin board illustrating uses of electronics in agriculture/agribusiness and renewable natural resources.

Relevant Web Sites

All About Circuits
www.allaboutcircuits.com

Public Broadcasting Service (PBS), information on transistors and diodes
www.pbs.org/transistor/index.html

Self-Evaluation

A. Multiple Choice. Select the best answer.

1. Static electricity was described as early as
   a. 10,000 B.C.
   b. 5000 B.C.
   c. 600 B.C.
   d. A.D. 1600

2. Archaeologists have found evidence of electro-plating dating to
   a. A.D. 1607
   b. A.D. 1240
   c. 4000 B.C.
   d. thousands of years ago

3. The conductance (G) of a 100-ohm resistor is
   a. 0.01 siemen
   b. 0.10 siemens
   c. 10 siemen
   d. 100 siemens

4. The resistance (R) of a conductor with a conductance of 300 siemens is
   a. 0.0033 ohm
   b. 0.0010 ohm
   c. 0.0013 ohm
   d. 0.0001 ohm

5. Unless discharged after power is turned off, a serious electric shock can occur from a
   a. transistor
   b. rheostat
   c. potentiometer
   d. capacitor

6. Which substance is not used in semiconductors?
   a. carbon
   b. germanium
   c. silicon
   d. titanium

7. Devices that interact with light are called
   a. chips
   b. optoelectric
   c. silicon saddles
   d. transistors

8. LED stands for
   a. low exterior density
   b. limited-emission device
   c. lithium emulsion drawdown
   d. light-emitting diode

9. Alternating current can be converted to direct current with
   a. a transformer
   b. an oscilloscope
   c. a rectifier
   d. an optoelectric device

10. Data are processed in a computer by
    a. bitmaps
    b. chips
    c. disks
    d. transformers

B. Matching. Match the terms in column I with those in column II.

Column I

1. lodestone
2. Thales
3. Charles du Fay
4. Alessandro Volta
5. potentiometer
6. resistance
7. henry (H)
8. hertz (Hz)
9. binary
10. digits

Column II

    a. discovered static electricity

    b. opposite of conductance

    c. variable resistor

    d. unit of inductance

    e. one cycle per second

    f. observed that like charges repel

    g. magnetized iron oxide

    h. 0 to 9

    i. developed the first battery

    j. 0 or 1

C. Completion. Fill in the blanks with the word or words that make the following statements correct.

1. What does each of the following color bands on a resistor tell you?
   a. first band _____
   b. second band _____
   c. third band _____
   d. fourth band _____

2. Computers may be classified by capacity and size as _____ and _____.

3. _____ are used on animals to indicate body temperatures, breeding status, and disease problems.

4. Two uses of electronics in agriculture are _____ and _____.

D. Brief Answer. Briefly answer the following questions.

1. What is an electrical meter? Name two types and explain how they differ.

2. What is inductance? Explain how it works to maintain current flow.

3. What information does an oscilloscope provide in a visual display?

4. What is a diode? What is a popular use of diodes?

5. What are the main functions of a power supply?

UNIT 36

Electric Motors, Drives, and Controls

Objective

To install, maintain, and utilize motors and controls.

Competencies to be developed

After studying this unit, you should be able to:

- List advantages of electric motor power.
- Discuss the types of electric motors.
- List factors to consider when selecting motors.
- Discuss mounts and drives for motors.
- Select and use motor controls.
- Maintain motors and controls.

Materials List

- Pencil
- Pad
- Calculator

Terms to Know

- ball bearings
- needle bearings
- bushing
- frame
- clutch
- drive train
- type
- induction
- centrifugal
- horsepower
- duty rating
- temperature rise
- pulley
- step pulleys
- adjustable pulley
- variable-speed pulley
- sprockets
- transmission
- gear
- sensor
- fusestat
- actuator
- float switch
- sump hole
- manual
- override
- pilot light

continued

Electric motors provide the power for most portable power tools, household appliances, shop power equipment, stationary farm equipment, processing equipment, and conveying equipment (Figure 36-1). They drive massive machinery for mining and manufacturing, trains, and water and oil pumps. Electric motors are also essential components of portable machines such as autos, trucks, tractors, earth movers, boats, airplanes, mowers, tillers, and combines. They start gasoline and diesel engines, provide power for accessories, and pump fluid for hydraulic systems (Figure 36-2).

Important information on electricity, electronics, and motors is presented in Unit 33 ("Electrical Principles and Wiring Materials"), Unit 34 ("Installing Branch Circuits"), and Unit 35 ("Electronics in Agriculture"). Students should review those before proceeding with this one.

SELECTING ELECTRIC MOTORS

Electric motors should be of the proper type, size, and capacity for the job. Economical and efficient operation depends upon selection of the correct motor for the job.

FIGURE 36-1 Electric motors have become a mainstay of modern technology.

FIGURE 36-2 Electric motors supply power for a feed mill.

Advantages of Electric Motors

Electric motor use in domestic and agricultural settings is increasing. Electric motors are frequently chosen over other types of power because they are:

- Adaptable—motors can be used in almost any location, including underwater.
- Automatic—motors are easily controlled by automatic devices.
- Compact—a small unit develops a relatively large amount of power.
- Dependable—electric motors selected specifically for the job seldom give trouble.
- Economical—a small motor operating at less than 25¢ per hour can do the work of several humans.
- Efficient—electric motors have efficiency ratings ranging up to 95 percent.
- Long endurance—electric motors frequently run trouble-free for 20 to 30 years.
- Low-maintenance—if protected from dust and dirt, motors require little or no maintenance.
- Quiet—motors are generally quieter than the machinery they drive.
- Safe—when properly installed, maintained, and used, motors are very safe to operate.
- Simple to operate—very little training is needed to operate most motors.

Factors in Selecting the Correct Electric Motor

For most users, the motor comes with the equipment, building, or facility. Therefore, every motor has been selected by a manufacturer, engineer, or electrician. However, a knowledge of motor selection factors is useful when motors are replaced or equipment is being designed or made on the job.

Power Supply. What electrical power is available? Is it 120 volts or 230 volts? Single-phase or three-phase? Motors larger than ½ horsepower run more efficiently and cause less voltage drop problems if operating with 230 volts. Additionally, the use of three-phase current further reduces such problems. Circuits are categorized by voltage. The two categories most often used in the SU are the 110 or 220 single phase. Because the voltage may vary from 110 volts to 115, the circuit may be called a 110-, a 115-, or a 120-volt circuit. Similarly, a 220-volt circuit may be called a 220-, a 230-, or a 240-volt circuit. The terms are used interchangeably.

New Wiring. Would the benefit of updated or modified wiring override the cost of new wiring from 230-volt or three-phase motors? Many farms, ranches, and businesses do not have three-phase current. However, 230-volt single-phase current is usually in the building, an adjacent building, or on a central meter pole on the property.

Power Requirements. How many horsepower are needed to run the machine? Is the machine easy to start? If not, the motor must have high torque starting capabilities. What speed is needed? Motors typically run at 1150, 1750, or 3450 rpm. The motor speed should match or be adaptable to that required by the machine. When replacing motors, the characteristics of the replacement motor should be as near as possible to those of the original motor. If the machine is being designed or rigged up, use a motor as large as or larger than that used on similar commercial equipment.

Base, Bearings, and Frame. Slotted holes in motor bases help to accommodate bolts and mounting. Similarly, mounting plates may have long slots to permit tightening of the drive belt. Further, holes in the base may have rubber mounts to cushion the motor against vibration or shock.

The better motors have ball bearings rather than bushings for shaft support. A **ball bearing** is a set of hardened steel balls in a container or cage that keeps a shaft centered in a frame (Figure 36-3). **Needle bearings** are steel rollers in a cage to keep a shaft centered and useful when extreme pressure is involved. Finally, a **bushing** is a sleeve in a frame that keeps a

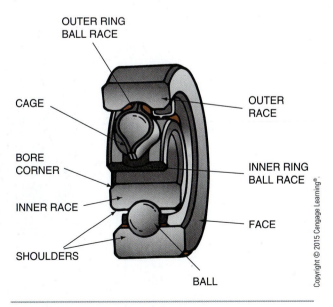

Copyright © 2015 Cengage Learning®.

FIGURE 36-3 A ball-bearing assembly.

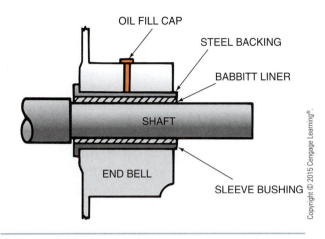

FIGURE 36-4 A bushing-type bearing.

shaft centered (Figure 36-4). Modern sleeve bearings are of a type called sintered bearings. This type bearing is usually made of bronze. The sintering process allows the bronze to be porous. This allows oil and lubricating materials to seep into the pores, allowing the bearings to be constantly lubricated.

Ball bearings or needle bearings are generally permanently lubricated with grease. However, bushings must be lubricated with small amounts of oil at regular intervals. The oil is generally stored in a felt wick or oil-impregnated bushing material.

Ball bearings add initial cost to a motor but have major advantages over bushings. They can be permanently lubricated and mounted in any position. They also create less friction, withstand more shaft pressure, and rarely need replacing.

The motor **frame** is the housing or case that contains the other motor components. The frame must be designed to permit the motor to be mounted appropriately; permit adequate ventilation and cooling; and support the bearings, shaft, and other components (Figure 36-5).

Starting and Running Characteristics. Electric motors are generally connected directly to the load, with no clutch in the drive train. Therefore, they must have sufficient torque to start. A **clutch** disengages a power source from its load. A **drive train** is the collective components that transfer power from a power source to a load.

Ease of starting of a load will dictate which types of motors are suitable. Examples of loads that are easy to start are fans, saws, and equipment with free-running lightweight shafts and components. Examples of hard-starting loads are compressors, conveyers, augers, and grinders that must compress air, move weight, or cut material as they start.

Motor **type** refers to the way current flows in the motor and accessories it has to help it start or run. Type includes a name for starting and a name for running, unless they are the same. For instance, an induction motor that has a capacitor to help in starting, but not when running, is called a "capacitor-start,

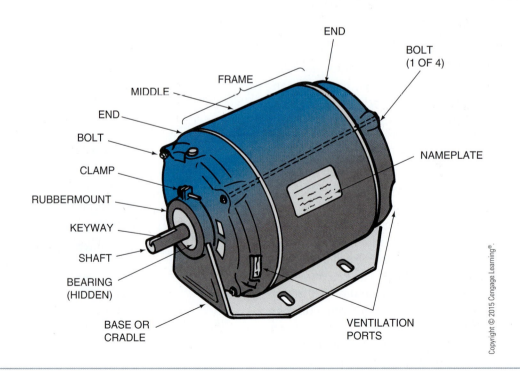

FIGURE 36-5 Exterior parts of a motor.

induction-run" motor. A capacitor manages the flow of current to provide increased motor torque. **Induction** is a type of motor where electricity from the power source does not flow directly to the armature. Motors typically use a centrifugal switch that clicks closed, or turns on, when the motor slows down to stop, and clicks open, or turns off, the capacitor or other booster when the motor approaches its running speed on startup. **Centrifugal** refers to the outward force caused by a spinning object.

Several types of motors have a high starting torque. Some draw up to six times as much current when starting under a heavy load as they do when running. Such motors cause temporary voltage drop, which causes lights in the circuit or building to dim until the motor gets up to speed. Since motor selection is complex, an electrician or engineer should be consulted before purchasing an expensive motor. A selection table may also be used to gain a general understanding of motor types (Figure 36-6).

INTERPRETING NAMEPLATE INFORMATION

The nameplate on a motor provides valuable information (Figure 36-7). The model and serial number enable the manufacturer to identify the motor and are useful for warranty service or other communications with the supplier. Type, horsepower, rpm, and duty ratios indicate the capabilities of the motor.

One **horsepower** is the force needed to lift 33,000 pounds 1 foot in 1 minute, or 550 pounds 1 foot in 1 second. The formula for calculating horsepower is as follows:

$$\text{Horse-power} = \frac{\substack{\text{Weight} \\ \text{(in pounds)}} \times \substack{\text{Distance} \\ \text{(in feet)}} \times \substack{\text{Time} \\ \text{(in seconds)}}}{550}$$

or

$$HP = \frac{W(lb) \times D(ft) \times T(sec)}{550}$$

An efficient motor will generate 1 horsepower of force while using about 800 watts or 0.8 kilowatt (kW) of electricity. This is about the work done by eight to ten men.

Volts, amps, hertz, and phase indicate the type of electricity that must be supplied for the motor to run. **Duty rating** refers to the percent time a motor may run without overheating. **Temperature rise** indicates how warm the motor can get without causing internal damage. Other data may be provided on motor nameplates, depending on the manufacturer.

MOTOR DRIVES

Motors may be coupled directly to the equipment they drive, but the speed of the two must be matched. Direct drives include flexible-hose, flange, cushion-flange, and flexible shaft couplings. The first three types are likely to be used to couple motors to water pumps. The flexible shaft is frequently used to couple a motor with a portable tool, such as a grinder.

Belts

The use of belts, gear boxes, and chains permits the motor to run at one speed and the machine at another. Transmissions and variable-speed pulley configurations permit the choice of two or more speeds or variable speeds. Since belts are flexible and quiet, they are used in most motor drive systems. Belts are reinforced with nylon or wire cords. They are classified according to shape and size. Popular shapes include the V belt, the multi-V belt, the micro-V belt, and the flat belt (Figure 36-8). The V refers to the shape of the cross section of the belt. The wedge shape provides more surface area that comes in contact with the pulley, thus providing more "grip." Belt length should be determined by measuring around the pulleys, including the space between the motor and the machine.

Pulleys

A **pulley** is a device attached to a shaft to carry a belt. The pulley is locked to the motor shaft key by a set screw. Pulleys may be single, step, adjustable, or variable speed (Figure 36-9). **Step pulleys** are several sizes of pulleys fused together. They permit changing of equipment speeds by moving the belt to a different level on the pulley. Such a move must be accompanied by repositioning the pulley on the mating device unless a second step pulley is part of the system. After moving the belt, the belt must be retightened. An **adjustable pulley** is one with a movable side locked in place with a nut or set screw. The effective diameter of the pulley is changed by repositioning the movable side. A **variable-speed pulley** is a pulley with a movable side that can be moved while the equipment is running. Such devices permit the changing of speeds of the machine over a certain range frequently and easily with the machine running.

Load Type	Motor Type (1)	Starting Ability (Torque) (2)	Starting Current (3)	Size HP (4)	Electrical Power Requirements Phase (5)	Voltage (6)	Speed Range (7)	Reversible (8)	Relative Cost (9)	Other Characteristics (10)	Typical Uses (11)
	(a) Shaded-Pole Induction	Very low. ½ to 1 times running torque.	Low.	.035–.2 kW (1/20–1/4 hp)	Single	Usually 120	900 1200 1800 3600	No	Very Low	Light duty, low in efficiency.	Small fans, freezer blowers, arc welder blower, hair dryers.
	(b) Split-Phase	Low. 1 to 1½ times running torque.	High. 6 to 8 times running current.	.035–.56 kW (1/20–3/4 hp)	Single	Usually 120	900 1200 1800 3600	Yes	Low	Simple construction.	Fans, furnace blowers, lathes, small shop tools, jet pumps.
Easy Starting Loads	(c) Permanent-Split, Capacitor-Induction	Very low. ½ to 1 times running torque.	Low.	.035–.8 kW (1/20–1 hp)	Single	Single voltage 120 or 240	Variable 900–1800	Yes	Low	Usually custom-designed for special application.	Air compressors, fans.
	(d) Soft-Start	Very low. ½ to 1 times running torque.	Low. 2 to 2½ times running current.	5.6–40 kW (7½–50 hp)	Single	240	1800 3600	Yes	High	Used in motor sizes normally served by 3-phase power when 3-phase power not available.	Centrifugal pumps, crop dryer fans, feed grinder.
	(e) Capacitor-Start, Induction-Run	High. 3 to 4 times running torque.	Medium. 3 to 6 times running current.	.14–8 kW (1/6–10 hp)	Single	120–240	900 1200 1800 3600	Yes	Moderate	Long service, low maintenance, very popular.	Water systems, air compressors, ventilating fans, grinders, blowers.
	(f) Repulsion-Start, Induction-Run	High. 4 times running torque.	Low. 2½ to 3 times running current.	.14–16 kW (1/6–20 hp)	Single	120–240	1200 1800 3600	Yes	Moderate to High	Handles large load variations with little variation in current demand.	Grinders, deep-well pumps, silo unloaders, grain conveyors, barn cleaners.
	(g) Capacitor-Start, Capacitor-Run	High. 3½ to 4½ times running torque.	Medium. 3 to 5 times running current.	.4–20 kW (½–25 hp)	Single	120–240	900 1200 1800 3600	Yes	Moderate	Good starting ability and full-load efficiency.	Pumps, air compressors, drying fans, large conveyors, feed mills.
Difficult Starting Loads	(h) Repulsion-Start, Capacitor-Run	High. 4 times running torque.	Low. 2½ to 3 times running current.	.8–12 kW (1–15 hp)	Single	Usually 240	1200 1800 3600	Yes	Moderate to High	High efficiency, requires more service than most motors.	Conveyors, deep-well pump, feed mill, silo unloader.
	(i) Three-Phase, General Purpose	Medium. 2 to 3 times running torque.	Low-med. 3 to 4 times running current.	.4–300 kW (½–400 hp)	Three	120–240 240–480 or higher	900 1200 1800 3600	Yes	Very Low	Very simple construction, dependable, service-free.	Conveyors, dryers, elevators, hoists, irrigation pumps.

FIGURE 36-6 Selection table of electric motor characteristics.

Source: Brown, R. H. and Henderson, G. E. (1982). Electric Motors: Selection/Protection/Drives. American Association for Vocational Instructional Materials.

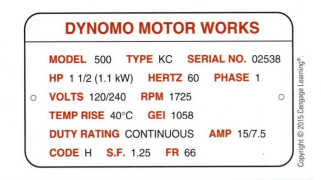

Copyright © 2015 Cengage Learning®.

FIGURE 36-7 A motor's nameplate provides a summary of critical information.

Pulley-Speed Ratio

It is important to understand the relationship between pulley size and belt speed. As the diameter of a motor pulley turning at a given rpm is increased, the resulting belt speed increases proportionally as does the speed of the driven machine. Therefore, the effect of changing the pulley size on the motor or the driven machine can be determined with a mathematical formula:

$$\text{Motor rpm (S)} \times \text{Motor pulley diameter (D)} = \text{Machine rpm (SV)} \times \text{Machine pulley diameter (DV)}$$

The formula can be shortened to:

$$S \times D = S' \times D'$$

where S = Speed of motor, D = Diameter of motor pulley, S' = Speed of machine, and D' = Diameter of machine pulley.

If a motor has a rated speed of 1,750 rpm, what size of pulley should be installed on the motor to drive a drill press with a 4-inch pulley at 975 rpm? The problem is solved as follows:

$$S \times D = S' \times D'$$
$$1{,}750 \times D = 875 \times 4$$

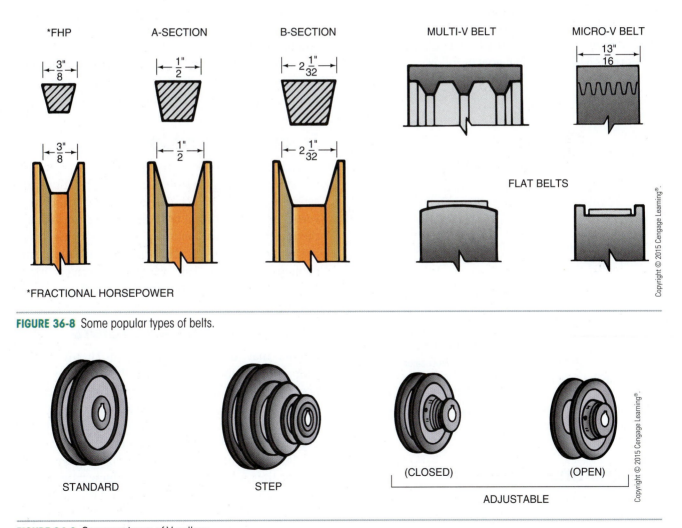

FIGURE 36-8 Some popular types of belts.

FIGURE 36-9 Common types of V pulleys.

$$1{,}750 \times D = 3{,}500$$

$$D = \frac{3{,}500}{1{,}750}$$

$$D = 2$$

Again, suppose the motor was rated at 1,150 rpm and came with a 3-inch pulley. What size of pulley should you install on the drill press to run it at 575 rpm? The solution is:

$$S \times D = S' \times D'$$

$$1{,}150 \times 3 = 575 \times D'$$

$$3{,}450 = 575 \times D'$$

$$\frac{3{,}450}{575} = D'$$

$$6 = D'$$

The same formula is used to calculate the speed of a machine if the speed of motor and pulley sizes are known. What is the speed of a saw with a 3-inch pulley if it is driven by a motor with a 4-inch pulley turning at 1,750 rpm? The calculation goes as follows:

$$S \times D = S' \times D'$$

$$1{,}750 \times 4 = S' \times 3$$

$$7{,}000 = S' \times 3$$

$$\frac{7{,}000}{3} = S'$$

$$2{,}333 = S'$$

From this formula and examples, one can make several observations:

1. To increase the speed of a machine, either increase the size of the motor pulley or decrease the size of the machine pulley.
2. To decrease the speed of a machine, either decrease the size of the motor pulley or increase the size of the machine pulley.

The formula applies also to gear sizes and motor and machine speeds when gears are used.

Chains and Sprockets

Drive chains function as belts except they are less flexible and cannot slip. They may be the roller type, like a bicycle chain, or link type. They travel on **sprockets**,

which are wheels with spokelike teeth. Chains and sprocket drives require regular maintenance such as cleaning, oiling, and repair. Their main advantage is strength and positive, nonslip power transfer.

Transmissions

A **transmission** is a gear box that permits two or more choices of speed (Figure 36-10). A **gear** is a wheel with teeth that mesh with teeth on another wheel to transfer power. Gears are available in many sizes, shapes, and tooth types (Figure 36-11). Some gears are designed for shifting while the shafts are turning, when a clutch is used. Others require the machinery to be stopped before the gears can be shifted.

ELECTRIC MOTOR CONTROLS

Electric motors are controlled by monitors, switches, and protective devices. These items enable electric motors to run continuously, on a timed basis, or when conditions call for them to run. They can be started and stopped by humans or controlled by switches activated by sensors. A **sensor** is a device that receives and responds to a signal, such as light or pressure.

Electrical Service

Electricity flows to the user from power plant to local transformer through a system of lines, substations, and switches (Figure 36-12). The local transformer converts the current to the type and voltage needed, and the meter measures how many kilowatts of electricity are used (Figure 36-13). The service distribution panel contains fuses or circuit breakers to protect circuits going to all electrical appliances (Figure 36-14).

Wiring Circuits for Motors

Motors may have an independent circuit directly from the distribution panel, or they may contain a power cord that plugs into an outlet (Figure 36-15). Whichever method is used, the circuit must be properly grounded (Figure 36-16) and be protected from overload and overheating by fuses or circuit breakers.

Fuses may be the cartridge type or plug type. They may be the regular type or the slow-blow type that can stand a temporary overload. The **fusestat** is a fuse with threads in its base that permit the fuse to be used only in circuits that match its capacity (Figure 36-17).

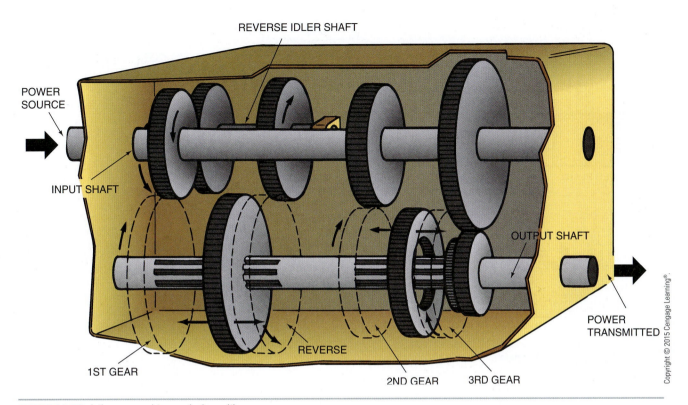

REVERSE IDLER SHAFT

POWER
SOURCE

INPUT SHAFT

OUTPUT SHAFT

POWER
TRANSMITTED

1ST GEAR

REVERSE

2ND GEAR   3RD GEAR

Copyright © 2015 Cengage Learning®.

FIGURE 36-10 A three-speed transmission with reverse.

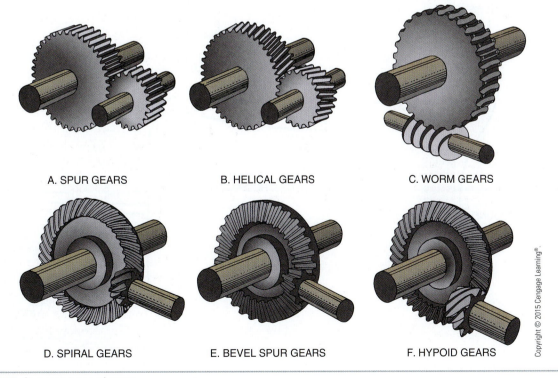

A. SPUR GEARS

B. HELICAL GEARS

C. WORM GEARS

D. SPIRAL GEARS

E. BEVEL SPUR GEARS

F. HYPOID GEARS

Copyright © 2015 Cengage Learning®

FIGURE 36-11 Common types of gears.

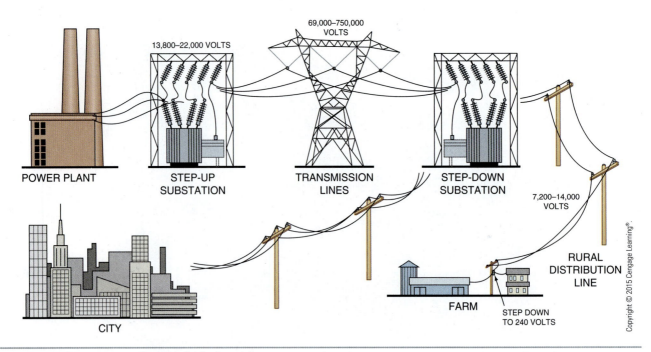

FIGURE 36-12 A power generation and distribution system.

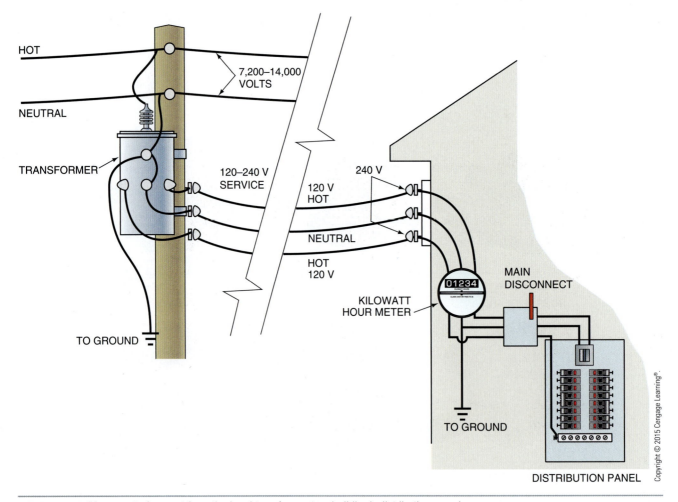

FIGURE 36-13 Movement of current from the local transformer to a building's distribution panel.

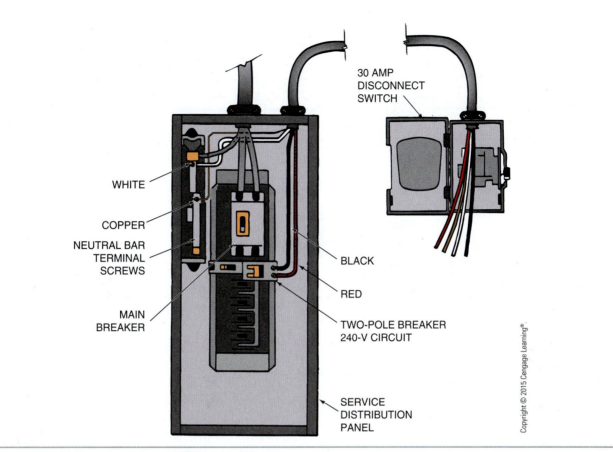

WHITE

COPPER

NEUTRAL BAR
TERMINAL
SCREWS

MAIN
BREAKER

30 AMP
DISCONNECT
SWITCH

BLACK

RED

TWO-POLE BREAKER
240-V CIRCUIT

SERVICE
DISTRIBUTION
PANEL

Copyright © 2015 Cengage Learning®.

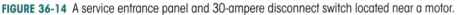

FIGURE 36-14 A service entrance panel and 30-ampere disconnect switch located near a motor.

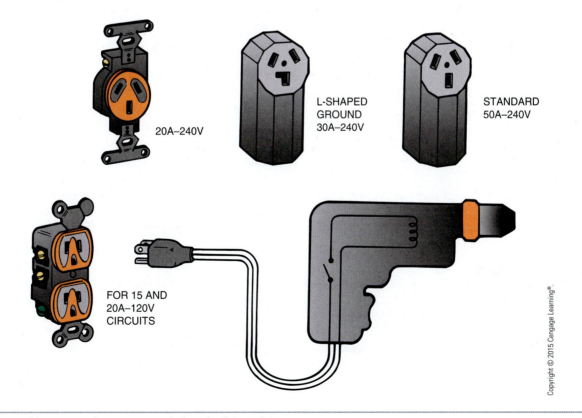

20A–240V

L-SHAPED
GROUND
30A–240V

STANDARD
50A–240V

FOR 15 AND
20A–120V
CIRCUITS

Copyright © 2015 Cengage Learning®.

FIGURE 36-15 Many motors have power cords that plug into outlets.

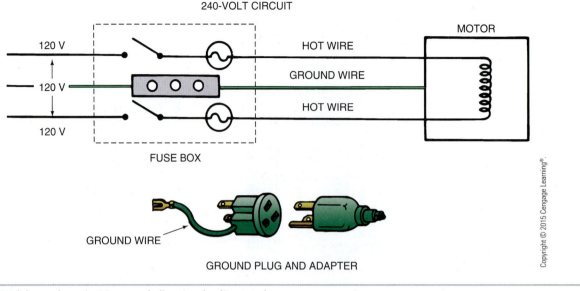

FIGURE 36-16 It is very important to ground all motor circuits properly.

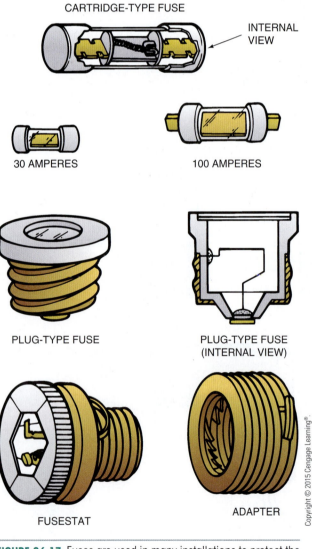

FIGURE 36-17 Fuses are used in many installations to protect the wiring circuit.

Circuit breakers protect the circuit, as well as fuses. They are more expensive than fuses, but serve as a switch in the circuit and can be reset if they trip due to overload. Ground-fault circuit interrupters (GFCIs) add a safety feature for the user. In the event of a faulty circuit that may shock the individual, the GFCI will interrupt or stop the flow of current before it causes injury (Figure 36-18).

Motor Protection

Fuses and circuit breakers do not provide adequate protection for motors. Motors need overload devices that trip and can be reset if overloaded. These may be mounted on the motor or may be part of the switch assembly (Figure 36-19).

Overload devices frequently utilize a bimetallic strip that bends when overheated and opens the circuit, which in turn trips the motor switch. When the bimetallic element cools, the protector can be reset, and the motor will function again (Figure 36-20). This device works because two different metals, most often copper and steel, are bonded together to create the working part of the switch. Because copper conducts heat much faster than steel, the copper expands more rapidly and bends the strip, which then opens or closes a circuit.

Motor Switches

Switches can be safely used only in the hot wires in a circuit. Never attach neutral or ground wires to switches. A switch generally has one, two, or three pairs of wire terminals.

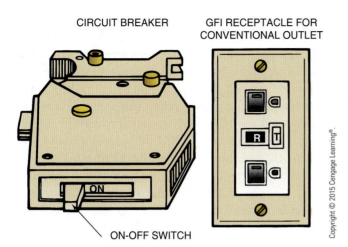

FIGURE 36-18 Circuit breakers and ground-fault circuit interrupters improve a circuit's safety and convenience.

Types of Switches. Switches may be of the toggle, rotary, push-button, trigger, or magnetic type. All of the switches are useful on motors of about 1 horsepower or less. However, magnetic switches, which close contact points by electromagnetic solenoids, are recommended for large motors. Such switches close rapidly and can be built to carry more current than the other types (Figure 36-21). Such switches are expensive but durable.

Reversing Switches. Some machines must run backward as well as forward. A plumber's pipe threading machine is an example. The drum-type switch is useful for this purpose, even when three pairs of wires are needed for three-phase motors (Figure 36-22).

Sensors and Other Switch Actuators

There are dozens of types of sensors and switch actuators. An **actuator** is a device responsible for placing another device in motion. Some of the more common ones will be discussed here.

Float Switch. A **float switch** is turned on and off by the action of a float. Float switches are used

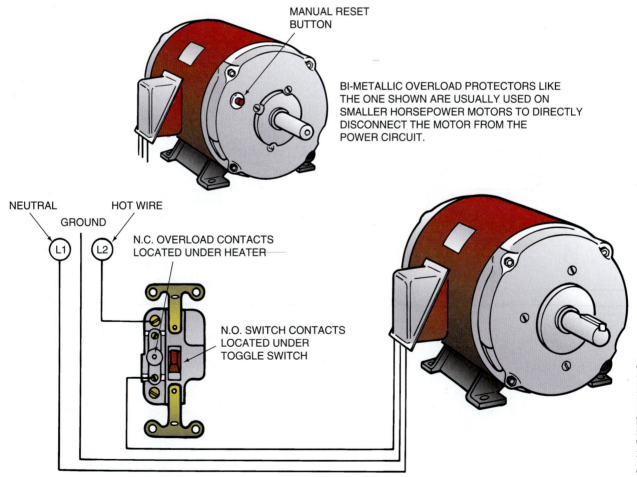

FIGURE 36-19 Motor overload protection devices.

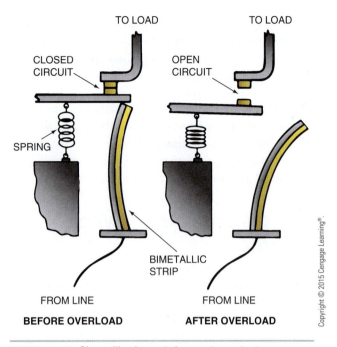

FROM LINE — **BEFORE OVERLOAD**

FROM LINE — **AFTER OVERLOAD**

FIGURE 36-20 Bimetallic elements in a motor protector.

on sump pumps, livestock water tanks, and other devices, where liquid levels must be controlled. A **sump hole** is a hole where unwanted water drains and is removed by a pump. As water accumulates and rises to a certain level, a float actuates the switch that turns on the pump. When the pump lowers the water level and the float drops to a certain point, the pump motor is switched off. Circuits with such automatic switching devices also need a manual or override switch for control while repairing equipment (Figure 36-23). **Manual** means by hand, and **override** means "to prevail over or take precedence over."

Pilot Light. A **pilot light**, or indicator light, glows when a circuit is energized. These lights are installed in circuits such as attic fans, sump pumps, and burners on kitchen stoves in order to indicate when electrical motors and electronic devices are operating (Figure 36-24).

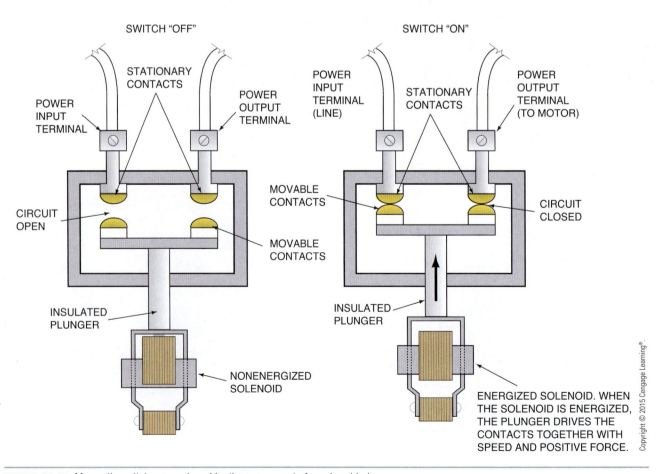

FIGURE 36-21 Magnetic switches are closed by the movement of a solenoid plunger.

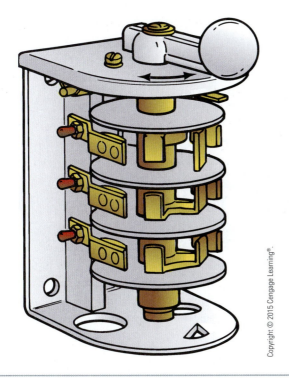

FIGURE 36-22 A drum switch is useful for motors that must be able to run forward and backward.

Timer. A **timer** is a device that activates a switch after a preset time lapse. Timers are used on stove ovens, clothes dryers, interior and exterior lights, slide projectors, water pumps, mist propagation systems, automatic plant watering systems, and others. Timers may have a manual start and a timed shut-off cycle, and some models can be set to numerous combinations of on/off cycles within a 24-hour period or longer (Figure 36-25).

Heat Sensor. **Heat sensors** are switches that respond to temperature changes. They control the electrical flow to oven burners, hot water heaters, boiler equipment, and motors in hot air and liquid systems. They can be designed to turn the current off when the temperature rises to a certain degree and on again when it falls to a certain degree. Other heat sensors may work in reverse. Many heat sensors use a **thermocouple** to detect heat changes and actuate switches. Thermocouples may use a bimetallic element to sense temperature changes. Others use mercury, which expands rapidly in a tube as temperature rises, and activates the switch (Figure 36-26).

Flow Switch. A **flow switch** reacts to the movement of a gas or liquid. Some are mounted on water lines to indicate when a sprinkler or other liquid system is running. Others are mounted in air ducts to indicate when fans are running (Figure 36-27). Flow switches could also be used to start fans if used to sense the absence of natural air movement in a system.

Pressure Switch. A **pressure switch** responds to changes in liquid, gas, or mechanical pressure. For instance, pressure switches on engines activate oil pressure gauges or warning lights (Figure 36-28). Pressure switches on water pumps for homes, farms, and ranches start and stop the water pump to maintain pressure between 20 and 50 pounds per square inch (psi) in the water tank. Similarly, air compressors utilize air pressure switches to keep the air in the tank within a narrow pressure range. Pressure switches may also be used to monitor the movement of livestock

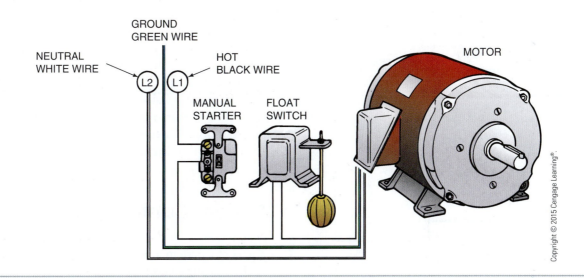

FIGURE 36-23 Float-activated switches are useful for controlling liquid levels.

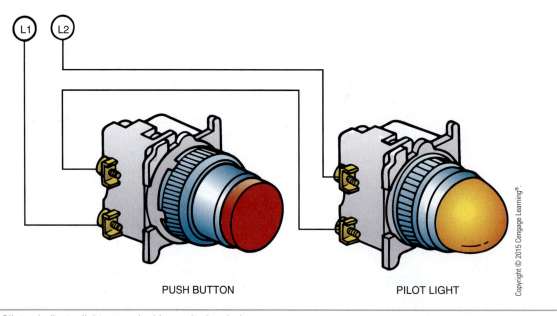

FIGURE 36-24 Pilot or indicator lights are valuable monitoring devices.

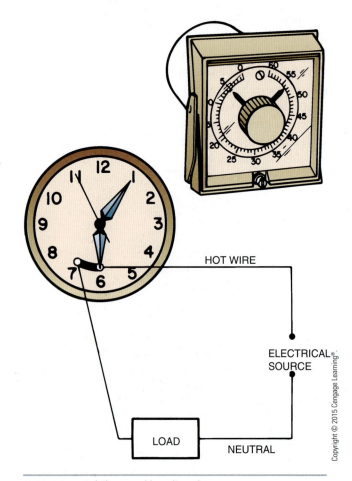

FIGURE 36-25 A timer and how it works.

through gates, chutes, and milking facilities and actuate silo unloaders, feeders, and augers.

Gas Analyzer. A **gas analyzer** is a device that determines the content of gases such as air and automotive exhaust. An **air pollution detector** is a type of gas analyzer that senses and presents a warning when air has unacceptable amounts of pollutants (Figure 36-29). On the other hand, an automotive gas analyzer has a set of meters that indicate the proportion of various gases in the exhaust, and guides the carburetor adjustment process.

Solid-State Devices. Many motor controls are solid-state electronic devices. They are very small and compact. They do not have moving parts to wear out, and are quite reliable. For these and other reasons, solid-state sensors and switches are rapidly replacing the traditional ones. Solid-state sensors and switching devices are discussed in Unit 35 ("Electronics in Agriculture").

System Controls

Many motors do not operate independently. They are a part of a system and contribute to a larger operation such as heating and moving air or water, operating an assembly line, or controlling traffic lights in a city.

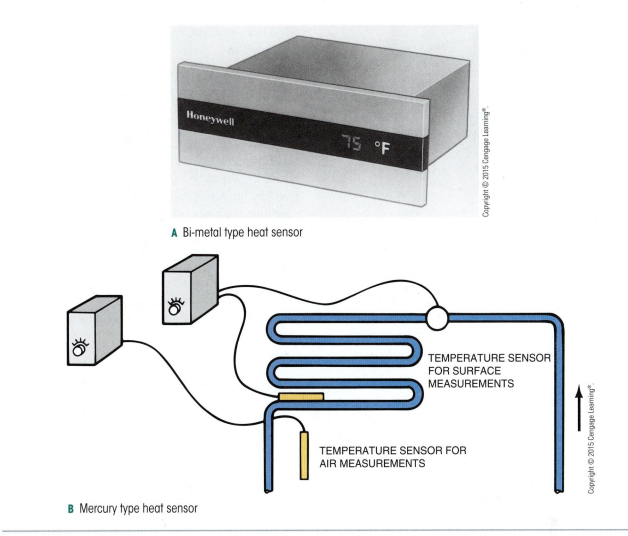

A Bi-metal type heat sensor

TEMPERATURE SENSOR
FOR SURFACE
MEASUREMENTS

TEMPERATURE SENSOR FOR
AIR MEASUREMENTS

B Mercury type heat sensor

FIGURE 36-26 Examples of heat sensors.

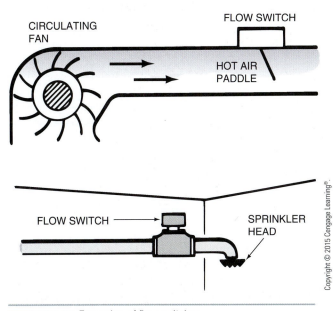

CIRCULATING
FAN

FLOW SWITCH

HOT AIR
PADDLE

FLOW SWITCH

SPRINKLER
HEAD

FIGURE 36-27 Examples of flow switches.

OIL
PRESSURE

NORMAL

FIGURE 36-28 Oil pressure gauge.

The oil-fired, forced-hot-air furnace is a good example of a system with many interrelated sensors and motor controls. It has only two motors, but many controls (Figure 36-30).

The oil pump motor in the furnace illustrated in Figure 36-30 pumps oil through a nozzle as a highly combustible mist and supplies air to the fire. The fan

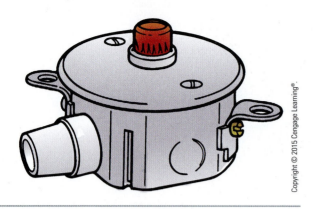

FIGURE 36-29 An air pollution detector.

motor drives a squirrel-cage fan, which draws cool air from the cool-air return duct and forces the heated air into the living spaces of the building. The following explanation illustrates how many controls direct the action of these two motors to make a furnace work safely and automatically. Briefly, the controls work as follows:

1. The system gets its power from a circuit *A*.
2. Disconnect switch *B* is installed on or near the furnace.
3. Room thermostat *C* senses temperature drop and signals the system to start.
4. Oil burner *D* switches on, and the fire causes the unit to heat up.
5. When limit control *E* heats up to its upper-limit setting, it signals fan motor *G* to start.
6. Fan *G* pulls cold air in from cool-air return duct *I* and pushes warmed air through duct *H* for distribution.
7. When heat warms the space and raises the temperature several degrees, thermostat *C* is "satisfied" and shuts off oil burner *D*.
8. Fan *G* continues to run because there is heat in the furnace.

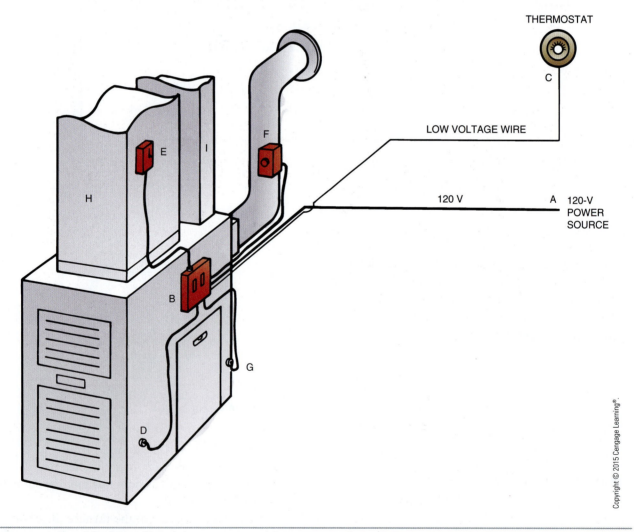

FIGURE 36-30 The oil-fired, forced-hot-air furnace uses an elaborate system of controls. The main text identifies each of the parts labeled here as A through I.

9. When air being pushed through *H* drops in temperature to the "lower setting," limit control *E* shuts off the current to fan motor *G*. The cycle is complete.

There are several additional safety and protective devices in the system. First, oil burner motor *D* and fan motor *G* have overload protectors that stop them if they get too hot or draw too much current due to overload. Second, if the flue or chimney gases get too hot, stack control *F* will shut down the system. Third, there is a switch in the limit control *E* that will shut down the system if the hot air duct gets too hot. Finally, there is a switch in the system to manually turn on the fan to circulate air continuously if desired, without generating heat.

Items *C*, *E*, and *F* systems are heat sensors that use bimetallic elements. These sensors measure and respond to temperature changes (Figure 36-31).

MAINTAINING MOTORS, DRIVES, AND CONTROLS

Most motors have permanently lubricated bearings. If not, they have oil filler tubes, which require only a few drops of electric motor oil periodically. Do not apply oil more frequently or in larger amounts than specified in the owner's manual. Over-oiling can be a problem just as under-oiling. Motors must be kept free of dust, dirt, and moisture.

Shafts need periodic greasing or oiling. When chains are used, they need to be oiled regularly. Additionally, they need to be cleaned in solvent and reoiled periodically to decrease wear caused by dirt and grit.

Belts must be kept clean and dry. Belt tension must be adjusted periodically to prevent slipping and compensate for wear. Whenever the motor is moved, care must be taken to assure proper alignment of belts and pulleys, as well as sprockets and chains.

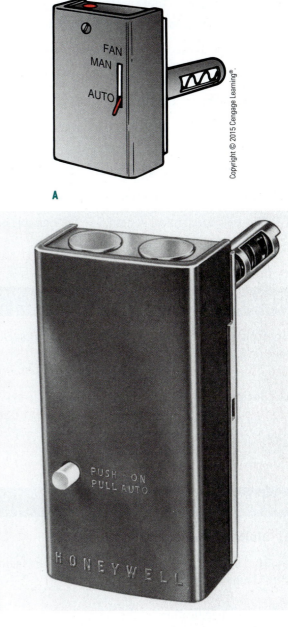

FIGURE 36-31 A fan/limit control (A) and a furnace stack control (B) for an oil-burning furnace.

SUMMARY

Electric motors provide low-cost, efficient, and flexible power. Motors are found in all facets of agriculture, agribusiness, and renewable natural resources. Sensors and controls are readily used to incorporate motors into mechanical systems. Motors are low in initial cost, economical to operate, and easy to maintain.

Student Activities

1. Define the Terms to Know in this unit.

2. Make a survey of your home, and write down the name of every device that has an electric motor.

3. Study the nameplate on a motor, and write down all the information provided. Discuss your findings in class.

4. Collect and make a display of parts and discarded belts, chains, and pulleys. Label the items.

5. Organize a class project in which each student brings in examples of sensors, protective devices, and switches. Organize these and label the items for teaching aids.

6. Choose one type of motor. Do research on how the motor starts, runs, and is designed. Report to the class.

7. Choose a mechanical system, like the example of a furnace in the text, and describe how the system uses sensors, motors, and controls.

Relevant Web Sites

Hyperphysics, exploration of concepts in physics, Department of Physics and Astronomy, Georgia State University, page on Electricity and Magnetism
http://hyperphysics.phy-astr.gsu.edu/hbase/hph.html

The Society of Women Engineers, Activities Center-Lesson Plans, Mechanical-Pulleys
http://technologystudent.com/gears1/geardex1.htm

TechnologyStudent.com, Gears and Pulleys

Self-Evaluation

A. Multiple Choice. Select the best answer.

1. The most reliable factor to use when replacing a motor is
   a. time required
   b. specifications of the original motor
   c. personal choice
   d. cost of motors

2. Another name for the case on a motor is
   a. bearing
   b. frame
   c. shield
   d. undercarriage

3. Components that transfer power from motor to machine make up the
   a. clutch assembly
   b. clutch housing
   c. drive train
   d. transmission

4. The device that disconnects a capacitor after a motor is up to speed is a
   a. centrifugal switch
   b. repulsor
   c. sensor
   d. voltage booster

5. A motor running at 1,750 rpm with a 3-inch pulley will drive a machine with a 2-inch pulley at
   a. 830 rpm
   b. 1,240 rpm
   c. 2,630 rpm
   d. 3,500 rpm

6. The formula for calculating horsepower is W (lb) × D (ft) × T (sec) divided by
   a. 10
   b. 50
   c. 500
   d. 550

7. One type of direct motor drive is
   a. belt and pulleys
   b. differential
   c. flexible shift
   d. transmission

8. The best pulley for changing speeds frequently is
   a. adjustable
   b. standard
   c. step
   d. variable-speed

9. The device that permits shifting of gears is
   a. transmission
   b. sensor
   c. differential
   d. clutch

10. Motors are best protected from burnout by
    a. circuit breakers
    b. fuses
    c. overload protectors
    d. all of these

B. Matching. Match the terms in column I with those in column II.

Column I

1. gear box
2. bimetallic
3. magnetic switch
4. reverses motor
5. float switch
6. flow switch
7. pressure switch
8. pollution detector
9. solid-state
10. limit control

Column II

a. electromagnetism
b. reacts to liquid levels
c. controls air compressors
d. use in liquid lines
e. controls furnace fans
f. electronic switches
g. gas analyzer
h. drum switch
i. heat sensor
j. transmission

C. Completion. Fill in the blanks with the word or words that make the following statements correct.

1. Calculate the unknown in the following belt systems:
   a. Motor speed is 3,450 rpm, motor pulley is 3½" in diameter, machine pulley is 3½" in diameter, machine speed is _____ rpm.
   b. Motor speed is 3,450 rpm, machine speed is 1,725 rpm, machine pulley is 3" in diameter, motor pulley must be _____.

2. To increase the speed of a machine, either _____ the size of the motor pulley or _____ the size of the machine pulley.

3. To decrease the speed of a machine, either _____ the size of the motor pulley or _____ the size of the machine pulley.

D. Brief Answer. Briefly answer the following questions.

1. Why are belts used in most motor drive systems?

2. What are the advantages of ball bearings over bushings for shaft support in motors?

3. Thermocouples may detect heat changes and activate switches in one of two ways. What are these?

4. What maintenance do shafts, chains, and belts need?

PLUMBING, HYDRAULIC, AND PNEUMATIC SYSTEMS

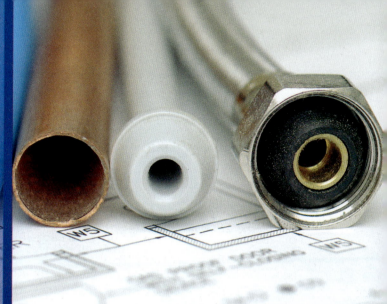

UNIT 37
Plumbing

Objective

To identify plumbing materials and perform basic plumbing procedures.

Competencies to be developed

After studying this unit, you should be able to:

- Identify plumbing tools.
- Identify and select pipe.
- Identify common pipe fittings.
- Assemble pipe.
- Maintain water systems.

Materials List

- Common pipe fittings:
  a. bushing
  b. 90° elbow
  c. 45° elbow
  d. coupling
  e. reducer
  f. cap
  g. street elbow
  h. plug
  i. tee
  j. Y
  k. union

continued

Terms to Know

- plumbing
- pipe
- fitting
- tubing
- bench yoke vise
- chain vise
- chain wrench
- tubing cutters
- ratchet die stock
- flaring tool
- flaring block
- pipe stand
- nominal
- ID
- OD
- black pipe
- standard pipe
- extra-heavy
- double-extra-heavy
- K
- L
- M
- compression fitting
- plastic
- polyethylene (PE)
- polyvinyl chloride (PVC)

continued

Materials list, *continued*

    l.   floor flange

   m. pipe nipples

- Examples of steel, copper, and plastic pipe
- All materials in the bill of materials shown in Figure 37-32

The term **plumbing** means installing and repairing water pipes and fixtures, including pipes for handling wastewater and sewage. The term came from the Latin, *Plumbum*, which is the name for lead. This stems from the use of lead in water pipes in times past. At one time pipes were actually made from lead. Later it was used to seal joints in iron or steel pipes. In more modern times, people discovered that lead was poisonous and has been banned from use in water pipes. In agricultural mechanics, all work done with pipe and pipe fittings is considered to be a part of plumbing. **Pipe** refers to rigid tube-like material. A **fitting** is a part used to connect pieces of pipe or to connect other objects to pipe. Tubing is also used in plumbing installations. **Tubing** generally refers to pipe that is flexible enough to bend.

TOOLS FOR PLUMBING

Many tools used for plumbing have been discussed in previous units. The following additional tools are also commonly required when working with pipe and tubing.

Tools useful for holding pipe and fittings are the **bench yoke vise** and the **chain vise** (Figure 37-1). The most useful tools for turning pipe and fittings are the **chain wrench** and pipe wrench (Figure 37-2). The hacksaw, file, pipe cutter, reamer, and **tubing cutters** are useful for cutting and smoothing pipe and tubing (Figure 37-3). Taps and dies are used for cutting threads in pipe. A **ratchet die stock** is used to turn the die.

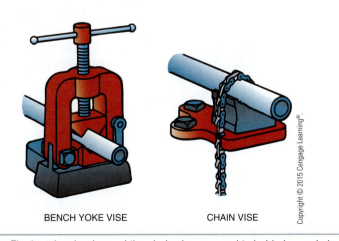

BENCH YOKE VISE   CHAIN VISE

Copyright © 2015 Cengage Learning®.

FIGURE 37-1 The bench yoke vise and the chain vise are used to hold pipe and pipe fittings.

Terms to know, *continued*

- chlorinated polyvinyl chloride (CPVC)
- acrylonitrite-butadiene-styrene (ABS)
- adaptors
- bushings
- reducers
- elbow (ell)
- tee
- Y
- coupling
- union
- plug
- cap
- nipple
- pipe dope
- Teflon tape
- faucet
- valve
- O ring
- packing
- faucet washer
- seat
- diaphragm
- aerators
- float valve assemblies
- flush valves

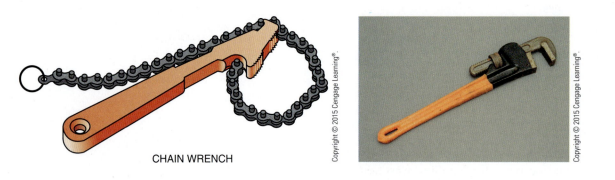

CHAIN WRENCH

Copyright © 2015 Cengage Learning®.

Copyright © 2015 Cengage Learning®.

FIGURE 37-2 The chain wrench and the pipe wrench are used to turn pipe and pipe fittings.

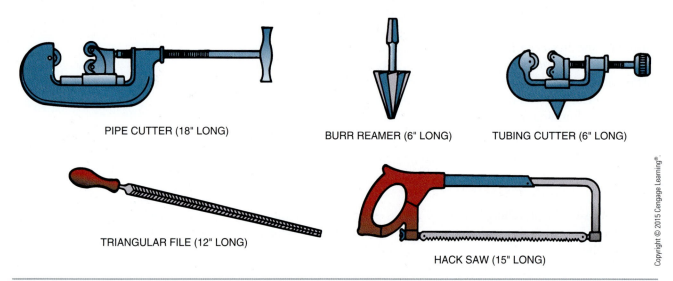

PIPE CUTTER (18" LONG)

BURR REAMER (6" LONG)

TUBING CUTTER (6" LONG)

TRIANGULAR FILE (12" LONG)

HACK SAW (15" LONG)

Copyright © 2015 Cengage Learning®.

FIGURE 37-3 Tools used for cutting and smoothing pipe and tubing.

The ends of soft copper tubing may be flared to fit bell-shaped fittings (Figure 37-4). Copper tubing is cut to length with a tubing cutter. The pressure and rolling action of the cutter wheel create a burr on the inside of the tubing at the cut (Figure 37-5). This burr is removed using the burr remover on the tubing

© iStockphoto/Chris Price

FIGURE 37-5 A tube cutter is used to cut tubing to length.

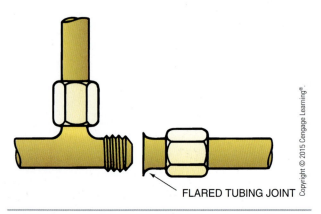

FLARED TUBING JOINT

Copyright © 2015 Cengage Learning®.

FIGURE 37-4 The ends of soft copper tubing may be flared to fit bell-shaped fittings.

cutter. The **flaring tool** and **flaring block** are used to hold the tubing and to form a flared end on the tubing (Figure 37-6A). The tool is turned into the end of the tube to create a flare (Figure 37-6B). This end is fastened against bell-type fittings by a flare nut to create a gas- or liquid-tight seal.

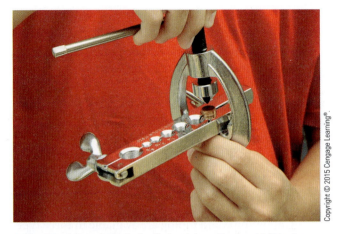

A The tube is inserted into the flaring tube with about 1/2 inch protruding. Then turn the lever down until the end of the tube is flared.

B The flare should be smooth and even all the way around.

FIGURE 37-6 A flaring tool is used to flare the ends of soft copper tubing.

FIGURE 37-7 A pipe vise can hold pipe for cutting or threading.

A tripod-type **pipe stand** with a vise makes an excellent work area for cutting, reaming, and threading pipe (Figure 37-7).

IDENTIFYING, SELECTING, AND CONNECTING PIPE

The **nominal**, or identifying, size of pipe is generally based on the inside diameter. Inside diameter is abbreviated **ID** and outside diameter, **OD**. Common sizes of pipe for home and farm applications (ID, in inches) are ¼, ⅜, ½, ¾, 1, 1¼, 1½, 2, 2½, 3, 4, and 6.

Steel Pipe

Steel pipe may be purchased as black pipe or galvanized pipe. At one time, steel pipe was commonly used in water systems. Now, it is used very little and has been replaced by copper and plastic pipe. Steel pipe is expensive, retains residue more easily, and may not last as long as copper or plastic pipe. **Black pipe** is painted black and has little resistance to rusting. Galvanized pipe is coated with zinc inside and outside. It resists rust for many years.

> **CAUTION!**
>
> The zinc coating on galvanized pipe gives off toxic fumes when welded or heated with a torch. Special precautions must be taken when using galvanized steel pipe for construction.

Steel pipe of each nominal size has a specified wall thickness and is classified as **standard pipe**. Special pipe may be ordered with thicker walls. This pipe is classified as **extra-heavy** or **double-extra-heavy**. It is generally used for construction rather than plumbing.

The outside diameter of standard, extra-heavy, and double-extra-heavy pipe is the same. The additional wall thickness of the extra-heavy and double-extra-heavy pipe results in reduced inside diameters (Figure 37-8). Because the outside diameter is consistent, standard pipe fittings can be used with all three types of pipe.

Copper Tubing and Pipe

Pipe made of copper is frequently referred to as tubing. It may be purchased in the soft, annealed form or the rigid form. For convenience, this unit refers to the soft form as tubing and the rigid form as pipe. The tubing

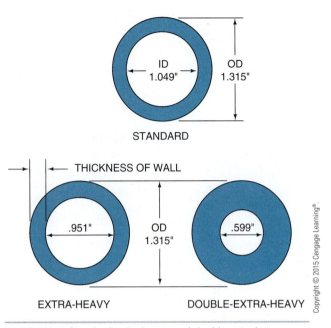

STANDARD

THICKNESS OF WALL

EXTRA-HEAVY   DOUBLE-EXTRA-HEAVY

FIGURE 37-8 Standard, extra-heavy, and double-extra-heavy varieties of steel pipe all have the same outside diameter for a given size of pipe.

can be bent around irregular parts of buildings. This makes it fast and easy to install. However, tubing does not have the neat appearance of rigid pipe. Therefore, it is used primarily in hidden spaces or where appearance is of little concern.

Copper is frequently preferred to steel for water lines because it resists corrosion, is easy to handle, and has been known to withstand freezing without breaking. Some disadvantages of copper are its high initial cost, high degree of expansion, and the bad taste created if the water is acid. Slightly acid water also reacts with copper to leave green stains in sinks and tubs.

Copper pipe and tubing are available in three types based on wall thickness: K, L, and M (Figure 37-9). Type **K** has the thickest wall, type **L** has medium wall thickness, and type **M** has the thinnest wall. Standard copper fittings are used for all three types. Generally, type L is specified by plumbing codes as the minimum acceptable size for use in buildings. Type K is generally required when the pipes are to be buried.

Flexible copper tubing may be joined by flaring the ends of the tubing and connecting them with bell-type fittings. Very small tubing can also be joined with

Standard Water Tube Size	Actual Outside Diameter	Nominal Wall Thickness		
		Type K	**Type L**	**Type M**
3/8	.500	.049	.035	.025
1/2	.625	.049	.040	.028
5/8	.750	.049	.042	. . . .
3/4	.875	.065	.045	.032
1	1.125	.065	.050	.035
1 1/4	1.375	.065	.055	.042
1 1/2	1.625	.072	.060	.049
2	2.125	.083	.070	.058
2 1/2	2.625	.095	.080	.065
3	3.125	.109	.090	.072
3 1/2	3.625	.120	.100	.083
4	4.125	.134	.110	.095
5	5.125	.160	.125	.109
6	6.125	.192	.140	.122
8	8.125	.271	.200	.170
10	10.125	.338	.250	.212
12	12.125	.405	.280	.254

Note: All measurements are in inches. Type M is not made in 5/8" size.

FIGURE 37-9 Sizes and wall thicknesses of copper tubing and pipe.

Copyright © 2015 Cengage Learning®

STEP 1

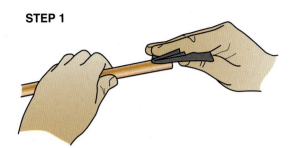

CLEAN THE OUTSIDE OF THE COPPER TUBE
WITH FINE STEEL WOOL OR EMERY CLOTH
TO A BRIGHT FINISH.

STEP 2

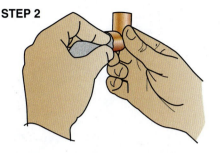

CLEAN THE INSIDE OF THE COPPER FITTING
WITH FINE STEEL WOOL OR EMERY CLOTH
TO A BRIGHT FINISH.

STEP 3

APPLY A THIN, EVEN COATING OF FLUX TO
THE OUTSIDE OF THE COPPER TUBE.

STEP 4

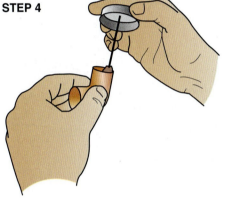

APPLY A THIN, EVEN COATING OF FLUX
TO THE INSIDE OF THE FITTING.

STEP 5

STEP 6

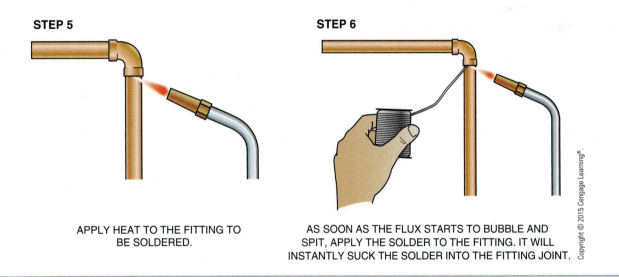

APPLY HEAT TO THE FITTING TO
BE SOLDERED.

AS SOON AS THE FLUX STARTS TO BUBBLE AND
SPIT, APPLY THE SOLDER TO THE FITTING. IT WILL
INSTANTLY SUCK THE SOLDER INTO THE FITTING JOINT.

Copyright © 2015 Cengage Learning®.

FIGURE 37-10 Six simple steps for soldering or sweating copper tubing.

compression fittings. A **compression fitting** grips the pipe by compressing a special collar with a threaded nut. This procedure does not require the use of a flaring tool or soldering torch. Figure 37-10 provides a brief review of sweating or soldering copper tubing.

Plastic Pipe

Plastic pipe is used extensively in water systems. Plastic is the easiest type of pipe to install. The term **plastic** is used here to include all pipe made from synthetic

STEP 1

CUT PIPE WITH FINE-TOOTH HANDSAW OR HACKSAW.

STEP 2

WITH A PIECE OF CLOTH, CLEAN THE INSIDE OF THE FITTING AND THE OUTSIDE OF THE PIPE.

STEP 3

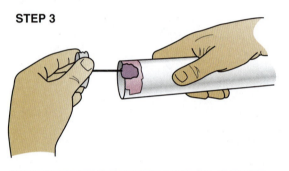

SWAB CONTACT AREAS WITH A SPECIAL CLEANER. APPLY A FULL, EVEN COATING OF CEMENT TO THE INSIDE OF THE FITTING AND TO THE OUTSIDE OF THE PIPE.

STEP 4

SLIP FITTING OVER PIPE, GIVING IT A QUARTER OF A TURN; HOLD FOR A FEW SECONDS UNTIL GLUE SETS.

Copyright © 2015 Cengage Learning®.

FIGURE 37-11 Steps involved in cementing plastic pipe.

materials. Such materials have long been used for cold-water lines. Some plumbing codes now permit the use of certain plastic pipe materials for hot-water lines. Most plastic pipe is fastened together by the simple procedure shown in Figure 37-11. The following procedure describes how to cement plastic pipe.

Procedure

1. Use a fine-tooth saw or special cutter to cut the pipe so the end is square.

2. Remove the rough or ragged edges using a knife or special cutter.

3. Clean the inside of the fitting and the outside of the pipe with a cloth and a special plastic cleaner (sometimes called primer).

4. Swab plastic cement on the outside of the pipe and in the fitting where the parts will join.

5. Press the fitting onto the pipe and give it a quarter of a turn. The fitting will be difficult to remove after a few seconds. It will set permanently shortly thereafter.

Plastic pipe is made from several different materials. **Polyethylene (PE)** has been in use for several decades. It is used extensively for cold-water lines.

PE pipe is usually black, somewhat flexible, and is available in rolls. It is popular for direct burial, pump installations, and surface water lines. PE pipe is assembled by pushing the pipe over the grooved section of the fittings. The pipe is secured to the fitting using a rustproof stainless steel clamp. There is no need for cement or pipe sealing compounds. Plastic pipe is easily removed by sawing with a hacksaw. Also, there is a special cutter for plastic pipe that cuts smoothly and evenly (Figure 37-12).

Polybutylene is another type of plastic. Between 1978 and 1995, polybutylene pipe was used extensively in homes and other buildings because it was inexpensive and easy to install (Figure 37-13). However, polybutylene is no longer used because of pipe failures that may have been due to chemical reactions between

FIGURE 37-14 Polybutylene pipe can be identified by the stainless steel or copper clamps used to fasten it.

FIGURE 37-12 A special cutter for plastic pipe makes a clean, even cut.

FIGURE 37-13 In the past, polybutylene pipe the small gray pipe, was widely used for cold-water lines in underground and surface areas, but it is no longer being used in new installations.

the polybutylene and minerals in some water supplies, which resulted in the pipes and fittings becoming brittle and weak. Many buildings still have this type of plumbing and have had no problems, probably due to the type of water being piped through the system (Figure 37-14).

Pipes composed of **polyvinyl chloride (PVC)** and **chlorinated polyvinyl chloride (CPVC)** are rigid and generally white or gray in appearance. PVC pipe is used for cold-water lines, and CPVC is approved by some plumbing codes for hot-water lines. Figure 37-15 shows plastic pipe being used for the vent and drain system for a tub, basin, and toilet. PVC pipe is especially easy to install. It is lightweight, may be cut with any saw, and is assembled using a liquid cement.

Acrylonitrite-butadiene-styrene (ABS) is used for sewerage and underground applications. Tests seem to indicate that it will last almost indefinitely. Manufacturers reportedly guarantee the material for 50 years. It is assembled by cementing the pipe to the fittings. The light weight of the pipe and the ease of assembly make plastic pipe especially attractive for sewerage installations where large diameters are required. This type of pipe has largely replaced cast-iron pipe.

Selecting Pipe Material

Each kind of pipe has advantages and disadvantages. When making repairs, it is generally easier to repair systems with the kind of pipe used in the original installation. When planning a new system, however, it is important to consider the relative merits of each kind of pipe (Figure 37-16).

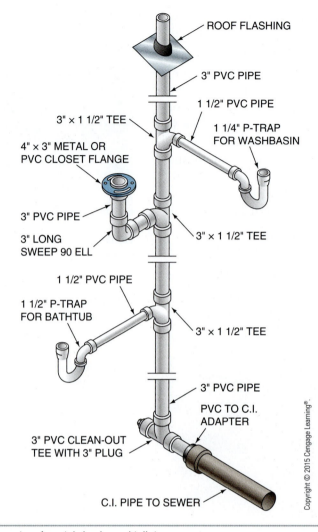

ROOF FLASHING

3" PVC PIPE

1 1/2" PVC PIPE

3" × 1 1/2" TEE

1 1/4" P-TRAP FOR WASHBASIN

4" × 3" METAL OR PVC CLOSET FLANGE

3" PVC PIPE

3" LONG SWEEP 90 ELL

3" × 1 1/2" TEE

1 1/2" PVC PIPE

1 1/2" P-TRAP FOR BATHTUB

3" × 1 1/2" TEE

3" PVC PIPE

PVC TO C.I. ADAPTER

3" PVC CLEAN-OUT TEE WITH 3" PLUG

C.I. PIPE TO SEWER

Copyright © 2015 Cengage Learning®.

FIGURE 37-15 Plastic vent and drain system for a tub, basin, and toilet.

IDENTIFYING PIPE FITTINGS

Pipe fittings are used to connect pipe. It is important to learn the names of basic pipe fittings in order to plan plumbing systems, order plumbing materials, and communicate with fellow workers about the job. The names of the fittings are the same for all kinds of pipe (Figure 37-17).

Fittings used to connect pipes of different types (such as copper tubing to steel pipe) are called **adaptors**. **Bushings** and **reducers** are used to connect pipes of different sizes (such as ½-inch steel pipe to ¾-inch steel pipe). Pipes coming together from different directions may be connected by an elbow, tee, or Y. The **elbow (ell)** is used where a single line changes direction; a **tee** or a **Y** is used where a line splits in two directions. A **coupling** is used to connect two pieces of similar pipe. A **union** also connects two pieces of similar pipe. An advantage

of the union is that it permits opening the line without cutting the pipe. A pipe fitting may be closed with a **plug**; a pipe end may be closed with a **cap**.

Short pieces of steel pipe are hard to thread on the job. Therefore, they are manufactured and sold in various lengths with both ends threaded. A short piece of pipe is called a **nipple**. Whenever fittings are threaded onto steel pipe, it is necessary to use **pipe dope** or **Teflon tape** to make a watertight seal. Teflon tape is also recommended for sealing threads when gas will be used in the lines.

MAINTAINING WATER SYSTEMS

Properly installed water systems are generally quite durable. However, certain types of water react with steel and copper and may shorten the life of pipe, fittings, faucets, valves, and pumps unless the proper

Factors to Consider	Copper		Plastic
	Type K (Heavy Duty)	Type L (Standard)	
Underground soil corrosion—Probable life expectancy (1)	40–100 yrs. under most conditions	30–80 yrs. under most conditions	Experience indicates durability is satisfactory under most soil conditions.
	14–20 yrs. in high sulfide conditions. May be less than 10 yrs. in cinders.	12–14 yrs. in high sulfide conditions. May be less than 10 yrs. in cinders.	
Resistance to corrosion inside pipe	Normally very resistant. May penetrate rapidly in water containing free carbon dioxide.		Very resistant
Resistance to deposits forming inside pipe	Subject to lime scale and encrustation from suspended materials.		Resistant, but occasional deposits will form.
Effect of freezing	Will stand mild freezes.		PE—will stand some freezing. PVC—will stand mild freezes.
Safe working pressures (lbs. per sq. in.)	Adequate for pressures developed by small water systems.		Working pressures at 73°F PE PVC 80 180 to to 160 600
Resistance to puncturing and rodents	Resistant to both		PE—Very limited resistance to puncture and rodents PVC—Resistant
Effect of sunlight	No effect		PE—Weakens with prolonged exposure PVC—High resistant
Effect on water flavor	Very acid water dissolves enough copper to cause off flavor.		Little effect
Lengths available	Soft temper: 60-ft.–100-ft. coils up to 1″ diameter 60-ft. coils above 1″ diameter Hard temper: 12- and 20-ft. lengths		PE—usually in 100-ft. coils or longer PVC—usually in 20-ft. lengths

Source: American Association for Vocational Instructional Materials

FIGURE 37-16 Characteristics of copper and plastic piping materials.

Factors to Consider	Copper		Plastic	
	Type K (Heavy Duty)	Type L (Standard)		
Comparative Weight (Approx. lbs. per foot) Inside diameter				
$1/2$	.34 lb.	.29 lb.	.06 lb.	
$3/4$	.64	.46	.09	
1	.84	.66	.14	
$1 1/4$	1.04	.88	.24	
$1 1/2$	1.36	1.14	.30	
2	2.06	1.75		
$2 1/2$	2.93	2.48		
Ease of bending	Soft temper bends readily, will collapse on short bends. Hard temper difficult to bend except for slight bends over long lengths		PE—Bends readily, will collapse on short bends PVC—Rigid. Bends on short bends long radius	
Conductor of electricity	Yes		No	
Comparative cost index Pipe size (in.)			PE	PVC
$1/2$	21	17	3	3
$3/4$	39	28	5	4
1	54	42	9	5
$1 1/4$	66	59	14	7
$1 1/2$	87	74	18	10
2	130	110	27	15

(1) Derived from studies reported by Dennison, Irving A. and Romanoff, Melvin, "Soil-corrosion Studies, 1946 and 1948; Copper Alloys, Lead and Zinc," Research paper RP2077, Vol. 44, March 1950 and "Corrosion of Galvanized Steel in Soils," Research paper 2366, Vol. 49, No. 5, 1952. National Bureau of Standards, U.S. Dept. of Commerce.

FIGURE 37-16 (*Continued*)

water conditioners are used. A **faucet** is a device that controls the flow of water from a pipe or container. A **valve** is a device that controls the flow of water in a pipe.

A green stain beneath the faucet in a tub or sink generally indicates a reaction between the water and copper pipes. Reddish water coming from a faucet generally means a rusting condition caused by a reaction between water and steel pipes or tanks.

The green stain or red water problems can generally be corrected by a water treatment system. Such systems are expensive but may be worth the investment. Money spent for water treatment may be offset by savings on repairs to the plumbing system.

Repairing Faucets

Most modern faucets are maintenance free. However, many farms and homes still have the old-type faucets. Normal maintenance of these older water systems includes the occasional tightening of packing nuts and material or changing faucet washers or O rings to stop faucets from dripping. An **O ring** is a piece of rubber shaped like an O, which fits into a groove on a shaft or in a housing. It prevents liquids from passing between a rotating shaft like a faucet stem and the surrounding hole. **Packing** is a soft, slippery, and wear-resistant material. It, too, is used to seal the space between a moving shaft and the surrounding area. A **faucet washer** is a

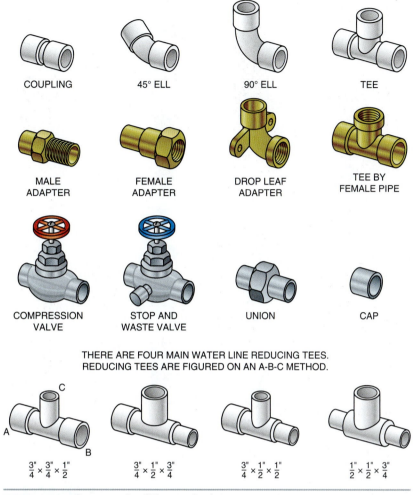

COUPLING   45° ELL   90° ELL   TEE

MALE ADAPTER   FEMALE ADAPTER   DROP LEAF ADAPTER   TEE BY FEMALE PIPE

COMPRESSION VALVE   STOP AND WASTE VALVE   UNION   CAP

THERE ARE FOUR MAIN WATER LINE REDUCING TEES.
REDUCING TEES ARE FIGURED ON AN A-B-C METHOD.

$\frac{3"}{4} \times \frac{3"}{4} \times \frac{1"}{2}$ $\frac{3"}{4} \times \frac{1"}{2} \times \frac{3"}{4}$ $\frac{3"}{4} \times \frac{1"}{2} \times \frac{1"}{2}$ $\frac{1"}{2} \times \frac{1"}{2} \times \frac{3"}{4}$

FIGURE 37-17 Common pipe fittings and valves.

rubberlike part that creates a seal when pressed against a metal seat. A **seat** is a nonmoving part that is designed to seal when a moving part is pressed against it.

Toilet tanks and automatic livestock waterers have shut-off valves that need periodic service. Typically the problem is a failure to shut off completely. The replacement of O rings, diaphragms, or rubber seats generally corrects the problem. A **diaphragm** is a flat piece of rubber that creates a seal between a part moving a short distance, and sealing against a nonmoving part.

The basic parts of valves and single faucets are the handle, stem, packing nut, packing, bonnet, washer, washer screw, seat, and body (Figure 37-18). Mixing faucets that control the flow of hot and cold water with one handle have a different design (Figure 37-19).

Aerators mix air with the water as it comes from the faucet to create a smooth stream. Aerators have a screen to catch particles of rust or dirt. This screen and other parts of the aerator must be removed and cleaned occasionally to work properly.

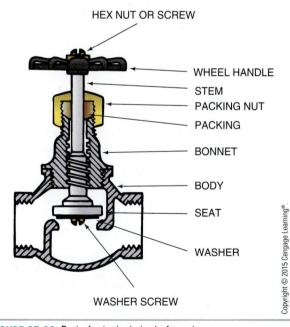

HEX NUT OR SCREW
WHEEL HANDLE
STEM
PACKING NUT
PACKING
BONNET
BODY
SEAT
WASHER
WASHER SCREW

Copyright © 2015 Cengage Learning®.

FIGURE 37-18 Part of a typical single faucet.

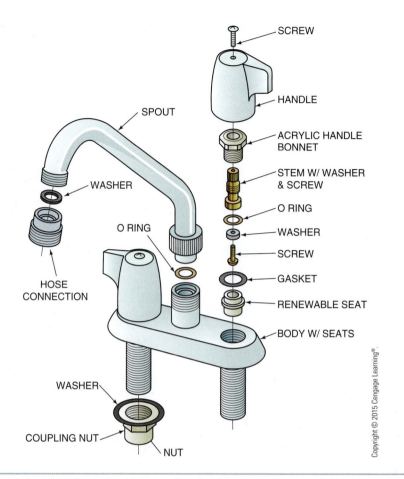

FIGURE 37-19 Parts of a washer-type faucet used in a kitchen or laundry sink.

Stopping Leaks Around Faucet Stems. To repair a faucet that leaks around the stem the following procedure is suggested.

Procedure

1. Tighten the bonnet or packing nut.

2. If the stem still leaks, turn off the water-line valve, and unscrew the bonnet or packing nut and the stem (Figures 37-20 and 37-21).

3. Replace the packing or O ring and reassemble the faucet (Figure 37-22).

4. Turn on the water to the faucet.

5. Tighten the packing nut or bonnet just enough to prevent leaking.

CAUTION!

Over-tightening the packing nut or bonnet will make the faucet hard to turn on and off.

FIGURE 37-20 Older-type faucet with the stem removed.

Faucet spouts that swing may leak at the base. To correct this problem simply tighten the nut that attaches the spout to the faucet body. If necessary, replace the O ring or packing on the spout.

Loose handles are tightened by prying off or unscrewing the plug in the handle and tightening

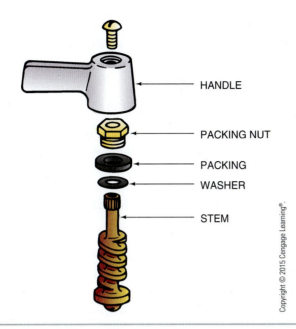

FIGURE 37-21 On some faucets, the handle must be removed to get to the packing nut. The stem may be removed after the packing nut is unscrewed.

HANDLE

PACKING NUT

PACKING

WASHER

STEM

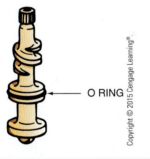

O RING

FIGURE 37-22 Some faucet stems may have an O ring instead of packing.

the handle screw. If the handle is broken or slips, a replacement handle can generally be purchased. Most faucet parts can be purchased from a hardware store or a plumber.

Stopping Faucets from Hammering or Dripping.

A common problem with faucets and valves is a hammering noise when the water is running slowly. This may be caused by worn threads on the stem or a loose washer (Figure 37-23). If the problem is worn threads, the stem or entire faucet may need replacing. The problem may be reduced or corrected by tightening the packing nut.

The more likely cause of a hammering faucet is a loose washer. This is corrected by removing the stem and tightening the washer. It is generally advisable to replace the washer while the faucet is open. When

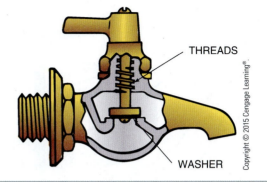

THREADS

WASHER

FIGURE 37-23 Worn stem threads or a loose washer may cause a faucet to make a hammering noise when water is running through it slowly.

replacing washers, be sure to use one of the correct size, shape, and style (Figure 37-24). The size is determined by inside diameter. When tightening the screw, draw it down snugly but not so tightly that the washer is forced out of shape (Figure 37-25).

Replacing Seats.

The seats of faucets may become eroded by chemical reaction with the water. This may happen within weeks if the faucet is permitted to drip.

FIGURE 37-24 It is important to use a washer of the correct size and shape.

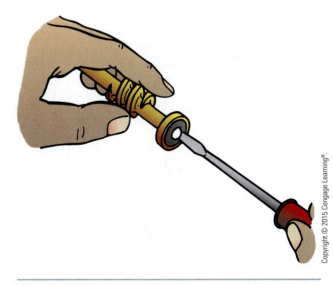

FIGURE 37-25 A faucet washer should be tightened snugly but not forced out of shape.

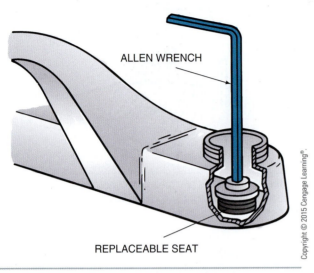

FIGURE 37-26 Many faucets have replaceable seats.

The seats in many faucets can be replaced using an Allen wrench (Figure 37-26).

Repairing Concealed Faucets. Bathtub faucets are generally inside the wall with only the handles in view. To repair the faucet, the handle must be removed first. A large, deep socket is then used to unscrew the bonnet (Figure 37-27). The bonnet and

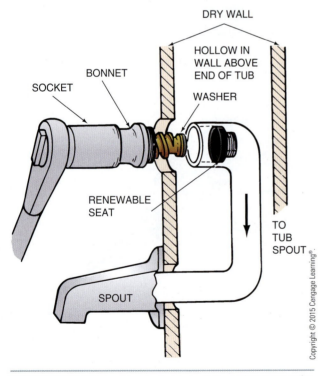

FIGURE 37-27 A large, deep socket wrench is used to remove the bonnet from a bathtub faucet.

stem are removed as an assembly. The washer or seat is then serviced as described previously.

Repairing Float Valve Assemblies

Float valve assemblies are used to control the water flow to flush toilets. Another important use is in livestock watering tanks with automatic filling devices. Float assemblies consist of a body with a diaphragm or washer valve and a float that is lifted by the water as the tank fills (Figure 37-28).

The action of the rising float puts pressure on the valve when the water in the tank reaches a specified level. This causes the valve to shut off the flow of incoming water. When the water level drops in the tank, the float drops, opens the valve, and permits water to enter until the tank refills.

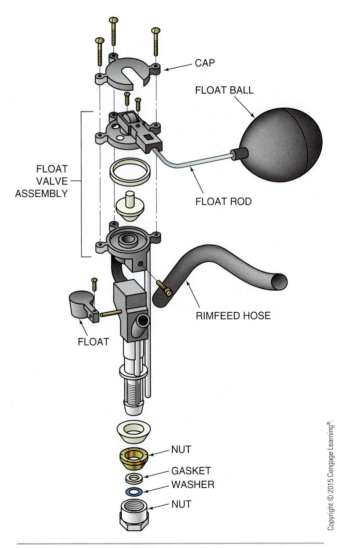

FIGURE 37-28 Parts of a float valve assembly used in toilet tanks and livestock watering tanks.

Repairs to float valve assemblies may involve tightening the nut or replacing a gasket to stop leaks at the point where the assembly passes through the bottom of the tank; or, a washer or diaphragm in the valve may need to be replaced to stop a flooding condition caused by the valve not shutting off completely. Some systems use a hollow ball for a float. If the ball develops a leak, it will fill with water and no longer float. In this case, the float and its rod are unscrewed from the assembly and replaced.

Repairing Flush Valves

Flush valves in toilet tanks control the water used in the flushing process (Figure 37-29). The valves frequently fail to seat properly and permit water to flow continuously from the tank to the toilet bowl (Figure 37-30). This causes the filler valve to remain open and water to flow in continuously until an adjustment is made. Such a condition, if not corrected, can cause a well to be pumped dry or a septic system to be flooded.

In most cases the problem is caused by the chain or linkage getting twisted or caught on another part. This can be corrected by simply flipping the trip lever up and down. If this does not permit the valve to reseal, then the chain flapper or ball seal, and other parts,

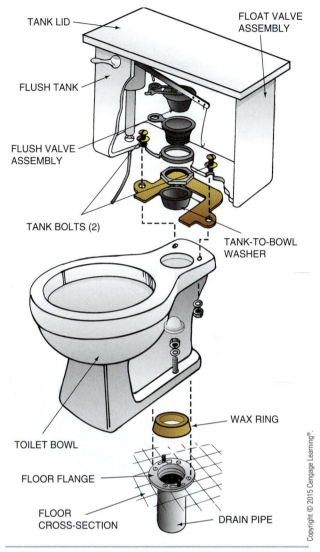

FIGURE 37-30 Toilet tank and bowl assembly.

should be cleaned of all rust, slime, or other deposits. If this does not correct the problem, replace the flapper and other parts as required.

Maintaining Septic Systems

Some kind of treatment plant is required for all human waste from bathrooms, kitchens, and laundry facilities if diseases are to be controlled. Proper treatment plants will provide conditions for bacteria to decompose solid waste and act upon paper, water, and other materials that have been contaminated by harmful microorganisms. The decomposition of solids in carbon dioxide and water, the filtration of water by soil particles, and oxidation by air will render sewage relatively free of disease-causing organisms in an effective treatment system. As a further precaution, municipal systems

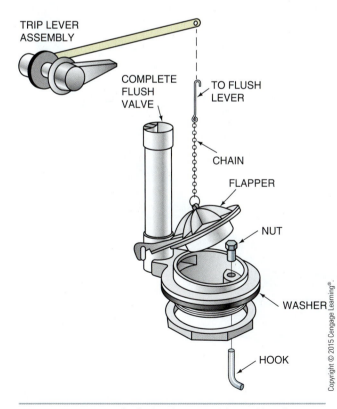

FIGURE 37-29 Parts of a flush valve assembly used in toilets.

and "package systems" for isolated institutions and housing developments carefully meter exact amounts of chlorine into the effluent (liquid) from the process. The chlorine will kill any harmful bacteria present.

The septic system of a house or business located in a city or other suburban area will be connected to a street sewer line and feed into a municipal sewage treatment plant. Such systems are maintained by the municipal government or other public agency. It is the property owner's responsibility to keep the sewer lines open from the fixtures in the building out to the connection at the street.

The system can accommodate human waste, toilet paper, garbage if shredded by a garbage disposal, and that is about all! Hair, excessive soap, grease, cotton, cloth, rubber, diapers, toys, and most other household objects are likely to plug up sewer lines and require the services of a plumber or specialized equipment to

reopen the lines. Sometimes, minor stoppages can be opened with chemical drain openers. Exterior lines may also become clogged with tree roots and require the use of a power drain tape or "snake" to cut the roots and restore the drain to good service.

Individual homes outside of cities and towns must have a septic tank to decompose the sewage and field drains or leach fields to distribute and filter the effluent (Figure 37-31). Such systems start with a septic tank near the house, where solids decompose and settle to the bottom as sludge and liquid rises to the top. As more sewage enters the septic tank, the accumulated water flows around a baffle and out to the distribution box. The distribution box has two or more field drains flowing out to the leach fields. Some localities may also permit round dry wells laid up with building blocks without mortar. Such wells have earthen floors and a concrete lid to prevent cave-in. Large gravel or stones

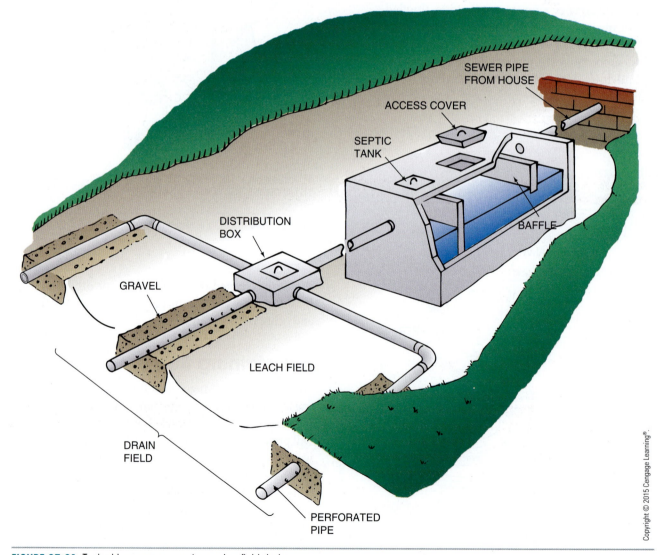

FIGURE 37-31 Typical home sewer system using field drains.

Copyright © 2015 Cengage Learning®.

about 2 inches in diameter are used to backfill around the structure to further extend the liquid storage area before it drains away through the surrounding soil for filtration and purification.

When field drains are used, a distribution box near the septic tank distributes the liquid to the various field drains, and gravel beds under the drains serve as a reservoir for liquids until they seep away into the soil. Under certain conditions, the drain lines may be placed close to the surface so water can evaporate into the air. When field drains are designed for evaporation, their locations can be observed by the tell-tale plushness of grass over the lines due to the abundance of water in the immediate area.

A properly designed septic system with only appropriate materials being flushed away will be trouble-free for many years. However, depending on the conditions of design, use, and soil conditions, some maintenance may be required. Sludge should be removed from the septic tank periodically and the system inspected by a professional plumber to be sure all components are functioning properly. Once the system is put into service, it is important to keep it in continuous service or the solids will dry on the surfaces of the perforated pipes, building blocks, and stones, and clog the system. Further, any accumulation of surface water with a sewerage odor signals the need for professional attention to the septic system.

Since the water from the septic tank is still contaminated, it is essential to place the septic tank as far as possible from the well. Specifications for such systems are determined by local health authorities, and systems must be installed according to state and local health codes.

SUMMARY

Many of the modern conveniences we enjoy are due to plumbing. Water is brought into buildings and wastewater and waste material are taken away through plumbing systems. Several materials such as steel, copper, and plastic are used in modern plumbing systems. If proper techniques are used and the correct procedures are followed, these materials will give many years of service.

Student Activities

1. Define the Terms to Know in this unit.

2. Sketch and label each pipe fitting illustrated in the unit.

3. Repair a faucet that leaks at the stem.

4. Repair a faucet that makes a hammering noise when the water is running slowly.

5. Repair a dripping faucet.

6. Repair the float mechanism in a water tank.

7. Complete the plumbing exercise in pipe fitting.

Plumbing Exercise in Pipe Fitting

Purpose: The purpose of the pipe fitting exercise is to obtain experience in assembling a watertight unit using different types of plumbing materials. Skills to learn include the following:

- Using various types of adapters and fittings
- Cutting pipe materials to length
- Reaming and cleaning pipe ends
- Threading and making galvanized connections
- Making cemented plastic connections
- Soldering copper connections
- Making polyethylene connections using clamps.

CAUTION!

This exercise involves the use of chemicals and extreme heat. Therefore, follow directions carefully.

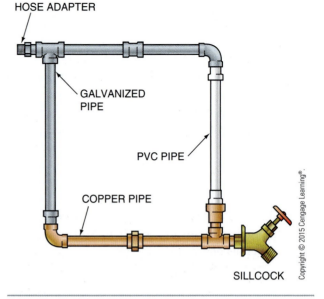

FIGURE 37-32 Layout of the finished pipe system for the plumbing exercise.

Procedure

1. Measure and cut pipe to length, as specified in the bill of materials.
2. Ream and clean ends of pipe.
3. Thread galvanized pipe.
4. Apply pipe joint compound to threads.
5. Assemble galvanized pipe and hose adapter as shown (Figure 37-32).
6. Working with one connection at a time, apply cement to PVC pipe and its fitting; then assemble quickly using a slight turning motion.
7. Assemble polyethylene pipe, using insert connectors and clamps.
8. Apply soldering flux to copper pipe and fittings.
9. Assemble and solder copper joints.
10. Cool copper parts before proceeding.
11. Assemble unit as shown, attaching the threaded sillcock. Make use of the union on the copper pipe for assembly purposes.
12. Attach a garden hose and test for leaks.

Relevant Web Sites

Do It Yourself®, Plumbing web page
http://doityourself.com/plumbing

This Old House®, Plumbing web page
www.thisoldhouse.com/toh/plumbing

Self-Evaluation

A. Multiple Choice. Select the best answer.

1. Extra-heavy steel pipe has
   a. thicker walls than standard pipe
   b. smaller ID than standard pipe
   c. the same OD as standard pipe
   d. all of these

2. Galvanized pipe is coated with
   a. zinc
   b. paint
   c. oil
   d. Galvanoleum

3. Galvanized pipe may be hazardous when
   a. used with acidic water
   b. mixed with black pipe in the system
   c. exposed to soil
   d. cut with a torch or welded

4. Pipe that is available in both flexible and rigid form is
   a. steel
   b. PVC
   c. copper
   d. none of these

5. The heaviest copper pipe is type
   a. K
   b. M
   c. L
   d. Z

6. A kind of flexible pipe that is preferred for long runs of outdoor cold-water lines is
   a. copper
   b. PE
   c. PVC
   d. steel

7. Green stain in a tub indicates
   a. poor quality finish on the tub
   b. reaction of water and steel
   c. reaction of water and copper
   d. reaction of water and plastic

8. Faucets may drip or leak due to faulty
   a. washers
   b. packing
   c. O rings
   d. any of these

9. A common problem with toilet tanks and automatic livestock waterers is
   a. incomplete shutoff
   b. green stain
   c. dripping
   d. all of these

10. The easiest water or sewer system to install uses pipe made of
    a. cast iron
    b. copper
    c. plastic
    d. steel

B. **Matching.** Match the terms in column I with those in column II.

Column I
   1. steel pipe
   2. copper pipe
   3. Teflon tape
   4. plastic pipe
   5. ID
   6. elbows
   7. tee and Y
   8. cap and plug
   9. coupling and union
   10. nipple

Column II
   a. nominal pipe size
   b. changes direction of single line
   c. connect several pipes
   d. used on steel pipe threads
   e. connect similar pipe in line
   f. short length of pipe
   g. used to close pipe and fittings
   h. connected by threading
   i. connected by cementing
   j. connected by soldering

C. **Completion.** Fill in the blanks with the word or words that make the following statements correct.

   1. A faucet that leaks around the stem may be corrected by
      a. _____
      b. _____

   2. Water reacting with steel pipe may be _____ in color.

   3. Two problems that may cause a faucet to make a hammering noise when water runs through it slowly are
      a. _____
      b. _____

   4. Three parts of a faucet are
      a. _____
      b. _____
      c. _____

   5. The water level in a flush tank or automatic livestock waterer is controlled by a _____.

   6. The valve that stops water from entering a flush tank or an automatic livestock waterer is shut off by the action of a _____.

D. **Brief Answer.** Briefly answer the following questions.

   1. What are the advantages and disadvantages of using copper pipe in water lines?

   2. How is polyethylene (PE) pipe assembled?

   3. What is acrylonitrite-butadiene-styrene (ABS) pipe used for? How long does it last?

   4. What does normal maintenance of older water systems include?

   5. How does one access the washer or seat in a bathtub faucet?

   6. If the flush valve on a toilet fails to seat properly, what steps should be tried before resorting to replacement of parts?

UNIT 38
Irrigation Technology

Objective

To select, install, and maintain a soil irrigation system.

Competencies to be developed

After studying this unit, you should be able to:

- State the benefits of irrigation.
- Select an irrigation system.
- Use soil moisture sensors.
- Discuss cost factors involved with irrigation.

Materials List

- Discarded 1-liter plastic soda bottles (five)
- Cheesecloth (2 square feet)
- Samples of different soils (five)
- Twistees (five)

Terms to Know

- transpiration
- precipitation
- irrigation
- pesticides
- acre-inch
- seasonal demand
- surface irrigation
- subirrigation
- trickle irrigation (drip irrigation)
- emitters
- wetted zones
- humidistat
- solenoid valve
- microtubing
- sprinkler irrigation
- sprinkler heads (nozzle)
- moisture sensors
- tensiometer
- gypsum blocks with meter
- neutron probe

IRRIGATION PLANNING

Plants take up water through their roots, use some for maintenance and growth, and give some off through **transpiration** (Figure 38-1). Natural **precipitation** can be supplemented by human-controlled watering called **irrigation**.

Water may be applied with sprinklers to simulate rain. Water can also be applied by flooding the surface of the land. Water may also be applied directly to the root zone in the soil.

Benefits of Irrigation

Irrigation provides many benefits to crops and crop growers, including the following:

- Water may be added to supplement precipitation.
- Water can be added when precipitation does not match crop needs.
- Plant nutrients may be applied efficiently in irrigation water.
- Pesticides may be applied in irrigation water. A **pesticide** is a substance used to control pests that decrease yields, or damage or destroy desirable plants.
- Irrigation water can reduce frost damage.
- Irrigation equipment can be used to distribute liquid manure.
- Irrigation may be used to cool crops on excessively hot days.

Points to Consider before Installing an Irrigation System

Costs for irrigation equipment and installation are high. Irrigation management requires considerable expertise, too, and operating costs are high for fuel, water, labor, and other items. Some questions that should be asked when considering an irrigation system are as follows:

- Should I irrigate?
- Will normal precipitation provide good profits?
- Will irrigation improve yields? Profits?
- Can I afford the cost?
- Can I get the money?
- What are the energy requirements?
- Do I have sufficient time, labor, and management expertise?
- What are the risks involved?
- Can I get sufficient water?
- Are my soils suitable for irrigation?
- What problems might I encounter?
- Who can help design the system?

WATER REQUIREMENTS

There is no use considering an irrigation system if there is not enough water for efficient crop production. Several steps are necessary to make this determination. A basic measurement of irrigation water use is a unit called an acre-inch. An **acre-inch** is the amount of water needed to cover one acre (43,560 square feet) with water 1 inch deep. Assuming no water is running off or seeping into the soil, it takes 27,000 gallons to cover an acre with an inch of water. Therefore, we say one acre-inch equals 27,000 gallons or it takes 27,000 gallons of irrigation water to provide an acre-inch of water to the soil.

Seasonal Demand. Scientists have determined the number of acre-inches of water taken from the soil in one season by efficiently growing crops in various climates. This is referred to as **seasonal demand**. The figures for some popular crops are presented in Figure 38-2. For example, alfalfa growing in southern Arizona requires 44 acre-inches per year, but it needs only 26 acre-inches per year if growing in New England. Crops in Arizona need more water than in New England because the hot dry climate causes more transpiration and evaporation.

The seasonal demand for one acre of alfalfa in southern Arizona would be 27,000 × 44 or 1,188,000 gallons of water. In New England it would be 27,000 × 26 or 702,000 gallons per acre. In many areas, annual precipitation provides most of the water needed for crops. However, if the precipitation does not occur when the crop needs it, supplemental irrigation is needed. Therefore, irrigation systems may be installed primarily to function during times of maximum or

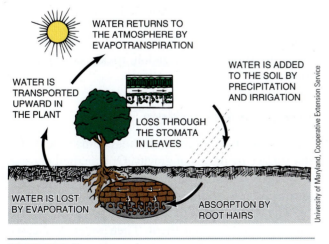

WATER RETURNS TO THE ATMOSPHERE BY EVAPOTRANSPIRATION

WATER IS ADDED TO THE SOIL BY PRECIPITATION AND IRRIGATION

WATER IS TRANSPORTED UPWARD IN THE PLANT

LOSS THROUGH THE STOMATA IN LEAVES

WATER IS LOST BY EVAPORATION

ABSORPTION BY ROOT HAIRS

University of Maryland, Cooperative Extension Service

FIGURE 38-1 The water cycle.

Agricultural Areas											
Crop	Lower California	Central Valley California	Washington	Southern Arizona	Western Colorado	Texas High Plains	Western Kansas	Northern Great Plains	Arkansas-Mississippi Bottoms	Piedmont Area	New England Area
(Inches per season)											
Alfalfa	36.0	40.0	34.0	44.0	27.0	45.0	39.0	23.0	39.0	—	26.0
Barley	—	—	—	—	—	—	—	16.0	—	—	—
Brome grass	—	—	—	—	—	—	31.0	26.0	—	—	—
Corn	—	26.0	—	—	20.0	32.0	26.0	17.0	23.0	23.0	17.0
Cotton	25.0	26.0	—	41.0	—	20.0	—	—	—	19.0	—
Grain sorghum	—	15.0	—	25.0	—	22.0	22.0	—	21.0	20.0	—
Grapefruit	—	—	—	48.0	—	—	—	—	—	—	—
Grapes	—	—	—	20.0	—	—	—	—	—	—	—
Oranges	27.0	—	—	39.0	—	—	—	—	—	—	—
Orchard	—	—	—	—	20.0	—	—	—	—	—	—
Pasture or hay	34.0	36.0	—	—	23.0	—	—	—	—	36.0	—
Potatoes	—	18.0	20.0	24.0	—	—	—	18.0	—	—	12.0
Small grain	—	—	—	23.0	15.0	27.0	27.0	14.0	—	20.0	—
Snap beans	—	—	—	—	—	—	—	—	—	—	9.0
Sugar beets	—	—	28.0	43.0	—	36.0	27.0	23.0	—	—	—
Tomatoes	18.0	20.0	—	—	—	—	—	—	—	—	14.0

FIGURE 38-2 Acre-inches of water needed by various crops for efficient production in various regions of the United States.

Copyright © 2015 Cengage Learning®.

peak demand by the crop. For a system to fulfill its purpose, sufficient water must be available to meet the peak demand or daily peak-use rate.

Peak-use rates needed by certain crops in various areas are presented in Figure 38-3. For example, alfalfa in California's San Joaquin Valley requires 0.25 acre-inch, or $0.25 \times 27,000 = 6,750$ gallons per acre per day, for peak use. In New York State, the same crop requires $0.20 \times 27,000 = 5,400$ gallons per day. To meet the most demanding needs for irrigation, the system must be capable of supplying peak-use demands every day for extended periods of time.

Sources of Water

Water may be used from streams, rivers, lakes, ponds, reservoirs, irrigation district pipes or canals, or deep wells. Before a decision is made to tap any of these sources, the state, county, and local water laws and ordinances must be checked carefully.

IRRIGATION SYSTEMS

Water may be applied by using one of four systems. These are surface, subirrigation, trickle, and sprinkler irrigation.

Surface Irrigation

The term **surface irrigation** means adding water to the soil surface by means of gravity. One method consists of flooding an entire area. The area must be level to assure even distribution of water by gravity.

The other method is to plant crops in rows or beds with furrows between them. Then water is released from supply ditches to the furrows. It then flows the

	Agricultural Areas										
Crop	Washington Columbia Crop Basin	Calif. San Joaquin Valley	Texas Southern High Plains	Arkansas- Mississippi Bottoms	Nebraska Eastern Part	Colorado Western Part	Wisconsin State	Indiana State	Piedmont Plateau	Virginia Coastal Plains	New York State
	(Inches per Day)										
Corn	0.27	0.26	0.30	0.24	0.28	0.23	0.30	0.30	0.22	0.18	0.20
Alfalfa	0.25	0.25	0.30	0.24	0.27	0.23	0.30	0.30	0.25	0.22	0.20
Pasture	0.29	0.32	0.30	0.22	0.29	0.23	0.30	0.30	0.25	0.22	—
Grain	0.21	0.17	0.15	0.15	0.26	0.22	0.25	—	0.16	—	—
Sugar Beets	0.26	0.22	—	—	0.26	0.20	0.25	—	—	—	—
Cotton	—	0.22	0.25	0.18	—	—	—	—	0.21	—	—
Potatoes	0.29	0.24	—	—	0.26	0.22	0.20	0.25	0.18	0.18	0.18
Deciduous Orchards	—	0.21	—	—	—	0.18	0.30	0.25	0.25	0.22	0.20
Citrus Orchards	—	0.19	—	—	—	—	—	—	—	—	—
Grapes	—	0.18	—	—	—	—	—	0.25	0.20	—	—
Annual Legumes	—	—	0.18	0.28	—	—	—	—	—	—	—
Soybeans	—	—	—	0.19	0.27	—	0.25	0.30	0.18	—	—
Shallow Truck	—	—	—	—	—	—	0.20	0.20	0.14	—	0.18
Medium Truck	—	—	—	0.12	—	—	0.20	0.20	0.14	0.16	0.18
Deep Truck	—	—	—	—	—	—	0.20	0.20	0.18	—	0.18
Tomatoes	—	—	—	—	—	0.22	0.20	0.20	0.21	0.18	0.18
Tobacco	—	—	—	—	—	—	—	0.25	0.18	0.17	—
Rice	—	—	—	0.17	—	—	—	—	—	—	—

Copyright © 2015 Cengage Learning®

FIGURE 38-3 Inches of water per day needed by various crops for efficient production in various regions of the United States.

length of the field by gravity. Water seeps to the root zone under furrows and rows as it goes (Figure 38-4).

Subirrigation

The addition of water to the soil below the surface is called **subirrigation**. The moisture reaches the root zone area while keeping the surface dry. The open-ditch method of subirrigation has water seeping into the root zone laterally from ditches or canals filled with water (Figure 38-5). The other type adds water to the root zone through perforated underground pipes or hoses.

Trickle Irrigation

The term **trickle irrigation** or **drip irrigation** refers to the application of water onto or below the surface of soil through small tubes or pipes. Water is released slowly

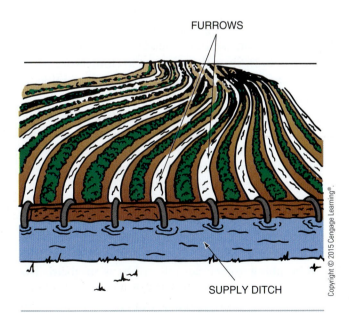

FURROWS

SUPPLY DITCH

Copyright © 2015 Cengage Learning®

FIGURE 38-4 Surface irrigation by flooding furrows.

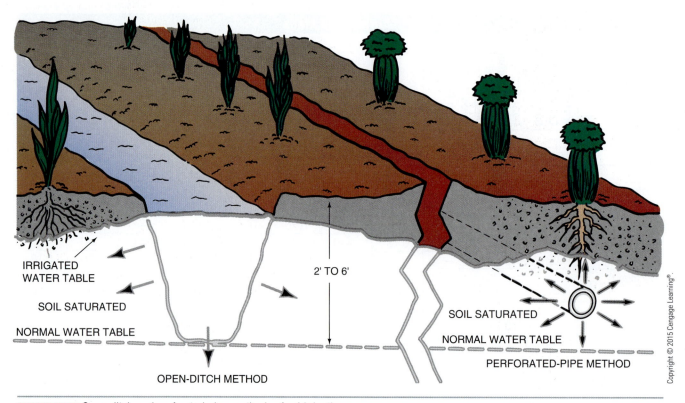

IRRIGATED
WATER TABLE

SOIL SATURATED

NORMAL WATER TABLE

2' TO 6'

SOIL SATURATED

NORMAL WATER TABLE

PERFORATED-PIPE METHOD

OPEN-DITCH METHOD

Copyright © 2015 Cengage Learning®.

FIGURE 38-5 Open-ditch and perforated-pipe methods of subirrigation.

through special nozzles called **emitters**. Emitters create small or local watered areas called **wetted zones** as the water moves through the soil (Figure 38-6). Trickle irrigation works especially well for watering individual plants in the field and container-growing plants (Figure 38-7).

Trickle irrigation systems require a water source, pump, check valve, gages, filters, control valves, main lines, submain and/or lateral lines, and emitters (Figure 38-8). Since the openings in emitters are small, filters should be placed in the system in several locations (Figure 38-9).

The use of solenoid valves permits the use of timers, humidistats, and other electronic devices to turn the system on and off. A **humidistat** measures humidity or moisture in the air. A **solenoid valve** is a valve actuated by an electromagnet.

Emitters may be inserted into PVC pipes at any location (Figure 38-10). They are quick and economical to install. Emitters are also available in the form of hollow lead devices attached to the end of thin, flexible, plastic **microtubing**, sometimes referred to as spaghetti. The hollow lead devices are weights at the end of the tubes that help hold the microtubing in place. The spaghetti system is especially handy for container irrigation.

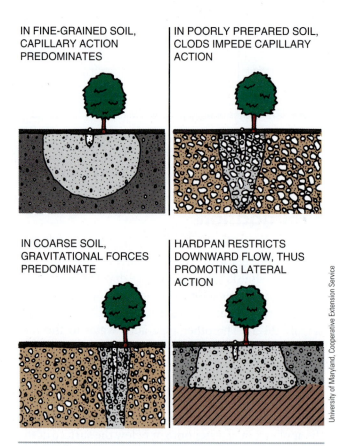

IN FINE-GRAINED SOIL, CAPILLARY ACTION PREDOMINATES

IN POORLY PREPARED SOIL, CLODS IMPEDE CAPILLARY ACTION

IN COARSE SOIL, GRAVITATIONAL FORCES PREDOMINATE

HARDPAN RESTRICTS DOWNWARD FLOW, THUS PROMOTING LATERAL ACTION

University of Maryland, Cooperative Extension Service

FIGURE 38-6 Wetted zones by soil types.

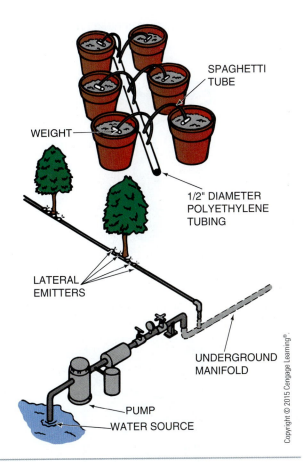

FIGURE 38-7 Trickle irrigation systems are suitable for field, orchard, nursery, greenhouse, shadehouse, and container-growing areas.

Sprinkler Irrigation

Sprinkler irrigation refers to dispersion of water in droplets to simulate rain. **Sprinkler heads** or **nozzles** are devices that break water into droplets and spread them evenly over soil, other media, or crops (Figure 38-11). Sprinkler irrigation is popular for lawns, golf courses, nurseries, field crops, vegetables, orchards, citrus groves, and wherever cooling or wetting is desired (Figure 38-12).

There are many types of sprinkler irrigation systems. All utilize the same basic equipment as trickle irrigation systems, except sprinkler nozzles can throw water great distances depending on the nozzle type, pipe size, and water pressure. Irrigation may be done by covering large areas with many nozzles or a relatively small area by moving the nozzle(s) as the job progresses (Figures 38-13 and 38-14). Special equipment may be added to the system for adding fertilizers and chemicals (Figure 38-15).

Irrigation Wells and Pumps

Since irrigation requires huge volumes of water, special wells are frequently drilled in the center of the area to irrigate. This procedure may save costs of pumping and piping from sources some distance from the crop. Such wells must be designed and constructed with expert assistance (Figure 38-16). Similarly, the

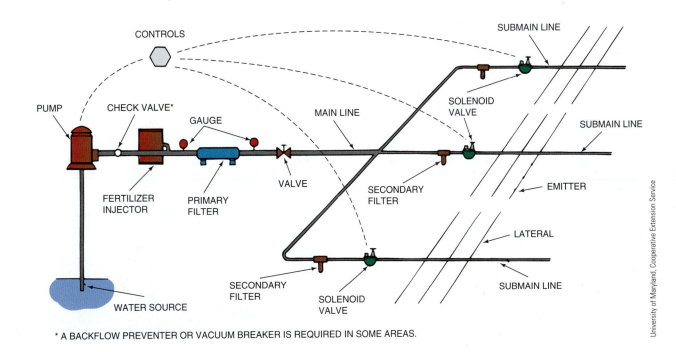

* A BACKFLOW PREVENTER OR VACUUM BREAKER IS REQUIRED IN SOME AREAS.

FIGURE 38-8 Trickle irrigation components.

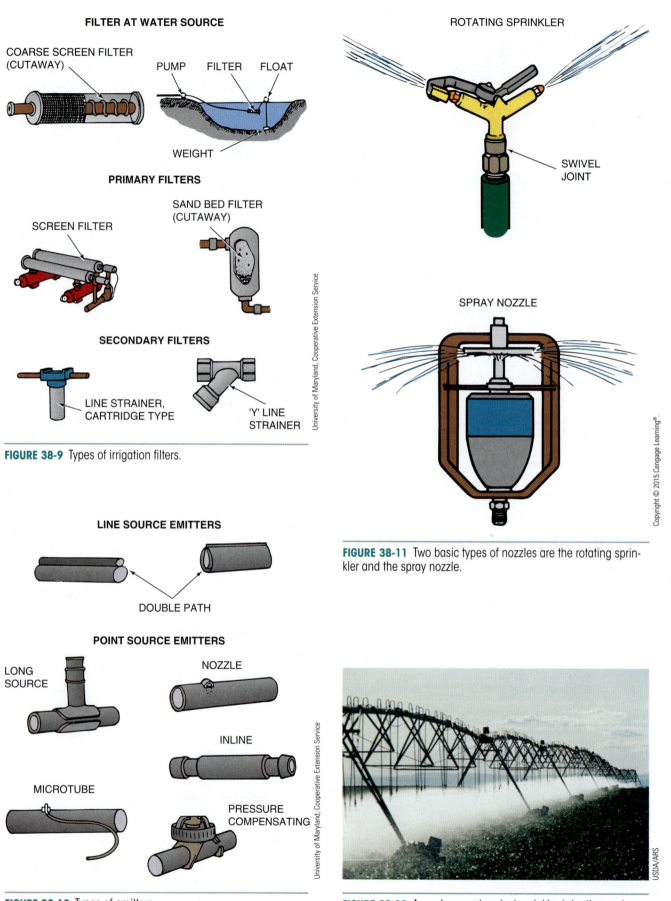

FILTER AT WATER SOURCE

COARSE SCREEN FILTER
(CUTAWAY)

PUMP   FILTER   FLOAT

WEIGHT

PRIMARY FILTERS

SCREEN FILTER

SAND BED FILTER
(CUTAWAY)

SECONDARY FILTERS

LINE STRAINER,
CARTRIDGE TYPE

'Y' LINE
STRAINER

University of Maryland, Cooperative Extension Service

FIGURE 38-9 Types of irrigation filters.

LINE SOURCE EMITTERS

DOUBLE PATH

POINT SOURCE EMITTERS

LONG
SOURCE

NOZZLE

INLINE

MICROTUBE

PRESSURE
COMPENSATING

University of Maryland, Cooperative Extension Service

FIGURE 38-10 Types of emitters.

ROTATING SPRINKLER

SWIVEL
JOINT

SPRAY NOZZLE

Copyright © 2015 Cengage Learning®

FIGURE 38-11 Two basic types of nozzles are the rotating sprinkler and the spray nozzle.

USDA/ARS

FIGURE 38-12 A modern center-pivot sprinkler irrigation system.

Type of System	Maximum Slope (Percent)	Water Application Rate (Inches/Hour)		Shape of Field	Field Surface Conditions	Max. Height of Crop (Feet)	Labor Required (Hrs./A)	Size of Single System (Acre)	Cooling and Frost Protect.	Adaptable To		Liquid Animal Waste Distribution
		Min.	Max.							Chemical Application	Fertilizer Application	
Multi-Sprinkler Permanent	No limit	.05	2	Any shape	No limit	No limit	.05–.10	1 or more	Yes			Not recom.
Hand-moved: Portable set	20	.10	2	Rectangular			.5–1.5	1–40	No			Yes
Solid set	No limit	.05	2	Any shape			.2–.5	1 or more	Yes			Not recom.
Tractor-moved: Wheel-mounted	10	.10	2		Smooth enough for safe tractor operation		.2–.5	20–50				Yes
Self-moved: Side-wheel-roll	10	.10	2	Rectangular	Reasonably smooth	4	1–3	20–80				
Side-moved	10	.10	2			4–6	1–3					
Self-propelled: Center-pivot	20	.20	1.5	Circular, square, or rectangular	Clear of obstructions, path for towers	8–10	.05–.15	20–200	No	Yes	Yes	
Lateral-move	20	.20	1			8–10	.05–.15	10–100				
Single-Sprinkler Hand-moved	20	2.5	2	Any shape	Safe operation of tractor		.5–1.5	20–40				Not recom.
Tractor-moved: Wheel-mounted	5–15	2.5	2		Safe operation of tractor	No limit	.2–.4	20–40				
Self-propelled	No limit	2.5	1	Rectangular	Lane for winch and hose		.1–.3	20–40				
Broom-Sprinkler Tractor-moved	5	2.5	1	Any shape	Safe operation of tractor	8–10	2–5	20–40				Yes
Self-propelled	5	2.5	1	Rectangular	Lane for boom and hose	8–10	.1–.5	20–40				

FIGURE 38-13 Factors to consider in selecting a sprinkler irrigation system.

Source: American Association for Vocational Instructional Materials.

Spacing (Feet)	1	2	3	4	5	6	7	8	9	10	11	12	13	14	15	16	17	18	19	20
20 × 30	0.24	0.48	0.72	0.97	1.21	1.45	1.70	1.94												
20 × 30	.16	.32	.48	.64	.80	.96	1.13	1.29	1.45	1.61	1.77	1.93	2.10							
20 × 40	.12	.24	.36	.48	.60	.72	.85	.97	1.09	1.21	1.33	1.45	1.57	1.70	1.81	1.93	2.06	2.18		
20 × 50	.10	.19	.29	.39	.48	.58	.68	.77	.87	.97	1.06	1.16	1.26	1.36	1.45	1.55	1.64	1.74	1.84	1.94
20 × 60	.08	.16	.24	.32	.40	.48	.56	.64	.72	.81	.88	.97	1.05	1.13	1.21	1.29	1.37	1.45	1.53	1.61
25 × 25	.15	.31	.46	.62	.77	.93	1.09	1.24	1.40	1.55	1.70	1.86	2.02							
30 × 30	.11	.21	.32	.43	.54	.64	.75	.86	.97	1.07	1.18	1.29	1.39	1.50	1.61	1.72	1.83	1.93	2.04	2.15
30 × 40		.16	.24	.32	.40	.48	.56	.64	.72	.81	.89	.97	1.05	1.13	1.21	1.29	1.37	1.45	1.53	1.61
30 × 50		.13	.19	.26	.32	.38	.45	.52	.58	.64	.71	.77	.84	.90	.97	1.03	1.09	1.16	1.22	1.29
30 × 60		.11	.16	.21	.27	.32	.37	.43	.48	.54	.59	.64	.70	.75	.81	.86	.91	.97	1.02	1.07
40 × 40		.12	.18	.24	.30	.36	.42	.48	.54	.60	.66	.72	.78	.84	.90	.96	1.02	1.09	1.14	1.20
40 × 50		.10	.14	.19	.24	.29	.34	.39	.43	.48	.53	.58	.63	.68	.73	.78	.82	.87	.92	.97
40 × 60			.12	.16	.20	.24	.28	.32	.36	.40	.44	.48	.52	.56	.60	.64	.68	.72	.77	.81
40 × 80			.09	.12	.15	.18	.21	.24	.27	.30	.33	.36	.39	.42	.45	.48	.51	.54	.57	.61
50 × 50			.12	.15	.19	.23	.27	.31	.35	.39	.43	.46	.50	.54	.58	.62	.66	.70	.73	.77
50 × 60			.10	.13	.16	.19	.22	.26	.29	.32	.35	.39	.42	.45	.48	.52	.55	.58	.61	.64
50 × 70				.11	.14	.17	.19	.22	.25	.28	.30	.33	.36	.39	.41	.44	.47	.50	.52	.55
50 × 80				.10	.12	.14	.17	.19	.22	.24	.27	.29	.31	.34	.36	.39	.41	.44	.46	.48
60 × 60				.11	.13	.16	.19	.21	.24	.27	.30	.32	.35	.38	.40	.43	.46	.48	.51	.54
60 × 70					.11	.14	.16	.18	.21	.23	.25	.28	.30	.32	.34	.37	.39	.41	.43	.46
60 × 80					.10	.12	.14	.16	.18	.20	.22	.24	.26	.28	.30	.32	.34	.36	.38	.40
70 × 70							.14	.16	.18	.20	.22	.24	.25	.28	.30	.31	.33	.36	.37	.39
70 × 80							.12	.14	.16	.17	.19	.21	.23	.24	.26	.28	.29	.31	.33	.34
70 × 90								.12	.14	.15	.17	.18	.20	.22	.23	.25	.26	.28	.29	.31
80 × 80								.12	.14	.15	.17	.18	.20	.21	.23	.24	.26	.27	.29	.30
80 × 90								.11	.12	.13	.15	.16	.17	.19	.20	.21	.23	.24	.26	.27
80 × 100								.10	.11	.12	.13	.15	.16	.17	.18	.19	.21	.22	.23	.24
100 × 100										.10	.11	.12	.13	.14	.14	.15	.16	.17	.18	.19
120 × 132																	.10	.11	.12	.12

Table heading: **Gallons per Minute from Each Sprinkler**

Source: University of Maryland, Cooperative Extension Service.

FIGURE 38-14 Average water application rates in inches per hour for sprinklers spaced at different distances apart and for water flowing at different rates.

high cost of pumping, piping, and components dictates the need for a system engineered according to the specific needs (Figure 38-17).

MANAGING IRRIGATION

Some areas have such a dry climate that nearly all water requirements must be met through irrigation. Others have sufficient rainfall generally, but need supplemental irrigation occasionally. In all situations, it is important to apply water in accordance with crop needs. Such needs will peak when crops are growing rapidly and producing seed (Figure 38-18).

Similarly, it is important to place the water where it does the most good. Some crops, such as azaleas, have shallow root systems and receive most of their water within several inches of the surface. Other

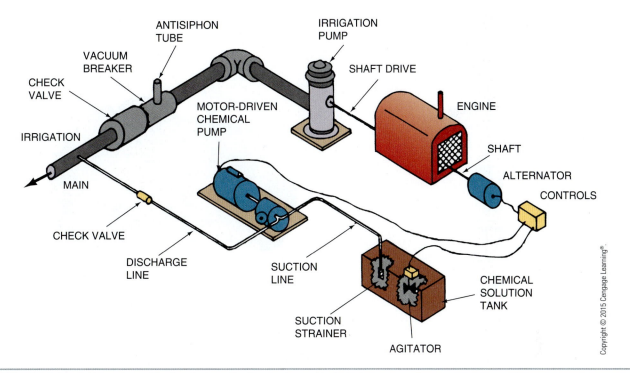

Copyright © 2015 Cengage Learning®.

FIGURE 38-15 A sprinkler system with chemical application equipment included.

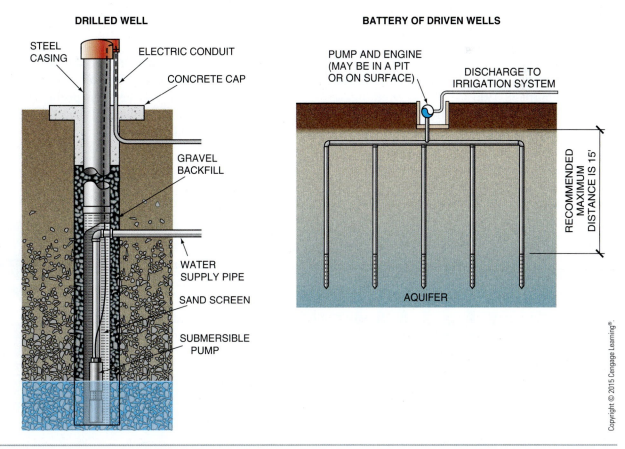

Copyright © 2015 Cengage Learning®.

FIGURE 38-16 Cross sections of two types of irrigation wells.

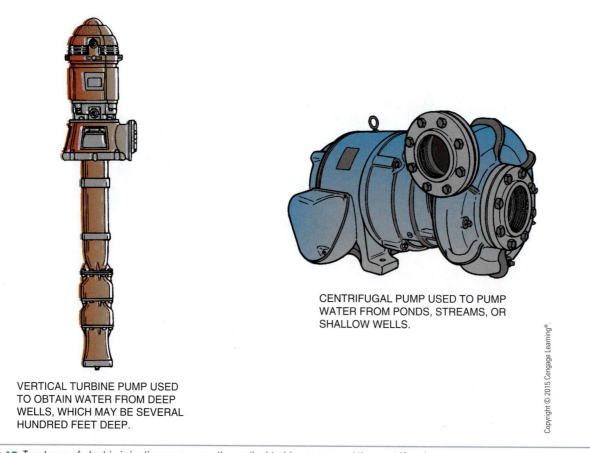

VERTICAL TURBINE PUMP USED
TO OBTAIN WATER FROM DEEP
WELLS, WHICH MAY BE SEVERAL
HUNDRED FEET DEEP.

CENTRIFUGAL PUMP USED TO PUMP
WATER FROM PONDS, STREAMS, OR
SHALLOW WELLS.

Copyright © 2015 Cengage Learning®.

FIGURE 38-17 Two types of electric irrigation pumps are the vertical turbine pump and the centrifugal pump.

crops, like alfalfa, may put roots down 10 or 15 feet, and can survive long dry periods. Most plants fall in between (Figure 38-19).

For maximum plant growth, water conservation, and profit, water should be applied only as needed.

Moisture sensors are devices placed in the soil to indicate the moisture content. The **tensiometer**, **gypsum blocks with meter**, and **neutron probe** are types of moisture sensors. Their use and placement are illustrated in Figure 38-20.

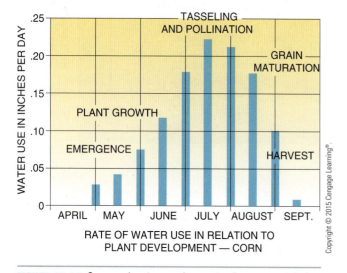

Copyright © 2015 Cengage Learning®.

RATE OF WATER USE IN RELATION TO
PLANT DEVELOPMENT — CORN

FIGURE 38-18 Seasonal water requirements of corn correlated with the plant's growth stage.

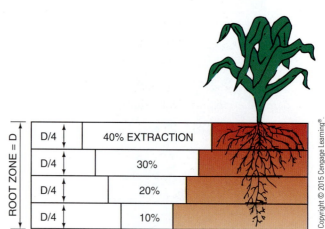

Copyright © 2015 Cengage Learning®.

FIGURE 38-19 Moisture use pattern.

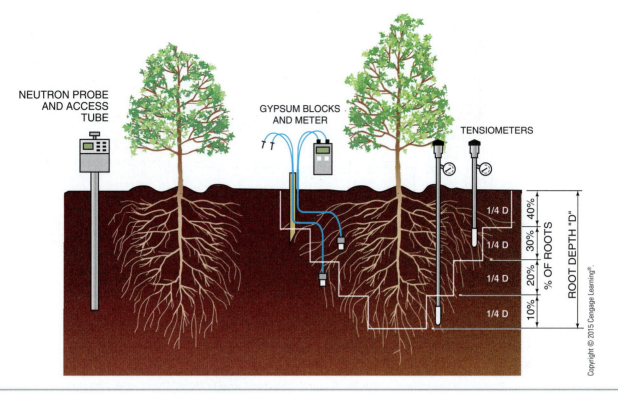

FIGURE 38-20 Plant rooting pattern and moisture sensor placement for three types of moisture sensors.

SUMMARY

Crop production can be increased and ornamental landscapes can be enhanced with water management practices. Irrigation has its place in a productive agriculture. For those needing this practice, the benefits of expert advice should more than offset the cost and should lead to satisfying results.

Student Activities

1. Define the Terms to Know in this unit.

2. Set up a soil percolation demonstration, as shown in Figure 38-21. Present the demonstration to your class.

3. Do research using a variety of soils and add explanatory charts to the soil percolation tests described in Figure 38-21. Enter research and findings in your school's science fair.

4. Read an encyclopedia article on irrigation practices in ancient Egypt, China, Babylonia, or other early civilization. Report your findings to the class.

5. Research the details of one type of irrigation system, and report your findings to the class.

6. Design and install an irrigation system for the school greenhouse or grounds, or for your lawn.

7. Choose a plot of ground and plan an irrigation system. This may be your home lawn or garden, or it could be a plot of ground on the school campus. Plan for a water source, conduct a percolation test, determine the types of piping and sprinklers to be used, and calculate the amount of water required. Refer to the tables and concepts in Unit 38 of the text.

8. Have your instructor evaluate your plan. Have you followed the correct procedures? Is the plan workable?

SOIL PERCOLATION EXPERIMENT

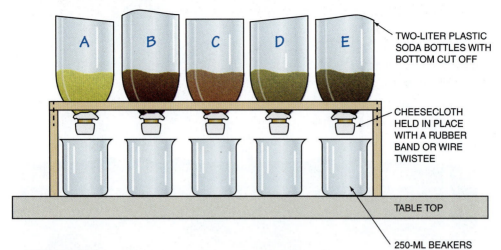

TWO-LITER PLASTIC SODA BOTTLES WITH BOTTOM CUT OFF

CHEESECLOTH HELD IN PLACE WITH A RUBBER BAND OR WIRE TWISTEE

TABLE TOP

250-ML BEAKERS

CONSTRUCTION

1. Cut two pieces of 1 × 6 × 6" long and one piece 1 × 6 × 24" long.
2. Fasten the boards together with nails or screws as shown in the diagram.
3. Measure the top of the soda bottles and choose the proper size drill bit to bore 5 evenly spaced holes in the top board.
4. Cut the bottoms out of five two-liter plastic soda bottles.
5. Place the bottles in the hole like in the diagram above.
6. Cover the opening of the bottles with a piece of cheesecloth and secure with a twisted wire or rubber band.
7. Place jars or beakers under each bottle. They should hold about a pint or 250 ml. Make sure all of the jars or beakers are the same size.
8. Mark bottles A through E with a marking pen or label.

DOING THE SOIL PERCOLATION EXPERIMENT

1. Select five samples of soil of 250 milliliters each.
    SAMPLE A – pure sand
    SAMPLE B – an organic potting soil
    SAMPLE C – a clay soil
    SAMPLE D – soil from a field where row crops are grown continuously
    SAMPLE E – soil from your yard
2. Crumble all the soil particles to their smallest size and, spread the soil out and let it dry.
3. Pour each sample into one of the one-liter bottles and record the bottle number and the type soil.
4. Add 250 ml. of water to each sample and record the time to the minute and second when you add the water to each bottle.
5. Record the instant the first drop of water flows into each beaker.
6. Compute the time in minutes and seconds for water to start draining from each.
7. Record the amount of water in each beaker after the last one stops dripping.
8. Record all the data you observe.
9. Rank your samples from 1 (fastest) to 5 (slowest) in releasing water.
10. Rank your samples from 1 (best) to 5 (worst) in water-holding capacity.
11. What are your observations and conclusions?
12. What are the implications of your findings on selection of drainage systems?

Copyright © 2015 Cengage Learning®.

FIGURE 38-21 Soil percolation experiment.

Relevant Web Sites

Clemson Extension, South Carolina Irrigation Pages, Irrigation Equipment
www.clemson.edu

IrrigationTutorials.com, Irrigation Design—Landscape Sprinkler System Design Tutorial
www.irrigationtutorials.com/sprinkler00.htm

UC IPM Online, The UC Guide to Healthy Lawns, Irrigation Design
www.ipm.ucdavis.edu/TOOLS/TURF/SITEPREP/irrdes.html

Self-Evaluation

A. Multiple Choice. Select the best answer.

1. The gallons of water required to provide one acre-inch of water on five acres is
   a. 27,000
   b. 50,000
   c. 100,000
   d. 135,000

2. Peak use is the maximum amount of water needed for
   a. one day
   b. the season
   c. the year
   d. none of these

3. The irrigation system that adds water by flooding furrows is
   a. surface
   b. subirrigation
   c. sprinkler
   d. trickle

4. Valves to start and stop irrigation can be controlled automatically with
   a. solenoids
   b. laddenvalves
   c. filters
   d. emitters

5. Plants require most water
   a. at germination
   b. during early growth
   c. during growth and seed formation
   d. near harvest

6. Emitters are used to
   a. release water slowly into the soil
   b. release water rapidly into the soil
   c. water the tops of plants
   d. flood the furrows with water

7. Which is not a moisture sensor?
   a. gypsum block and meter
   b. neutron probe
   c. philometer
   d. tensiometer

B. Matching. Match the terms in column I with those in column II.

Column I

1. humidistat
2. solenoid valve
3. wetted zones
4. surface irrigation
5. transpiration
6. 1 acre-inch
7. emitter
8. trickle irrigation
9. irrigation nozzle
10. subirrigation

Column II

a. local watered areas
b. 27,000 gallons
c. trickle irrigation
d. activated by an electromagnet
e. measures humidity
f. microtubing
g. creates droplets
h. water from below
i. water given off by plants
j. adding water by using gravity

C. Completion. Fill in the blanks with the word or words that make the statement correct.

1. The number of acre-inches of water taken from the soil in one season by efficiently growing crops in various climates is called _____ _____.

2. _____ irrigation works especially well for watering individual plants in a field.

3. The spaghetti system is especially handy for _____ irrigation.

4. It is important to apply water in accordance with _____ _____.

D. Brief Answer. Briefly answer the following questions.

1. What are two methods of surface irrigation? How do they differ?

2. What are two methods of subirrigation? How do they differ?

3. What is sprinkler irrigation? Where is sprinkler irrigation popular?

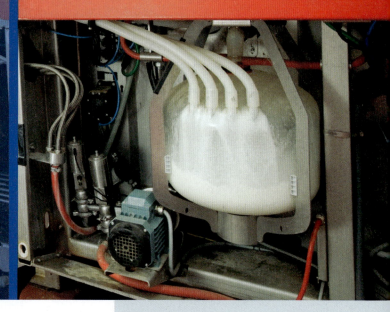

UNIT 39

Hydraulic, Pneumatic, and Robotic Power

Objective

To maintain and use fluid and robotic power in agricultural applications.

Competencies to be developed

After studying this unit, you should be able to:

- Compare hydraulic and pneumatic systems.
- Identify basic theories that apply to fluid dynamics.
- Describe fluid power principles.
- Discuss fluid characteristics.
- Identify major components of fluid systems.
- Use and maintain fluid power equipment.
- Discuss some concepts of robotics.

Materials List

- Bulletin-board materials
- Publications with photos of robotics
- Reference materials on hydraulics and pneumatics
- Camera
- Materials to demonstrate the principles of fluid dynamics

Terms to Know

- fluid power
- hydraulics
- pneumatics
- force
- pressure
- Bourdon tube
- manometer
- viscosity index
- additive
- viscosity improver
- antifoam
- corrosion inhibitor
- rust inhibitor
- antiscuff
- extreme-pressure resistor
- Society of Automotive Engineers (SAE)
- SAE rating
- American Petroleum Institute (API)
- API rating
- double-action cylinder
- positive displacement
- eccentric
- air compressors
- fluid coupling
- impeller

continued

Fluid power is an integral part of agricultural and industrial technology. **Fluid power** is work done utilizing liquids and gases to transfer force. The use of liquids to transfer force is called **hydraulics**. The use of air or other gases to transfer force is called **pneumatics**. Together they comprise fluid power.

The use of hydraulic power and pneumatic power is extensive in agriculture/agribusiness and the management of renewable natural resources. Pneumatics is used extensively in repair shops to clean parts, drive tools, and operate paint sprayers (Figure 39-1). It is used in milking operations to open and close gates and to direct livestock movement. It is also used in food-processing plants to control items on production lines by opening and closing various devices.

Hydraulics permit the use of pistons to generate linear power and the use of hydraulic motors to generate rotary power for equipment (Figure 39-2). Hydraulic power is especially useful in remote parts of automobiles and tractors (Figures 39-3 and 39-4). Vehicles such as trucks, earthmovers, boats, and airplanes rely heavily on hydraulic systems.

PRINCIPLES OF FLUID DYNAMICS
Nature of Hydraulic Fluids and Air

Both liquids and gases are considered fluids. However, they have different characteristics:

1. Hydraulic fluids cannot be compressed; air can be.
2. Hydraulic fluids must have complete circuits, with the fluid returning to a system reservoir; air can be pumped in from the atmosphere, used

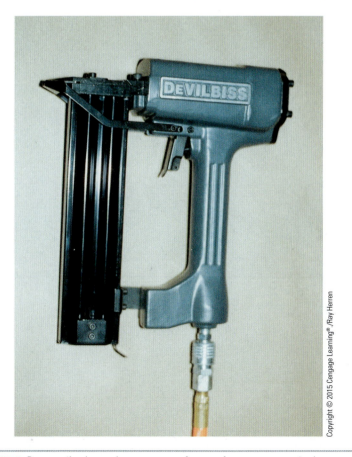

Copyright © 2015 Cengage Learning®./Ray Herren

FIGURE 39-1 Pneumatics is used as a source of power for many power tools.

Copyright © 2015 Cengage Learning®./Ray Herren

FIGURE 39-2 This tree-harvesting machine uses several different hydraulic systems.

to move components, and be expelled back out into the atmosphere.

Force and Pressure Defined

To understand fluid power, both force and pressure must be defined. **Force** is the pushing or pulling action of one object on another. Force usually causes objects to move. Force is measured in pounds (lb).

On the other hand, **pressure** is force acting upon an area. For instance, air pressure in a tire may be 30 pounds of pressure per square inch of tire surface. Pressure is measured in pounds per square inch (psi).

Pascal's Law

Pressure applied to a confined fluid is transmitted undiminished to every portion of the surface of the containing vessel. This is *Pascal's law*, which may be shortened to: Pressure on a liquid in a container is transferred equally to all surfaces (Figure 39-5). Therefore, pressure exerted on a master cylinder piston exerts equal force on all wheel cylinders (Figure 39-6).

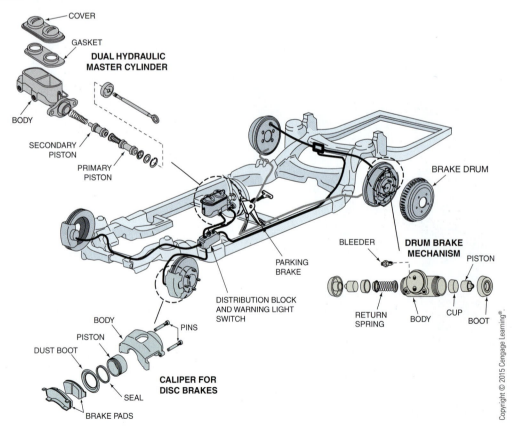

Copyright © 2015 Cengage Learning®.

FIGURE 39-3 Automotive braking systems use fluid to transfer the driver's foot power applied to the master cylinder to piston power on the brake shoes and pads.

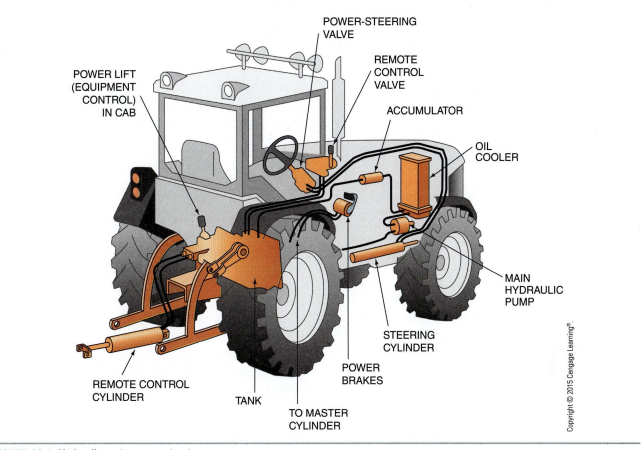

POWER-STEERING VALVE

REMOTE CONTROL VALVE

ACCUMULATOR

POWER LIFT (EQUIPMENT CONTROL) IN CAB

OIL COOLER

MAIN HYDRAULIC PUMP

STEERING CYLINDER

POWER BRAKES

REMOTE CONTROL CYLINDER

TANK

TO MASTER CYLINDER

Copyright © 2015 Cengage Learning®.

FIGURE 39-4 Hydraulic systems on a tractor.

Also, pressure on a fluid is equal to the force divided by the area:

$$P = \frac{F}{A}$$

where P = Pressure (in psi)
F = Force (in lb)
A = Area to which force is applied (in si)

Therefore, if F = 100 pounds, A = 1 square inch, then P = 100 pounds per square inch.

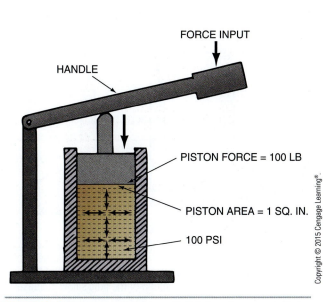

FORCE INPUT

HANDLE

PISTON FORCE = 100 LB

PISTON AREA = 1 SQ. IN.

100 PSI

Copyright © 2015 Cengage Learning®.

FIGURE 39-5 Pressure exerted on a fluid is transferred equally to all surfaces of the fluid's container.

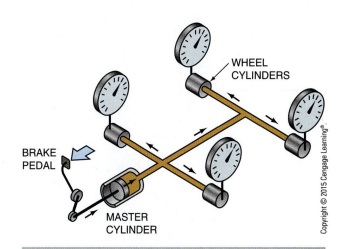
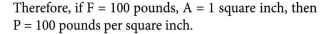

WHEEL CYLINDERS

BRAKE PEDAL

MASTER CYLINDER

Copyright © 2015 Cengage Learning®.

FIGURE 39-6 Pressure transfer in the hydraulic brake system of an automobile.

The output or resulting pressure transferred in a system can be changed or manipulated by using a different size piston. For instance, if 100 psi of pressure is exerted on a piston with one square inch area, the output would be only 50 psi on a piston with one-half square inch area. But, the pressure exerted on a piston of 2 square inches would be 200 pounds (Figure 39-7). Therefore, it can be concluded that P × A (input) = P'/A' (output). In Figure 39-6 we observe: 100 × 1/2 = 50 in cylinder B, 100 × 1 = 100/1 in cylinder C, and 100 × 1 = 200/2 in cylinder D.

Boyle's Law

A scientist by the name of Robert Boyle made a significant discovery in 1662 about the behavior of gases. Boyle's law states that the volume of gas is inversely proportional to its pressure. That means, as pressure increases volume decreases, and as pressure decreases volume increases. It is like a basketful of dry leaves. Press them with your foot and the leaves decrease in volume. Lift your foot and take the pressure off and the leaves bounce back to their original volume. It is this rebound tendency of gases that makes air a good medium for rubber tires, pressurized water tanks, and aerosol cans.

Bernoulli's Principle

Daniel Bernoulli made an important discovery in 1738 about fluids. Bernoulli's principle states that when a fluid flows through a pipe, pressure remains constant unless the diameter of the pipe changes. As the diameter of the pipe decreases, the pressure decreases (Figure 39-8).

Measuring Pressure

A **Bourdon tube** is a tube bent into a circular pattern that tends to straighten as internal fluid pressure increases. The outward movement at the end of the loop moves a pointer. The resulting pressure is read directly from a scale on the face of the meter (Figure 39-9).

A **manometer** registers pressure as well as vacuum. It uses a glass U-shaped tube filled with water or mercury. Pressure pushes and vacuum pulls the liquid column. The pressure or vacuum is read from the scale behind the liquid column (Figure 39-10).

Viscosity Index

The viscosity of a liquid refers to its tendency to flow. A **viscosity index** measures a fluid's tendency to flow at a given temperature. As a fluid becomes colder, it becomes thicker and tends to flow less. As it becomes hotter, it becomes thinner and flows faster.

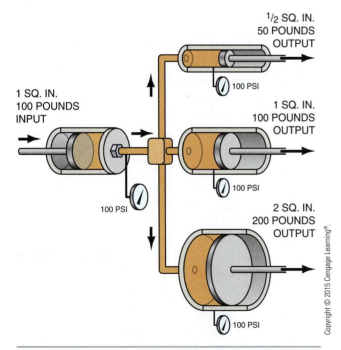

½ SQ. IN.
50 POUNDS
OUTPUT

100 PSI

1 SQ. IN.
100 POUNDS
INPUT

1 SQ. IN.
100 POUNDS
OUTPUT

100 PSI

100 PSI

2 SQ. IN.
200 POUNDS
OUTPUT

100 PSI

Copyright © 2015 Cengage Learning®.

FIGURE 39-7 The output force in a hydraulic system can be increased by changing the size of the piston diameter on the output side of the system.

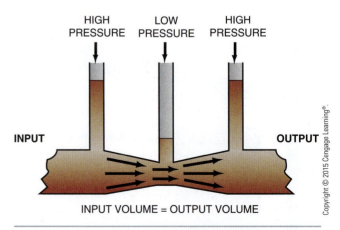

HIGH
PRESSURE

LOW
PRESSURE

HIGH
PRESSURE

INPUT

OUTPUT

INPUT VOLUME = OUTPUT VOLUME

Copyright © 2015 Cengage Learning®.

FIGURE 39-8 Bernoulli's Principle states that as the diameter of pipe containing a flowing fluid increases, the pressure at that point decreases; likewise, the speed of the fluid's flow increases at that point.

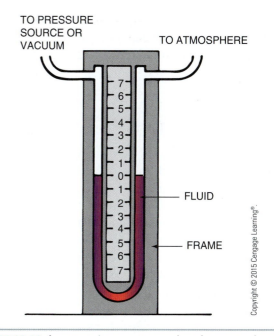

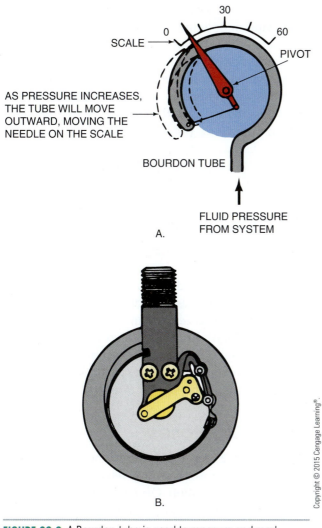

FIGURE 39-9 A Bourdon tube is used to measure and read hydraulic pressure A. The diagram in B illustrates how the tube is used in a pressure gauge.

FIGURE 39-10 A manometer measures and reads pressure by movement of a liquid column that is subject to the forces of atmospheric pressure and a vacuum.

- **Extreme-pressure resistor**: an additive to prevent splitting of molecules and oil breakdown under high pressure

Fluid Ratings

SAE Rating. Most fluids and lubricating oils are rated by the **Society of Automotive Engineers (SAE)**. The **SAE rating** on containers of fluids and oils is a viscosity rating. Fluids and oils must have appropriate viscosity for the climate and the conditions. For instance, fluids used in vehicles in Alaska need different SAE ratings than those used in Florida. SAE ratings of fluids generally range from 5 to 90. SAE 10 flows easily at 0°F, while SAE 40 is too stiff for good lubrication. However, when the engine is operating at 180°F, SAE 10 is too runny, but SAE 40 works fine. The solution is to add viscosity improvers to make the fluid flow like SAE 10 at 0°F and like SAE 40 at 180°F. Such a fluid would be rated SAE 10W-40.

API Rating. The **American Petroleum Institute (API)** rates oils by the type of service they can tolerate. Oils receive **API ratings** of SA, SB, SC, SD, or SE for gasoline engines, and CA, CB, CC, CD, or CE for the more demanding diesel engine service. In general,

Fluid Additives

An **additive** is a material added to a fluid to change its characteristics. Some types of additives used to improve fluids used in hydraulic systems include:

- **Viscosity improver**: a chemical added to make fluids more stable with temperature changes
- **Antifoam**: a chemical that reduces foaming in fluids
- **Corrosion inhibitor**: a chemical that reduces or prevents corrosion
- **Rust inhibitor**: a chemical that reduces or prevents rusting
- **Antiscuff**: a chemical that helps fluids polish moving parts

the further down the alphabet, the better (within the S or C group) the oil. However, it is always very important to use the fluid prescribed by the equipment manufacturer.

HYDRAULIC SYSTEMS

Hydraulic systems must have a reservoir, a pump, control valves, a check valve, one or more filters, lines, and one or more cylinders. The pump draws the fluid from the reservoir and pushes it through lines and valves to a piston in a cylinder. The pressure on the piston moves the piston and connecting apparatus to move the object. A valve then closes and holds the fluid against the piston. When the object should be lowered, the valve is opened and the piston pushes the fluid back into the reservoir.

Some hydraulic systems are built to apply pressure in either direction. A cylinder that can exert pressure by pushing or pulling is called a **double-action cylinder** (Figure 39-11).

Hydraulic Pumps

Hydraulic pumps may be of three types: gear, vane, or piston. Within the gear category are standard and eccentric subcategories. Some gear-type pumps and all piston-type pumps are positive-displacement pumps. **Positive displacement** means that for each revolution an exact volume of fluid will be pumped.

Standard Gear Pump. A standard gear-type pump consists of two gears that mesh with each other and turn in opposite directions. As they turn, fluid is drawn into one side, trapped between the gear teeth and housing, and expelled out the other side (Figure 39-12). This is a positive-displacement pump.

Eccentric Gear Pump. An eccentric gear pump operates on the same principle as a standard gear-type pump. However, its gears move in an **eccentric**, or off-center, pattern, which creates the vacuum and pressure action necessary to move fluids (Figure 39-13).

Vane Pump. A vane-type pump consists of spring-loaded paddles (vanes) on an off-center hub. As the hub turns, its body moves closer to the housing wall during part of its revolution and farther away during another part of it. The vanes maintain a seal with the wall and move the fluid through the pump (Figure 39-14). The vane-type pump is also useful as an air pump.

Piston Pump. Piston-type pumps can be purchased for moving air or liquids. Pumps that increase pressure on air are called **air compressors**. Some are two-stage air compressors. A *two-stage air compressor* compresses a large cylinder of air into a small cylinder, and then a second piston compresses that air into a tank.

It may be recalled that liquids cannot be compressed. Therefore, piston-type hydraulic pumps are positive-displacement pumps. They are relatively expensive, but capable of pumping fluids under great pressure.

Fluid Coupling

Many applications call for a gradual and smooth transfer of power from an engine to the vehicle. Automobile transmissions use fluid couplings for this purpose. A **fluid coupling** utilizes fluid to transfer power from one turning impeller to drive another, as the fluid spins (Figure 39-15). An **impeller** is a fan-shaped rotor. The action of a fluid coupling can be demonstrated using a motor-driven fan (or *air pump*) to force air through a tube and drive another fan (or *turbine*).

Hydraulic Hoses

Hydraulic hoses carry fluid from one component to another. They must be strong, flexible, and resistant to pressure and abrasion. Generally, the inner layer or inner tube is made of tough rubber. The intermediate layer or layers are composed of synthetic fibers or wire to withstand high pressure. The outer cover is also rubber, for reducing vibration and providing resistance to abrasion, weather, dirt, and moisture (Figure 39-16).

Hydraulic Valves

A system of valves is necessary to direct the flow of fluids from pump to cylinders or motors and back to the reservoir. Such valves may determine the routing of fluid or may simply control the direction of flow.

Pressure Regulators. A **pressure regulator** is a valve that holds pressure in a line at a given value. In hydraulic systems, pressure regulators determine the amount of pressure that can be exerted on the system before fluid is routed back to the reservoir. This provision protects the system from damage caused by overloading. Therefore, it also protects operators from injury that could result from rupturing hoses or exploding parts (Figure 39-17).

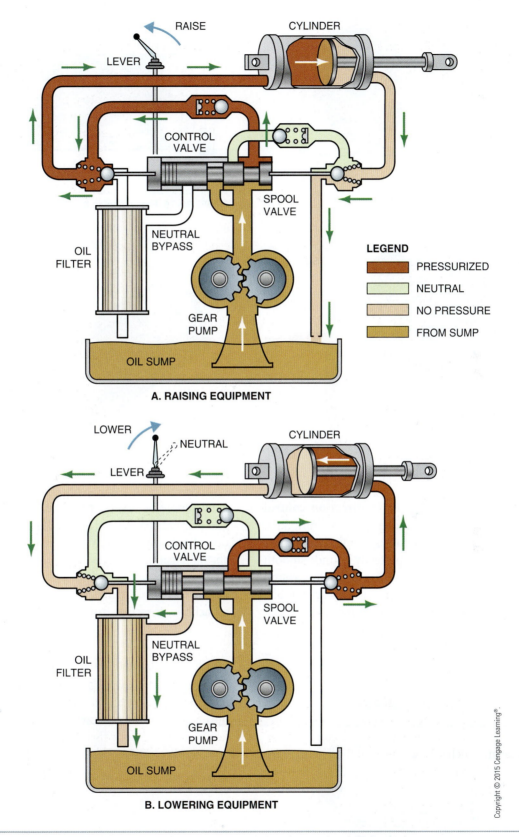

A. RAISING EQUIPMENT

B. LOWERING EQUIPMENT

FIGURE 39-11 A hydraulic system showing activity when equipment is being (A) raised and (B) lowered.

Copyright © 2015 Cengage Learning®.

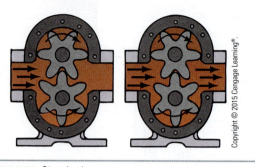

FIGURE 39-12 Standard gear pump.

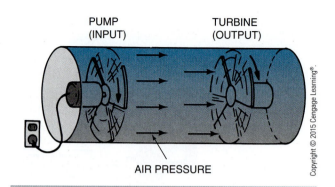

FIGURE 39-15 Fans set up to function as a fluid coupling.

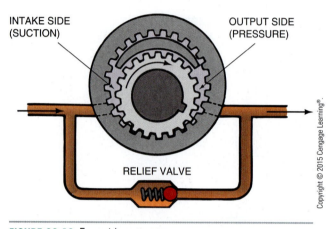

FIGURE 39-13 Eccentric gear pump.

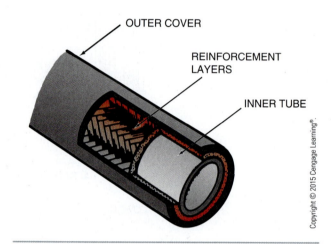

FIGURE 39-16 Composition of a hydraulic hose.

Direction Control Valves. A **direction control valve** permits fluid to flow in one direction only. Such valves are also called **check valves**. Some check valves consist of a housing with a perfectly round, smooth ball that sits on an opening with a carefully ground seat. Fluid coming in under pressure can lift the ball and flow around it. However, if the pressure drops or is overcome by reverse pressure, the ball drops, seals, and prevents backflow (Figure 39-18). Such valves can operate in any position if the ball is installed under spring tension.

Equipment Control Valves. A person operating a backhoe, trencher, front-end loader, or aerial bucket

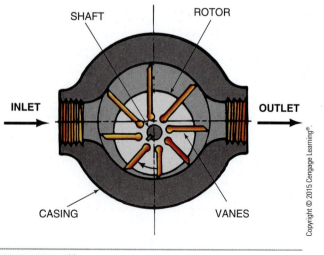

FIGURE 39-14 Vane pump.

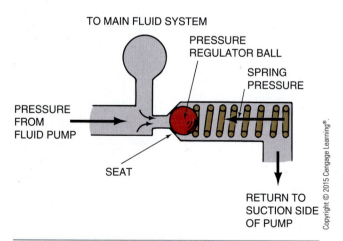

FIGURE 39-17 A pressure regulator protects both the system and the operator.

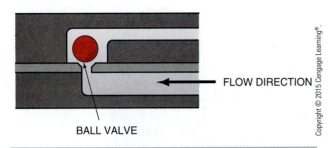

FIGURE 39-18 A check valve (direction control valve) with a ball-type mechanism.

controls the machine by pushing or pulling several levers. These levers operate the equipment control valves.

The most common type of equipment control valve is the spool valve. A **spool valve** is a valve with one inlet and several outlets, all controlled by a sliding spool. The spool is constructed so its wheels can block the flow of fluid and its axle can permit fluid to flow through. The position of the sliding spool determines the outlet (and equipment) that receives the flow and pressure of fluid at any given time (Figure 39-19).

FILTERING SYSTEMS

Hydraulic system parts are machined and operated at very close tolerance. **Tolerance** means leeway or acceptance of variation. Therefore, the smallest speck of dust, dirt, grit, or metal can damage the moving parts of a hydraulic system. A good system of filters must be in place and properly maintained at all times. Most fluid filters are made of special paper folded into pleats that encircle the core. The fluid enters the outside chamber, passes through the paper filter, and leaves the filter through a center tube.

Full-Flow Filter System. A **full-flow system** directs all fluid leaving the pump through the filter. As long as the filter is changed as needed, the system works well. However, if the filter becomes clogged, the unfiltered fluid will bypass the filter and flow throughout the system. If not corrected promptly, the system will be damaged (Figure 39-20).

Bypass Filter System. A **bypass system** is one that filters only a part of the fluid in any given pass through. However, since the fluid circulates frequently in an active system, all fluid is eventually filtered. This system can utilize a filter with less flow capacity than a full-flow system (Figure 39-21).

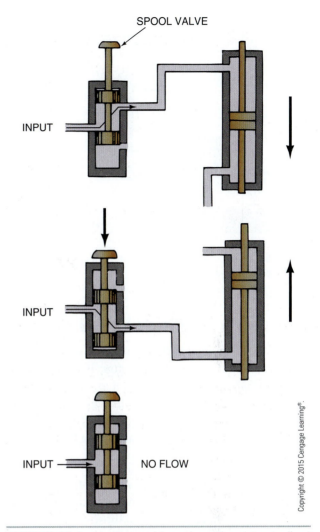

FIGURE 39-19 Spool valves enable equipment operators to direct the power and movement of hydraulic devices.

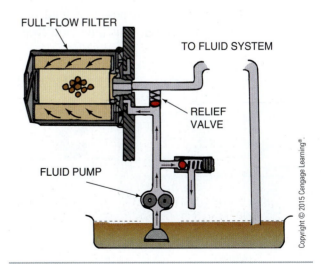

FIGURE 39-20 All fluid flows continuously through the filter of a full-flow system.

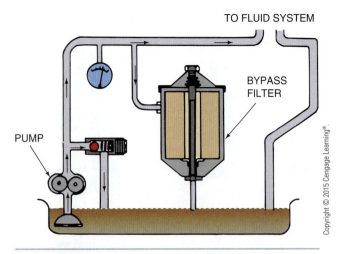

Copyright © 2015 Cengage Learning®.

FIGURE 39-21 In a bypass system, only part of the fluid is filtered during any single pass through.

Hydraulic Cylinder and Motors

Cylinders. A **hydraulic cylinder** is a tank with a fluid inlet and a close-fitting piston. If it is a *double-action cylinder*, it has a fluid outlet as well. The piston has a connecting rod extending through the end of the cylinder to which movable equipment is attached. The piston has a rubber ring called an **O ring** to maintain a fluid seal (Figure 39-22).

Hydraulic Motors. A **hydraulic motor** is a motor that receives its power from moving fluid. Its construction and operation are similar to those of a gear pump. Fluid, under pressure from the pump, enters the inlet side of the motor, drives the gears as it passes through, and exits under low pressure to return to the reservoir (Figure 39-23).

Although the hydraulic motor is simple in design, it can convert the tremendous power of fluid under pressure into a rotary power. This quiet and flexible

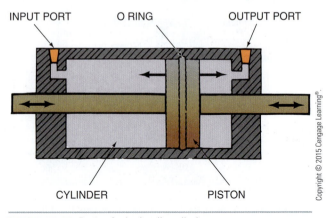

Copyright © 2015 Cengage Learning®.

FIGURE 39-22 Parts of a hydraulic cylinder.

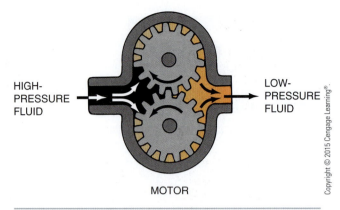

Copyright © 2015 Cengage Learning®.

FIGURE 39-23 A hydraulic motor.

power source is especially useful in remote parts of engine-driven machines.

PNEUMATIC SYSTEMS

Pneumatic systems are relatively simple, compact, and have great flexibility. Small, electric-motor-driven vane pumps and compressors can be very portable. Larger compressors are stationary and are driven by large electric motors or gasoline or diesel engines. Service of such units involves frequently draining condensed water from air tanks and using moisture-removing devices in the lines.

Regulators and valves for pneumatic systems are similar in function to those for hydraulic systems. However, air pressure regulators are generally adjustable (Figure 39-24). Pneumatic cylinders frequently are single-action. These rely on an internal spring to return the piston to its home position when air pressure is released (Figure 39-25). Additional information on pneumatic systems is provided in Unit 29 ("Selecting and Applying Painting Materials").

ROBOTICS

A **robot** is a mechanical device that is capable of performing human tasks. Robots are used extensively in industry. They are becoming increasingly important in agriculture/agribusiness and the management of renewable natural resources.

The term **robotics** refers to the study and application of the technology of robots. Robots may be distinguished from other machines in that they:

- are freely computer programmable.
- can do a variety of tasks.
- have three-dimensional freedom of motion.
- are equipped with grippers or tools.

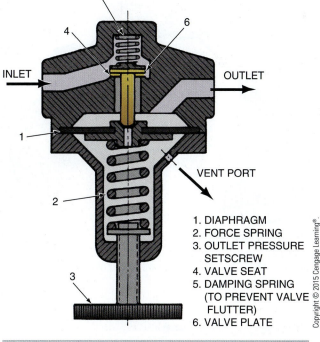

1. DIAPHRAGM
2. FORCE SPRING
3. OUTLET PRESSURE SETSCREW
4. VALVE SEAT
5. DAMPING SPRING (TO PREVENT VALVE FLUTTER)
6. VALVE PLATE

FIGURE 39-24 An air pressure regulator.

COMPONENTS OF A SINGLE-ACTING CYLINDER

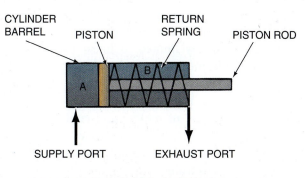

SINGLE-ACTING CYLINDER IN EXTENDED POSITION

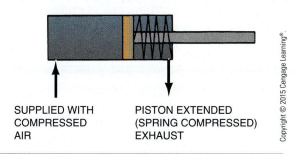

FIGURE 39-25 A single-action air cylinder.

Robotic Functions

Robots can be built to perform many tasks faster and more accurately than humans can. They are precise in movement and repeat tasks in exactly the same way for long periods of time. Robots are especially adaptable to the following tasks:

- arranging parts
- handling parts
- distributing items
- positioning tools and workpieces
- moving tools in predetermined patterns
- gripping, directing, and assembling
- fastening, attaching, and detaching

Robotic Movements

Robots are built to manage circular and linear motion. Circular motion is referred to as **rotation**, and linear motion is called **translation**.

Robots have one or more translational and/or rotational axes (*axes* is the plural of *axis*). An **axis** is a real or imaginary straight line around which a body rotates. The more axes a robot has, the more motions it can perform. Each axis provides the robot with one **degree of freedom**. A robot has as many degrees of freedom as it has axes.

Cartesian Work Area. A robot with three translational axes can perform translational tasks in a space with box-like characteristics. A box-like work space is known as a **Cartesian work area** (Figure 39-26).

Cylindrical Working Area. The work space of the robot in the previous example can be changed drastically by changing the axis of the base. If the translational axis is changed to a rotary axis, the working area is **cylindrical**, or in the shape of a cylinder. However, since the robot now works around itself, there is a **hollow** (space) in the center where the robot cannot function (Figure 39-27).

Hollow Sphere Working Area. A ball-shaped working area is known as a **hollow sphere**. It can be

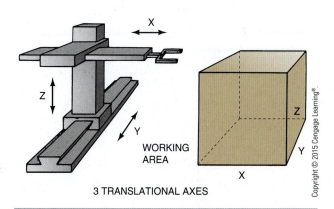

3 TRANSLATIONAL AXES

WORKING AREA

FIGURE 39-26 The Cartesian work area of a three-translational robot.

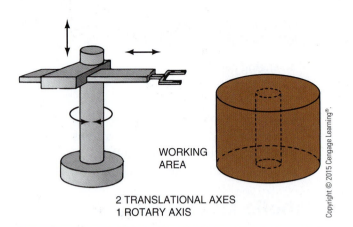

WORKING AREA

2 TRANSLATIONAL AXES
1 ROTARY AXIS

FIGURE 39-27 The cylindrical work area of a robot with two translational axes and one rotational axis.

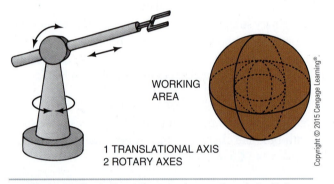

WORKING AREA

1 TRANSLATIONAL AXIS
2 ROTARY AXES

FIGURE 39-28 The hollow sphere work area of a robot with two rotational axes and one translational axis.

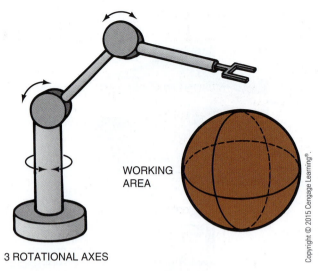

WORKING AREA

3 ROTATIONAL AXES

FIGURE 39-29 The solid sphere work area of a robot with three rotational axes.

Robotic Power

Robotic structures are powered by pneumatic, hydraulic, or electrical drives. Both pneumatic and hydraulic power may be utilized with cylinders and motors. However, since air is compressible, the movements of air-driven equipment are not as precise as that of hydraulic equipment. For this reason, pneumatic drives are generally limited to linear point-to-point functions. This means that the robot operates in a straight line from one point to another.

Electric power is the choice for many robotic applications. The variety of motors and controls is great, providing an almost infinite number of combinations in equipment design. Electric motors are powerful, but compact, and electronic controls are tiny. They lend themselves well to computer-controlled operations. The advantages of compactness, ease of getting power to the motor, ease of maintenance, and high reliability all contribute to the advantages of electric power for robotics.

achieved with two rotary axes and one translational axis. Here the robot is also working around itself and creates a hollow in the center of the work area (Figure 39-28).

Solid Sphere Working Area. A solid sphere is like a solid ball. A **solid sphere work area** can be achieved with three rotational axes on a robot. With this arrangement the robot can work in any part of a round pattern to form a solid sphere work area (Figure 39-29).

SUMMARY

Hydraulic and pneumatic systems provide flexible power in remote areas. Compressed air is useful as a propellant, it can drive motors, and it can actuate cylinders. Pressurized liquids provide considerable force when driving hydraulic motors and permit excellent control over hydraulic cylinders.

Hydraulic, pneumatic, and electrical devices are used to operate robots. Robots are performing many tasks faster, cheaper, and better than what can be done by human hands. The use of robots and the expansion of the robotics industry will provide new career opportunities in the future.

Student Activities

1. Define the Terms to Know in this unit.

2. Develop a bulletin board illustrating the use of hydraulics and pneumatics in agricultural settings.

3. Make a collage of robots in action.

4. Work with your agricultural mechanics instructor and/or with a science teacher to prepare and present a class demonstration on Boyle's Law.

5. Present a class demonstration to illustrate the truth of Bernoulli's Principle.

6. Demonstrate one or more applications of Pascal's Law.

7. Prepare and present a report on the use of additives in hydraulic fluids.

8. Do a shop project that requires the servicing of a hydraulic system.

9. Make a shop project that utilizes hydraulics or pneumatics.

10. Visit a firm that uses robots for production. Take pictures of the robots, and report to your class about how the robots are designed, how they function, and how they are maintained.

Relevant Web Sites

Engineering.com, Pascal's Law
http://www.engineering.com/Library/ArticlesPage/tabid/85/ArticleID/214/Pascals-Law.aspx

NASA Glenn Research Center, The Beginner's Guide to Aeronautics, Animated Gas Lab, Boyle's Law
www.grc.nasa.gov/WWW/K-12/airplane/aboyle.html

Self-Evaluation

A. Multiple Choice. Select the best answer.

1. Fluid power is
   a. hydraulics
   b. pneumatics
   c. pneumatics and hydraulics
   d. pneumatics, hydraulics, and electricity

2. Linear power is provided directly by
   a. pistons in cylinders
   b. fluid couplings
   c. motors
   d. rotational robot axes

3. Force acting on an area is
   a. compression
   b. horsepower
   c. measured in pounds
   d. pressure

4. Important laws of force and pressure in hydraulics were formulated by
   a. Bernoulli
   b. Bourdon
   c. Boyle
   d. Pascal

5. Pressure (in psi) = force (in pounds) divided by
   a. area (in square inches)
   b. distance (in inches)
   c. number of pistons
   d. time (in seconds)

6. The volume of a gas is inversely proportional to its
   a. weight
   b. temperature
   c. pressure
   d. area

7. When fluid flows through a pipe and encounters a decrease in pipe diameter, pressure in the narrower area
   a. cannot be predicted
   b. decreases
   c. increases
   d. remains the same

8. Which is not a type of hydraulic pump?
   a. vane
   b. ratchet
   c. piston
   d. gear

9. The full-flow system refers to the use of
   a. valves
   b. spools
   c. lines
   d. filters

10. A hollow sphere working area can be achieved with
    a. three translational axes
    b. two translational axes and one rotational axis
    c. two rotational axes and one translational axis
    d. none of the above

B. **Matching.** Match the terms in column I with those in column II.

Column I
1. Bourdon tube
2. manometer
3. viscosity
4. rust inhibitor
5. SAE rating
6. API rating
7. spool valve
8. quick coupling
9. Cartesian
10. translational

Column II
a. additive
b. service rating
c. measures fluid pressure
d. controls hydraulic equipment
e. measures pressure
f. tendency to resist flowing
g. uses a sliding collar
h. boxlike work area
i. linear action
j. viscosity rating

C. **Completion.** Fill in the blanks with the word or words that make the following statements correct.

1. In addition to SA and CA, three API oil service ratings are _____, _____, and _____.

2. Two types of gear pumps are the _____ _____ type and the _____ type.

3. The piston in a hydraulic cylinder typically contains a(n) _____ to create a moving seal between the piston and cylinder wall.

4. Three important maintenance procedures for hydraulic systems are _____, _____, and _____.

5. A disadvantage of pneumatics in robots is _____.

D. **Brief Answer.** Briefly answer the following questions.

1. What property of gases makes air a good medium for rubber tires, pressurized water tanks, and aerosol cans? Who first discovered that gases behave in this way?

2. Name six types of additives used to improve fluids in hydraulic systems.

3. Why is proper maintenance of filters so important in a hydraulic system?

4. What four aspects of robots distinguish them from other machines?

5. Why is electric power the preferred choice for robotics?

Section 13
CONCRETE AND MASONRY

UNIT 40
- Concrete and Masonry

UNIT 40

Concrete and Masonry

Objective

To mix and place concrete and use masonry materials.

Competencies to be developed

After studying this unit, you should be able to:

- Identify tools used for concrete work.
- Select ingredients for mixing concrete.
- Make a workable masonry mix.
- Prepare forms for concreting.
- Insulate concrete floors.
- Pour concrete.
- Finish concrete.
- Calculate concrete and block for a job.
- Lay masonry block.

Materials List

- Portland cement
- Sand
- Gravel
- Mortar mix
- Wall reinforcement
- Lumber for forms
- Stakes
- Bull float
- Hand float

continued

Terms to Know

- masonry
- portland cement
- concrete
- sand
- mortar
- finishing lime
- fine aggregate
- gravel
- coarse aggregate
- silt
- clay
- washed sand
- cement paste
- ratio
- mix
- workable mix
- form
- footer
- footing
- moisture barriers
- vapor barriers
- insulation
- construction joint
- control joint
- reinforced concrete
- air-entrained concrete
- screeding

continued

Materials list, *continued*

- Edging tool
- Grooving tool
- Push broom
- Oil for forms
- Level, 4 foot
- Mason's trowel
- Corner poles
- Wheelbarrow
- Mortar pan
- Striking tool
- Rakes
- Shovels
- Quart jar with lid

Terms to know, *continued*

- bull float
- broom finish
- curing
- masonry units
- laying block
- mortar bed
- cores
- hollow core block
- stretcher block
- ears
- sash block
- jamb block
- course
- frost line
- corner pole

Concrete and masonry construction is basic to most farm and agribusiness facilities. **Masonry** refers to anything constructed of brick, stone, tile, or concrete units set or held in place with portland cement. **Portland cement** is a dry powder made by burning limestone and clay, and then grinding and mixing to an even consistency.

Concrete and mortar are two construction materials with portland cement as a component. **Concrete** is a mixture of stone aggregates, sand, portland cement, and water that hardens as it dries. **Sand** consists of small particles of stone. **Mortar** is a mixture of sand, portland cement, water, and finishing lime. **Finishing lime** is a powder made by grinding and treating limestone. Like concrete, mortar is a mixture of materials mixed with water that hardens as it dries.

Nearly all buildings have concrete and masonry components somewhere in their construction. The ability to use masonry materials is therefore essential to many maintenance and repair operations.

MASONRY CONSTRUCTION

Masonry construction has been used for thousands of years. Its basic materials come from the earth and are plentiful in most areas. The pyramids of Egypt, the Coliseum of ancient Rome, the temples of the Mayan Indians, and the skyscrapers of modern America all demonstrate the use of masonry construction. Many masonry structures have lasted for hundreds and even thousands of years (Figure 40-1). Masonry construction has most of the advantages of concrete. However, the hardness, durability, and moisture-resistant characteristics of the masonry units and mortar will determine the characteristics of the masonry structure.

CONCRETE CONSTRUCTION

Concrete has many advantages for farm use and in other agriculture enterprises. Concrete is:

- fireproof
- insect and rodent proof

FIGURE 40-1 Mayan temples are a good example of masonry construction that has lasted hundreds of years.

- decay resistant
- highly storm resistant
- wear resistant
- waterproof
- strong
- attractive
- easy to make on the job without expensive equipment
- available locally
- low in original and maintenance costs
- sanitary and easy to keep clean
- easily broken up and used as fill material when the structure becomes obsolete

The strength and durability of concrete depend upon several factors: (1) the strength of the stone particles, (2) the proportion of stone particles by size, (3) the type of portland cement, (4) the purity of the water, (5) the uniformity of the mixture, and (6) the procedures used in placing, finishing, and curing the concrete.

Many types of stone found in fields, streams of water, and solid rock formations are suitable for making concrete. Occasionally the correct type and mixture of stone particles may be found in stream beds or gravel deposits. Usually, however, stone particles must be made by crushing stone and grading it to size. The correct proportion of small particles, called sand or **fine aggregate**, and large particles, called **gravel** or **coarse aggregate**, can then be mixed to meet the requirements of specific jobs (Figure 40-2). In recent years different materials have been used as aggregates. For example, a fine glass-like powder called fly ash has been used as an aggregate in concrete. This material is a by-product produced by huge electric

FIGURE 40-2 Aggregate comes in a variety of sizes.

power generators that use coal as a fuel source. As the coal burns, fly ash is generated in the gases that emerge as a result of the combustion. These power plants are found all across the country and produce millions of tons of fly ash every year. This material can actually replace the portland cement portion of the concrete. It can produce concrete that is less expensive and is stronger. In addition, it provides a use for a material that for a long time was simply hauled to the landfill.

Other materials that can be used as aggregate include recycled, ground glass; blast furnace slag; and recycled concrete. Some glass does not recycle well for use as glass products and is ground up and used as a sand substitute for sand in concrete. A blast furnace is a huge furnace that smelts iron ore. When the iron is separated from the stone, through a tremendous heating process, a rock-like substance called slag is left (Figure 40-3). This material can substitute for gravel in concrete. Recycled concrete is ground up concrete from building demolitions where concrete slabs are left or from roads that are to be resurfaced (Figure 40-4).

To make good concrete, aggregates of various sizes should fit together to form a fairly solid mass. The stone particles must be clean and free of clay, silt, chaff, or any other material. **Silt** is composed of intermediate-sized soil particles. **Clay** consists of the smallest group of soil

FIGURE 40-3 Slag is a by-product of the iron smelting process. It is used as aggregate in concrete.

FIGURE 40-4 Broken concrete from demolished slabs or roots can be ground up and recycled as concrete aggregate.

particles. Both silt and clay particles are too small for use as aggregates in concrete and decrease the quality of concrete if present.

Sand sold commercially for use in concrete generally contains very little clay or silt. If the clay or silt content is above an acceptable level when the material comes from the quarry, the sand is flushed with water to remove the clay or silt (Figure 40-5). The material is then referred to as **washed sand**.

Testing Sand for Silt or Clay Content

Sand from streambeds or other farm areas may be suitable for making concrete. However, when sand from such places is being considered, a test for clay and silt must be made. The procedure follows.

FIGURE 40-5 Sand from a quarry is washed to remove clay and silt.

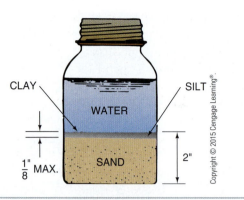

FIGURE 40-6 This simple test will determine the content of materials being considered for use in concrete.

Procedure

1. Fill a 1-quart glass jar to a depth of 2 inches with the sand to be tested (Figure 40-6).

2. Add water until the jar is ¾ full.

3. Screw on a lid and shake the mixture vigorously for 1 minute to mix all particles with the water.

4. Shake the jar sideways several times to level the sand.

5. Place the jar where it will not be disturbed for 1 hour for a silt test or 12 hours for a clay and silt test.

6. After 1 hour, measure the thickness of the silt layer on top of the sand.

7. If the layer is more than ⅛ inch thick, the sand is not suitable for use in concrete unless the silt is removed by washing.

8. If the layer is not ⅛ inch thick in 1 hour, let the mixture stand for 12 hours. Remeasure the layer(s) that have settled on the sand.

9. If the silt plus clay layer exceeds ⅛ inch, wash the sand before using it in concrete.

MIXING CONCRETE

Cement paste is made by mixing portland cement and clean water in precise proportions. The reaction between the cement and the water is a chemical one, and the strength of the concrete is directly affected by the proportions of the materials. As a result, precision in mixing portland cement and water is as important as precision in mixing ingredients when baking a cake.

CAUTION!

A mistake in measuring will ruin the product.

All sand has some water attached to its particles. This water must be estimated and an allowance made for it when deciding how much water to use in the concrete mix. Sand is described as "damp," "wet," or "very wet." The more moisture there is in the sand, the less water is added from other sources when mixing concrete or mortar. To obtain a more durable concrete to withstand severe weather conditions and/or traffic, add less water (Figure 40-7).

Intended Use of Concrete	For Maximum Aggregate Size of	Water (Gallons) Added to 1 Cu Ft of Cement if Sand Is:			Suggested Mixture for 1 Cu Ft Trial Batches*		
		Damp	Wet (Average Sand)	Very Wet	Cement Cu Ft	Fine Cu Ft	Coarse Cu Ft
						Aggregates	
Mild Exposure	1½ in	6¼	5½	4¾	1	3	4
Normal Exposure	1 in	5½	5	4¼	1	2¼	3
Severe Exposure	1 in	4½	4	3½	1	2	2¼

*Mix proportions will vary slightly depending on the gradation of aggregate sizes.

Note: Batches may be made by using 100 pounds (1 cwt) of each ingredient where 1 cubic foot is indicated.

FIGURE 40-7 Ratio of ingredients for mixing concrete.

Copyright © 2015 Cengage Learning®

For Maximum Aggregate Size of:	Suggested Mixture for 1 Cwt Trial Batches*		
	Cement Cwt** or Cu Ft	Aggregates	
		Fine Cu Ft	Coarse Cu Ft
3/4 in	1	2	2 1/4
1 in	1	2 1/4	3
(preferred mix) 1 1/2 in	1	2 1/2	3 1/2
(optional mix) 1 1/2 in	1	3	4

*Mix proportions will vary slightly depending on the gradation of the aggregates. A 10-percent allowance for normal wastage is included in the fine and coarse aggregate values.

**1 cwt of cement equals about 1 cubic foot.

FIGURE 40-8 Materials needed to make trial batches of concrete with separated fine and coarse aggregates.

The ratio of cement to fine aggregate (sand) and coarse aggregate (gravel) is important. However, the exact proportion will vary with the makeup of each aggregate (Figure 40-8). A correct mixture will assure that (1) each particle of sand and gravel is covered with cement paste, and (2) each particle is bound to others when the cement paste dries and hardens.

The proportion of cement to fine and coarse aggregate is known as the **ratio**. The ratio is expressed as a three-digit number called a **mix**. For example, a 1-2-3 mix has one part cement, two parts fine aggregate, and three parts coarse aggregate. Once the mix for a job has been determined, a table can be used to help estimate the amount of material needed for a cubic yard of concrete (Figure 40-9). A rule of thumb is that the resulting concrete will be about two-thirds the combined volume of the cement and aggregate used in the mix.

It should be noted that portland cement is sold in bags containing 94 pounds, or exactly 1 cubic foot, of cement. A 1-2-3 mix would consist of 1 cubic foot of cement (one bag), 2 cubic feet of sand, and 3 cubic feet of coarse aggregate. For the purpose of mixing, one bag of cement may be considered as 100 pounds or 1 hundredweight (1 cwt). This is especially useful when the aggregates being used are measured by weight rather than volume. Proportions may be based on either weight or volume.

The term **workable mix** refers to the consistency of the wet concrete after the various ingredients have been mixed together. A workable mix has the following characteristics:

- All aggregates are clean.
- All portland cement is mixed with water to form a cement paste. No dry powder is present.

Mix Recommended for Maximum Aggregate Size of:	Cement Cwt or Cu Ft	Materials Needed to Make One Cu Yd of Concrete			
		Aggregates			
		Fine		Coarse	
		Cu Ft	Lb	Cu Ft	Lb
3/4 in 1-2-2 1/4	7	17	1,550	19.5	1,950
1 in 1-2 1/4-3	5.5	15.5	1,400	21	2,100
(preferred mix) 1 1/2 in 1-2 1/2-3 1/2	5.25	16.5	1,500	23	2,300
(optional mix) 1 1/2 in 1-3-4	4.5	16.5	1,500	22	2,200

Note: 1 cwt of cement or aggregate equals approximately 1 cubic foot.

FIGURE 40-9 Materials needed to make 1 cubic yard of concrete.

- Every particle of aggregate is covered with cement paste.
- Aggregates are distributed evenly throughout the mix.
- No more than the recommended amount of water is added.
- No lumps are present.
- The mixture has uniform color and consistency.
- The mixture can be mixed, moved, and placed with a shovel or spade.

ESTIMATING MATERIALS FOR A JOB

It is important to know how to estimate the amount of concrete needed for a job. If the estimate is inaccurate, the job will cost more than is necessary and ingredients may be wasted. If insufficient material is available, the job must be stopped and then started again at a considerable increase in cost and time (Figure 40-10). The following example will show how to determine the amounts of materials needed for a concrete floor 4 inches thick, 14 feet wide, and 24 feet long.

Procedure

1. Determine the cubic yards of concrete needed by multiplying the thickness of the floor by its width and length in feet: Cu Ft = T' × W' × L'. In this case,

thickness equals 4 inches, which must be converted to $\frac{1}{3}$ foot. Therefore, $\frac{1}{3}' \times 14' \times 24' = 112$ cubic foot. There are 27 cubic feet in 1 cubic yard. Thus 112 cubic feet ÷ 27 = 4.15 cubic yards.

2. Select the ratio of the mix needed for the job. In this example, a 1:2¼:3 ratio is selected.

3. Determine the amount of each ingredient needed to make 1 cubic yard of concrete. According to Figure 40-9, 1 cubic yard of 1:2¼:3 mix requires: 5.5 cwt (550 pounds) of cement, 1,400 pounds of fine aggregate, and 2,100 pounds of coarse aggregate.

4. Multiply each of those figures by the number of cubic yards in the project. In this case, multiply each ingredient by 4.15:

   4.15 × 550 = 2,282.5 lb of cement

   4.15 × 1,400 = 5,810 lb of fine aggregate

   4.15 × 2,100 = 8,715 lb of coarse aggregate

NOTE

If the concrete is to be purchased in ready-mixed form, the estimate is complete once it is determined how many cubic yards are needed. In this case, 4.15 plus a little extra in case the estimate is too low might come to 4.5 cubic yards. If the base is not leveled carefully, 5 cubic yards could be ordered and any excess used in another location.

PREPARING FORMS FOR CONCRETE

A **form** is a metal or wooden structure that confines the concrete to the desired shape or form until it hardens. Preparing forms for concrete can be as simple as nailing a board in place or as complex as constructing forms for the side of a highway bridge. On farms, around homes, and in landscape projects, form construction is usually required for walks, floors, feedlots, steps, ornamental projects, and building footers. A **footer**, or **footing**, is the concrete base that provides a solid, level foundation for brick, stone, or block walls. Footers, walks, and concrete slabs are generally constructed with 2" × 4" lumber

FIGURE 40-10 One of the most critical parts of a concrete construction job is calculating the correct amount of concrete needed.

Photo courtesy of Portland Cement Association

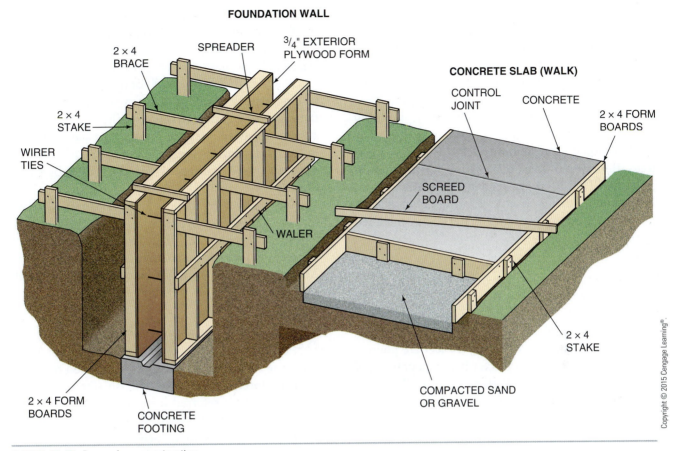

FIGURE 40-11 Proper form construction.

nailed to 2" × 4" stakes driven into the soil. Forms for walls are generally constructed by nailing ¾-inch plywood to 2" × 4" materials (Figure 40-11). Forms may also be purchased that are made from aluminum plate and struts.

When constructing forms, the following points may be helpful:

- Use soft, clean, straight lumber.
- Sharpen stakes evenly so they can be driven in straight.
- Place stakes about 30 inches apart along the outside of forms for 4-inch-thick concrete. Place the stakes closer when the concrete is more than 4 inches thick.
- Use a transit or level to adjust the height of forms for the desired slope or "fall" of the slab.
- Drive nails through the form and into, but not through, the stakes.
- Be sure the stakes do not stick up above the top of the forms. If they do, saw them off so they are level with or tapered down from the form.

CAUTION!

Do not saw into the top of the form.

- Construct the inside surfaces of the forms to create the desired shape of the finished concrete. For example, if you want a rounded extension on the front of a concrete step, the shape of the extension is determined by the inside of the form.
- Brush used motor oil on wood surfaces that will be touched by concrete. This is a low-cost product that will prevent the wood from sticking to the concrete and permits easy removal of the forms.

INSULATING CONCRETE FLOORS

In any concrete or masonry structure, consideration should be given to moisture barriers and insulation. **Moisture barriers**, also known as **vapor barriers**,

Copyright © 2015 Cengage Learning®.

prevent moisture from passing through. In climates with either high or low temperatures, energy costs can be reduced with proper insulation. **Insulation** retards heat movement. Further, moisture barriers and insulation will make floors more comfortable, and waterproofed walls will prevent groundwater and precipitation from entering the building (Figure 40-12).

Additional information on insulating concrete floors and masonry walls is provided in Unit 41 ("Planning and Constructing Agriculture Structures"). Further, a table showing the insulation values of common building materials is included there.

MAKING JOINTS

Construction joints are needed when slabs of concrete larger than about 10 feet by 10 feet are poured. A **construction joint** is a place where one pouring of concrete stops and another starts. A **control joint** is a planned break that permits concrete to expand and contract without cracking. Control joints are placed at equally spaced intervals in the slab.

Control joints are made by placing a wood or galvanized metal "key" on the form (Figure 40-13). A beveled 1" × 2" wood strip is adequate for most slabs. A saw kerf in the key prevents breakage of the concrete when moisture expands the wood strip. The groove left in the concrete after the form is removed provides an indentation into which the next

section locks. Before the second section is poured, the first section is coated with a curing compound. This prevents the second section from sticking to the first.

Another popular type of control joint is made by placing asphalt material between sections to absorb the expansion and contraction of the two slabs. This system, however, does not prevent one slab from rising or falling during freezing and thawing, which may leave an uneven surface where the two slabs come together.

Sidewalks need shallow grooves or joints across the surface at 3-foot intervals to control cracking due to expansion. When joints are present, the concrete will crack under the joint if stress occurs. A crack under the joint will not create an unsightly appearance.

REINFORCING CONCRETE

Concrete slabs may be greatly strengthened by using steel reinforcing rods or wire mesh (Figure 40-14). Such concrete is referred to as **reinforced concrete**. Any concrete slab that will carry vehicles such as tractors, loaders, or machinery must be reinforced. Otherwise, money is wasted on extra thicknesses of concrete that are still likely to crack under stress. Concrete posts and other thin concrete structures must be reinforced with steel to withstand normal stresses.

Concrete reinforcing rods are made of steel and have a rough, pebbly surface. This surface prevents

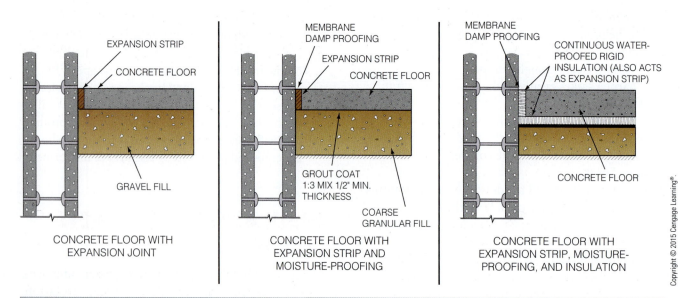

FIGURE 40-12 Moisture-proofing and insulating concrete joints.

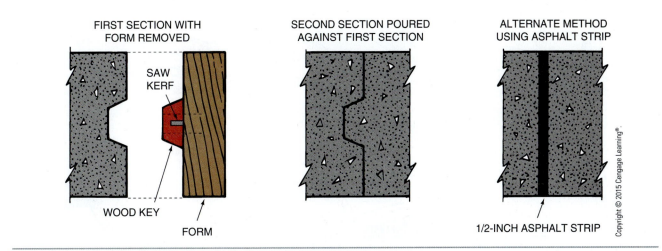

FIGURE 40-13 Control joints.

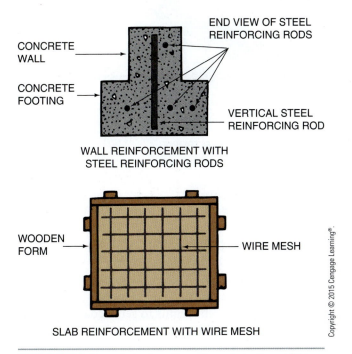

FIGURE 40-14 Concrete is strengthened greatly by the addition of steel rods or wire mesh.

them from slipping when embedded in concrete. They are classified according to diameter and are available in diameters from ¼ to 1 inch and over. They may be purchased in 20-, 40-, or 60-foot lengths.

Two or more bars may be joined with wire or by welding to make longer units. These should be wired so that they lap one another by at least a foot. A rule of thumb is to lap bars by 24 times the diameter. Therefore, a ½-inch bar should be lapped 12 inches and a ¾-inch bar lapped 18 inches. Rods should be placed at least ¾ inch from all surfaces of the concrete. For most applications, 1½ inches or more is preferred.

Rods may be placed in concrete slabs in a cross-sectional pattern. Such reinforcement is stronger if the bars are wired together where they cross one another. Reinforcing bars should be free of rust, dirt, oil, or other materials that will reduce adhesion by the concrete.

Concrete slabs are generally reinforced with wire mesh or fabric. Wire fabric is steel wire welded to form a cross-sectional pattern. It is generally available in a 6-by-6-inch pattern and consists of No. 6-, 8-, or 10-gauge wire. Both wire mesh and reinforcing rods are more effective if they are not rusted.

Wire fabric is first rolled out on a flat surface to eliminate curves in the material. It is then laid inside the concrete forms. Multiple pieces are wired together to form a continuous piece. Where two or more widths are needed, they should overlap each other by at least one and a half squares. The pieces should be tied securely at regular intervals.

The wire fabric should be supported by stones to keep it off the base material. As the concrete is being poured, it is important to pull the wire up until it is approximately centered in the slab. This ensures maximum reinforcement.

Finally, ready-mixed concrete may be ordered with tiny air bubbles trapped throughout the mixture. Such concrete is known as **air-entrained concrete**. It is stronger and more resistant to acid, salt, and frost than regular concrete.

POURING, FINISHING, AND CURING CONCRETE

When pouring concrete, common tools such as shovels, spades, and rakes are needed to move and spread the material. Other tools are needed to finish the concrete

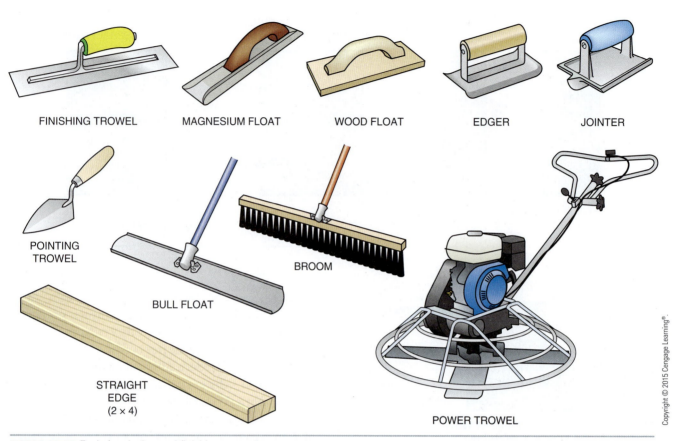

FINISHING TROWEL   MAGNESIUM FLOAT   WOOD FLOAT   EDGER   JOINTER

POINTING TROWEL

BULL FLOAT

BROOM

STRAIGHT EDGE (2 × 4)

POWER TROWEL

FIGURE 40-15 Tools for placing and finishing concrete.

(Figure 40-15). Since concrete is so heavy, it is best to move it downhill with a chute and across level surfaces with a wheelbarrow. This permits the concrete to be pushed, pulled, or lowered rather than lifted. Move a spade up and down in the concrete to help it settle in close to the forms.

Concrete starts to harden about 15 minutes after it is mixed. Thus, it is very important to have all materials on hand before mixing starts.

Before pouring concrete, the stone base should be wetted down with water so that water does not move from the concrete into the dry base. Loss of water from the mixture will weaken the concrete.

Screeding, Floating, and Finishing

After spreading concrete, a straight 2" × 4" or 2" × 6" plank is moved back and forth across the forms to strike off the excess concrete and create a smooth and level surface. This process is called **screeding** (Figure 40-16). When the screeding is finished, a smooth board attached to a handle is pushed and pulled across the surface. This process is called floating, and the tool is called a **bull float**. The process brings the fine aggregate and cement

paste to the surface for a smooth finish. Small areas may be finished with wooden or magnesium floats. Hand floats, edging tools, jointing tools, and trowels are used to finish the edges and/or place grooves or joints in the concrete (Figure 40-17). All of these procedures must be done before the concrete begins to set.

After the concrete begins to set, a coarse bristled broom may be pulled across it to create a rough surface

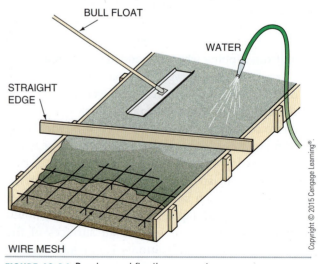

BULL FLOAT

WATER

STRAIGHT EDGE

WIRE MESH

FIGURE 40-16 Pouring and floating concrete.

Copyright © 2015 Cengage Learning®.

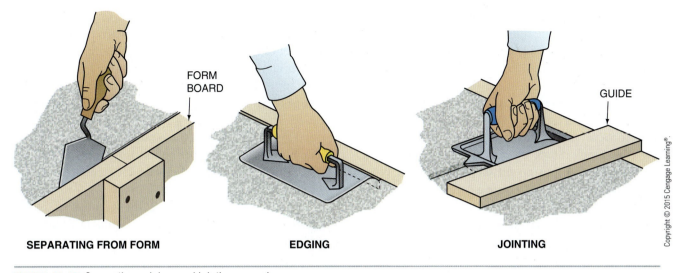

| SEPARATING FROM FORM | EDGING | JOINTING |

FORM BOARD

GUIDE

Copyright © 2015 Cengage Learning®

FIGURE 40-17 Separating, edging, and jointing procedures.

and improve footing or traction (Figure 40-18). This is called a **broom finish**. A very rough, pebbly finish may be obtained by washing away some of the cement paste, leaving the aggregate exposed. A smooth finish is created by troweling after the concrete starts to harden.

When an extremely smooth surface is desirable, the concrete is rubbed with a metal trowel. Generally the trowel is dipped in water and worked over the semihardened surface. This action brings the cement paste to the surface. Excessive troweling will leave the surface weak and easily damaged by frost and chemicals.

Curing

Concrete must dry slowly or it will crack, crumble, and break up long before its intended lifetime has passed. The proper drying of concrete is called **curing**. To cure concrete, it must be protected from drying air, excessive

© Jernej5m/Dreamstime.com

FIGURE 40-18 A broom finish creates a rougher finish that will improve footing.

heat, and freezing temperatures for several days after it is poured.

Concrete is kept moist by covering it with plastic or canvas to prevent the water from evaporating. The surface may also be covered with straw, sawdust, or other insulating material and sprinkled with water occasionally to keep it moist. Insulating materials help protect the concrete from air that is too hot or too cold as well as preventing moisture from escaping. Wooden forms help to protect concrete from drying out. Therefore, forms and other materials used for curing concrete should be left in place for up to a week until curing is complete.

LAYING MASONRY UNITS

Blocks made from concrete, cinders, or other aggregates are used extensively for agricultural buildings. Such blocks are called **masonry units**. They are referred to as "block" in both singular and plural. Masonry units are held together with mortar. The process of mixing mortar, applying it to block, and placing the block to create walls is called **laying block**.

Types of Blocks

Concrete and cinder block are available in a variety of sizes and shapes. Standard block are $15\frac{5}{8}$ inches long and $7\frac{5}{8}$ inches high. When laid with $\frac{3}{8}$-inch mortar joints, they cover an area 16 inches long and 8 inches high. Standard block are available in 4-, 6-, 8-, 10-, and 12-inch widths. The 4-, 8-, and 12-inch sizes are used most frequently.

Block are wider when viewed from the top than when viewed from the bottom. This is because the block is poured into forms that are tapered slightly to permit the block to be removed easily. Block are usually laid with the thicker part up to provide a larger area for the mortar bed. A layer of mortar is called a **mortar bed**.

Block are generally manufactured with hollow spaces called **cores**. They are available with two or three cores per block. Such block are referred to as **hollow core block**. Some block, called **stretcher block**, have slight extensions on the ends called **ears**. When laid end-to-end, the ears create a core. Stretcher block are used in straight wall sections. Corner block have one flat end to create attractive walls at corners. Block with special grooves can be laid to receive window (**sash block**) and door (**jamb block**) parts so the openings are attractive and secure (Figure 40-19).

Estimating Block

As stated before, standard block are 8 inches or $\frac{2}{3}$ foot high and 16 inches or $1\frac{1}{3}$ foot long when they are laid

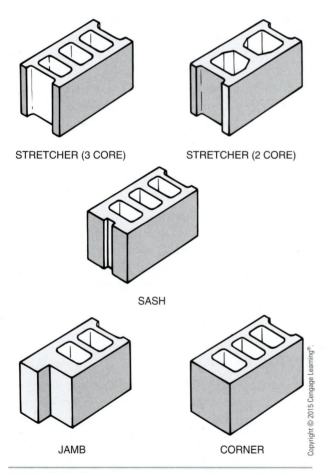

STRETCHER (3 CORE)   STRETCHER (2 CORE)

SASH

JAMB   CORNER

Copyright © 2015 Cengage Learning®.

FIGURE 40-19 Stretcher block are used in straight wall sections; sash block hold windows; jamb block hold doors; corner block are used wherever the ends of block are in view.

with a $\frac{3}{8}$-inch mortar joint. One foot, or 12 inches, is $\frac{3}{4}$ the length of one block. Therefore, when estimating the number of block needed for a job, the length of the wall in feet can simply be multiplied by $\frac{3}{4}$. This gives the number of block needed for one row, which is called a **course**. Similarly, 1 foot is $\frac{12}{8}$ or $\frac{3}{2}$ of the height of a block. Therefore, the height of the wall can be multiplied by $\frac{3}{2}$ to determine the number of courses needed. The number of block per course is then multiplied by the number of courses to obtain the number of block needed for the wall.

For example, to estimate the number of block needed to build a wall 12 feet long and 8 feet high, the procedure is as follows:

12' (length in feet) × $\frac{3}{4}$ = 9 block per course

8' (height in feet) × $\frac{3}{2}$ = 12 courses

9 (block per course) × 12 (courses) = 108 block needed for the wall

When estimating block for buildings, this procedure is used for each of the four walls. The estimate is then decreased by the area of the doors and windows.

> **NOTE**
>
> When planning a building, it is important that the dimensions between corners, windows, and doors utilize full- or half-length block. Otherwise, large numbers of block must be cut, a procedure that is expensive and time consuming and that results in an unattractive appearance. Figure 40-20 has no cut block.

Constructing Footers

A footer or footing is a continuous slab of concrete that provides a solid, level foundation for block and brick walls. Footers should be at least as deep as the wall is wide (Figure 40-21). A wall that is 8 inches wide rests on a footer that is at least 8 inches deep. The width of the footer should be at least twice that of the wall (or more where there is poor soil support). Hence, the 8-inch wall would rest on a footer 16 inches wide. In localities where temperatures drop to 32°. For lower, footers must be placed below the frost line to prevent damage by freezing in cold weather. The **frost line** is the maximum depth that the soil freezes in cold weather.

When starting a building or wall, a trench is dug by backhoe or shovel. The bottom of the trench should

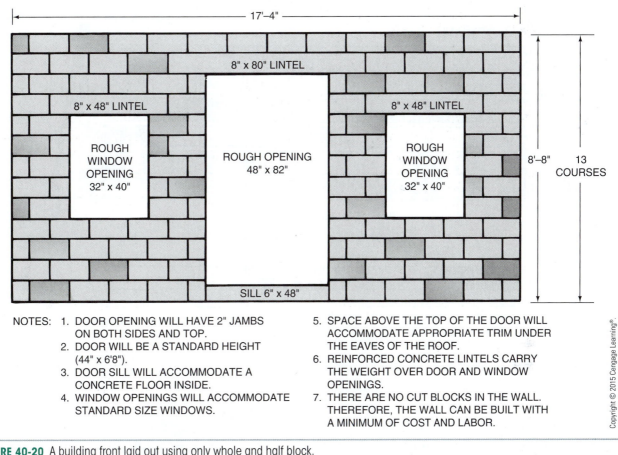

17'-4"

8" x 80" LINTEL

8" x 48" LINTEL

8" x 48" LINTEL

ROUGH
WINDOW
OPENING
32" x 40"

ROUGH OPENING
48" x 82"

ROUGH
WINDOW
OPENING
32" x 40"

8'-8"   13 COURSES

SILL 6" x 48"

NOTES:
1. DOOR OPENING WILL HAVE 2" JAMBS ON BOTH SIDES AND TOP.
2. DOOR WILL BE A STANDARD HEIGHT (44" x 6'8").
3. DOOR SILL WILL ACCOMMODATE A CONCRETE FLOOR INSIDE.
4. WINDOW OPENINGS WILL ACCOMMODATE STANDARD SIZE WINDOWS.
5. SPACE ABOVE THE TOP OF THE DOOR WILL ACCOMMODATE APPROPRIATE TRIM UNDER THE EAVES OF THE ROOF.
6. REINFORCED CONCRETE LINTELS CARRY THE WEIGHT OVER DOOR AND WINDOW OPENINGS.
7. THERE ARE NO CUT BLOCKS IN THE WALL. THEREFORE, THE WALL CAN BE BUILT WITH A MINIMUM OF COST AND LABOR.

Copyright © 2015 Cengage Learning®.

FIGURE 40-20 A building front laid out using only whole and half block.

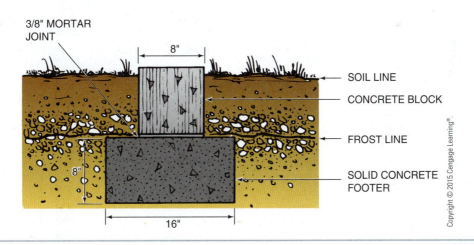

3/8" MORTAR JOINT

8"

SOIL LINE

CONCRETE BLOCK

FROST LINE

SOLID CONCRETE FOOTER

8"

16"

Copyright © 2015 Cengage Learning®.

FIGURE 40-21 The footer should be as thick as the wall resting on it is wide. The footer should also be at least twice as wide as the wall. Concrete footers are placed below the frost line to provide a solid base for masonry walls.

be the width of the footer and as deep as the intended bottom of the footer. Stakes are driven into the bottom of the trench so their tops are level with the top of the intended footer. Concrete is then poured and leveled to the height of the stakes. The stakes are pulled out and the holes filled after the concrete is leveled but is still workable.

Mixing Mortar

Before block can be laid, the mortar must be mixed to a working consistency. Premixed mortar can be purchased that requires only the addition of water. Mortar can also be made at the job site by mixing various combinations of portland cement, finishing lime, sand, and water. The

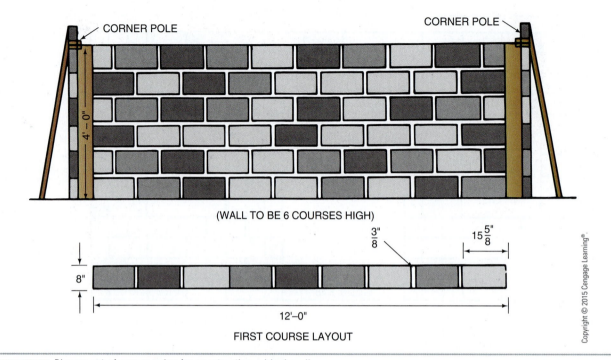

FIGURE 40-22 Placement of corner poles for constructing a block wall.

different combinations yield different characteristics, such as greater strength, higher waterproofness, and greater ease of handling. Unfortunately, no combination has all the ideal qualities.

Small batches of mortar may be mixed with a hoe in a mortar pan or wheelbarrow. Jobs requiring more than a wheelbarrow or two of mortar can be mixed more thoroughly and efficiently in a motor- or engine-driven mixer.

Block must be laid with mortar of the correct wetness. If the mortar is too stiff, it will not bond tightly to the block. If it is too thin, it will be squeezed out of the joint by the weight of the block. In such a case, the joint will be less than ⅜ inch thick and the block must be relaid.

Laying Block

To ensure that block are laid straight and plumb, usually a corner pole is set at each corner of the building (Figure 40-22). A **corner pole** is a straight piece of wood or metal held plumb by diagonal supports. It is used to support the lines to which the wall is built. This process is described further in the procedure that follows.

An alternate method is to lay block corners using a level to guide the construction. Since the level has bubble vials placed in different positions, it can be used to plumb the block as well as to level them. The corners

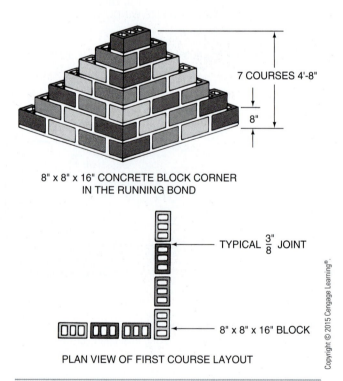

FIGURE 40-23 Construction of corners for block walls.

are built, and then the areas between them are filled with courses of stretcher block using a line to keep the courses straight (Figure 40-23).

The following procedure for laying block is also illustrated in Figure 40-24.

1. SPREAD THE MORTAR BED FOR THE FIRST COURSE.

2. LEVEL THE BLOCK.

3. APPLY MORTAR TO THE EARS OF THE STRETCHER BLOCK. NOTE THE DOWNWARD WIPING OF THE TROWEL.

4. LAY THE STRETCHER BLOCKS AND CHECK THE FIRST COURSE TO BE SURE IT IS LEVEL.

5. CHECK THE FIRST COURSE TO BE SURE IT IS PLUMB.

6. CHECK THE FIRST COURSE TO SEE THAT THE BLOCKS FORM AN EVEN, STRAIGHT LINE.

7. APPLY THE MORTAR BED FOR THE SECOND COURSE.

FIGURE 40-24 Procedure for laying block.

8. LAY REINFORCEMENT WIRE IN
THE MORTAR BED.

9. LAY BLOCK TO THE LINE TO ENSURE THE COURSES
ARE STRAIGHT. THE LINE IS POSITIONED TO BE
LEVEL ALONG THE TOP EDGE OF THE BLOCK.

10. REMOVE EXCESS MORTAR.

11. CUT BLOCK BY MAKING A GROOVE WITH
MODERATE BLOWS OF MASON'S HAMMER.

12. FOLLOW BY A SHARP RAP ON THE
EDGE OF THE WEB.

13. MEASURE A CORNER TO BE SURE
THAT EACH COURSE IS 8 INCHES HIGH.

FIGURE 40-24 (Continued)

14. FINISH A FLUSH OR SMOOTH JOINT WITH A PIECE OF BROKEN BLOCK.

15. STRIKE A JOINT WITH A SLED RUNNER JOINTER.

FIGURE 40-24 *(Continued)*

Procedure

1. Spread a layer of mortar, called a mortar bed, on the footer.

2. Position the block on the mortar bed so that its outside corner is exactly where the outside corner of the wall should be. Level the block by first placing the level across the block and then lengthwise along the block.

3. Turn several stretcher block on end and apply mortar to the ears with a wiping or swiping stroke of the trowel.

4. Lay several stretcher block in place by working away from the end or corner block.

5. Use the end of the trowel handle to tap the block until each block is plumb.

6. Make sure the course is in a straight line.

7. Apply a mortar bed on top of the first course in preparation for the second course.

8. If extra strength is needed in the wall, install reinforcement in the mortar bed.

9. As the block laying progresses, cut off excess mortar with the trowel.

10. Use a line to keep the courses straight. The line is positioned to be level along the top of the block.

11. When a block must be cut, use a mason's hammer and make multiple strikes along the line of cut.

12. To complete the cut, make one sharp strike on the web.

13. Check the height to be sure each new course is an additional 8 inches high.

14. After the mortar dries and hardens slightly, finish the joint by rubbing it with a broken piece of block.

15. If a joint other than a flush joint is desired, use a jointer to compress the mortar and create a watertight joint (Figure 40-25). Tools are available to create joints that are concave, V-shaped, flush, or raked.

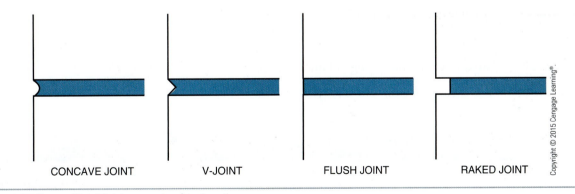

CONCAVE JOINT   V-JOINT   FLUSH JOINT   RAKED JOINT

FIGURE 40-25 Types of joint finishes used on block walls.

Walls that are above ground level may be damaged by wind pressure. This is especially true before the mortar cures. Therefore, walls higher than 8 feet should be braced on both sides at 8- to 10-foot intervals (Figure 40-26). The braces are removed after the wall is strengthened by curing and by the installation of floor or roof parts.

Mortar joints must be protected from freezing for several days until they are cured. This is generally done by covering the wall with plastic or canvas when temperatures are below 32°F.

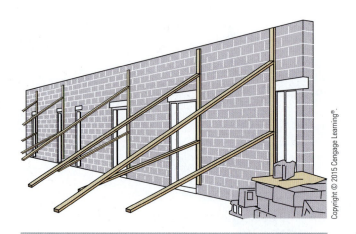

FIGURE 40-26 The temporary bracing of this wall will prevent wind damage until additional building parts provide adequate support.

SUMMARY

Concrete and masonry construction has been used for thousands of years. It is one of the most durable building materials used. Strength and durability depend on using the proper ratio of components and using the correct procedures for mixing, applying, and curing. Following the basic rules of concrete and masonry construction will result in a structure capable of lasting thousands of years.

Student Activities

1. Define the Terms to Know in this unit.

2. Examine samples of sand and gravel to observe the variety of particle sizes.

3. Conduct a clay and silt test on a sample of sand.

4. Prepare the forms for a small concrete project. Mix and pour the concrete. Finish and cure the concrete.

5. Visit a ready-mix concrete plant to observe the equipment and techniques.

6. Help pour concrete for a project using ready-mix concrete.

7. Pour a footer.

8. Estimate the block needed for a wall or building.

9. Construct a small block wall.

10. Protect concrete from freezing and drying out until it is cured.

11. Construct a wooden float for use on future concrete projects.

Relevant Web Sites

Cement & Concrete Basics
www.cement.org/basics/concretebasics_concretebasics.asp

The History of Concrete and Cement
http://inventors.about.com/library/inventors/blconcrete.htm

Historical Timeline of Concrete
www.auburn.edu/academic/architecture/bsc/classes/bsc314/timeline/timeline.htm

Concrete Calculator
www.calculator.net/concrete-calculator.html

Self-Evaluation

A. Multiple Choice. Select the best answer.

1. Concrete is a mixture of sand, gravel, water, and
   a. clay cement
   b. finishing cement
   c. finishing lime
   d. portland cement

2. Mortar is a mixture of portland cement, sand, water, and
   a. aggregate
   b. clay
   c. finishing cement
   d. finishing lime

3. The strength and durability of concrete are dependent upon the
   a. purity of water
   b. proportion of stone particles by size
   c. type of cement
   d. all of these

4. A concrete slab that is 6 inches deep by 10 feet wide by 20 feet long will contain
   a. 2 cubic yards of concrete
   b. 2.5 cubic yards of concrete
   c. 3.7 cubic yards of concrete
   d. 5.0 cubic yards of concrete

5. Forms for concrete slabs are usually made of
   a. 2" × 4"
   b. 2" × 8"
   c. 2" × 10"
   d. none of these

6. To prevent forms from sticking to the concrete, they are treated with
   a. fat
   b. oil
   c. paint
   d. wax

7. Concrete is reinforced with
   a. air bubbles
   b. aluminum wire
   c. steel bars
   d. wood fibers

8. Concrete is cured by
   a. covering with plastic
   b. protecting from wind
   c. sprinkling with water
   d. all of these

9. A standard-sized block when laid will cover an area
   a. 8 inches high by 16 inches long
   b. 4 inches high by 16 inches long
   c. 8 inches high by 12 inches long
   d. none of these

10. Courses of block are laid in a straight line by using a
    a. center point
    b. line
    c. plumb bob
    d. sighting tool

B. **Matching.** Match the terms in column I with those in column II.

Column I

1. sand
2. gravel
3. 1:2:3
4. air
5. drying
6. footing
7. raked joint
8. flat end
9. ears
10. flush joint

Column II

a. course aggregate
b. cement/sand/gravel
c. curing
d. foundation for walls
e. jointing tool
f. fine aggregate
g. entrained
h. corner block
i. stretcher block
j. broken block

C. **Completion.** Fill in the blanks with the word or words that make the following statements correct.

1. A quart jar with 2 inches of sand for concrete mixtures should not yield more than a _____ inch layer of silt.

2. The amount of water added to a mix is important because _____.

3. A concrete mixture with uniform color and consistency and the correct proportion of ingredients is known as a _____.

4. A block wall is plumbed with a _____.

5. Concrete that is already mixed when it arrives at the job site is called _____.

6. Moisture is prevented from entering or passing through a concrete slab by installing a _____ _____.

7. Heat loss through concrete floors may be reduced by installing _____.

D. **Brief Answer.** Briefly answer the following questions.

1. When constructing forms, how far apart should stakes be placed?

2. When are construction joints needed? What is a control joint and what is its purpose?

3. As concrete is poured over wire fabric, what should one do to ensure maximum reinforcement?

4. During the curing of concrete, it must be protected from what? How long should coverings, insulating material, and wooden forms be left in place?

5. What are the size requirements of a footer?

Section 14

AGRICULTURAL STRUCTURES

UNIT 41
- Planning and Constructing Agricultural Structures

UNIT 42
- Aquaculture, Greenhouse, and Hydroponics Structures

UNIT 43
- Fence Design and Construction

UNIT 41

Planning and Constructing Agricultural Structures

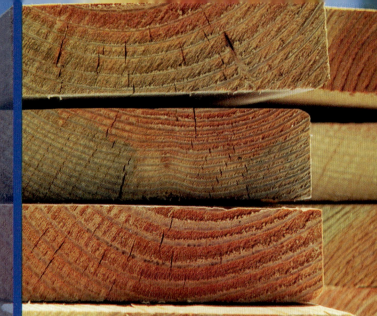

Objective

To plan and construct agricultural structures.

Competencies to be developed

After studying this unit, you should be able to:

- Discuss appropriate considerations for site planning.
- Determine storage space requirements for farm machinery.
- Identify major types of buildings used in agricultural settings.
- Name major building parts.
- Describe special features used to make buildings waterproof and wind-resistant.
- Construct agricultural buildings.
- Insulate agriculture structures.
- Tie basic knots and hitches.

Materials List

- Drawing supplies
- Heavy paper for templates
- Reference material on agricultural buildings
- Materials for model agricultural building
- Rope

Terms to Know

- rafter
- girder
- truss
- pole buildings
- post-and-girder
- post-frame
- rigid-frame
- plywood
- veneer
- subfloor
- sheathing
- interior grade
- exterior grade
- oriented strand board (OSB)
- pressure-treated
- acid copper chromate (ACC)
- ammoniacal copper arsenate (ACA)
- chromated copper arsenate (CCA)
- ridge
- footing pad
- girts
- purlins
- gusset

continued

Buildings of appropriate type, size, and function are an important asset to any business. Agricultural businesses use buildings for protecting field machinery, storing crops, keeping animals and animal products, milking cows, selling commodities, processing crops and animal products, manufacturing commodities, and numerous other activities (Figure 41-1).

PLANNING AGRICULTURAL BUILDINGS

Planning saves time and money. Careful consideration is needed when planning buildings for farm, ranch, horticulture, hydroponics, agribusiness sales, aquaculture, or any other agricultural enterprise. Building size, type, design, and placement are all important.

Inadequate buildings can result in crop loss, machinery deterioration, inadequate livestock production, loss of food and related commodities, human discomfort, and wasted energy.

Farmstead Layout and Building Site Selection

Generally, one cannot start from the beginning and build a farmstead. However, as one adds buildings, it is useful to apply some principles of efficient farmstead layout. Some of these are:

- Place the farmstead in a well-drained area.
- Since high winds come mostly from the northwest, plant a windbreak on the northwest side of the farmstead.
- Place the electrical meter pole so it will be as close as possible to all buildings that require a lot of electrical power.
- Place the livestock facilities downwind from the house to reduce livestock odors around the house.
- Face buildings to the south or east for maximum heat and light in the winter and shade in the summer.

© JorgHackemann/Shutterstock.com

FIGURE 41-1 A modern farmstead.

- Position all buildings so they can be enlarged or expanded.
- Provide a circle to permit traffic to get in and out easily.
- Hand-surface or gravel the main traffic areas.
- Provide proper drainage away from each building to avoid polluting streams and wells (Figure 41-2).

As many as possible of these principles should be incorporated into building plans. An efficient farmstead is attractive as well as productive (Figure 41-3).

Building Use and Size

Most buildings are large, expensive, and specialized. This means professional help is needed for planning the buildings, and probably constructing them. For instance, a milking parlor and free-stall operation must be very carefully planned to be efficient and cost-effective (Figure 41-4).

However, some buildings can be planned adequately by the owner or manager and family. Machine storage can be adequately planned with guides on space requirements and building details (Figures 41-5 and 41-6). When machinery storage is built, adequate facilities for machinery and farmstead maintenance should be included (Figure 41-7).

Some buildings are relatively simple in design and construction. Some buildings such as calf barns, sheep and lambing sheds, horse barns, and roadside stands

DO NOT BUILD IN A HOLE!

THE BEST SITE IS REASONABLY HIGH AND WELL DRAINED. DRAINAGE CAN ALSO BE IMPROVED BY GRADING, DITCHING, AND/OR INSTALLING SUBSURFACE DRAINS.

SLOPE THE GROUND AWAY FROM THE BUILDING.

DO NOT DRAIN DIRECTLY INTO WATERWAYS.

DIVERSION TERRACE, 50' OR MORE AWAY FROM THE BUILDING.

SLOPE 1' IN 30'

INTERCEPTOR TRENCH

SEEPAGE FROM HILL

FIGURE 41-2 Building sites should be well drained, but not pollute water sources.

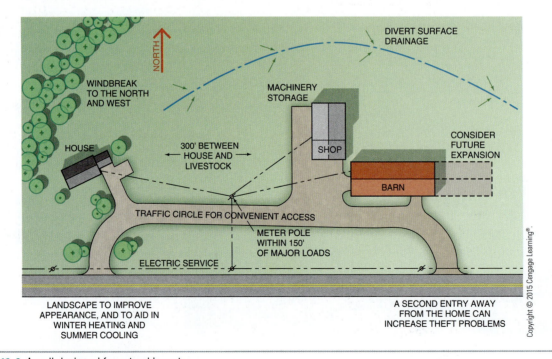

NORTH

DIVERT SURFACE DRAINAGE

WINDBREAK TO THE NORTH AND WEST

MACHINERY STORAGE

HOUSE

SHOP

CONSIDER FUTURE EXPANSION

300' BETWEEN HOUSE AND LIVESTOCK

BARN

TRAFFIC CIRCLE FOR CONVENIENT ACCESS

METER POLE WITHIN 150' OF MAJOR LOADS

ELECTRIC SERVICE

LANDSCAPE TO IMPROVE APPEARANCE, AND TO AID IN WINTER HEATING AND SUMMER COOLING

A SECOND ENTRY AWAY FROM THE HOME CAN INCREASE THEFT PROBLEMS

FIGURE 41-3 A well-designed farmstead layout.

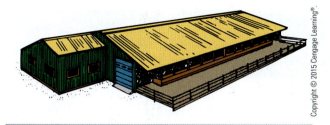

FIGURE 41-4 A free-stall dairy operation.

Copyright © 2015 Cengage Learning®

are constructed by the farm family. In such cases, plans should be obtained from the state agricultural college or USDA (Figure 41-8).

Building Types and Shapes

Buildings can be built in almost any shape, but certain shapes have advantages over others. Similarly, some shapes are easier and less expensive to build. For instance, straight gable-type roofs are less expensive to build than hip roofs. For small buildings, shed roofs are less expensive than gable roofs. Still, each roof type has its place for certain uses (Figure 41-9).

Types of Trusses. In the past, buildings have been limited in width, without interior posts, due to the limited strength of rafters and girders. A **rafter** is a single timber that supports a section of the roof. A **girder** is a timber that carries the weight of floors and interior walls. However, engineers have designed large, lightweight trusses capable of spanning over 100 feet without posts. A **truss** is a rigid framework (Figure 41-10).

Types of Construction. Buildings may be classified by their construction design. For instance, buildings are known as **pole buildings** if they are supported by poles erected in the soil. If they rest upon posts and rely on heavy girders to carry their sides and roof, they are called **post-and-girder** or **post-frame** buildings. Buildings that have heavy steel frames are called **rigid-frame** buildings.

Types of Coverings. Buildings are also classified according to their coverings. They may be known as steel-, aluminum-, or wood-sided buildings.

CONSTRUCTING AGRICULTURAL BUILDINGS

Reference is made to the previous units in this text. Most of the earlier units provide information that is helpful in constructing buildings. Therefore, the reader should be prepared to review previous units if needed.

BUILDING MATERIALS

Lumber. Construction lumber comes in standard sizes ranging from 1" to 6" thick and from 2" to 12" wide. Lengths typically run from 6" to 16" and longer. It is important to remember that thickness and width are stated as nominal dimensions, but when lumber is dressed or planed, the actual thickness is ¼" less than nominal thickness and width is ½" less than nominal width. Actual length is the same as nominal length. The dimensions of frequently used lumber are displayed in Table 11 in Appendix B. There are also many other tables with information useful for constructing buildings in Appendix B, as well as in this unit.

Plywood. **Plywood** is made of **veneer** or thin sheets of wood glued with alternate layers perpendicular to each other. It is used extensively for subfloors, wall sheathing, and roof sheathing. **Subfloor** refers to the first layer of flooring. **Sheathing** refers to the first exterior layer of a wall or roof.

Plywood is graded according to the quality of the exterior layers as well as the glue's ability to tolerate moisture. If the glue weakens in dampness, the plywood is classified as **interior grade**. If the plywood can remain strong and usable when exposed to weather, it is classified as **exterior grade**.

The top and bottom layers of a sheet of plywood are classified with a letter from A to D. A letter designation is provided for both outside layers. The higher the letter, the better the grade. For instance, a piece of plywood with an AD grade has one side that is smooth and paintable and one side with knots or knotholes up to 2½ inches across grain (Figure 41-11). Plywood is sold in 4' × 8' sheets of ¼", ⅜", ½", ⅝", ¾", and 1" thicknesses.

Oriented Strand Board. Plywood is being replaced in many applications such as decking, walls, and subflooring by a relatively new product called **oriented strand board (OSB)**. OSB is made from small round logs that have been chipped into strands up to 6" long and 1" wide. The strands are oriented to give the most strength, combined with waterproof resin, and subjected to intense heat and pressure. The strands and resins are molded into sheets and trimmed to a standard 4' × 8' sheet.

Machine	Length (Ft.)	Width (Ft.)	Area (Sq. Ft.)	Height (Total if Above 8 Ft.)	Machine	Length (Ft.)	Width (Ft.)	Area (Sq. Ft.)	Height (Total if Above 8 Ft.)
Row-crop types,					Combine, one-man,				
1-plow	10½	5*	53		5-ft cut	19	10	190	10
2- to 3-plow	1½	6*	69		6-ft cut	21	11	231	10
4- to 5-plow	13	6½*	85		8-ft cut	24	11	264	11
6-plow	15	8*	112	12	12-ft cut	24	12	288	13
General-purpose types,					Combine, two-man, 14-ft. cut				
2- to 3-plow	11	6*	66		(with header removed)	22½	11	248	13
4- to 5-plow	12½	6½*	82		Combine, self-propelled,				
6-plow or larger	14	8*	112		7-ft cut	18	9	162	
Moldboard plow,					9-ft cut	18	10	180	12
2-bottom	6	3½	21		12-ft cut	23½	13½	317	13
3-bottom	9	5	45		14-ft cut	21	15½	326	13
4-bottom	12	6½	78		24-ft cut	30	24	720	13
5-bottom	15	8	120		Cotton picker,				
Disk plow,					1-row	19	10	190	13
2-bottom	11½	6½	75		2-row	20	11	220	13
3-bottom	15	8	120		Cotton stripper, tractor mounted				
4-bottom	17	10	170		1-row	12	2	24	10
Disk harrow, single (with end sections folded)					2-row	20	8	160	10
15-ft	5½	9	50		Cane loader	29	9	261	12
18-ft	6	10	60		Corn harvester	22	7	154	8½
21-ft	6	12	72		Ensilage harvester, 1-row	13	7	91	
Disk harrow, tandem,					Ensilage cutter & blower	12	5½	66	
7-ft	10½	7	74		Forage crop blower	10-13	6	60	
8-ft	11	8	88		Corn picker, drawn,				
9-ft	11	9	99		1 row	15	9½	143	8½
10-ft	1½	10	115		2 row	15	11½	173	8½
Lister,					Corn picker, mounted,				
2-row	9½	8½	81		2-row	20½	8½	174	8
4-row	11	16	176		(with elevator removed)	19	5½	105	
Corn cultivator,					Corn sheller	22	8	176	11
2-row	8	8	64		Peanut picker	17	6	102	
4-row	8½	12	102		Peanut shaker	6	6	36	
Field cultivator	10½	15	158		Potato digger,				
Field cultivator					1-row	17	5	85	
(quack digger) 8-ft	10	9½	95		2-row	18	8½	153	
Spring tooth harrow—					Tobacco harvester	16	9½	152	2-14
per section	5	5	25		Field forage harvester-hay, grass silage and corn silage (with delivery spout removed)	13½	8	108	
Spike tooth or smoothing harrow—lean against wall					**Haying Machinery**				
Rotary hoe	7	7½	53		Mower,				
Corrugated roller—8 ft	4	9½	38		6-ft	7½	6½	49	
Stalk cutter	4	6	24		7-ft	7½	7½	56	8½
Sub-soiler	9	4½	41		Side-delivery rake	13	11	143	
Planting and Seeding Machinery					Dump rake—10 foot	5	11½	58	
Grain drill,					Hay loader	13	8	104	10
12 × 6	7	8	56		Field hay chopper	12	9	108	9
20 × 6	7	13	91		Pickup baler	17	13	221	8
23 × 6	9	16	144		**Hauling Machinery**				
13 × 7	9	12½	113		Manure spreader	15½	6	93	
24 × 7 press drill	10	15½	155		Tractor trailer	18	7	126	
16 × 10 press drill	8½	15	128		Truck, 1 1/2-ton	21	8	168	8
Corn or Cotton planter,					Wagons:				
2-row (with hitch)	10	8	80		Wagon with box bed	15	6	90	
(without hitch)	5	8	40		Wagon with 14-ft hay rack	16	7-8	112	
4-row (with hitch)	12	14	168		Wagon and tractor—tractor backed over tongue	25	7-8	175	
(without hitch)	6	14	84		**Miscellaneous**				
Potato planter,					Lime spreader, 8-ft	4	10½	42	
1-row	7	4	28		Manure loader	14	6	84	
2-row	11	8	88		Sprayers:				
Transplanter	8	5	40		Orchard	8	6	48	
Harvesting Machinery					Weed	5	8	40	
**Grain windrower,					Cut off (buzz) saw	4	4	16	
6-ft	12	5	60		Rotary cutter				
8-ft	14	5	70		(field shredder)	6½	9½	62	
12-ft	17½	7½	131						

FIGURE 41-5 Approximate space requirements for farm machinery.

Source: American Association for Vocational Instructional Material

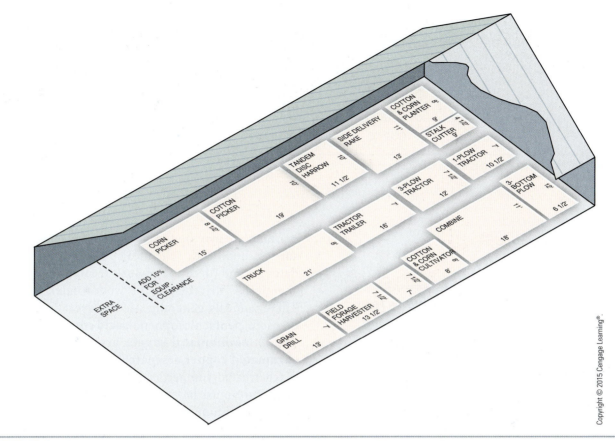

FIGURE 41-6 Using paper templates to plan machinery storage and determine building size.

FIGURE 41-7 Machinery shed with agricultural mechanics shop.

OSB is as strong as plywood and less expensive (Figure 41-12). It is now the standard material used in construction projects where inexpensive panels are needed such as roof decking. It is water- and moisture-resistant, and holds screws and nails very well.

Pressure-Treated Lumber. Most species of lumber will rot if exposed to weather for long periods of time. Wood is even more susceptible to rotting if it remains in contact with the earth.

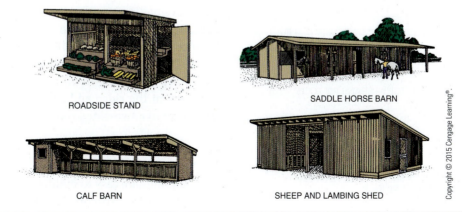

ROADSIDE STAND

CALF BARN

SADDLE HORSE BARN

SHEEP AND LAMBING SHED

FIGURE 41-8 Buildings frequently built by the farm family.

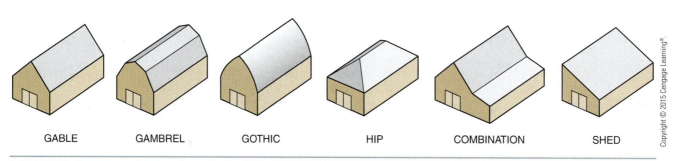

GABLE   GAMBREL   GOTHIC   HIP   COMBINATION   SHED

FIGURE 41-9 Common roof designs for agricultural buildings.

BELGIAN (DOUBLE OR TRIPLE FINK)—SPANS UP TO 80'.

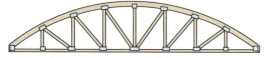

BOWSTRING—GENERALLY USED FOR SPANS FROM 40' TO 120'.

FINK (W)—A POPULAR, EFFICIENT DESIGN FOR SPANS OF 20' TO 50'.

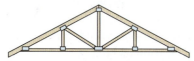

HOWE—SPANS 20' TO 50' FOR HEAVIER CEILING LOADS THAN CARRIED BY THE FINK TRUSS.

SCISSORS—USED WHERE HIGH CENTER CLEARANCE IS DESIRED. SPANS 20' TO 40'.

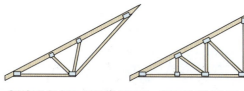

SINGLE SLOPE (MONO PITCH)—SPANS 20' TO 35'.

PRATT—SPANS 20' TO 60'. FOR USE WITH OR WITHOUT CEILINGS.

FIGURE 41-10 Common truss designs.

Lumber may be purchased that has been pressure-treated with chemicals to prevent insect damage and decay. **Pressure-treated** means the chemical is driven into the wood under pressure. The chemicals generally used to treat lumber are creosote, pentachlorophenol, **acid copper chromate (ACC)**, **ammoniacal copper arsenate (ACA)**, and **chromated copper arsenate (CCA)** (Figure 41-13). Since treated lumber is toxic to plants and animals, care must be exercised when choosing pressure-treated lumber for use in agricultural settings (Figures 41-14A and B).

Roofing and Siding. Steel and aluminum sheets are used extensively to cover agricultural buildings. Most are sold in 24" and wider widths and lengths of 8', 12', or 16'. Steel sheets are strong, but subject to rust. They are galvanized or coated with a layer of zinc equal to 1.25 ounces per square foot. However, better grades of sheet steel are coated with 2 ounces of zinc per square foot (Figure 41-15). Modern steel roofing sheets may also be coated with baked-on enamel. This gives a very durable finish.

Aluminum sheeting will not rust in normal weather, but it is expensive, thin, and easily damaged. Therefore, buildings to be covered with aluminum must have nailing boards with closer spacings than for steel. When using steel or aluminum, the manufacturer's instructions must be followed carefully as to type of nails, nail placement and spacing, overlapping of sheets, use of wicking to improve water tightness, and treatment at the ridge and gables. **Ridge** refers to the high point of a two-sided roof.

Pole Buildings

Pole buildings are easy to build, fast going up, economical, and flexible. To construct a pole building, a set of plans must be secured. In many locations the plans

Copyright © 2015 Cengage Learning®.

Veneer Grades	
Symbol	**Description**
A	Smooth, paintable. Not more than 18 neatly made repairs—boat, sled, or router type, and parallel to grain—permitted. May be used for natural finish in less demanding applications. Synthetic repairs permitted.
B	Solid surface. Shims, circular repair plugs, and tight knots to 1 inch across grain permitted. Some minor splits permitted. Synthetic repairs permitted.
C Plugged	Improved C veneer, with splits limited to ⅛ inch width and knotholes and borer holes limited to ¼ × ½ inch. Allows some broken grain. Synthetic repairs permitted.
C	Tight knots to 1½ inch. Knotholes to 1 inch across grain and some to 1½ inch if total width of knots and knotholes is within specified limits. Synthetic or wood repairs. Discoloration and sanding defects that do not impair strength permitted. Limited splits allowed. Stitching permitted.
D	Knots and knotholes to 2½ inch width across grain and ½ inch larger within specified limits. Limited splits allowed. Stitching permitted. Limited to interior, exposure 1, and exposure 2 panels.

American Plywood Association, *Grades and Specifications,* Page 5, August 1987.

FIGURE 41-11 Grades of plywood.

Courtesy of Russell & Herder Advertisers

FIGURE 41-12 Oriented strand board (OSB) is strong and inexpensive. It is used for a lot of different construction uses.

have to be approved to ensure that they meet local building codes. After carefully laying out the building dimensions and digging the holes, pressure-treated poles are set in holes with footing pads. A **footing pad** is a concrete slab or flat stone in the bottom of a hole to prevent a pole from sinking.

Girders are bolted to the poles and rafters are added. Then **girts**, or horizontal side nailers, and pressure-treated skirt planking are added. **Purlins**, or longitudinal roof nailers, are added, and finally the roofing and siding are installed (Figure 41-16).

Care must be taken to brace the building carefully. Wood braces and plywood or metal gussets are used for this purpose. A **gusset** is a piece of wood or metal used to reinforce a joint (Figure 41-17). Windows and skylights may be installed in the structure (Figure 41-18).

Post-Frame Buildings

Post-frame buildings may be built with posts extending to the floor level or to the square. The **square** is the top of the wall. If posts go to the square, it is easy to add a concrete floor and truss roof. The walls and roof may then be insulated and siding added (Figure 41-19).

If a building has a wooden floor, the posts are leveled and cut off at 1' to 3' above the ground. Heavy girders are placed on the posts and floor joists are nailed to them. A **joist** is the timber used to support the flooring. Construction of the walls proceeds after the subfloor is installed. Trusses are generally used to give the building stability and support the roof.

Rigid-Frame Buildings

The use of formed steel for framing parts is called rigid-frame construction. It permits very long roof spans with relatively flat roofs. Rigid-frame buildings must be set on and bolted to extremely strong concrete posts or floors. The downward pressure of roof loads tends to exert an outward pressure on the base of the steel frame.

The frame, sheet coverings, and all accessories are prepunched and ready for assembly upon arrival at the site. Erection is fast and easy for an experienced crew. Steel- and aluminum-covered buildings can be equipped with various styles of doors, windows, ventilators, skylights, gutters, spouting, and other special features. They can be insulated and finished with interior walls for office, manufacturing, warehousing, processing, storage, or other uses (Figure 41-20).

Copyright © 2015 Cengage Learning®.

	Creosote and Creosote Solutions	Pentachloropheno	Acid Copper Chromate (ACC)	Ammoniacal Copper Arsenate (ACA)	Chromated Copper Arsenate (CCA)
Round Poles as structural members					
Southern Pine, Ponderosa Pine	7.5	0.38	NR	0.6	0.6
Red Pine	10.5	0.53	NR	0.6	0.6
Coastal Douglas-Fir	9.0	0.45	NR	0.6	0.6
Jack Pine, Lodgepole Pine	12.0	0.60	NR	0.6	0.6
Western Red Cedar, Western Larch, Inter Mountain Douglas-Fir	16.0	0.80	NR	0.6	0.6
Posts					
Sawn four sides as structural members; all softwood species	12.0	0.60	NR	0.60	0.60
Lumber					
All softwood species in contact with soil	10.0	0.5	0.62	0.40	0.40
Not in contact with soil	8.0	0.4	0.25	0.25	0.25
Plywood					
In contact with soil	10.0	0.5	0.62	0.40	0.40
Not in contact with soil	8.0	0.4	0.25	0.25	0.25
Foundation	NR	NR	NR	0.60	0.60
Greenhouses					
Above ground	NR	NR	NR	0.25	0.25
Soil contact	NR	NR	NR	0.40	0.40
Structural posts	NR	NR	NR	0.60	0.60

1 As recommended by the American Wood Preservers Association.
Standard C 16-82.
NR—Not Recommended

Source: American Plywood Association

FIGURE 41-13 Recommended minimum preservative retention (in pounds per cubic feet) for wood use on farms.

Woodframe Buildings

Woodframe buildings have walls constructed of vertical 2" × 4" boards placed 16" on center (OC). Homes, garages, and other one- to three-story buildings with finished interior walls are generally of the woodframe type. These buildings have floors nailed to the floor joists that are carried by the concrete or block walls or **foundation** and one or more centerline girders. They rely on solid masonry foundations and one or more posts to bear the weight of the structure, provide anchorage against wind, and prevent moisture from entering below grade. **Below grade** means below the surface level of the surrounding land. Interior partitions are made of studs and drywall or paneling. They help support upper stories. Woodframe buildings may also be built on concrete slabs in well-drained areas or on posts or masonry pillars in wet areas.

Laying Out a Foundation. The exact location of corners is necessary for all buildings. In addition, level lines must be stretched along the outer edges of the foundation or posts to establish their placement and elevations. **Elevation** means relative height and position from sea level. Such lines are attached to **batter boards**, which are level boards placed 8' to 12' beyond the foundation. The procedure for laying out a building is provided in Figure 41-21.

A Pressure-treated wood is labeled according to use.

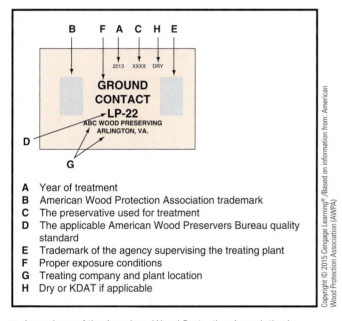

A Year of treatment
B American Wood Protection Association trademark
C The preservative used for treatment
D The applicable American Wood Preservers Bureau quality standard
E Trademark of the agency supervising the treating plant
F Proper exposure conditions
G Treating company and plant location
H Dry or KDAT if applicable

B A mock-up of the American Wood Protection Association's stamp and its interpretation.

FIGURE 41-14 Labeling used for pressure treated wood.

Excavation. To **excavate** is to dig or remove earth. After a building is laid out, a backhoe and front-end loader are used to dig the trenches for the footer and foundation and remove the soil from the basement area. A **footer** or footing is a continuous block of concrete under a foundation.

The excavation is made several feet beyond the outside of the layout lines and as deep as the basement subgrade. **Subgrade** means the level where stone begins for a concrete floor or topsoil begins when grading off land. After the basement excavation is completed, the trenches are dug for the footer. The footer should be at least as deep as the foundation width and 1½ times as wide. However, engineering specifications and local building codes should be consulted (Figure 41-22).

Footers and Foundations. Footers should be poured with concrete made to correct specifications and reinforcement details. The foundation may be poured with solid concrete or laid up with masonry block. Block laying is discussed in detail in Unit 40 ("Concrete and Masonry").

Walls must be specifically treated to be waterproof or water may seep through. The joint where the footer and foundation meet and the foundation wall must be plastered with waterproof cement or special masonry sealer and topped with an asphalt finish. A tile or PVC pipe drainage system should be installed around the total foundation in a stone bed with stone cover above the footing. The system must have a suitable outlet or pump to carry away water that gets against the foundation (Figure 41-23).

Walls and Floors. There are several ways of integrating the floors and walls in a woodframe house. With the modern, platform, and Western methods of building, floor structure and the subflooring are constructed before the wall of the next story is built. This provides a safe working area to construct the walls. Wall and partition framing consisting of 2" × 4" studs with 2" × 4" plates at the top and bottom are laid out and nailed together on the floor. The wall unit is then tilted up and into place, braced, and nailed. After the wall framing is in place, exterior sheathing and siding or brick veneer is installed (Figure 41-24).

When a two- or three-story building is desired, one or more floors and wall systems are built on the first story. Figure 41-25 shows some additional details of construction. Trusses may be substituted for rafters and ceiling joists. Solid plywood sheathing is generally used instead of diagonal sheathing and eliminates the need for diagonal bracing. Also, plywood is frequently used instead of diagonal subflooring. Figure 41-26 shows detailed treatment of eaves of the building. The **eave** is the overhang.

Source: American Association for Vocational Instructional Material

Roofing materials	Probable life expectancy due to weathering		Fire resistance	Hail resistance	Resistance to heat from sun	Resistance to strong wind	Maintenance	Minimum roof slope with ordinary lap (in. per ft.)	Approximate wt. per sq. ft. (100 sq. ft.)** (lbs.)	Corrosion resistance to salt and industrial fumes	Relative cost***
	Years until first rust (approx.) (years)	Probable life before roof rusts through* (years)									
Galvanized Steel (all types)			Highly resistant	Good	Excellent. If painted white is effective for reflecting heat.	Good if well fastened.	If painted with metallic zinc paint when rust starts to appear, life is increased.	Corrugated: 4; V-crimp: 3	80 / 85 / 100	Poor	Low to medium
Gage 29 .0172 — with coating 1.25 / 2.00	15 / 23	40 / 50+									
Gage 28 .0187 1.25 / Gage 26 .0217 2.00	15 / 23	45 / 50+									
Aluminum Thickness generally available (inches): .019, .0215, .024, .032	50+ years		Good	Fair with .019 inch unless used with solid decking. Good with heavier sheets.	Excellent. Also excellent for heat reflection.	Fair. Good if well fastened	.019 inch punctures easily. Requires no painting. Sheets must be well nailed.	4	29 / 36.5 / 55.2	Poor	Medium
Wood Shingles	8–30 yrs.		Burn readily	Good	Fair	Fair	May damage from strong winds.	6	200	Good	High
Asphalt Shingles	7–25 yrs. Less durable in warmer regions.		Will burn after heat builds up.	Fair	Fair	Fair. Good if well laid.	May damage from strong winds or hail.	5	250 to 300	Good	Low to medium
Asphalt Roll Roofing: Mineral—surfaced, selvage edge	About the same as asphalt shingles		Will burn	Fair	Fair	Fair	May damage from strong winds or hail.	2- to 4-inch lap: 4	100 to 140	Good	Low
Smooth—surfaced	Much less than mineral-surfaced shingles or selvage edge		Will burn	Fair	Fair	Fair	May damage from strong winds or hail.	2- to 4-inch lap: 4	65	Good	Low
Asbestos-Cement Shingles and Corrugated sheets	50+ yrs.		Good	Good	Excel.	Good	May be damaged from flying objects.	5 / 3	260 to 300 / 300	Good	Medium to high

*Tests under rural conditions are not complete. Estimates were derived by projecting rates of corrosion in rural areas and correlating the data with tests already completed under more corrosive conditions. Farms located near industrial areas, or near saltwater, will get much less service. Painting with zinc paint when rust first appears will prolong the life of the steel.

**The different types of roofing vary in weight per square foot according to the thickness and density of the material.

***Cost for roofing material only. Costs will vary with the weights of roofing and also with the location in which it is purchased.

****Standard thicknesses specified by "United States Galvanized Sheet Gauge." Sheets imported from other countries may come in under other gauge standards that are not of the same thickness as the U.S. gauge.

FIGURE 41-15 Characteristics of roofing materials.

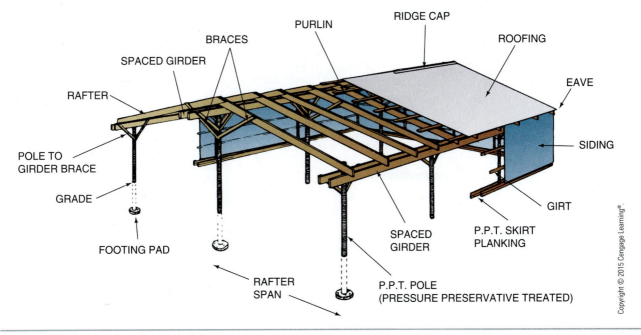

FIGURE 41-16 Components of a pole building.

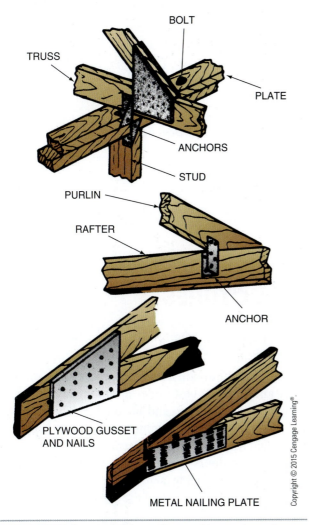

FIGURE 41-17 Securing girders, rafters, and trusses to posts or poles.

LAYING OUT STAIRS AND RAFTERS

Stairs. Stairs may be needed in original construction or for replacement purposes. When replacing old stairs, remove the trim and treads first. The **tread** is the part of the stair you step on. Remove the **riser** or back of the stair step. Finally, remove the **stair stringer** or **stair horse**, which is the structure that supports the

NOTE: LARGE FIELD EQUIPMENT MAY REQUIRE DOORWAYS 16' TO 24' WIDE, AND SMALL AIRPLANES MAY NEED 40' WIDE DOORS.

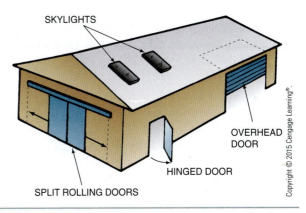

FIGURE 41-18 Doors, windows, skylights, and vents add to the utility of the building.

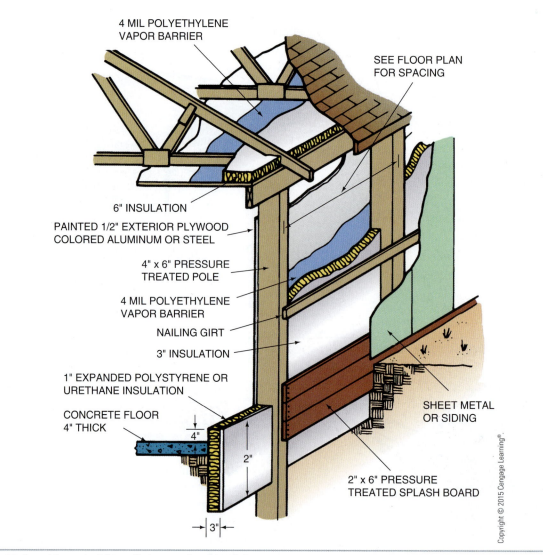

4 MIL POLYETHYLENE
VAPOR BARRIER

SEE FLOOR PLAN
FOR SPACING

6" INSULATION

PAINTED 1/2" EXTERIOR PLYWOOD
COLORED ALUMINUM OR STEEL

4" x 6" PRESSURE
TREATED POLE

4 MIL POLYETHYLENE
VAPOR BARRIER

NAILING GIRT

3" INSULATION

1" EXPANDED POLYSTYRENE OR
URETHANE INSULATION

CONCRETE FLOOR
4" THICK

4"

2"

3"

SHEET METAL
OR SIDING

2" x 6" PRESSURE
TREATED SPLASH BOARD

Copyright © 2015 Cengage Learning®.

FIGURE 41-19 A post-frame building with truss roof and insulation.

stairs. Be careful not to break the stair horse, and use it as the pattern for cutting replacement parts.

Horses for stairs in new construction must be laid out and sawed out of 2" × 10" or wider material. The framing square is used to lay out and mark the cuts for the bottom and top of the horse and each tread and riser along the way (Figure 41-27).

Rafters. There are many kinds of rafters. For straight roofs, trusses are now used in the construction of most new agricultural buildings. Rafter layout and cutting may be needed for small buildings or special roof shapes. The layout of a rafter is discussed in Figure 41-28.

INSULATING AGRICULTURE STRUCTURES

Energy conservation and moisture control are of major concern in most buildings. Workers' health, efficiency, and safety are generally improved by comfortable temperatures and humidity. Similarly, plants and animals grow and produce best under optimum conditions of temperature, humidity, and air quality. These are all easier to control when buildings have appropriate insulation and moisture or vapor barriers.

Concrete floors may be protected from moisture penetration by placing a moisture-proof membrane between the fill material and the concrete. The surface

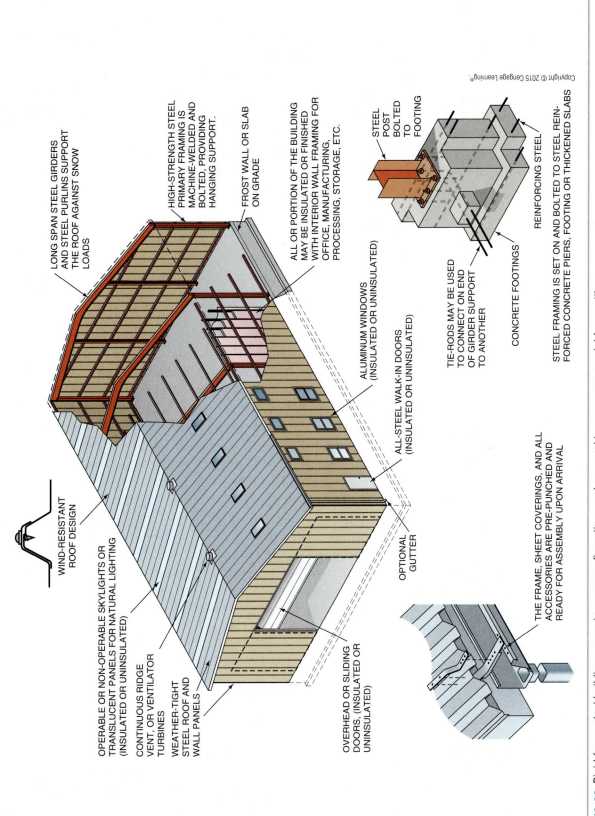

LONG SPAN STEEL GIRDERS AND STEEL PURLINS SUPPORT THE ROOF AGAINST SNOW LOADS

HIGH-STRENGTH STEEL PRIMARY FRAMING IS MACHINE-WELDED AND BOLTED, PROVIDING HANGING SUPPORT.

FROST WALL OR SLAB ON GRADE

ALL OR PORTION OF THE BUILDING MAY BE INSULATED OR FINISHED WITH INTERIOR WALL FRAMING FOR OFFICE, MANUFACTURING, PROCESSING, STORAGE, ETC.

ALUMINUM WINDOWS (INSULATED OR UNINSULATED)

ALL-STEEL WALK-IN DOORS (INSULATED OR UNINSULATED)

OPTIONAL GUTTER

OVERHEAD OR SLIDING DOORS, (INSULATED OR UNINSULATED)

WEATHER-TIGHT STEEL ROOF AND WALL PANELS

CONTINUOUS RIDGE VENT, OR VENTILATOR TURBINES

OPERABLE OR NON-OPERABLE SKYLIGHTS OR TRANSLUCENT PANELS FOR NATURAL LIGHTING (INSULATED OR UNINSULATED)

WIND-RESISTANT ROOF DESIGN

STEEL POST BOLTED TO FOOTING

TIE-RODS MAY BE USED TO CONNECT ON END OF GIRDER SUPPORT TO ANOTHER

CONCRETE FOOTINGS

REINFORCING STEEL

STEEL FRAMING IS SET ON AND BOLTED TO STEEL REIN-FORCED CONCRETE PIERS, FOOTING OR THICKENED SLABS

THE FRAME, SHEET COVERINGS, AND ALL ACCESSORIES ARE PRE-PUNCHED AND READY FOR ASSEMBLY UPON ARRIVAL

Copyright © 2015 Cengage Learning®.

FIGURE 41-20 Rigid-frame steel buildings come in many configurations and are used in many commercial farm settings.

Building Layout

1. Lay out the baseline at the front edge or side of the building by driving stakes in the ground (A&B below).

2. Create a baseline by driving a nail in the top of the corner stake (A below). This will serve as a reference point.

3. Lay out the building length along the baseline from the reference point (Stake A) and set corner Stake B. Drive a nail in the top of the stake at the exact length of the building and move it to achieve the proper distance.

4. To square the end wall, measure 18' along the base line from Stake B and set a temporary stake. The point 30' from this temporary stake and 24' from Stake B is perpendicular to the base line.

Set a temporary stake at this point. This is what is called the 3, 4, 5, rule. Any of the measurements that meet this ration will work. For example, 30', 40', and 50' can be used also.

5. Measure the width of the building along this line and set the third corner Stake C. Drive a nail in top of the stake at the exact width of building.

6. From Stake C measure the building length. From nail in Stake A measure building width. Square this corner by using the 3, 4, 5, rule for this corner as in step 4.

7. Check the layout to make sure it is square. This means that all corners are perpendicular or at 90 degrees. The diagonal measurements from B to D and A to C should be exactly the same. Adjust nails or stakes B, C, D as necessary.

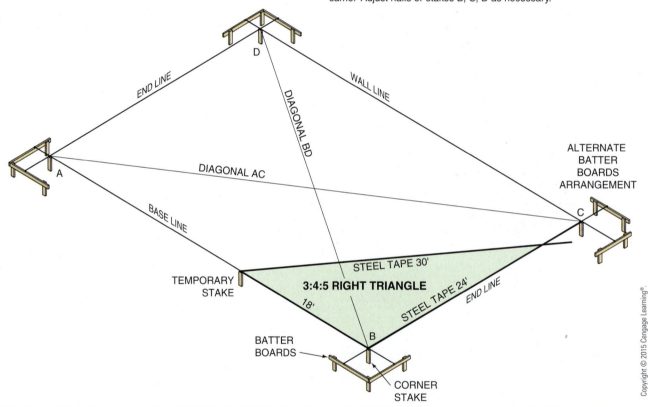

FIGURE 41-21 Building layout procedures.

of the fill must be relatively smooth to avoid puncturing the film or a thin coat of concrete is recommended over the fill. Four-mil polyethylene film is available in large sheets and makes a fast, low-cost, and effective moisture barrier. For small floor areas, 55-lb roll roofing material overlapped 6 inches with hot asphalt or bituminous cement in the laps makes a satisfactory vapor barrier. The vapor barrier should be carefully placed across the fill material and up the side walls before the concrete is placed.

Outside surfaces of foundations should be sealed with special masonry products that penetrate the pores and expand as they dry. Further protection against moisture and mechanical damage may be ensured by using a waterproof cement coating followed by a final coat of asphalt. The exteriors of masonry walls can be

Copyright © 2015 Cengage Learning®.

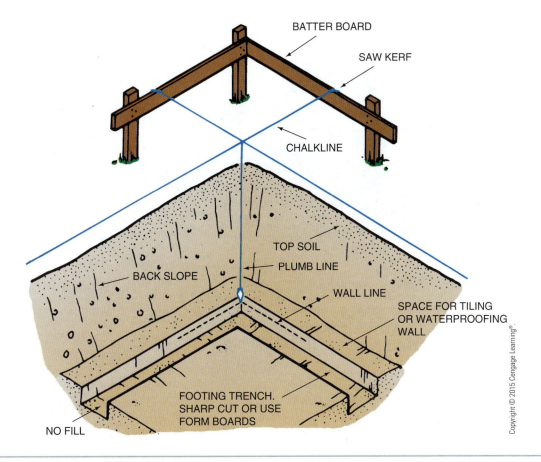

FIGURE 41-22 Excavation for a basement and footing.

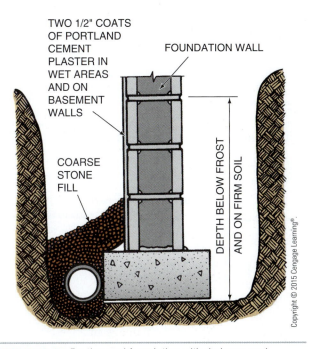

FIGURE 41-23 Footing and foundation with drainage and waterproofing.

protected from driving rains by appropriate waterproof paints.

In heated buildings, floors will be warmer and energy conservation better if solid insulation material is placed between the floor and the side walls or foundation. Further, if energy costs are high, consideration should be given to placing solid insulation under the slab.

Masonry and concrete materials have varying insulation values according to the material they contain. For instance, a heavy, high-density concrete block 8" wide may have a low insulation value or resistance (R) of only .98, but a light-aggregate block 8" wide may have an R value as high as 2.3. By filling the cores of such light aggregate block with appropriate insulation material, the R value of the wall may be increased to nearly 7.5. Materials typically used to fill masonry block cores or make concrete with high-insulation value include perlite, vermiculite, and cellulose (Figure 41-29).

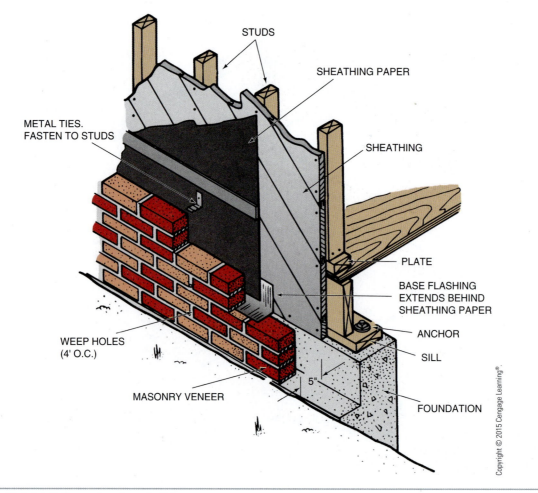

STUDS

SHEATHING PAPER

METAL TIES.
FASTEN TO STUDS

SHEATHING

PLATE

BASE FLASHING
EXTENDS BEHIND
SHEATHING PAPER

ANCHOR

SILL

WEEP HOLES
(4' O.C.)

5"

FOUNDATION

MASONRY VENEER

Copyright © 2015 Cengage Learning®

FIGURE 41-24 Wall detail of a modern frame building.

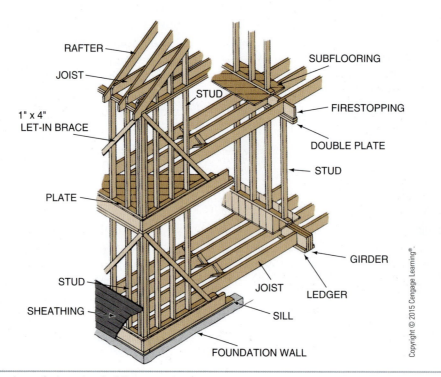

RAFTER

JOIST

SUBFLOORING

STUD

FIRESTOPPING

1" x 4"
LET-IN BRACE

DOUBLE PLATE

STUD

PLATE

GIRDER

STUD

JOIST

LEDGER

SHEATHING

SILL

FOUNDATION WALL

Copyright © 2015 Cengage Learning®

FIGURE 41-25 Framing members of a two-story building.

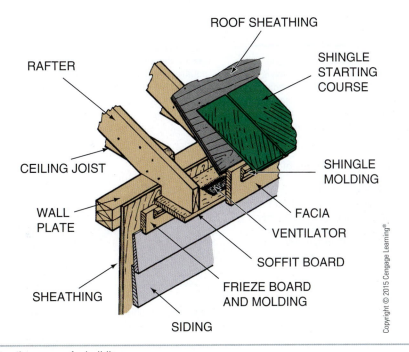

ROOF SHEATHING

RAFTER

SHINGLE
STARTING
COURSE

CEILING JOIST

SHINGLE
MOLDING

WALL
PLATE

FACIA

VENTILATOR

SOFFIT BOARD

FRIEZE BOARD
AND MOLDING

SHEATHING

SIDING

Copyright © 2015 Cengage Learning®.

FIGURE 41-26 Finishing the eaves of a building.

Exterior masonry or metal walls may also be insulated by building a 4" wood or metal framed wall inside and placing roll or batt insulation in the wall. Woodframe buildings typically have stud walls with 2" or 3" batt or roll insulation between 2" × 4" studs with an R value of 7 or 11, respectively, or 6" insulation between 2" × 6" wall studs with an R value of 19. Ceilings typically have a minimum of 6" insulation or more for an R value of 19 to 30. Popular insulation materials for this purpose include fiber, mineral, fiberglass, and slag (Figure 41-30).

Metal and pole buildings may be insulated with rigid insulating materials or fiberglass with a polyethylene vapor barrier. Special attention must be given to the high volume of moisture given off by livestock and appropriate fans must be installed to keep the humidity under control. Further, the ability of birds to pick apart and damage fiberglass and other soft insulating materials must be considered in buildings where birds cannot be excluded.

In all cases, a proper vapor barrier must be installed or the insulation may absorb moisture and lose its insulating ability. Moisture may also lead to the decay of wood structures and the rusting of steel components. Vapor barriers generally consist of asphalt-impregnated paper, aluminum-coated wood or paper products, or polyethylene film. These are generally installed on the warm side of walls, but should be installed according to local climatic conditions, building codes, and manufacturers' recommendations.

ROPE WORK IN AGRICULTURAL SETTINGS

Ropes are used for numerous purposes, such as to secure bundles, lift objects, drag materials, secure livestock, and rig boats. The practice of splicing is seldom used today, and rope halters for livestock are generally cheaper to buy than make. However, one should know how to tie knots that hold and yet can be untied.

Common Knots and Hitches

Square Knot. The same basic knots are useful in construction as for other agricultural settings. The granny knot is often used, but will pull out and is, therefore, unsafe. The **square knot** is easy to tie, easy to untie, and will not pull out under tension. It is the preferred knot for attaching ropes of the same size to each other. The **sheet bend** is good for fastening ropes of different sizes.

Laying Out Stairs

Although there are many ways to build stairways, only one method is discussed here to illustrate the principles involved. A staircase has a total rise and run, because it travels both horizontally and vertically; but it is also divided into a series of steps, each of which has a rise and run. The width of each tread is its run and the height of each riser makes up the rise of each step.

Determining the width of treads and height of risers

The "magic number" in stair building is 17. This is because the width of the tread and the height of the riser should add together to form 17 inches, or approximately this amount. (Some stair builders try to stay between $16\frac{1}{2}$ and $17\frac{1}{2}$ inches.) Any great departure from 10-inch treads and 7-inch risers is uncomfortable for the average person and may be dangerous.

In laying out a stairway, the total rise is usually a fixed figure. It is important to divide this distance into a number of steps of equal rise. The procedure usually followed is first to determine total rise in inches and divide this distance by 7 in order to determine how many steps there should be. This answer is determined only to the closest whole number. Then divide the total rise by this number and work it out to the closest $\frac{1}{16}$ of an inch.

Example: Total rise is 54 inches.
$\frac{54}{7} = 7\frac{5}{7}$ or 8 (probable number of steps).
$\frac{54"}{8} = 6\frac{6}{8}$, or $6\frac{3}{4}"$ (height of each riser)
Width of tread should be $17" - 6\frac{3}{4}" = 10\frac{1}{4}"$
(Can vary from $9\frac{3}{4}"$ to $10\frac{3}{4}"$)

Laying out the horses

The side members, which support the treads, are called "horses." There is usually one less tread than the number of risers as figured above. It may be desirable to draw a sketch and number the risers in order to avoid making an error in laying out the stair horses. Such a sketch is shown below.

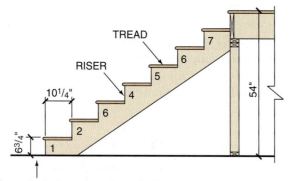

After calculating the number of steps and the height of each riser, draw a sketch and number the risers.

Place the square along the 2" x 10" with the amount of rise on the tongue and the width of tread on the blade of the square. It is more convenient to use clips or wooden fence to avoid the necessity of reading the square each time it is applied.

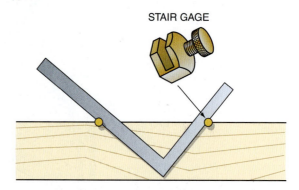

STAIR GAGE

Mark lines as shown in the sketch below where the pieces are to be cut out of the horses. Finally, make a correction at the lower end to shorten the first riser by an amount equal to the thickness of the tread. This amount is usually $1\frac{1}{4}"$ or $1\frac{1}{16}"$ depending upon whether nominal 2" or $1\frac{1}{4}"$ lumber is used. The one correction at the lower end takes care of the top step also as it lowers the entire horse by one tread thickness.

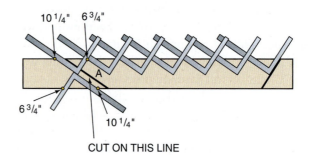

CUT ON THIS LINE

Lay out the stairs horses with the square, making correction at A for the thickness of the stair treads.

FIGURE 41-27 Laying out stairs.

Copyright © 2015 Cengage Learning®

Laying out a gable-roof rafter

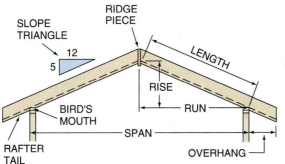

Fig. 1 Standard terms are applied to gable roof buildings and rafters.

Since gable rafters meet in midair, the height of the plate is the only fixed point and it is not necessary to determine the total rise. The only information needed is the total run and the slope, or inches of rise per foot of run. The run is determined by measuring the span from plate to plate (Fig. 1) and dividing by two. Rate of rise is obtained from the slope triangle. In this case the total run is 11 feet and the slope is 6 inches of rise per foot of run.

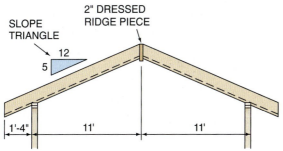

Fig. 2 Total run of a rafter for this building is 22/2" or 11'.

Using a framing square, put a marking line parallel to the edge of the rafter and 2" from the bottom edge. Using 12" on the blade and the number of inches rise per foot of run (6" in this case) on the tongue, lay out the lower end of the rafter first. Mark the lower plumb cut first and then measure with the square in position the desired amount of overhang as a horizontal distance and locate the corner of the bird's mouth on the marking line. Mark out the bird's mouth with a horizontal line where the rafter is to rest on top of the plate and vertical line where it is to contact the edge of the plate.

Calculate the length of the rafter as follows: Find the rafter table on the square and under the 6" mark, since the slope is 6" rise per foot of run, find the figure opposite "Length of Common Rafters per Foot of Run." This figure is 13.42, which means that for each foot of run, the rafter is 13.42" long. Calculate the length for the number of feet of run as follows:

Calculation: 13.42" x 11 = 147.62 inches
$$\frac{147.62"}{12} = 12' 3.62" \text{ or}$$
12' 3 5/8" (.625 = 5/8")

Measure out 12' 3 5/8" with a steel tape from the corner of the bird's mouth to a point on the marking line near where the upper plumb cut is to be made. Place the square with the numbers 6 and 12 on the marking line, with the number 12 even with the length mark (Fig. 4). Then measure back along the square a distance equal to half the thickness of the ridge piece. This is 3/4" for a standard-dressed 2" ridge piece, or 3/8"+

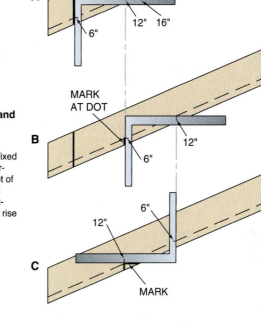

Fig. 3 Lay out lower end of gable rafter as follows:
(A) Place the square so that 6" and 12" are on the marking line. Mark lower plumb cut and place a dot opposite 16" on the blade of the square to indicate length of overhang.
(B) Slide the square up to the rafter and mark the plumb line for the bird's mouth even with the dot.
(C) Rotate the square and mark the level line for the bird's mouth.

for a 1" ridge piece. Draw the line for the plumb cut through this corrected length mark and the rafter is ready to cut. If no ridge piece is used, the upper plumb cut is made to intersect the marking line at the full measured length.

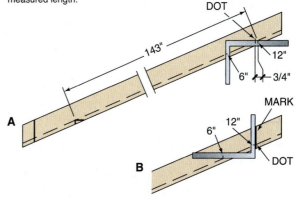

Fig. 4 Lay out upper end of gable rafter as follows:
(A) Measure the calculated length (12' 3 5/8") along the marking line from the corner of the lower bird's mouth and make a dot. Place the square with 6" and 12" on the marking line and with 12" mark even with the dot. Measure back along the blade of the square a distance of 3/4" and make a second dot.
(B) Rotate the square and draw a plumb line through the second dot.

Copyright © 2015 Cengage Learning®

FIGURE 41-28 Laying out a gable-roof rafter.

Material	*R Value per Inch of Thickness	*R Value for Thickness Shown
Insulation, rigid fiberglass	4.00	
Mineral fiber, resin binder	3.45	
Polystyrene—cut cell surface	4.00	
Polystyrene—smooth skin surface	5.26	
Polystyrene—molded bead	3.57	
Polyurethane	6.25	
Polyisocyanurate foam, plastic core, and foil each face	7.2	
Insulation—batt fiber type, mineral glass, slag		
2–2.75 in		7
3–3.50 in		11
6.5 in		19
Insulation, fill Perlite	2.70	
Vermiculite	2.27	
Cellulose	3.13	
Miscellaneous brick, 4 in		.44
Gyp board, ½ in		.45
Particle board, ⅝ in		.82
Plaster, ½ in		.10
Wood, hard	0.9	
Wood, soft	1.2	
Wood, plywood	0.8	

*R = Resistance to heat transfer

FIGURE 41-29 Insulation values of popular building materials.

Bowline. The **bowline** is a good knot for creating a loop that will not slip or close up under tension. It is easy to tie and untie.

Clove Hitch and Two Half Hitches. The **clove hitch** and **two half hitches** are quick to tie and easy to untie. They are used for fastening ropes to posts.

Timber Hitch. The **timber hitch** can be used as a single hitch or multiple hitch to lift or drag poles.

Barrel Sling. The **barrel sling** is useful for rigging barrels and similar objects for lifting. Care must be taken to be certain the rope is across the center of the bottom and crosses on the sides at equal distances around the barrel.

The knots and hitches described here are illustrated in Figure 41-31.

FIGURE 41-30 Blown-in insulation completely seals and insulates the ceiling below.

Whipping Rope Ends

Most rope will unravel at the end unless treated in some way. Rope ends may be **whipped** by tightly coiling strong cord around the end for a distance of 1/2" or 3/4", depending on the diameter of the rope. Also friction tape or plastic electrical tape may be tightly wrapped around rope ends for temporary use. Nylon rope ends may be fused into a solid end by burning with a flame.

SUMMARY

Planning, constructing, and maintaining buildings and structures are important parts of agriculture. This unit addressed some principles of planning and types of structures. While some details of construction are presented, the scope of the book does not permit sufficient detail to proceed without professional advice, plans, and information on local, state, and national building codes.

Related information is presented in previous sections of the text. The unit that follows provides information on some specialized structures.

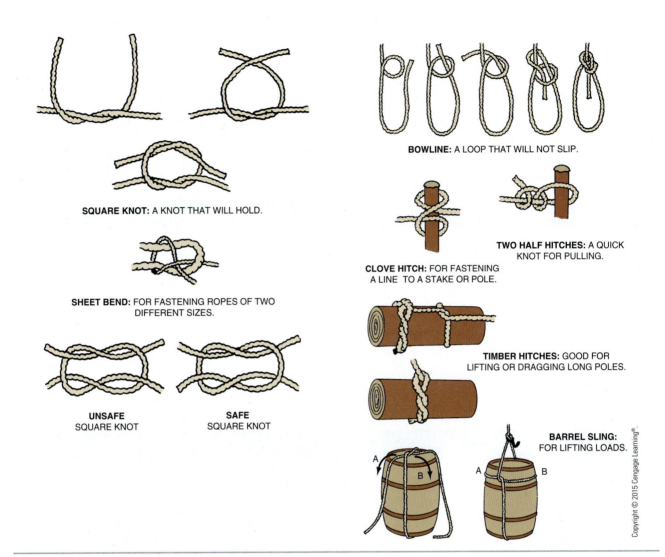

FIGURE 41-31 Useful knots and hitches.

Student Activities

1. Define the Terms to Know in this unit.

2. Visit a farm or other agricultural business. Study the building layout and discuss the advantages and disadvantages of the various building locations.

3. Design and draw a farmstead plan showing electrical meter pole, buildings, windbreak, and roads. Report the advantages of your plan to the class.

4. Study Figures 41-5 and 41-6. Draw templates to scale on heavy paper for the major pieces of machinery on a farm with which you are familiar. Cut out the templates and use them to design a proposed machinery storage building to accommodate the machinery.

5. Choose an agricultural enterprise, such as corn, potatoes, poultry, or beef cattle, in which you have special interest. Research the building requirements for the enterprise, and describe in detail the buildings you would recommend for production and/or storage. Report your findings to the class.

6. Build a model of some agricultural structure of your choice.

7. Work with a group of classmates to plan and build a model farmstead to scale.

8. Display the model described in activity #7 in a school showcase, library, and/or local fairs and shows. Consider it for entry in the school science fair.

9. Make a display board showing all the knots described in this unit.

10. Plan and construct a fence. Refer to Appendix C, and Unit 17 for basics of drawing.

Relevant Web Sites

American Wood Protection Association
www.awpa.com

APA—The Engineered Wood Association
www.apawood.org

Dow Building Solutions
building.dow.com

Self-Evaluation

A. Multiple Choice. Select the best answer.

1. Prevailing winds generally blow from the
   a. east
   b. north
   c. northwest
   d. west

2. The sun's warmth in winter can be captured best if the building faces
   a. north
   b. northwest
   c. south
   d. west

3. The electrical meter pole for a farmstead should be placed
   a. in a central location
   b. beside the house
   c. close to the highway
   d. at the building using the most electricity

4. Farm building planning should be done by
   a. farm family
   b. farm operator
   c. professional farm or building planner
   d. cooperatively by all the above

5. The item that permits a building to span over 100 feet without posts is a
   a. truss
   b. stud
   c. rafter
   d. girder

6. The actual thickness of 1" × 6" dressed lumber is
   a. ⅛" less than nominal
   b. ¼" less than nominal
   c. ½" less than nominal
   d. same as nominal

7. An example of a grade of plywood is
   a. AC
   a. S2S
   a. pressure-treated
   a. laminated

8. Which is not a material used to pressure-treat lumber?
   a. chromated copper arsenate
   b. creosote
   c. pentachlorophenol
   d. sodium biarsenate

9. A building foundation rests upon a
   a. wall
   b. tile drain
   c. girder
   d. footing

10. A woodframe building has walls constructed of
    a. beams
    b. girders
    c. purlins
    d. studs

B. Matching. Match the items in column I with those in column II.

Column I

1. ACA
2. granny knot
3. galvanized-coated
4. plywood
5. high point of roof
6. joist
7. rigid-frame
8. horse
9. square
10. whipping

Column II

a. made of veneer
b. 1.25 or 2.0 oz. coating
c. ridge
d. supports the floor
e. stair stringer
f. used for pressure treatment
g. safe knot
h. used on rope ends
i. unsafe knot
j. support of steel buildings

C. Completion. Fill in the blanks with the word or words that will make the following statements correct.

1. Outside surfaces of foundations are best waterproofed by sealing with a product that _____ and expands as it dries.

2. Foundation walls may be further protected after being sealed by using _____ and/or _____.

3. The ability of a certain material to restrict the flow of heat and be useful for insulation is referred to as _____ value.

4. Vapor barriers are important in agriculture buildings for
   a. _____
   b. _____
   c. _____
   d. _____

5. Specify the knot or hitch recommended for each of the following:
   a. creating a nonslip loop _____
   b. fastening livestock lead ropes _____
   c. lift poles _____
   d. lift barrels _____
   e. joining ropes of equal size _____

D. Brief Answer. Briefly answer the following questions.

1. What is oriented strand board? How does it compare with plywood for construction projects?

2. What is the purpose of pressure-treating? Why should care be taken when selecting pressure-treated lumber for use in agricultural settings?

3. Although aluminum will not rust in normal weather, it has disadvantages when used for roofing. What are they? What should be done to improve the strength of the aluminum roof when building?

4. Identify the parts of stairs.

5. When insulating metal and pole buildings designed for livestock, what are two special concerns to keep in mind?

UNIT 42

Aquaculture, Greenhouse, and Hydroponics Structures

Objective

To build and maintain structures for aquaculture, greenhouse, and hydroponics management.

Competencies to be developed

After studying this unit, you should be able to:

- Recognize structures used in aquaculture, greenhouse, and hydroponics technology.
- Maintain common structures used in aquaculture, greenhouse, and hydroponics technology.
- Construct simple aquaculture production facilities.
- Erect a small greenhouse.
- Build hydroponic units.

Materials List

- Protective goggles and coveralls
- General shop tools
- Specific construction materials are listed for Projects 52 to 54 in Appendix A

Terms to Know

- aquaculture
- open aquaculture systems
- fingerling
- microorganisms
- dissolved oxygen
- rotary air pumps
- diffuser tube
- air stones
- paddle-wheel aerator
- ammonia
- water quality
- submersible pumps
- centrifugal head pumps
- air-lift pumps
- net cages
- closed aquaculture systems
- controlled-environment structures
- biological filters
- raceways
- fish feeders
- settling tank
- greenhouses
- translucent
- greenhouse effect
- shadecloth

continued

CAUTION!

Read and observe all safety precautions when working with electrical equipment in wet areas.

ENVIRONMENTAL STRUCTURES USED IN AQUACULTURE

Types of Systems

Aquaculture. **Aquaculture** production systems have been developed to improve the production of aquatic plants and animals used for food, fiber, and recreation. These systems center around bodies of water such as lakes, streams, ponds, or artificial tanks that support the growth and development of economic aquatic species.

Open Aquaculture Systems. **Open aquaculture systems** are managed with relatively low densities of plants or animals. With low densities, an acceptable balance between the inputs (food, air, or nutrients) and outputs (waste products) may be maintained and pollution of the growing system can be minimized. For example, in the spring, ponds are stocked with small fish called **fingerling** and managed so that the final number of adult fish in the fall will not pollute the natural ecosystem of bacteria and fungi (**microorganisms**) that detoxify the fish waste products.

As growers continue to increase the number, or density, of fish in the tank, cage, or pond, it is necessary to add inputs, such as feed and oxygen, or remove the accumulated fish wastes by flushing or water treatments. Such systems are called semi-open aquaculture systems and are managed like open aquaculture systems until the fish densities start to limit growth (Figure 42-1).

FIGURE 42-1 Net pens or cages assist producers in managing fish produced in lakes.

When feed is added to the fish population, an increase in the consumption of oxygen in the water, by both fish and microorganisms, results. Air must then be mixed with the water to increase the levels of **dissolved oxygen** so that the fish can continue to grow. Air is pumped into the water using **rotary air pumps** or air compressors. The air is then spread or diffused by thin-walled **diffuser tubes** or **air stones** that break the airstream into fine bubbles (Figure 42-2). An alternative method is to use a **paddle-wheel aerator**, driven by a motor or tractor, that splashes water into the air for additional oxygen to be incorporated (Figure 42-3).

As additional feed is consumed, fish produce more toxic waste products such as **ammonia**, a gas that is a by-product of decomposition. Wastes may be removed by microorganisms that live along the bottom and sides of the ponds or by adding new water into the system. Many ponds are flushed with 50 gallons of fresh water/acre/minute to assure good **water quality** that is free of excessive pollutants. Many growers have designed ponds to use the natural elevation drops to maintain water flows. However, most growers must use pumps to force water flow. Pumps include **submersible pumps** with submersible motors that can be totally immersed in water, **centrifugal head pumps** driven by auxiliary motors, and **air-lift pumps** that use compressed air to lift water.

Fish can be grown in **net cages** or pens to assist in feeding and harvest (Figure 42-4). However, because the fish are generally forced into greater density by this method, care must be taken to make sure water quality is maintained within the net cages.

Recirculating systems or **closed aquaculture systems** are now being developed in many parts of the world where good water quality or land resources are limited. Because of the increased cost of construction, these systems are maintained in **controlled-environment structures**. Here temperature, light, and other requirements are managed to maximize fish growth. In this system, fish are grown indoors within tanks. Water flows through production tanks where the fish are held; then, it flows through waste treatment facilities or **biological filters** in which waste products are detoxified and recycled back to the production tanks (Figure 42-5). Management includes maintaining the optimum temperature, dissolved oxygen levels, and feed rations for the animals or plants (Figure 42-6).

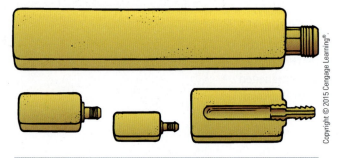

Copyright © 2015 Cengage Learning®.

FIGURE 42-2 Ceramic air stones break the airstream into small bubbles.

© iStockphoto/Muammer Mujdat Uzel

FIGURE 42-3 Mechanical aerators mix air and water to increase the amount of dissolved oxygen.

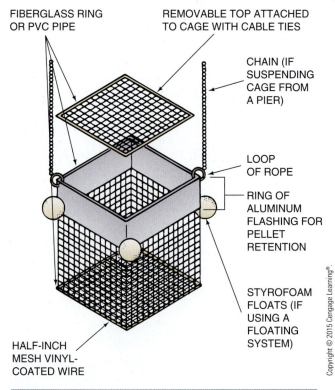

FIBERGLASS RING OR PVC PIPE

REMOVABLE TOP ATTACHED TO CAGE WITH CABLE TIES

CHAIN (IF SUSPENDING CAGE FROM A PIER)

LOOP OF ROPE

RING OF ALUMINUM FLASHING FOR PELLET RETENTION

STYROFOAM FLOATS (IF USING A FLOATING SYSTEM)

HALF-INCH MESH VINYL-COATED WIRE

Copyright © 2015 Cengage Learning®.

FIGURE 42-4 Net cages or pens are constructed to minimize water movement.

design components include the tank and tank equipment, air supply, waste treatment, and building.

Tanks may be constructed of fiberglass, polypropylene, epoxy-coated steel, aluminum, or plastic liners supported by wood or steel frames. Most structures average 4' to 6' deep and can be constructed above or below ground level. Round tanks are generally the most economical shape (cost/gallon). Rectangular **raceways** (Figure 42-7), or cages (Figure 42-8), provide economical use of floor space but the corners encourage fighting among fish, which stunts overall growth in some species. Use of liners provides the cheapest material cost for initial construction. However, liners must be replaced periodically and create difficulty in connecting equipment to the necessary plumbing fixtures.

The auxiliary equipment necessary to maintain the growing parameters of the fish is based on pounds of fish or density in the system. For example, a 6,000-gallon system may support 1,200 pounds of fish or more. The density for this system is 0.2 pound of fish/gallon. Fish are typically fed bulk feed with automatic **fish feeders** that release preset pounds of fish food during the day, or they may be hand fed (Figure 42-9). With a feed conversion rate of 1.5 pounds of feed per pound of fish, the system should produce 4,800 pounds of fish per year.

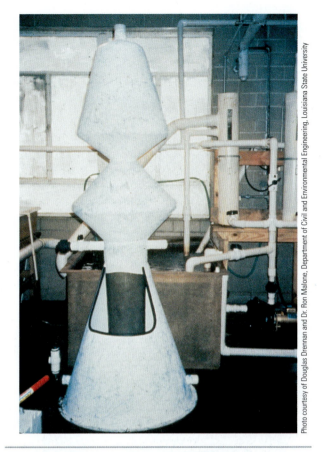

FIGURE 42-5 Biological filters provide surface area for microorganisms to grow and detoxify waste materials.

FIGURE 42-6 Water quality measurements must be taken every day to maintain optimum growing conditions.

Construction of Indoor Aquaculture Systems

The design of an indoor, recirculating aquaculture production model must include equipment and materials that can operate in constant contact with water and high humidity. Electrical units, switches, and motors must be installed with proper safety precautions. Major

FIGURE 42-7 Rectangular raceways provide economical use of space.

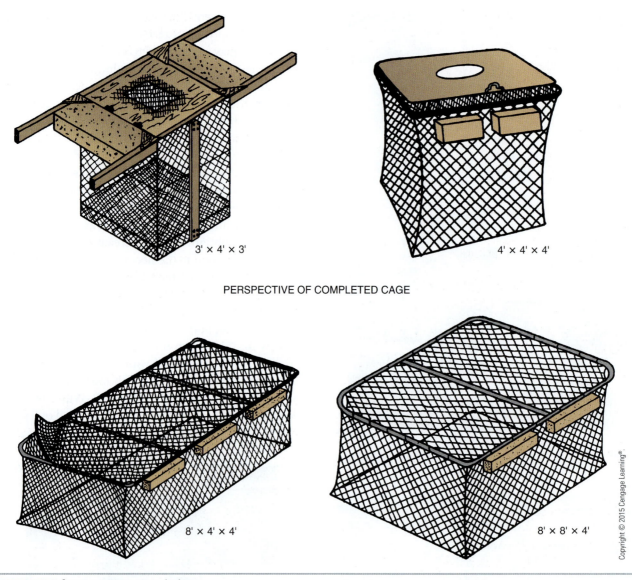

PERSPECTIVE OF COMPLETED CAGE

FIGURE 42-8 Some common cage designs.

FIGURE 42-9 Fish feeders deliver feed on a timed schedule to increase the feed conversion ratio.

Dissolved oxygen may be maintained in water by rotary air pumps, which bubble air into the water through a diffuser tube system. Air blowers are typically engineered to provide about 1 pound of oxygen infused into the water per pound of food added (Figure 42-10).

A waste treatment component may settle the solid wastes produced by the fish by pumping water sufficient for about seven tank changes per day through a **settling tank**. The effluent passes through a biological filter, which contains the growing surface covered with natural microorganisms that detoxify the wastes and ammonia before the water is recycled to the production tank. The biological filter consists of inert material such as fine gravel, pea gravel, coarse sand, plastic beads, or open-cell foam that provides large amounts

FIGURE 42-10 Tanks of oxygen are used to maintain dissolved oxygen levels during the transportation of fingerlings.

of surface area on which the microorganisms grow. The size of the biological filter is determined by:

- Estimating the ammonia produced from the feed added to the system.
- The surface area per cubic foot of the selected material.
- The necessary surface area of microorganisms to remove the ammonia to safe levels.

Plumbing fixtures and pipes are typically constructed from PVC Schedule 40 grade materials. Copper and iron may be released from other plumbing materials and can become toxic to fish raised in tanks made from these metals.

Climate control in the building is determined by local climate and insulation installed during construction. General building materials must withstand the corrosive humidity and ammonia produced by the system. Recommended materials include fiberglass sheets, Lexan sheets, plastic, galvanized tin sheets with vapor barrier, or treated lumber. Care must be taken that no condensation on the building materials comes in contact with the fish water.

General maintenance for such systems requires maintaining the integrity of the water tanks, pump and blower maintenance, and routine flushing of the settling tank. Plans for constructing a closed system aquaculture model are provided in Appendix A, Project 52.

GREENHOUSES

Greenhouse Types and Orientation

Greenhouses are specialized agricultural structures that provide a controlled environment for the production of both plants and animals where either light or heat generated by sunlight creates optimal growth (Figure 42-11). The greenhouse structure simply supports a transparent or **translucent** covering that allows between 70 to 90 percent of the sunlight to enter. Such coverings include glass, plastic films, fiberglass sheets, and Lexan materials.

Each type of covering has special advantages. Glass is the most durable but often the highest in initial cost. Fiberglass can last up to 15 years but must be cleaned and repaired about every 5 years. Lexan materials have been recently introduced with a 15-year maintenance-free period, but have been reported to be susceptible to hail or other mechanical damage. Plastic films (4 to 6 mm) are economical and are the most commonly used materials on Quonset-style greenhouses (Figure 42-12). The expected durability of commercial grades of plastic is 2 to 3 years before ultraviolet radiation destroys the integrity of the plastic. Most

FIGURE 42-11 Greenhouse production is a major industry in many countries.

FIGURE 42-12 Plastic-covered, Quonset-style greenhouses are economical for plant production.

commercial operations use special ultraviolet-resistant grades of plastics that are more expensive, but last 4 to 5 years before needing replacement. To conserve energy, double layers of film are often used. Equipment needed to inflate and separate the film layers includes a squirrel-cage blower, a damper to adjust blower pressure, and flexible tubes to transfer air between sidewall and end wall sections.

As sunlight enters the greenhouse, the light is absorbed by objects, which then re-radiate the energy as heat waves, even on cloudy days. These low-energy waves cannot pass back through the coverings and accumulate inside. This buildup of heat is called the **greenhouse effect**. During the summer months, heat must be removed by ventilation or the greenhouse must be shaded to limit the penetration of sunlight. Shading can be applied as a dilute white paint on the outside of the greenhouse that reflects sunlight, or as a woven black **shadecloth**, which blocks between 30 and 70 percent of the sunlight. Shading materials are also installed inside the greenhouse to reflect the solar heat to the outside.

The support work in greenhouses is usually aluminum, galvanized steel pipe, treated wood, or plastic pipe.

Since greenhouses may be low in cost per square foot to build, they are used in a variety of settings from plant production to aquaculture production to recreational activities.

Structural Form

A greenhouse with a straight wall and a gable roof is common and has advantages in framing and in space utilization. Post and beam, post and truss, and arches are used to form the gable structure. The part-circle or Quonset frame is constructed with rolled sections of aluminum or steel. The Quonset design makes better use of the frame material than gable does, but in some applications the curved sidewall limits space use.

Freestanding greenhouses can be truss-rafter or sawtooth, gothic, or Quonset in style (Figure 42-13). These shapes support the translucent coverings and provide several alternative methods for ventilation. Vent openings are placed in the top peaks of truss-rafter- or sawtooth-type greenhouses so that as the hot air rises, it is vented to the outside. The gothic- or Quonset-style greenhouses are usually covered with single sheets of plastic that stretch from side to side. Vent openings and fans are located on opposite end walls. The fans draw air through the greenhouse and exhaust the heated air outside.

Lean-to or supported greenhouses are connected to existing structures with the translucent covering having a southern exposure. Vent openings and exhaust fans are located on the opposite end wall. Lean-to greenhouses are located on the southern exposure of support buildings to provide maximum exposure to sunlight.

FIGURE 42-13 Freestanding greenhouses are built following several different construction styles.

Greenhouse Construction

The structural design of a greenhouse must provide safety from wind, snow, and crop load damage while permitting maximum light transmission. Therefore, opaque framing members should be as small as possible but still provide adequate strength to resist expected loads over the planned life of the greenhouse.

Design loads include dead loads (weight of the permanent equipment carried by the frame), live loads (items used for a short period of time), and loads from wind and snow. For example, a polyethylene-covered house will have a lighter dead load than a glass house. If hanging baskets are used for a three-month growing period in the greenhouse, this (live load) would have to be considered in the structural design (Figure 42-14).

Site Selection

Consideration should be given to site selection when constructing a greenhouse. Orientation of the building and site drainage area are two important factors that can make a difference in greenhouse crop productivity. An east–west facing greenhouse will transmit more light than a north–south one. The site should be level and should provide adequate drainage of surface water from the area. Additionally, zoning laws, accessibility, water quality, and space for future expansion should be considered.

Foundation

The foundation is the connection between the ground and the building. The footing should be set below frost or to a minimum depth of 24 inches below ground surface. Footings should always be placed on level undisturbed soil. Individual pier footings are less expensive than a continuous wall and can be used if they fit the load and soil conditions. A continuous foundation wall should be set on poured concrete footing.

Floor

A gravel floor or ground cloths are most often used in production and wholesale structures. For a retail greenhouse or one that has high traffic flow, a gravel floor with concrete traffic aisles works well and still provides drainage and weed control (Figure 42-15).

Porous concrete makes a good floor surface for greenhouses, because it allows water to pass through. This avoids the puddles or standing water that is common on regular concrete floors. If solid concrete is used, floors should be sloped to drains that are connected to underground pipe to carry the water away from the site.

Environmental Equipment

A greenhouse is constructed and operated to provide an acceptable plant environment that will contribute to a profitable enterprise. Light, temperature, ventilation, and water are factors that directly affect a plant's productivity.

Heating

Heating systems in greenhouses range from large boilers to small unit heaters, and they use a variety of energy sources. Wood, coal, LP gas, natural gas, fuel oil, electricity, and solar are the most commonly used

FIGURE 42-14 The weight of hanging baskets must be considered in planning the construction of greenhouses.

FIGURE 42-15 Greenhouses are built with gravel floors and concrete traffic aisles.

energy sources. The selection of heating equipment depends on the size and type of greenhouse operation. Additionally, the availability, economics, and dependability of the energy source should be considered.

The method of heat distribution depends on the type of heating system installed. When boilers are used, the heat is usually generated in an external furnace and transferred to the greenhouse through hot water pipes. With freestanding heaters, the complete heat transfer is located in the greenhouse using a forced hot air system. A convection tube or horizontal airflow fans distribute the heat through the greenhouse to maintain a constant temperature. Several factors must be considered when calculating what size heater is needed to heat a greenhouse structure:

- type of greenhouse covering
- exposed surface area
- inside temperature desired
- lowest average outside temperature expected during heating season
- wind speed
- condition of covering

Cooling and Ventilation

Estimates of cooling requirements for greenhouses are also based on acceptable temperature differences between inside and outside air. The most common methods of cooling include the use of vents, fans, shading, evaporative pad systems, and fogging systems.

Fans and Vents

Depending on the size of the greenhouse, one or more fans are placed on one side (end) of the greenhouse, from which the greenhouse air is exhausted (Figure 42-16).

On the opposite side (end) of the greenhouse, louvered vents are installed to permit airflow from the outside into the greenhouse when exhaust fans are operating. Ideally, air that enters the greenhouse should move through and exit the greenhouse in 1 minute. Fans are rated on how much air they can move per cubic foot. To calculate the required fan size, it is necessary to first determine the volume of air contained in the greenhouse.

Evaporative Cooling

Evaporative cooling systems are installed on the opposite side (end) of the greenhouse from the exhaust fans. The system consists of a water reservoir, a pump, pipes, and a water return system. The water is pumped to the top of the **evaporative cooling pads** through an overhead pipe. Water drips through the pad while the exhaust fans pull outside air through the wet pads, which cools the air moving through the greenhouse (Figure 42-17).

FIGURE 42-16 Fans are used to cool and exhaust greenhouses.

FIGURE 42-17 Water drips through a cooling pad to cool the temperature inside a greenhouse.

Copyright © 2015 Cengage Learning®.

Copyright © 2015 Cengage Learning®.

Excess water is drained into a sump, where it is stored until, once again, it is filtered and pumped to the top of the cooling pad. Aspen or formed paper pads are commonly used in the evaporative cooling system. Chemicals are placed in the water reservoir to control algae growth in the tank and on the pads.

Fogging Systems

High-pressure fog cooling is based on the theory that evaporating water takes heat from the air. In this system, a fine fog is emitted in the greenhouse by forcing water through small nozzles. The system requires a water supply that is free of silt, sand, or debris that might cause obstruction of the emitters.

Environmental Controls

Nearly all controls for the greenhouse can be automated. Most modern greenhouses use thermostats that are calibrated around the desired greenhouse temperature. A thermostat consists of a sensor and a switch to control equipment operation (Figure 42-18).

Control can be improved by using more than one thermostat, each set to operate at a different temperature and to control a part of the supply. A common approach uses one single-stage and one double-stage thermostat. For example, using a single-stage thermostat with the heating set at 60°F, the daytime temperature in the greenhouse might rise to 70°F. As the sun sets and the temperature drops to 59°F, the thermostat contacts close, causing the heat to come on. As the temperature increases to 63°F, the thermostat contacts open, causing the heating system to shut down, thus maintaining a relatively constant temperature.

If a ventilation or two-stage thermostat is set at 70°F and the temperature rises to 69°F, stage-one contact of the thermostat closes and the ventilation shutter opens. If the temperature continues to rise to 71°F, the second-stage contacts of the thermostat close and the exhaust fan begins to pull air through the greenhouse to cool the temperature.

Electricity

An adequate electric power supply and distribution system should be provided to serve the environmental control and mechanization needs of the greenhouse. Early in plan development, contact the local supplier to determine the availability and cost of power, and the best service drop location. Once this is done, the distribution system can be planned.

To determine service drop size, the size and number of motors and other electrical components should be known. The distribution system within the greenhouse will have to meet the National Electrical Code and any local codes. Watertight boxes, UF wire, and ground-fault interrupters are often required (see Section 11).

Water

Plants require an adequate supply of moisture for optimum growth. By supplying an adequate but regulated amount of moisture, it is possible to control the growth and flowering of plants. A correctly designed water system will satisfy daily water requirements with a minimum of manual labor (Figure 42-19).

The volume of water required will depend on the area to be watered, the crop grown, the weather conditions, the time of year, and whether the heating or

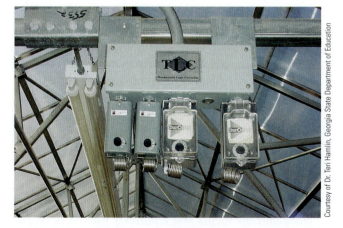

FIGURE 42-18 Thermostats are used to control the amount of heating and cooling applied to the inside of a greenhouse.

FIGURE 42-19 A correctly designed water system will satisfy daily water requirements with a minimum of manual labor. This is a mist irrigation system.

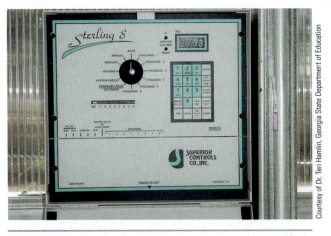

Courtesy of Dr. Teri Hamlin, Georgia State Department of Education

FIGURE 42-20 Water systems are timed to come on at specified times and are controlled by electronic devices.

ventilating system is operating. Most greenhouse irrigation systems are regulated by timers that water the plants at certain preset intervals. Examples are misting systems for propagation areas and drip irrigation for potted plants. Twenty-four hours and seven days are common two cycles or interval lengths for time clocks, which open and close contacts at preset times by inserting appropriate trippers, riders, or pins in a dial (Figure 42-20).

Plans for building small greenhouses with PVC pipe are shown in Appendix A, Project 54.

HYDROPONICS

Hydroponics is a production technology in plant science that attempts to maximize growth by stimulating nutrient uptake (Figure 42-21). It literally means "water culture," a process in which the necessary plant nutrients are dissolved in water. Water with nutrients added—called a **nutrient solution**—contains all the required elements for plant growth. Therefore, it is not necessary for plants to be planted in soil. Water flow can be maintained so that the dissolved oxygen levels in the nutrient solution are optimal for plant growth.

Hydroponic systems are designed to maintain the nutrient solution in contact with the root systems. This requires a stock tank, pump system, irrigation manifold, plants, and drainage system.

All plumbing should be constructed of PVC plastic pipe and fiberglass because of the corrosive nutrient solution. Copper, steel, or lead pipes affect the nutrient composition of the solution or release toxic compounds. Submersible pumps or centrifugal head pumps used to circulate the nutrient solution must be manufactured to handle saltwater solutions. All electrical equipment and utilities must be installed with safety precautions outlined for wet environments.

© Mack2happy/Dreamstime.com

FIGURE 42-21 Hydroponic crops are grown using a nutrient solution containing all the essential salts a plant absorbs through its roots.

Hydroponic systems can be installed at ground level or at tabletop level. They operate by either maintaining a constant solution around the roots or by flooding the roots periodically with nutrient solution.

The plants are maintained in various types of bags, troughs, or tubes. Large plants, like tomato and cucumber, are planted directly into bags filled with **vermiculite**, which is an artificial soil medium made from mica. In this **bag culture** system, the bags are laid directly on the ground, punched with holes, and planted with two plants per bag. Trickle irrigation emitters are placed into the bags and the nutrient solution dripped continuously. A drainage system returns the effluent to the stock tank where it is pumped back into the irrigation system (Figure 42-22).

In a **trough culture** system, trenches are built and filled with vermiculite or crushed gravel. Plants are placed into the media and a pump is used to flood the trough. When the pump turns off, the nutrient solution drains back to the stock tank. The pump is on a timed cycle that prevents the trough media from becoming dry.

FIGURE 42-22 Plants are grown in tubes where trickle irrigation delivers nutrients and water.

Plants can also be held in small-diameter gutters or tubes with nutrient solution circulating continuously. This **tube culture** maintains a thin stream of nutrient solution in contact with the root system. A plan for constructing this system is provided in Appendix A, Project 35.

Hydroponic systems have been demonstrated to produce the highest production levels of any plant culture system. However, the grower must maintain an intensive inspection schedule that monitors both the irrigation system and the nutrient solution.

The optimum growing conditions of light and temperature are maintained by the environmental controls in the greenhouse structure.

SUMMARY

Aquaculture units are used to produce fish and other aquatic animals for human food. The systems are as varied as the conditions and type of fish being grown. Whether the units are large outside ponds or systems within a building, they all have components that have to be constructed and properly maintained. Similarly, greenhouses provide controlled environments for plants and come in a wide array of different types of structures. Constructing and maintaining all types of greenhouses require the proper skills and techniques that can be learned through a quality agricultural mechanics program.

Student Activities

1. Define the Terms to Know in this unit.

2. Describe how waste materials are detoxified in a pond or aquarium.

3. Examine a wholesale catalog for aquaculture farmers.

4. Research the permits or requirements for starting an aquaculture industry in your state.

5. Contact the local waste treatment facility and understand how waste products generated in your community are detoxified.

6. Determine what aquatic plants and animals are being grown in your state and locality.

7. Start an aquaculture library for your school.

8. Sketch and label the major structural components of a greenhouse.

9. Outline a detailed procedure for erecting a plastic-covered greenhouse.

10. Determine what vegetables in your local supermarket are being produced in hydroponic systems.

11. Construct models of aquaculture, greenhouse, and hydroponics systems.

Relevant Web Sites

Southern Regional Aquaculture Center
http://srac.msstatc.cdu

United States Aquaculture Society
https://www.was.org

Virginia Cooperative Extension, articles on Home Hydroponics and other greenhouse topics
www.ext.vt.edu

West Virginia University Extension Service, articles on Greenhouse Construction, Planning and Building a Greenhouse, and more
https://extension.wvu.edu/

Self-Evaluation

A. **Multiple Choice.** Select the best answer.

1. Aquaculture production includes
   a. growing fish for food
   b. growing plants for aquariums
   c. growing fish for recreational fishing
   d. all of the above

2. Farmers can increase the amount of fish they grow by
   a. lowering the water level
   b. decreasing the dissolved oxygen levels
   c. increasing the dissolved oxygen levels
   d. slowing the water flow

3. Fish wastes are detoxified by
   a. birds
   b. small fish
   c. microorganisms
   d. soil

4. Closed aquaculture systems
   a. are expensive to construct
   b. are easy to operate
   c. discharge large amounts of water
   d. are used in ponds

5. Greenhouses built with plastic film coverings
   a. are more expensive to construct than glass greenhouses
   b. must be recovered with new plastic every 2 to 3 years
   c. cannot handle heavy snow loads
   d. cannot be exposed to the sun

6. The "greenhouse effect" occurs
   a. on sunny summer days
   b. on cloudy summer days
   c. on sunny winter days
   d. during all of the above days

7. A greenhouse should be built
   a. on the southern exposure of a building
   b. where the afternoon sun is blocked
   c. in a low area where water collects
   d. on the northern exposure of a building

8. Heat is generated in the greenhouse from
   a. propane heaters
   b. electrical heaters
   c. sunlight
   d. all of the above

9. Dissolved oxygen in a hydroponic nutrient solution is provided by
   a. the stock tank
   b. the submersible pump
   c. the circulating water in contact with air
   d. none of the above

10. Greenhouse structures are used for
    a. plant production
    b. animal production
    c. aquaculture production
    d. all of the above

B. **Matching.** Match the word or phrase in column I with the correct word or phrase in column II.

Column I

    1. air-lift pump

    2. hydroponics

    3. biological filter

    4. raceway

    5. gothic style

    6. shadecloth

    7. wind brace

    8. aquaculture

    9. nutrient solution

    10. ammonia

Column II

    a. oval fish tank

    b. waste product

    c. freestanding greenhouse

    d. triangle brace or greenhouse

    e. production of fish

    f. mixture of plant nutrients

    g. "water culture"

    h. water pump using bubbles

    i. light reflecting cover

    j. mixture of microorganisms

C. **Completion.** Fill in the blanks with the word or words that will make the following statements correct.

    1. Closed aquaculture systems require farmers to add extra feed, dissolved oxygen, and remove _____ products in order to grow fish.

    2. Copper plumbing is not used in aquaculture systems because high levels of copper are toxic to _____.

    3. Dissolved oxygen is increased in water by mixing water with _____.

    4. Diffuser tubes and _____ _____ are used to generate small bubbles that help mix the air in water.

    5. The microorganisms in the _____ _____ or growing on the bottom of a pond detoxify waste products.

    6. A controlled-environment structure, like a _____, is used to grow plants.

    7. Glass, plastic films, and fiberglass sheets are _____ coverings that let sunlight enter the greenhouse.

    8. The "greenhouse effect" causes the temperature in the greenhouse to _____.

    9. A supported greenhouse is best built on the _____ side of the house.

    10. A hydroponic system recirculates the _____ _____ that provides the root system with water, oxygen, and nutrients.

D. **Brief Answer.** Briefly answer the following questions.

    1. In closed aquaculture systems, how is the size of the biological filter determined?

    2. Explain how the greenhouse effect works.

    3. On which side of the house should a lean-to or supported greenhouse be built? Why?

    4. In hydroponic systems, what makes planting in soil unnecessary?

    5. How are large plants maintained in hydroponic systems? What is such a system called?

UNIT 43

Fence Design and Construction

Objective

To identify the main types of fences and explain how fences are constructed.

Competencies to be developed

After studying this unit, you should be able to:

- Discuss why humans have built fences for thousands of years.
- Identify problems associated with the development of barbwire.
- Identify the various types of fences used today.
- Discuss the advantages and disadvantages of different types of fences.
- Discuss the various types of materials used in modern fences.
- Explain the safety rules used in fencing.
- Explain the proper methods used in building fences.

Materials List

- Fence charger
- Wooden post
- PVC post
- Steel post
- Insulators
- Lightning arrester
- Fence stretcher

continued

Terms to Know

- rails
- barbwire
- landscape fence
- polyvinyl chloride (PVC)
- charger
- lightning arrester
- joule
- ribbon wire
- galvanizing
- high-tensile wire
- stays
- pullout
- stretch indicator

Materials list, *continued*

• Samples of the following:
  Barbwire
  Vinyl fencing material
  Electric fence wire
  Ribbon wire
  Fencing staples
  High-tensile wire
  Woven wire
  Chain-link fencing

EARLY USES OF FENCES

Humans were building fences long before recorded history. They probably started building them to protect themselves from animals or other humans. When people began to grow crops, they built fences to keep animals from eating the plants. These first primitive fences were built of brush with sharp thorns to deter animals from entering the fields. Later on, when animals were domesticated, this type of fence was used to keep animals in. In fact, in some parts of the world even today, sophisticated hedges are grown around fields to either keep livestock in or animals out of the area.

A very early fence-building material was stones, piled or stacked in rows. In many places, stones were readily available and were so durable that the fences have lasted thousands of years (Figure 43-1). In the New England area of the United States, stone fences remain that were built 300 years ago. Farmers picked up the stones from their fields and built neat fences by stacking the stones along the borders.

Another material used in early American fencing was wooden **rails**. Wood was plentiful as vast forests were cleared, so it was readily available for use in fence construction. Most often a durable wood such as oak,

© Anneka/Shutterstock.com

FIGURE 43-1 Stone fences were among the earliest types of fences built by humans.

cedar, or chestnut was used. The trees were felled and cut into logs, and the logs were split lengthwise into rails. The rails were then stacked in a crisscross manner that could wind for a very long distance (Figure 43-2). These fences were very effective in controlling livestock.

The invention of barbwire in the 1860s and 1870s brought about a real revolution in fencing, because of the lack of fencing materials on the plains in the West. **Barbwire** was simply sharp wire barbs woven into two strands of smooth wire. The sharp barbs kept the animals from pushing through the fence. Many types of wire were tried, and collectors today display many of these early designs. Barbwire fences are used extensively today.

Traditionally, cattle had been raised on free range where they were allowed to roam freely in search of grazing. As the western part of the United States became more populated, there was less room for cattle to roam free and there was a great need for fencing to limit areas where cattle could graze. Ironically, barbwire was first used to keep animals out rather than to keep them within the fence. Many of the settlers moving West were farmers who grew crops rather than cattle. If cattle were allowed to range anywhere they wished, the crops of the farmers could be destroyed. The use of the new barbwire fences caused problems because cattle could become entangled in the wire and suffer injury. In addition, cattle were moved to market in long cattle drives over land to a rail head, where they could be loaded onto rail cars and shipped to the population centers in the East. The use of fences meant that the cattle drives were severely limited in where they could go. Disputes over the use of fences led to what became known as range wars, where battles were fought between cattle ranchers and farmers. Eventually, the courts ruled in favor of the farmers who used fencing. With the further settling of the West by farmers, the issue was settled.

MODERN FENCES

In our modern era, fences are widely used. There are various reasons for building fences. These include landscaping, privacy, boundary marking, keeping animals in, and keeping animals out. Modern fences are made of a wide variety of materials that differ depending on the type of fence and the desire of the landowner.

Landscape fences are built for aesthetic purposes and must be attractive (Figure 43-3). Most of these fences can be divided into two basic types, wood and masonry, although a few landscape fences are constructed of wrought iron. Masonry fences are built from stone, bricks, or other masonry materials. They are expensive and time-consuming to build, but they are very durable and are considered to be permanent. They can serve other purposes as well, such as acting as retaining walls to hold soil on a hillside.

Wooden fences are built from boards that have been treated to prevent decay. Some of the older types of treatment were toxic, so care should be taken when repairing old fences that may have been treated with toxic substances. Some wooden fences are constructed of woods such as redwood, cypress, or cedar that are naturally resistant to decay. However, none of these woods will last as long as pressure-treated wood. These fences come in a variety of different designs and layouts that depend on the overall landscape scheme and the tastes of the landowners. Wooden fences may also

FIGURE 43-2 Early American settlers built fences out of split wooden rails.

FIGURE 43-3 Landscape fences add beauty to the home or other areas.

FIGURE 43-4 Vinyl fences are effective, durable, and attractive.

be used for horse pastures. Boards are generally considered to be safer for horses than many types of wire fences.

The newest type of fencing material is vinyl (plastic). The main ingredient is **polyvinyl chloride (PVC)**, the substance used in the manufacture of water pipes. Originally designed for horse fencing, the fences are now used in a variety of ways, such as for landscaping. Vinyl is initially more expensive than wood, but it is stronger, is more flexible, and lasts many times longer (Figure 43-4). In fact, some companies guarantee that their vinyl fencing will last a lifetime. Both posts and rails are made from this material and, as with most other types of building materials, vinyl fencing comes in different grades. The best type contains an ingredient that counteracts the effects of ultraviolet sunrays. Without this ingredient, vinyl will deteriorate in the sun. Posts are generally set in concrete and the rails or vinyl boards attached with screws. Some rails are fitted into the post and locked into place without any hardware. This prevents injury to horses if they push the fence down. It also makes a very attractive fence.

Electric Fences

For many years, livestock producers have used electric fences consisting of one or more strands of bare wire that carry an electrical current. When livestock come in contact with the wire, they feel a small electrical shock. The shock is not dangerous, but it is strong enough to feel uncomfortable. This trains the animals to remain within the boundaries of the fence. The advantages of electric fences are that they are less expensive, take less time to install, and are more portable than other forms of fencing. Disadvantages are that weeds growing up to the wire can ground the current flow, and that the fence must have access to an electrical outlet. Some fences use a battery, but batteries have to be replaced or recharged on a regular basis.

The parts of an electric fence include the **charger**, **lightning arrester**, posts, wire, and insulators (Figure 43-5). The charger is a type of transformer that steps a household current of 120 volts up to several thousand volts. As explained in Unit 33, an electric current must have voltage and amperage. Voltage is the amount of electric pressure, and amperage is the amount of electron flow along the conductor. The higher voltage is needed to push the current along the wire; in addition, the higher the voltage, the fewer problems with weeds. Low amperage and the pulsation of the charger ensure safety. Pulsation is the amount of time the electric current flows along the wire.

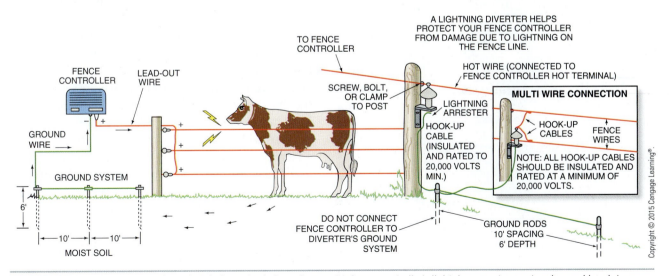

FIGURE 43-5 The components of an electric fence include a charger (or fence controller), lightning arrester, posts, wire, and insulators.

The charger sends the current in short time segments, such as $\frac{1}{10}$ of a second. Chargers are sometimes rated in joules. A **joule** is a measurement of energy calculated by multiplying amps × volts × time. Care should be taken in using the joule rating alone to choose a charger. Keep in mind that the joule rating can be influenced by the amount of time, volts, or amps, or any combination of the three.

A lightning arrester is a device that prevents lightning from running on a wire fence. During an electrical storm, a tremendous amount of current can run along a wire. This can not only destroy the charger, but can also kill any animal or human that comes in contact with the lightning-charged fence. A simple ground is not sufficient, because the current will be grounded and the fence will not work. A lightning arrester allows only the proper amount of current to flow along the fence wire. Any large surge of current is diverted into the ground.

Posts for electric fences can be much smaller than posts for other types of fences. Remember that the post only needs to support one to three strands of light wire. The wire is usually a 12-gauge, smooth, zinc-coated wire that is attached to the post by insulators. A variation is the use of a fabric ribbon that has a much smaller wire attached to it. The **ribbon wire** is much easier to see at a distance, is cheaper, and can be rolled up and handled more easily. Insulators are made of plastic with holes for driving nails or staples to secure them to the post (Figure 43-6). The plastic prevents the wire from contacting the post. In wet weather, even a wooden post can act as a ground that will disrupt the current flow.

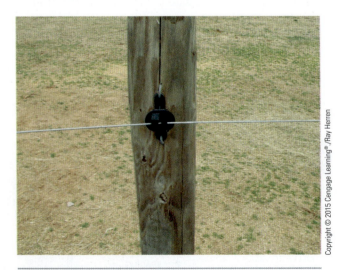

Copyright © 2015 Cengage Learning®./Ray Herren

FIGURE 43-6 Insulators prevent the post from grounding out the electric current flowing through the fence wire (note the split in the post due to incorrect placing of the staples).

Wire Fences

By far the most common types of fences are those constructed of wire. Wire fences are relatively inexpensive when compared with wood or vinyl fences. Also, when properly installed, they last a long time and are effective for penning livestock. Basically there are two types of wire fencing: strands and woven wire. Strand fences include barbwire and high-tensile wire.

Wire used for fencing material has several characteristics that should be considered. The size of the wire (the diameter) is measured by gauge. The higher the gauge number, the smaller the wire. For example, a 12-gauge wire is 0.105 inches or 0.267 centimeters in diameter. A 7-gauge wire is 0.177 inches or 0.450 centimeters in diameter. The heavier the wire is, the more holding power it will have. A fence used to hold large animals, such as cattle, needs a larger gauge wire than a fence intended to hold smaller animals, such as sheep. Also, in areas where the climate is humid, rusting occurs as the protective coating wears off the wire. Larger gauge wires will last longer as this process takes place.

Wires also have different types of protective coatings. Fence wire is made from steel because steel is strong and relatively inexpensive. A disadvantage of steel is that it rusts or corrodes. To combat this process, wires are coated with materials such as copper or zinc that do not rust. Keep in mind that the coating only delays rusting and eventually the coating wears away and the wire begins to rust. By far the most common coating is zinc, which is applied by a process called **galvanizing**. This is the same coating used on sheets of steel roofing to prevent rusting. The thickness of the zinc coating, which determines how long the wire is protected from rusting, is divided into classes. Class I galvanizing refers to the lightest coating, and class III refers to the thickest coating. A 7-gauge wire with a class I coating has 18.3 ounces of coating per square foot of surface area; a class III coating has 24.4 ounces of coating per square foot of surface area. Obviously, the thicker the coating is, the longer the wire will last. Most farm supply stores carry class I wire.

Barbwire was the first type of wire to be used on a large scale for containing livestock. It is still widely used today and comes in a variety of types. Lightweight barbwire can be as small as 16 gauge; heavier wire is 12 gauge. There are also differences in the type and number of barbs on the wire. Some barbs are spaced further apart than others. In addition, some wire has barbs with 4 points; other wire has barbs with only 2 points (Figure 43-7).

FIGURE 43-7 There are several types of barbwire. This type has 4 points.

The effectiveness of barbwire fences depends on several factors. The wire has to be stretched to the proper tightness. If the wire is too tight, it can break; if it is too loose, animals might push through it. The number of strands in the fence is also an important factor in the fence's ability to hold livestock. Most livestock producers use at least 5 strands with the strands spaced 8 to 10 inches apart.

High-Tensile Wire Fences. Another type of stranded fence, called high-tensile or New Zealand fencing, makes use of wires that are stronger than conventional barbwire. In fact, **high-tensile wire** can be stretched almost twice as tightly as barbwire without breaking. Because the strands are smooth, as opposed to having barbs, there is less risk of animal injury (Figure 43-8). High-tensile fences are usually constructed of 4 to 8 strands of wire. Sometimes fewer strands are used and an electrically charged wire is added to the fence. High-tensile fences are more expensive than barbwire, but they last longer because of the heavier coating of zinc on the wires, and they have less potential for causing harm to livestock.

One problem with high-tensile wire fences is keeping the wires tight. In areas where the temperature falls below zero, the wires contract during the winter and expand during the summer. This causes problems with consistency in wire tension that can be solved in two ways. Springs can be used to allow the fence to expand or contract as the temperature fluctuates. A ratcheting device that allows the wire to be tightened or loosened as needed can also be used (Figure 43-9). Obviously this creates extra work in maintaining the fence.

Woven Wire Fences. Woven wire fences consist of wires woven together to form a fence that resembles a net. This type of fencing material is sometimes referred to as net wire. The fence net is made by wrapping the vertical wires (called **stays**) around the horizontal wires to secure the wires together. Some woven wires have separate wires tied in a knot at the juncture of the stays and the horizontal wires. These joints also act as hinges to give the wire flexibility (Figure 43-10).

There are several types of woven wire, which vary by the diameter of the wires, the spacing of the wires, and the height of the woven wires. Heights commonly vary from 28 inches to 8 feet or more. The height of the fence depends on the type of livestock that is confined. Taller woven wire is used to keep animals such as deer and elk out of an area. A three- or four-digit number

FIGURE 43-8 A properly built high-tensile wire fence. This one is also an electric fence.

FIGURE 43-9 A ratcheting type device is used to maintain tension on high-tensile wires.

FIGURE 43-10 Woven wire of this type has joints that serve as hinges to add flexibility to the wire.

indicating the number of wires that run horizontally (lengthwise) and the height of the wire in inches identifies the type of wire. For example, the number 828 refers to a woven wire that has 8 horizontal wires and is 28 inches tall. Some manufacturers add other numbers to indicate the **pullout** (the distance between the stay wires) and the gauge. For example, 1047-6-9 refers to wire that has 10 horizontal wires, is 47 inches tall, has a pullout of 6 inches, and is made of 9-gauge wires. Woven fencing materials are also classified by weight as light, medium, heavy, and extra heavy, depending on the size of the wires in the fence.

Chain-Link Fences

Another type of woven wire is a chain-link fence consisting of crimped vertical wires that are linked together very much like the links in a chain. These fences are used primarily to fence in yards and as security fences. Although they are more attractive and stronger, chain-link fences are the most expensive of the woven wire fences. Steel tubes are used at the bottom and the top of the fence, so the posts must be closely spaced and placed in concrete. Cost usually limits the use of this type fence to smaller areas.

Selecting Posts

Selecting the proper type and size of posts is just as important as the type and size of wire used to build the fence. The most commonly used posts are made of wood, vinyl, or steel. Each has advantages and disadvantages.

Wooden posts are probably used most often. They are relatively inexpensive and, if properly preserved, last for many years. If the fence is made of boards, a square post is usually used. If it is made of wire, round posts are generally used. Round posts are less expensive, because they do not have to be sawed into a square shape (Figure 43-11). Small trees are simply harvested when they reach the appropriate size for fence posts. The size can range from about 2 inches in diameter to more than 6 inches in diameter. The wood most commonly used for fence posts is southern pine. It is fast-growing and absorbs preserving solution very well. Once the trees are cut into logs, the bark is peeled off and the posts are cut to the proper length. They are then dried to the correct moisture content and are pressure-treated with a preservative that prevents decay for up to 30 years.

Steel posts are more expensive than wooden posts but they are easier to install. Generally they are driven into the ground with a post driver, so there is no need to dig a posthole. Wire is attached by tying it to the post

FIGURE 43-11 Round posts are less expensive because they do not have to be sawed into a square shape.

FIGURE 43-12 Wire is attached to steel posts by a tie wire.

(Figure 43-12). Vinyl posts are used with vinyl fencing, but they can also be used much like wooden fences when installing wire.

CONSTRUCTING FENCES

No matter what materials are used to build a fence, there are guidelines that must be followed to build a fence that will suit the intended purpose, last for years, and require minimum maintenance. As with any project, time spent in planning will eliminate problems during construction and prevent future problems. Skimping on materials and taking shortcuts in construction will usually result in a poor fence.

The most critical guidelines to follow are those related to safety. Following are safety guidelines for constructing fences.

- Make sure the fence line is cleared of all brush before construction begins. This prevents falls and tangled wire.
- Wear protective clothing such as eye protection, leather gloves, and steel-toe boots.
- Never stand in the direct line of a wire that is being stretched. If the wire breaks, severe injury can occur.
- Never use a tractor to stretch wire. Too much power can cause the wire to break and injure people in the area.
- Stand clear of any power machines, such as power posthole diggers.
- Keep the work area free of clutter. Tools should be kept orderly and out of the way. Wire can become tangled in clutter, and people can trip over tools that are scattered around.

Setting the Corner Posts

Once the fence line has been cleared of all brush, the corner posts should be set to establish the size and shape of the fenced-in area. All fence lines are then lined up on the corner posts. This means that the corners should be carefully placed and constructed to provide anchoring for the fence. Remember that all wires will be stretched around the corner posts. Generally the wire is placed inside the posts along the fence line but is wrapped around the outside of the corner posts (Figure 43-13).

The corner posts should be the largest posts in the fence—at least 6 to 8 inches in diameter and up to 8 feet in height. To set a corner post, begin by digging a hole that is approximately $\frac{1}{3}$ the length of the post. For example, if the fence is to be $5\frac{1}{2}$ feet high, the corner posts should be about 8 feet long to allow $2\frac{1}{2}$ feet to be in the ground and $5\frac{1}{2}$ feet above the ground. Dig the hole 12 to 18 inches in diameter, and place the post in the corner position. Pour properly mixed concrete into the hole and let it set up for about 48 hours. When all the corner posts are set, a heavy string or cord can be run between them. This will serve as a guideline for setting all of the posts on the fence line. Using the string as a guideline, dig holes for the brace posts about 6 feet from each side of the corner post (Figure 43-14). When these posts have been placed in the holes, nail a brace between the brace posts and the corner posts. Fence posts can be used for braces. Notches are cut in the upright posts to provide a place to nail the braces. Carefully line the brace posts up with the string and concrete them in. To further strengthen the corners, run at least two loops of wire between the top of the corner post and the bottom of each brace post. Place a

FIGURE 43-13 All wire is stretched around the fence posts. Generally the wire is on the inside of the posts on the fence line but around the outside of the corner.

Courtesy of *Progressive Farmer* magazine, February, 2004

FIGURE 43-14 A strong string or cord is used to line up the posts in a straight line.

strong stick or metal rod between the strands and twist until the wire is tight. This should provide a strong corner for the fence.

Setting the Fence Line

After the corner posts are properly set and braced, the posts in the fence line are carefully lined up using the string stretched between the corners. To keep the wires tight, the posts in the fence line must be in a straight line. Many pastures are shaped so that it would be impossible for the fence to run in perfectly straight lines from corner to corner. If the line varies from straight, a brace must be constructed that is very much like those used at the corners. Also, if the fence line is long, bracing is necessary about every 200 to 300 feet to help stretch the fence and keep the wire tight (Figure 43-15).

Copyright © 2015 Cengage Learning®./Ray Herren

FIGURE 43-15 Bracing is necessary at least every 200 to 300 feet in order to keep the fence tight. This is a properly constructed fence brace.

For wire fences, posts should be set about 15 feet apart. If the posts are small or the ground is not level, the posts should be set closer together. Postholes are dug with a hand digger, a digger run by a small gasoline engine, or a digger mounted on a tractor. Metal posts and some wooden posts can be driven into the ground using a post driver that operates off the hydraulic system of a tractor. As the posts are set into the ground, they should be plumbed with a level. Having the posts straight will aid greatly in keeping the wires tight and properly aligned. They are usually tamped into place using a strip of wood or metal rod that packs the soil around the post.

Stretching and Securing the Wire

Wires must be stretched to the proper tension. Woven wire has **stretch indicators** in the form of a bend or crimp in the wire (Figure 43-16). When the wire is stretched to the proper tension, the bend straightens out to a degree indicated by the wire manufacturer. A ratcheting wire stretcher is often used to stretch the wire. It can be anchored to the corner post or to a tractor parked in the proper location (Figure 43-17). Remember that a tractor is used only as an anchor for stretching wire. Never attempt to stretch the wire by pulling it with a tractor. Not only is there danger that the wire will be stretched too tightly, but serious bodily injury can also occur if the wire breaks or suddenly comes loose.

Once the wire is properly stretched, it is secured to wooden posts with staples. These fasteners may have slight barbs on them to help prevent pullout. The staple should be driven in across the grain and not with the

Copyright © 2015 Cengage Learning®./Ray Herren

FIGURE 43-16 The bend in the wire is a stretch indicator.

FIGURE 43-17 Proper wire tension may be obtained using a fence stretcher. Do not use a tractor to stretch fence wire.

FIGURE 43-18 This gate serves a functional purpose and is not intended to be attractive. Note the proper bracing of the gate posts.

grain of the post. This prevents the post from splitting and helps to hold the staple more securely. The staple should be driven in only until the wire touches the post. Driving the wire into the wood only weakens it. Wire is fastened to steel posts by wrapping it around the wire and then around the post. The wire fits into notches in the post that prevent the wire from slipping down the post. When installing barbwire or high-tensile wire, space the wires carefully on the posts to ensure that the wires are an equal distance apart on the posts.

Gates

Obviously, to be useful, fences must have a means of entrance and exit. Therefore, a gate must be constructed somewhere along the fence line. Care should be taken in planning where the gate should be located. Consider where access to the enclosed area is needed the most and what needs to move through the gate. Will the gate be used for livestock, humans, trucks, tractors, large machinery, or all of these? The gate should be of sufficient size to allow access for the largest object that is to be moved through it. At the same time, the gate should not be larger than necessary.

People have been creative in the materials they use to build gates. Often, they look like objects of art. Wheels from wagons or horse-drawn implements are common in some places. Strips of metal can be shaped into interesting designs and carefully built wooden fences can be attractive. Many people buy ready-made gates composed of wire stretched over a frame of metal tubing. Of course, there are some places where looks are not important in the gate design (Figure 43-18). Whether the gate is designed to be attractive or merely functional, it must be properly constructed and hung in order to operate properly and last for many years.

Posts on either side of the gate should be properly braced in much the same manner as the corner posts. A lot of stress will be on the gateposts from the stretched wire. The post on which the gate is hung will need to be of sufficient size to carry the weight of the gate. Hinges used in wooden gateposts are usually the screw hook and strap type (see Figure 7-25) or some variation of it. These hinges are easy to install, operate very well, and are strong enough to last for many years with little maintenance.

SUMMARY

Fences have been used for thousands of years for a variety of purposes. Today there are a wide variety of fences designed for beauty, for protection, for use as a boundary, or to keep animals in or out of an area. Properly constructed fences built with the correct materials can last for many years. Today's building materials offer a broad choice of materials to use in fencing. Planning and careful layout will result in fences that are both attractive and functional.

Student Activities

1. Define the Terms to Know listed at the beginning of this unit.

2. Design a fence for an open area near your home. Describe the purpose for the fence, what materials are to be used, and a rationale for using those particular materials. Complete a bill of materials list for everything you will need to construct the fence, and calculate a final cost.

3. Make a list of the different types of gates you see in your area. Which are the most attractive? Which are the most functional?

4. Locate some well-constructed fences in your area. Take pictures and share with the class the reasons you consider the fences to be well-made.

Relevant Web Sites

Red Brand
www.redbrand.com

Stay-Tuff Fence Manufacturing
www.staytuff.com

Self-Evaluation

A. Multiple Choice. Select the best answer.

1. Early American fencing for controlling live-stock was made with logs split into rails and stacked
   a. in neat rows
   b. around small, circular enclosures
   c. end-to-end
   d. in a crisscross manner

2. Vinyl will deteriorate if exposed to
   a. pesticides
   b. moisture
   c. sun
   d. petroleum-based products

3. Livestock producers use at least five strands of barbwire spaced
   a. 2 to 3 inches apart
   b. 10 to 12 inches apart
   c. 8 to 10 inches apart
   d. 6 to 8 inches apart

4. Of the woven wire fences, chain-link fences are the most
   a. expensive
   b. effective
   c. unattractive
   d. all of these

5. The wood most commonly used for fence posts is
   a. cedar
   b. redwood
   c. birch
   d. Southern pine

6. From each side of the corner posts, brace posts should be placed
   a. 2 feet
   b. 6 feet
   c. 10 feet
   d. 12 feet

7. To keep wires tight, it is necessary to
   a. set the posts in the fence line in a straight line
   b. brace long fence lines every 200 yards
   c. plumb posts with a level as they are set into ground to keep them straight
   d. all of the above

B. Matching. Match the items in column I with those in column II.

Column I

1. masonry fences
2. charger
3. vinyl fences
4. lightning arrester
5. a "come along"
6. stretch indicator
7. joule
8. pullout
9. pulsation
10. high-tensile wire

Column II

a. a bend or crimp that straightens to indicate proper tension
b. steps up a household current of 120 volts to several thousand
c. amount of time the electric current flows along the wire
d. guaranteed to last a lifetime
e. controls amount of current flowing along the fence wire
f. used for retaining walls
g. can be very tightly stretched
h. amps × volts × time
i. used for stretching wire
j. distance between stay wires

C. Completion. Fill in the blanks with the word or words that will make the following statements correct.

1. The first primitive fences were built of _____ to deter criminals from entering the fields.
2. In electric fencing, a higher voltage reduces problems with _____.
3. The higher the gauge number is, the _____ the wire.
4. If a fence is made of boards, a _____ post is used, and if it is made of wire, _____ posts are used.
5. Never stand in the _____ _____ of a wire that is being stretched.

D. Brief Answer. Briefly answer the following questions.

1. What brought about the invention of barb-wire? How was it first used?
2. What are some advantages and disadvantages of electric fences?
3. What is the most common coating used to prevent rusting of wire? Explain how the thickness of the coating is categorized.
4. What does the number 840-8-12 on woven wire fencing indicate?
5. Explain the procedure for setting a corner post.

APPENDICES

CONTENTS

Project 1: CONCRETE FLOAT

Plan for Concrete Float

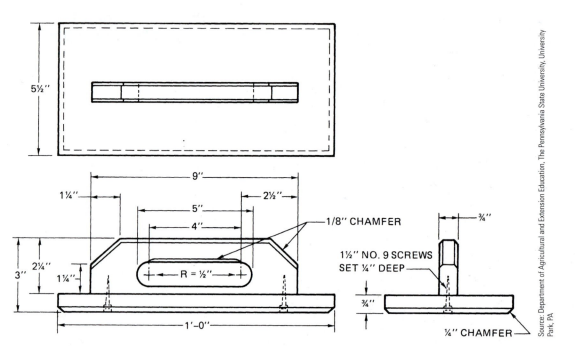

Source: Department of Agricultural and Extension Education, The Pennsylvania State University, University Park, PA

Construction Procedure for Concrete Float

1. Lay out and cut base to the dimensions given in the plan drawing.

2. Chamfer the base edges as shown in the plan.

3. Lay out and cut the outside dimensions of the handle.

4. Locate position of two 1-inch holes in the handle.

5. Bore the two 1-inch holes in the handle and saw out the piece between the holes.

6. Chamfer the edges of the handle as noted in the plan.

7. Drill pilot holes for the screws.

8. Center the handle on the base; locate and drill all screw holes.

9. Fasten handle to the base using glue and two 2½-inch No. 9 wood screws.

10. Sand the float and apply one coat of linseed oil.

Bill of Materials

Materials Needed	Quantity	Dimensions	Description or Use
Lumber (hard)	1 piece	¾" × 5½" × 1'0"	Base
Lumber (hard)	1 piece	¾" × 2¼" × 9"	Handle

Project 2: **PUSH STICK**

Plan for Push Stick

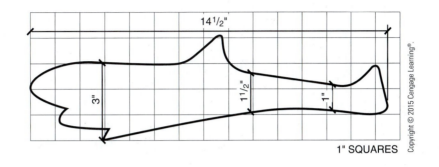

Copyright © 2015 Cengage Learning®.

1" SQUARES

Construction Procedure for Push Stick

1. Surface the 1-inch board to ¾-inch thickness with a hand plane or sander.

2. Lay off 1-inch squares on a piece of heavy paper or cardboard, and sketch an outline of the push stick, as shown in the drawing.

3. Cut out the pattern to use as a template.

4. Lay out the push stick by marking around the template.

5. Cut out the push stick by sawing along the outside edges of the mark with a band saw, jig saw, coping saw, or compass saw.

6. Sand and finish.

Bill of Materials

Materials Needed	Quantity	Dimensions	Description or Use
Lumber	1 piece	1" × 3" × 12"	—
Unsurfaced			
or			
Surfaced	1 piece	¾" × 3" × 12"	—
or			
Plywood	1 piece	½" × 3" × 12"	—

Project 3: **MITER BOX**

Plan for Miter Box

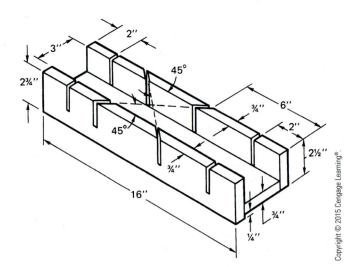

Copyright © 2015 Cengage Learning®.

Construction Procedure for Miter Box

1. Cut a kiln-dried piece of ¾-inch hard wood to 49 inches length.

2. Joint one edge of the board.

3. Rip the board to 3¹⁄₁₆ inches wide.

4. Joint the board to 3 inches wide.

5. Saw board into three pieces 16 inches long.

6. Rip pieces to ¹⁄₁₆ inch wider than is shown in the materials list. Then set the jointer to remove ¹⁄₁₆ inch and joint to exact width.

7. Lay off, bore, and countersink holes for wood screws. Space a hole 1" from each end on each side. Space the other hole in the middle on each side. Fasten side pieces to bottom piece with three wood screws on each side.

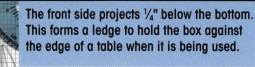

NOTE

The front side projects ¼" below the bottom. This forms a ledge to hold the box against the edge of a table when it is being used.

8. Lay off and saw a square cut in the side pieces 2 inches from the ends, with a miter box saw. Saw both side pieces at the same time.

9. Lay off and saw 45-degree cuts in the side pieces 6 inches from the right end, as shown in the drawing. These cuts serve as a guide for the saw when square or miter cuts are desired. In some instances, it may be desirable to cut other angles in the miter box to fit a particular situation. If so, the desired angle cut may be substituted for one of the 45-degree cuts.

Bill of Materials

Materials Needed	Quantity	Dimensions	Description or Use
Lumber (hard)	1 piece	¾" × 3" × 16"	Bottom
Lumber (hard)	1 piece	¾" × 2¾" × 16"	Front
Lumber (hard)	1 piece	¾" × 2½" × 16"	Back
Wood Screws	6	No. 8 × 2" Flat-head	Attach

Project 4: **GUN RACK**

Plan for Gun Rack

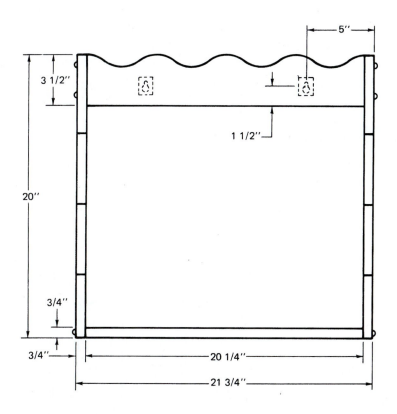

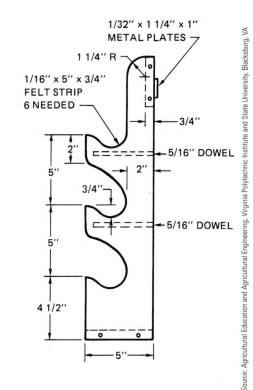

Source: Agricultural Education and Agricultural Engineering, Virginia Polytechnic Institute and State University, Blacksburg, VA

Construction Procedure for Gun Rack

1. Cut a piece of heavy paper 3½" × 20¼". Fold the paper in half end-to-end and cut a pattern for the back.

2. Draw the pattern topline onto a piece of stock ¾" × 3½" × 20¼".

3. Use a piece of heavy paper 5" × 20" to cut a pattern for the sides.

4. Cut the topline of the back using a coping, jig, band, or bayonet saw.

5. Cut two 5" × 20" end pieces from ¾-inch stock.

6. Use the pattern to draw the gun supports on the end pieces.

7. Lay out the ¾" × 3½" cut on each piece to receive the back.

8. Saw out the ends.

9. Cut a ¾" × 5" × 20¼" piece for the bottom.

10. Carefully drill holes and install dowel pins in the ends.

11. Carefully and thoroughly sand all surfaces, being careful not to round edges where the joints are made.

12. Glue and screw all parts together.

13. Apply a suitable finish.

14. Use stick-on felt strips in the curves where the guns are cradled.

15. Install hangers or plan to hang the rack by two nails or screws.

Bill of Materials

Materials Needed	Quantity	Dimensions	Description or Use
Lumber, surfaced	1 piece	$\frac{3}{4}" \times 3\frac{1}{2}" \times 20\frac{1}{4}"$	Back
Lumber, surfaced	2 pieces	$\frac{3}{4}" \times 5" \times 20"$	Ends
Lumber, surfaced	1 piece	$\frac{3}{4}" \times 5" \times 20\frac{1}{4}"$	Bottom
Lumber, surfaced	2 pieces	$\frac{5}{16}" \times 4\frac{1}{4}"$	Dowels
Screws	6 each	No. $8 \times 1\frac{1}{2}"$	Round or flat-head
Metal Plates	2 each	$\frac{1}{32}" \times 1\frac{1}{4}" \times 1"$	Hangers

Project 5: **FLICKER/WOODPECKER HOUSES**

Plan for Flicker/Woodpecker Houses

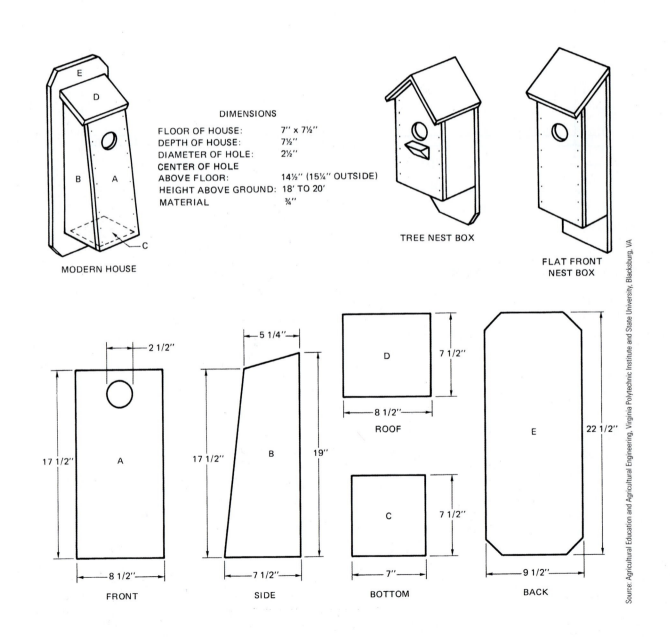

DIMENSIONS

FLOOR OF HOUSE:	7'' x 7½''
DEPTH OF HOUSE:	7½''
DIAMETER OF HOLE:	2½''
CENTER OF HOLE ABOVE FLOOR:	14½'' (15¼'' OUTSIDE)
HEIGHT ABOVE GROUND:	18' TO 20'
MATERIAL	¾''

MODERN HOUSE

TREE NEST BOX

FLAT FRONT NEST BOX

FRONT

SIDE

ROOF

BOTTOM

BACK

Source: Agricultural Education and Agricultural Engineering, Virginia Polytechnic Institute and State University, Blacksburg, VA

The flicker will nest readily in a well-placed birdhouse of the proper dimensions. Roughen the interior of the nest box to assist the young woodpeckers to reach the entrance hole. Cover the bottom with sawdust so that the mother bird can shape the nest for eggs and the young birds. Material may be ½ inch instead of ¾ inch thick, in which case the width of the front and roof will be 8 inches.

Construction Procedure for Flicker/Woodpecker Houses

1. Select pine, redwood, or other lumber that will withstand weather without paint. Rough lumber is generally preferred and will provide a rustic appearance. Planed lumber should be used if the house is to be painted for decorative purposes, but this is not recommended for most species except Martin houses.

2. Cut all parts from a ¾" × 9½" × 8" board.

3. Cut the sides to the proper angle.

4. Bevel the top edge of the front to conform to the angle of the sides.

5. Bevel the upper edge of the roof to conform to the back and sides.

6. Nail the front, sides, and bottom together.

7. Nail the back to the sides and bottom. (Center the assembly on the back for a good appearance. Leave 1 inch at bottom and 2½ inches at the top.)

8. Attach the roof with two hinges or nail it lightly so the roof is easily removed for periodic cleaning of the nest area.

Bill of Materials (for modern house)

Materials Needed	Quantity	Dimensions	Description or Use
Lumber (rough and rustic)	1	¾" × 8½" × 17½"	Front (A)
"	2	¾" × 7½" × ⅑"	Sides (B)
"	1	¾" × 7" × 7½"	Bottom (C)
"	1	¾" × 7½" × 8½"	Roof (D)
"	1	¾" × 9½" × 22½"	Back (E)
Box nails or finishing nails		4d or 6d	Attach

Project 6: **NESTING AND DEN BOXES**

Plan for Nesting and Den Boxes

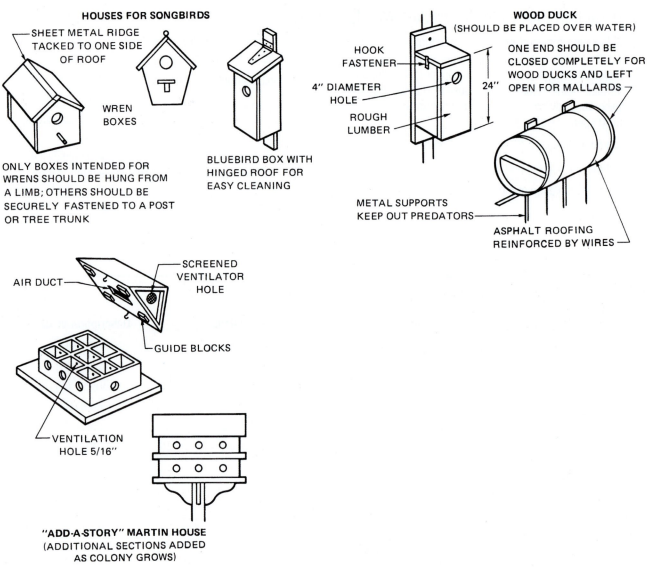

HOUSES FOR SONGBIRDS

SHEET METAL RIDGE TACKED TO ONE SIDE OF ROOF

WREN BOXES

ONLY BOXES INTENDED FOR WRENS SHOULD BE HUNG FROM A LIMB; OTHERS SHOULD BE SECURELY FASTENED TO A POST OR TREE TRUNK

BLUEBIRD BOX WITH HINGED ROOF FOR EASY CLEANING

WOOD DUCK
(SHOULD BE PLACED OVER WATER)

HOOK FASTENER

4" DIAMETER HOLE

ROUGH LUMBER

24"

ONE END SHOULD BE CLOSED COMPLETELY FOR WOOD DUCKS AND LEFT OPEN FOR MALLARDS

METAL SUPPORTS KEEP OUT PREDATORS

ASPHALT ROOFING REINFORCED BY WIRES

AIR DUCT

SCREENED VENTILATOR HOLE

GUIDE BLOCKS

VENTILATION HOLE 5/16"

"ADD-A-STORY" MARTIN HOUSE
(ADDITIONAL SECTIONS ADDED AS COLONY GROWS)

Based on information from Maryland Department of Natural Resources, Public Information Service

Construction Procedure for Nesting and Den Boxes

Construction procedures will vary with the nesting or den box being constructed. However, the procedure outlined for Project 5, Flicker/Woodpecker Houses, may be helpful.

Bill of Materials

Bills of materials are not provided with these nesting and den boxes due to the variety of types pictured. Nesting and den boxes should be planned from the information provided. An appropriate bill of materials may then be developed for the plan.

Dimensions for Each Species

Species	Diameter of Entrance	Floor of Cavity	Depth of Cavity	Entrance above Floor
Bluebird	1½"	5" × 5"	8"	6"
Chickadee	1⅛"	4" × 4"	8"–10"	6"–8"
Titmouse	1¼"	4" × 4"	8"–10"	6"–8"
Nuthatches	1¼"	4" × 4"	8"–10"	6"–8"
House Wren	⅞"	4" × 4"	6"–8"	1"–6"
Carolina Wren	1⅛"	4" × 4"	6"–8"	1"–6"
Crested Flycatcher	2"	6" × 6"	8"–10"	6"–8"
Flicker	2½"	7" × 7½"	16"–18"	14"–16"
Purple Martin	2½"	6" × 6"	6"	1"
Tree Swallow	1½"	5" × 5"	6"	1"–5"
Sparrow Hawk	3"	8" × 8"	12"–15"	9"–12"
Barn Owl	3"	8" × 8"	12"–15"	9"–12"

Project 7: **SMALL ENGINE STAND—WOOD**

Plan for Small Engine Stand—Wood

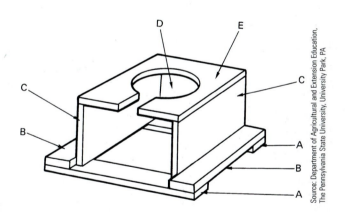

Source: Department of Agricultural and Extension Education, The Pennsylvania State University, University Park, PA

Construction Procedure for Small Engine Stand—Wood

1. All wooden joints should be made with glue and nails, or glue and screws.

2. Use pieces "A" and "B" to construct the frame base as shown in the drawing, making sure the base is square.

3. Fasten the two "C" pieces into position. One end of each "C" piece should be flush to the rear edge of the base frame; the front ends will fall short of the front edge of the base frame by ½ inch.

4. Place the "D" piece or back of the stand between the sides of the stand and flush with the back ends.

5. Determine the center of the top plywood piece "E" and draw a 7-inch diameter circle around it. Cut a 7-inch hole and a 2-inch front slot (centered on the front) before fastening the top to the stand.

6. For the front of the stand a safety door, not shown, should be constructed from ½-inch material. It may be fitted to the front and secured in place by small hooks and eyes or hinges.

7. All wooden parts should be painted with an enamel paint. Recommended colors are grey, green, or blue.

Bill of Materials

Materials Needed	Quantity	Dimensions	Description or Use
Lumber, No. 2 White Pine, surfaced	2 pieces	$\frac{3}{4}" \times 2\frac{1}{4}" \times 16"$	Front and back of base frame (A)
Lumber, No. 2 White Pine, surfaced	2 pieces	$\frac{3}{4}" \times 2\frac{1}{4}" \times 12"$	Sides of base frame (B)
Lumber, No. 2 White Pine, surfaced	2 pieces	$\frac{3}{4}" \times 5\frac{1}{2}" \times 11\frac{1}{2}"$	Sides of stand (C)
Lumber, No. 2 White Pine, surfaced	1 piece	$\frac{3}{4}" \times 5\frac{1}{2}" \times 10"$	Back of stand (D)
Plywood, Good one side	1 piece	$\frac{3}{4}" \times 11\frac{1}{2}" \times 11\frac{1}{2}"$	Top of stand (E)

Project 8: **SADDLE RACK**

Plan for Saddle Rack

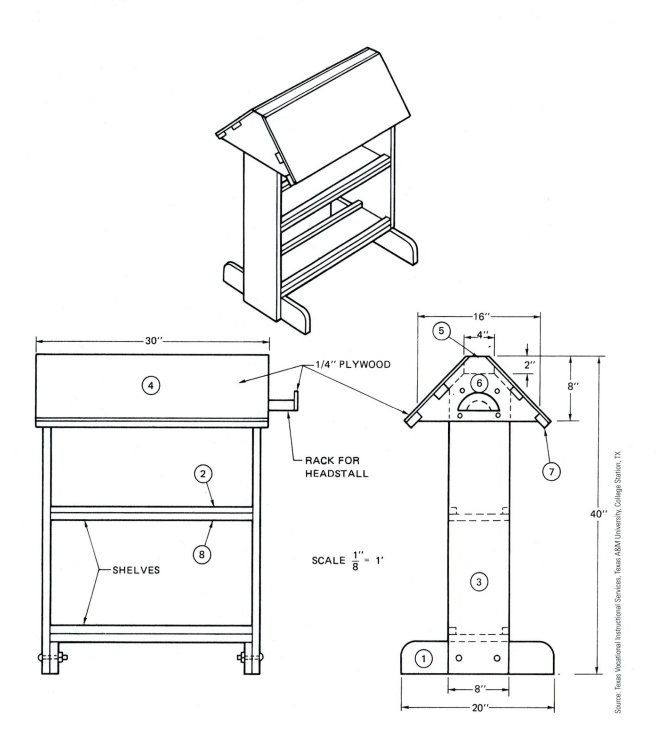

1/4" PLYWOOD

RACK FOR
HEADSTALL

SHELVES

SCALE $\frac{1}{8}" = 1'$

30"

16"

4"

2"

8"

40"

8"

20"

Source: Texas Vocational Instructional Services, Texas A&M University, College Station, TX

Construction Procedure for Saddle Rack

1. Cut two pieces 1" × 8" × 16" for the top ends. Set the boards on edge and mark the center on the top edge. Mark locations 1 inch out in both directions from the centers. Draw a line from each of these points to the lower corners on each end of the boards. Saw the ends to these lines.

2. Cut the ridge to length and joint two corners so its shape matches the top of the end pieces.

3. Cut all other pieces to size as listed in the bill of materials, except the shelves and cleats. These should be cut about 1 inch longer than indicated.

4. Use glue and nails or screws to assemble the top.

5. Use glue and $\frac{5}{16}$-inch carriage bolts with washers to attach the sides to the base pieces.

6. Use glue and eight $\frac{1}{4}$-inch carriage bolts with washers to attach the sides to the top. Be very careful to attach the ends so the unit is straight and sits squarely on its feet on a flat surface.

7. Attach the cleats to the shelf pieces with glue and nails.

8. Measure the distance from inside to inside of the two ends at the top of the rack. Cut the length of the shelf and cleat assemblies to this dimension.

9. Install the shelf and cleat assemblies with glue and 10d finishing nails or No. 8 × 2½" wood screws.

10. Smooth all surfaces and apply a suitable finish.

Bill of Materials

Materials Needed	Quantity	Dimensions	Description or Use
Lumber	2 pieces	1" × 4" × 20"	Base (1)
Lumber	4 pieces	1" × 1" × 26"	Cleats (2)
Lumber	2 pieces	1" × 8" × 40"	End Supports (3)
Lumber	2 pieces	¼" × 12" × 30"	Top (4)
Lumber	1 piece	2" × 4" × 30"	Center (Ridge) (5)
Lumber	2 pieces	1" × 8" × 16"	Ends (6)
Lumber	4 pieces	1" × 2" × 30"	Ribs (7)
Lumber	2 pieces	1" × 8" × 26"	Shelves (8)
Carriage bolts	12	$\frac{5}{16}$" × 2½"	—
Wood screws	20	#8 × 2½"	—

Project 9: **SAWHORSE—LIGHTWEIGHT**

Plan for Sawhorse—Lightweight

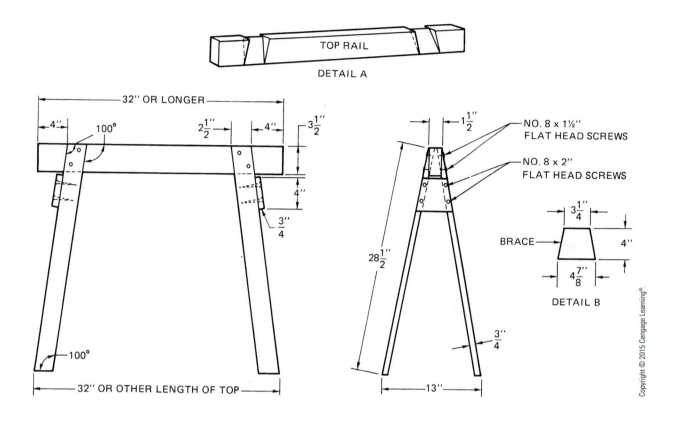

Construction Procedure for Sawhorse—Lightweight

1. Cut the top rail from a surfaced 2 × 4.

2. Cut one leg from clear lumber so the finished size is ¾" × 2½" × 28½" with ends cut at 100-degree angles as shown in the drawing. Tough wood such as oak or yellow pine is recommended.

3. Use the first leg as a pattern to cut the other three legs.

4. Mark "top" on one of the 1½-inch edges of the top rail.

5. Mark the top of the rail 4 inches in from each end and square a line across the 1½-inch surface at each mark. (See Detail A for illustration of steps 5–8.)

6. Use a T-bevel to draw a line at 100 degrees from each end of each line down the sides of the rail.

7. Lay the leg pattern against each line and draw a second line parallel to each. The lines for all four legs are now drawn along the sides of the top rail. These should be such that when legs are attached, the outside toe of each leg will be in line with the end of the top rail.

8. Draw a line down the center of the top of the rail between each pair of lines.

9. Use a handsaw to make multiple cuts inside the sets of parallel lines as the first step in cutting a dado for each leg. The depth of the cuts is ¾ inch on the top of the rail and 0 inch on the bottom of the rail.

10. Use a wood chisel to cut out the dados. Smooth the bottom of each dado with a coarse file.

11. Drill and countersink holes in the legs for wood screws.

12. Cut out two leg braces ¾" × 4" wide by 3¼" and 4⅞". See Detail B.

13. Drill and countersink for the wood screws.

14. Assemble the sawhorse with glue and screws.

15. If the sawhorse rocks when placed on a flat surface, place a thin block of wood on the floor against the leg. Draw a line at the top of the block and one on each side and on each edge of the leg. Repeat the process on the other three legs. Be careful that the sawhorse does not rock during this procedure. Resaw the legs on the lines so established.

16. Smooth the ends of the legs on top of the rail.

17. Apply a suitable finish.

Bill of Materials

Materials Needed	Quantity	Dimensions	Description or Use
Lumber, surfaced	1 piece	1½" × 3½" × 32"	Top rail
*Lumber, surfaced	4 pieces	¾" × 2½" × 29"	Legs
Lumber, surfaced	2 pieces	¾" × 4" × 5"	Braces
Wood screws	8	No. 8 × 1½"	Flat-head, legs
Wood screws	8	No. 8 × 2"	Flat-head, braces

*Use clear lumber, preferably hardwood.

Project 10: **TOOL-SHARPENING GAUGE**

Plan for Tool-Sharpening Gauge

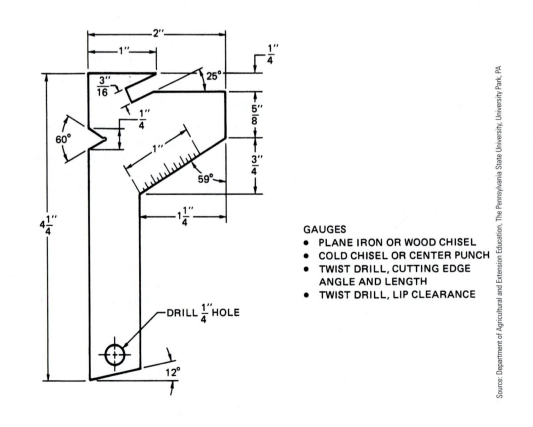

GAUGES
- PLANE IRON OR WOOD CHISEL
- COLD CHISEL OR CENTER PUNCH
- TWIST DRILL, CUTTING EDGE ANGLE AND LENGTH
- TWIST DRILL, LIP CLEARANCE

Source: Department of Agricultural and Extension Education, The Pennsylvania State University, University Park, PA

Construction Procedure for Tool-Sharpening Gauge

1. Measure and scribe outline on the stock using a scratch awl.

2. Cut out tool gauge with snips.

3. Use snips and a flat file to cut the 25-degree wood chisel slot.

4. Use a taper file to cut the 60-degree cold chisel vee.

5. Position and drill a ¼-inch hole at the narrow end of the gauge.

6. Measure and scribe a 1-inch rule by ¹⁄₁₆-inch graduations.

7. Cut rule indicator marks with a cold chisel or awl.

8. Smooth all edges with a flat file.

Bill of Materials

Materials Needed	Quantity	Dimensions	Description or Use
Sheet metal, galvanized, 24 gauge	1	2" × 4¼"	—

Project 11: **FEED SCOOP**

Plan for Feed Scoop

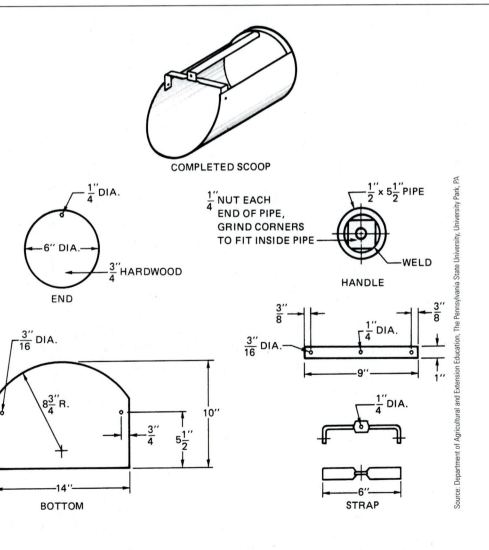

COMPLETED SCOOP

END

HANDLE

BOTTOM

STRAP

Source: Department of Agricultural and Extension Education, The Pennsylvania State University, University Park, PA

Construction Procedure for Feed Scoop

1. Lay out and cut the sheet metal and wood end.
2. Cut and bore holes in the strap iron, $\frac{1}{8}" \times 1" \times 9"$.
3. Cut the light steel tubing, $\frac{1}{2}" \times 5\frac{1}{2}"$ long for the handle.

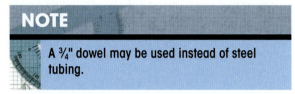

NOTE
A $\frac{3}{4}"$ dowel may be used instead of steel tubing.

4. Bend the bottom sheet metal to fit the end.
5. Drill holes for screws and handle bolt.
6. Screw the metal to the wood end.

NOTE
Screw-shank or other improved nails may be used in place of screws.

7. Heat and bend the iron strap.
8. Attach the iron strap to the sheet metal using rivets.
9. Weld nuts into ends of the tubing.

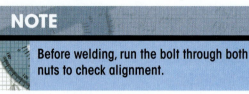

NOTE
Before welding, run the bolt through both nuts to check alignment.

10. Attach the handle with $\frac{1}{4}" \times 6\frac{1}{2}"$ bolt.
11. Finish and paint the handle, bracket, and wood end.

Bill of Materials

Materials Needed	Quantity	Dimensions	Description or Use
Sheet metal, galvanized, 24 gauge	1	$10" \times 14"$	Bottom
Lumber, hardwood	1	$\frac{3}{4}" \times 6" \times 6"$	End
Mild steel, hot rolled	1	$\frac{1}{8}" \times 1" \times 9"$	Strap
Steel tubing	1	$\frac{1}{2}"$ I.D. $\times 5\frac{1}{2}"$	Handle
Hardware			
Stove bolt	1	$\frac{1}{4}" \times 6\frac{1}{2}"$	—
Nut	2	$\frac{1}{4}"$	—
Rivet	2	$\frac{3}{16}"$	—
Screw	6	$\#6 \times \frac{3}{4}"$	Pan head
Paint (nonlead)	$\frac{1}{2}$ pint	—	For end, handle, and hand bracket

Project 12: **STAPLE**

Plan for Staple

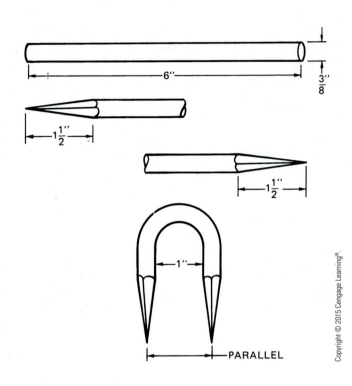

Construction Procedure for Staple

1. Cut a piece of ⅜-inch round rod 6-inches long.

2. Heat and put a long, round point on one end.

3. Heat and put a long, square point on the other end.

4. Bend to form a staple:

    a. One inch width between staple legs

    b. Legs of equal length

Bill of Materials

Materials Needed	Quantity	Dimensions	Description or Use
Steel, round hot or cold rolled	1	⅜" × 6"	—

Project 13: **COLD CHISEL**

Plan for Cold Chisel

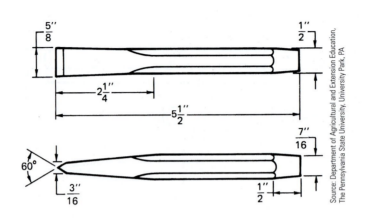

Source: Department of Agricultural and Extension Education, The Pennsylvania State University, University Park, PA

Construction Procedure for Cold Chisel

1. Heat 2¼ inches of one end of the stock to a uniform cherry red color.

2. Place one side against the anvil face. Using drawing blows, work to shape rapidly starting at the end and working back to 2¼-inch taper.

3. Finish to ³⁄₁₆-inch thickness and ⅝-inch width at the tip. Keep area hot (from a dull red to a cherry red color) when working it.

4. Anneal by heating to cherry red and cooling slowly (12 to 24 hours) in lime or sand.

5. File and polish the forged faces. Do not grind.

6. Temper with water (practice on an old cold chisel).

   a. Heat 2 inches to 3 inches of the tip to a uniform cherry red color.

   b. Cool ¾" to 1" in water until drops cling to the tip when it is removed from the water.

   c. Move the tip to avoid cracks at the water line.

   d. Quickly remove scale using a steel brush or

   e. Observe the color changes and quench lower ¼ inch on purple color. Color order is light straw, dark straw, brown, purple, dark blue, and light blue.

7. Grind the cutting edge to a 60-degree angle. Use a tool gauge to check the angle.

8. Chamfer the opposite end to approximately ½" by ⁷⁄₁₆" to prevent mushrooming.

Bill of Materials

Materials Needed	Quantity	Dimensions	Description or Use
Tool steel	1	½" × 5½"	Octagonal

Project 14: **DRAWBAR HITCH PIN**

Plan for Drawbar Hitch Pin

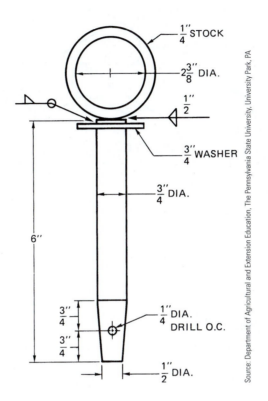

Source: Department of Agricultural and Extension Education, The Pennsylvania State University, University Park, PA

Construction Procedure for Drawbar Hitch Pin

1. Bend $\frac{1}{4}$" × 8" round stock around 2-inch pipe.

2. Cut the pin to length.

3. Drill a $\frac{1}{4}$-inch hole $\frac{3}{4}$ inch from one end of the pin.

4. Shape the end as indicated in the plan.

5. Weld the washer to the other end of the pin.

6. Place handle on pin and weld.

7. Remove slag and clean with a steel brush.

Bill of Materials

Materials Needed	Quantity	Dimensions	Description or Use
Hot rolled steel, M1020	1	$\frac{1}{4}$" × 8"	Handle
C1042 cold rolled or C1045			
hot rolled steel, round	1	$\frac{3}{4}$" × 6"	Pin
Washer, flat	1	$\frac{3}{4}$"	—
Hair pin	1	$\frac{1}{4}$" × 2"	Safety pin

Project 15: **HAY HOOK**

Plan for Hay Hook

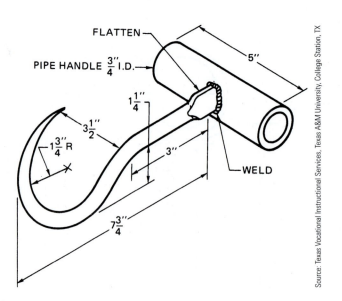

Source: Texas Vocational Instructional Services, Texas A&M University, College Station, TX

Construction Procedure for Hay Hook

1. Heat one end of the rod and forge it into a long, slender point.

2. Finish shaping the point with a grinder.

3. Reheat the pointed end and shape it to the dimensions given in the plan drawing.

4. Heat the other end and forge a flat area about ¼ inch thick and in line with the curved section.

5. Cut a piece of ¾-inch ID steel pipe 5-inches long.

6. Weld the hook to the center of the handle, being careful to keep the hook square to the handle.

7. Grind and sand all surfaces smooth and round for a comfortable feel.

Bill of Materials

Materials Needed	Quantity	Dimensions	Description or Use
Pipe, steel	1	¾" ID × 5"	Handle
Rod	1	⁷⁄₁₆" dia × 12"	Hook

Project 16: **FOOT SCRAPER**

Plan for Foot Scraper

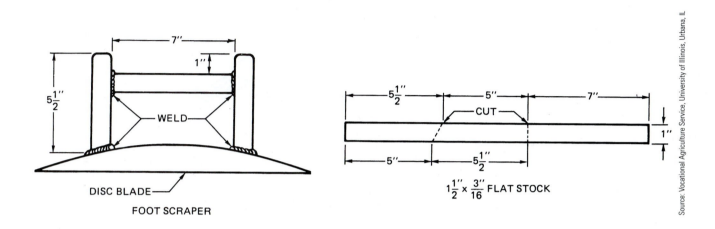

DISC BLADE

FOOT SCRAPER

$1\frac{1}{2}'' \times \frac{3}{16}''$ FLAT STOCK

Source: Vocational Agriculture Service, University of Illinois, Urbana, IL

Construction Procedure for Foot Scraper

1. Mark and cut the three pieces of flat stock with a hacksaw or torch.

2. Square the ends of the 7-inch piece and round the upper ends of the upright pieces on grinder.

3. Place the scraper piece on the edge of the welding table with the angled ends extending over the edge.

4. Clamp firmly and weld the crosspiece to the uprights.

5. Fit the angled ends of the uprights to the contour of the disc blade by grinding if necessary.

6. Clean the disc blade with a wire brush or portable grinder.

7. Weld the scraper to the disc blade carefully. Use low heat. Since the disc blade is high-carbon steel, either preheat or use a low-hydrogen electrode.

Bill of Materials

Materials Needed	Quantity	Dimensions	Description or Use
Disc blade	1	15" diameter	Base
Band, steel	2	$\frac{3}{16}'' \times 1\frac{1}{2}'' \times 5\frac{1}{2}''$	Legs
	1	$\frac{3}{16}'' \times 1\frac{1}{2}'' \times 7''$	Scraper

Project 17: **SMALL ENGINE STAND—METAL**

Plan for Small Engine Stand—Metal

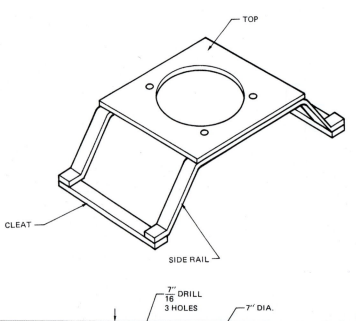

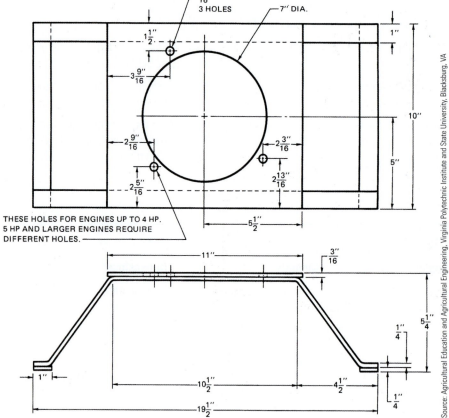

THESE HOLES FOR ENGINES UP TO 4 HP.
5 HP AND LARGER ENGINES REQUIRE
DIFFERENT HOLES.

Source: Agricultural Education and Agricultural Engineering, Virginia Polytechnic Institute and State University, Blacksburg, VA

Construction Procedure for Small Engine Stand—Metal

1. Locate and mark the midpoint of the two side pieces.

2. Measure and mark points 5½ inches in both directions from the midpoint.

3. Measure and mark points 1 inch from both ends of the side pieces.

4. Bend the side pieces so they are 19½ inches from end to end and 4¹³⁄₁₆ inches from the bottom to the upper surface. Place them side-by-side and be sure their shapes are identical.

5. Weld the cleats to the sides at 90-degree angles.

6. Lay out and center-punch for all holes in the top.

7. Drill the ⁷⁄₁₆-inch holes.

8. Use a divider to lay out the 7-inch diameter hole.

9. Drill a hole inside the 7-inch circle to insert a metal cutting bayonet saw blade and cut out the circle; or, use a cutting torch to make the cut.

10. Weld the top to the sides.

11. Apply a suitable finish.

Bill of Materials

Materials Needed	Quantity	Dimensions	Description or Use
Band, steel	2	¼" × 1" × 24"	Side rails
Band, steel	2	¼" × 1" × 10"	Cleats
Plate, steel	1	³⁄₁₆" × 10" × 11"	Top

Project 18: **SHOP STAND—ADJUSTABLE**

Plan for Shop Stand—Adjustable

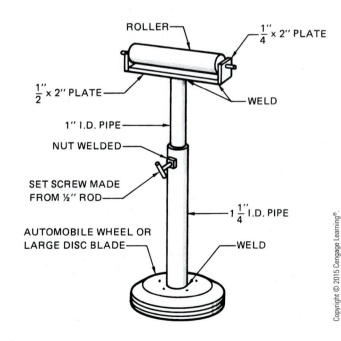

Copyright © 2015 Cengage Learning®.

Construction Procedure for Shop Stand—Adjustable

1. Obtain a rim of suitable size and weight to support the height of the stand needed. A large steel disc or other heavy metal object will also work well.

2. Determine the length of pipes needed to provide the desired height of the stand.

3. Drill a $\frac{5}{8}$-inch hole 1-inch from one end of the external post. Weld a $\frac{1}{2}$-inch nut over the hole. Make a setscrew from $\frac{1}{2}$-inch cold rolled round stock or use a standard bolt with a cross handle welded to it.

4. Weld the external post to the rim.

5. Obtain a standard roller or make one from 2-inch wood stock or $1\frac{1}{2}$-inch ID pipe 12 inches long.

6. Cut the base plate $\frac{1}{4}$-inch longer than the roller.

7. Lay out and drill bearing holes in the bearing ends to hold the top of the roller slightly above the tops of the ends.

8. Weld the bearing ends with the roller in place.

9. Apply a suitable finish.

Bill of Materials

Materials Needed	Quantity	Dimensions	Description or Use
Steel automobile or truck rim		According to need	Base
Pipe, black steel	1	$1\frac{1}{4}$" ID × *A	External post
Pipe, black steel	1	1" ID × *B	Internal post
Band, steel	1	$\frac{1}{2}$" × 2" × *C	Base plate
Band, steel	2	$\frac{1}{4}$" × 2" × 2"	Bearing ends
Wood or metal or rubber	1	Approx. 2" dia × 12"	Roller

*A—Determined by the desired minimum height of the stand minus the height of the base and roller.

*B—Same length as A.

*C—Same length as the roller plus $\frac{1}{4}$".

Note: Standard 1" ID pipe will fit well in standard $1\frac{1}{4}$" ID pipe. 2" ID pipe will fit well in $2\frac{1}{2}$" ID pipe.

Project 19: **STEEL POST DRIVER**

Plan for Steel Post Driver

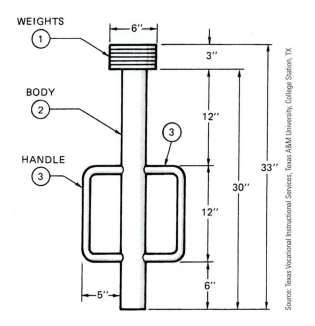

Source: Texas Vocational Instructional Services, Texas A&M University, College Station, TX

Construction Procedure for Steel Post Driver

1. Cut six pieces for weights and weld them to form a mass 3 inches thick; 6" × 6", or 6" in diameter.

2. Weld the weight mass securely to the body.

3. Bend the two handles so they are identical.

4. Weld the handles to the body.

5. Apply a suitable finish.

Bill of Materials

Materials Needed	Quantity	Dimensions	Description or Use
Flat or plate steel	6	½" × 6" × 6" or 6" dia	Weights
Pipe	1	3" ID × 30"	Body
Round stock	2	½" × 20"	Handles

Project 20: **EXTENSION CORD STAND**

Plan for Extension Cord Stand

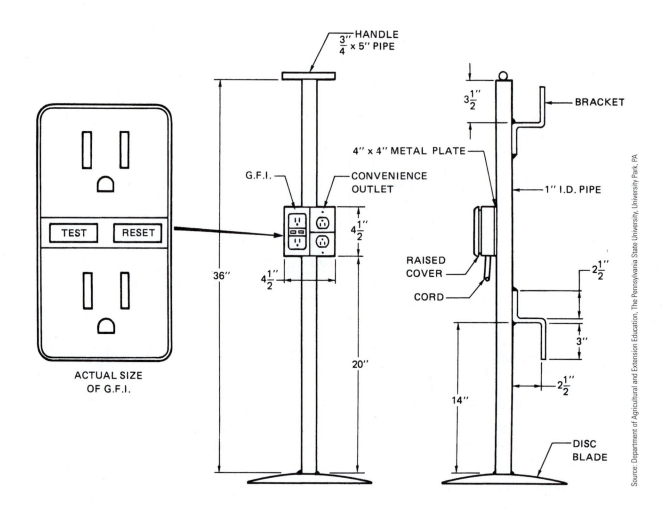

Source: Department of Agricultural and Extension Education, The Pennsylvania State University, University Park, PA

Construction Procedure for Extension Cord Stand

1. Cut the handle 5 inches long.

2. Cut the bracket pieces 8 inches long.

3. Shape the brackets to hold the electrical cable.

4. Cut 1-inch pipe to 36 inches length.

5. Weld the pipe to the disc blade, using a low hydrogen electrode.

6. Weld the handle and brackets to the pipe.

7. Drill ⅛" × 4" × 4" steel plate for electrical boxes; weld the plate to the pipe.

8. Bolt the electrical box to the plate.

9. Install one end of conductor into the box.

10. Install the ground-fault interrupter (GFI).

11. Install the duplex receptacle so it is protected by the GFI.

12. Add the cover plate.

13. Connect the male plug to the electrical conductor.

14. Apply a suitable finish.

Bill of Materials

Materials Needed	Quantity	Dimensions	Description or Use
Pipe, black iron	1	1" ID × 36"	Upright
Pipe, black iron	1	¾" × 5"	Handle
Band, iron	2	⅛" × ¾" × 8"	Brackets
Disc blade	1	12" dia	Base
Plate, steel	1	⅛" × 4" × 4"	Outlet box base
Outlet box, square	1	4" square, 1½" deep	—
*Ground fault interrupter	1	—	—
Duplex receptacle	1	—	—
Grounded cap, male plug	1	—	—
Raised cover	1	—	—
Box connector	1	¾"	—
Solderless connectors	3	—	—
Cable	100 feet	—	Neoprene or plastic covered portable cable, AWG
			12-2 stranded conductor with ground

*Substitute a duplex receptacle if the circuit is already protected by a ground-fault interrupter.

Project 21: **JACK STAND—AUTOMOBILE**

Plan for Jack Stand—Automobile

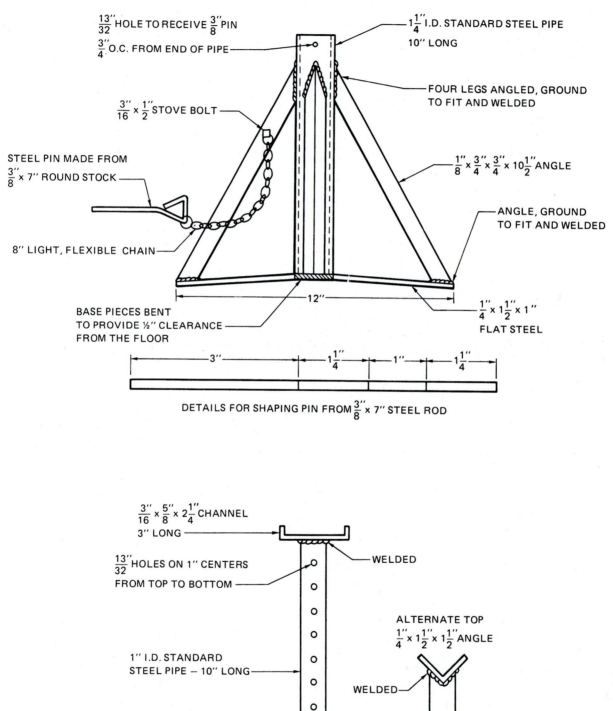

$\frac{13}{32}''$ HOLE TO RECEIVE $\frac{3}{8}''$ PIN

$\frac{3}{4}''$ O.C. FROM END OF PIPE

$1\frac{1}{4}''$ I.D. STANDARD STEEL PIPE
10'' LONG

$\frac{3}{16}'' \times \frac{1}{2}''$ STOVE BOLT

FOUR LEGS ANGLED, GROUND
TO FIT AND WELDED

STEEL PIN MADE FROM
$\frac{3}{8}'' \times 7''$ ROUND STOCK

$\frac{1}{8}'' \times \frac{3}{4}'' \times \frac{3}{4}'' \times 10\frac{1}{2}''$ ANGLE

8'' LIGHT, FLEXIBLE CHAIN

ANGLE, GROUND
TO FIT AND WELDED

12''

$\frac{1}{4}'' \times 1\frac{1}{2}'' \times 1''$
FLAT STEEL

BASE PIECES BENT
TO PROVIDE ½'' CLEARANCE
FROM THE FLOOR

3'' $1\frac{1}{4}''$ 1'' $1\frac{1}{4}''$

DETAILS FOR SHAPING PIN FROM $\frac{3}{8}'' \times 7''$ STEEL ROD

$\frac{3}{16}'' \times \frac{5}{8}'' \times 2\frac{1}{4}''$ CHANNEL
3'' LONG

$\frac{13}{32}''$ HOLES ON 1'' CENTERS
FROM TOP TO BOTTOM

WELDED

1'' I.D. STANDARD
STEEL PIPE — 10'' LONG

ALTERNATE TOP
$\frac{1}{4}'' \times 1\frac{1}{2}'' \times 1\frac{1}{2}''$ ANGLE

WELDED

Copyright © 2015 Cengage Learning®.

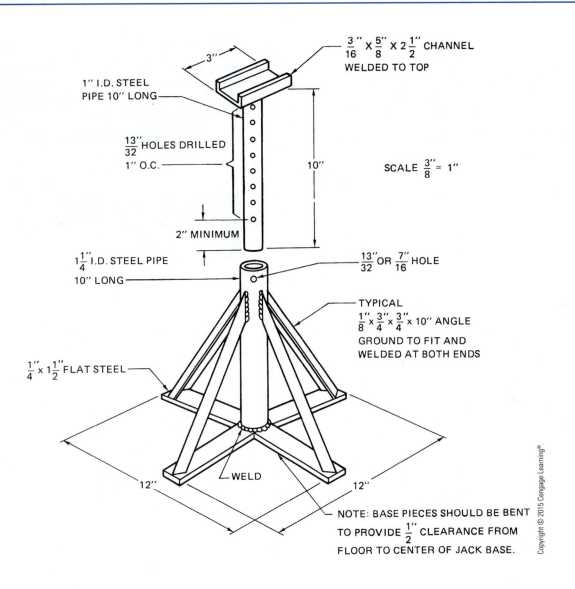

$\frac{3}{16}'' \times \frac{5}{8}'' \times 2\frac{1}{2}''$ CHANNEL
WELDED TO TOP

1" I.D. STEEL
PIPE 10" LONG

$\frac{13}{32}''$ HOLES DRILLED
1" O.C.

10"

SCALE $\frac{3}{8}'' = 1''$

2" MINIMUM

$1\frac{1}{4}''$ I.D. STEEL PIPE
10" LONG

$\frac{13}{32}''$ OR $\frac{7}{16}''$ HOLE

TYPICAL
$\frac{1}{8}'' \times \frac{3}{4}'' \times \frac{3}{4}'' \times 10''$ ANGLE
GROUND TO FIT AND
WELDED AT BOTH ENDS

$\frac{1}{4}'' \times 1\frac{1}{2}''$ FLAT STEEL

WELD

12"

12"

NOTE: BASE PIECES SHOULD BE BENT
TO PROVIDE $\frac{1}{2}''$ CLEARANCE FROM
FLOOR TO CENTER OF JACK BASE.

Copyright © 2015 Cengage Learning®.

Construction Procedure for Jack Stand—Automobile

1. Cut all pieces to length.

2. Weld two short base pieces to the longer piece to form a 90-degree cross. Grind welds smooth and slightly round off the exposed ends.

3. Bend the pieces of the base slightly so there will be ½-inch clearance between the floor and the center of the cross.

4. Ream the inside of the ends of the external post.

5. Center-punch ¾ inch from one end of the external post.

6. Scribe a mark from end-to-end down the center axis of the interior post.

7. Center-punch at 1-inch intervals on the line from end-to-end of the interior post.

8. Place the exterior post in a drill press vise or clamp it in a vee block and drill a 1¾₃₂-inch hole through both walls of the pipe.

9. Without unclamping the exterior post, slide the interior post into the exterior post and run the drill through the exterior post and on through both walls of the interior post. Drill all holes in the interior post in like fashion.

10. Weld the exterior post to the center of the base so the assembly sits with the post perpendicular to the floor.

11. Grind the legs so they fit the post and base. Tack-weld the legs to the base and to the post, being careful to keep the parts and spacings equal and the post perpendicular. Weld all joints completely.

12. Weld the top to the interior post.

13. Bend a loop in the pin and attach the chain to the pin.

14. Drill a $\frac{3}{16}$"-inch hole $1\frac{1}{2}$ inches from the top of one leg and anchor the chain with a $\frac{3}{16}$" × $\frac{1}{2}$" stove bolt.

15. Chip all welds, and remove all burrs.

16. Apply a suitable finish.

NOTE

All welds in this project must be high quality and a full-size pin must pass through all four pipewalls when stand is in use. For added safety, insert a hardened bolt in the first hole above the external post.

CAUTION!

Maximum capacity is one-half ton.

Bill of Materials

Materials Needed	Quantity	Dimensions	Description or Use
Band, steel	1	$\frac{1}{4}$" × $1\frac{1}{2}$" × $\frac{1}{2}$"	Base
Band, steel	2	$\frac{1}{4}$" × $1\frac{1}{2}$" × $5\frac{1}{4}$"	Base
Angle, steel	4	$\frac{1}{8}$" × $\frac{3}{4}$" × $\frac{3}{4}$" × 10"	Legs
Pipe, steel	1	$1\frac{1}{4}$" ID × 10"	External post
Pipe, steel	1	1" ID × 10"	Internal post
Channel, steel	1	$\frac{3}{16}$" × $\frac{5}{8}$" × $2\frac{1}{2}$" × 3"	Top
Round stock	1	$\frac{3}{8}$" dia × 7"	Pin, hardened
Chain	1	Light weight, 8"	Pin chain
Bolt, stove	1	$\frac{3}{16}$" × $\frac{1}{2}$"	Chain fastener

Project 22: JACK STAND—TRACTOR

Plan for Jack Stand—Tractor

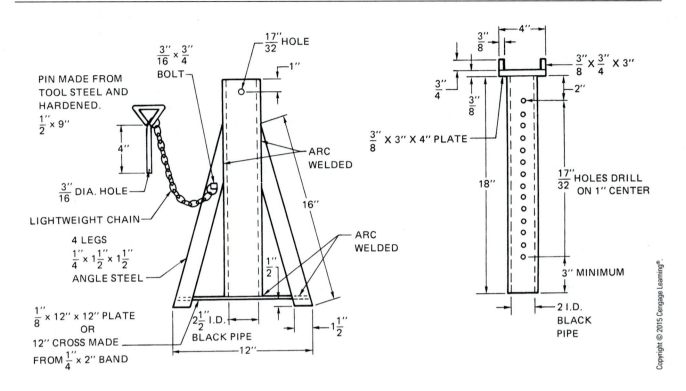

Copyright © 2015 Cengage Learning®

Construction Procedure for Jack Stand—Tractor

1. Cut all pieces to length.

2. Make a cross by welding two pieces of ¼" × 6½" band to a piece of ¼" × ⅕" band to form a 90-degree cross. Grind the welds smooth.

3. Cut all four ends with two 45-degree angles so the ends will fit inside the legs.

4. Ream the inside of the ends of the external post.

5. Center-punch 1 inch from one end of the external post.

6. Scribe a mark from end-to-end down the center axis of the interior post.

7. Center-punch 2 inches from one end and at 1-inch intervals on the line from that end to a point 3 inches from the other end of the interior post.

8. Place the exterior post in a drill press vise or clamp it in a vee block and drill the $1\frac{7}{32}$-inch hole through both walls of the pipe.

9. Without unclamping the exterior post, slide the interior post into the exterior post and run the drill through the exterior post and on through both walls of the interior post. Drill all holes in the interior post in like fashion.

10. Weld the exterior post to the center of the base so the assembly sits with the post perpendicular to the floor.

11. Grind the legs so they fit the post. Tack-weld the legs to the base and to the post, being careful to keep the parts and spacing equal and the post perpendicular. Weld all joints completely.

12. Weld the top on the interior post.

13. Bend a loop in the pin, harden the pin, and attach the chain to the pin.

14. Drill a ³/₁₆-inch hole 1½ inches from the top of one leg and anchor the chain with a ³/₁₆" × ¾" stove bolt.

15. Chip all welds and remove all burrs.

16. Apply a suitable finish.

Bill of Materials

Materials Needed	Quantity	Dimensions	Description or Use
Band, steel	1	$\frac{1}{4}$" × 2" × 15"	Base
Band, steel	2	$\frac{1}{4}$" × 2" × $6\frac{1}{2}$"	Base
Angle, steel	4	$\frac{1}{4}$" × $1\frac{1}{2}$" × $1\frac{1}{2}$" × 16"	Legs
Pipe, steel	1	$2\frac{1}{2}$" ID × 18"	External post
Pipe, steel	1	2" ID × 18"	Internal post
Plate	1	$\frac{3}{8}$" × 3" × 4"	Top base
Plate	2	$\frac{3}{8}$" × $\frac{3}{4}$" × 3"	Top rails
Round stock	1	$\frac{1}{2}$" dia × 9"	Pin, hardened
Chain	1	Light weight, 12"	Pin chain
Bolt, stove	1	$\frac{3}{16}$" × $\frac{3}{4}$"	Chain fastener

Project 23: **CAR RAMP**

Plan for Car Ramp

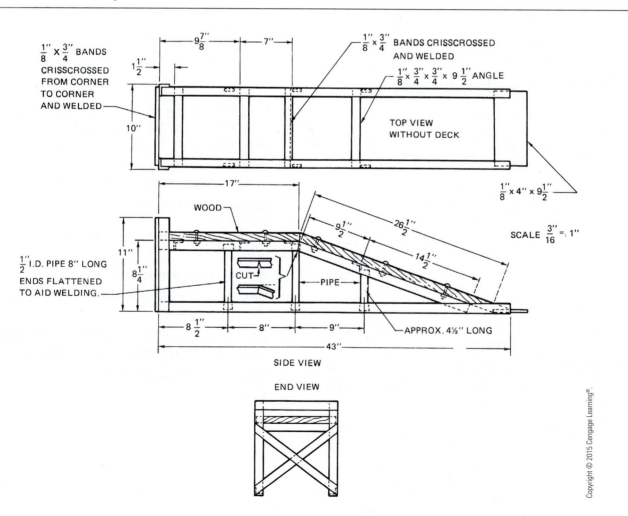

SIDE VIEW

END VIEW

Copyright © 2015 Cengage Learning®

Construction Procedure for Car Ramp

1. Cut to length, exactly as specified, the bottom runners, top rails, cross supports, corner standards, toe plate, stop bar, and posts.

2. Assemble the parts and tack-weld one side frame.

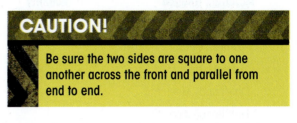

> **CAUTION!**
>
> **Sides must be welded so the corner standard will be on the outside of the runner and rail. When fully assembled, runners will have one flat side down and one flat side out. Rails have one flat side up and one flat side out.**

The following procedure is suggested:

a. Lay a runner and rail on a flat surface parallel to each other. One leg of each points toward the opposite piece and the other leg of each points upward.

b. Slip one leg of a standard under the ends of the runner and rail.

c. Lay two 8-inch posts in place between the runner and rail.

d. Mark the standard where the rail intersects. The top of the rail should be 8¼ inches from the bottom of the runner.

e. Starting at the standard, measure out the rail 17 inches and lay out and cut a notch in the rail to permit a 25-degree bend.

f. Place the rail in a vise and bend it to close the notch.

g. Clamp the runner, standard, and rail assembly to a flat metal surface. Square the joints carefully and tack-weld.

h. Place the 8-inch posts in position and tack-weld.

i. Tack-weld the rail to the runner at the bottom of the ramp.

j. Position the 4½-inch post as shown in the plan. If it is too long, shorten it as needed. Tack-weld.

3. Assemble and tack-weld the opposite side frame.

> **CAUTION!**
>
> **Make the unit match the first one except be careful to set it up so it faces the first one.**

4. Install and tack-weld cross members.

> **CAUTION!**
>
> **Be sure the two sides are square to one another across the front and parallel from end to end.**

5. Recheck the unit to be sure it is square and parallel. Install and tack-weld the toe plate, stop bar, and cross braces.

6. Make permanent welds at all points.

> **CAUTION!**
>
> **Select a welding sequence that minimizes distortion.**

7. Chip all welds. Prime and paint the unit.

8. Measure the unit for the exact cutting of the deck pieces.

9. Cut, drill holes, and install the deck pieces.

> **CAUTION!**
>
> **Cut angles on the ends of the ramp pieces so they fit well for maximum strength at the floor and where they meet at the top of the incline. Drill bolt holes in the frame and deck so there is room to apply nuts when bolts are installed.**

10. Prime and paint the deck parts.

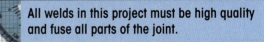

> **NOTE**
>
> **All welds in this project must be high quality and fuse all parts of the joint.**

> **CAUTION!**
>
> **Maximum capacity is one-half ton.**

Bill of Materials

Materials Needed	Quantity	Dimensions	Description or Use
Angle, steel	2	$\frac{1}{8}" \times \frac{3}{4}" \times \frac{3}{4}" \times 43"$	Bottom runners
Angle, steel	2	$\frac{1}{8}" \times \frac{3}{4}" \times \frac{3}{4}" \times 41"$	Top rails
Angle, steel	4	$\frac{1}{8}" \times \frac{3}{4}" \times \frac{3}{4}" \times 9\frac{1}{2}"$	Cross supports
Flat (band), steel	2	$\frac{1}{8}" \times \frac{3}{4}" \times \frac{3}{4}" \times 11"$	Corner standards
Flat (band), steel	1	$\frac{1}{8}" \times 4" \times 9\frac{1}{2}"$	Toe plate
Flat (band), steel	4	$\frac{1}{8}" \times \frac{3}{4}" \times 12"$ (approx)	Diagonal braces
Flat (band), steel	1	$\frac{1}{8}" \times \frac{3}{4}" \times 10"$	Stop bar
*Pipe, standard, black steel	4	$\frac{1}{4}"$ ID $\times 8"$	Posts
Pipe, standard, black steel	2	$\frac{1}{4}"$ ID $\times 4\frac{1}{2}"$ (approx)	Ramp posts
*Plywood, exterior	1	$\frac{3}{4}" \times 9\frac{3}{4}" \times 17"$ (approx)	Deck
Plywood, exterior	1	$\frac{3}{4}" \times 9\frac{3}{4}" \times 26\frac{1}{2}"$ (approx)	Ramp deck
Bolts, carriage or stove	10	$\frac{1}{4}" \times 1\frac{1}{2}"$	Ramp fasteners

*Flatten ends to improve welds. Angles ($\frac{1}{8}" \times \frac{3}{4}" \times \frac{3}{4}"$) may be used in place of pipe.

*Sawed lumber may be used if it is placed with the grain running across the ramp, or sheet steel may be welded on for decking.

Project 24: **CREEPER**

Plan for Creeper

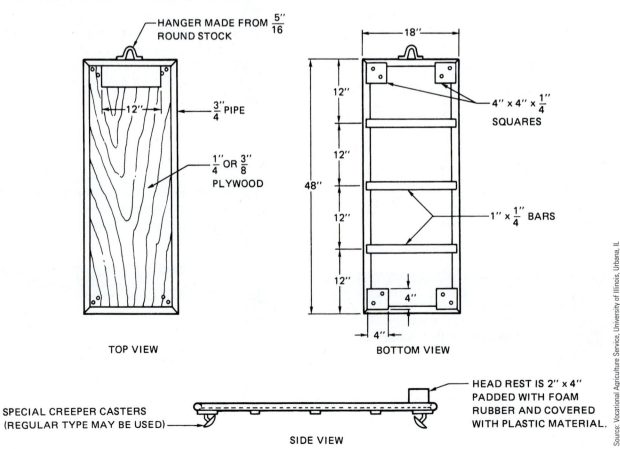

HANGER MADE FROM $\frac{5}{16}"$ ROUND STOCK

$\frac{3}{4}"$ PIPE

12"

$\frac{1}{4}"$ OR $\frac{3}{8}"$ PLYWOOD

TOP VIEW

18"

12"

12"

48"

12"

12"

4"

4"

4" $\times$ 4" $\times \frac{1}{4}"$ SQUARES

1" $\times \frac{1}{4}"$ BARS

BOTTOM VIEW

SPECIAL CREEPER CASTERS (REGULAR TYPE MAY BE USED)

HEAD REST IS 2" $\times$ 4" PADDED WITH FOAM RUBBER AND COVERED WITH PLASTIC MATERIAL.

SIDE VIEW

Source: Vocational Agriculture Service, University of Illinois, Urbana, IL

Construction Procedure for Creeper

1. Cut all metal pieces to size.

2. Assemble and weld the rails, bars, plates, and hanger.

3. Cut the deck to size.

4. Install the deck and casters using bolts with lock washers. Saw off excess bolt material.

5. Apply a suitable finish.

6. Construct and install the headrest using wood screws.

Bill of Materials

Materials Needed	Quantity	Dimensions	Description or Use
*Pipe, black steel	2	$\frac{3}{4}$" ID × 48"	Side rails
*Pipe, black steel	2	$\frac{3}{4}$" ID × 18"	End rails
Band, steel	3	$\frac{1}{4}$" × 1" × 17"	Bars
Band or plate, steel	4	$\frac{1}{8}$" or $\frac{1}{4}$" × 4" × 4"	Square plates
Round stock, steel	1	$\frac{5}{16}$" × 4"	Hangers
Plywood	1	$\frac{1}{4}$" or $\frac{3}{8}$" × 15" × 46"	Deck
Lumber	1	2" × 4" × 12"	Headrest
Foam	1	As needed	Headrest
Plastic cover	1	As needed	Headrest
Bolt, carriage	8	$\frac{1}{4}$" × 1"	Deck and caster fasteners
Lock washers	8	$\frac{1}{4}$"	Deck and caster fasteners
Creeper casters	4	As needed	Casters
Wood screws	3	#8 × 1$\frac{1}{4}$"	Headrest

*Frame may also be constructed of hard wood or angle steel.

Project 25: **SAWHORSE—PIPE**

Plan for Sawhorse—Pipe

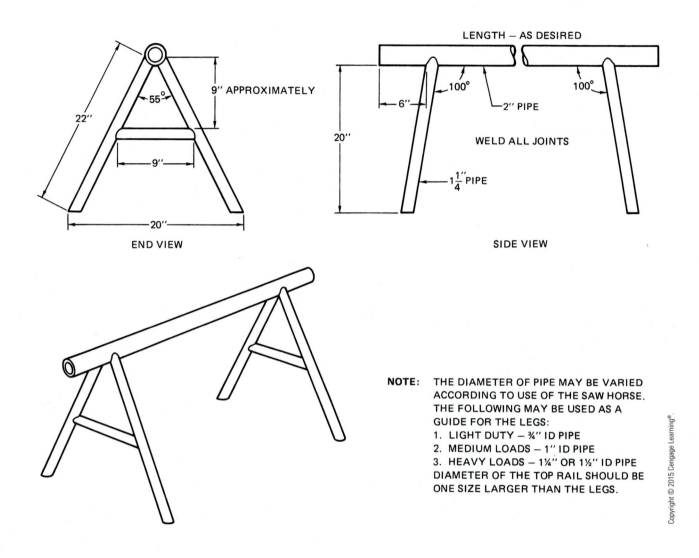

END VIEW

55°
22″
9″ APPROXIMATELY
9″
20″

SIDE VIEW

LENGTH – AS DESIRED
100°
6″
2″ PIPE
100°
20″
WELD ALL JOINTS
$1\frac{1}{4}''$ PIPE

NOTE: THE DIAMETER OF PIPE MAY BE VARIED ACCORDING TO USE OF THE SAW HORSE. THE FOLLOWING MAY BE USED AS A GUIDE FOR THE LEGS:
1. LIGHT DUTY – ¾″ ID PIPE
2. MEDIUM LOADS – 1″ ID PIPE
3. HEAVY LOADS – 1¼″ OR 1½″ ID PIPE
DIAMETER OF THE TOP RAIL SHOULD BE ONE SIZE LARGER THAN THE LEGS.

Copyright © 2015 Cengage Learning®.

Construction Procedure for Sawhorse—Pipe

1. Cut all pipe to length.
2. Grind the ends of the legs to fit the top rail.

NOTE

Welding will be easier if the end of each leg is flattened slightly where it will be fitted to the top rail.

3. Tack-weld the legs to the top rail.
4. Adjust the leg positions so the sawhorse stands squarely on a flat surface and all spacings are correct.
5. Weld the legs permanently.
6. Slightly flatten the ends of the braces and grind them to fit the legs.
7. Weld the braces in place.
8. Grind the bottoms of the legs so they are flat on the floor.

Bill of Materials

Materials Needed	Quantity	Dimensions	Description or Use
Pipe, black steel	1	$1\frac{1}{4}$" ID × 36"	Top rail
Pipe, black steel	4	1" ID × 20"	Legs
Pipe, black steel	2	1" ID × 9"	Braces

Project 26: **WORKBENCH BRACKETS**

Plan for Workbench Brackets

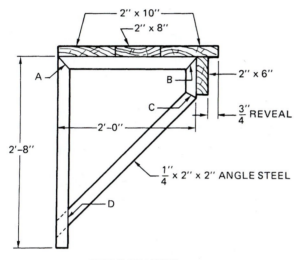

ANGLE BRACKET

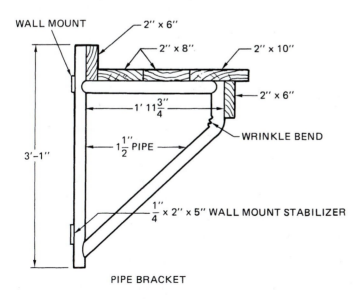

PIPE BRACKET

GENERAL

THREE BRACKETS ARE USUALLY REQUIRED FOR A 10- OR 12- FOOT BENCH. TWO MAY BE SUFFICIENT FOR AN 8- FOOT BENCH.

BOLT BRACKETS TO OR THROUGH THE WALL.

BOLT ALL WOOD PIECES TO BRACKETS, USING 3/8" CARRIAGE BOLTS. PLANKS SHOULD BE JOINTED FOR A COMBINED TOTAL BENCH WIDTH TO PROVIDE A 3/4" REVEAL ON THE FRONT. EDGES OF PLANKS MAY BE GLUED AND CLAMPED TO PROVIDE A CONTINUOUS BENCH TOP. HOLES ON TOP ARE COUNTERBORED 1 INCH AND FILLED WITH DOWEL PLUGS.

Source: Vocational Agriculture Service, University of Illinois, Urbana, IL

Construction Procedure for Workbench Brackets

Angle Iron Bracket

1. Cut two standards each with a 45-degree angle for the upper joint. Cut the horizontal support with a 45-degree angle on each end.

2. Cut a 30-degree wedge out at point C.

3. Square and tack-weld A.

4. Heat and bend at C to create correct fits at B and D.

5. Tack-weld B and D, being careful to keep A and B square.

6. Weld all joints.

Pipe Bracket

1. Cut pipes to length, make wrinkle bend in one standard, and weld together.

2. Cut, drill, and weld on wall brackets.

Bill of Materials

Materials Needed	Quantity	Dimensions	Description or Use
ANGLE BRACKET (EACH):			
Angle, steel	2	$\frac{1}{4}$" × 2" × 2" × 2'8"	Legs
Angle, steel	1	$\frac{1}{4}$" × 2" × 2" × 2'0"	Horizontal support
Or			
PIPE BRACKET (EACH):			
Pipe, black	2	$1\frac{1}{2}$" ID × 3'1"	Standards
Pipe, black	1	$1\frac{1}{2}$" ID × 1'10"	Horizontal support
Band, steel	2	$\frac{1}{4}$" × 2" × 5"	Wall mounts

Project 27: **WORKBENCH BRACKET—ALTERNATE PLAN**

Plan for Workbench Bracket—Alternate Plan

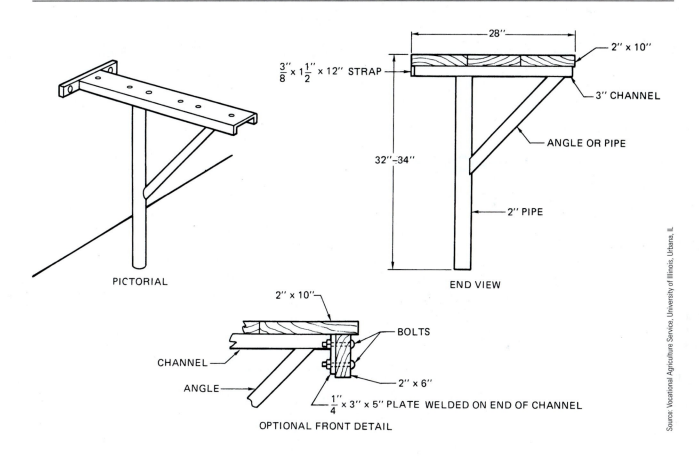

PICTORIAL

END VIEW

OPTIONAL FRONT DETAIL

Source: Vocational Agriculture Service, University of Illinois, Urbana, IL.

Construction Procedure for Workbench

Bracket—Alternate Plan

1. Cut all pieces to length.

2. Drill $\frac{7}{16}$-inch holes in the bracket.

3. Drill holes in the top as needed.

4. Round the corners of the channel that will protrude forward.

5. Weld the bracket to the top.

6. Tack-weld the top to the leg and square the assembly.

7. Grind the ends of the brace to fit. Tack-weld the brace.

8. Weld all joints permanently.

9. Apply a suitable finish.

Bill of Materials

Materials Needed	Quantity	Dimensions	Description or Use
Channel, steel	1	3" × 27"	Top
Pipe, black steel	1	2" ID × 32"	Leg
Pipe, black steel	1	1" ID × 15"	Brace
Band, steel	1	$\frac{3}{8}$" × $1\frac{1}{2}$" × 12"	Bracket

Project 28: **HAND CART**

Plan for Hand Cart

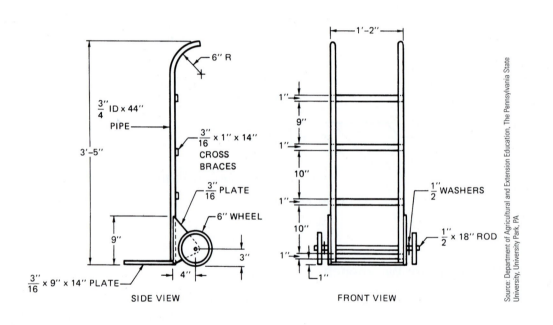

SIDE VIEW   FRONT VIEW

Source: Department of Agricultural and Extension Education, The Pennsylvania State University, University Park, PA

Construction Procedure for Hand Cart

1. Measure and cut the pipe to the dimension given in the bill of materials.

2. Bend the handles to a 6-inch radius.

3. Cut the axle supports using an oxyacetylene torch. Grind to identical shape.

4. Drill ½-inch holes with centers ½ inch from the point of the axle supports. Insert the axle.

5. Cut the four cross braces and weld in place.

6. Tack-weld the base in position, handles in air.

7. Position the axle supports with the wheels mounted so that the base and wheels are level with the floor. Tack-weld in place.

8. Drill the axle for the ⅛-inch cotter pins to keep wheels on the cart and weld the axle in place.

9. Weld the ½-inch washers to the axle on the inside of each wheel to serve as stops.

10. Complete welding all weld joints.

11. Paint the finished product with metal primer and a finish coat of enamel.

NOTE

If cart is to be used as a bag cart, add fenders by welding ⅛" × 2" × 6" steel plate to the pipe over each wheel.

Bill of Materials

Materials Needed	Quantity	Dimensions	Description or Use
Wheels, heavy duty	2	6" × 1.50	For ½" axle
Round, cold rolled	1	½" × 18"	Long axle*
Plate, steel	1	³⁄₁₆" × 9" × 14"	Base
Pipe, black	2	¾' × 44"	Long Handles
Plate, steel	2	³⁄₁₆" × 4" × 9"	For axle supports
Band, steel	4	³⁄₁₆" × 1" × 14"	Cross braces
Washers	4	½"	—
Cotter pins	2	⅛" × 1"	—
Metal primer	1 pint	—	—
Enamel	1 pint	—	—

*Axle length may vary with different wheels.

Project 29: **WELDING STAND—ACETYLENE**

Plan for Welding Stand—Acetylene

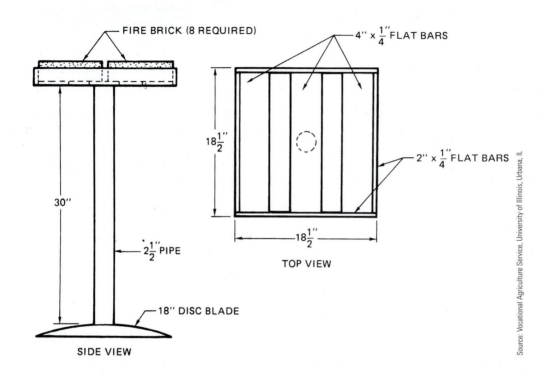

FIRE BRICK (8 REQUIRED)

4" × $\frac{1}{4}$" FLAT BARS

18$\frac{1}{2}$"

30"

2" × $\frac{1}{4}$" FLAT BARS

2$\frac{1}{2}$" PIPE

18$\frac{1}{2}$"

TOP VIEW

18" DISC BLADE

SIDE VIEW

Source: Vocational Agriculture Service, University of Illinois, Urbana, IL

Construction Procedure for Welding Stand—Acetylene

1. Set two rows of four firebricks each on a flat surface. Check to see if they form a bed 18" × 18". If they do not, determine the changes needed in the plan to accommodate your bricks. Modify the bill of materials accordingly.

2. Cut all pieces to length.

3. Tack-weld the sides on the outside corners.

4. Tack-weld the bottom pieces to the sides.

5. Invert the table on a flat surface and weld all joints from the underside.

6. Weld the post to the center bottom piece.

7. Weld the post to the disc base using a low-hydrogen electrode.

8. Apply a suitable finish.

9. Add the firebrick.

Bill of Materials

Materials Needed	Quantity	Dimensions	Description or Use
Disc blade	1	18" diameter	Base
Pipe, steel	12	2½" ID × 30"	Post
Flat (band) steel	3	¼" × 4" × 18"	Bottom
Flat (band) steel	2	¼" × 2" × 18½"	Sides
Flat (band) steel	2	¼" × 2" × 18"	Ends

Project 30: **FARROWING CRATE**

Plan for Farrowing Crate

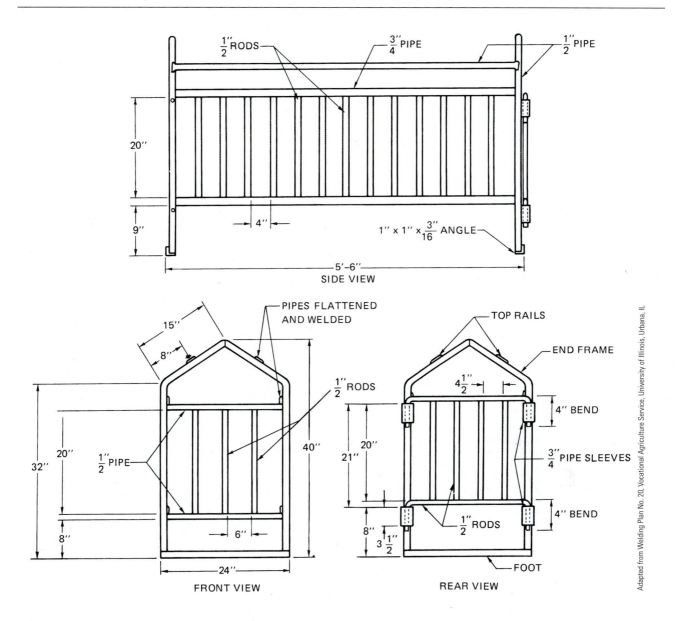

SIDE VIEW

FRONT VIEW

REAR VIEW

Adapted from Welding Plan No. 20, Vocational Agriculture Service, University of Illinois, Urbana, IL

Construction Procedure for Farrowing Crate

1. Cut the pieces to be used in the end frame halves and feet.

2. Measure in 15 inches from one end and bend the four half frames to match each other.

3. Saw or grind the ridge angle on the half frames.

4. Weld the half frames at the top and weld them to the feet to form the ends.

5. Cut and weld the front rails and slats in place.

6. Cut and weld the side rails and slats in place.

7. Set up the ends and weld the side rails to the ends.

8. Make up the rear gate.

9. Cut four 3½-inch lengths of a ¾-inch ID pipe for the sleeves and weld them into place.

10. Apply a suitable finish.

Bill of Materials

Materials Needed	Quantity	Dimensions	Description or Use
Pipe, steel	2	½" ID × 5'6"	Top side rails
	4	¾" ID × 5'4"	Upper and lower side rails
Rod, steel	26	½" × 20"	Vertical slats for sides
Pipe, steel	4	½" ID × 4'	Half of end frame
	4	¾" ID × 3½"	Sleeves
	2	½" ID × 22"	Top and bottom rails of front
Rod, steel	7	½" × 20"	Vertical slats for ends
	2	½" × 32"	Top and bottom rails of rear gate
Angle, steel	2	³⁄₁₆" × 1" × 1" × 24"	Feet

Project 31: **FEED/SILAGE CART**

Plan for Feed/Silage Cart

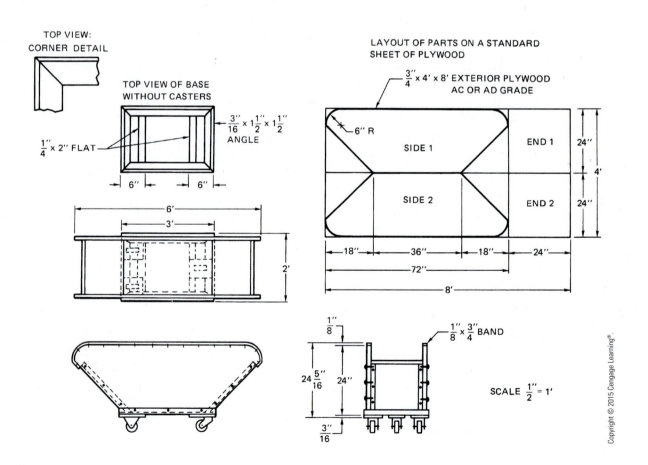

Copyright © 2015 Cengage Learning®.

Construction Procedure for Feed/Silage Cart

1. Using a ¾" × 4' × 8' sheet of plywood, cut the sides by cutting two pieces 2' × 6'. Then lay out the other lines and finish cutting out the sides. See plan for most efficient layout.

2. Cut the ends by sawing the remaining piece in half.

3. Cut two pieces of ³⁄₁₆" × 1½" × 1½" angle 36" long and two pieces of ³⁄₁₆" × 1½" × 1½" angle 24" long with 45-degree angles on all ends. Weld them into a rectangular frame to form the base.

4. Cut two ¼" × 2" bands 21" long.

5. Place the base on a flat surface. Weld the caster supports into place according to the size of the base of the casters.

6. Bolt the casters to the base using flat-head stove bolts passing downward through countersunk holes and through the base of each caster.

7. Cut four pieces of ⅛" × 1" × 1" angle 22" long. Drill three ⁹⁄₃₂-inch holes in each leg of each piece. Holes should be spaced so they are 1 inch to 1½ inches from each end with one in the middle. Holes must be placed so the bolts do not collide.

8. Notch the ends of the sides so they fit down into the base. Bolt the sides to the base using ¼" × 1¼" flat-head stove bolts passing through countersunk holes in the frame and then through the wood sides. Install ¼-inch flat washers and draw them slightly into the wood. Saw off any excess bolt length.

9. Using the four pieces of angle, bolt the ends to the sides at each corner. The angle steel and bolt nuts should be on the exterior of the ends and interior of the sides.

10. Cut a piece of ¾-inch plywood to drop in to form the bottom. Small notches may be needed to clear the nuts at the bottom of the sides. The bottom must fit tight to prevent feed from running through. Caulk any cracks that can permit feed to pass through.

11. Cut two pieces of ⅛" × ¾" band 90" long. Drill and countersink ½ inch from each end and at 6-inch intervals for No. 6 screws.

12. Shape the bands to fit the top of the sides.

13. Apply suitable finishes to all wood and metal parts.

14. Install the bands on the top of each side of the cart.

Bill of Materials

Materials Needed	Quantity	Dimensions	Description or Use
Plywood, exterior AC or better recommended	2	¾" × 2' × 6'	Sides
	2	¾" × 2' × 2'	Ends
	1	¾" × 2' × 3'	Bottom
Steel, angle	2	³⁄₁₆" × 1½" × 1½" × 36"	Sides of base
	2	³⁄₁₆" × 1½" × 1½" × 24"	Ends of base
	4	⅛" × 1" × 1" × 22"	Corner angles
Steel, band	2	⅛" × 1" × 90"	Edge bands
	2	³⁄₁₆" × 2" × 21"	Caster supports
Bolts, carriage, with nuts and lock washers	2	4¼" × 1¼"	Bolt plywood to angles
Bolts, stove, flat-head	6	¼" × 1¼"	Bolt sides to base
Bolts, stove, flat-head	12	⅜" × 1"	Attach casters
*Caster, straight	2	6" or 8" dia	Rigid wheels on one end
Caster, swivel	1	6" or 8"	Swivel wheel in center on other end
Screws, wood, flat	30	No. 6 × 1"	Fasten edge bands

*Two 6", 8", or 10" wheels with supports and axle may be used in place of the straight casters. A metal support is then fabricated on the other end to level the cart on the swivel caster.

Project 32: **HOIST FRAME—PORTABLE**

Plan for Hoist Frame—Portable

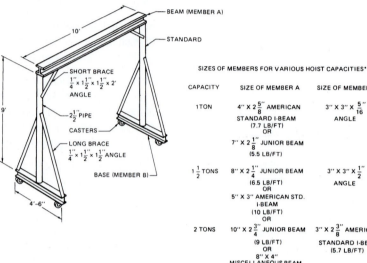

SIZES OF MEMBERS FOR VARIOUS HOIST CAPACITIES*

CAPACITY	SIZE OF MEMBER A	SIZE OF MEMBER B
1 TON	4" X 2$\frac{5}{8}$" AMERICAN STANDARD I-BEAM (7.7 LB/FT) OR 7" X 2$\frac{1}{8}$" JUNIOR BEAM (5.5 LB/FT)	3" X 3" X $\frac{5}{16}$" ANGLE
1$\frac{1}{2}$ TONS	8" X 2$\frac{1}{4}$" JUNIOR BEAM (6.5 LB/FT) OR 5" X 3" AMERICAN STD. I-BEAM (10 LB/FT) OR	3" X 3" X $\frac{1}{2}$" ANGLE
2 TONS	10" X 2$\frac{3}{4}$" JUNIOR BEAM (9 LB/FT) OR 8" X 4" MISCELLANEOUS BEAM (10 LB/FT) OR 6" X 3$\frac{3}{8}$" AMERICAN STD. I-BEAM (12.5 LB/FT)	3" X 2$\frac{3}{8}$" AMERICAN STANDARD I-BEAM (5.7 LB/FT)

* Beams are described in terms familiar to structural steel dealers. Name of beam, dimensions, and weight per foot must all be specified in order to get the correct member. Some alternate sizes are given in case first beam listed (most economical) is not available.

Source: Vocational Agriculture Service, University of Illinois, Urbana, IL

Construction Procedure for Hoist Frame—Portable

NOTE

Be sure the casters used are designed to carry the combined weight of the frame, hoist, and load for which the hoist is designed.

1. Cut all pieces to length.

2. Lay the standards and member A on a level floor. Square and tack-weld them.

3. Check each joint between the standards and member A, and tack-weld the braces into place. Weld all joints permanently.

4. Set the assembly up on the B members.

5. Check for squareness and parallelism between the B members, then tack-weld the standards to the B members.

6. Check for squareness and weld the braces between the standards and B members.

7. Weld all joints permanently.

8. Attach casters to the B members.

CAUTION!

Bill of materials is for a hoist with a maximum capacity of 1 ton.

Bill of Materials

Materials Needed	Quantity	Dimensions	Description
*1 beam, standard, steel	1	4" × 2⅝" × 10' (7.7 lb/ft)	Beam (member A)
Pipe, steel, standard	2	2½" × 9'	Standards
Angle, steel	2	⁵⁄₁₆" × 3" × 3" × 4'6"	Base (member B)
Angle, steel	4	¼" × 1½" × 1½" × 6'6"	Long braces
Angle, steel	2	¼" × 1½" × 1½" × 2	Short braces

*A 7" × 2⅛" × 10' (5.5 lb/ft) junior beam should be cheaper, if available, and may be substituted. See plan for specifications of beams for capacity greater than one ton.

Project 33: **UTILITY TRAILER**

Plan for Utility Trailer

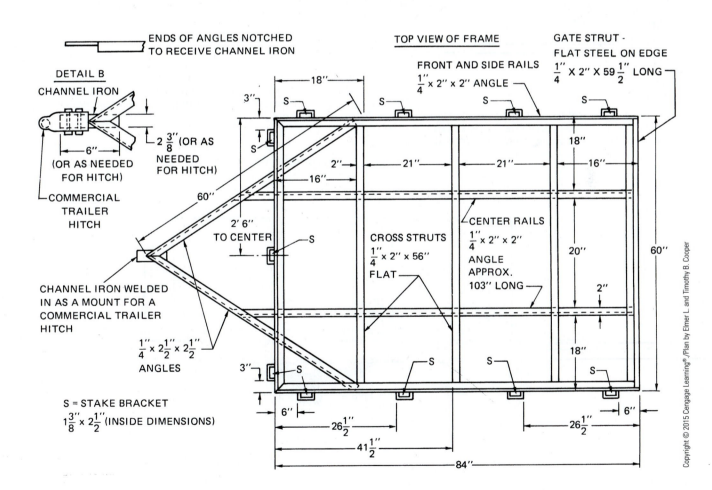

Copyright © 2015 Cengage Learning®./Plan by Elmer L. and Timothy B. Cooper

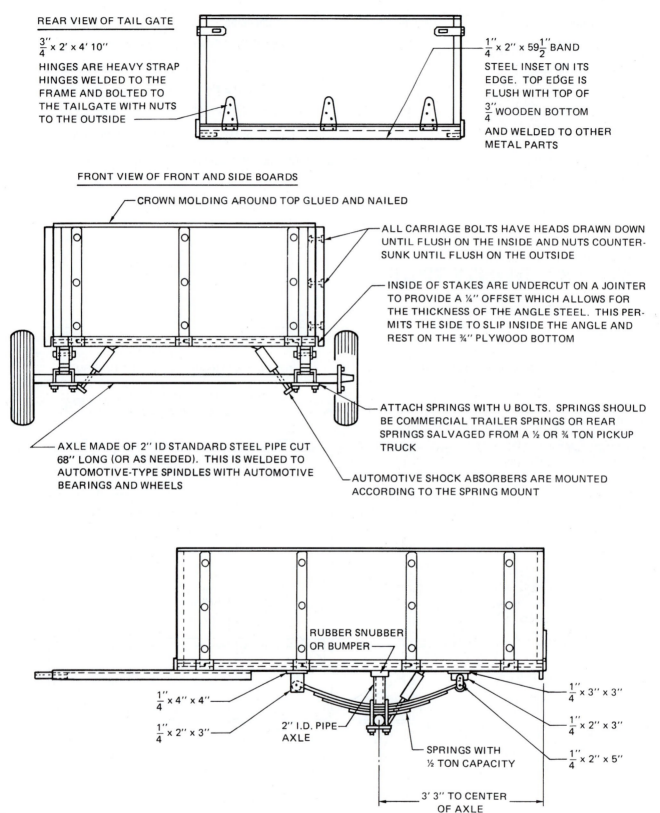

REAR VIEW OF TAIL GATE

$\frac{3''}{4}$ x 2' x 4' 10''

HINGES ARE HEAVY STRAP HINGES WELDED TO THE FRAME AND BOLTED TO THE TAILGATE WITH NUTS TO THE OUTSIDE

$\frac{1''}{4}$ x 2'' x 59$\frac{1''}{2}$ BAND STEEL INSET ON ITS EDGE. TOP EDGE IS FLUSH WITH TOP OF $\frac{3''}{4}$ WOODEN BOTTOM AND WELDED TO OTHER METAL PARTS

FRONT VIEW OF FRONT AND SIDE BOARDS

CROWN MOLDING AROUND TOP GLUED AND NAILED

ALL CARRIAGE BOLTS HAVE HEADS DRAWN DOWN UNTIL FLUSH ON THE INSIDE AND NUTS COUNTER-SUNK UNTIL FLUSH ON THE OUTSIDE

INSIDE OF STAKES ARE UNDERCUT ON A JOINTER TO PROVIDE A ¼'' OFFSET WHICH ALLOWS FOR THE THICKNESS OF THE ANGLE STEEL. THIS PERMITS THE SIDE TO SLIP INSIDE THE ANGLE AND REST ON THE ¾'' PLYWOOD BOTTOM

ATTACH SPRINGS WITH U BOLTS. SPRINGS SHOULD BE COMMERCIAL TRAILER SPRINGS OR REAR SPRINGS SALVAGED FROM A ½ OR ¾ TON PICKUP TRUCK

AXLE MADE OF 2'' ID STANDARD STEEL PIPE CUT 68'' LONG (OR AS NEEDED). THIS IS WELDED TO AUTOMOTIVE-TYPE SPINDLES WITH AUTOMOTIVE BEARINGS AND WHEELS

AUTOMOTIVE SHOCK ABSORBERS ARE MOUNTED ACCORDING TO THE SPRING MOUNT

RUBBER SNUBBER OR BUMPER

$\frac{1''}{4}$ x 4'' x 4''

$\frac{1''}{4}$ x 2'' x 3''

2'' I.D. PIPE AXLE

SPRINGS WITH ½ TON CAPACITY

$\frac{1''}{4}$ x 3'' x 3''

$\frac{1''}{4}$ x 2'' x 3''

$\frac{1''}{4}$ x 2'' x 5''

3' 3'' TO CENTER OF AXLE

Copyright © 2015 Cengage Learning®./Plan by Elmer L. and Timothy B. Cooper

Construction Procedure for Utility Trailer

1. Cut the two side rails to length with a 45-degree cut in the front end. Cut $\frac{1}{4}$ inch off the bottom leg in the rear.

2. Cut the front rail with 45-degree cuts on both ends.

> **CAUTION!**
>
> Be careful to lay out the cuts with the parts as they will be positioned on the completed trailer.

3. Weld the front and side rails together.

4. Cut the cross struts and weld them in place.

5. Cut the gate strut and weld it in place. Be careful to position the gate strut so it extends $\frac{3}{4}$ inch above the top edge of the back cross strut to protect the edge of the $\frac{3}{4}$-inch plywood bottom.

6. Make up and install the stake brackets. Drill a $\frac{3}{16}$-inch hole in the face of each bracket that is located adjacent to each corner for insertion of wood screws into the stakes.

7. Install the tongue pieces and hitch.

8. Install the center rails and weld them at all contact points.

9. Make up the spring and shock absorber mounts and shackles according to the springs being used. Place the spring mounts so the axle will be 3 to 4 inches slightly to the rear of center to provide appropriate balance in the completed trailer.

10. Make up the axle and spindle assembly to allow approximately 2 inches clearance between the body and tires.

11. Decide if the axle should be mounted above or below the springs to obtain the desired clearance to permit loading of the springs and to provide the desired height of the trailer to the towing vehicle.

12. Mount the axle assembly and shock absorbers.

13. Cut the plywood bottom to fit the frame and install it temporarily. Carefully lay out and drill six $\frac{1}{4}$-inch holes through the plywood floor and into the frame for carriage bolts to anchor the floor.

14. Cut 11 stakes for the front and sides. Cut a $\frac{1}{4}$" × 2" rabbet on the flat side of one end of each stake. The resulting end should fit into the brackets and the inside of the stake should be in line with the inside of the metal side rail. Chamfer ($\frac{1}{4}$ inch) the appropriate three edges on the opposite end of each stake to provide a finished appearance.

15. Cut and install the side boards with the stakes in their brackets so the side boards fit inside the side rails.

16. Cut and install the front board between the side boards.

17. Tack-weld the T-hinges to the gate strut so the hinge pins are in an absolutely straight line and parallel to the gate strut edge.

18. Cut and install the back gate. Install one bolt in each of the outside hinges. Move the gate up and down to be sure it does not bind. Install a bolt in the center hinge. Check again for binding. Weld all hinges permanently.

19. Install corner hasps. The front corners may be secured with rigid corner brackets rather than hasps if desired.

20. Drill a $\frac{1}{2}$-inch hole through the floor near each front corner for drainage.

21. Remove all wooden parts for painting on all sides and edges.

22. Chip all welds thoroughly. Smooth all rough areas and prepare all parts for painting.

23. Install commercial fenders if desired.

24. Prime all wood and metal parts with suitable primers.

25. Apply two coats of machinery enamel to all parts.

26. Re-install all wooden parts. Install No. 10 screws in the stakes to secure the sides.

> **CAUTION!**
>
> Trailer is designed for maximum load of half ton.

Bill of Materials

Materials Needed	Quantity	Dimensions	Description or Use
Angle, steel	2	$\frac{1}{4}$" × 2" × 2" × 7'	Side rails
	1	$\frac{1}{4}$" × 2" × 2" × 5'	Front rail
	2	$\frac{1}{4}$" × 2" × 2" × 8'7" (approx)	Center rails
	2	$\frac{1}{4}$" × 2$\frac{1}{2}$" × 2$\frac{1}{2}$" × 5'	Tongue rails
Band (flat), steel	4	$\frac{1}{4}$" × 2" × 4'8"	Cross struts
	1	$\frac{1}{4}$" × 2" × 4'11$\frac{1}{2}$"	Gate strut
	1	$\frac{1}{8}$" or $\frac{1}{4}$" × 2" × 4'	Materials for pockets (or use 11 commercial pockets) 1$\frac{3}{8}$" × 2$\frac{1}{2}$" inside dimensions
	2	$\frac{1}{4}$" × 4" × 4"	Front spring mount base
	4	$\frac{1}{4}$" × 2" × 3"	Front spring mount
	2	$\frac{1}{4}$" × 3" × 3"	Rear spring mount base
	4	$\frac{1}{4}$" × 2" × 3"	Rear spring mount
	4	$\frac{1}{4}$" × 2" × 5"	Rear spring shackle
Channel, steel	1	2$\frac{3}{8}$" wide × 6" long (or as needed) for the trailer hitch	Hitch mount
Trailer hitch	1	Commercial	Ball type, heavy duty
Pipe, steel, standard	1	2" ID × 5'8" (approx)	Axle
Springs, leaf type	2	Rated to carry $\frac{1}{2}$ ton (rear truck or trailer type)	Springs with matching base mounts and U bolts
Spindles, automotive type	2	To carry $\frac{1}{2}$ ton	With matching wheels and tires
Shock absorbers	2	As needed	Automotive type
Fenders	2	As needed	Commercial boat or utility trailer fenders
*Plywood, exterior type AC or better	1	$\frac{3}{4}$" × 5' × 7'	Bottom
	2	$\frac{3}{4}$" × 2' × 7'	Side boards
	2	$\frac{3}{4}$" × 2' × 4'10"	Front and back boards
Lumber	11	1$\frac{1}{2}$" × 2$\frac{1}{2}$" × 2'1"	Stakes
Crown molding	2	$\frac{1}{2}$" × $\frac{3}{4}$" × 7'	Top of side boards
	2	$\frac{1}{2}$" × $\frac{3}{4}$" × 5'	Top of front and gate boards
T hinge, steel	3	8" heavy duty	Back gate
Hasps, steel	4	4" or as needed	Corner fasteners
Bolts, carriage	33	$\frac{1}{4}$" × 2$\frac{1}{2}$" with hexnuts and small washers	Stake bolts
Wood screws	12	$\frac{1}{4}$" × 1$\frac{1}{4}$"	Hinge bolts
Wood screws	11	#10 × 1$\frac{1}{4}$"	Stake brackets

*Plywood may be special ordered in five-foot widths, and lengths other than eight feet. Local prices should be explored to determine the most economical sizes to purchase. The trailer is designed with a five-foot width to accommodate a large lawn tractor with a four-foot mower. All dimensions for plywood pieces must be adjusted according to the dimensions of the actual frame.

Project 34: **AQUACULTURE MODEL—CLOSED SYSTEM**

Plan for Aquaculture Model—Closed System (Biological Filter)

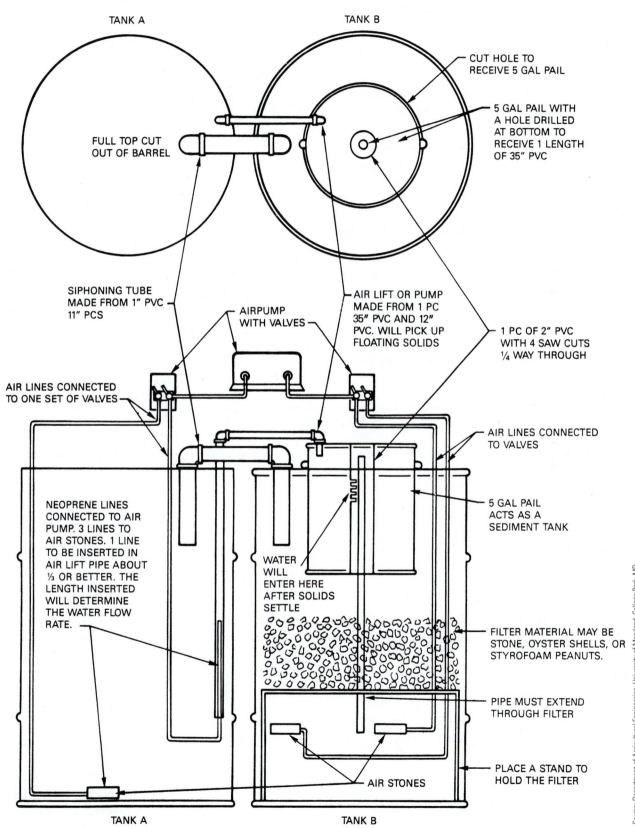

TANK A

TANK B

CUT HOLE TO RECEIVE 5 GAL PAIL

5 GAL PAIL WITH A HOLE DRILLED AT BOTTOM TO RECEIVE 1 LENGTH OF 35" PVC

FULL TOP CUT OUT OF BARREL

SIPHONING TUBE MADE FROM 1" PVC 11" PCS

AIRPUMP WITH VALVES

AIR LIFT OR PUMP MADE FROM 1 PC 35" PVC AND 12" PVC. WILL PICK UP FLOATING SOLIDS

1 PC OF 2" PVC WITH 4 SAW CUTS ¼ WAY THROUGH

AIR LINES CONNECTED TO ONE SET OF VALVES

AIR LINES CONNECTED TO VALVES

NEOPRENE LINES CONNECTED TO AIR PUMP. 3 LINES TO AIR STONES. 1 LINE TO BE INSERTED IN AIR LIFT PIPE ABOUT ⅓ OR BETTER. THE LENGTH INSERTED WILL DETERMINE THE WATER FLOW RATE.

5 GAL PAIL ACTS AS A SEDIMENT TANK

WATER WILL ENTER HERE AFTER SOLIDS SETTLE

FILTER MATERIAL MAY BE STONE, OYSTER SHELLS, OR STYROFOAM PEANUTS.

PIPE MUST EXTEND THROUGH FILTER

PLACE A STAND TO HOLD THE FILTER

AIR STONES

TANK A

TANK B

Source: Department of Agricultural Engineering, University of Maryland, College Park, MD

Construction Procedure for Aquaculture Model—Closed System

1. Cut the top out of tank A to hold fish.

2. Cut section out of tank B to support the 5-gallon sedimentation tank by the top rim.

3. Drill a hole in center of sediment tank and insert 35 inches of 1-inch PVC pipe. The top of this pipe will determine the water height in the 5-gallon pail.

4. Prepare a bag of filter material (crushed small gravel, oyster shell, or glass beads) so that the water that runs down the sediment pipe will flow out below the filter matrix.

5. Position both barrels together and construct an air-lift pump using the ½-inch PVC pipe. An air line will be inserted into the bottom of the lift in tank A so that water flows up and over into the outer ring of the 5-gallon sediment tank.

6. Cut a 12-inch piece of 2-inch PVC pipe and cut several saw slits about 4 inches from the end. This slips over the 1" center PVC pipe and acts as a sleeve to calm the water flow.

7. Construct a siphon tube using 1-inch PVC pipe as illustrated to return the clean water from tank B to tank A.

8. Connect the air pump to control valves so that two lines with 2" air stones can be placed in tank B under the biological filter. Two other lines are used in tank A, one for the air-lift and one with a 2-inch air stone suspended in the tank.

Operation of the Closed Aquaculture Model

It will take about three to four weeks for the microorganisms to become established in the biological filter. This can be aided by adding 1 ounce of feed every two days until fish are added.

Fish are kept in tank A. The air-lift pump moves water into the sediment tank in tank B. It flows into the inner chamber, down the 1-inch PVC stand pipe and up through the biological filter. The siphon returns the water back to tank A.

If fish are not eating, the water quality might have deteriorated: stop feeding, clean sediment tank, exchange 10 percent of the water, and observe the fish behavior.

Bill of Materials

Materials Needed	Quantity	Dimensions	Description or Use
55 gal. drums	2	—	—
5 gal. pail	1	—	—
½" PVC 35"	2 pieces	—	—
½" PVC 12"	1 piece	—	—
90° PVC ½" elbows	2	—	—
1" PVC 11"	3 pieces	—	—
90° PVC 1" elbows	2	—	—
2" PVC 12"	1 piece	—	—
Neoprene tubing ¼"	15 ft	—	—
Air pump with 2 outlets	1	—	—
Air stones	3	—	—
Two-way valves	2	—	—
Filter material	1 bag	—	—
Stand	1	—	(to hold filter material off bottom of barrel)

Project 35: **HYDROPONICS MODEL—TUBE CULTURE**

Plan for Hydroponics Model—Tube Culture

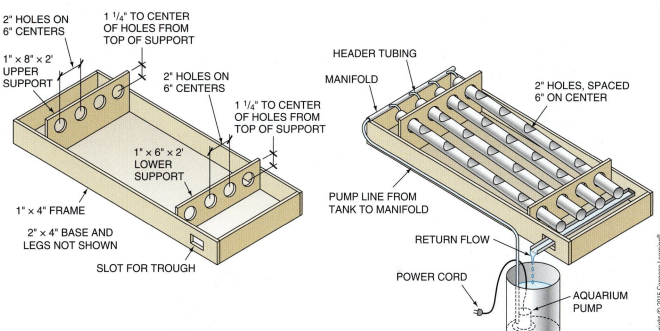

2" HOLES ON 6" CENTERS

1 1/4" TO CENTER OF HOLES FROM TOP OF SUPPORT

1" × 8" × 2' UPPER SUPPORT

2" HOLES ON 6" CENTERS

1 1/4" TO CENTER OF HOLES FROM TOP OF SUPPORT

1" × 6" × 2' LOWER SUPPORT

1" × 4" FRAME

2" × 4" BASE AND LEGS NOT SHOWN

SLOT FOR TROUGH

HEADER TUBING

MANIFOLD

2" HOLES, SPACED 6" ON CENTER

PUMP LINE FROM TANK TO MANIFOLD

RETURN FLOW

POWER CORD

AQUARIUM PUMP

Copyright © 2015 Cengage Learning®

Construction Procedure for Hydroponics Model—Tube Culture

1. Construct a 2' × 4' base frame (1" × 4" material).

2. Drill 2¼-inch holes, 1½ inches to center from the top edge of the 1" × 8" and 1" × 6" supports. Space four holes across each member on 6-inch centers.

3. Attach supports within the framework 6 inches from the upper end and 10 inches from the lower end.

4. Cut 2-inch PVC pipe into four 3-feet-6-inch lengths. Insert into framework and fasten with small wood screws through top of the wooden supports.

5. Drill 2-inch holes along the top axis of each pipe, spaced 6 inches on center to hold plants.

6. Saw a 2½-inch length of PVC pipe in half (end-to-end). Glue a 2-inch cap on one end and install as drain trough at the lower end of the tubes.

7. Place submersible pump in pail, attach tubing, and adjust length to reach manifold connection.

8. Attach a 1-inch manifold along the upper framework. Cap one end and connect to other tubing.

9. Drill manifold and install irrigation header tubes.

10. Friction fit 2-inch PVC caps to upper ends of tubes. Drill ¼-inch hole in each and insert header tubes.

11. Apply a suitable finish. Observe all safety precautions when working with electrical equipment in a wet environment.

Bill of Materials

Materials Needed	Quantity	Dimensions	Description or Use
Wood	12'	1" × 4"	Frame
Wood	2'	1" × 8"	Upper support
Wood	2'	1" × 6"	Lower support
PVC pipe Sch. 40	20'	2"	Growing tube
PVC cap	5, 1	2", 1"	Caps
Plastic pipe	2'	1"	Manifold
Plastic tubing and fittings	10'	$\frac{5}{16}$"	Supply line
Header tubing	4'	$\frac{1}{4}$"	Header tubing
Submersible pump	1	Aquarium size	Pump
Reducing coupling	1	1" <> $\frac{5}{16}$"	Connection
Plastic pail	1	5 gallon	Stock tank
Miscellaneous: wood screws paint and nails	—	—	—

Project 36: **GREENHOUSE—PIPE FRAME**

Plan for Greenhouse—Pipe Frame

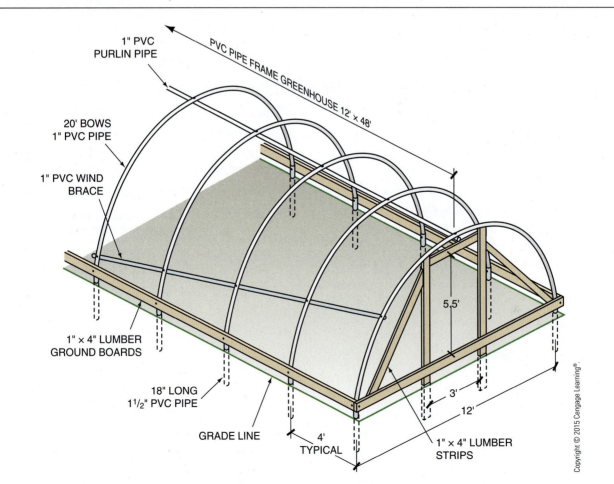

Copyright © 2015 Cengage Learning®

Construction Procedure for Greenhouse—Pipe Frame

1. Prepare a level site and ensure adequate drainage.

2. Square the structure. Drive or set in concrete the 18" × 1½" PVC standards so only 4 inches are visible.

3. Drive or set remaining standards on 4 feet centers down each side until only 4 inches are visible. The tops of all standards should be level.

4. Mark four 1-inch PVC pipes with a line 6 inches from each end and at the midpoint of each pipe.

5. Place the 1" × 4" ground boards around the outside of all the standards.

6. Insert one end of a 1-inch PVC pipe into the first corner standard to the 6" mark and drill through the ground board, bow, and standard. Insert a ¼" bolt and tighten.

7. Insert the opposite end of the 1-inch PVC pipe into the corner standard across from the first one and secure as described in #6.

8. Complete installation of all bows.

9. Cut 1" PVC pipe 12 feet long and mark every 4 feet. Hold this purlin up to the inside of the arched bows. Drill and insert ¼-inch bolts with purlin marks aligned with the center of all bows. Saw off any protruding bolts to prevent head injuries.

10. Take 1-inch PVC pipe and bolt end about 4 feet high on the inside of end bow. Allow the pipe to triangulate down to the ground board. Bolt the pipe to all bows that it passes and to the ground board. This is a wind brace and is an important member of the greenhouse. Be sure bows are plumb.

11. Construct the end walls and door frames as illustrated.

12. Drape plastic across the framework and staple the leading edge to the door frame. Proceed down each side, pulling the plastic tight to the ground board. Roll the edge of the plastic several times and staple the roll to the wooden members. Finish with the trailing end attached to the opposite door frame.

13. Doors, hinges, and utilities are constructed according to standard practices.

Bill of Materials

Materials Needed	Quantity	Dimensions	Description or Use
Wood	4	1" × 4" × 12'	Ground board
Wood	4	2" × 4" × 8'	Door frame
Plywood, exterior grade	2	4' × 8' A/C	Doors
PVC pipe Sch. 40	7	1" × 20'	Bows and braces
PVC pipe Sch. 40	8	1½" × 18"	Standards
#9 Wire	10	#9 × 10'	Straps
Hex bolts and nuts	12	¼" × 2"	Attach bows
Plastic film	1	20' × 30'	6 mil thick
Miscellaneous: Door hinges, staples, nails, and paint	—	—	—

Project 37: **10-FRAME LANGSTROTH BEEHIVE (FOR ¾" THICK LUMBER)**

Plan for 10-Frame Langstroth Beehive (for ¾" Thick Lumber)

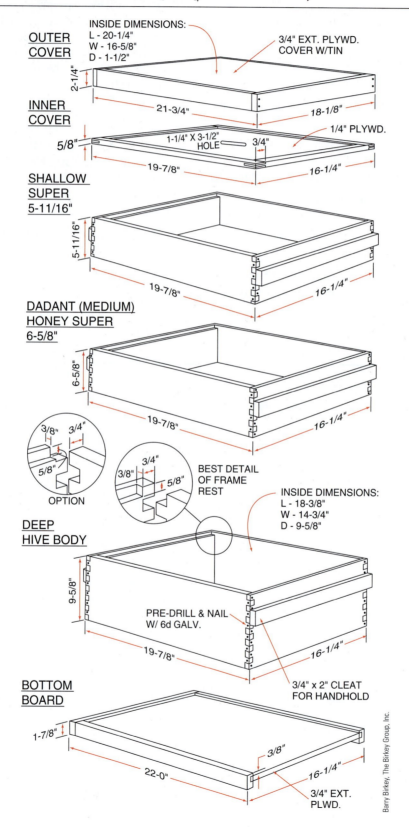

OUTER COVER

INSIDE DIMENSIONS:
L - 20-1/4"
W - 16-5/8"
D - 1-1/2"

3/4" EXT. PLYWD. COVER W/TIN

2-1/4"

21-3/4"   18-1/8"

INNER COVER

5/8"

1-1/4 X 3-1/2" HOLE   3/4"

1/4" PLYWD.

19-7/8"   16-1/4"

SHALLOW SUPER 5-11/16"

5-11/16"

19-7/8"   16-1/4"

DADANT (MEDIUM) HONEY SUPER 6-5/8"

6-5/8"

19-7/8"   16-1/4"

3/8"   3/4"
5/8"
OPTION

3/4"
3/8"   5/8"

BEST DETAIL OF FRAME REST

INSIDE DIMENSIONS:
L - 18-3/8"
W - 14-3/4"
D - 9-5/8"

DEEP HIVE BODY

9-5/8"

PRE-DRILL & NAIL W/ 6d GALV.

19-7/8"   16-1/4"

BOTTOM BOARD

1-7/8"

3/4" x 2" CLEAT FOR HANDHOLD

3/8"

22-0"   16-1/4"

3/4" EXT. PLWD.

Barry Birkey, The Birkey Group, Inc.

The species of wood used to make a beehive can vary depending upon what is available in your area. The minimum thickness should not be less than ¾ inch. If you are using standard dimensional lumber, you can use 138(¾" × 7¼") for both shallow and medium super, and 1 × 12(¾" × 11¼") for the deep hive body.

Start by cutting the boards to length. For fronts and backs, cut them a smidgen over 16¼ inches. For sides, cut a smidgen over 19⅞ inches.

Now that you have the joint cut and the boards cut to finished size, cut the ⅝" × ⅜" rabbet on the 16¼-inch boards stopping just short of the box joint pin at each end. (Chisel these square after the boards are assembled.) Note detail of frame rest at left. Predrill holes for nails in each pin.

Assemble boxes with glue and nail each pin with a 6d galvanized nail. Attach 1 × 2 handholds with screws and glue. Attach metal rabbets on the frame rest notch. Fill any holes and paint all surfaces, both outside and inside and top and bottom edges, with two coats of paint.

Project 38: DADANT TYPE FRAMES (FOR LANGSTROTH HIVE)

Plan for Dadant Type Frames (for Langstroth Hive)

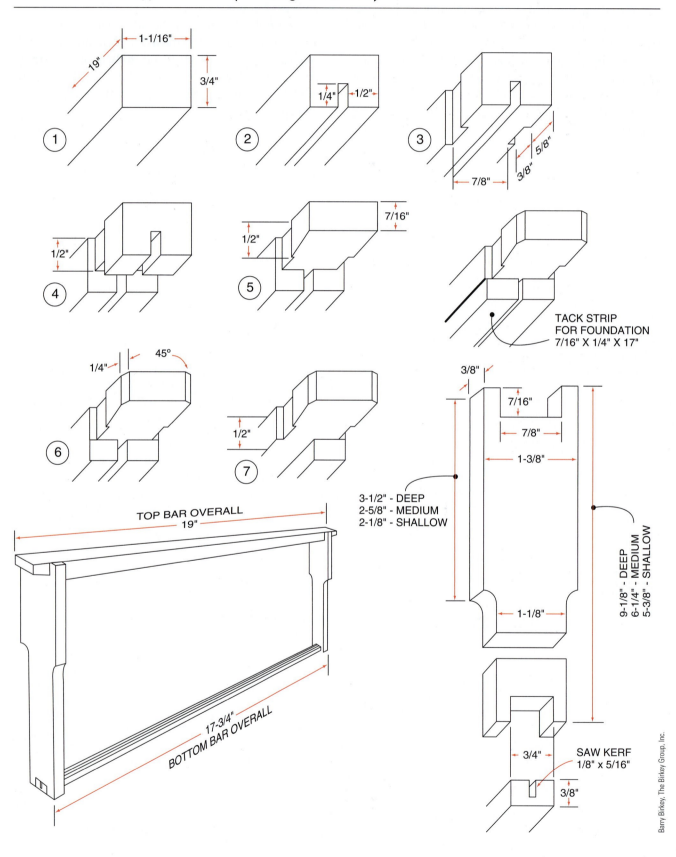

1
19"   1-1/16"   3/4"

2
1/4"   1/2"

3
7/8"   3/8"   5/8"

4
1/2"

5
1/2"   7/16"

TACK STRIP
FOR FOUNDATION
7/16" X 1/4" X 17"

6
1/4"   45°

7
1/2"

TOP BAR OVERALL
19"

17-3/4"
BOTTOM BAR OVERALL

3/8"   7/16"
7/8"
1-3/8"

3-1/2" - DEEP
2-5/8" - MEDIUM
2-1/8" - SHALLOW

9-1/8" - DEEP
6-1/4" - MEDIUM
5-3/8" - SHALLOW

1-1/8"

3/4"

SAW KERF
1/8" x 5/16"

3/8"

Barry Birkey, The Birkey Group, Inc.

Project 39: **3-FRAME OBSERVATION BEEHIVE (USING OAK LUMBER)**

Plan for 3-Frame Observation Beehive (Using Oak Lumber)

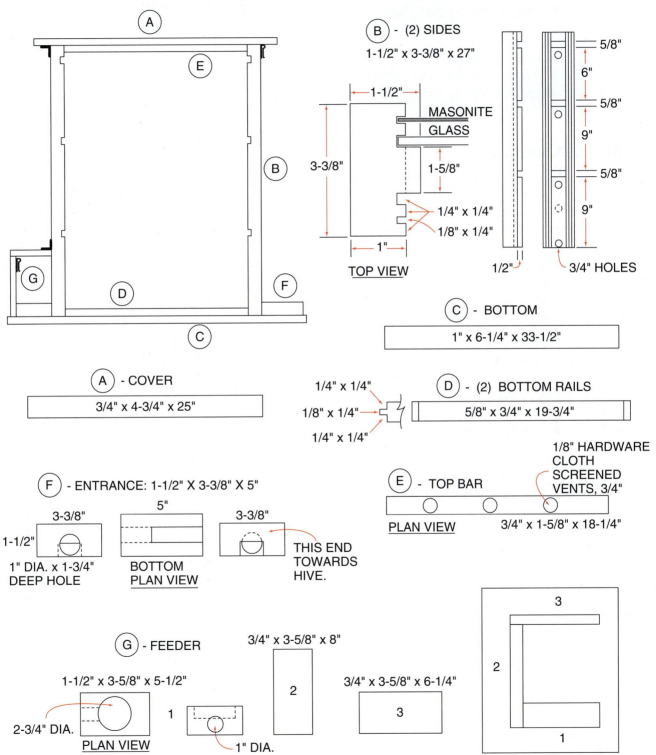

B - (2) SIDES
1-1/2" x 3-3/8" x 27"

MASONITE
GLASS

1-1/2"

3-3/8"

1-5/8"

1/4" x 1/4"
1/8" x 1/4"

1"

TOP VIEW

5/8"
6"
5/8"
9"
5/8"
9"

1/2"   3/4" HOLES

C - BOTTOM
1" x 6-1/4" x 33-1/2"

A - COVER
3/4" x 4-3/4" x 25"

1/4" x 1/4"
1/8" x 1/4"
1/4" x 1/4"

D - (2) BOTTOM RAILS
5/8" x 3/4" x 19-3/4"

F - ENTRANCE: 1-1/2" X 3-3/8" X 5"

3-3/8"
1-1/2"
1" DIA. x 1-3/4"
DEEP HOLE

5"
BOTTOM
PLAN VIEW

3-3/8"
THIS END
TOWARDS
HIVE.

E - TOP BAR

PLAN VIEW

1/8" HARDWARE
CLOTH
SCREENED
VENTS, 3/4"

3/4" x 1-5/8" x 18-1/4"

G - FEEDER

1-1/2" x 3-5/8" x 5-1/2"

2-3/4" DIA.
PLAN VIEW

1

1" DIA.

3/4" x 3-5/8" x 8"

2

3/4" x 3-5/8" x 6-1/4"

3

3

2

1

Barry Birkey, The Birkey Group, Inc.

Additional Material:

NOTE

Depending on your use and needs of the observation hive, several kinds and thicknesses of glass can be used. For a lighter-weight hive, use Plexiglass; if long-term clarity is desired, use glass. You can reverse the two dado widths for glass and panel if a thinner glass panel is used such as ⅛-inch glass and ¼-inch panel.

Two pieces of ¼-inch Plexiglass approx. 19⅝" × 27"

Two pieces of ⅛-inch hardboard or luan 19⅝" × 27"

Two 3-inch brass hinges

Two brass latches that can be locked

1-inch clear tubing (length as required)

One mason jar with holes drilled in the cover

Construction Notes

For making sides, start with a piece of ⁶⁄₄ oak stock about 8 inches in width. Cut dados for frame rests first and then rip the stock into two pieces, 3¾ inches wide. Next, adjust fence to 2⅞ inches and cut dado, then turn the stock around and cut the other dado that holds the panels. Now set the fence to 2½ inches and cut the dado that holds the glass. Go back and widen the first dados to the thickness of the panel material you are using (e.g., luan, birch plywood, hardboard).

- The height of part (2) for the feeder can vary depending on the jar used. Adjust to fit (pint, quart, mason, etc.). An entrance feeder can be modified and used in place of feeder bottom "G" so there's tin under the jar instead of wood. Do not move the hive with the sugar water in the jar. Put an empty one on first.

- On one of the sides, drill the bottom vent hole about halfway up the first frame space. This prevents a conflict when attaching the feeder top.
- If you use screws for attaching members, they can be installed from bottom so that none show. The top bar screws get covered with the latch and hinge.
- Install a nice brass pull on top to act as handle and rout the edges to give a nice profile.
- Apply petroleum jelly to the edge of the glass before installing to prevent the bees from propolising the glass to the frame. If they do propolise, use a hair drier on the glass edge to soften.

Project 40: **HONEY EXTRACTOR—USDA MODEL**

Plan for Honey Extractor—USDA Model

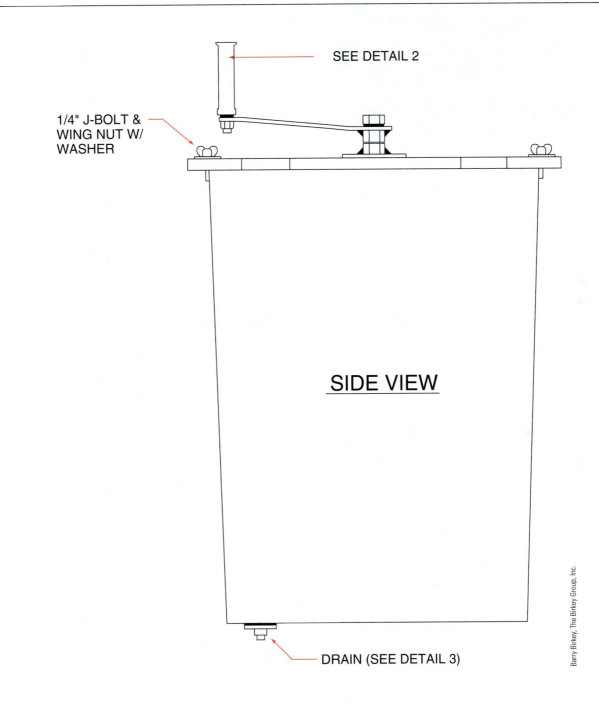

SEE DETAIL 2

1/4" J-BOLT &
WING NUT W/
WASHER

SIDE VIEW

DRAIN (SEE DETAIL 3)

Barry Birkey, The Birkey Group, Inc.

NOTE

The honey extractor is used to spin honey out of the combs after they have been uncapped. This unit will handle four combs at once. Caution should be used when extracting from a new or very old combs as they are subject to breakage. These combs should be partially emptied on one side, then turned and partially emptied on the other side, repeating this procedure until all honey is removed.

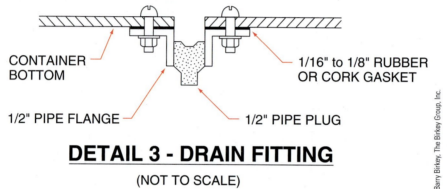

CONTAINER BOTTOM

1/16" to 1/8" RUBBER OR CORK GASKET

1/2" PIPE FLANGE

1/2" PIPE PLUG

DETAIL 3 - DRAIN FITTING

(NOT TO SCALE)

Barry Birkey, The Birkey Group, Inc.

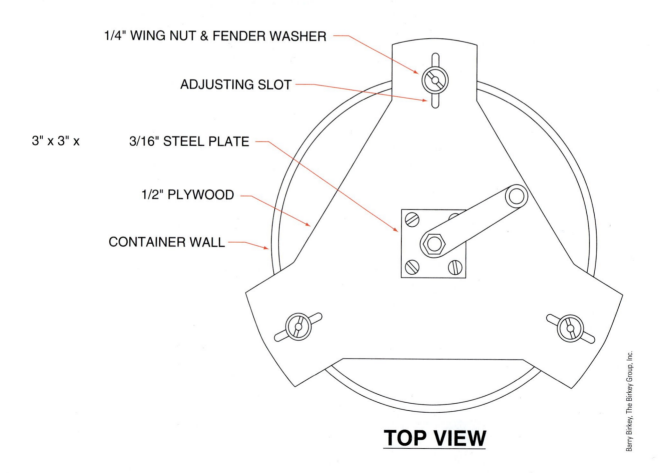

1/4" WING NUT & FENDER WASHER

ADJUSTING SLOT

3" x 3" x

3/16" STEEL PLATE

1/2" PLYWOOD

CONTAINER WALL

TOP VIEW

Barry Birkey, The Birkey Group, Inc.

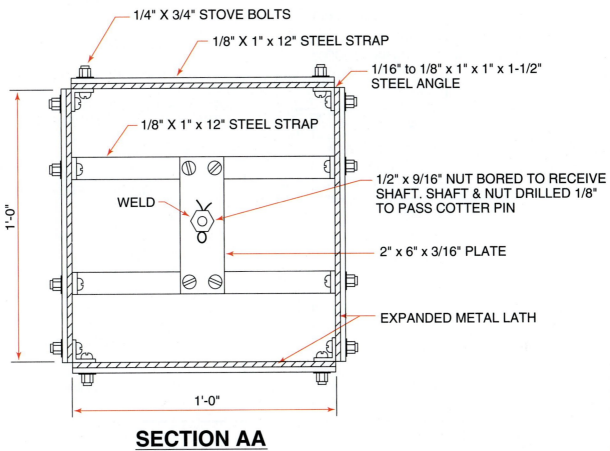

1/4" X 3/4" STOVE BOLTS

1/8" X 1" x 12" STEEL STRAP

1/16" to 1/8" x 1" x 1" x 1-1/2" STEEL ANGLE

1/8" X 1" x 12" STEEL STRAP

1/2" x 9/16" NUT BORED TO RECEIVE SHAFT. SHAFT & NUT DRILLED 1/8" TO PASS COTTER PIN

WELD

2" x 6" x 3/16" PLATE

EXPANDED METAL LATH

1'-0"

1'-0"

SECTION AA

Barry Birkey, The Birkey Group, Inc.

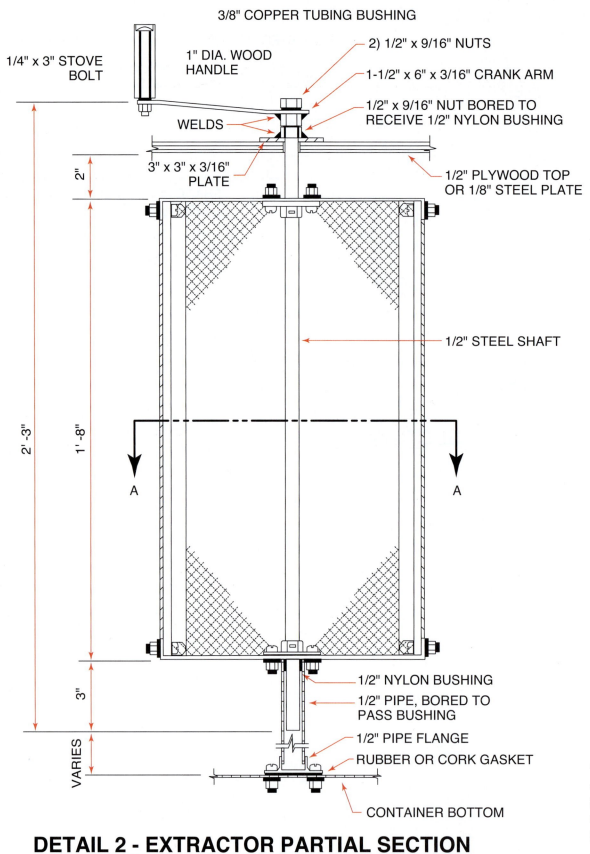

3/8" COPPER TUBING BUSHING

2) 1/2" x 9/16" NUTS

1-1/2" x 6" x 3/16" CRANK ARM

1/2" x 9/16" NUT BORED TO
RECEIVE 1/2" NYLON BUSHING

1" DIA. WOOD
HANDLE

1/4" x 3" STOVE
BOLT

WELDS

3" x 3" x 3/16"
PLATE

1/2" PLYWOOD TOP
OR 1/8" STEEL PLATE

2"

1/2" STEEL SHAFT

2' -3"

1' -8"

A

A

3"

1/2" NYLON BUSHING

1/2" PIPE, BORED TO
PASS BUSHING

VARIES

1/2" PIPE FLANGE

RUBBER OR CORK GASKET

CONTAINER BOTTOM

DETAIL 2 - EXTRACTOR PARTIAL SECTION

Barry Birkey, The Birkey Group, Inc.

APPENDIX B
Data Tables

Table 1

Weights and Measures

Liquid/Dry Measure

16 fluid ounces or 1 pound (water)	1 pint
2 pints	1 quart
4 quarts	1 gallon
8 quarts	1 peck
4 pecks	1 bushel
1 bushel	1¼ cu ft
2 barrels	1 hogshead
31½ gallons (US)	1 barrel
7.48 gallons (US)	1 cu ft
1 gallon (US)	231 cu in
1 inch of water per acre	27,154 gallons
1 inch of water per hectare	67,885 gallons

(1 hectare = 2.5 acres)

Weight Measure

Gram	15.432 grains
Gram	0.0353 ounce
Kilogram	2.2046 pounds
Kilogram	0.0011 ton (Short)
Met. Ton	1.1025 ton (Short)
Grain	0.064 gram
Ounce	28.35 grams
16 ounces	1 pound
Pound	453.5 grams
100 pounds	1 hundredweight
20 hundredweight	1 ton
Ton (Short)	907.18 kilograms
Ton (Short)	0.907 met. ton
Ton (Short)	2,000 pounds

1 gallon of water weighs 8.34 pounds

1 gallon of milk weighs 8.6 pounds

Temperature

°C	= (°F – 32) × 5/9
°F	= (9/5 °C) + 32

Linear Measure

12 inches	1 foot
3 feet	1 yard
5½ yards (16½ feet)	1 rod
320 rods	1 mile
1,760 yards (5,280 feet)	1 mile

Square Measure

144 sq in	1 sq ft
9 sq ft	1 sq yd
30¼ sq yd	1 sq rod
160 sq rods	1 acre (43,560 sq ft)
640 acres	1 sq mi
1 sq mi	1 section

Cubic Measure

1,728 cu in	1 cu ft
27 cu ft	1 cu yd
128 cu ft (8' × 4' × 4')	1 cord
1' × 1' × 1"	1 bd ft

Land Measure

To find number of acres:

Divide no. of sq ft by 43,560

Divide no. of sq yds by 4,840

(continued)

Table 1

Weights and Measures, *continued*

Divide no. of sq rods by 160

Divide no. of sq chains by 10

Height of Tree or Building

To estimate the height of a tree or building:

1. Measure the height (H_1) of a nearby object that is vertical.
2. Measure the length of the shadow (S_1) cast by that object.

3. Measure the length of the shadow (S_2) cast by the tree or building.
4. Length of shadow (S_2) of tree or building times the height (H_1) of the object divided by length of shadow (S_1) of the object equals the height (H_2) of the tree or building ($H_2 = S_2 \times H_1 \times S_1$)

Copyright © 2015 Cengage Learning®.

Table 2

Metric Conversions and Measurements

Comparison of the International Metric System and the English System of Measurement

1 centimeter = 0.3937 inches

1 inch = 2.54 centimeters

1 foot = 30.48 centimeters

1 meter = 39.37 inches

1 meter = 100 centimeters

1 meter = 1.094 yards

1 meter = 1,000 millimeters

1 millimeter = 0.001 meter

1 yard = 0.9144 meter

1 mile = 1,609.344 meters

1 kilometer = 1,000 meters

1 kilometer = 0.62137 miles

1 sq centimeter = 0.155 sq inches

1 sq decimeter = 100 cu centimeters

1 cu centimeter = 0.061 cu inches

1 cu decimeter = 1,000 cu centimeters

1 cu meter = 100 liters

1 fluid ounce = 29.54 milliliters

1 liter = 1,000 cu centimeters

1 liter = 1.057 quarts

1 gallon = 3.785 liters

1 gram = 15.43 grains

1 ounce = 28.35 grams

1 kilogram = 1,000 grams

1 kilogram = 2.205 lb

1 pound = 7,000 grains

1 pound = 0.4536 kilograms

1 kilogram = 1,000 milliliters

1 kilogram = 1 liter

Approximate Conversion of Common Units

U.S. to Metric

LENGTH

1 inch .... = 25.0 millimeters (mm)

1 foot .... = 0.3 meter (m)

1 yard .... = 0.9 meter

1 mile .... = 1.6 kilometers (km)

AREA

1 sq in .... = 6.5 sq centimeters (cm²)

1 foot² .... = 0.09 sq meter (m²)

1 sq yd² .... = 0.8 sq meter (m²)

1 acre ....= 0.4 hectare

1 mi² .... = 2.6 km²

*1 hectare equals 10,000 sq meters

Metric to U.S.

LENGTH

1 mm .... = 0.04 inch

1 m .... = 3.3 feet

1 m .... = 1.1 yards

1 km .... = 0.6 mile

AREA

1 sq centimeter (cm²) .... = 0.16 in²

1 sq meter (m²) .... = 11.0 ft²

1 sq meter (m²) .... = 1.2 yd²

1 ha .... = 2.5 acres

1 km² .... = 0.39 mi²

Approximate Conversion of Common Units

U.S. to Metric	Metric to U.S.

MASS

U.S. to Metric	Metric to U.S.
1 grain = 64.8 milligrams (mg)	1 mg = 0.015 grain
1 ounce (dry) = 28.0 grams (g)	1 g = 0.035 ounce
1 pound = 0.45 kilogram (kg)	1 kg = 2.2 pounds
1 short ton = 9.071 kg	1 metric ton = 1.102 tons (short)

VOLUME

U.S. to Metric	Metric to U.S.
1 cubic inch (in^3) = 16.0 cubic centimeters (cm^3)	1 cubic centimeter (cm^3) = 0.06 inch^3
1 cubic foot (ft^3) = 0.03 cubic meter (m^3)	1 cubic meter (m^3) = 35.0 feet^3
1 cubic yard (yd^3) = 0.76 m^3	1 ml = 1.3 yards^3
1 teaspoon = 5.0 milliliters (ml)	1 ml = 0.2 teaspoon
1 tablespoon = 15.0 ml	1 ml = 0.07 tablespoon
1 fl. ounce = 30.0 ml	1 ml = 0.03 ounce
1 cup = 0.24 liter (l)**	1 l = 4.2 cups
1 pint = 0.47 l	1 l = 2.1 pints
1 quart (liq.) = 0.95 l	1 l = 1.1 quarts
1 gallon (liq.) = 0.004 m^3	1 m^3 = 264.0 gallons
1 peck = 0.009 m^3	1 m^3 = 113.0 pecks
1 bushel = 0.04 m^3	1 m^3 = 28.0 bushels

**1 liter equals 1 cubic decimeter (dm^3)

Source: The Maryland State Department of Education, Division of Vocational Technical Education, Baltimore, MD

Table 3

Approximate Dilution Ratios and Proportions

Dilution of Liquid Pesticides at Various Concentrations

Dilution	1 Gal	3 Gal	5 Gal	15 Gal
1–100	2 tbs + 2 tsp	½ cup	¾ cup + 2 tsp	2 cups + 6½ tbs
1–200	4 tsp	¼ cup	6½ tbs	1 cup + 3⅓ tbs
1–400	2 tsp	2 tbs	3 tbs	½ cup + 1 tbs
1–800	1 tsp	1 tbs	1 tbs + 2 tsp	4 tbs + 2 tsp
1–1,000	¾ tsp	2¼ tsp	1 tbs + 2 tsp	4 tbs

NOTE: 1 gal = 16 cups or 256 tbs or 768 tsp; 1 cup = 16 tbs; 1 tbs = 3 tsp

Water	Quantity of Material					
100 gal	1 lb	2 lb	3 lb	4 lb	5 lb	6 lb
25 gal	4 oz	8 oz	12 oz	1 lb	1¼ lb	1½ lb
5 gal	3 tbs	1½ oz	2½ oz	3¼ oz	4 oz	5 oz
1 gal	1 tsp	2 tsp	1 tbs	4 tsp	5 tsp	2 tbs

Equivalent quantities of liquid materials (emulsion concentrates, etc.) for various quantities of water based on pints per 100 gallons.

(continued)

Table 3

Approximate Dilution Ratios and Proportions, *continued*

Water			Quantity of Material			
100 gal	½ pint	1 pint	2 pints	3 pints	4 pints	5 pints
25 gal	2 fl. oz	4 fl. oz	8 fl. oz	12 fl. oz	1 pint	1¼ pints
5 gal	1 tbs	1 fl. oz	2 fl. oz	2½ fl. oz	3 fl. oz	4 fl. oz
1 gal	½ tsp	1 tsp	2 tsp	3 tsp	4 tsp	5 tsp

Copyright © 2015 Cengage Learning®.

Table 4

Cubic Air Content of a Greenhouse

FORMULA: $\dfrac{E + R}{2} \times W \times L = $ CUBIC CONTENT

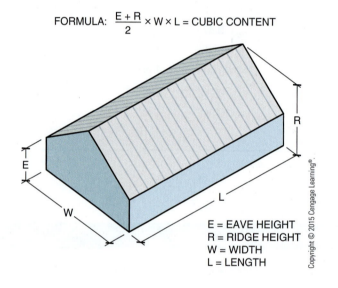

E = EAVE HEIGHT
R = RIDGE HEIGHT
W = WIDTH
L = LENGTH

Copyright © 2015 Cengage Learning®.

Table 5

Weights per Bushel of Commodities

Commodity	Weight per Bushel (Pounds)
FIELD CROPS	
Barley	48
Cottonseed	30
Corn, ear	70
Corn, shelled	56
Oats	32
Sorghum seed	56
Soybeans	60
Wheat	60
LEGUMES	
Lespedeza, common	25
Lespedeza, Korean	40
Lespedeza, Kobe	25
Clover, crimson	60
Clover, white	60
Alfalfa	60
Vetch	60
GRASSES	
Bermuda	40
Dallis	15
Bahia	15–17
Fescue	10–30
Orchard	14
Rye	53
FRUITS	
Apples	48
Peaches	48
Pears	50
VEGETABLES	
Potatoes, Irish	56
Potatoes, Sweet	50–55
Cabbage	50
Cucumbers	48
Okra	32–36
Green beans	30
English peas	30
Dried peas	60
Turnips	54

Table 6

Moisture Content of Grains for Long-Term Storage

Crop	Maximum Moisture Content (%)
Wheat	12
Oats	13
Barley	13
Grain sorghum	12
Shelled corn	13
Soybeans	11
Rice	12

Copyright © 2015 Cengage Learning®.

Table 7

Quick Reference Fraction to Decimal Equivalents

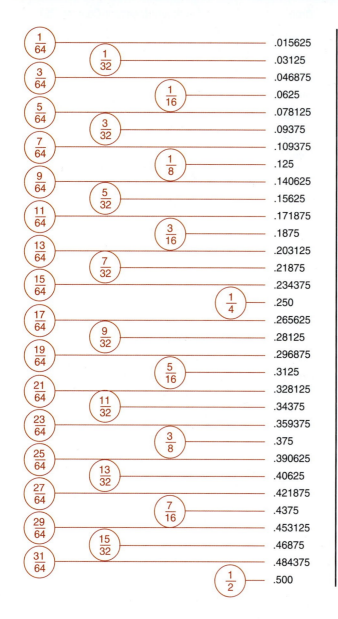

Fraction			Decimal
1/64			.015625
	1/32		.03125
3/64			.046875
		1/16	.0625
5/64			.078125
	3/32		.09375
7/64			.109375
		1/8	.125
9/64			.140625
	5/32		.15625
11/64			.171875
		3/16	.1875
13/64			.203125
	7/32		.21875
15/64			.234375
		1/4	.250
17/64			.265625
	9/32		.28125
19/64			.296875
		5/16	.3125
21/64			.328125
	11/32		.34375
23/64			.359375
		3/8	.375
25/64			.390625
	13/32		.40625
27/64			.421875
		7/16	.4375
29/64			.453125
	15/32		.46875
31/64			.484375
		1/2	.500

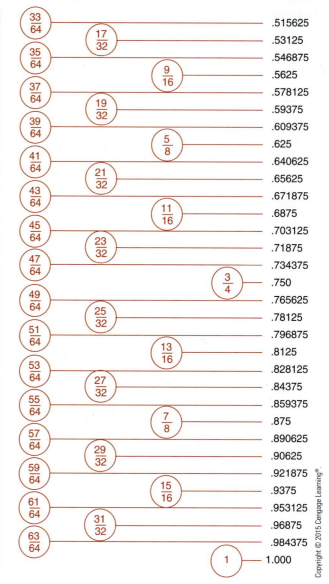

Fraction			Decimal
33/64			.515625
	17/32		.53125
35/64			.546875
		9/16	.5625
37/64			.578125
	19/32		.59375
39/64			.609375
		5/8	.625
41/64			.640625
	21/32		.65625
43/64			.671875
		11/16	.6875
45/64			.703125
	23/32		.71875
47/64			.734375
		3/4	.750
49/64			.765625
	25/32		.78125
51/64			.796875
		13/16	.8125
53/64			.828125
	27/32		.84375
55/64			.859375
		7/8	.875
57/64			.890625
	29/32		.90625
59/64			.921875
		15/16	.9375
61/64			.953125
	31/32		.96875
63/64			.984375
		1	1.000

Copyright © 2015 Cengage Learning®

Table 8

Nail Data

Penny	Length (in)	Gauge No. of Shank			Approx. No. per Lb.		
		Common	**Box**	**Finish**	**Common**	**Box**	**Finish**
2d	1	15	15½	16½	830	1,010	1,351
3d	1¼	14	14½	15½	528	635	807
4d	1½	12½	14	15	316	473	584
6d	2	11½	12½	13	168	236	309
8d	2½	10¼	11½	12½	106	145	189
10d	3	9	10½	11½	69	94	121
16d	3½	8	10	11	49	71	90
20d	4	6	9	10	31	52	62
30d	4½	5	—	—	24	—	—
40d	5	4	—	—	18	—	—
60d	6	2	—	—	11	—	—

Copyright © 2015 Cengage Learning®.

Table 9

Wood Screw Information

Wood Screw Size	Shank Hole Size	Pilot Hole Size		Auger Bit Sizes by 16th for Counterbore Hole
	All Woods	**Hardwood**	**Softwood**	
	Nearest Fractional Size Drill	**Nearest Fractional Size Drill**	**Nearest Fractional Size Drill**	
0	1/16	1/32	—	—
1	5/64	1/32	1/32	—
2	3/32	3/64	1/32	3
3	7/64	1/16	3/64	4
4	7/64	1/16	3/64	4
5	1/8	5/64	1/16	4
6	9/64	5/64	1/16	5
7	5/32	3/32	1/16	5
8	11/64	3/32	5/64	6
9	3/16	7/64	5/64	6
10	3/16	7/64	3/32	6
11	13/64	1/8	3/32	7
12	7/32	1/8	7/64	7

Standard screw lengths in inches are ³⁄₁₆, ¼, ³⁄₈, ½, ⅝, ¾, ⅞, 1, 1¼, 1½, 1¾, 2, 2¼, 2½, 2¾, 3, 3½, and up.

Copyright © 2015 Cengage Learning®.

Table 10

Grades of Hardwood Lumber*

Grades	Clear cut** (%)	Minimum Dimensions (W × L)	Principal Farm Uses
Firsts	91.7	6" × 8'	Flooring, trim, furniture, implement & tool handles
Seconds	83.3	6" × 8'	
Selects	91.7	4" × 6'	
No. 1 Common	66.7	3" × 4'	Permanent structures (barns, pens, feeders, troughs, fences)
No. 2 Common	50.0	3" × 4'	
Sound & Wormy	33.3	3" × 4'	
No. 3A Common	33.3	3" × 4'	Temporary structures (sheds, fences, feeders, crates)
No. 3B Common	25.0	3" × 4'	

*Adopted by National Hardwood Lumber Association.
**Percentage of face surface that is free of all defects.

Source: National Hardwood Lumber Association

Table 11

Lumber Dimensions

Type of Lumber	Rough Size (in) (Sawmill Cut)	Actual Size (in) (S × S)*	Bd ft per ft of Length	Amount to Add to Cover**
Square Edge Boards	1 × 4	¾ × 3½	⅓	1/10
	1 × 6	¾ × 5½	½	1/16
	1 × 8	¾ × 7¼	⅔	1/16
	1 × 10	¾ × 9¼	⅚	1/20
	1 × 12	¾ × 11¼	1	1/24
Framing	2 × 4	1½ × 3½	⅔	—
	2 × 6	1½ × 5½	1	—
	2 × 8	1½ × 7¼	1⅓	—
	2 × 10	1½ × 9¼	1⅔	—
	2×12	1½ × 11¼	2	—
	4×4	3½ × 3½	1⅓	—
	6 × 6	5½ × 5½	3	—

*Surfaced on all four sides.
**Amount to add to total surface to allow for surfacing and matching of lumber. (Does not include carpenter's waste.)

Copyright © 2015 Cengage Learning®.

Table 12

Wood Measure

A board foot of lumber = A Unit 1' long × 1' wide × 1" thick or its equivalent

1 U.S. cord of wood = A Pile 4' wide × 4' high × 8' long or its equivalent

1 cord foot of wood = 4' wide × 4' high × 1' long or its equivalent

Board feet = Number of pieces of wood × thickness in inches × width in inches × length in feet ÷ 12

Table 13

Positions of Framing Square for Some Common Angles

Position of		For Angle at	
Body (inches)	Tongue (inches)	Body (degrees)	Tongue (degrees)
12	1$\frac{1}{16}$	5	85
12	2$\frac{1}{8}$	10	80
12	3$\frac{3}{16}$	15	75
12	4$\frac{3}{8}$	20	70
12	5$\frac{9}{16}$	25	65
12	6$\frac{15}{16}$	30	60
12	8$\frac{3}{8}$	35	55
12	10$\frac{1}{16}$	40	50
12	12	45	45

Table 14

Recommended RPM for Wood Turning

Stock Dia. (inches)	Rounding Stock	General Cutting	Finishing
1	900–1,400	2,400–3,000	3,000–4,400
2	800–1,200	2,000–2,700	2,700–4,000
3	700–1,050	1,800–2,400	2,400–3,500
4	600–900	1,600–2,100	2,100–3,000
5	550–750	1,400–1,800	1,850–2,600
6	500–650	1,200–1,550	1,600–2,200
7	450–550	1,000–1,300	1,350–1,650
8	400–500	800–1,050	1,100–1,400
9	350–450	600–700	750–1,150
10	300–400	400–500	600–950

Copyright © 2015 Cengage Learning®.

Table 15

Abrasive Paper Grades

Grit	Equiv. "O" Series	General Description	Remarks
600	—	Super	Range of abrasive papers used for wet sanding
500	—	Fine	
400	10/0		
360	—		
320	9/0		
280	8/0	Very	Used for dry sanding all finishing undercoats
240	7/0	Fine	
220	6/0		
180	5/0	Fine	For final sanding of bare wood; Good for smoothing old paint
150	4/0		
120	3/0	Medium	Use for general wood sanding; Good for first smoothing of old paint, plaster patches
100	2/0		
80	1/0		
60	½	Coarse	For rough-wood sanding
50	1		
40	1½		
36	2	Very	Too coarse for pad sanders. Heavy machines and high speed recommended
30	2½	Coarse	
24	3		

Copyright © 2015 Cengage Learning®.

Table 16

Recommended RPM of High-Speed Drills in Various Metals*

Drill Size**	Wrought Iron, Low-Carbon Steel	Medium-Carbon Steel	High-Carbon Tool Steel	Cast Iron	Aluminum & Brass
1/16	5,000–6,700	4,000–4,900	3,000–3,600	4,200–6,100	12,000–18,000
1/8	2,500–3,350	2,100–2,450	1,500–1,800	2,100–3,000	6,000–9,000
3/16	1,600–2,200	1,400–1,600	1,000–1,200	1,400–2,000	4,000–6,000
1/4	1,200–1,700	1,000–1,200	750–900	1,100–1,500	3,000–4,500
5/16	1,000–1,300	850–950	600–725	850–1,200	2,400–3,600
3/8	800–1,100	700–800	500–600	700–1,000	2,000–3,000
7/16	700–950	600–700	425–525	600–875	1,700–2,500
1/2	600–850	450–600	375–450	525–750	1,500–2,250
9/16	550–750	425–550	340–400	475–675	1,325–2,000
5/8	500–650	400–500	300–350	425–600	1,200–1,800
11/16	450–600	375–450	275–325	375–550	1,100–1,650
3/4	400–550	350–400	250–300	350–500	1,000–1,500
13/16	375–500	325–375	235–275	325–460	900–1,375
7/8	350–450	300–350	225–350	300–425	850–1,275
15/16	325–425	275–325	210–235	275–400	800–1,200
1	300–400	250–300	200–225	250–375	750–1,125

*Reduce RPM one-half for carbon drills.
**For number and letter drills, use speed of nearest fractional-sized drill.

Copyright © 2015 Cengage Learning®.

Table 17

Data for Screw Extractors

Screw or Bolt (in)	Twist Drill (in)	Extractor (No.)	Pipe (in)
$3/16$–$1/4$	$5/64$	1	—
$1/4$–$5/16$	$7/64$	2	—
$5/16$–$7/16$	$5/32$	3	—
$7/16$–$9/16$	$1/4$	4	$1/8$
$9/16$–$3/4$	$17/64$	5	$1/4$
$3/4$–1	$13/32$	6	$3/8$
1–$1\frac{1}{8}$	$17/32$	7	$1/2$
$1\frac{3}{8}$–$1\frac{3}{4}$	$13/16$	8	$3/4$
$1\frac{3}{4}$–$2\frac{1}{8}$	$1\frac{1}{16}$	9	1
$2\frac{1}{8}$–$2\frac{1}{2}$	$1\frac{5}{16}$	10	$1\frac{1}{4}$
$2\frac{1}{2}$–3	$1\frac{9}{16}$	11	$1\frac{1}{2}$
3–$3\frac{1}{2}$	$1\frac{15}{16}$	12	2

Copyright © 2015 Cengage Learning®.

Table 18

Sheet Metal Riveting Data

Metal Gauge	Rivets			Drill, Punch (in)	Rivet Set
	Weight	Dia. (in)	Length (in)		
28	14 oz.	0.109	$3/16$	$7/64$	#8
26	1 lb.	0.112	$13/64$	$1/8$	#7
24	2 lb.	0.144	$17/64$	$9/64$	#5
22	$2\frac{1}{2}$ lb.	0.148	$9/32$	$5/32$	#4
20	3 lb.	0.160	$5/16$	$11/64$	#4

Copyright © 2015 Cengage Learning®.

Table 19

Self-Tapping Sheet Metal Screw Data (Pointed Type)

Screw No.	Screw Lengths*	Metal Gauge	Drill Size No.	Inches
4	$1/4$–$3/4$	29–22	42	$3/32$
6	$3/8$–1	29–20	38	$7/64$
8	$3/8$–$1\frac{1}{2}$	26–18	32	$1/8$
10	$3/8$–2	24–18	29	$9/64$
12	$1/2$–2	24–18	25	$5/32$
14	$1/2$–2	24–18	12	$3/16$

*Standard screw lengths in inches are: $1/4$, $3/8$, $1/2$, $5/8$, $3/4$, $7/8$, 1, $1\frac{1}{4}$, $1\frac{1}{2}$, and 2. This column shows range of screw lengths available—example, for No. 4, lengths are $1/4$, $3/8$, $1/2$, $5/8$, and $3/4$.

Copyright © 2015 Cengage Learning®.

Table 20

Fluxes for Soft Soldering

Metal	Flux	Chemical Name
Aluminum	Special compounds	
Brass, copper	Cut acid, rosin, sal ammoniac	Zinc chloride, colophony, ammonium chloride
Galvanized iron	Muriatic acid	Hydrochloric acid
Iron, steel	Cut acid, sal ammoniac	Zinc chloride, ammonium chloride
Tin	Rosin, cut acid	Colophony, zinc chloride

Copyright © 2015 Cengage Learning®.

Table 21

Melting Points of Soft Solders

Tin (%)	Lead (%)	Liquifies at (°F)*
63	37	360°F
50	50	415
45	55	437
40	60	459
33	67	621

*First melting point for each type is 360°F.

Copyright © 2015 Cengage Learning®.

TABLE 23

Hacksaw Blades

Teeth (per in)	Stock	Material
14	1" & over	cast iron, machine steel, brass, copper, aluminum, bronze, slate
18	¼" to 1"	annealed toolsteel, high-speed steel, rails, bronze, copper, aluminum
24	⅛" to ¼"	iron, steel, drill rod, brass & copper tubing, wrought-iron pipe, conduit, trim
32	⅛" & less	same as for 24 teeth per in

Copyright © 2015 Cengage Learning®.

Table 22

Recommended Colors for Tempering Various Tools*

Degrees Fahrenheit	Color	Kind of Tool
430	Yellow	Scrapers, lathe cutting, tools, hammers
470	Straw	Punches, dies, hacksaw, blades, drills, taps, knives, reamers
500	Brown	Axes, wood chisels, drifts, shears
530	Purple	Cold chisels, center punches, rivet sets
560	Blue	Screwdrivers, springs, gears, picks, saws
700	Grey	Soft (annealed), must harden again

Copyright © 2015 Cengage Learning®.

*Because of the difference in the quality of steel used in tool manufacture, the color recommendations will not always apply. E.g., It may be necessary to cool a cold chisel when at a straw or blue color instead of purple to get the desired temper.

Table 24

Fuel-Oil Mixtures

Mix.	U.S. Gallons		Imperial Gallons		Metric	
	Fuel (gal)	Oil (oz)	Fuel (gal)	Oil (oz)	Petrol (l)	Oil (l)
16:1	1	8	1	10	4	0.250
	3	24	3	30	12	0.750
	5	40	5	50	20	1.250
	6	48	6	60	24	1.500
24:1	1	5.33	1	6.4	4	0.160
	3	16	3	19.2	12	0.470
	5	26.66	5	32	20	0.790
	6	32	6	38.4	24	0.940
32:1	1	4	1	5	4	0.125
	3	12	3	15	12	0.375
	5	20	5	25	20	0.625
	6	24	6	30	24	0.750
50:1	1	2.5	1	3	4	0.080
	3	8	3	9	12	0.240
	5	13	5	15	20	0.400
	6	15.5	6	18.5	24	0.480

Copyright © 2015 Cengage Learning®.

Table 25

Electrodes for Farm Welding

Type of Work	Electrode		Current & Polarity (S-straight) (R-reversed)	Welding Position	Penetration	Characteristics	Uses
	AWS. No	NEMA Color					
Mild Steel	E6010	None	DC-R	All	Deep	Digging affects it; leaves rough, rippled surface; slow burn off; a strong weld with much spatter.	All mild-steel welding; $\frac{1}{8}$" very good for holes and cutting; best farm electrode.
	E6011	Blue spot	All	All	Deep		
	E6012	White spot	All	All	Medium	Melted metal is gummy.	Fill in gaps and poor fit ups; medium arc keeps sagging metal from closing the circuit.
	E6013	Brown spot	All	All	Shallow		Very good for down welds on thin sheet metal.
	E6013	Brown spot	All	All	Shallow	General-purpose, easy to use; less burn through owing to shallower penetration.	All mild steel; for extra strength, lay a heavy first bead or two light beads.
	E6014	Brown spot	All	All	Shallow	Cross between E6013 and E6024 easy to use; fast burn off.	Second best for farm; same uses as above.
	E6024	Yellow spot	All	Flat, Horizontal	Medium	Fast deposit; slag removes itself if amperage is right and edges not pinned by poor electrode motion.	All downhand welds, in fillets and horizontal; $\frac{3}{32}$" good as spray rod on vertical down welds.
	E6027	Silver spot	All	Flat, Horizontal	Medium	Less undercut than E6024; better weld; slag not as easily removed; less penetration on rusty metal.	Same as E6024, but see Characteristics (to the left).
Low-Alloy Metal	E7016	(See Characteristics.)	AC, DC-R	All	Varies	Requires clean surface*; NEMA code; green group, orange spot, blue end.	All difficult welds; rather hard for beginners to use.

(continued)

Table 25

Electrodes for Farm Welding, *continued*

Type of Work	Electrode		Current & Polarity (S-straight) (R-reversed)	Welding Position	Penetration	Characteristics	Uses
	AWS. No	NEMA Color					
Non Machinable Cast Iron	No AWS; non-machinable cast iron	Orange end	AC, DC-S	All	Medium	Requires clean surface*; low heat needed reduces cracking of weld or work; do not permit work to become dark red.	For shielded arc welds if not to be machined; hold close arc, but coating must not touch molten metal; intermittent beads not over 3" long; peen lightly after each bead; cool and clean before next bead.
Cast Iron	No AWS; cast Iron	Orange spot	AC, DC-R	All	Medium	Same as above, except machinable.	All cast iron, machinery gears, housings, parts, mold boards, etc.
Stainless Steel	No AWS; s. steel 308-16	Yellow group & end	All	All	Medium	Requires short arc, electrode held 15° in direction of travel.	High-to-medium carbon-alloy steel and most nonferrous metals; auto bumpers, mold boards, etc.
Hard Surfacing			AC, DC-R	All	Light	For all hard surfacing; requires clean surface*; not for joining parts; build up worn parts with a strength rod, then hard surface; for wear resistance, lay straight or wavy bead not over ¾" wide; remove slag before new bead.	Metal-to-metal wear: building-up gives moderate hardness to resist shock and abrasion.
			AC, DC-S	All	Light		
			AC, DC-S	All	Light		
			AC, Arc torch	All	Light		Metal in rocky soil: build-up for resistance to impact and severe abrasion; for mild or carbon steel, low-alloy or high-manganese steel.
							Metal in sandy soil: to resist any kind of abrasion, mild impact; for carbon, alloy, or manganese steel.
							Knife edges: fine-grain alloy powder applied with carbon arc gives smooth, abrasion-resistant surface.

*This means that surfaces must be cleaned of rust, grease, oil, moisture, and other foreign matter by wire brushing or grinding.

Copyright © 2015 Cengage Learning®.

Table 26

Electrical Wire Sizes for Copper at 115–120 Volts

Minimum Allowable Size of Conductor | COPPER up to 200 Amperes, *115-120 Volts*, Single Phase, Based on 2% Voltage Drop

Length of Run in Feet — Compare size shown below with size shown to left of double line. Use the larger size.

Load in Amps	Types R,T, TW	Types RH, RHW, THW	Covered Conductors	30	40	50	60	75	100	125	150	175	200	225	250	275	300	350	400	450	500	550	600	650	700
5	12	12	10	12	12	12	12	12	12	12	10	10	10	10	8	8	8	6	6	6	6	4	4	4	
7	12	12	10	12	12	12	12	12	12	10	10	8	8	8	8	6	6	6	6	4	4	4	4	4	3
10	12	12	10	12	12	12	12	10	10	8	8	8	6	6	6	6	4	4	4	4	3	3	2	2	2
15	12	12	10	12	12	10	10	10	8	6	6	6	4	4	4	4	4	3	2	2	1	1	1	0	0
20	12	12	10	12	10	10	8	8	6	6	4	4	4	4	3	3	2	2	1	1	0	0	00	00	00
25	10	10	10	10	10	8	8	6	6	4	4	4	3	3	2	1	1	0	0	00	00	000	000	000	
30	10	10	10	10	8	8	8	6	4	4	4	3	2	2	1	1	1	0	00	00	000	000	000	4/0	4/0
35	8	8	10	10	8	8	6	6	4	4	3	2	2	1	1	0	0	00	00	000	000	4/0	4/0	4/0	250
40	8	8	10	8	8	6	6	4	4	3	2	2	1	1	0	0	00	00	000	000	4/0	4/0	250	250	300
45	6	8	10	8	8	6	6	4	4	3	2	1	1	0	0	00	000	000	4/0	4/0	250	250	300	300	
50	6	6	10	8	6	6	4	4	3	2	1	1	0	0	00	00	000	000	4/0	4/0	250	250	300	300	350
60	4	6	8	8	6	4	4	4	2	1	1	0	00	00	000	000	4/0	250	250	300	300	350	400	400	
70	4	4	8	6	6	4	4	3	2	1	0	00	00	000	000	4/0	250	300	300	350	400	400	500	500	
80	2	4	6	6	4	4	3	2	1	0	00	00	000	000	4/0	250	300	300	350	400	400	500	500	600	
90	2	3	6	6	4	4	3	2	1	0	00	000	000	4/0	4/0	250	300	350	400	500	500	500	600	600	
100	1	3	6	4	4	3	2	1	0	00	000	000	4/0	4/0	250	250	300	350	400	500	500	500	600	600	700
115	0	2	4	4	4	3	2	1	0	00	000	4/0	4/0	250	300	350	400	500	500	600	600	700	700	750	
130	00	1	4	4	3	2	1	0	00	000	4/0	4/0	250	300	300	400	500	500	600	600	700	750	800	900	
150	000	0	2	4	2	1	1	0	000	4/0	4/0	250	300	350	350	500	500	600	700	700	800	900	900	1M	
175	4/0	00	2	3	2	1	0	00	000	4/0	250	300	350	400	400	500	600	700	750	800	900	1M			
200	250	000	1	2	1	0	00	000	4/0	25	0300	350	400	500	500	500	600	700	750	900	1M				

Source: National Food and Energy Council, Inc., Columbia, Missouri

Table 27

Electrical Wire Sizes for Aluminum at 115–120 Volts

Minimum Allowable Size of Conductor · ALUMINUM up to 200 Amperes, *115-120 Volts*, Single Phase, Based on 2% Voltage Drop

| In Cable, Conduit, Earth | | | Overhead in Air | Length of Run in Feet | | | | | | | | | | | | | | | | | | | | | |
| Load in Amps | Types R,T,TW | Types RH,RHW,THW | Covered Conductors | Compare size shown below with size shown to left of double line. Use the larger size. | | | | | | | | | | | | | | | | | | | | | |
				30	40	50	60	75	100	125	150	175	200	225	250	275	300	350	400	450	500	550	600	650	700
5	12	12	10	12	12	12	12	12	10	10	8	8	8	8	6		6	6	6	4	4	4	4	3	3
7	12	12	10	12	12	12	12	10	10	8	8	6	6	6	6	4	4	4	4	3	3	2	2	2	1
10	12	12	10	12	12	10	10	8	8	6	6	6	4	4	4	4	3	3	2	2	1	1	0	0	0
15	12	12	10	12	10	8	8	8	6	4	4	4	3	3	2	2	2	1	0	0	00	00	00	000	000
20	10	10	10	10	8	8	6	6	4	4	3	3	2	2	1	1	0	0	00	00	000	000	4/0	4/0	4/0
25	10	10	10	8	8	6	6	4	4	3	2	2	1	1	0	0	00	00	000	000	4/0	4/0	250	250	300
30	8	8	10	8	6	6	6	4	3	2	2	1	0	0	00	00	00	000	4/0	4/0	250	250	300	300	350
35	6	8	10	8	6	6	4	4	3	2	1	0	0	00	00	000	000	4/0	4/0	250	300	300	350	350	400
40	6	8	10	6	6	4	4	3	2	1	0	0	00	00	000	000	4/0	4/0	250	300	300	350	350	400	500
45	4	6	10	6	6	4	4	3	2	1	0	00	00	000	000	4/0	4/0	250	300	300	350	400	400	500	500
50	4	6	8	6	4	4	3	2	1	0	00	00	000	000	4/0	4/0	250	300	300	350	400	400	500	500	600
60	2	4	6	6	4	3	3	2	0	00	00	000	4/0	4/0	250	250	300	350	350	400	500	500	600	600	700
70	2	2	6	4	4	3	2	1	0	00	000	4/0	4/0	250	300	300	350	400	500	500	600	600	700	700	750
80	1	2	6	4	3	2	1	0	00	000	4/0	4/0	250	300	300	350	350	500	500	600	600	700	750	800	900
90	0	2	4	4	3	2	1	0	00	000	4/0	250	300	300	350	400	400	500	600	600	700	750	800	900	1M
100	0	1	4	3	2	1	0	00	000	4/0	250	300	300	350	400	400	500	600	600	700	750	900	900	1M	
115	00	0	2	3	1	1	0	00	000	4/0	300	300	350	400	500	500	600	600	700	800	900	1M			
130	000	00	2	2	1	0	00	000	4/0	250	300	350	400	500	500	600	600	700	800	900	1M				
150	4/0	000	1	2	0	00	00	000	250	300	350	400	500	500	600	700	800	900	1M	600					
175	300	4/0	0	1	0	00	000	4/0	300	350	400	500	600	60	750	800	900	700							
200	350	250	00	0	00	000	4/0	250	300	400	500	600	600	900	900	700	750								

Source: National Food and Energy Council, Inc., Columbia, Missouri

Table 28

Electrical Wire Sizes for Copper at 230–240 Volts

Minimum Allowable Size of Conductor — COPPER up to 400 Amperes, *230-240 Volts*, Single Phase, Based on 2% Voltage Drop

In Cable, Conduit, Earth			Overhead in Air	Length of Run in Feet																				
			Covered Conductors	Compare size shown below with size shown to left of double line.																				
Load in Amps	Types R, T, TW	Types RH, RHW, THW		50	60	75	100	125	150	175	200	225	250	275	300	350	400	450	500	550	600	650	700	750	800
5	12	12	10	12	12	12	12	12	12	12	12	12	12	12	10										
7	12	12	10	12	12	12	12	12	12	12	12	10	10	10	10	10	10	10	8	8	8	8	8	6	6
10	12	12	10	12	12	12	12	12	10	10	10	10	8	8	8	8	8	8	8	6	6	6	6	6	6
15	12	12	10	12	12	12	10	10	10	8	8	8	6	6	6	8	6	6	6	6	4	4	4	4	4
20	12	12	10	12	12	10	10	8	8	8	6	6	6	6	4	6	4	4	4	4	3	3	3	3	2
25	10	10	10	12	10	10	8	8	6	6	6	6	4	4	4	4	4	4	3	3	2	2	2	1	1
30	10	10	10	10	10	10	8	6	6	6	4	4	4	4	4	4	3	3	2	2	1	1	1	0	0
35	8	8	10	10	10	8	8	6	6	4	4	4	4	3	3	3	2	2	1	1	1	0	0	0	00
40	8	8	10	10	8	8	6	6	4	4	4	4	3	3	2	2	2	1	1	0	0	0	00	00	00
45	6	8	10	10	8	8	6	6	4	4	4	3	3	2	2	2	1	1	0	0	00	00	00	000	000
50	6	6	10	8	8	6	6	4	4	4	3	3	2	2	1	1	1	0	0	00	00	00	000	000	000
60	4	6	8	8	8	6	4	4	4	3	2	2	1	1	1	1	0	0	00	00	000	000	000	4/0	4/0
70	4	4	8	8	6	6	4	4	3	2	2	1	1	0	0	0	00	00	000	000	000	4/0	4/0	4/0	250
80	2	4	6	6	6	4	4	3	2	2	1	1	0	0	00	00	00	000	000	4/0	4/0	4/0	250	250	300
90	2	3	6	6	6	4	4	3	2	1	1	0	0	00	00	00	000	000	4/0	4/0	250	250	300	300	300
100	1	3	6	6	4	4	3	2	1	1	0	0	00	00	000	000	000	4/0	4/0	250	250	300	300	350	350
115	0	2	4	6	4	4	3	2	1	0	0	00	00	000	000	000	4/0	4/0	250	250	300	300	350	350	400
130	00	1	4	4	4	3	2	1	0	0	00	00	000	000	4/0	4/0	4/0	250	300	300	350	350	400	400	500
150	000	0	2	4	4	3	1	0	0	00	000	000	4/0	4/0	4/0	4/0	250	300	300	350	400	400	500	500	500
175	4/0	00	2	4	3	2	1	0	00	000	000	4/0	4/0	250	250	250	300	350	350	400	500	500	500	600	600
200	250	000	1	3	2	1	0	00	000	000	4/0	4/0	250	250	300	300	350	400	400	500	500	600	600	600	700
225	300	4/0	0	3	2	1	0	00	000	4/0	4/0	250	300	300	350	350	400	500	500	500	600	600	700	700	750
250	350	250	00	2	1	0	00	000	4/0	4/0	250	300	300	350	350	400	500	500	600	600	700	700	750	800	900
275	400	300	00	2	1	0	00	000	4/0	250	250	300	350	350	400	400	500	600	600	700	700	750	800	900	1M
300	500	350	000	1	1	0	000	4/0	4/0	250	300	350	350	400	500	500	500	600	700	700	800	900	900	1M	
325	600	400	4/0	1	0	00	000	4/0	250	300	300	350	400	500	500	500	600	700	700	800	900	900	1M		
350	600	500	4/0	1	0	00	000	4/0	250	300	350	400	400	500	500	600	600	700	750	900	900	1M			
375	700	500	250	0	0	00	4/0	250	300	300	350	400	500	500	600	600	700	750	800	900	1M				
400	750	600	250	0	00	000	4/0	250	300	350	400	500	500	500	600	600	700	800	900	1M					

Conductors in overhead spans must be at least No. 10 for spans up to 50 feet and No. 8 for longer spans.

Source: National Food and Energy Council, Inc., Columbia, Missouri

Table 29

Electrical Wire Sizes for Aluminum at 230–240 Volts

Minimum Allowable Size of Conductor				ALUMINUM up to 400 Amperes, *230-240 Volts*, Single Phase, Based on 2% Voltage Drop																					
In Cable, Conduit, Earth			Overhead in Air	Length of Run in Feet																					
Load in Amps	Types R,T,TW	Types RH, RHW, THW	Covered Conductors	Compare size shown below with size shown to left of double line.																					
				50	60	75	100	125	150	175	200	225	250	275	300	350	400	450	500	550	600	650	700	750	800
5	12	12	10	12	12	12	12	12	12	12	10	10	10	10	8	8	8	8	6	6	6	6	6	4	4
7	12	12	10	12	12	12	12	12	10	10	10	8	8	8	8	6	6	6	6	4	4	4	4	4	4
10	12	12	10	12	12	12	10	10	8	8	8	8	6	6	6	6	4	4	4	4	3	3	3	2	2
15	12	12	10	12	12	10	8	8	8	6	6	6	4	4	4	4	3	3	2	2	2	1	1	1	0
20	10	10	10	10	10	8	8	6	6	6	4	4	4	4	3	3	2	2	1	1	0	0	0	00	00
25	10	10	10	10	8	8	6	6	4	4	4	4	3	3	2	2	1	1	0	0	00	00	00	000	000
30	8	8	10	8	8	8	6	4	4	4	3	3	2	2	2	1	0	0	00	00	00	000	000	00	4/0
35	6	8	10	8	8	6	6	4	4	3	3	2	2	1	1	0	0	00	00	000	000	000	4/0	4/0	4/0
40	6	8	10	8	6	6	4	4	3	3	2	2	1	1	0	0	00	00	000	000	4/0	4/0	4/0	250	250
45	4	6	10	8	6	6	4	4	3	2	2	1	1	0	0	00	00	000	000	4/0	4/0	250	250	250	300
50	4	6	8	6	6	4	4	3	2	2	1	1	0	0	00	00	000	000	4/0	4/0	250	250	300	300	300
60	2	4	6	6	6	4	3	2	2	1	0	0	00	00	00	000	4/0	4/0	250	250	300	300	350	350	350
70	2	2(a)	6	6	4	4	3	2	1	0	0	00	00	000	000	4/0	4/0	250	300	300	350	350	400	400	500
80	1	2(a)	6	4	4	3	2	1	0	0	00	00	000	000	4/0	4/0	250	300	300	350	350	400	500	500	500
90	0	2(a)	4	4	4	3	2	1	0	00	00	000	000	4/0	4/0		250	300	300	350	400	400	500	500	500
100	0	1(a)	4	4	3	2	1	0	00	00	000	000	4/0	4/0	250		300	300	350	400	400	500	500	600	600
115	00	0(a)	2	4	3	2	1	0	00	000	000	4/0	4/0	250	300	350	400	500	500	600	600	600	700	700	
130	000	00(a)	2	3	2	1	0	00	000	000	4/0	250	250	300	300	400	500	500	600	600	700	700	750	800	
150	4/0	000(a)	1	2	2	1	00	000	000	4/0	250	250	300	300	350	500	500	600	600	700	750	800	900	900	
175	300	4/0(a)	0	2	1	0	00	000	4/0	250	300	300	350	400	400	600	600	700	750	800	900	900	1M		
200	350	250	00	1	0	00	000	4/0	250	300	300	350	400	400	500	600	700	750	900	900	1M				
225	400	300	000	1	0	00	00	4/0	250	300	350	400	500	500	500	600	700	750	900	1M	1M				
250	500	350	000	0	00	000	4/0	250	300	350	400	500	500	500	600	700	750	900	1M						
275	600	500	4/0	0	00	000	4/0	250	300	400	400	500	500	600	600	750	900	1M							
300	700	500	250	00	00	000	250	300	350	400	500	500	600	600	800	900	1M	700							
325	800	600	300	00	000	4/0	250	300	400	500	500	600	600	700	900	1M	750								
350	900	700	300	00	000	4/0	300	350	400	500	600	600	700	750	900	800									
375	1M	700	350	000	000	4/0	300	350	500	500	600	700	700	800	1M	900									
400		900	350	000	4/0	250	300	400	500	600	600	700	750	900	900										

Conductors in overhead spans must be at least No. 10 for spans up to 50 feet and No. 8 for longer spans.

Source: National Food and Energy Council, Inc., Columbia, Missouri

Table 30

Average Daily Water Requirements of Farm Animals	
Horse	10–12 gallons
Beef Cattle	8–12 gallons
Dairy Cows (dry)	8–12 gallons
Dairy Cows	12–20 gallons
Hogs	4–11 gallons
Sheep	1–2 gallons
Layers (100)	5 gallons
Broilers (100)	1–10 gallons

Copyright © 2015 Cengage Learning®

Table 31

Steel and Wrought-Iron Pipe Data*

Nom. Pipe Size	Out-side Dia.	Standard		Extra Strong		Threads per Inch	Tap Drill Size	Hole Size**	Length of Thread	Dist. Pipe Goes into Fitting
		Inside Dia.	Weight lb per ft	Inside Dia.	Weight lb per ft					
$\frac{1}{8}$	0.405	0.269	0.24	0.215	0.31	27	$\frac{5}{16}$	$\frac{7}{16}$	$\frac{13}{32}$	$\frac{1}{4}$
$\frac{1}{4}$	0.540	0.364	0.42	0.302	0.53	18	$\frac{7}{32}$	$\frac{9}{16}$	$\frac{5}{8}$	$\frac{3}{8}$
$\frac{3}{8}$	0.675	0.493	0.57	0.423	0.74	18	$\frac{19}{32}$	$\frac{11}{16}$	$\frac{5}{8}$	$\frac{3}{8}$
$\frac{1}{2}$	0.840	0.622	0.85	0.546	1.09	14	$\frac{23}{32}$	$\frac{7}{8}$	$\frac{13}{16}$	$\frac{1}{2}$
$\frac{3}{4}$	1.050	0.824	1.13	0.742	1.47	14	$\frac{15}{16}$	$1\frac{1}{16}$	$\frac{7}{8}$	$\frac{1}{2}$
1	1.315	1.049	1.68	0.957	2.17	$11\frac{1}{2}$	$1\frac{3}{16}$	$1\frac{11}{32}$	$1\frac{1}{32}$	$\frac{9}{16}$
$1\frac{1}{4}$	1.660	1.380	2.27	1.278	3.00	$11\frac{1}{2}$	$1\frac{15}{16}$	$1\frac{11}{16}$	$1\frac{1}{16}$	$\frac{5}{8}$
$1\frac{1}{2}$	1.900	1.610	2.72	1.500	3.63	$11\frac{1}{2}$	$1\frac{25}{32}$	$1\frac{15}{18}$	$1\frac{1}{16}$	$\frac{11}{16}$
2	2.375	2.067	3.65	1.937	5.04	$11\frac{1}{2}$	$2\frac{3}{16}$	$2\frac{7}{16}$	$1\frac{1}{8}$	$\frac{3}{4}$
$2\frac{1}{2}$	2.875	2.469	5.79	2.323	7.66	8	$2\frac{5}{8}$	$2\frac{15}{16}$	$1\frac{5}{8}$	$\frac{7}{8}$
3	3.500	3.068	7.58	2.900	10.25	8	$3\frac{1}{4}$	$3\frac{9}{16}$	$1\frac{11}{16}$	$\frac{15}{16}$

*All measurements are stated in fractional or decimal inches.
**Size of hole to drill to insert pipe through walls, floors, and other materials.

Copyright © 2015 Cengage Learning®

Table 32

Standard Malleable Fittings

Nominal Size (in)	General Dimensions (in)					
	A	**B**	**C**	**D**	**E**	**F**
1/8	11/16	11/16	15/16	—	1	13/16
1/4	13/16	3/4	1 1/16	1	1 3/16	15/16
3/8	15/16	13/16	1 3/16	1 1/8	1 7/16	1
1/2	1 1/8	7/8	1 5/16	1 1/4	1 5/8	1 1/8
3/4	1 5/16	1	1 1/2	1 1/2	1 7/8	1 5/16
1	1 1/2	1 1/8	1 11/16	1 11/16	2 1/18	1 7/16
1 1/4	1 3/4	1 5/16	1 15/16	2 1/16	2 7/16	1 11/16
1 1/2	1 15/16	1 7/16	2 1/8	2 5/16	2 11/16	1 7/8
2	2 1/4	1 11/16	2 1/2	2 13/16	3 1/4	2 1/4
2 1/2	2 11/16	1 15/16	2 7/8	3 1/4	3 7/8	—
3	3 1/16	2 3/16	3 3/16	3 11/16	4 1/2	—

Copyright © 2015 Cengage Learning®

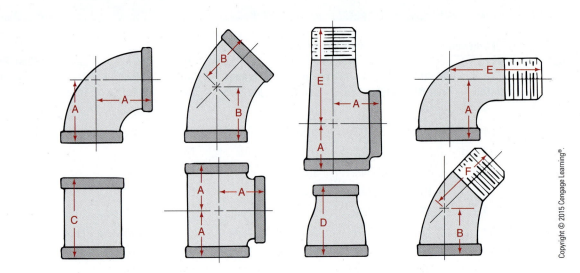

Copyright © 2015 Cengage Learning®

Table 33

Recommended Mixtures for Concrete

Uses of Concrete	Water (gal) for 1 Sack Cement—Sand				Trial Batch 1 Sack Cement plus		Space Trial Batch Fills (ft³)	Materials for 1 yd 3**		
	Dry	Damp	Wet	Very Wet	Sand (ft³)	Gravel (ft³)		Cement (sack)	Sand (ft³)	Gravel (ft³)
Acid-, alkali-resistant; dairy, creamery floors	5*	4¾	4½	4¼	1¾	2	3.38	8	14	16
Medium wear; reinforced; watertight; floors, tanks, etc.	6*	5½	5	4¼	2¼	3	4.32	6¼	14	19
Medium wear, indoor, underground, no water pressure	7*	6¼	5½	4¾	2¾	4	5.4	5	14	20

*Called 5-, 6-, or 7-gallon pastes, indicating that there are 5, 6, or 7 gallons of water to each sack of cement, including the amount of water in the sand.

**To calculate materials for 1 cu ft, divide these figures by 27.

Table 34

Approximate Materials Required for One Cubic Yard of Concrete

Proportions of Concrete or Mortar			Quantities of Materials		
Sacks Cement	Cu Yd Sand	Cu Yd Gravel	Cement	Sand	Gravel
1	1.5	—	15.5	0.86	—
1	2.0	—	12.8	0.95	—
1	2.5	—	11.0	1.02	—
1	3.0	—	9.6	1.07	—
1	1.5	3	7.6	0.42	0.85
1	2.0	2	8.2	0.60	0.60
1	2.0	3	7.0	0.52	0.78
1	2.0	4	6.0	0.44	0.89
1	2.5	3.5	5.9	0.55	0.77
1	2.5	4	5.6	0.52	0.83
1	2.5	5	5.0	0.46	0.92
1	3.0	5	4.6	0.51	0.85
1	3.0	6	4.2	0.47	0.94

Copyright © 2015 Cengage Learning®.

Table 35

Areas Covered by One Cubic Yard of Mixed Concrete

Depth, in	Sq ft	Depth, in	Sq ft	Depth, in	Sq ft
1	324	4¾	68	8½	38
1¼	259	5	65	8¾	37
1½	216	5¼	62	9	36
1¾	185	5½	59	9¼	35
2	162	5¾	56	9½	34
2¼	144	6	54	9¾	33
2½	130	6¼	52	10	32½
2¾	118	6½	50	10¼	31½
3	108	6½	48	10½	31
3¼	100	7	46	10¾	30
3½	93	7¼	45	11	29½
3¾	86	7½	43	11¼	29
4	81	7¾	42	11½	28
4¼	76	8	40	11¾	27½
4½	72	8¼	39	12	27

Copyright © 2015 Cengage Learning®.

Table 36

Pigments for Coloring Concrete

Color Desired	Pigment to Use
Blues	Cobalt oxide
Browns	Brown oxide of iron
Buffs	Synthetic yellow oxide of iron
Greens	Chromium oxide
Reds	Red oxide of iron
Grays and slate effects	Black iron oxide or carbon black (never use common lamp black)

Copyright © 2015 Cengage Learning®.

APPENDIX C
Supervised Agricultural Experience

Supervised Agricultural Experience (SAE) is an integral part of an agricultural educational program. It is where you plan, propose, conduct, document, and evaluate your programmatic experiential learning activities. It is where you apply and test what you have learned in your classes and FFA experiences. An SAE will help you become well versed in recordkeeping for your own portfolio. It will also aid in your personal leadership development and in the strategic planning process for teams, groups, organizations, and programs.

TERMS TO KNOW

- Supervised Agricultural Experience (SAE)
- Foundational SAE
- Ownership/Entrepreneurship SAE
- Placement/Internship SAE
- Research SAE
- scientific process
- School-based Enterprise SAE
- Service Learning SAE
- improvement project
- journal
- portfolio
- recordkeeping
- degree program
- proficiency awards

OBJECTIVES
Competencies to be developed

After completing this unit, you should be able to:

- Define SAE.
- List and explain the types of SAE.
- Plan an SAE program.
- Propose an SAE program.
- Conduct an SAE program.
- Document an SAE program.

- Evaluate an SAE program.
- Apply proper recordkeeping skill.
- Participate in youth leadership opportunities to create a well-rounded SAE.
- Produce in a local program of activities using a strategic planning process.
- Participate in a local program of activities using a strategic planning process.

Supervised Agricultural Experience (SAE) is a program of experiential learning activities conducted outside of the regular agricultural education class time. The student-led, instructor-supervised, work-based learning experience is designed to help students develop and apply the knowledge and skills learned in an agricultural education classroom and/or laboratories.[1] The SAE used to be called the Supervised Occupational Experience Program (SOEP), and the focus was on the occupation. These days the focus is on entrepreneurship and experiences rather than the occupation or the actual career.

The SAE takes place under the direction of your agricultural education teacher. Many times individual projects (e.g., showing pigs and raising a plant in the greenhouse) are considered to be SAE, but the intent of a quality SAE is for it to be conducted as a series of career-related experiences completed during your enrollment in agricultural education.

SAE OPTIONS

SAE projects can be completed in a variety of categories, always beginning with the foundational SAE, and then transitioning to usually one of the following based on career plans and interests:

Ownership/entrepreneurship
Placement/internship
Research: Experimental, analysis, or invention
School-based enterprise, and
Service learning

Foundational SAE—Get It Started

The career-related **Foundational SAE** is where an SAE begins. "Every Foundational SAE will provide experiences in five components as a graded part of each of your agricultural education courses." The foundational SAE results in the development of a plan to begin an SAE. Agricultural education instructors assist students with the plan and working through concepts and activities to achieve the following five components of a foundational SAE.

1. Career exploration and planning according to your interests
2. Employability (leadership) skills for college and career readiness
3. Financial management and planning
4. Workplace safety
5. Agricultural literacy[2]

Ownership/Entrepreneurship—Work for Self

In the **Ownership/Entrepreneurship SAE**, you plan, implement, operate, and assume all or some of the financial risks in a productive or service activity or agriculture, food, or natural resource–related business.[3] If you have this type of SAE, you own the materials and other inputs, and you keep financial records to determine return on investment. Some examples of SAE's experiential learning activities that can make up an SAE program are listed next. A much more comprehensive list can be found at the end of this unit.

Examples:

1. Growing grapes
2. Raising bees
3. Growing an acre of corn

Placement/Internship—Work for Someone Else

If you work for someone else on a farm or ranch, in an agricultural business, or in a verified nonprofit organization providing a "learning-by-doing" environment, your SAE qualifies for the **Placement/Internship SAE** type. This type of experience may be paid or nonpaid.[4] Not everyone reading this can have an ownership/entrepreneurship SAE, but almost everyone has the opportunity to seek out a paid or unpaid "working for someone else" experience for the purpose of learning and applying your agricultural knowledge and skills. Some examples are listed next, but many more examples are provided at the end of this unit.

Examples:

1. Working in a flower shop
2. Working on Saturdays at a local stable
3. Working at a grocery store

Research: Experimental, Analysis, or Invention—Solve a Problem

The agriculture industry is becoming a highly scientific field full of problems to be solved by you and your classmates. Problems range from world hunger to climate change and even social ideals and political policies related to the agriculture industry. A **Research SAE** helps you prepare for a long and productive career solving these important issues for our society. The research SAE involves a program of extensive activities where you plan and conduct experiments or other forms of scientific evaluation using the scientific process. There are three types of research SAEs: experimental, analytical, and invention.[5]

Experimental Research SAE. According to the National Council for Agricultural Education, the experimental research SAE involves an extensive activity where the student plans and conducts a major agricultural experiment using the scientific process. The **scientific process** is a way of answering scientific questions through observations, making assumptions, experimenting, and drawing conclusions based on your analysis. The purpose of the experiment is to provide students firsthand experience in verifying, learning, or demonstrating scientific principles in agriculture, discovering new knowledge, and using the scientific process. In an Experimental Research SAE, there is a hypothesis and a control group, and variables are manipulated.[6]

Analytical Research SAE. In an analytical research SAE, you choose a real-world agriculture, food, or natural resource-related problem that is not amenable to experimentation and design a plan to investigate and analyze the problem. You will gather and evaluate data from a variety of sources and then produce some type of finished product. The product may include a marketing display or marketing plan for a commodity, product, or service; a series of newspaper articles; a land-use plan for a farm; a detailed landscape design for a community facility; an advertising campaign for an agribusiness; and so forth. A student-led Analytical Research SAE is flexible enough so that it could be used in any type of agricultural class, provides valuable experience, and contributes to the development of critical thinking skills.[7]

Invention Research SAE. In the Invention Research SAE, you identify a need in an agriculture, food, or natural resource–related industry and perform research and analysis in order to solve a problem or increase efficiency by developing/adapting a new product or service to the industry. The student plans, documents, and develops his or her innovation through the iterative processes of design, prototyping, and testing with the goal of creating a marketable product or service.

Examples:

1. Comparing the effect of various plant food on plant growth
2. Demonstrating the impact of different levels of soil acidity on plant growth
3. Determining if different rations of feed improve growth rate of pigs

School-Based Enterprise SAE— Manage It at School

The **School-based Enterprise SAE** is an entrepreneurial operation managed by you. Your operation, however, takes place in a school setting that provides not only facilities but also goods and services that meet the needs of an identified market.[8] To give you the most educational value, this type of SAE should replicate the real world of work as much as possible. This type of SAE is usually cooperative in nature and management decisions are made by you in "cooperation" with your teacher. Activities in this type of SAE "may include, but are not limited to: cooperative livestock raising, school gardens and land labs, production greenhouses, school based agricultural research, agricultural equipment fabrication, equipment maintenance services, or a school store."[9]

Service Learning SAE—Plan a Service Project

Service learning is one of the highest forms of leadership. Our goal should always be to leave things better than we found them. The goal of the **Service Learning SAE** is to make the community in which you live better. This SAE type combines community service activities with structured reflection.

As part of this type of SAE, you can become involved in the development of a needs assessment, planning goals, creating objectives and budgets, implementing the activity, and promoting and evaluating a chosen project. The project could be in support of a school, community organization, religious institution, or nonprofit organization. You would be responsible for raising necessary funds for the project (if funds are needed). The project must stand by itself and not be part of an ongoing chapter project or community fundraiser. The project must be somewhat challenging and require the awesome leadership you possess.[10]

Examples:

1. Test water wells in community for contamination.
2. Design a web page for your FFA chapter or for a local organization.
3. Design and install a landscape plan for a church.

PLAN AN SAE PROGRAM

Planning is an important part of any team, business, project, or program. Your SAE program isn't any different. To properly plan, you will need to understand the steps in planning, as well as the guidelines for planning, the SAE.

Steps in Planning the SAE

Step One　　Identify your career interests in agriculture. Your SAE program, experiential learning activities, and the career you choose someday need to be something about which you can get excited. If you like the SAE or your future career choice, you are more likely to stick with it and be successful.

Step Two　　Review the job responsibilities of career interest areas you may have. You might like the idea of a particular project or career, but if you don't like specific tasks involved, you might need to choose another project or program. For example, if you like the idea of a placement SAE as a veterinarian technical assistant, but you don't want to clean out cages, you may have made a poor choice.

Step Three   Identify SAE programs of interest by interviewing friends who have an SAE or by viewing suggestions found in various resources. Suggestions for certain experiential learning activities or projects are at the end of this unit.

Step Four   Develop a timeline for your SAE program. In other words, which projects will happen first, second, third, and so on? What is the completion date for each activity? Electronic portfolios, recordkeeping systems, or simple calendar systems can be used for this step.

Step Five   Building on step four, build a long-range plan for the SAE program. Remember, projects or activities happen in a shorter time span, but an SAE program, which is necessary to be competitive for FFA proficiency and Star awards, happens over a longer frame and grows in scope and diversity.

Step Six   Develop the first-year (annual) plan. You have to actually start now. After you know the long-range plan and the timeline for the different experiential learning activities/projects, you are ready to put the first-year plan together. Again, think about completion dates and specific strategies for reaching those goals.

Step Seven   Replan on a regular basis. Part of good planning is reviewing your activities in light of your plan and then adjusting your plan. This should happen fairly often, but not so often that you are always planning instead of completing tasks.[11]

Guidelines to Planning SAE

The steps outlined previously should get you moving in the right direction, but the following guidelines will help ensure a successful SAE.

1. Plan for year-round experiences.
2. Include ownership and/or placement projects at home, school, or in the community.
3. Identify a number of improvement projects and supplementary skills for the year.
4. Develop a budget.
5. Plan ownership projects with some form of profit.
6. Explore different locations to gain desired experiences.
7. Discuss the SAE program with parents/guardians.
8. Plan the scope of the program to earn enough profit in order to qualify for advanced FFA degrees.
9. Provide for a variety of activities and experiences.
10. Increase the SAE scope annually.
11. Choose SAE program experiences that relate to career interest areas.[12]

Parts of an Annual SAE Plan

A yearlong SAE program plan will keep you on track, and it will help you make decisions about your SAE. The annual plan consists of a calendar, description of projects, budget, improvement projects, and supplementary skills. For entrepreneurship/ownership projects, prepare a description of the size/scope of your enterprise, the location of your projects, the nature of the business or enterprise, partners involved, methods of marketing, facilities needed, and months involved in the project. For placement projects, detail the location, beginning and ending dates, and pay schedule. It is also important to include a budget and to keep up with income and receipts as the year progresses. You will also want to include a description of improvement projects you wish to complete and supplemental skills you would like to develop. An **improvement project** is an activity that improves the appearance, convenience, efficiency, safety, or value of a home, farm, ranch, agribusiness, or other agriculture facility. Your annual plan should include specific activities of the improvement project as well as hours of labor committed and an estimate of costs.[13]

Propose an SAE Program

Following your SAE plan, it is a good practice to write a statement justifying your plans with a proposal. This proposal should explain why you chose the SAE you did as well as discuss why your plan is laid out the way it is. Your proposal should reiterate the importance of completion dates on your timeline. The SAE proposal should also detail specific agricultural or leadership concepts that you hope to achieve as a result of your SAE program.

CONDUCT AN SAE PROGRAM

The first step to conducting a quality SAE program is getting started. All of the planning and decision making can take time, but it is worth it. Once you've started, you are going to have a great time, but you have some work ahead of you as well. Specifically, you will need to become skilled at documenting your SAE program as well as evaluating it.

Document an SAE Program

Documenting your SAE is important because it gives you an opportunity to see the planning come to life. Documenting your SAE makes it permanent record from which you can learn and improve your project. It shows the problems encountered and solved, which applies to future activities and provides us with confidence to continue in our SAE.

Calendar. Using a calendar can help with more than planning. You can use a calendar, such as the one on your smartphone, to document hours invested, skills learned, and even tasks performed.

Journal or Portfolio. An alternative to using a simple calendar system is to document your SAE with a journal or portfolio, paper or electronic. A paper system could be kept using a notebook or ledger, but these days an electronic system is much easier to use. Many states have an online journal or portfolio that allows you to keep up with important documentation.

A **journal** is "a daily written record of (usually personal) experiences and observations, or a ledger in which transactions have been recorded as they occurred."[14] A **portfolio**, usually kept online, houses all of your work (skills, projects, activities, service, finances, etc.) compiled over a period of time. Like many other tools discussed in this chapter, it is helpful for improving your project. It is also advisable to utilize your portfolio for recordkeeping as well.

APPLY PROPER RECORDKEEPING SKILLS

Recordkeeping is the process of keeping a journal or portfolio of everything you have done. As stated previously, you will need to make notes in your SAE whenever you do or learn something new, and you need to document time and money spent on your activities, projects, and program. The skill of recordkeeping will serve you well in your career.[15]

At its core, recordkeeping is quite simple. For every project, you will need to capture the following for all activities or items: the date, the name or description, the hours you spent, and expenses or income that resulted. Be sure to add up the hours you've invested in all activities. Time is money, and each one of your experiences represents new knowledge and skills you've developed. You will also add up expenses and income to determine your profit and loss (P&L). It's really that simple. You can use state-endorsed, preprogrammed spreadsheets or websites to determine your P&L and to track your hours invested or keep records on paper. It's up to your personal preference, available resources, and the wishes of your advisor or business partner.

Learning to keep records for and through your SAE has many benefits. Keeping accurate records can help you determine if you made or lost money. It can help you keep others from cheating you out of what you have earned. Your knowledge and experience in recordkeeping will help you determine which parts of the business are doing well and which parts are not. Becoming proficient at recordkeeping will also help you make management decisions, document your net worth for loans, prepare tax returns, and plan for future events. Being a good recordkeeper for your SAE will also help document your activities for FFA recognitions and degree purposes. Good records can also protect you legally and help you plan a budget for the following year.[16]

Evaluate an SAE Program

When your teacher/adviser comes to visit you and your SAE, he or she is there to answer any questions you may have. The teacher is also there to help you evaluate your SAE. The evaluation looks different for different types of SAEs. Lesson 8 from the *SAE Handbook* resource available online (http://harvest.cals.ncsu.edu /site/WebFile/IIB8.pdf) breaks down specific components of SAE evaluation. They are listed next.

Entrepreneurship/Ownership:

1. Accuracy of records
2. Neatness of records
3. Dates of records
4. Net income (total income minus total expenses)
5. Are good management practices being used?
6. Efficiency factors (yield per acre, number of offspring raised, etc.)
7. Improvements made since the last observation
8. Cleanliness of facilities
9. Customer satisfaction
10. What skills were learned

Placement/Internship:

1. Satisfaction of the employer
2. Number of hours worked

3. Accuracy of records
4. Neatness of records
5. Currency of records
6. Does student report to work on time?
7. Ability to get along with the other workers
8. Ability to get along with customers
9. Skill in performing the expected tasks
10. Attitude of the student

Research:

1. Was the scientific process used?
2. How well was the job done?
3. What was learned?
4. What was the reaction of the beneficiaries of the service-learning project?
5. Were records kept on the activity, and what shape are they in?

Exploratory:

1. How many different activities were conducted?
2. How many hours were involved?
3. What was learned?
4. Accuracy of records
5. Neatness of records
6. Dates of records

School-based Enterprise:

1. How many hours were worked?
2. What was the quality of the work done?
3. What was the scope or size of the activity?
4. Accuracy of records
5. Neatness of records
6. Dates of records

Service Learning:

1. How many hours were involved?
2. How well was the job done?
3. What was learned?
4. What was the reaction of the beneficiaries of the service-learning project?
5. Were records kept on the activity, and what shape are they in?[17]

ENHANCING AN SAE WITH OTHER OPPORTUNITIES

An SAE in not only financial recordkeeping but also a record of skills, knowledge, credentials, certifications, experiences, career planning, reflection, and leadership development.[18] A solid SAE will provide you the opportunity to receive FFA degrees and compete for different proficiency awards.

The FFA **degree program** is FFA's primary recognition program. An SAE helps students apply for each degree, and every degree has certain SAE requirements. There are five degrees of FFA membership as follows: Discovery, Greenhand, Chapter, State, and American. These five degree areas recognize you for your overall participation in FFA. Everything you do in FFA, when combined, helps you move toward a degree.[19]

Proficiency awards are awards that recognize students' excellence in SAE.[20] Developing your SAE into a proficiency award is a time-consuming but rewarding task. You should apply for the proficiency award in the specific area and career pathway for which you are the strongest and have the most experience. For instance, if you have worked for a livestock producer for three years and only raised goats for one year, it would be best to apply in the placement area for Animal Systems rather than Entrepreneurship for the same system. The career pathways for your SAE and proficiency awards follow:

1. Agribusiness Systems—the study of business principles, including management, marketing, and finance, and their application to enterprises in agriculture, food, and natural resources
2. Animal Systems—the study of animal systems, including life processes, health, nutrition, genetics, management, and processing, through the study of small animals, aquaculture, livestock, dairy, horses, and/or poultry
3. Biotechnology Systems—the study of data and techniques of applied science for the solution of problems concerning living organisms
4. Cluster Skills—the student will demonstrate competence in the application of leadership, personal growth, and career success skills necessary for a chosen profession while effectively contributing to society
5. Environmental Service Systems—the study of systems, instruments, and technology used in waste management and their influence on the environment
6. Food Products and Processing Systems—the study of product development, quality assurance, food safety, production, sales and service, regulation and compliance, and food service within the food science industry
7. Natural Resources Systems—the study of the management of soil, water, wildlife, forests, and air as natural resources

8. Plant Systems—the study of plant life cycles, classifications, functions, and practices, through the study of crops, turf grass, trees and shrubs, and/or ornamental plants

9. Power, Structure, and Technical Systems—the study of agricultural equipment, power systems, alternative fuel sources, and precision technology, as well as woodworking, metalworking, welding, and project planning for agricultural structures[21]

SUMMARY

An SAE is a program of experiential learning activities conducted outside of the regular agricultural education class time. There are different types of SAEs and related opportunities and rewards programs, so every student in agricultural education and FFA can learn and develop from the authentic experience. Recordkeeping and documentation of the valuable experiences are very important, and will serve you and your future career well.

EXAMPLES OF PROJECTS FOR YOUR SAE

Agricultural Mechanics Project Examples

- Build a patio for the home.
- Build frames for raised beds for gardeners.
- Build handicap ramps in local community.
- Build picnic tables/sell to schools and local community.
- Construct a utility building.
- Construct a hydro ram pump and calculate the efficiency and water delivery rate.
- Construct a wind-powered generator and show its applications to agriculture.
- Construct and sell birdhouses and feeders.
- Construct and sell lawn furniture made of PVC.
- Construct compost bins to sell.
- Construct concrete projects for the home or farm.
- Construct or recondition a welding project (such as a trailer and cooker) at home or in school-provided facilities.
- Construct prefabricated wooden fence panels to sell to local hardware and building supply stores.
- Construct spray rigs for four-wheelers.
- Construct and market woodworking projects (birdhouses, dog houses, etc.).

- Construct metal projects.
- Contract with local EMCs or power companies to remove bolts, wire, and so on from old power poles (sell copper for recycling).
- Contract with a school system to maintain and service lawn-care equipment.
- Cut out and paint lawn figures to sell.
- Electrical repair service.
- Install plumbing fixtures or a plumbing system in your own building.
- Lawn mower maintenance service.
- Make craft items from wood, metal, or concrete to sell at arts and crafts shows.
- Make personalized signs to sell.
- Paint the home, supervised by an agricultural education teacher.
- Placement in a parts store.
- Provide a poultry house maintenance preparation business.
- Provide custom-painted mailboxes and stands.
- Repair and rebuild damaged pallets for businesses.
- Start a chain saw basic-maintenance and service business.
- Start a custom-vehicle refurbishing or painting business.
- Start a detailing business for cleaning farm equipment on the farm (wash, wax, clean, and maintain).
- Start an equipment locating business and match folks with something for sale with folks who want to buy something.
- Start a farm equipment tire disposal business (turn old tires into livestock feeders).
- Start a farm-fence maintenance business (cleaning fencerows, and repairing).
- Start a farm-fencing company for custom work.
- Start a pallet manufacturing business.
- Start a small-engine repair service.
- Wire a home shop, utility room, barn, or treehouse.
- Work as an agricultural mechanic's aide.
- Work at a welding operation.
- Work at a building supply business.
- Work with a farm equipment dealer.
- Start an equipment-trailer fabrication business.

Agribusiness Sales and Service Project Examples

- Become an agricultural news consultant for local radio or newspapers.
- Conduct a study of commodity trading over a period of time.

- Conduct general home maintenance.
- Contract with the local Chamber of Commerce to conduct county tours for prospective businesses.
- Create a custom labor venture: mow pastures, remove undesirable weeds from crops, paint outbuildings, and so on.
- Design a computer application plan for an agricultural facility or program.
- Develop a marketing plan for an agricultural commodity.
- Fry pork rinds for local stores.
- Install electrical circuits or a wiring system at home.
- Job placement in food distribution, restaurant, and so on.
- Job placement with a local florist.
- Job-shadow agribusiness professionals, visit agribusinesses to interview personnel, take educational tours, and so on.
- Offer a custom-parts or supplies delivery business to farms in your county.
- Presell fresh meat to clients on a weekly basis.
- Presell fresh seafood to clients on a weekly basis.
- Presell fresh vegetables in family portions delivered weekly.
- Preserve food for home use.
- Process creamed corn in a food processing facility.
- Provide a co-op program for an agribusiness.
- Provide a custom barbecue service for the community.
- Provide custom feed for livestock (tap the organic, all-natural, no-chemical market).
- Provide a hand-weeding crew for local peanut/ vegetable farmers.
- Provide a sausage-making business at home; can be sold if regulations are met.
- Provide custom hay baling and/or hauling.
- Provide a farm sign business (manufacture, sell, install, and maintain).
- Provide livestock hauling.
- Provide small-engine maintenance and repair service.
- Provide systematic maintenance and service on outdoor power equipment at home or at school-provided facilities.
- Purchase and resell aerial photographs from the tax office to local landowners.
- Package fresh fruit or vegetable gift packs.
- Remove pesticide jugs monthly from farms and transport to landfill.
- Sell ready-to-freeze processed vegetables.

- Start a composting business by buying cow manure from local farmers, bagging for resale.
- Start a farm-sitting business for vacationing farmers.
- Start a kerosene route for homeowners (probably little demand in the summer time).
- Start an MSDS compliance business by compiling and maintaining current sheets for farms and businesses in your county.
- Start a recycling business (collecting and selling newspapers and plastics to recycling plants).
- Start an agricultural business promotion business (sell custom caps, T-shirts with farm or ag business names or logos to clients).
- Start an agricultural photography service (animals, equipment, barns, families, children with animals, show animals).
- Start a local farm produce sale paper and sell ads to farmers.
- Form a cooperative with other students and share in profits of a greenhouse crop.
- Write "how to" pamphlets to sell at local garden supply stores (e.g., How to Grow Tomatoes).
- Write news articles on agriculture or FFA for local newspapers.

Agriscience Project Examples

- Compare weight gain of chicks fed different feed rations.
- Conduct a plant growth and physiology experiment in school agriscience lab.
- Conduct a research project for agriscience fair (local and national).
- Conduct a research project on a specific career; set up a business plan, including expenses, possible income, and so on.
- Conduct a supervised control burn and assess plant growth in the area.
- Conduct food science experiments.
- Grow crops with different mechanical/chemical applications, fertilizer, growth regulator, and so on, and observe/report results.
- Monitor local air quality; record and report.
- Plant and maintain a research plot on different types of turf grasses.
- Plant raised beds and monitor the growth of the plants.
- Research project on how light intensity affects plant growth.
- Research project on how light quality affects plant growth.

- Research project on plant reproduction.
- Soil conservation project on private or public land.
- Study effect of fertilizer runoff into a stream or pond.
- Study effect of manure runoff into a stream or pond.
- Study effects of herbicide type and varying concentrations.
- Study temperature effects on worms' food consumption.
- Work with agencies involved in research (USDA, etc.).
- Conduct a plant-growth and mineral-deficiency experiment.

Alternative Animals Project Examples

- Provide a beehive rental service for farms and gardens.
- Raise a dog for show.
- Raise dairy goats.
- Raise dogs to sell.
- Raise fish in tanks or floating cages (research the rate of growth based on factors such as temperature and amount of feed given).
- Raise llamas.
- Raise market goats for show.
- Raise meat birds (chickens, turkeys, and ducks) to the desired weight and sell to customers.
- Raise meat goats.
- Raise mice, hamsters, or gerbils.
- Raise miniature cattle.
- Raise miniature horses.
- Raise quail or other game birds for flight and meat.
- Raise rabbits for pets or meat animals.
- Raise special breeds of dogs.
- Raise tropical fish in aquariums.
- Raise worms; collect and sell to bait stores.
- Start a crawfish farm.
- Start a cricket ranch.
- Start a dog exercising business for elderly folks or sick people.
- Start a dog obedience school.
- Start a fish bait farm (mealworms, golden grubs, etc.).
- Start a gopher-tortoise relocation service for landowners.
- Start a honey production business (would work well with hive rental).

- Start a pet-grooming business.
- Start a turtle farm (sell to pet stores and pond owners).
- Train sporting dogs (quail, rabbit, and retriever dogs).
- Work at a dog kennel.
- Work at a pet store.
- Work at a veterinary hospital.

Animal Science Project Examples

- Board horses.
- Build a backyard poultry research project.
- Contract finish swine.
- Develop a cow-calf operation.
- Develop a small swine operation.
- Develop a stocker cattle operation.
- Raise replacement heifers.
- Raise dairy replacement heifers.
- Produce feeder pigs.
- Provide a deer-processing service.
- Provide a home-animal care service.
- Provide a horse-training service.
- Provide a horseshoeing service.
- Provide a meat-processing service.
- Provide a poultry-processing service.
- Raise a beef heifer for show.
- Raise a horse for show.
- Raise a market hog for show.
- Raise a market steer for show.
- Raise breeding sheep for show.
- Raise breeding swine for show or breeding.
- Raise dairy heifers for show.
- Raise market lambs for show.
- Raise poultry for show.
- Start a small animal-care business.
- Start an Easter egg business.
- Work at a horse operation or stables.
- Work at a poultry-processing operation.
- Work in the egg industry (packaging and distribution).
- Work on a beef cattle operation.
- Work on a dairy operation.
- Work on a poultry operation.
- Work on a sheep operation.
- Work on a swine operation.
- Operate a pay-to-fish business.
- Provide fish-pond management.
- Raise catfish in cages.
- Raise fish in an aquaculture system.

- Raise fish in cages in a pond or other body of water.
- Care and incubation of hatching eggs.

Crops Project Examples

- Produce organic vegetables.
- Produce vegetables for decoration (Indian corn, mini pumpkins, gourds, etc.).
- Produce farm crops at home or at school-provided facilities.
- Produce forage crops at home or at school-provided facilities.
- Produce watermelons.
- Scout cotton or peanuts for producers.

Forestry Project Examples

- Bale and market pine straw.
- Buy unusable lumber from builders supply and building sites; grind up and chip for mulch to sell.
- Collect green pine cones (for seed in the fall).
- Collect/market natural supplies (i.e., pine cones, acorns, nuts, and corn shucks) to sell to craft stores.
- Container pine seedling production.
- Contract with a tree-removal service to cut firewood and remove fallen trees.
- Contract with local timber companies and landowners to maintain boundary lines by painting and chopping.
- Cut and sell firewood provided free by national forests and state and local parks.
- Cut and/or market firewood.
- Grow longleaf pine seedlings.
- Measure timber on school forestry plot; determine volume and establish a management plan.
- Provide a soil sampling service for farms and lawns.
- Purchase bulk pine bark from sawmill; bag and resell.
- Purchase seedlings from Georgia Forestry Commission; pot and grow out to sell.
- Remove damaged (from lightning strike, insect damage, or mechanical injury) trees for landowners.
- Start a custom forest herbicide application crew (must have forest commercial pesticide license).
- Start a forest tree planting business.
- Start an ornamental tree care service.

Horticulture Project Examples

- Adopt a community building for beautification.
- Adopt an area of the school campus for beautification.
- Collect and sell dry/preserved native plant materials (acorns, leaves, wiregrass), especially for floral design retail/wholesale.
- Collect, press, mount, and identify plants that are growing on campus.
- Construct a garden arbor.
- Construct backyard water gardens.
- Container gardening ornamental plants.
- Container gardening vegetables.
- Create and market custom floral designs.
- Develop a business making dried arrangements to sell.
- Grow herbs.
- Produce daylilies.
- Develop a park on public property.
- Entrepreneurship in floral design.
- Establish a community roadside wildflower planting.
- Establish a garden plot at home or at school; produce crops to market.
- Grow and sell mushrooms.
- Grow and sell produce crops.
- Grow greenhouse plants on rented school greenhouse/cold-frame space.
- Grow, harvest, and can or preserve fruits and vegetables.
- Grow organic cut flowers for farmers' markets.
- Horticulture therapy.
- Indoor plant rentals and care service for businesses and offices.
- Landscape maintenance.
- Landscape pruning enterprise.
- Native plant materials.
- Offer a shrub-care service (pruning, trimming, and cutting back shrubs; fertilization).
- Produce fruit crops (at home or school-provided facility—e.g., watermelons).
- Produce greenhouse crop (at home or school-provided facility—e.g., ferns).
- Produce perennials from seed.
- Produce turf grass (at home or school-provided facility).
- Propagate and market shrubs.
- Provide a fruit tree pruning service.
- Provide a mulching service for urban gardeners.
- Provide landscaping materials for local businesses (pine straw, rocks, etc.).

- Raise a trial garden plot on school grounds (similar to UGA); seed companies may donate seeds/plugs.
- Raise tomato seedlings and replant into one-gallon pots to sell.
- Rent indoor plants to teachers in your school.
- Rent houseplants to homeowners (care for plants and change plants weekly).
- Rent-A-Plant; rent plants for weddings, banquets, parties (e.g., ferns and tropicals).
- Start a commercial flower upkeep business; change hanging baskets, potted plants, and window boxes for businesses.
- Start a floral design business by creating table centerpieces to sell at farmers' markets, grocery stores, and vegetable stands.
- Start a garden photography business.
- Start a hydroponics vegetable business.
- Start a lawn irrigation installation business.
- Start a houseplant renovation business.
- Start a turfgrass establishment business (seedlings, sodding, hydroseeding, etc.).
- Start a vegetable transplant seedling business.
- Work at a florist.
- Work at a garden center.
- Work in a nursery business.

Natural Resources Project Examples

- Adopt a local stream to monitor water quality.
- Collect water runoff from school parking lot, and analyze for various pollution indicators.
- Collect, mount, and identify insects found on school campus.
- Conduct a research project on how to prevent deer damage to a home garden.
- Conduct a water-quality study on area lakes or streams.
- Conduct endangered-plant surveys for landowners.
- Construct deer stands to sell (portable and stationary).
- Construct duck-nesting boxes to sell to landowners.
- Construct turtle traps for pond owners (use this in conjunction with turtle farm as a source of breeding stock).
- Develop a backyard bird habitat.
- Develop a backyard wildlife habitat.
- Develop a schoolyard wildlife habitat.
- Develop and/or maintain a wildlife food plot on private or public land.

- Develop and/or maintain wetland area on private or public land.
- Measure land for the local FSA office.
- Monitor success rate of bluebird houses.
- Plan and develop a school nature trail.
- Plan and develop an outdoor classroom.
- Plant a butterfly garden at school.
- Provide a debris-removal service along rivers and streams; sell driftwood and other items to consumers.
- Provide pond fertilization and testing service.
- Provide custom dove shoots or quail hunts.
- Raise mallard or wood ducks to sell to pond owners.
- Raise popular games birds; sell them for meat and as taxidermy products.
- Start a bullfrog farm; sell fresh frog legs to local restaurants.
- Start a fish fingerling nursery (catfish, trout, bream, etc.).
- Start a red cockaded woodpecker relocation service.
- Start a rock store; sell for landscaping purposes (gravel, pebbles, stones, etc.).
- Start a wildlife food plot and native plant enhancement business for local landowners and hunting clubs.
- Trap nuisance animals.
- Provide nongame wildlife management.[22]

Search Terms

portfolio

Notes

1. National Council for Agricultural Education, *Supervised Agricultural Experience SAE for All: Student Guide,* p. 2.
2. Ibid., p. 6.
3. National FFA Organization, *Supervised Agricultural Experiences*, Accessed February 18, 2016. https://www.ffa.org/about/supervised-agricultural-experiences.
4. Ibid.
5. Ibid.
6. "Philosophy and Guiding Principles for Execution of the Supervised Agricultural Experience Component of the Total School Based Agricultural Education Program." *The Council: A National Partner for Excellence in*

Agriculture and Education. National Council for Agricultural Education. Web. 18 Feb. 2016. https://www.ffa.org/thecouncil/sae, pp. 2–3.

7. Ibid, p. 3.

8. Ibid.

9. Ibid, p. 3.

10. Ibid, p. 3

11. California Core Agriscience Lesson Library. Lesson 612a: *Planning your SAE program.* Accessed February 20, 2016. http://calaged. csuchico.edu/ResourceFiles/Curriculum /CoreAgriscience/CD_old/Lessons/612a.pdf

12. Ibid, p. 3.

13. Ibid, p. 4.

14. National FFA Organization, ed. *SAE Handbook, Lesson RK.3: What Is a Journal and How Do We Prepare One?* Accessed February 22, 2016. http://harvest.cals.ncsu.edu/site/WebFile/lp3 .pdf, p. 2.

15. National FFA Organization, ed. *Life Knowledge, Lesson MS.69: Record Keeping.* Accessed February 22, 2016. http://harvest.cals.ncsu.edu/site/ WebFile/MS69.PDF, p. 3.

16. National FFA Organization, ed. *SAE Handbook, Lesson 7: Why Do We Keep Records?* Accessed

February 23, 2016. http://harvest.cals.ncsu.edu/ site/WebFile/IIB7.pdf, p. 4.

17. National FFA Organization, ed. *SAE Handbook, Lesson 8: How will my lesson be evaluated?* Accessed February 23, 2016, http://harvest.cals. ncsu.edu/site/WebFile/IIB8.pdf, p. 3.

18. Philosophy and Guiding Principles for Execution of the Supervised Agricultural Experience Component of the Total School Based Agricultural Education Program, p. 4.

19. National FFA Organization, ed. *SAE Handbook, Lesson MS. 70: Proficiency awards and SAE.* Accessed February 23, 2016, http://harvest.cals. ncsu.edu/site/WebFile/IIB8.pdf, p. 3.

20. Ibid, p. 3.

21. National FFA Organization, ed. *National FFA Agricultural Proficiency Awards: A Special Project of the National FFA Foundation.* Accessed February 23, 2016, https://www.ffa.org /sitecollectiondocuments/prof_handbook .pdf, p. 11.

22. J. Ricketts. *Project Workshop.* SAE Handbook. Unpublished document for teacher training in the Republic of Georgia.

APPENDIX D
National FFA Career Development Events and Leadership Events

NATIONAL FFA CAREER DEVELOPMENT EVENTS

Agricultural Communications
Agricultural Sales
Agricultural Technology and Mechanical Systems
Agronomy
Dairy Cattle Evaluation and Management
Dairy Cattle Handlers Activity
Environmental and Natural Resources
Farm and Agribusiness Management
Floriculture
Food Science and Technology
Forestry
Horse Evaluation
Livestock Evaluation
Marketing Plan
Meats Evaluation and Technology
Milk Quality and Products
Nursery/Landscape
Poultry Evaluation
Veterinary Science

NATIONAL FFA LEADERSHIP EVENTS

Agricultural Issues Forum
Conduct of Chapter Meetings Leadership Development
Creed Speaking
Employment Skills
Extemporaneous Public Speaking
Parliamentary Procedure
Prepared Public Speaking

GLOSSARY / GLOSARIO

A

abuse to use wrongly, make bad use of, or misuse.

abusar utilizar injustamente, usa mal de una cosa, o maltratar.

ACA ammoniacal copper arsenate.

ACA arsenato de cobre amoniacal.

ACC acid copper chromate.

CCA cromato de cobre ácido.

acid copper chromate wood preservative.

cromato de cobre ácido preservativo de la madera.

acre-inch amount needed to cover one acre of land one inch deep in water.

pulgada acre cantidad de agua que se necesita para sumergir un acre de tierra con agua de una pulgada de profundidad.

acrylonitrite-butadiene-styrene (ABS) a type of plastic pipe used for sewerage and underground applications.

acrilonitrito-butadiene-estireno un tipo de tubo plástico que se usa para el alcantarillado y servicios subterráneos.

actuator device that places another device in motion.

impulsor dispositivo que da impulso a otro dispositivo, poniéndolo en movimiento.

adaptor a fitting used to connect pipes of different types.

adaptor un acoplador que se usa para conectar tubos de tipos diferentes.

additive a chemical or material introduced into a fluid to change one or more of the fluid's characteristics.

aditivo cambia las características de un fluido.

adhesive a sticky substance used to bind two materials.

adhesivo sustancia pegajosa que se usa para ligar dos materiales.

adjust to set a part or parts to function as designed.

ajustar encajar una pieza o unas piezas para que funcionen así como deben, según su diseño.

adjustable pulley pulley with a movable side.

polea ajustable polea cuyo canto es movible.

aerator device that mixes air with water as it comes from a faucet.

aireador aparato que mezcla el aire con el agua mientras sale e ésta de un grifo.

aerosol high-pressure container with a valve and spray nozzle.

aerosol recipiente de alta presión con una válvula y un atomizador.

agribusiness and agricultural production broad range of activities associated with agriculture.

agroindustria una gran variedad de actividades relativas a la agricultura.

agricultural mechanics selection, operation, maintenance, service, sale, and use of power units, machinery, equipment, structures, and utilities in agriculture.

mecánica agrícola elección, operación, mantenimiento, servicio, venta, y uso de motores, maquinaria, equipo, estructuras y herramientas de la agricultura.

agriculture enterprises involving the production of plants and animals, along with supplies, services, mechanics, products, processing, and marketing related to those enterprises.

agricultura empresas que comprenden la producción de plantas y animales, junto con los artículos, servicios, mecanismos, productos, el elaborar y la venta relativos a esas empresas.

agriscience the body of scientific concepts and skills underlying the growing of plants and animals for human use.

agrociencia los conceptos y oficios de la ciencia que sirven de base para cultivar plantas y criar animales para el uso de los humanos.

air atomization paint split into tiny droplets using compressed air.

atomización de aire pintura pulverizada en gotitas por usar aire comprimido.

air compressor pump that increases pressure on air.

compresor de aire bomba que aumenta la presión sobre el aire.

air-dried lumber sawed lumber separated with wooden strips and protected from rain and snow for six months or more.

madera de construcción secada al aire libre madera aserrada separada con listones y protegida de la lluvia y la nieve por seis meses o más.

air-entrained concrete ready-mixed concrete with tiny bubbles of air trapped throughout the mixture to strengthen it.

hormigón de oclusión de aire hormigón hecho con pequeñas burbujas de aire retenidas en toda la mezcla para fortalecerlo.

air filter a device or element on an engine carburetor that blocks dust or other solid particles from entering the engine along with the incoming air.

filtro de aire dispositivo en un carburador de motor que limpia el aire que entra.

air-fuel mixture the ratio of air to fuel that enters the combustion chamber of an engine.

mezcla de aire-carburante la proporción de aire carburante que entra en la cámara de combustión de un motor.

air-lift pump pumps or lifts water using rising air bubbles.

bomba elevadora de líquidos por aire comprimido saca o levanta el agua usando burbujas de aire que se levantan.

air pollution detector a gas analyzer that senses and presents a warning when air has an unacceptable amount of pollution.

detector de contaminación del aire un analizador de gas que detecta y alerta cuando el aire tiene una cantidad inaceptable de contaminación.

air stones ceramic tips that produce small air bubbles.

puntas de soplado puntas arcillosas que producen pequeñas burbujas de aire.

air tool a tool powered by compressed air.

herramienta aérea herramienta accionada por el aire comprimido.

air vane type of governor used on small engines.

aspa un tipo de manostato regulador usado en motores pequeños.

alkyd oil-base paint.

alquídica pintura derivada de petróleo.

Allen screw a screw with a six-sided hole in the head.

tornillo de cabeza allen un tornillo cuya cabeza consiste en un agujero hexagonal.

alloy mixture of two or more elemental metals.

aleación mezcla de dos o más metales elementales.

alternating current (AC) current that reverses its direction 60 times per second.

corriente alterna (CA) corriente que recorre ya en un sentido, ya en otro, sesenta veces por segundo.

alternator device that produces alternating electrical current but usually equipped with diodes to change current to direct current.

alternador dispositivo que produce la corriente eléctrica alterna pero usualmente equipado con diodos para convertir la corriente en corriente continua.

aluminum tough, light, and durable metal; an element associated with high-quality paint; a paint color used as a focal color on waste containers.

sluminio metal fuerte, liviano y durable; un elemento asociado con pintura de alta calidad; color de pintura usado como color central en recipientes para desechos.

American Petroleum Institute (API) an organization that rates oils by the types of service they can tolerate.

Instituto Americano de Petróleo [API por sus siglas en inglés]) una asociación que clasifica los petróleos según el tipo de servicio que pueden tolerar.

American Welding Society (AWS) an organization that supports education in welding processes and that developed a system of numerical classification of electrodes.

Sociedad Americana de Soldadura (AWS) una organización que apoya la instrucción en los procesos de soldadura y que ha realizado un sistema de clasificación numérica de electrodos.

ammeter measures current in amperes.

amperímetro mide la corriente en amperios.

ammonia nitrogen waste product of animals.

amoníaco desecho de nitrógeno producido por animales.

ammoniacal copper arsenate wood preservative.

arsenato de cobre amoniacal preservativo de la madera.

amp See *ampere*.

amperio V. *amperio*.

amperage a measure of the rate of flow of current in a conductor.

amperaje una medida de velocidad del movimiento de la corriente en un conductor.

ampere (A) (amp) the unit of measure for amperage.

amperio (A) (amp) la unidad de medida para el amperaje.

amplifier circuit used to increase the amplitude of a signal.

amplificador circuito que se usa para aumentar la amplitud de una señal.

analog meter instrument with a graduated scale and a pointer.

medidor análogo instrumento que tiene una regla graduada y un índice.

anchor hole See *pilot hole*.

agujero de áncora V. *guía*.

anneal to cool steel slowly so as to make it soft and malleable.

recocer dejar enfriar lentamente el acero para que se haga maleable y suave.

annual rings pattern in a tree caused by the hardening and disuse of tubes in the woody parts.

capa cortical diseño en un árbol que resulta del endurecimiento y desuso de las partes leñosas.

antifoam an additive that reduces a fluid's tendency to foam.

agente antiespumante reduce la tendencia a hacer espuma.

antifoam agents additives added to engine oil to prevent foam from forming.

agentes de antiespuma aditivos que se añaden al petróleo del motor para evitar la formación de espuma.

antiscuff an additive that helps polish moving parts.

antipatinador ayuda en bruñir las partes móviles.

anvil heavy steel object used to help bend, cut, and shape metal.

yunke objeto pesado de acero empleado para ayudar en remachar, embutir y cortar el metal.

API See *American Petroleum Institute*.

IAP V. *Instituto Americano de Petróleo*.

API rating a rating given to an oil based on its ability to withstand a certain level of engine performance.

IAP rating una estimación de la habilidad del aceite del motor para resistir cierto nivel de funcionamiento de motor.

apparatus equipment necessary to carry out a function.

aparato equipo necesario para llevar a cabo una función.

aquaculture production of aquatic plants and animals.

piscicultura producción de plantas y animales acuáticos.

arbor a rod-shaped length of steel upon which one or more wheels (such as a grindstone or saw) turn.

mandril vástago metálico en que una o más ruedas (como una muela o sierra) giran.

arc a continuous discharge of electricity through the air across a gap in a circuit.

arco la descarga de electricidad por el espacio entre polos y armadura.

arc welder a machine that produces current for welding.

transformador para soldadura por arco una máquina que produce la corriente eléctrica para soldadura.

arc welding See *soldadura manual por arco metálico*.

soldadura por arco V. *soldadura por arco protegido*.

armature the rotating part of a motor. Also the iron core portion of a magneto.

armadura la parte de rotación del motor. También la parte del núcleo de hierro de una máquina magnetoeléctrica.

armored cable flexible metal sheath with individual wires inside.

cable armado forro flexible de metal que tiene alambres individuales adentro.

ax stone a small abrasive stone used for sharpening an ax.

piedra afiladura una piedra pequeña y abrasiva empleada para afilar un hacha.

asphalt a tarry substance, derived from various chemical refining processes, that produces a black paint pigment with excellent hiding power.

asfalto una sustancia alquitranada, derivada de varios procedimientos químicos de refinamiento que produce un pigmento negro con un poder excelente de cubrimiento.

assembly the joining of parts together.

ensamblaje la unión de partes.

attraction the action of two magnets that pull together when opposite poles are placed near each other.

atracción acción de dos imanes que se atraen cuando los polos opuestos se juntan uno al otro.

axis a straight line around which a body rotates.

eje varilla sobre la que gira un cuerpo.

B

backfire a loud snap or popping noise in a gas torch that generally blows out the flame.

retorno de llama un ruido de explosión o un chasquido fuerte del soplete oxídrico, que generalmente sopla la llama.

backsaw a saw that has very fine teeth and a stiff metal back.

cola de zorro una sierra de dientes muy finos y de una banda dura de metal.

back-stepping series of short welds each made in the backward direction while the bead progresses forward.

soldadura de retroceso una serie de soldaduras cortas, cada una hecha en dirección de retroceso mientras el cordón sigue adelante.

bag culture plant culture with artificial media contained in bags fed with a nutrient solution.

cultivo sin suelo, bolso para plantas cultivo vegetal con medios artificiales contenidos en bolsas y alimentado con una solución nutritiva.

balance weight equally distributed on all sides of center.

equilibrio el peso distribuido igualmente por todos lados del centro.

ball bearings sets of hardened steel balls in a container.

cojinete de bolas juego de bolas duras de acero en un contenedor (que se apoya un eje).

band saw power tool with saw teeth on a continuous blade or band.

sierra de cinta tronzadora de trozos máquina herramienta cuyos dientes son partes de una hoja continua o una cinta.

barbwire type of fence wire that consists of sharp barbs woven into two strands of smooth wire.

alambre de púas tipo de alambrado de lenguetas afiladas que se teje con dos filamentos.

barrel sling procedure for attaching rope to barrels.

nudo de barril procedimiento de atar cuerda a los barriles.

base metal main piece of metal or main component.

metal de base pieza principal de metal o componente mayor.

batter board board placed to carry level guide lines.

tabla marcadora para establecer la línea de excavación tabla colocada para llevar líneas de guía a nivel.

battery produces electricity by chemical action.

acumulador produce la electricidad con acción química.

bayonet saw See *sabre saw.*

sierra de sable V. *sierra de sable.*

bead continuous and uniform line of filler metal.

cordón raya continua y uniforme de masilla de metal de aporte.

below grade below surface level of land.

por debajo de la superficie por debajo de la superficie de la tierra.

belt sander power tool with a moving sanding belt.

lijadora de cinta máquina–herramienta con una cinta móvil para lijar.

bench brush handheld brush used to clean tables, benches, and power equipment.

cepillo de banco de trabajo cepillo de manija que se usa para limpiar mesas, bancas y máquinas de potencia.

bench top saw a stationary circular saw with either a tilting arbor or a tilting table.

sierra de mesa una sierra circular fija con mesa o árbol inclinado.

bench stone sharpening stone designed to rest on a bench.

piedra de banco piedra afiladora, hecha para apoyarse sobre un banco.

bench yoke vise a tool useful for holding pipe.

tornillo de banco una herramienta útil para sujetar tubería.

bevel a sloping edge.

bisel un borde pendiente.

bill of materials a listing of materials with specifications that are needed in a project.

lista de materiales una lista de los materiales y sus especificaciones que se necesitan para un proyecto.

binary having two parts or components.

binario que tiene dos partes o componentes.

binary system contains two digits, 0 and 1.

sistema de numeración binaria tiene solamente dos dígitos, cero y uno.

biological filters colony of microorganisms that detoxify waste products.

filtros biológicos colonia de microorganismos que desintoxican los desechos.

biosensor device that measures processes of the body.

biosensor dispositivo que mide los procesos del cuerpo.

biscuit A thin oval-shaped piece of wood that is used to join wood together.

cuña pieza oval de madera que se usa para juntar madera.

bit one digit, one number, or a yes/no answer to a question.

bitio un dígito, un número o una respuesta sí o no a una pregunta.

black and yellow stripes stripes used in safety color coding to indicate radiation hazards.

cintas negras y amarillas usadas en códigos de color de seguridad para indicar peligros de radiación.

black pipe steel pipe painted black that has little resistance to rusting.

tubo de hierro negro tubo de acero, pintado negro, que tiene poca resistencia a oxidarse.

blade guide metal blocks or wheels that support the blade of a band saw.

guías para la sierra bloques o ruedas de metal que sostienen la cinta de una cierra de cinta tronzadora de trozos.

bleeding the lines removing gas pressure from all lines and equipment.

purgar las líneas quitar la presión de gas de todas las líneas y el equipo.

blind cut cut made by piercing a hole with a sabre saw blade or by slowly lowering a portable handsaw blade into the material.

picadura inferior corte hecho perforando un hueco con una hoja de sierra de sable o bajando lentamente la hoja de un serrucho dentro del material.

block a mass of metal that houses engine cylinders and pistons.

bloque una masa de metal que contiene cilindros y pistones del motor.

blowby compression leakage past pistons.

oclusión de aire pérdida o escape de compresión más allá de los émbolos.

blue the safety color used for signs of warning or caution.

azul el color de seguridad que se usa para señales de peligro o aviso.

board foot amount of wood equal to a board 1 inch thick, 1 foot wide, and 1 foot long, or 144 cubic inches in volume.

pie de madera cantidad de madera igual a un madero de una pulgada de espesor, un pie de ancho y un pie de largo, o sea de 144 pulgadas cúbicas de volumen.

bolt threads straight threads that can be cut the entire length of a rod.

Roscas de perno filamentos que van de un extremo al otro [y] pueden cortarse a todo lo largo de una varilla.

border (border line) heavy, solid, black line drawn close to the outer edges of paper used for drawing plans.

orladura marginal; filete de encuadramiento línea gruesa negra y continua, dibujada cerca de los bordes de papel usado para planos.

bore to make or drill a hole.

taladrar hacer un agujero penetrando o atravesando.

boring tool any device used for drilling a hole.

herramienta barrenada cualquier aparato empleado para taladrar un agujero.

bottom dead center (BDC) piston at its lowest point (point nearest the crankshaft).

punto muerto inferior (PMI) cuando el émbolo está en su punto más bajo (punto más cerca al cigüeñal).

bottoming tap cuts threads to the bottom of a blind hole.

macho de abrir roscas corta roscas [de tornillo] hasta el fondo de un agujero ciego.

Bourdon tube measures fluid pressure.

manómetro de Bourdon mide la presión de los fluidos.

bowline useful as a nonslip loop in a rope.

nudo marinero útil como lazo antideslizante en una cuerda.

bows a framework that supports the greenhouse covering.

arcos armazón que sostiene el cobertizo de un invernadero.

Boy Scouts of America an international organization for boys dedicated to developing character and training for the responsibilities of adult life.

Niños Exploradores de la América una organización internacional para niños dedicada a desarrollarles el carácter y a entrenarlos para las responsabilidades de la vida adulta.

branch circuit electrical wiring, switches, and outlets that extend out from a service entrance panel.

circuito derivado cableado, interruptores y toma de corriente que extienden desde un panel de acometida de energía.

brass mixture of copper and zinc.

latón una mezcla de cobre y cinc.

braze welding bonding with metals and alloys that melt at or above 840 °F when capillary action does not occur.

soldeo oxigás con metal de aportación unir piezas con metales y aleaciones que se funden a los 840 °F o más, cuando no ocurre la acción capilar.

brazing bonding with metals and alloys that melt at or above 840 °F when capillary action occurs.

soldadura fuerte unir piezas con metales y aleaciones que se funden a los 840 °F o más, cuando sí ocurre la acción capilar.

break line solid, zigzag line used to show that the illustration stops but the object does not.

filete tremente con zigzag una línea continua en zigzag usada para mostrar el límite de la ilustración aunque todavía continúe el objeto.

brick set wide chisel used for breaking masonry units.

cincel de ladrillería buril ancho empleado para quebrar los elementos de albañilería.

broom finish rough surface formed by pushing a broom over concrete before it hardens to improve footing or traction.

acabado de escoba superficie áspera formada pasando una escoba sobre el hormigón antes de que se endurezca para mejorar el lecho de cimentación o la fricción.

bull float smooth board attached to a long handle used to smooth out newly poured concrete.

llana para hormigón madero liso ligado a un mango largo, que se usa para extender el hormigón que se acaba de derramar.

burn-through the unwanted creation of a hole in metal during the welding process. It is caused by improper techniques.

quemar un agujero creación no deseada de un agujero en el metal durante el procedimiento de soldadura. Es causado por técnicas incorrectas.

bushing a fitting used to connect pipe to pipe fittings or larger sizes. Also, a device used to fill the space between a shaft and a larger hole in a blade or wheel. Also, a sleeve in a frame.

aislador de entrada ajuste empleado para conectar tubos a los accesorios para tubos o los tubos de tamaños más grandes. También un dispositivo empleado para llenar el vacío entre el eje y el agujero en una rueda. También un manguito de un bastidor.

business work done for profit.

negocio trabajo hecho para ganancia.

butt hinge a square or rectangular hinge that mounts flush or flat with the surface (usually on the edge) of a door.

bisagra un gozne cuadrado o rectangular que se monta llano con la superficie.

butt joint joint formed by placing two pieces end-to-end in line or at a 90-degree angle.

empalme plano ensambladura formada uniendo dos piezas por sus extremos o a un ángulo de 90 grados.

butt weld bead laid between two pieces of metal set edge-to-edge or end-to-end.

costadura transversal soldada a tope cordón soldado entre dos piezas de metal puestas por sus extremos o por sus bordes.

bypass system a filter system in which some of the fluid bypasses the filter.

válvula de desvío parte del sistema de fluidos que desvía el filtro.

C

cadmium metal used for plating of fasteners and other steel products for rust-resistance.

cadmio metal empleado en el revestimiento de cierres y otros productos de acero para que sean resistentes al orín.

calcium a low-quality paint pigment.

calcio pigmento de pintura de mala calidad.

caliper an instrument used to measure the diameter or thickness of an object.

calibre un instrumento que se usa para medir el diámetro o espesor de un objeto.

cam a raised area on a shaft.

leva un área elevada en un eje.

camshaft a rotating shaft with raised areas that open and close valves.

árbol [eje] de levas/árbol de distribución eje giratorio con áreas levantadas que abren y cierran válvulas.

cap a fitting used to close a pipe end.

tapa de recubrimiento un encastre usado para tapar la boca de un tubo.

capacitance ability of a device to store electrical energy in an electrostatic field.

capacitancia la capacidad de un dispositivo de almacenar la energía eléctrica en un campo electrostático.

capacitor two conductors separated by an insulator.

capacitor dos conductores separados por un aislador.

capillary action rising of the surface of a liquid at the point of contact with a solid.

acción capilar el subir la superficie de un líquido al punto de contacto con un sólido.

cap screw a hex-head screw that is threaded over its entire length and is generally 2 inches or shorter.

perno de máquina un tornillo de cabeza hexagonal que está roscado a todo lo largo de su cuerpo y generalmente mide 2 pulgadas o menos.

carbon arc torch a device that holds two carbon sticks and produces a flame from the energy of an electric welder.

linterna de arco de carbón un dispositivo que tiene dos barras de carbono y produce una llama que proviene de la energía de una soldadora eléctrica.

carbonizing flame a flame with an excess of acetylene.

llama carbonizada una llama con exceso de acetileno.

carburetor provides fuel and air to the engine in appropriate portions and volume.

carburador proporciona el combustible y el aire al motor en proporciones y volumen adecuados.

cardiopulmonary resuscitation (CPR) a first-aid technique to provide oxygen to the body and to circulate blood when a victim's breathing and heartbeat stop.

resucitación cardiopulmonar una técnica de primeros auxilios para suministrarle oxígeno al cuerpo y circulación a la sangre cuando la respiración y el latido del corazón de la víctima han parado.

carriage bolt a threaded fastener with a round head over square shoulders.

perno de carrocería cierre roscado con una cabeza redonda encima de hombros cuadrados.

Cartesian work area a work area with a boxlike shape and no hollow space at the center.

área de trabajo cartesiano en forma de caja y sin espacio vacío en el centro.

casting a mold that holds molten metal.

colada matriz que guarda el metal fundido.

caulk material that stretches, compresses, and rebounds to maintain a tight seal between materials as they expand and contract.

calafateado materia que se estira, se comprime y regota para mantener los materiales fuertemente sellados entre sí mientras se expanden y se contraen.

CCA chromated copper arsenate.

ACC arsenato de cobre cromado.

cement paste a paste made by mixing portland cement and clean water in exact proportions.

pasta de cemento una masa hecha por mezclar el cemento portland y agua limpia en proporciones exactas.

center line a long-short-long line used to indicate the center of a found object on a diagram or plan.

línea central un punto de referencia dibujado en forma de raya-punto-raya, usado para indicar el centro de un objeto hallado en un diagrama o plano.

center punch steel punch with a sharp point.

sacabocados perforadora de acero con punzón agudo.

centrifugal directed outward from the center.

centrífugo dícese de la fuerza que se aleja del centro, causado por un objeto que da vueltas.

centrifugal head pumps self-contained unit powered by an auxiliary motor.

bombas centrífugas unidad autosuficiente impulsada por un motor auxiliar.

CFM See *cubic feet per minute.*

pcm V. *pie cúbico por minuto.*

chain vise a tool useful for holding pipe.

prensa de cadena herramienta útil para fijar tubería.

chain wrench wrench that utilizes a chain to grip and turn pipe.

llave de cadena tornillo portátil para tubería con una cadena que permite fijar el tubo y tornearlo.

chalking a process whereby paint gradually washes off in the rain to stay bright.

ayesamiento un proceso para que permanezca fresca la pintura que gradualmente se quita con la lluvia.

chalk line a cotton cord with chalk applied used to create long, straight lines.

cordel entizado una cuerda de algodón entizado, que se usa para hacer líneas largas y derechas.

charger device designed to provide the correct amount of current on an electric fence.

el cargador dispositivo que provee la cantidad correcta de corriente para un alambrado eléctrico.

check valve See *direction control valve.*

válvula de retención V. *válvula de control de dirección.*

chip tiny electronic processing circuit in a computer.

chip circuito pequeñito del procesamiento electrónico empleado en un computador.

chip breaker a device on a planer that breaks up chips as they are cut from a board.

cortavirutas un dispositivo en un cepillo mecánico que rompe las virutas a medida que se cortan de la tabla.

chipping hammer a hammer with a sharp edge and/or point used to remove slag from a welding bead.

martillo ancelador martillo con un corte o una punta afilados que se usa para extraer escoria de un cordón de soldadura.

chlorinated polyvinyl chloride (CPVC) plastic pipe suitable for hot water lines.

cloruro de polivilino clorado (CPVC) tubo plástico adecuado para líneas de agua caliente.

chromated copper arsenate wood preservative.

arsenato de cobre cromado preservativo de madera.

chuck a device with jaws that open and close to receive and hold bits or other objects.

mandril dispositivo con quijadas que abren y cierran para asegurar brocas y otros objetos.

circuit an electrical source and wires connected to a light, heater, or motor.

circuito conjunto de conductores y cables eléctricos conectados a una luz, un filamento o un motor.

circuit breaker a switch that trips and breaks the circuit when more than a specified amount of current passes through it.

ruptor de circuito un interruptor que interrumpe y corta el circuito cuando más de la cantidad especificada de corriente pasa por él.

claw the cup of a milking machine.

distribuidor filtrante el gancho o garfio de la ordeñadora.

clay the smallest group of soil particles.

arcilla el grupo más pequeño de partículas de tierra.

cleanup assignment sheet a sheet that assigns a certain number of students to clean a specific area.

hoja para asignar [trabajos de] limpieza una hoja que asigna a cierto número de estudiantes para limpiar un área específica.

cleanup skills checklist a checklist used to evaluate the performance of each student in cleaning his or her assigned area.

lista de comprobación para habilidades en limpieza una lista utilizada para evaluar el desempeño de cada estudiante en la limpieza de su área asignada.

cleanup wheel cleaning system that utilizes a chart shaped like a wheel.

rueda para la limpieza un sistema que utiliza un gráfico conformado como una rueda.

clearance hole shank hole.

agujero de holgura agujero del mango.

clinch to bend a nail over and drive the flattened end down into the material.

remachar curvar un clavo y machacar la cabeza o punta allanada en el material.

clockwise in the direction that a clock's hands move.

en el sentido de las agujas del reloj de derecha a izquierda.

closed aquaculture systems intensive production systems using oxygen, feed, and waste treatment to maximize economic return.

sistemas cerrados de cultivos marinos sistemas de producción intensivos que usan el oxígeno, alimentación y tratamiento de residuos para obtener al máximo el rendimiento económico.

cloth fabric material used in some measuring tapes.

tela un tejido usado en algunas cintas métricas.

clove hitch knot useful for tying livestock.

ballestrinque nudo útil para atar ganado.

clutch a device used to engage and disengage a power source from its load.

embarque dispositivo que se usa para meter y sacar una fuente de energía de su carga.

coarse aggregate gravel; larger particles of stone used in concrete.

árido grueso grava; partículas más grandes de piedra que se usan en el hormigón.

cold chisel a piece of tool steel shaped, tempered, and sharpened to cut mild steel when driven with a hammer.

cortafrío una herramienta de acero labrada, templada y afilada para cortar el acero suave cuando se clava con un martillo.

combination square type of square that combines many tools in one.

escuadra de combinación tipo de escuadra que combina muchos instrumentos en uno.

combustion burning.

combustión acción de quemar.

commutator the part of a motor that changes the flow of current at appropriate times.

colector de delgas la parte de un motor que cambia la corriente en el momento apropiado.

compass saw a small saw used for cutting circles in wood.

sierra de compás una sierra pequeña que se usa para cortar los círculos en madera.

compatible going together with no undesirable reactions.

compatible que se llevan bien sin reacciones indeseables.

compress to reduce in volume by pressure.

comprimir reducir el volumen de algo haciendo presión.

compressed air air pumped under high pressure that may be carried by pipe, tubing, or reinforced hose.

aire comprimido aire inyectado bajo alta presión que puede ser llevado por medio de tubo o de manguera reforzada.

compression fitting a fitting that grips copper or steel tubing by compressing a special collar with a tapered, threaded nut.

ajuste de compresión un ajuste que agarra entubado de cobre o acero comprimiendo una abrazadera especial con una tuerca afilada y roscada.

compression ratio the ratio of the combined volume of the cylinder and the combustion chamber at the beginning of the compression stroke (when the piston is all the way down) to the volume of the combustion chamber at the end of the compression stroke (when the piston is all the way up).

índice [relación, grado] de comprensión la relación del volumen combinado del cilindro y la cámara de combustión al principio del golpe de compresión (cuando el pistón está totalmente abajo) con el volumen de la cámara de combustión al final del golpe de compresión (cuando del pistón está completamente arriba).

compression stroke movement of an engine piston to squeeze or compress the air-fuel mixture.

carrera de émbolo movimiento del émbolo de motor para comprimir la mezcla carburante.

computer-aided design (CAD) a computer software program that aids in the designing of projects ranging from cabinets to large buildings.

el diseño ayudado de la computadora programación almacenada en una computadora que ayuda a diseñar projectos desde armarios hasta edificios grandes.

concave hollow or curved in.

cóncavo hueco o curvado hacia el interior.

concrete a mixture of stone aggregate sand, portland cement, and water that hardens as it dries.

hormigón una mezcla de áridos de piedra, arena, pórtland y agua, que se fragua mientras seca.

condenser stores and releases current to boost current in the primary circuit.

condensador almacena y suelta la corriente para aumentar la corriente del circuito primario.

conductance (G) ability of a material to carry electrons.

conductancia (G) capacidad de un material para llevar electrones.

conductor any material that will permit electrons (electricity) to move through it.

conductor cualquier material que permite que los electrones (electricidad) pasen por él.

conduit metal tube with individual insulated wires inside.

tubo aislante tubo de metal que contiene los cables individuales aislados adentro.

construction joint place where one pouring of concrete stops and another starts.

junta de construcción lugar en donde un bloque de hormigón termina y otro empieza.

contaminant any material that does not belong in a substance.

contaminador cualquier material que no debe ser parte de la sustancia.

continuity connectedness.

continuidad ser conectado.

continuity tester device used to determine if electricity can flow between two points.

probador de continuidad dispositivo empleado para comprobar si puede recorrer la corriente eléctrica entre dos polos.

continuous duty an electrical tool that can be used without any cooling-off period over a 6- or 8-hour day.

servicio continuo dícese de una herramienta que se puede usar todo un día de trabajo de 6 o a 8 horas.

control joint planned break that permits concrete to expand and contract without cracking.

junta de dilatación, junta de retracción una abertura o rotura intencional que permite la expansión y la contracción del hormigón sin que se rompa.

controlled-environment structures production facilities where temperature, light, and humidity are maintained for agricultural production.

instalaciones de ambiente controlado instalaciones de producción en donde se mantienen la temperatura, la luz y la humedad para la producción agrícola.

convex curved out.

convexo curvado hacia el exterior.

coolant liquid used to cool parts or assemblies.

refrigerante líquido usado para enfriar piezas o montajes.

coping saw a saw that has a very thin and narrow blade supported by a spring steel frame.

sierra bastidor una sierra que tiene una hoja cortante muy delgada y estrecha, sostenida por un bastido de muelle aceroso.

cordless tool containing a rechargeable battery pack to drive the unit when not plugged into an electrical outlet.

inalámbrica herramienta que incluye una batería recargable para hacer funcionar la unidad cuando no está conectada a un enchufe eléctrico.

core the hollow space in a masonry block.

núcleo hueco el hueco de un bloque de albañilería.

corner pole a straight piece of wood or metal held plumb by diagonal supports and used to support a line when laying block or brick.

poste de ángulo un poste recto de madera o metal aplomado por tirantes de extensión y empleado para sostener una línea cuando se están poniendo ladrillos o bloques.

corrosion reaction of metal to liquids and gases that causes them to deteriorate or break down.

corrosión una reacción del metal a líquidos y a gases que les causa deteriorarse o corroerse.

corrosion inhibitor a material or additive that reduces or prevents corrosion.

sustancia anticorrosiva un material o aditivo que reduce o evita la corrosión.

counterclockwise in the direction opposite to the direction that a clock's hands move.

en sentido opuesto al de las agujas del reloj de derecha a izquierda.

countersink tapered hole for a screw head.

contrafresadura bisel practicado en el borde de un agujero para recibir tornillos.

coupling a fitting used to connect two pieces of similar pipe.

acoplamiento un ajuste que se usa para unir entre sí dos piezas semejantes de entubado.

course a row of masonry units.

hilada de ladrillos una fila de elementos de albañilería.

CPR See *cardiopulmonary resuscitation*.

resucitación cardiopulminar V. *resucitación cardiopulminar*.

CPVC See *chlorinated polyvinyl chloride*.

CPVC V. *cloruro de polivilino clorado*.

crack the cylinder turn gas on and off quickly to blow dust from the opening.

destaponar un cilindro abrir el cilindro de gas y en seguida muy rápidamente cerrarlo para soplar el polvo de la abertura.

craftsperson skilled worker.

artesano obrero cualificado.

crankshaft a shaft with an offset projection that converts circular motion to reciprocal motion, or vice versa.

cigüeñal una leva con una proyección desviada que convierte el movimiento circular en el movimiento rectilíneo o vice versa.

crater a low spot in metal where the force of a flame has pushed out molten metal.

cráter depresión en el metal donde la fuerza de la llama ha abollado el metal fundido.

creosote popular wood preservative that is black in color.

creosota preservativo de madera popular que es de color negro.

crosscut to cut across the grain of wood.

cortadura en cruz cortar a través de la fibra de la madera.

crosscut saw saw with teeth cut and filed to a point, used to cut across the grain of wood.

sierra de cortadura en cruz sierra de dientes cortados y afilados en punta, usada para cortar a través de la fibra de madera.

crown the part of a tool that receives the blow of a hammer.

cabeza parte de la herramienta que recibe el golpe del martillo.

cubic feet per minute (CFM) amount of air output by the compressor in a spray-painting system.

pies cúbicos por minuto la cantidad de aire que sale por un compresor de pintura por pulverización.

cuprinol popular wood preservative used when plants will be near the wood.

cuprinol preservativo de madera popular que se usa cuando hay plantas cerca de la madera.

curing proper drying of concrete.

fraguar método correcto de secar y endurecer la masa de hormigón.

cutter head a device on a jointer or planer that holds the blades.

portacuchillas un dispositivo que sujeta las cuchillas en un cepillo mecánico de banco o en una acepilladora.

cutting tool a tool used to cut, chop, saw, or otherwise remove material.

herramienta de cortadura una herramienta usada para cortar, tronchar, serrar o remover material de otro modo.

cycle all the events that take place as an engine takes in air and fuel, compresses the air-fuel mixture, burns the fuel, and expels the burned gasses. Also the current produced by one complete turn of a generator's armature.

ciclo todos los sucesos que tienen mientras el motor toma aire y combustible, comprime la mezcla carburante, quema el combustible y expele los gases quemados. También la corriente producida por una vuelta de la armadura del generador.

cylinder long round tank with extremely thick walls built to hold gases under great pressure. Also, engine cavity containing a piston.

cilindro cámara larga y redonda con paredes sumamente gruesas, construida para contener los gases bajo gran presión. También, la cavidad del motor que contiene el émbolo.

cylinder hone tool for restoring cylinders.

rectificador de cilindros herramienta usada para restaurar los cilindros.

cylindrical work area a work area in shape of a cylinder, containing a hollow space in the center.

área cilíndrica de trabajar área de trabajar en forma de cilindro con un espacio vacío en el centro.

D

dado square or rectangular groove in a board.

dado ranura cuadrada o rectangular en un madero.

dado head special blade that is adjustable to cut kerfs from ⅛ to ¾ inch in one pass.

fresa rotativa de ranurar hoja especial que se ajusta para cortar cortes de ⅛ a ¾ de pulgadas en una pasada.

dado joint rectangular groove cut in a board with the end or edge of another board inserted.

mortaja filete ranura rectangular cortada de un madero en cuyo borde otro listón o madero está encajado.

decibel (dB) the standard unit of sound.

decibelio (dB) unidad estándar de sonido.

degree program an educational line of study that leads to the completion of a degree such as an Associate, Bachelor, Master, or Doctorate.

programa educativo un plan de estudios que tiene de objetivo obtener un título universitario como un grado de asociado, licenciatura, maestría o doctorado.

degrees of freedom a way in which a body may move, based on its having a certain axis of rotation; the total degrees of freedom a body has is equal to its number of axes.

grado/medida de movimiento una manera en la que un cuerpo puede moverse, basada en cierto número de ejes de rotación que pueda tener; el total de grados de movimiento de un cuerpo tiene como equivalente su número de ejes.

detergents additives to engine oil that clean engine parts and suspend tiny contaminants in the oil.

detergentes aditivos que se añaden al aceite del motor; limpian las partes del motor y suspenden los contaminantes minúsculos en el aceite.

diameter the distance across the center of a circle or round object.

diámetro la distancia a través del centro de un círculo o de objeto redondo.

diaphragm a flat piece of rubber that creates a seal between moving and nonmoving parts.

diafragma una pieza llana de hule que forma un sello entre piezas móviles e inmóviles.

die an instrument used to cut threads onto a rod or bolt.

troquel instrumento usado para cortar roscas en un biela o un perno.

die stock handle used to turn a die.

terraja de anillo mango empleada para tornear un troquel.

dielectric insulated part of a capacitor.

dieléctrico pieza aislada de un capacitor.

diffuser tube porous underwater tube for producing small bubbles.

difusor tubo poroso que va debajo del agua y produce pequeñas burbujas.

digging tool any device used to move soil or rock.

herramienta de cava dispositivo empleado para remover tierra o piedra.

digital using digits.

digital que usa números.

digital computer electronic processor using digital techniques.

computadora digital máquina electrónica de proceso de datos que usa técnicas digitales.

digital meter device providing a direct numerical readout.

medidor digital dispositivo que da presentación numérica directa.

digits the numbers 0 through 9.

dígitos los números 0 a 9.

dimension measurement of length, width, or thickness.

dimensión medida de lo largo, lo ancho y lo alto.

dimension line solid line with arrowheads at the ends to indicate the length, width, or height of an object or part.

línea de dimensión línea continua con puntas de flecha en los extremos para indicar lo largo, lo ancho y lo alto de un objeto o de una pieza.

diode simplest type of semiconductor.

diodo tipo más sencillo del semiconductor.

direct current (DC) current that flows in one direction continuously.

corriente continua (CC) corriente que corre en una dirección continuamente.

direction control valve (check valve) restricts fluid movement to one direction.

válvula de regulación restringe el movimiento de fluidos a una sola dirección.

disc sander tool with sanding grit on a revolving plate.

esmeriladora de disco herramienta que tiene arenisca de grano en una placa giratoria.

discharge loss of power from a battery.

descarga pérdida de potencia del acumulador.

dissolved oxygen level of oxygen dissolved in water used by aquatic plants, animals, and microorganisms.

óxigeno disuelto índice de oxígeno disuelto en agua que se usan plantas, animales y microorganismos acuáticos.

distilled free from ions or other impurities.

destilado libre de iones o de otras impurezas.

dividers two sharp steel legs connected on one end that are used to make arcs and circles.

compás de división dos patas agudas de acero conectadas en un extremo que se emplean para dibujar arcos y círculos.

double-action cylinder works two ways.

cilindro de doble-acción que funciona en dos sentidos.

double-cut file file that has teeth laid out in two directions.

lima de doble corte lima cuyas aristas cortantes son talladas en dos direcciones.

double-extra-heavy pipe steel pipe with walls thicker than standard and extra heavy pipe.

tubo pesado extra doble tubo de acero cuyas paredes son más gruesas que las de los tubos estándares y extra gruesos.

double-insulated motor a motor whose electrical parts are shielded or separated from the user by special insulation inside the motor and by its insulating plastic motor housing.

motor doble-mente-aislado motor con piezas eléctricas que son aisladas o separadas del usuario por un aislamiento especial dentro del motor y por su carcasa plástica de motor aislante.

dovetail joint a joint formed by interlocking parts of two pieces.

ranura en cola de milano ensambladura formada por las partes entrelazadas de dos piezas.

dowel a cylindrical piece of wood.

pasador pieza cilíndrica de madera.

doweling jig a device designed to guide the drilling of properly aligned dowel holes in wood.

plantilla para poner en posición espigas de montaje un dispositivo designado para guiar la perforación de huecos adecuadamente alineados para espigas de montaje en madera.

draw filing a filing procedure done by placing the file at a 90-degree angle to the metal.

limar a rectángulo una acción de limar realizada por poner la lima a un ángulo de 90 grados al metal.

draw the temper modify the temper of steel to render it softer.

revenir modifica el templado de acero para renderlo más blando.

drawing a picture or likeness made with a pencil, pen, chalk, crayon, or other instrument.

dibujo figura o retrato hecho con lápiz, pluma, tiza, crayola u otro instrumento.

dress to remove material and leave a smooth surface.

raboter quitar el material y dejar una superficie lisa.

drill any rotating device used for making a hole.

taladro cualquier dispositivo giratorio empleado para hacer un agujero.

drill press stationary tool used for making holes in metal and other materials.

taladradora radial herramienta fijada empleada para hacer agujas en el metal u otros materiales.

drip irrigation See *trickle irrigation*.

irrigación por infiltración V. *irrigación por infiltración*.

drive train the collective set of components used to transfer power from a power source to a load.

tren de propulsión conjunto de piezas que transfieren la potencia desde la punta de potencia hasta la carga.

driving tool a tool used to move another tool or object.

herramienta de impulso una herramienta que se emplea para poner en moción otra herramienta u objeto.

drop cloth material used to protect floors, furniture, shrubbery, and the like from falling paint.

lona protectora tela que se usa para proteger los pisos, las muebles, los arbustos y otras cosas semejantes de las gotitas de pintura que se caen.

dry-element filter an engine air filter that uses dry fibrous material to prevent dust and other solid particle from entering an engine's interior along with the needed air.

filtro de elemento seco un filtro de aire de motor que se usa el material seco y fibroso que limpia el aire.

drywall screw a hardened steel self-threading screw.

tornillo de muro de piedra un tornillo auto-roscante de acero colado.

dual having or consisting of two.

dual de dos.

dual-element filter an engine air filter involving the use of both a dry and oil foam filter.

filtro de elemento doble un filtro de aire de motor que se usa los dos, un filtro seco y un filtro de aceite de goma.

ductile able to be bent slightly without breaking.

dúctil que puede estirarse sin romperse.

duplex receptacle double receptacle wired so that both outlets are on the same circuit.

receptáculo dúplex receptáculo doble que se conecta para que las dos tomas del receptáculo corran en el mismo circuito.

dust mop large rectangular-shaped mop that is used to clean up small dust particles from the floor.

trapeador para polvo trapeador grande en forma rectangular que se usa para limpiar partículas pequeñas de polvo del piso.

dust pan small handheld pan used to collect dust, dirt, or other trash for deposit in a waste container.

pala para recoger la basura pala pequeña de mano usada para recoger polvo, tierra u otra basura para ponerlos en un recipiente para desperdicios.

duty cycle the proportion of time a motor or welder can run without overheating.

ciclo de servicio la proporción de tiempo en que se puede marchar un motor o soldador sin calentar demasiado.

duty rating percent of time a motor may run without overheating.

relación de trabajo el porcentaje de tiempo en que se puede marchar un motor sin calentar demasiado.

dysfunctional not working the way it should.

disfuncional que no funciona en la manera a que debe.

E

ear short extensions on the ends of masonry block.

asa salientes cortos que sobresalen de los extremos de ladrillos.

eave the overhang on the roof of a building.

alero parte inferior del tejado de un edificio.

eccentric off-center.

excéntrico muy alejado del centro.

efficiency the ability to produce with minimum waste of time, energy, and materials.

de buen rendimiento que puede producir dentro del mínimo de tiempo, energía y gasto.

efficient able to produce with minimum waste of time, energy, and materials.

de buen rendimiento que puede producir dentro del mínimo de tiempo, energía y gasto.

elbow (ell) a fitting used where a single pipe line changes direction.

codo un ajuste que es empleado donde la tubería cambia de sentido.

electrical metallic tubing (EMT) a type of bendable conduit in which electrical wires are encased.

tubos mecánicos de electricidad un tipo de tubo flexible en el cual alambres eléctricos se encierran.

electrical meter a device that measures the activities of electrons.

electrómetro un dispositivo que mide las actividades de electrones.

electricity a form of energy that can produce light, heat, magnetic force, and chemical changes.

electricidad forma de energía que puede producir luz, calentura, magnetismo y reacciones químicas.

electrode in arc welding, a metal welding rod that is coated with flux and used with an electric welder.

electrodo una varilla de soldadura de metal, cubierta con pasta para soldar y usada con una soldadora eléctrica.

electrode holder a spring-loaded device with insulated handles used to grip welding electrodes.

portaelectrodo un dispositivo de válvula de descarga de muelle que tiene mangos aislantes y que se usa para fijar electrodos de soldadura.

electrolyte acid solution in a battery.

electrólito solución acídico en un acumulador.

electromagnet a core of magnetic material surrounded by a coil of wire through which an electric current is passed to magnetize the core.

electroimán un núcleo de materia magnética rodeado por una bobina de alambre a través de que pasa una corriente eléctrica para imantar el núcleo.

electromagnetic induction conversion of low-voltage current to high-voltage current with a coil.

inducción electromagnético la conversión de corriente de baja tensión a la de alta tensión usando una bobina.

elevation relative height above or below sea level.

elevación altura relativa sobre el nivel del mar.

EMF electromotive force—the amount of force produced by an electrical device.

fuerza electromotriz (*EMF* por sus siglas en inglés) la cantidad de fuerza producida por un dispositivo eléctrico.

emitters nozzles that release water slowly.

emisores boquillas que emiten el agua lentamente.

employment work done for which one is paid by the hour, day, week, month, or year.

empleo trabajo hecho, por lo cual se paga a uno por hora, día, semana, mes o año.

EMT See *electrical metallic tubing.*

ETM V. *tubos mecánicos de electricidad.*

enamel paint with a gloss or semigloss finish.

pintura esmaltada pintura con acabado brillante o semi brillante.

end marking the color-coded marking on the end of a welding electrode.

marca al extremo [de electrodos de soldadura] marcas con colores en código al extremo de un electrodo de soldadura.

end nailing nailing through the thickness of one piece and into the end of another piece.

clavar contrahilo clavar por el espesor de una pieza y por el borde de otra pieza.

end walls wall construction at the end of greenhouses supporting doors, fans, and vents.

pared del extremo pared en el extremo del invernadero que sostiene las puertas, los abanicos y las rejillas de ventilación.

English system a system of measure based on measurements of common objects.

sistema inglesa es un sistema de medida basado en medidas de objectos ordinarios.

entrance head a waterproof conduit connection where an electrical entrance cable is connected to a building.

cabeza de entrada un tubo impermeable de conexión donde una línea eléctrica se conecta a un edificio.

entrance panel a device that distributes incoming electrical current to various circuits in a building.

panel de entrada dispositivo que distribuye corriente eléctrica a varios circuitos en un edificio.

epoxy a synthetic paint material with exceptional adhesion and wear-resistant qualities.

epoxi un material sintético de pintura que tiene adhesión excepcional y cualidades resistentes al desgaste.

evaporative cooling pads porous excelsior or material kept moist to help cool incoming air into the greenhouse.

almohadillas evaporativas de enfriamiento excelsior poroso u otro material que se mantiene húmedo para enfriar el aire entrante al invernadero.

excavate to dig.

excavar cavar.

exhaust burned gases removed by the motion of a piston.

escape gases quemados traslados por la moción de un émbolo.

exhaust stroke movement of a piston that expels burned gases from a cylinder.

tiempo de escape movimiento del émbolo que expulsa los gases quemados del cilindro.

extension line solid line showing the exact area specified by a dimension.

línea de extensión línea continua que representa el área exacta que se especifica una dimensión.

exterior grade a grade of plywood suitable for outdoor use.

cualidad para exterior madera contra-chapada apropiada para el uso al aire libre.

exterior paint paint able to withstand moisture and outside weather conditions.

pintura exterior pintura que puede aguantar la precipitación y el clima.

external combustion engine an engine in which the combustion takes place outside the engine as in a steam engine.

motor de combustión exterior un motor en que la combustión ocurre afuera del motor como en un motor de vapor.

extinguish to put out a fire by cooling it, smothering it, or removing fuel from it.

extinguir hacer que cese un fuego por congelar, apagar o quitar el combustible.

extractor an assembly that removes moisture and oil from compressed air.

extractor dispositivo que extrae el agua y aceite del aire comprimido.

extra-heavy pipe pipe with thicker than standard walls.

tubería extra gruesa tubería cuyas paredes son más gruesas que las estándares.

extreme-pressure resistor an additive that reduces oil breakdown under high pressure.

lubricante para presión extrema reduce la descomposición del aceite.

eye a piece of metal bent into a small circle.

ojo trozo de metal remachado, formando un círculo pequeño.

F

face surface that is intended for use. Also tapered section of a valve head.

cara superficie que se intenta usar.

cara de válvula también la parte ahusada de la cabeza de válvula.

farmstead layout the efficient arrangement of buildings on a farm.

disposición de las instalaciones de la finca arreglo de los edificios en una finca.

fastener a device used to hold two or more pieces of material together.

sujetador, broche un dispositivo que se usa para juntar dos tejidos.

faucet device that controls the flow of water from a pipe or container.

grifa dispositivo que controla el flujo del agua de un tubo o contenedor.

faucet washer rubberlike part that creates a seal when pressed against a metal seat.

anillo obturador pieza de tipo de hule que forma una obturación cuando aprieta fuerte a una hoja de metal.

feather to sand so that a chipped edge is tapered.

recubrimiento solapado lijar hasta que se haga ahusado un filo desportillado.

ferrous metal that comes from iron ore.

ferroso dícese del metal que es de mineral de hierro.

ferrule a metal collar fitted on a handle to prevent splitting of the wood.

virola una abrazadera que encaja sobre un mango para evitar el hendimiento de la madera.

FFA the students organization for students studying high school agricultural education. This part of the program teaches leadership skills and provides motivation to learn.

Granjeros Futuros de América organización de estudiantes para los estudiantes que estudian la agricultura. Esta parte del programa enseña las habilidades de mando y da motivo para aprender.

field magnetic force around an electric wire or iron core.

campo fuerza magnética que rodea un hilo eléctrico o un núcleo de hierro.

field use location where repairs and other operations are done.

estación de reparación y servicio técnico sitio en donde se realizan las reparaciones y otras operaciones.

50-50 solder solder composed of 50 percent tin and 50 percent lead.

soldadura mitad y mitad soldadura compuesta de 50 por ciento de estaño y 50 por ciento de plomo.

filament special metal element in a vacuum that produces light when electricity flows through it.

filamento elemento especial de metal en un vacío que produce luz al pasar la corriente.

file flat, round, half-round, square, or three-sided piece of metal with fine teeth.

lima pieza de metal plana, redonda, de media caña, de cuatro cuartes o triangular, tallada de dientes muy finos.

file card a tool used for cleaning a file.

ficha una herramienta que se usa para limpiar una lima.

filler rod metal in the form of a long, thin rod used to add to or fill joints when brazing and welding.

varilla de metal de aportación metal en la forma de una válvula larga y estrecha empleada para llenar junturas al soldar.

fillet weld a weld placed in a joint created by a 90-degree angle.

soldadura de filete soldadura hecha en una unión formada por un ángulo recto.

filter converts pulsating DC voltage to smooth voltage.

filtro convierte corriente continua modulada en voltaje uniforme.

fine aggregate sand and other small particles of stone.

árido fino arena y otras partículas pequeñas de piedra.

fingering dividing and clustering of the bristles of a paintbrush.

fingering dividir y poner en grupitos las cerditas de una brocha.

finish chemical layer that protects the surface of a material.

acabado capa química que protege la superficie de un material.

finishing lime powder made by grinding and treating limestone for use in masonry materials.

cal apagada polvillo que se hace al pulverizar y tratar el cal; se usa en materiales de albañilería.

finishing sander tool with a small sanding pad driven in a forward-backward or circular pattern.

enarenador de acabado herramienta con placa pequeña de lijar, operada en movimientos circulares o hacia adelante, hacia atrás.

fire flame; to make a spark jump across an air gap.

fuego llama; hacer que brinque una chispa a través de un espacio de aire.

fire triangle the three conditions—fuel, heat, and oxygen—that must be present to produce a fire.

triángulo de fuego las tres condiciones—combustible, oxígeno y calor—que tienen que ser presentes para producir un fuego.

fish feeders built feed holders calibrated to disperse feed at timed intervals.

canales de alimentación para los peces construidos y calibrados para disponer alimentación a intervalos determinados.

fitting a part used to connect pieces of pipe or to connect other objects to pipe.

ajuste una pieza empleada para conectar pedazos de tubería o para conectar otros objetos a la tubería.

fixture a base or housing for a lightbulb, fan motor, or other electrical device.

instalación un fundamento o cárter para una bombilla, un ventilador u otro dispositivo eléctrico.

flammable capable of burning easily.

inflamable que puede encenderse fácilmente.

flammable materials cabinets cabinets specially made to reduce the risk of fire caused by flammable materials.

gabinetes para materiales inflamables gabinetes hechos especialmente para reducir el riesgo de fuego causado por materiales inflamables.

flaring block tool used to hold tubing when flaring.

bloque abocinadora herramienta empleada para expandir la boca de la tubería en ensanchando.

flaring tool tool used to expand the opening in tubing to a bell shape.

herramienta abocinadora herramienta empleada para expandir la boca de la tubería para que tome la forma de una campana.

flashback burning inside an oxyfuel torch that causes a squealing or hissing noise.

retroceso de la llama lo ardiendo dentro de la llama de soplete que causa un ruido chirriante o silbante.

flat finish dull or without shine.

acabado mate sin relieve o sin lustre.

flat-head screw a screw that has a flat head with a tapered underside designed to fit down into the material being secured.

tornillo de cabeza plana un tornillo que tiene cabeza llana con cara inferior ahusada, hecha para encajar el material que se asegura.

flat nailing two flat pieces nailed to each other.

clavar sobre costero dos piezas planas clavadas una a la otra.

float switch a switch turned on and off by the action of a float.

interruptor de flotador un interruptor encendido y apagado por la acción de un flotador.

float valve assembly a device used to control liquid flow into a reservoir, such as a flush tank in a toilet.

montaje de válvula de flotador un dispositivo empleado para controlar el flujo de líquido a un embalse, como el tanque de un lavabo.

floor broom broom that is used primarily to sweep floors.

escoba para pisos escoba que se usa principalmente para barrer pisos.

flow switch reacts to the movement of a gas or liquid.

controlador de circulación de fluídos reacciona al movimiento de un gas o un líquido.

fluid coupling uses fluid to transfer motion.

acoplamiento fluido usa fluidos para transferir movimiento.

fluid power transfer of force with liquids to gases.

potencia fluida transmisión de la fuerza de líquidos a los gases.

fluorescent light light tube that glows as a result of electricity flowing through gases.

lámpara fluorescente tubo ligero que se enciende a causa de la corriente eléctrica que corre por los gases.

flush plate a thin piece of hardware fastened across a joint to provide support.

placa para embutir a línea recta dispositivo delgado asegurado a través de una ensambladura para sostenerla.

flush valve used to control water used in the flushing process.

válvula de chorreo aparato empleado para regular el agua durante el proceso de chorrear.

flux material that removes tarnish or corrosion, prevents corrosion from developing, and acts as an agent to help solder spread over metal.

fundente material que quita la empañadura o corrosión, evita que ocurra la corrosión y actua como agente para que la soldadura pueda esparcirse sobre el metal.

flux-core solder See *hollow-core solder*.

soldadura con núcleo de fundente V. *soldadura con núcleo de fundente.*

flywheel a relatively heavy wheel attached to an engine to make it run more smoothly.

volante una rueda relativamente pesada que se sujeta a un motor para hacerlo funcionar sin dificultad.

focal color a color used to draw attention to a particular area or to warn of a danger. See also *vista green*.

color central es un cierto color que acera atención a un área particular o advierte de un peligro.

folding rule rigid rule of 2 to 8 feet in length that folds into a compact unit.

metro plegable una regla rígida de 2 a 8 pies de largo que se dobla para ponerse compacta.

footer (footing) a continuous slab of concrete that provides a solid, level foundation for block and other masonry.

zócalo un bloque continuo de hormigón que sirve como cimientos sólidos y planos para el bloque y otra albañilería.

footing concrete base under a wall.

zócalo fundamento de hormigón debajo de una pared.

footing pad concrete or stone under a vertical pole.

zapata del soporte hormigón o piedra bajo un poste vertical.

force pushing or pulling action.

fuerza acción de empujar o jalar.

form a metal or wooden structure that contains and shapes concrete until it hardens.

molde una estructura de metal o de madera que contiene y moldea el hormigón hasta que se fragüe.

formulate to put together according to a formula.

formular recetar conforme a una fórmula.

foundation block or concrete wall supporting a building.

cimientos bloque o pared de hormigón que sostiene un edificio.

foundational SAE the beginning part of an SAE where the student develops a detailed plan for conducting a Supervised Agricultural Experience.

fundacional SAE el inicio de una SAE donde el estudiante desarrolla un plan detallado para dirigir una experiencia agrícola supervisada.

four-cycle engine See *four-stroke-cycle engine*.

motor de ciclo de cuatro tiempos un motor que tiene cuatro tiempos por ciclo.

4-H a youth organization administered through the Cooperative Extension Service.

4 H una organización para los jovenes que se administra para el servicio de extensión cooperativo.

four-stroke-cycle engine an engine with four strokes per cycle.

motor de ciclo de cuatro tiempos un motor que tiene cuatro tiempos por ciclo.

frame the housing or case for motor components.

bastidor el cárter o armazón para componentes del motor.

framing square (carpenter's square) a flat square with a body and tongue.

escuadra un instrumento plano con cuerpo y lengüeta, usado para seguir ensambladuras.

freestanding construction-style greenhouse requiring no support structures.

autoportante invernadero de tipo de construcción que no requiere armazones para sostenerlo.

frost line the maximum depth to which the soil freezes in a given locality.

profundidad de penetración del congelamiento la profundidad máxima a la que se congela el suelo en determinada localidad.

fuel any material that will burn.

combustible cualquier material que se quema.

full-flow system a filter system in which all fluid flows through the filter.

sistema de flujo completo en que todo fluido recorre por un filtro.

full scale a drawing the same size as the object it represents.

tamaño natural una representación del mismo tamaño que el objeto a que representa.

fuse plug or cartridge containing a strip of metal that melts when more than a specified amount of current passes through it.

fusible enchufe o cartucho que contiene una chapa metálica que se funde al pasar una cantidad excesiva de corriente eléctrica.

fusestat a fuse with threads that permit it to be used only in circuits that match its capacity.

fustato un fusible con filamentos que le permiten emplearse solamente en circuitos de la misma capacidad.

fusion joining by melting.

fusión unir por fundición.

fusion welding joining parts by melting them together.

soldadura por fusión unir piezas por fundirlas.

G

galvanized coated with zinc for rust resistance.

galvanizado cubierto con una capa de cinc para protección contra la corrosión.

gas any fluid substance that can expand without limit.

gas cualquier sustancia fluida que puede expandirse sin límite.

gas analyzer device that determines the content of gases.

analizador de gas aparato que determina el contenido de gases.

gas forge gas-burning unit used to heat pieces of metal and temper tools.

forjador de gas quemador de gas empleado para fundir piezas de metal y templar las herramientas.

gas furnace gas-burning unit used to heat objects.

horno de gas quemador de gas empleado para calentar objetos.

gas metal arc welding (GMAW) a unique method of infusing molten metal into a base metal, using a gaseous shield around the molten puddle.

asfixié con gas soldadura de arco de metal método único que hace una infusión de metal fundido en metal usando un revestimiento gaseoso alrededor del charco fundido.

gas tungsten arc welding (GTAW) a method used to weld very light, thin metal, using a tungsten electrode and a gaseous shield.

soldaduras por arco de tungsteno con gas método que se usa para soldar el metal muy ligero y delgado usando un electrodo de tungsteno y un revestimiento gaseoso.

gauge a device used to determine the dimension of materials or space; a device used to measure and indicate pressure in a hose, pipe, or tank.

medida un dispositivo empleado para medir la dimensión de materiales o del vacío; un dispositivo empleado para medir e indicar la presión en una manga, un tubo o un tanque.

gear wheel with teeth that mesh to another wheel.

engranaje rueda dentada que penetra entre los dientes de otra rueda.

generator a device that produces direct current.

generador un dispositivo que produce corriente continua.

GFCI See *ground-fault circuit interrupter.*

IFT V. *interruptor de la falta de tierra.*

girder carries floors and interior walls.

viga sostiene pisos y paredes interiores.

girt horizontal side nailers in a building.

riostra vigas de rigidez horizontales en un edificio.

glazing compound special puttylike material used to install window glass.

masilla materia especial de tipo de yeso usada para instalar los cristales en los bastidores de las ventanas.

Global Positioning System (GPS) a system that uses a group of satellites to determine the exact location of a point anywhere in the world.

sistema GPS un sistema que usa un grupo de satélites para determinar el sitio exacto de un punto en todas partes del mundo.

gloss paint with a shiny finish.

pintura brillante pintura con un acabado brillante.

glue sticky liquid used to hold things together.

pegamento líquido pegajoso que se usa para unir o pegar cosas.

governor speed control device.

regulador dispositivo que controla la velocidad de algo.

GPS See *Global Positioning System.*

GPS V. *sistema GPS.*

gradations numbers and lines stamped or painted on measuring devices.

gradaciones números y líneas embutidos o pintados sobre los dispositivos de medida.

grade amount of drop or fall in the land.

grado medida del declive del terreno.

grain natural lines on lumber caused by the annual growth rings in a tree.

grano rayas o estrías en los maderos causadas por la capa cortical del árbol.

graph paper paper laid out in squares of equal size.

papel cuadriculado papel en el cual son imprimidas cuadrículas.

gravel particles of stone larger than sand, also called *coarse aggregate.*

gravilla partículas de piedra más grande que arena, también se llama árido grueso.

gray a neutral color used in the safety color system as a restful background color.

gris un color neutro que se usa en el sistema de colores de seguridad para crear una sensación tranquila.

green the safety color used to indicate the presence of safety equipment, safety areas, first-aid, and medical practice.

verde el color de seguridad que se usa para indicar la presencia de equipo de seguridad, áreas de seguridad, primeros auxilios y tratamiento médico.

greenhouse structure that provides light and heat from solar energy.

invernadero una construcción que proporciona luz y calor que provienen de la energía solar.

greenhouse effect the buildup of heat waves accumulated inside a translucent structure or greenhouse.

efecto invernadero el aumento de ondas de calor acumuladas dentro de una construcción translúcida o un invernadero.

grinding wheel abrasive cutting particles formed into a wheel by a bonding agent.

rueda de muela abrasiva partículas abrasivas para cortar que son formadas para hacerse una rueda, por un agente de encolamiento.

grinders tools that cut metal only with rigid grinding wheels instead of flexible discs.

trituradoras para metal herramientas que cortan metal solamente con ruedas rígidas de trituración hechas de metal, en vez de discos flexibles.

grit cutting particles.

partículas arenosas duras partículas cortantes.

ground board wooden or aluminum construction member for attaching plastic or fiberglass coverings on greenhouses.

ground board pieza de construcción de madera o aluminio que sirve para atar cubiertas plásticas o de fibra de vidrio a los invernaderos.

ground clamp connector used to attach a cable, wire, or object to a ground source.

conector a tierra conectador usado para conectar un cable, alambre u objeto a una fuente en tierra.

ground (gee) clip a device by which groundwires are attached to electrical boxes.

conector a tierra dispositivo por lo cual los cables de toma de tierra son conectados a cajas eléctricas.

ground-fault circuit interrupter (GFCI) a device that cuts off the electricity if even very tiny amounts of current leave the normal circuit.

interruptor de circuito de falla un dispositivo que interrumpe la corriente eléctrica, incluso si cantidades muy pequeñas de corriente eléctrica salgan del circuito normal.

grounding making an electrical connection between a piece of equipment and the earth.

conectar con tierra establecer una conexión eléctrica entre un aparato y la tierra.

group marking the marking on a package of welding electrodes that indicates the type of electrodes within the package.

marca de grupo marco sobre un paquete de electrodos de soldadura que indica el tipo de electrodos adentro del paquete.

gusset a plate or bracket used to reinforce a wood or metal joint.

esquinero chapa de madera o metal empleada para reforzar una juntura.

gypsum blocks with meter a type of moisture meter that involves the use of two or more gypsum blocks buried in the soil.

bloques de yeso un tipo de metro de humedad que usa dos o más bloques de yeso enterrado en la tierra.

H

hacksaw a device that holds a blade designed for cutting metal.

sierra de arco un dispositivo que tiene una hoja creada para cortar metal.

hammer a type of tool for driving other implements and hardware; also, a rapid striking action that gives a power drill more power.

martillazo acción de golpear.

hand stone a sharpening stone designed to be held in the hand.

piedra de aguzamiento de mano piedra de aguzamiento hecho para sujetarse en la mano.

hand tool a tool operated by hand to do work.

herramienta manual una herramienta manejada a mano para hacer una obra.

handsaw a saw used to cut across boards or to rip boards and panels.

sierra de mano una sierra usada para cortar a través de maderos o hender maderos y tableros.

hands-on experience experience that comes from actually doing something rather than reading or hearing others tell about it.

capacitación práctica experiencia que se gana haciendo algo, en vez de aprenderlo leyendo u oyendo de otros cómo se hace.

hard drive a rigid internal or external component capable of storing vast quantities of computer data.

disco duro disco rígido empleado para almacenaje.

hardpan impervious clay.

arcilla dura arcilla impermeable.

hardware fasteners; objects made from metal.

ferretería sujetadores; objetos hechos de metal.

hardwood lumber from a deciduous tree.

madera dura madera de un árbol de hoja caduca.

hasp a flat, hingelike device used for attaching a padlock on a door or gate.

cierre dispositivo llano que se emplea para poner un candado en una puerta o un portón.

head the flat part of a valve; the enlarged part on top of a nail or screw; a cylinder cover containing the spark plug and combustion chamber.

cabeza la parte llana de la válvula.

culata parte superior del cilindro, conteniendo la bujía y la cámara de combustión.

head gasket seal between the head and the cylinder block.

junta de culata juntura entre la culata y el bloque del cilindro.

heat type of energy that causes the temperature of an object or environment to rise.

calor forma de energía que resulta en el aumento de la temperatura.

heat sensor switch that responds to temperature change.

ruptor térmico interruptor que responde al cambio de la temperatura.

heel the place where the tongue and body of a framing square meet.

talón lugar donde se unen la lengüeta y la parte principal de una escuadra.

henry (H) unit of inductance.

henrio unidad de inductancia.

hertz (Hz) one cycle per second.

hertz (Hz) un ciclo por segundo.

hex head a bolt or nut having six sides.

cabeza hex un pestillo o una tuerca con una cabeza hexagonal.

hidden line a series of dashes that indicates the presence of unseen edges.

línea oculta una serie de rayas que indican la presencia de bordes ocultos.

hiding power the ability of a material to create color and mask out the presence of colors over which it is spread.

poder cubriente la capacidad que tiene una materia para crear color y recubrir los colores en los cuales se extiende.

high-speed drill bit a twist drill bit made and tempered especially for drilling metal.

barrena de alta velocidad una broca helicoidal hecha y templada específicamente para taladrar metal.

high-tensile wire type of very strong, smooth wire used for fencing.

alambrado de alta tensión tipo de alambre liso y muy fuerte que se usa para cercos y vallas.

high-tension wire high-voltage wire in a secondary circuit.

alambre de alta tensión alambre de alto voltaje en un circuito secundario.

hinge an object that pivots and permits a door or other attached object to swing back and forth or up and down.

bisagra un objeto que da vueltas sobre su eje y permite que una puerta u otro objeto gire de acá para allá o de arriba abajo.

holding tool a tool used to grip wood, metal, plastic, and other materials.

herramienta fijadora herramienta empleada para agarrar madera, metal, plástico y otros materiales.

hollow a space in the center where the robot cannot function.

hueco espacio en el centro donde el robot no puede funcionar.

hollow-core block masonry block with two or three holes per block.

núcleo de aire bloque de albañilería con dos o tres cavidades por bloque.

hollow-core solder (flux-core solder) solder with flux inside.

soldadura con núcleo de fundente soldadura envolviendo el fundente que está en el centro.

hollow-ground a saw blade with teeth wider at the points than at the base.

afilado con cara cóncava una hoja de sierra con dientes más anchos de los puntos que de la base.

hollow sphere work area a work area in the shape of a hollow ball, with a hollow area at the center.

esfera hueca área de trabajo en la forma de una bola hueca.

horizontal flat or level.

horizontal llano o a nivel.

horizontal band saw has a blade that saws parallel to the ground.

aserradora de banda horizontal tiene una cuchilla que asierra paralelamente con el suelo.

horizontal shaft engine an engine with a crankshaft that lies crossways for normal operation.

motor de eje horizontal un motor que tiene un cigüeñal que yace transversalmente para funcionamiento normal.

horizontal weld a weld that is made from left to right or right to left on horizontally positioned metal.

soldadura horizontal soldadura que es hecha de izquierda a derecha o de derecha a izquierda en un métal que se situa horizontal.

horsepower force needed to lift 550 pounds one foot in one second.

caballo de vapor la potencia necesario para levantar 550 libras un pie por segundo.

hose flexible line that carries gases or liquids.

manga tubo largo flexible que lleva gases o líquidos.

hot wires See *positive wires.*

alambres de alta tensión V. *alambres de alta tensión.*

humidistat an instrument that measures air humidity.

higrostato un instrumento que mide la humedad del aire.

hydraulic cylinder tank with inlet and piston.

cilindro hidráulico tanque con válvula de admisión y émbolo.

hydraulic motor receives power from moving fluid.

motor hidráulico recibe potencia de fluido en movimiento.

hydraulics use of liquids to transfer force.

hidráulica uso de líquidos para transferir potencia.

hydroponics a plant production technology that uses water as a growing medium.

hidroponía (cultivo hidropónico) tecnología de producción de las plantas que usa el agua como el instrumento para cultivar.

hydroponics production of plants using dissolved nutrients in a recirculating system.

hidropónica cultivación de plantas usando alimentos nutritivos disueltos en un sistema de recirculación.

I

ID inside diameter.

DI diámetro interior.

ignition a spark igniting an air-fuel mixture.

encendido una chispa encendiendo una mezcla carburante.

ignition spark hot electrical arc across an air gap.

chispa de encendido arco eléctrico caliente a través de un espacio de aire.

impeller fan-shaped rotor.

propulsor rueda de aletas.

improvement project an activity that improves the appearance, convenience, efficiency, safety, or value of a home, farm, ranch, agribusiness, or other agriculture facility.

actividad de mejoramiento proyecto que mejora la belleza, conveniencia, seguridad, valor o eficacia y que se aprende fuera de las clases o laboratorios programados normalmente.

impurity any material other than the base metal.

impureza cualquier materia que no es el metal principal.

inclined plane a surface at an angle to another surface; one of the six simple machines.

plano inclinado una superficie de un ángulo a otra superficie; una de las seis máquinas sencillas.

inductance (L) ability to draw energy and store it in a magnetic field.

inductancia (L) capacidad de sacar energía y almacenarla en un campo magnético.

induction motor where power does not flow directly to the armature.

inducción motor a que la potencia no corre directamente a la armadura.

inductor a device that provides inductance.

inductor un dispositivo que sirve como fuente de inductancia.

infrared light sensors used to detect heat or motion.

detector de rayos infrarrojos usado para detectar calor o movimiento.

inside micrometer telescoping gauge used to measure inside surfaces of hollow objects.

micrómetro interior medida telescópica empleada para medir las caras interiores de objetos huecos.

insulation retards heat movement.

aislación/aislamiento retrasa el movimiento del calor.

insulator material that provides great resistance to the flow of electricity.

aislador material que da gran resistencia al paso de la electricidad.

intake stroke engine process of taking fuel and air into the combustion chamber.

carrera de admisión el proceso motriz de tomar combustible y aire a la cámara de combustión.

integrated circuit a miniature electronic circuit.

circuito integrado un circuito electrónico pequeño.

interior grade a grade of plywood suitable for indoor use only.

cualidad para interior madera contrachapada apropiada solamente para uso dentro de la casa.

interior paint paint that will not hold up if exposed to weather.

pintura interior pintura que se desgasta cuando está expuesta a condiciones climáticas.

internal combustion engine device that burns fuel inside a cylinder to create a force that drives a piston.

motor de combustión interna dispositivo que quema el combustible dentro de un cilindro para crear una fuerza que impulsa el émbolo.

iron ferrous; an element associated with high-quality paint.

hierro ferroso; un elemento asociado con pintura de buena calidad.

irrigation human-controlled water system applied to plants and soils.

irrigación abastecimiento de agua controlado por los seres humanos que se aplica a los cultivos y los suelos.

item separate thing.

artículo cosa distinta.

ivory a soothing neutral beige or off-white color used as a focal color.

color marfil color beige tranquilizante neutro o blancuzco usado como color focal.

J

jamb block a concrete block used to frame a door.

bloque de jamb un bloque de hormigón que se usa para hacer el armazón de una puerta.

jet or seat a hole shaped to receive the needle and control the flow of fuel.

chicler un agujero formado para recibir el agujo y controlar el paso de combustible.

jigsaw (scroll saw) a saw that cuts very short curves by reciprocal action.

sierra de vaivén una sierra que corta en curvas muy cortas por movimiento alternativo.

jobbers-length drill bit a drill bit of longer than standard length.

taladradora para agujos profundos taladradora cuyo mandril es más largo que lo estándar.

joint the place where two pieces of linked material come together.

juntura sitio en donde se juntan dos piezas.

ensambladura sitio en donde dos piezas de madera se juntan.

jointer a machine with rotating knives used to straighten and smooth edges of boards; a masonry tool used to shape joints.

garlopa para igualar y lisar las superficies de madera.

palustre una herramienta de albañilería empleado para formar junturas.

joist a timber used to support the floor or ceiling of a building.

viqueta una viga que se usa para sostener el suelo o el techo de un edificio.

joule a measurement of energy calculated by multiplying amps × volts × time.

julio medida de energía que se calcula multiplicando amperios × voltios × tiempo.

journal an accurate record of events and observations, such as weight gain in livestock, personal accomplishments, data collection, etc.

diario un registro preciso de los acontecimientos y las observaciones, como el aumento de peso del ganado, los logros personales, la recolección de datos, etc.

K

K classification of thick wall copper pipe or tubing.

K clasificación de tubos o cañerías de pared gruesa de cobre gruesa.

kerf opening in board made by a saw; opening in steel made by oxyfuel cutting.

aserradura corte hecho en un madero por una sierra.

corte abertura hecha en el acero por el soldeo con soplete oxiacetilénico.

keyhole saw a saw designed for starting in holes.

serrucho de punta una sierra creada para empezar en agujeros.

kick starter foot-operated lever and cog mechanism to crank an engine.

pedal de arranque mecanismo de palanca y rueda dentada en que pisa para arrancar el motor.

kiln a huge oven that heats boards at a steady rate to remove the moisture slowly.

horno de secado un horno grande que seca las tablas a un ritmo constante para extraerles la humedad lentamente.

kiln dried lumber that has been dried by heat in a special oven called a kiln.

secado en horno dícese de los maderos que han sido secados con calor en un gran horno especial que se llama horno de secado.

kilo one thousand.

kilo mil.

kilowatt-hour the use of 1,000 watts for one hour.

kilovatio-hora el uso de 1,000 vatios por una hora.

knockout partially punched impression in electrical boxes.

cortacircuitos un molde en las cajas eléctricas parcialmente perforado.

L

L classification of medium wall copper pipe or tubing.

L clasificación de tubos o cañerías de pared media de cobre.

labor to struggle or work hard to keep running.

funcionar con dificultad luchar o avanzar penosamente para que siga en marcha.

lag screw (lag bolt) a screw with coarse threads designed for use in structural timber or lead anchors.

tirafondo un tornillo con roscas gruesas creados para usarse en maderos de construcción o áncoras de plomo.

laminate to fasten two or more flat pieces together with an adhesive.

contrachapear unir dos chapas planas o más con un adhesivo.

land level handheld level used for sighting level points.

escuadra de agrimensor nivel usado para averigüar si la tierra es plana u horizontal a ciertos puntos.

landscape fence fence constructed to provide beauty to a landscaping scheme.

cercado paisajista cercado diseñado para proveer belleza a un paisaje

lap joint joint formed by fastening one member face-to-face on another member of an assembly.

juntura imbricante juntura formada por sujetar una pieza cara-a-cara a otra pieza, de una ensambladura.

lapping compound gritty material used for lapping in valves.

compuesto para pulir materia arenosa usada para lapear las válvulas.

lapping in grinding valves to fit the seat for a perfect seal.

lapear las válvulas esmerilar las válvulas para encajar el asiento para hacer una obturación perfecta.

laser intense beam of light.

láser rayo de luz intensa.

latex water-based paint.

látex pintura de base de agua.

laying block the process of mixing mortar, applying it to masonry block, and placing the block to create walls.

tender bloque el proceso de mezclar mortero, aplicándolo a bloques de albañilería y colocando los ladrillos o piedras para crear paredes.

layout a plan, map, or pattern for future operations.

trazado un plano, mapa o dibujo para operaciones futuras.

layout tool a tool used to measure or mark wood, metal, and other materials.

instrumento para trazar un instrumento usado para medir o marcar madera metal y otras materiales.

lead heavy metal element that stays in the body once it is ingested; one component of solder.

plomo elemento metal pesado que permanece en el cuerpo una vez que es ingerido; un componente de soldadura.

leader line solid line with an arrow used with an explanatory note to point to a specific feature.

línea de guía línea continua con una flecha usada con comentario explanatorio para señalar una característica específica.

leaner a greater proportion of air and a lesser proportion of fuel in an air-fuel mixture.

mezcla pobre una proporción más grande de aire y una proporción menor de combustible en la mezcla carburante.

LED See *light-emitting diode.*

DEL V. *diodo emiso de luz.*

level a device used to determine if an object has the same height at two or more points. See *spirit level.*

nivel un instrumento empleado para determinar si un objeto tiene la misma altura entre dos puntos o más. V. *nivel de burbuja.*

light-emitting diode (LED) produces light when subjected to a current flow.

diodo emiso de luz (DEL) produce luz al sujetarse a la corriente eléctrica.

lightning arrester device installed on an electric fence that directs large surges of current created by lightning into the ground.

para-rayos dispositivo que se instala en un alambrado eléctrico para dirigir descargas eléctricas producidas por los rayos a tierra.

line small-diameter material stretched tightly between two or more points. See also *purge the lines.*

cordón flexible material de diámetro que se puede estirar tensamente entre dos puntos o más. V. *purgar las líneas.*

linear in a straight line.

lineal de línea recta.

loading the brush the process of dipping the bristles of a brush into paint.

preparar la brocha el proceso de zambullir las cerdas en la pintura.

lodestone magnetized iron oxide.

magnetita piedra de imán de óxido de hierro negro.

M

M classification of thin wall copper pipe or tubing.

M clasificación de tubos o cañerías de pared delgada de cobre.

machine bolt a fastener with a square or hex head on one end and threads on the last inch or so on the other end.

perno hecho a torno cierre de cabeza cuadrada o hexagonal de un extremo y roscado de la última pulgada del otro extremo.

machine screw a small bolt with an accompanying hex nut.

tornillo a máquina un pestillo con una tuerca hexagonal.

magnesium a low-quality paint pigment.

magnesio pigmento de pintura de mala calidad.

magnetic field area around an iron core or electric wire influenced by the presence of magnetism.

campo magnético región alrededor del núcleo de hierro o alambre eléctrico sometida a la influencia de magnetismo.

magnetic flux lines of magnetic force that occur in patterns between poles of a magnet.

flujo magnético rayos de fuerza magnética que ocurren en patrones de campo entre los polos de un imán.

magnetism a force that attracts or repels iron or steel.

magnetismo una fuerza que atrae o repele el hierro o acero.

magneto produces electricity by magnetism.

magneto produce electricidad por el magnetismo.

mainframe a very large-capacity computer.

computadora principal una computadora muy grande de mucha capacidad.

maintenance doing the tasks that keep a machine in good condition.

mantenimiento hacer las tareas para el cuidado de una máquina, preservándola en buen estado.

malleable workable.

maleable que puede labrarse.

malleable cast iron combination of a cast iron core and a ductile metal outer layer.

fundición maleable mezcla de núcleo de arrabio y metal dúctil.

manifold device with openings to serve two or more units.

tubería multiple dispositivo con aberturas que puede acometer a dos unidades o más.

manometer measures fluid pressure and vacuum.

manómetro mide las presiones de los líquidos y de lo neumático.

manual by hand.

manual a mano.

margin outer edge of a valve head.

margen borde externo de la cabeza de válvula.

mask to cover so paint will not touch.

ocultar cubrir para que no se manche con pintura.

masonry anything constructed of brick, stone, tile, or concrete units held in place with portland cement.

albañilería cualquier cosa construida de unidades de ladrillo, piedra, azulejo u hormigón, unidas por cemento pórtland.

masonry units block made from concrete or cinders.

unidades de albañilería bloques hechos de hormigón o escoria.

measuring tape See *tape*.

cinta métrica V. *cinta métrica*.

mechanic a person specifically trained to perform mechanical tasks.

mecánico persona especializada en las obras o tareas mecánicas.

mechanical of or having to do with a machine, mechanism, or machinery.

mecánico de o que tiene que ver con una máquina, un mecanismo o la mecánica.

mechanics the branch of physics dealing with motion and the action of forces on bodies or fluids.

mecánica el ramo de la física que trata del movimiento y la acción de las fuerzas sobre cuerpos o líquidos.

member a component of a project such as a piece of wood.

miembro componente de un proyecto como una pieza de madera.

metal inert gas (MIG) gas such as argon, carbon dioxide, or helium that is used in welding.

gaz inertede metal gas como argón, dióxido de carbono o helio que se usa en la soldadura.

meter a device that measures electricity, gas, or liquid that passes through a system.

metro un dispositivo que mide la corriente eléctrica, el gas o el líquido que pasa por un sistema.

meter (m) a metric unit of linear measure equalling 39.37 inches.

metro (m) unidad métrica de medida longitudinal que es igual a 39.37 de pulgadas.

metric system a decimal system of measures and weights.

sistema métrico sistema decimal de medidas y pesas.

microtubing small tubes used in an irrigation system.

microtubo tubos pequeños que se usan en una sistema de irrigación.

microcomputer self-contained small computer.

microcomputadora computadora pequeña y autosuficiente.

micrometer (telescoping gauge) instrument used to measure outside surfaces of round objects.

micrómetro usado para medir las superficies de objetos redondos.

microorganisms colony of microscopic bacteria and fungi.

microorganismos colonia de bacterias y mohos microscópicos.

microprocessor central processing unit (CPU) of small computer.

microprocesador unidad central de proceso (CPU) de una computadora pequeña.

milliampere thousandth of an ampere.

miliamperio milésima parte de un amperio.

millimeter (mm) thousandth of a meter.

milímetro (mm) milésima parte de un metro.

minicomputer intermediate-capacity computer.

minicomputadora computadora de capacidad intermedia.

miter an angle.

inglete un ángulo.

miter box a device used to cut molding and other narrow boards at angles.

caja de ingletes dispositivo empleado para cortar molduras y otras tablas estrechas a ángulos.

miter gauge adjustable, sliding device to guide stock into a saw at the desired angle.

medida de inglete dispositivo ajustible, móvil, para guiar materia prima a una sierra al ángulo determinado.

miter joint joint formed by cutting the ends of two pieces at a 45-degree angle.

inglete ensambladura formada por cortar los bordes de dos piezas a un ángulo de 45 grados.

miter saw motor-driven circular blade for cross cutting angle cuts that is fed down into the material.

sierra inglete cuchilla circular motarizada para hacer cortes transversales angulares que se impulsa directamente hacia el material.

mix the ratio of materials in concrete or mortar.

mezcla la ración de materias en el hormigón o el mortero.

moisture barrier prevents movement of water, steam, or vapor.

barrera de humedad impide el movimiento de agua o vapor.

moisture sensor a device that measures soil moisture.

sensor de humedad dispositivo que mide la humedad de la tierra.

molding head a device that holds knives to shape wood into moldings of various types.

cabeza de limadora un dispositivo que sujeta cuchillas que moldean la madera en molduras de varios tipos.

momentum turning force of the flywheel and other moving parts that carries an engine through non-power strokes.

velocidad fuerza giratoria del volante y otras partes móviles que lleva un motor por carreras no motrices.

mortar a mixture of portland cement, finishing lime, water, and sand.

mortero una argamasa de pórtland, cal apagada en polvo, agua y arena.

mortar bed a layer of mortar.

lecho de mortero capa de mortero.

multimeter combination volt-ohm-milliammeter.

aparato medidor multiple una combinación voltio-ohmio-miliamperímetro.

mushroom to spread out over an edge.

esparcirse como hongo desbordar un borde.

mushroomed a spread or pushed-over condition caused by being struck repeatedly.

tomar forma de seta una condición de rebosar o hacer caer causada por ser golpeada repetidamente.

N

nail a fastener with a single point that is driven into the material it holds.

clavo/puntilla un sujetador con una sola punta que se inserta en el material que va a sujetar.

nail set punchlike tool with a cupped end.

embutidera herramienta de tipo perforador con un extremo en forma de bocina.

nap soft, woolly, threadlike surface.

lanilla superficie de lana finos, suave.

National Electrical Manufacturers Association (NEMA) the group that developed the system of color coding electrodes.

Asociación Nacional de Fabricantes Electrotécnicos (ANFE) el grupo que desarrolló el sistema de codificar electrodos por asignarlos colores específicos.

neat's-foot oil light yellow oil obtained by boiling the toe bones and shinbones of cattle and used to soften and preserve leather.

aceite de pata de vaca aceite amarillo ligero que se obtiene hirviendo las pezuñas y las tibias de ganado y que se usa para suavizar y preservar el cuero.

needle a long, tapered shaft.

aguja eje largo y afilado.

needle bearings sets of steel rollers in a cage to keep a shaft centered.

cojinete de agujas rodillos de acero en un cojinete en el cual se apoya y gira un eje.

negative polarity See *straight polarity.*

polaridad negativa V. *polaridad negativa.*

neoprene tough rubberlike material that is resistant to moisture and rotting.

neopreno material duro de tipo de hule que es resistente a la humedad y al pudrimiento.

net cages pens or confinement structures for herding fish.

trasmallos redes o estructuras de restricción para cazar peces.

neutral flame flame with a balance of acetylene and oxygen.

llama neutra llama con un equilibrio de acetileno y oxígeno.

neutral wires conductors with white insulation that carry current from an appliance back to the source.

hilos neutros conductores con aislamiento blanco que llevan la corriente desde un aparato eléctrica hasta el origen.

neutron probe type of moisture sensor.

sonda de neutrones un tipo de sensor de humedad.

new wood wood that has never been painted or sealed.

madera virgen madera que nunca ha sido pintada ni chapada.

new work original construction, including new wood.

obra original construcción original, inclusive la madera nueva.

nipple a short piece of steel pipe threaded at both ends.

boquilla de unión un trozo de tubo de acero roscado en ambos extremos.

noise duration the length of time a person is exposed to a sound.

duración de ruido la duración de tiempo que una persona se expone al ruido.

noise intensity the amount of energy in the sound waves emitted by a particular source.

intensidad de ruido energía de las ondas acústicas.

nominal the size of material as used in the name for trade and building purposes.

nominal el tamaño del material como lo usado en la nomenclatura para fines comerciales y de construcción.

nonferrous metal that does not come from iron ore.

metal no férreo que no proviene del metal de hierro.

nonmetallic sheathed cable cable composed of copper or aluminum wires covered with paper, rubber, or vinyl for insulation and protection.

cable con cubierta de material no metálico cable compuesto de alambres de cobre o aluminio cubiertas con papel, hule o vinilo para aislamiento y protección.

north pole end of a magnet opposite the south pole.

polo positivo terminal del imán opuesto del polo negativo.

nozzle (sprinkler head) a device on a hose or pipe that directs the out flow of water or other liquid.

boquilla dispositivo en una manguera o un tubo que dirige el chorro de agua y otro liquido.

nut a hardware device with shoulders and a threaded hole in the center, used with a bolt to fasten two items together.

tuerca dispositivo roscado que tiene hombros.

nutrient solution dissolved nutrients and dissolved oxygen needed for plant growth.

solución nutritiva alimentos nutritivos disueltos y oxígeno disuelto necesarios para el crecimiento de plantas.

O

object line solid line in a drawing that shows visible edges of an object.

object line línea continua en un dibujo que representa los bordes visibles de un objeto.

occupation business, employment, or trade.

ocupación negocio, empleo o empresa.

occupational cluster group of related jobs.

grupo profesional grupo de empleos relacionados.

occupational division group of occupations or jobs within a cluster that require similar skills.

división profesional grupo de ocupaciones o trabajos dentro de un grupo que requieren oficios semejantes.

octane rating a rating of the ability of gasoline to prevent engine knock.

indice de octano una estimación de la abilidad de gas evitar golpeteo.

OD outside diameter.

DO diámetro exterior.

off-the-farm agricultural jobs those jobs requiring agricultural skills, but not regarded as farming or ranching.

trabajos agrícolos fuera de la finca aquellos trabajos que exigen técnicas agrícolas, pero que no se consideran cultivar la tierra o llevar una hacienda.

ohm a measure of the resistance of a material to the flow of electrical current.

ohmio medida de la resistencia de un material a la corriente eléctrica.

Ohm's Law the relationship between electric current (I), electromotive force (E), and resistance (R): $E = IR$.

la ley de Ohm la relación entre la corriente eléctrica (I), la fuerza electromotriz (E) y la resistencia (R): $E = IR$.

oil-based paint paint containing some type of oil as the vehicle.

colores al óleo pintura que contiene algún tipo de óleo como el vehículo.

oil-foam filter an engine air filter that involves the use of foam rubber coated with oil.

filtro de aceite de goma un filtro de aire de motor que se usa la goma cubierta de aceite.

old work wood that was previously painted or finished.

obra vieja maderos que fueron pintados o acabados anteriormente.

open aquaculture system natural water environment for growing plants and animals.

sistema abierto de piscicultura ambiente acuático natural para cultivar plantas y animales.

open circuit an electrical circuit in which there is at least one place where current cannot flow.

circuito abierto un circuito eléctrico en que hay por lo menos un sitio donde no puede pasar la corriente.

optoelectric device device designed to interact with light energy in the visible, infrared, and ultraviolet ranges.

dispositivo optoelectrónico dispositivo diseñado para actuar recíprocamente con energía de luz en las escalas visibles infrarrojas y ultravioletas.

orange safety color used to designate machine hazards, such as edges and openings.

anaranjado color de seguridad usado para señalar peligros de máquina, como bordes y aberturas.

orbital circular or egg-shaped pattern.

orbital diseño circular o de curva elíptica.

orbital sander a finishing sander that travels in a circular pattern.

enarenador orbital un enarenador de acabado que se mueve en una manera circular.

oriented strand board (OSB) a building material made from chips of wood. The chips are oriented to provide the maximum strength.

la tabla orientada de hilo un material de construcción hecha de virutas de madera. Las virutas se orientan para suministrar la fuerza máxima.

O ring a piece of rubber shaped like an O that fits in a groove on a shaft and prevents liquids from passing.

junta tórica pieza de hule en forma de la O que cabe en una ranura sobre un eje y previene que pasen líquidos.

oscillator nonrotating device for producing alternating current.

oscilador aparato que no gira, empleado para producir corriente alterna.

oscilloscope provides visual display of frequency, duration, phase, wave form, and amplitude of signals.

osciloscopio da representación visual de la frecuencia, duración, fase, ondulación y amplitud de sintonías.

oval head screw a screw with a head that extends both above and below the surface of the material being held.

tornillo de cabeza ovalada un tornillo con cabeza que asoma encima de y está hincada debajo de la superficie del material a que sujete.

overhaul complete disassembly with cleaning and reconditioning or replacement of most moving parts.

reparar totalmente; poner en buen estado desmontaje completo con limpieza y arreglo o reemplazo de la mayoría de las piezas móviles.

override take precedence over.

dominar tener prioridad sobre algo.

overspray spray paint that goes on objects other than the one that is supposed to be painted.

exceso de rocío pintar o cubrir con pintura objetos que no se iban a pintar.

overvoltage protection circuit circuit protection device.

circuito de protección de máxima tensión protección contra sobretensiones.

ownership/entrepreneurship SAE a student program through which a student owns an enterprise such as livestock, crops, etc.

la propiedad/espíritu empresarial SAE un programa estudiantil a través del cual un estudiante posee una empresa como ganado, cosechas, etc.

oxidation combining with oxygen.

oxidación lo que resulta al combinarse algo con el oxígeno.

oxidation inhibitors oil additive that prevents the oil from breaking down because of reactions with oxygen.

inhibidores de oxidación aditivo que se le añade al aceite para que impide que el aceite se descomponga a causa de reacciones con el oxígeno.

oxide the product resulting from oxidation of metal.

óxido el producto que resulta de la oxidación de metal.

oxidizing flame flame with an excess of oxygen; hottest type of oxyfuel flame.

llama oxidante llama con demasiado oxígeno; tipo de llama oxifuel.

oxyacetylene oxygen and acetylene combined.

oxiacetilénico la combinación de oxígeno con acetileno.

oxyfuel the combination of nearly pure oxygen and a combustible gas to produce a flame.

oxifuel la combinación de oxígeno casi puro y un gas combustible para producir una llama.

oxyfuel cutting process in which steel is heated to the point when it burns and is removed in a thin line, or kerf.

soldeo con soplete oxiacetilénico proceso en que se caldea el acero hasta que se funda y se corte en un corte delgado.

oxygen gas in the atmosphere that is necessary to support combustion.

oxígeno el gas en la atmosfera que es necesario para sostener la combustión.

P

packing soft, slippery, wear-resistant material used to seal space between a moving and nonmoving part.

embalaje material suave, escurridizo, resistente al desgaste, que sirve para llenar el vacío entre las partes móviles e inmóviles.

pad metal for practicing welding.

placa de soldadura chapa de metal para entrenamiento en soldadura.

paddle-wheel aerator splashing paddle wheel used to mix air and water.

aireador de rueda de paletas álabe salpicante que se usa para mezclar aire y agua.

paint coating material consisting of pigment suspended in a vehicle.

pintura sustancia que consiste en pigmentos y un vehículo.

paint film material left after paint has dried.

película de pintura material residuo al secarse la pintura.

paint roller a cylinder that turns on a handle and is used to apply paint.

rodillo un cilindro que gira por medio de un mango y que se usa para pintar.

pallet a portable wooden platform for storing or moving cargo.

paleta plataforma portátil para almacenar o trasladar la carga.

pan head screw a screw that has a head that looks like a frying pan turned upside down.

tornillo de cabeza troncocónica un tornillo cuya cabeza es parecida a una sartén puesta boca abajo.

parallel having two edges or lines the same distance apart at all points.

paralelo dos bordes o líneas equidistantes entre sí a todos puntos.

parallel circuit a circuit that provides two or more paths for current flow.

circuito en paralelo proporciona dos senderos o más para el paso de corriente.

pass one bead of metal or other material.

pasada un cordón de metal u otra materia.

pattern a model or guide for something to be made.

muestra un modelo o guía para algo que se ha de hacer.

PE See *polyethylene.*

PE V. *polietileno.*

penny a unit of measure used to designate the length of most nails.

penique unidad de medida usada para determinar el longitud de la mayoría de los clavos.

pentachlorophenol a popular wood preservative sold in a nearly clear vehicle.

pentachlorofenol preservativo popular de la madera que se vende en un vehículo casi claro.

permanent magnet a piece of iron or steel that holds its magnetism.

imán permanente un trozo de hierro o acero imantado.

perpendicular at a 90-degree angle to some object.

perpendicular algo que forma ángulo recto con otro objeto.

pesticide substance used to control pests.

pesticida sustancia empleada para controlar los animales e insectos nocivos.

Phillips-head screw a screw with screwdriver slots in the shape of a plus sign.

tornillo de cabeza Phillips un tornillo con muescas en forma de cruz.

Phillips screwdriver a screwdriver with the tip shaped like a plus sign (+).

destornillador Phillips un destornillador con punta en forma de cruz (+).

photoconductive cell light-sensitive device.

célula fotoconductor dispositivo sensible a la luz.

photodiode light-sensitive device that controls current flow.

fotodiodo dispositivo sensible a la luz que controla el paso de la corriente.

phototransistor produces higher output than a photodiode.

fototransistor produce más energía que un fotodiodo.

photovoltaic cell (solar cell) converts light energy into electrical energy.

célula fotovoltaica (célula solar) transforma energía solar en energía eléctrica.

pictorial drawing a drawing that shows three views in one.

representación esquemática un dibujo que presenta tres vistas.

pierce to make a hole by pushing through.

agujerear hacer agujeros por perforar.

pigment a solid coloring substance suspended in a liquid or vehicle.

pigmento una sustancia colorante sólida que es suspendida en un líquido o vehículo.

pilot hole (anchor hole) a small hole drilled in material to guide the center point of larger drills; hole drilled to receive the threaded part of a screw.

guía (agujero de áncora) un agujero pequeño taladrado en el material para guiar el punto central de taladradoras más grandes; agujero taladrado para recibir la parte roscada de un tornillo.

pilot light glows when a circuit is energized.

lámpara indicadora está encendida cuando se activa un circuito.

pipe rigid tubelike material.

tubo material cilíndrico rígido.

pipe dope See *pipe joint compound.*

pegamento para tubería V. *compuesto para tuberías.*

pipe fitting See *fitting.*

ajuste de tubería V. *ajuste.*

pipe joint compound (pipe dope) substance applied to threaded fittings to prevent leakage at pipe connections.

compuesto para tuberías sustancia aplicada a los accesorios roscados para impedir el derrame de substancias por conexiones de tubería.

pipe stand work area for cutting and threading pipe.

soporte de tubería área de trabajo usado para cortar y roscar tubería.

piston a sliding cylinder fitting within a cylindrical vessel that receives the force of combusting fuel.

émbolo un ajuste cilíndrico móvil dentro de un recipiente cilíndrico que recibe la fuerza de carburante ardiendo.

piston ring completes the seal between a piston and cylinder wall.

aro del émbolo termina la junta entre el émbolo y la superficie del cilindro.

piston ring compressor used to force piston rings into their grooves.

compresor de aro de émbolo empleado para meter a la fuerza los aros del émbolo en las gargantas.

pivot to turn or swing on.

hacer giras dar vueltas o girar sobre algo.

placement/internship SAE a student experience where the student works with a business, farm, or other place in order to learn skills.

colocación/práctica SAE un experiencia estudiantil donde el estudiante trabaja con una empresa, granja u otro lugar para aprender habilidades.

plan reading reading or interpreting scale drawings.

leer los planes leer o interpretar los dibujos a escala.

planer a machine with turning knives that dress the sides of boards to a uniform thickness.

cepillo una máquina con cuchillas de tornear que cepillan los lados de los tableros a un espesor uniforme.

plasma a mass of charged particles that conduct electronic across a gap.

plasma un grupo de partículas cargadas que conducen electrones a través del vacío.

plasma arc cutting a procedure involving the use of electric arc and inert gas to cut nonferrous metals.

arco cortando de plasma procedimiento en que se usa un arco eléctrico y gas inerte para cortar metales no ferrosos.

plastic any of a group of synthetic materials made from chemicals and molded into objects.

plástica término que describe un grupo de materias sintéticas hechas de sustancias químicas y moldeado en objetos.

plastic wood a soft wood filler that dries and hardens very quickly.

madera plástica un relleno blando para maderos que se seca y se endurezca muy rápidamente.

Plastigage carefully designed material that flattens out uniformly when pressed.

Plastigage material diseñado cuidadosamente que se aplasta uniformemente al presionarlo.

plate 2″ × 4″ on the top or bottom of a stud wall.

viga horizontal 2″ × 4″ encima de o debajo del montante.

plates the conducting parts of a capacitor.

placas las piezas conductivas del capacitor.

play the flame to alternately move a flame into and out of an area to carefully control temperatures.

jugar la llama trasladar una llama alternativamente aquí dentro allá fuera del área para controlar las temperaturas cuidadosamente.

pliable able to move without breaking or separating if pushed or pulled.

flexible capaz de moverse sin partirse o separarse al ser empujado o arrastrado.

plow bolt a bolt with a square tapered head that is flush with the surface when installed.

perno de arado un perno con cabeza cuadrada ahusada que se pone embutido cuando instalado.

plug tap a device used to cut pipe threads in a hole.

macho de aterrajar un dispositivo que se emplea para aterrajar las roscas de tubo en un agujero o una apertura.

plumb vertical to the axis of the Earth or in line with the pull of gravity.

aplomo vertical al eje de la tierra o alineado con la fuerza de gravedad.

plumb bob a round, tapered piece of metal attached to a plumb line.

pesa de plomo un trozo de metal redondo, ahusado, atado a una plomada.

plumb line a string with a plumb bob attached.

plomada una pesa de plomo colgada de un hilo.

plumbing installing and repairing water pipes and fixtures.

fontanería la instalación y reparación de cañerías y conductos domésticos de agua.

plywood sheets of lumber made from veneer.

madera contrachapada tablas de maderos hechas de capas de madera terciada.

pneumatics use of air to transfer force.

neumática uso de aire para transmitir potencia.

polarity direction of electrical flow in a circuit.

polaridad dirección o sentido del flujo eléctrico de un circuito.

pole building building supported by erect poles in the soil.

pole building edificio apoyado de postes derechos en el suelo.

poles opposite ends of a magnet.

polos terminales opuestos del imán.

policy a plan of action or a way of management.

política/sistema un plan de acción o un modo de gerencia.

pollution detector gas analyzer that senses and warns when air is unacceptable.

detector de contaminación del aire analizador de gas que detecta y avisa cuando está inaceptable la contaminación del aire.

polyethylene (PE) plastic pipe used extensively for cold-water lines.

polietileno tubo plástico que se usa extensivamente para cañería de agua fría.

polyurethane a clear, durable, water-resistant finish.

poliuretano un acabado claro, duradero e impermeable al agua.

polyvinyl chloride (PVC) a relatively new type of plastic pipe suitable for interior plumbing.

cloruro de polivinilo un tipo relativamente recién de tubo plástico que es adecuado para instalación sanitaria dentro de la casa.

pop rivet a tubular rivet that is enlarged by pulling a ball-ended stem in it until the stem breaks, leaving the rivet secured.

roblón de salto un roblón tubular que se amplia por tirar del vástago con bola en su extremo, hasta romperse, dejando seguro al roblón.

poppet valve a valve that controls the flow of air and gases by moving up and down.

válvula de movimiento vertical una válvula que controla el movimiento de aire y gases por moverse verticalmente en sentido alternativo.

port a special hole in the cylinder wall of a two-cycle engine to permit gases to flow in or out of the cylinder.

portillo una abertura especial en la pared de cilindro de un motor de ciclo de dos tiempos que permite que los gases entren a y salgan del cilindro.

portable circular saw (power handsaw) a lightweight, motor-driven, round-bladed saw.

sierra circular portátil una sierra ligera, propulsada por motor y de hoja circular.

portfolio a collection of papers and other material that represents a body of work over a period of time.

portafolio una colección de documentos y otro material que representa un cuerpo de trabajo durante un periodo de tiempo.

portland cement dry powder made by burning limestone and clay followed by grinding and mixing.

polvo pórtland polvo seco obtenido por calcinación de caliza y arcilla, y en seguida pulverizándolo y mezclándolo.

positive displacement a design arrangement that ensures that a consistent volume of fluid is pumped during each revolution.

desplazamiento positivo un arreglo de un sistema que asegura que se bombee un volumen consistente de líquido durante cada revolución.

positive polarity See *reverse polarity*.

polaridad positiva V. *polaridad inversa*.

positive wires (hot wires) conductors with black, red, or blue insulation that carry current to an appliance.

alambres de alta tensión conductores con aislamiento negro, rojo o azul, que elevan la corriente a un aparato.

post-and-girder building building resting on posts with a heavy timber frame.

edificio de poste y viga edificio que descansa sobre postes con una armazón pesada de maderos.

post-frame another name for post-and-girder building.

post-frame otro nombre por edificio de poste y viga.

potentiometer (pot) variable resistor used to control voltage.

potenciómetro resistencia variable empleada para controlar la tensión.

pour-point depressants additive that allows oil to flow in cold weather.

poner efecto depresivo aditivo que permite al aceite del motor fluir en clima frío.

power handsaw See *portable circular saw*.

sierra mecánica portátil V. *sierra circular portátil*.

power machine a tool driven by electric motor, hydraulics, air, gas engine, or some force other than or in addition to human power.

motor mecánico una herramienta accionada por un motor eléctrico, por hidráulica, por motor de gas o por alguna fuerza distinta o en adición a la fuerza humana.

power stroke the engine process in which burning fuel expands rapidly but evenly to drive the piston down.

carrera motriz el proceso motriz en que el combustible ardiente se expansiona rápidamente pero uniformemente para apretar el émbolo.

power supply provides electricity to circuits.

suministro de electricidad suministra la corriente eléctrica a los circuitos.

power takeoff (PTO) a mechanical component of a tractor that provides rotary power to an employment (such as a mower).

toma de fuerza (TF) un componente mecánico de un tractor que le provee potencia rotatoria a un aparato (como un segadora).

power tool a tool operated by some source other than human power.

herramienta eléctrica una herramienta que es accionada por otra fuente que no es por la fuerza humana.

precipitation rain or snow.

precipitación lluvia o nieve.

precleaner a device that removes large particles from air entering an air cleaner.

filtro un dispositivo que, al entrar una limpiadora de aire, quita las partículas grandes al aire.

pressure force acting upon an area.

presión la fuerza ejerciéndose sobre un área.

pressure feed cup system a spray painting system that delivers paint under pressure to the mixing nozzle of the spray gun.

sistema de alimentación bajo presión un sistema de pistola atomizadora que entrega la pintura bajo presión a la boquilla de mezclado de la pistola.

pressure feed tank system a spray painting system that delivers paint under pressure to the mixing nozzle of the spray gun.

sistema de alimentación bajo presión un sistema de pistola atomizadora que entrega la pintura bajo presión a la boquilla de mezclado de la pistola.

pressure regulator maintains constant line pressure.

regulador de presión mantiene constante la presión de línea.

pressure switch responds to changes of liquid, gas, or mechanical pressure.

interruptor automático responde a cambios de presión líquida, gaseosa o mecánica.

pressure-treated chemicals injected under pressure.

manipulado bajo presión sustancias químicas inyectadas bajo presión.

primary circuit low-voltage circuit of an ignition system.

circuito primario circuito de baja tensión de un sistema de encendido.

primer a special paint used to seal bare wood and prepare metal surfaces for high-quality top coats.

esmalte primario una pintura especial usada para chapar madera bruta y preparar las superficies de metal para top coats de buena calidad.

procedure a method of doing things or a particular course of action.

procedimiento una sucesión de medidas para hacer algo; una línea particular de acción.

production agriculture the phase of agricultural industry that involves the actual production of plants and animals.

agricultura de producción es la fase de la industria agrícola que supone la producción real de la comida.

proficiency awards awards that are given to students based on superior Supervised Agricultural Experiences.

premios de competencia los premios que se conceden a los estudiantes con experiencias agrícolas supervisadas superiores.

profit income made from the sale of goods or services.

beneficios ingresos ganados de la venta de productos o servicios.

project special activity planned and conducted with the purpose of learning.

proyecto actividad especial planeado y conducido con el objetivo de aprender.

propellant the gas that propels paint toward the object being painted.

propulsor el gas que propela la pintura hacia el objeto que se ha de pintar.

protractor an instrument for drawing or measuring angles.

transportador un instrumento empleado para trazar o medir triángulos.

PTO See *power takeoff.*

TF V. *toma de fuerza.*

puddle a small pool of liquid metal.

mezcla goteo de metal líquido.

puddling to control the puddle.

proceso de reducción de oxidación para controlar el material en bruto.

pulley device attached to a shaft to carry a belt.

polea dispositivo atado a un eje para llevar una correa.

pullout distance between the stays on a woven fence.

sacador distancia entre los soportes de un alambrado de tejido.

purge the lines remove undesirable gases.

purgar las líneas quitar los gases no deseosos.

purlin longitudinal roof nailers.

correa claves longitudinales del techo.

purloins greenhouse construction member maintains distance between bows.

correa superior pieza de construcción del invernadero que mantiene la distancia entre los arcos.

purple the safety color used to designate radiation hazards.

morado el color de seguridad que se usa para señalar peligros de radiación.

push stick a wooden device with a notch in the end to push or guide stock on the table of a power tool.

varilla de guiar un instrumento de madera con una muesca en el extremo para empujar o guiar la materia prima que está en la mesa de una herramienta-máquina.

pushing the puddle creating a puddle of molten metal with a torch.

empujando el charco la creación de un charco de métal liquido con una lámpara de soldar.

putty soft material containing oils that keep it pliable over a long period of time, used to seal joints or fill holes.

masilla materia consistente en aceites que la mantienen flexible durante mucho tiempo; se usa para tapar junturas o tapar agujeros.

PVC See *polyvinyl chloride.*

CPV V. *cloruro de polivinilo.*

Q

quick coupling hose ends that snap together.

empalme automático bocas de manga que cierran de golpe.

R

rabbet a cut or groove at the end of a board made to receive another board and form a joint.

ranura un corte o hendidura en el borde del tablero hecho para que se encaje con otro tablero, formando una ensambladura.

rabbet joint a joint formed by setting the end or edge of one board into a groove in the end or edge of another board.

ensambladura de ranura una ensambladura formada por encajar el borde de un tablero en la ranura o borde de otro tablero.

raceways rectangular or oval fish tanks.

raceways peceras elípticas o rectangulares.

radial-arm saw a power circular saw that rolls along a horizontal arm.

sierra de brazo de movimiento radial una sierra circular mecánica que se roda a lo largo de un brazo horizontal.

rafter single timber supporting a roof section.

par una viga única que apoya una sección del techo.

rails lengths of wood split from logs used to make a wooden fence.

barras trozos de madera que se parte de troncos que se usan para hacer una alambrada de madera.

rasp a file with very coarse teeth.

escofina lima con dientes muy gruesos.

ratchet die stock a tool used for cutting threads on a rod or pipe.

trinquete una herramienta empleada para cortar agujas en una barra o tubo.

ratio proportion of one component to another by weight or volume.

proporción razón del peso o volumen de un componente a otro.

receptacle a device for receiving an electrical plug.

enchufe hembra un dispositivo que recibe un enchufe (macho).

reciprocal back and forth or opposite.

recíprico de acá para allá o al contrario.

reciprocate return; move back and forth.

reciprocar regresar; tener movimiento alterno.

reciprocating saw a saw with a stiff blade that moves back and forth.

sierra de movimiento alternativo sierra con una banda rígida que tiene movimiento alterno.

recondition to do what is needed to put back into good condition.

arreglar hacer lo que sea necesario para que se ponga de nuevo.

recordkeeping the process of keeping a journal or portfolio of everything you have done.

mantenimiento de datos el proceso de escribir en un diario o armar una carpeta de todo lo que se ha hecho.

rectifier converts AC to DC current.

rectificador convierte la corriente CA en la corriente CC.

red the safety color used to designate areas or items of danger or emergency.

rojo el color de seguridad usado para señalar áreas o artículos de peligro o de emergencia.

reducer a fitting used to connect pipes of different sizes.

reductor un ajuste empleado para conectar tubos de tamaños distintos.

reed valve a flat, flexible plate that permits air or liquid to pass in one direction but seals when the flow reverses.

válvula de lengüeta una placa llana flexible que permite que el aire o líquido pase en un sentido pero que cierre cuando pase en otro sentido.

regulator a device that keeps pressure at a set level or controls the rate of flow of a gas or liquid.

regulador un dispositivo que mantiene la presión a un nivel determinado o controla la velocidad del flujo de un gas o líquido.

reinforced concrete concrete slabs or structures that are strengthened with embedded steel rods or wire mesh.

hormigón armado bloques o armazones de hormigón que son armados entre su masa con alambres y barras de acero.

renewable natural resources resources provided by nature that can be replaced or renewed.

recursos naturales renovables recursos proporcionados por la naturaleza que pueden volver a llenarse o renoverse sí mismos.

repair to replace a faulty part or make it work correctly.

reparar reemplazar una pieza defectuosa o hacer que esta funcione correctamente.

represent to stand for or to be a sign or symbol of.

representar hacer presente algo o servir como señal de algo.

repulsion the action of two magnets that push away from each other when like poles are placed near each other.

repulsión la acción de dos imanes que se apartan mutuamente cuando los polos semejantes están colocados cerca el uno del otro.

research SAE a student experience where the student conducts and reports on a research experiment or topic.

investigación SAE una experiencia estudiantil en la cual el estudiante realiza y hace un informe sobre una investigación o un tema.

resistance any tendency of a material to prevent electrical flow.

resistencia cualquier tendencia de un material de impedir el flujo eléctrico.

resistivity the ability to prevent flow of electricity.

resistividad la capacidad de impedir el flujo de electricidad.

resistor device that offers specific resistance to a current.

reóstato dispositivo que ofrece resistencia específica a una corriente eléctrica.

reverse (positive) polarity (RP) DC current flowing in the opposite direction from straight polarity; to reverse the direction of current.

polaridad inversa (positiva) corriente continua que fluye en sentido contrario a la polaridad negativa.

invertir la polaridad invertir el sentido de la corriente.

reversible capable of running backward as well as forward.

reversible dícese de un aparato que es capaz de funcionar en los dos sentidos contrarios.

revolution one complete turn of 360°.

revolución una rotación, o un giro completo de 360°.

rheostat variable resistor used to control current.

reóstato resistencia variable empleado para controlar la corriente.

ribbon wire a small bare electric wire attached to a "ribbon"; it is used as a conductor in an electric fence.

alambre de cinta un alambre eléctrico pequeño sin recubrimiento unido a una "cinta"; se usa como conductor en una cerca eléctrica.

richer a mixture with an increased proportion of fuel to air.

mezcla rica una mezcla que tiene más combustible en proporción con aire.

ridge highest point of a roof.

caballete el lomo de un techo.

rig a piece of apparatus assembled to conduct an operation.

aparejo conjunto de cosas necesarias para conducir una operación.

right triangle a three-sided figure with one 90-degree angle.

rectángulo figura delimitada por tres líneas rectas que tiene un ángulo recto.

rigid-frame building with a steel frame.

de armazón rígida edificio con armazón de acero.

ring expander tool used to remove and install piston rings.

llave de aro herramienta empleada para instalar y quitar los aros de émbolo.

rip to cut the long way on a board or with the grain.

serrar al hilo cortar según la dirección de los hilos, o del grano.

rip fence a guide that helps keep work in a straight line with a saw blade.

guía de aserrar al hilo una guía que ayuda mantener los materiales alineados con la hoja de sierra.

ripsaw a saw with teeth filed to a knifelike edge and used to cut with the grain.

sierra de hender una sierra cuyos dientes son afilados a corte como cuchillo y que se emplea para cortar según la dirección de los hilos.

riser vertical part of a step.

contrahuella parte vertical del peldaño.

rivet to spread or shape by hammering; a fastening device held in place by spreading one or both ends.

remachar triturar o cambiar la forma por martillazos.

roblón un sujetador que se sujete por machacar o partir la punta.

robot mechanical device capable of humanlike movements.

robot aparato mecánico capaz de movimientos así como los de los seres humanos.

robotics study and application of technology to robots.

robótica estudio y aplicación de tecnología a los robot.

rod a round cylindrical piece of metal.

barra una pieza de métal redondo y cilíndrico.

roller cover a hollow, fabric-covered cylinder that slides onto the handle of a paint roller.

cilindro de rodillo un cilindro hueco y cubierto con tela que se introduce en el mango del rodillo para pintar.

root the deepest point in a weld.

raíz el punto más hondo de una soldadura.

root pass the first welding pass that is made in a joint.

pasada de raíz la primera pasada que se hace en una soldadura.

rope starter rope wrapped around a pulley for turning power to start an engine.

arranque de soga soga que corre por una polea para potencia de torno para poner en marcha un motor.

rot to decay or break down into other substances.

pudrir descomponer.

rotary air pumps high-volume, low-pressure air pumps.

bombas aspirantes giratorias bombas de aire de gran cantidad y presión baja.

rotation circular motion performed by a robot or other machine.

rotación movimiento circular que llevan a cabo un robot u otra máquina.

rough lumber lumber as it comes from the sawmill.

madera en bruto maderos así como llegan del aserradero.

rounded up or down going to the next higher or lower whole number.

redondear convertir la cantidad en el número completo que la precede o sigue.

round-head screw a screw that extends in an even curve above the surface of the material being held.

tornillo de cabeza redonda un tornillo que se extiende en una curva uniforme encima del material sujetado.

router a portable power tool used to shape the edge, cut a groove, or mill a design in a piece of wood.

fresadora una máquina-herramienta portátil empleada para dar forma al borde, cortar una ranura, aserrar un dibujo en un pedazo de madera.

rpm revolutions per minute.

rpm revoluciones por minuto.

run a sag in a coat of paint or other finish caused by applying too much paint or finish at one time.

curso hundimiento en una mano de pintura u otro acabado causado por aplicar demasiada pintura.

rust reddish brown or orange coating that results when iron reacts with air and moisture.

oxído capa castaña y anaranjada que resulta cuando el hierro reacciona con el aire y la humedad.

rust inhibitor a material or additive that reduces or prevents rusting.

antioxidante un material o aditivo que reduce o previene la oxidación.

S

saber saw (bayonet saw) a reciprocal saw used primarily for cutting curves or holes in wood, metal, cardboard, and similar materials.

sierra de sable una sierra de movimiento alternativo que se emplea mayormente para cortar curvas o agujeros en madera, metal, cartón y materiales semejantes.

saddle soap a mild soap used to clean, soften, and preserve leather.

jabón de aceite producto que se usa para limpiar, ablandar y curar el cuero.

SAE See *Society of Automotive Engineers.*

SIA V. *Sociedad de Ingenieros Automotrices.*

SAE rating a number assigned to an oil to represent its viscosity.

clasificación de SAE (siglas en inglés para *Society of Automotive Engineers*) un número asignado a un aceite para representar su viscosidad.

safe free from harm or danger.

seguro libre de daño o peligro.

safety freedom from accidents.

seguridad libre de accidentes.

safety color colors used as part of a standardized coding system according to which each color conveys a specific safety message. See also *blue, green, orange, red, white, white and black stripes,* and *yellow.*

colores de seguridad colores que se usan como parte de un sistema de codificación estandarizado en el cual cada color transmite un mensaje específico de seguridad. Ver también *azul, verde, naranja, rojo, blanco, amarillo y rayas blancas y negras.*

sag shifting of a large area of paint downward.

caída movedizo de una gran área de pintura hacia abajo.

sal ammoniac a cube or block of special flux for cleaning soldering coppers.

sal amoníaco un cubo o bloque de fundente especial para limpiar los soldadores.

sand small particles of stone.

arena partículas pequeñas de piedra.

sandblasting cleaning by sand particles thrown by compressed air.

limpiar con chorro de arena limpiar con partículas por el aire comprimido.

sash block a concrete block used to frame a window.

bloque de sash un bloque de hormigón que se usa para hacer el armazón de una ventana.

satin a wood or metal finish with low sheen or shine.

raso un acabado para madera o metal con poco lustre.

saturate to add a substance until the excess starts to run out.

empapar añadir una sustancia hasta que lo exceso empiece a rebosar.

sawhorse (trestle) a wood or metal bar with legs, used for temporary support of materials.

burro (caballete) madero o metal horizontal con patas empleado para el soporte temporario de materiales.

scale an instrument with all increments shortened according to proportion; numbers and gradations on measuring tools; a rigid steel or wooden measuring device; the size of a plan compared with that of the object it represents.

escala instrumento cuyos incrementos son graduados según de la proporción; números y gradaciones en las herramientas de medida; un dispositivo rígido de acero o madera; la representación de un plan relacionado a la dimensión del objeto que representa.

scale drawing a drawing that represents an object in exact proportion although the object is larger or smaller than the drawing itself.

dibujo a escala un trazado que representa un objeto en proporción exacta, aunque sea el objeto más grande o más pequeño que el trazado si mismo.

school-based enterprise SAE a Supervised Agricultural Experience that is conducted on the school campus.

educación escolar iniciativa SAE una experiencia agrícola supervisada que se lleva a cabo en la escuela.

scientific process a step-by-step process that a researcher uses to ensure that the research and outcome is correctly completed.

proceso científico un proceso desarrollado paso a paso que un investigador utiliza para garantizar que la investigación y sus resultados se han completado correctamente.

scoop shovel large shovel with a flat edge used to scoop up trash or other materials.

pala recogedora pala grande con borde plano que se usa para recoger basura u otros materiales.

scope size and complexity.

amplitud tamaño y complejidad.

score to scratch.

rayar raspar (a metal, madera, etc.)

scratch awl a sharply pointed tool with a wooden or plastic handle used to mark metal.

punzón marcador un instrumento con una punta muy aguda y un mango de madera o plástica, que se emplea para raspar metal.

screeding striking off excess concrete to create a smooth and level surface.

descantillar usar un escantillón para enrasar el hormigón exceso para crear una superficie lisa y plana.

screw a fastener with threads that bite into material as it is turned.

tornillo un sujetador con filetes que se introducen en el material cuando da vueltas.

screw hook and strap hinge a large hinge composed of a hook and a strap, used on gates.

gancho de tornillo y bisagra de correa una bisagra grande que se compone de un gancho y una tira que se usa en verjas.

screw plate See *tap and die set.*

dispositivo de roscado por fresa V. *dispositivo de roscado por fresa.*

screwdriver a turning tool with a straight tip, Phillips tip, or special tip.

destornillador una herramienta que da vueltas y que tiene punta llana, de Phillips o especial.

scriber a very small, metal, sharp-tipped marker.

punta de trazar un marcador muy pequeña, de metal y puntiaguda.

scroll saw See *jigsaw.*

sierra de contornear V. *sierra de vaivén.*

seal to apply a coating that fills or blocks the pores so that no material can pass through the surface.

impermeabilizar aplicar una capa que tapa los agrietos o poros para que un material no pueda penetrar la superficie.

sealer a coating used to fill pores and prevent material from passing through a surface.

producto de selladura una capa usada para tapar los poros y impedir que penetren materiales por la superficie.

seasonal demand amount of acre-inches of water for growing a crop.

demanda de temporada cantidad de acre-pulgadas de agua que se exige la cultivación de un cultivo determinado.

seat a stationary or nonmoving part that is designed to seal when a moving part is pressed against it.

asiento una pieza fija que es hecha a obturar cuando se la aprieta una pieza en movimiento.

secondary circuit high-voltage circuit of an ignition system.

circuito secundario circuito de alta tensión del sistema de encendido.

semiconductor a material with characteristics that fall between those of insulators and conductors.

semiconductor un material cuyas características caen entre los de los aisladores y de los conductores.

semigloss paint with a slight shine or gloss.

pintura semi-brillante pintura con un acabado de poco lustre o brilla.

sensor device that receives and responds to a signal.

sensor dispositivo que recibe y responde a una señal.

series circuit circuit that provides a single path for current flow.

circuito en serie da un sendero solo para el flujo de corriente eléctrica.

series-parallel circuit circuit that combines properties of series and parallel circuits.

circuito serie-paralelo combina propiedades de los circuitos en serie y los circuitos paralelos.

serrated notched.

dentado entallado.

service classification a rating of an engine oil's ability to withstand a certain level of engine performance.

clasificación de servicio una estimación de la habilidad del aceite del motor para resistir un cierto nivel de funcionamiento de motor.

service drop an assembly of electrical wires, connectors, and fasteners used to transmit electricity from a transformer to a service entrance panel.

conductor de bajada del poste un conjunto de alambres, conectadores y abrazaderas eléctricos usados para transmitir la electricidad de un transformador a una caja de línea de acometida de energía.

service entrance panel a box with fuses or circuit breakers where electricity enters a building.

caja de línea de acometida de energía una caja con fusibles o cortacircuitos donde la electricidad entra el edificio.

service learning SAE a Supervised Agricultural Experience where the student does a significant amount of community service.

servicio de aprendizaje SAE una experiencia agrícola supervisada donde el estudiante realiza una cantidad significativa de servicio a la comunidad.

set to drive the head of a nail below the surface.

embutir hincar la cabeza del clavo bajo la superficie.

settling tank low-pressure tank with slow water velocity that allows solid waste particles to settle to the bottom.

depósito de sedimentación depósito con velocidad lenta de agua que deja asentarse al fondo a las partículas sólidas residuales.

shadecloth woven plastic fabric that reflects 30 to 70 percent of the sunlight hitting it.

tela de sombras tejido plástico que refleja 30 a 70 por ciento de la luz del sol que lo golpea.

shank the nonthreaded part of a nail or screw.

pie la parte no roscada del tornillo o clave.

shank hole the hole provided for the shank of a screw.

agujero del mango el agujero a que se introduce el pie del tornillo.

shear to cut by action of opposed cutting edges.

cizallar cortar por medio de acción contraria de dos filos cortantes.

shears large scissorlike tools for cutting sheet metal and fabrics.

cizallas instrumentos grandes parecidos a tijeras, que se emplean para cortar chapas de metal y tejidos.

sheathing first exterior layer of wall or roof.

revestimiento primera capa exterior de pared o techo.

sheet bend knot for connecting ropes of different sizes.

nudo de escota nudo para conectar sogas de tamaños diferentes.

sheet-metal screw a screw that has threads wide enough to permit thin metal to fit between the ridges of the threads.

tornillo de chapa un tornillo cuyos filetes son bastante anchos para que chapa delgada de metal pueda encajar entre las partes salientes de los filetes.

shielded-metal arc welding (SMAW) (arc welding, stick welding) welding with electrical power as a source of heat and rods covered with flux which form a gaseous shield around the molten metal until it solidifies.

soldadura manual por arco metálico soldar con energía eléctrica como la fuente de caldeo y con electrodos cubiertos con fundente que forman campo gaseoso alrededor del metal fundido hasta que se frague.

shock the body's reaction to an electric current.

sacudida la reacción del cuerpo a una corriente eléctrica.

short circuit a condition that occurs when electricity flows back to its source too rapidly and blows fuses, burns wires, and drains batteries.

cortocircuito fenómeno que ocurre cuando la corriente eléctrica pasa al retroceso a su fuente rápidamente y funde los fusibles y los cables, y drena las baterías o pilas.

show box a box designed to hold equipment used in showing livestock.

caja para exhibición una caja designada para mantener equipo que se usa para exhibir ganado.

shroud to cover; a cover.

cubrir envolver.

cubierta lienzo para envolver.

siemen (S) unit for measuring conductance.

siemen (S) unidad para medir la conductancia.

silhouettes colored outlines of tools that make it easy to check for missing tools.

siluetas contornos coloreados de herramientas que facilitan ver qué herramientas faltan.

silicon a low-quality pigment. Also the material most commonly used for semiconductors.

silicio un pigmento de mala calidad. También el material lo más usado en los semiconductores.

silt a substance composed of intermediate-size soil particles.

légamo una sustancia compuesta de partículas de tamaño intermedio de tierra.

sine wave wave created by the flow of current through one cycle.

onda sinusoidal formada por el paso de corriente por un ciclo.

single-cut file a file with teeth going only in one direction.

lima de una corta una lima con dientes que va solo en una dirección.

single-line system single or independent line.

vía sencilla línea individua o independiente.

single-pole switch the one switch that controls one or more lights and/or outlets.

interruptor unipolar el único interruptor que controla una o más luces y/o tomas.

siphon feed cup system a spray painting system that creates a vacuum at the nozzle of the gun and utilizes atmospheric pressure to push the paint to the nozzle.

aparato de sifón un sistema de pistola atomizadora que crea un vacío en la boquilla de la pistola y utiliza la presión atmosférica para avanzar la pintura a la boquilla de mezclado.

60-cycle current electricity that alternates direction of flow 60 times per second.

corriente de ciclo 60 corriente eléctrica que cambia el sentido 60 veces por segundo.

sketch a rough drawing of an idea, object, or procedure.

croquis un trazado de una idea, objeto o procedimiento.

slag the material formed when burning steel combines with oxygen.

escoria el producto que se forma cuando acero ardiente combina con oxígeno.

slag box a container of water or sand placed to catch hot slag and metal from the cutting process.

depósito de escorias un recipiente de agua o arena colocado para recibir escoria y metal calientes del proceso de soldar.

sliding T bevel a device used to draw angles on boards or metal.

escuadra corrediza un instrumento usado para trazar ángulos sobre maderos o metal.

slotted-head screw a screw with one straight slot across the head.

tornillo de cabeza ranurada un tornillo con una sola ranura derecha a través de la cabeza.

slow-moving vehicle (SMV) a vehicle that moves no faster than 25 miles per hour. Such vehicles must have an SMV emblem—a reflective emblem consisting of an orange triangle with a red strip on each of the three sides—attached to its back end.

vehículo que se mueve lentamente (*SMV* por sus siglas en inglés) un vehículo que se mueve a no más de 25 millas por hora. Tal vehículo debe tener un emblema de vehículo de movimiento lento (SMV), un emblema reflector que consista en un triángulo anaranjado con una raya roja en cada uno de los tres lados—debe ir pegado a la parte de atrás del vehículo.

snipe a defect on planed wood caused by the feed tables being out of adjustment.

cavadura o desnivelación un defecto en madera cepillada que resulta cuando hay desajuste en las tablas al ser impulsadas [por el cepillo mecánico].

snips (shears) a large scissorlike tool for cutting sheet metal and fabrics.

cizallas gran instrumento parecido a tijeras, que se emplea para cortar chapa de metal y tejidos.

soapstone a soft, gray rock that shows up well when marked on most metals.

esteatita una piedra blanda y, gris que se muestra bien al marcarse sobre la mayoría de los metales.

Society of Automotive Engineers (SAE) an organization of engineers that sets standards for automotive components such as engine oil.

Sociedad de Ingenieros Automoviles una organización de ingenieros que establecen niveles por componentes automotores como aceite de motor.

softwood lumber from a conifer.

madera de coníferas madera de un arból conífera.

solar cell See *photovoltaic cell.* Converts light energy into electrical energy.

célula solar (*célula fotovoltáica*) transforma energía solar en energía eléctrica.

solder mixture of tin and lead.

soldadura mezcla de estaño y plomo.

soldering bonding with metals and alloys that melt at temperatures below 840 °F.

soldadura unir con metales y aleaciones que se funden a temperaturas menos de 840 °F.

soldering copper a tool consisting of a handle, steel shank, and copper tip used to heat metal for soldering.

soldador un instrumento consistente en mango, pierna de acero y punta de cobre usado para caldear el metal a que se ha de soldar.

solderless connector See *wire nut.*

conectador sin soldadura V. conectador de alambres.

solenoid valve a valve actuated by an electromagnet.

válvula de soleinoide válvula accionada por un electro-imán.

solid sphere work area a work area in the shape of a solid ball, with no hollow area at the center.

área de trabajo en forma de esfera sólida zona de trabajo en forma de una bola sólida, que no tiene un espacio hueco al centro de la bola.

solid state ignition a computerized engine ignition system that replaced the old point ignition.

encendido de estado sólido sistema computerizado de motor que reemplazó el punto de ignición anterior.

solidify to harden or change from a liquid into a solid.

solidificarse endurecerse o pasar de líquido a sólido.

south pole end of magnet opposite the north pole.

polo negativo terminal opuesto del polo positivo del imán.

species plants or animals with the same permanent characteristics.

especies conjunto de plantas o animales que tienen las mismas características permanentes.

speed indicator a device used to measure revolutions per minute (RPM) of a turning shaft or part.

contador de velocidad un dispositivo usado para medir las revoluciones por minutos (rpm) de un eje o pieza girándose.

spirit level a tool containing alcohol in a sealed, curved tube with a small air space or bubble, used to determine whether an object has the same height at two or more points.

nivel de burbuja, un instrumento que contiene alcohol en un tubo tapado y con una burbuja de aire, empleado para determinar si un objeto tenga la misma altura a dos o más puntos.

splitting the receptacle two receptacles in the duplex may be wired to two different circuits if the metal strap between the two screws is removed.

división del recipiente dos tomacorrientes en el dúplex pueden cablearse a dos circuitos diferentes si se quita la lámina metálica (de contacto) entre los dos tornillos.

spool valve a valve with one inlet and multiple outlets.

válvula de bobina una válvula con un orificio de entrada y múltiples orificios de salida.

spot marking the color-coded marking on the bare surface of the exposed metal rod of a welding electrode.

sistema de marcaje la marca en colores de código sobre la superficie sin revestir de la varilla de metal expuesta de un electrodo para soldar.

spray gun a device that uses compressed air to spray paint or other finish onto a surface.

pistola de pintura automizada dispositivo que usa aire comprimido para pintar con pistola u otro acabado a una superficie.

sprinkler head See *nozzle.*

cabezal rociador V. boquilla.

sprinkler irrigation dispersion of water in droplets.

riego por aspersión dispersión del agua en gotitas.

sprocket a wheel with spiked teeth.

catalina una rueda con dientes estacas.

square a device used to draw angles for cutting and to check the cuts for accuracy. Also the top of a wall.

escuadra instrumento que se emplea en trazar ángulos para cortar y para revisar la precisión de los cortes.

square head a head that has four equal sides.

cabeza cuadrada cabeza cuadrilátera de lados.

square knot safe knot for connecting two ropes of equal size.

nudo de envergue nudo seguro para conectar dos sogas de tamaños iguales.

stair horse See *stair stringer.*

zanca V. estructura que apoya una escalera.

stair stringer structure that supports stairs.

zanca estructura que apoya una escalera.

standard pipe steel pipe sold or used unless extra-heavy or double-extra-heavy is specified.

tubo estándar tubo de acero que se vende o se usa a menos que se especifique extragrueso o doble extragrueso.

staple a piece of wire with both ends sharpened and bent into a U or into a bracket shape with two legs of equal length.

grapa una pieza de alambre con ambos extremos afilados y combados para formar dos piernas del mismo longitud.

stationary having a fixed position.

estacionario mantiene una posición fija.

stays vertical wires on a woven wire fence.

soportes alambres verticales en una alambrada tejida.

steam cleaner a portable machine that uses water, a pump, and a burner to produce steam.

limpiadora de vapor una máquina portátil que utiliza el agua, una bomba y un quemador para producir el vapor.

steel screw a screw made of steel with a blued, galvanized, cadmium, nickel, chromium, or brass finish.

tornillo de acero un tornillo hecho de acero con un acabado pavonado, galvanizado, cadmio, de níquel, cromado o de latón.

stem long, round section of a valve.

vástago parte larga y cilíndrica de la válvula.

step pulley a pulley with several sizes.

motón una polea de varios tamaños.

stick welding See *soldadura manual por arco metálico*.

soldadura TIG V. *soldadura por arco protegido*.

stock a piece of material such as wood or metal.

materia prima un trozo de material como madera o metal.

stove bolt a round-head bolt with a straight screwdriver slot, threaded its entire length.

tornillo de estufa un perno de cabeza redonda ranurada, y roscado por su longitud entero.

straight polarity (SP) (negative polarity) DC current flowing in one direction, the opposite of reverse polarity.

polaridad negativa corriente continua pasando en una sola dirección, al contrario a la polaridad inversa.

strands small wires placed with others to form bundles to improve flexibility and conductivity.

filamentos cables pequeños que se colocan con otros para mejorar la flexibilidad y la conductividad.

strap hinge a long narrow hinge that mounts on the front surface of a door.

bisagra de correa una bisagra larga y estrecha que se pone en la parte delantera de una puerta.

stretch indicator a bend in the stays of a woven fence that indicates when the wire is stretched to the proper tension.

indicador de temple curva en los soportes de un alambrado de tejido que indica cuando el alambre está tensionado.

stretcher block a long concrete block used to construct straight wall sections.

bloque de stretcher un bloque largo de hormigón que se usa para construir una pared estrecha.

stringer bead a weld bead produced without weaving.

cordón longitudinal cordón producido sin pasada pendular.

stroke the movement of a piston from top to bottom or from bottom to top.

carrera el movimiento de un émbolo de arriba abajo o de abajo arriba.

subfloor first layer of flooring.

subpiso primera capa de suelo.

subgrade the level where stone begins for a concrete floor or topsoil begins when grading off land.

subgrado nivel donde la piedra comienza por el suelo del hormigón o comienza la capa cuando se nivela la tierra.

subirrigation water supplied below the ground surface.

riego subterráneo agua suministrado subterráneamente.

submersible pumps low-pressure, high-volume immersible pumps.

bombas sumergibles bombas de baja presión, y gran cantidad que se pueden sumergir.

sump hole where unwanted water drains and is removed by a pump.

pozo negro hoyo en que el agua superfluo se drena y de que se recoge por una bomba.

supercomputer enormous capacity computer.

supercomputadora computadora de capacidad enorme.

Supervised Agricultural Experience (SAE) activities of the student outside the agricultural class or laboratory done to develop agricultural skills.

Experiencia Agrícola Supervisada (*SAE*) por sus siglas en inglés) actividades que realiza el estudiante fuera de la clase de agricultura o del laboratorio para desarrollar habilidades en agricultura.

surface irrigation adding water to the soil by gravity.

riego por aspersión suministrar agua al suelo por la gravedad.

sweating process of soldering a piece of copper pipe into a fitting.

suelda por fusión el proceso de soldar un trozo de tubo de cobre a un ajuste.

switch a device used to stop the flow of electricity.

interruptor un dispositivo usado para interrumpir el flujo de una corriente eléctrica.

switching wires (traveler wires) a pair of wires attached to the light-colored screws on three-way or four-way switches.

circuitos selectores par de cables atado a los tornillos de colores claros en los interruptores de tres o cuatro direcciones.

T

table saw (bench saw) a stationary circular saw with either a tilting arbor or a tilting table.

sierra de mesa una sierra circular fija con mesa o árbol inclinado.

tack rag a rag dampened with solvent.

paño antipolvo trapo humedecido con disolvente.

tacking making a small weld to hold metal parts temporarily.

soldadura discontinua el hacer de una soldadura pequeña para unir piezas metales temporalmente.

tacky sticky.

pegadizo pegajoso.

tang tapered end on a file.

cola parte extremo ahusado de una lima.

tap a hardened, brittle, fluted tool used to cut threads into holes in metal.

macho de aterrajar instrumento endurecido, quebradizo y estriado que se usa para labrar la rosca en los agujeros de metal.

tap and die set (screw plate) a set of taps, dies, and handles used for making threads.

juegos de roscas con macho un juego de machos, troqueles y mangos usado para hacer filetes o roscados.

tape flexible measuring device that rolls onto a spool.

cinta métrica un instrumento flexible de medida que bobina a una bobina.

taper tap starts threads and completes them if the tap can extend through the material.

cortador macho de terrajadora empieza a cortar los filetes para los tornillos y los completa si el cortador puede extenderse a través del material.

tap wrench a device used to turn a tap.

giramacho un instrumento empleado para abrir o cerrar un grifo.

technician a person who has acquired the ability to perform skills unique to a certain area of expertise.

técnico una persona que ha conseguido la capacidad para desempeñar las habilidades que son únicas a una cierta área de perecia.

tee a fitting used where a pipe line splits into two directions.

unión en T un ajuste empleado donde la tubería se bifurca.

Teflon tape tape used on threaded fittings to prevent water or gas leakage at pipe connections.

cinta teflón cinta que se le aplica a los ajustes roscados para evitar escape de agua o gas de los conexiones de tubos.

telescoping gauge See *micrometer.*

micrómetro V. *micrómetro.*

temper to heat a piece of tool steel followed by controlled cooling so as to control the degree of hardness.

templar caldear una pieza de acero de herramientas y seguir con enfriamento controlado para controlar el grado de dureza.

temperature rise how warm the motor can get without causing internal damage.

elevación de temperatura la temperatura a que puede alcanzar un motor sin causar daño interno.

tensile strength the amount of tension or pull a weld can withstand.

resistencia a la tracción la tensión o fuerza máxima que una soldadura puede resistir.

tensiometer a type of moisture meter that uses water tension to indicate moisture content.

tensiometro un tipo de metro de humedad que se usa la tensión de agua para indicar el contenido de humedad.

thermocouple a special temperature sensor that can detect heat changes and activate switches.

termopar un sensor especial de temperatura que puede detectar cambios de calor y activar interruptores.

T hinge a hinge that is shaped like the letter T.

T bisagra una bisagra que es en forma de la letra T.

threads grooves of even shape and taper that wrap continuously around a shank or hole.

filetes ranuras de forma y estriamiento uniforme que se enrollan continuamente en un mango o agujero.

three-view drawing a drawing that shows three views or sides of an object in three separate representations.

representación esquemática un dibujo que muestra tres vistas o caras de un objeto en tres representaciones distintas.

three-way switch a switch that permits a light or receptacle to be controlled from two different locations by a pair of switches.

conmutador de tres direcciones un interruptor que permite que una luz o receptáculo sea controlado desde dos sitios distintos por un par de interruptores.

throat plate insert a plate in the table of a table saw where the blade rotates through the table.

inserción de placa de garganta una placa en la sierra circular estacionaria donde la cuchilla gira a través de la mesa.

throughput rate of flow of grain and straw through a combine.

rendimiento la tasa de flujo de granos y hechaduras o granzos por una segadora trilladora.

thyristor a semiconductor used to electronically control switching.

tiristor un semiconductor que se emplea para controlar electránicamente la interrupción de una corriente.

tiger saw See *reciprocating saw.*

sierra de movimiento alternativo V. *sierra de movimiento alternativo.*

tilting arbor a motor, belt, pulley, shaft, and blade assembly that tilts as a unit in a power saw.

árbol inclinable un conjunto de motor, cinta, polea, eje y banda que se inclina como grupo en una sierra de máquina.

tilting table a table that can be set at various angles.

mesa inclinable una mesa que se puede fijar a varias inclinaciones.

timber hitch knot for attaching rope to poles.

vuelta de braza nudo para atar la soga a los postes.

timer device that activates a switch for a preset time.

temporizador electrónico dispositivo que activa un interruptor programado a una hora específica.

tin a metallic element that is one component of solder.

estaño un componente de soldadura.

tinning bonding filler material to a base metal.

estañar uniendo materia de revestimiento a un metal de base.

tip cleaners rods with rough edges designed to remove soot, dirt, or metal residue from the hole in the tip of a torch.

limpiadoras de la boquilla del soplete varillas con puntos ásperos hechos para quitar residuos de metal, mugre u hollín de la cavidad de la boquilla del soplete.

titanium oxide a high-quality pigment used in many paints.

dióxido de titanio un pigmento de buena calidad que se usa en muchas pinturas.

title block the section of a drawing reserved for information about the drawing in general.

viñeta de título la parte de un dibujo reservada para información sobre el dibujo en general.

toe nailing driving a nail at an angle near the end of one piece and into the face of another piece.

clavar en cruz clavar un clavo a un ángulo cerca del borde de una pieza y por la cara de otra pieza.

tolerance acceptable variation.

tolerancia variación aceptable.

tongue-and-groove (T&G) board a board with a tonguelike edge on one side and a groove on the other which permits boards to be locked together.

ranura y lengüeta tablero con un borde de tipo lengüeta de un extremo y una ranura en el otro que permite que se encajen los tableros.

tool any instrument used in doing work.

herramienta cualquier instrumento empleado para trabajar.

tool fitting cleaning, reshaping, repairing, or resharpening a tool.

ajustar una herramienta limpiar, reformar, reparar o reafilar una herramienta.

tool steel steel with a specific carbon content that allows the tool to be annealed and tempered.

acero de herramientas acero con un índice específico de carbono que permite que se rocozca y se temple el acero.

top dead center (TDC) position of a piston when at its highest point (furthest from the crankshaft).

punto muerto superior posición del émbolo cuando está a su punto más alto (más lejano del cigüeñal).

torch an assembly that mixes gases and discharges them to support a controllable flame.

soplete un montaje que mezcla y descarga gases para sostener una llama controlable.

torque a twisting force; to twist.

momento de torsión una fuerza de torsión; torcer.

trade specific kinds of work or businesses, especially those that require skilled mechanical work.

industria empresas o negocios de tipo específico, especialmente aquellos que requieren una fuerza laboral experta.

transformer a device that converts high voltage from high-power lines to 230 volts for home and farm installations; device used to step current up or down.

transformador un aparato que transforma la tensión alta de las líneas de alta tensión en 230 voltios para instalaciones de casa y de finca; un aparato empleado para aumentar o reducir la corriente eléctrica.

transistor a three-element, two-junction device used to control electron flow.

transistor un dispositivo de tres elementos y dos empalmes que sirve para controlar el flujo de electrones.

transit a surveying instrument.

escala de agrimensura instrumento para medir tierras.

translation linear movement motion performed by a robot or other machine.

traslación un movimiento lineal realizado por un robot o por otra máquina.

translucent allowing sunlight to pass through.

translúcido que deja pasar la luz del sol.

transmission gear box that permits two or more choices of speed.

transmisión caja de cambios que permite dos o más cambios de velocidad.

traveler wires See *switching wires*.

cable flexible de ascensor V. *circuitos selectores*.

tread the horizontal surface of a stair step.

escalón la superficie horizontal del peldaño de una escalera.

trestle See *sawhorse*.

caballete V. *(burro) caballete*.

trickle irrigation (drip irrigation) tubes putting water onto or below the soil.

irrigación por infiltración tubería que lleva el agua sobre o por debajo de la tierra.

trim and shutter paint paint formulated to stay bright without chalking.

pintura para marcos y contravientos pintura formulada para permanecer brillante sin ponerse gredoso.

troubleshooting determining what causes a malfunction in a machine or process.

localizador de averías determina lo que causa un funcionamiento defectuoso de una máquina o proceso.

trough culture plant production where the root zone is periodically flooded with nutrient solution.

jardines NFT producción de cultivos en donde la zona radicular es inundada periódicamente con solución nutritiva.

true to restore to the original shape or condition.

rectificar restaurar a la primera forma o condicción.

trunnion a heavy cast-iron mechanism attached to the underside of a tablesaw or a band saw that mounts the carriage for the blade.

soporte giratorio un mecanismo pesado de hierro fundido que va unido a la parte de debajo de una sierra circular estacionaria o de una aserradora de banda que monta el soporte móvil para la cuchilla.

truss rigid framework.

cercha armazón rígida.

try square a tool used to try or test the accuracy of cuts that have been made.

escuadra de comprobación un instrumento utilizado para probar la precisión de los cortes de que se han hecho.

tube culture plant production by which the root zone is immersed in a constant flow of nutrient solution.

jardines tubulares producción de cultivos por la cual la zona radicular es sumergida en un flujo constante de solución nutritiva.

tubing pipe that is flexible enough to bend.

tubos flexibles que son bastante flexibles para doblar.

tubing cutter a tool used for cutting and smoothing pipe or tubing.

trefiladora de tubos una herramienta que se emplea para cortar y lisar tubos y cañería.

tungsten inert gas (TIG) a welding process involving the use of inert gas and a tungsten tipped electrode.

gas inerte de tungsteno un proceso de soldadura que usa gas inerte y un electrodo con contera de tungsteno.

turbine a high-speed rotary engine driven by water, steam, or other gases.

turbina un motor giratorio de alta velocidad, impulsado por agua, vapor u otros gases.

turning tool a tool used to turn nuts, bolts, or screws.

herramienta de tornear una herramienta empleada para tornear tuercas, pernos o tornillos.

two-cycle engine (two-stroke-cycle engine) an engine with two strokes per cycle.

motor de dos tiempos un motor que tiene dos carreras por ciclo.

two half hitches knot useful for tying livestock.

dos vueltas de cabo nudo útil para atar ganado.

type the way current flows in a motor.

tipo la manera en que pasa la corriente en un motor.

U

undercoater special paint used to prepare surfaces for high-quality top coats.

pintura de primera mano pintura especial que se usa para preparar las superficies para esmaltes protectores de buena calidad.

union a fitting that can be easily opened between two pieces of similar threaded pipe.

unión de tuerca un accesorio que puede abrirse fácilmente entre dos trozos de tubo roscado de la misma manera.

unique having no like or equal.

único no teniendo ni parecido ni igual.

U.S. customary system the English system of measurement that is used in the United States.

sistema de costumbre de los E.U. el sistema de medida que se usa en E.U.

V

valve a device that controls the flow of water or gas.

válvula un dispositivo que regula el flujo de agua o gas.

valve grinding tool a tool that rotates a valve back and forth when lapping.

rectificadora para machos de válvula gira de movimiento alternativo mientras esmerila.

valve guide a device that holds valve stems in alignment.

soporte de válvula sujeta los vástagos de las válvulas en alineación.

valve keepers (valve pin) transfers spring force to valve stem.

retenes de las válvulas transmite el empuje de muelle al vástago de la válvula.

valve spring compressor compresses valve spring to remove valve keepers.

compresor de muelle de válvula comprime el muelle de válvula para quitar los retenes.

valve stem clearance air gap between valve and push rod.

aire libre del vástago-guía de la válvula espacio de aire entre válvula e impulsador.

vapor barrier moisture barrier.

barrera de vapor barrera de humedad.

variable-speed motor a motor that has speed that can be controlled by the operator.

motor de velocidad regulable un motor que tiene velocidad que puede ser controlada por el operador.

variable-speed pulley pulley with a movable side that can be moved while the equipment is running.

polea de velocidad regulable polea con canto móvil que puede moverse mientras está puesto en marcha el equipo.

varnish clear, tough, water-resistant finish.

barniz acabado claro, duro e impermeable al agua.

vehicle a device for carrying something.

vehículo un instrumento para cargar algo.

veneer thin sheets of wood.

chapa de madera maderos muy delgados.

venturi a constriction in a carburetor that increases air flow into the carburetor.

venturi constricción en un carburador que aumenta el flujo del aire en el caburador.

vermiculite artificial planting media manufactured from heated mica.

vermiculita medios artificiales de planta fabricados de mica calentada.

vertical down weld a weld made by moving downward across the metal.

soldadura en descendente una soldadura realizada por mover hacia abajo a través del metal.

vertical racks permit storage of both long and short items.

estantes verticales permiten guardar artículos largos y cortos.

vertical shaft engine an engine with a vertical crankshaft for normal operation.

motor de eje vertical un motor con un cigüeñal rectolineo para funcionamiento normal.

vertical up weld a weld made by moving upward on the metal.

soldadura en ascendente una soldadura realizada por mover hacia arriba a través del metal.

video amplifier used to amplify pictorial signals.

amplificador de imagen se utiliza para amplificar señales pictoriales.

viscosimeter an instrument used to measure the rate of flow of a liquid.

viscosímetro un instrumento usado para medir el índice del flujo de un líquido.

viscosity the tendency of a fluid to resist flowing.

viscosidad la tendencia de un líquido a resistirse a fluir.

viscosity improver a chemical added to a fluid to make it more stable when subjected to temperature changes.

mejorador de viscosidad una sustancia química que se agrega a un líquido para hacerlo más estable cuando se le somete a cambios de temperatura.

viscosity index a measure of a fluid's tendency to flow at a given temperature.

indice de viscosidad una medida de la tendencia de fluido a fluir a una.

vista green special shade of green used as a focal color.

verde vista tono especial de verde que se usa como color de foco.

volt (V) the unit of measure for voltage.

voltio (V) unidad de medida para voltaje.

voltage a measure of electrical pressure.

voltaje una medida de presión eléctrica.

voltage drop loss of voltage as electricity travels through a wire.

caída de tensión pérdida de voltaje mientras corre la corriente eléctrica por un alambre.

voltage multiplier used to step up DC voltage.

multiplicador de tensión utilizado para aumentar la tensión CC.

voltage regulator controls voltage output.

regulador de tensión controla la salida de tensión.

voltmeter an instrument that measures voltage.

voltímetro un instrumento que mide voltaje.

volt-ohm-milliampere meter (VOM meter) a popular combination meter used in electronics.

medidor VOM un medidor combinado popular empleado en la electrónica.

W

warp to bend or twist out of shape.

alabear combar o torcear hasta que se pierda la forma o el pérfil original.

washed sand sand flushed with water to remove clay or silt.

arena lavada arena limpiada con agua para quitar tierra arcillosa y légamo.

washer a flat piece of hardware with a hole in the center, used to increase the holding power or prevent loosening of a bolt, nut, or screw.

arandela un aparato con agujera que se usa para aumentar la retención o evitar el aflojamiento de perno, tuerca o tornillo.

water-based paint paint containing water as the vehicle.

colores al agua pintura que contiene agua como su vehículo.

waterproof material that is sealed against the entrance of water.

impenetrable al agua materia hermética a que no puede penetrar el agua.

water quality levels of dissolved oxygen, pH, salts, and waste products that affect aquatic plants and animals.

calidad de agua niveles de oxígeno disuelto, el pH y desechos que afectan plantas y animales acuáticos.

watt (W) the unit of measure for wattage.

vatio (W) la unidad para medir vataje [potencia en vatios].

wattage a measure of energy available or work that can be done using one ampere at one volt.

vataje una medida de energía disponible o el trabajo que se puede hacer usando un amperio a un voltio.

watt-hour the use of one watt of electricity for one hour.

vatio-hora el uso de un vatio de corriente eléctrica durante una hora.

weaving moving an electrode sideways to create a wider bead.

pasada pendular pasando el electrodo a un lado para crear un cordón más ancho.

wedge a piece of wood or metal that is thick on one end and tapers down to a thin edge on the other; to pack tightly.

calzo una pieza de madera o metal que es gruesa de un lado y termina en ángulo agudo del otro.

weld to join by fusion; the seam created by fusion.

soldar unir por fusión.

soldadura la juntura formada por fusión.

welder a machine that welds.

soldadora una máquina que solda.

welder a person who welds.

soldador una persona que solda.

wetted zones watered areas.

zonas mojadas regiones mojadas.

whet to sharpen by rubbing on a stone.

afilar con piedra de agua afilar por frotar sobre una piedra.

whip end wrap on a rope.

látigo extremo cónico de una cuerda.

white the color used to mark traffic areas; color of insulation on neutral wires.

blanco el color que se usa para marcar áreas de tráfico; color de aislante de cables eléctricos neutros.

white and black stripes stripes used as traffic markings.

rayas blancas y negras rayas utilizadas como señalización de tráfico.

wind brace construction member forming a triangle brace that strengthens the greenhouse structure.

contraviento pieza de construcción que forma una riostra triangular que fortalece el armazón del invernadero.

wind-up starter a device that uses a lever to coil a spring for cranking an engine.

desvanado de puesta en marcha usa una palanca para desvanar una muella para arrancar un motor con la manivela.

wire nut an insulated solderless connector used to connect the ends of electrical wires.

conectador de alambre un conducto aislado sin soldadura usado para conectar los terminales de hilos eléctricos.

wood the hard, compact, fibrous material that comes from the stems and branches of trees.

madera el material fibroso, duro y compacto que viene de los tallos y ramos de los árboles.

wood filler a thick material used to fill the pores of open-grained woods.

auxiliar materia gruesa usada para llenar los poros de maderas porosas.

woodframe walls constructed of vertical $2'' \times 4''$s.

entramado paredes construidas de los $2'' \times 4''$.

wood preservatives liquids used on wood to prevent rotting and insect damage.

preservativos de madera sustancias líquidas usadas sobre la madera para evitar pudrimiento y daño de los insectos.

wood screw a screw with threads designed to bite into wood fibers and draw down when turned.

tornillo de rosca de madera un tornillo con filetes que penetran las fibras de madera y que se bajan al tornearse.

workable mix the consistency of wet concrete when the various ingredients are mixed together correctly.

mezcla buena la consistencia de la masa de hormigón cuando los ingredientes son mezclados correctamente.

wrist pin pin between a connecting rod and piston.

pasador de articulación pasador entre la biela y el émbolo.

Y

Y a fitting used where a pipeline splits in two directions at an angle of less than 90 degrees.

unión en Y un montaje empleado donde tubería se bifurca a un ángulo de menos de 90 grados.

yellow the safety color meaning caution, used to identify parts of machines such as wheels, levers, and knobs that control or adjust the machine.

amarillo el color de seguridad, que significa cuidado, usado para identificar piezas de máquinas como ruedas, volantes, palancas y bultos que controlan o ajustan la máquina.

Z

zinc element associated with high-quality paint; material used for galvanizing steel.

cinc elemento asociado con pintura de buena calidad; materia que se usa para galvanizar el acero.

INDEX